THE ELECTRICAL MACHINES PROBLEM SOLVER®

REGISTERED TRADEMARK

Staff of Research and Education Association,
Dr. M. Fogiel, Director

Research and Education Association
505 Eighth Avenue
New York, N. Y. 10018

THE ELECTRICAL MACHINES PROBLEM SOLVER®

Printed in the United States of America

Library of Congress Catalog Card Number 83-62280

International Standard Book Number 0-87891-551-6

WHAT THIS BOOK IS FOR

Students have generally found electrical machines a difficult subject to understand and learn. Despite the publication of hundreds of textbooks in this field, each one intended to provide an improvement over previous textbooks, students continue to remain perplexed as a result of the numerous conditions that must often be remembered and correlated in solving a problem. Various possible interpretations of terms used in electrical machines have also contributed to much of the difficulties experienced by students.

In a study of the problem, REA found the following basic reasons underlying students' difficulties with electrical machines taught in schools:

(a) No systematic rules of analysis have been developed which students may follow in a step-by-step manner to solve the usual problems encountered. This results from the fact that the numerous different conditions and principles which may be involved in a problem, lead to many possible different methods of solution. To prescribe a set of rules to be followed for each of the possible variations, would involve an enormous number of rules and steps to be searched through by students, and this task would perhaps be more burdensome than solving the problem directly with some accompanying trial and error to find the correct solution route.

(b) Textbooks currently available will usually explain a given principle in a few pages written by a professional who has an insight in the subject matter that is not shared by students. The explanations are often written in an abstract manner which leaves the students confused as to the application of the principle. The explanations given are not sufficiently detailed and extensive to make the student aware of the wide range of applications and different aspects of the principle being studied. The numerous possible variations of principles and their applications are usually not discussed, and it is left for the students to discover these for themselves while doing exercises. Accordingly, the average student is expected to rediscover that which has been long known and practiced, but not published or explained extensively.

(c) The examples usually following the explanation of a topic are too few in number and too simple to enable the student to obtain a thorough grasp of the principles involved. The explanations do not provide sufficient basis to enable a student to solve problems that may be subsequently assigned for homework or given on examinations.

The examples are presented in abbreviated form which leaves out much material between steps, and requires that students derive the omitted material themselves. As a result, students find the examples difficult to understand--contrary to the purpose of the examples.

Examples are, furthermore, often worded in a confusing manner. They do not state the problem and then present the solution. Instead, they pass through a general discussion, never revealing what is to be solved for.

Examples, also, do not always include diagrams/graphs, wherever appropriate, and students do not obtain the training to draw diagrams or graphs to simplify and organize their thinking.

(d) Students can learn the subject only by doing the exercises themselves and reviewing them in class, to obtain experience in applying the principles with their different remifications.

In doing the exercises by themselves, students find that they are required to devote considerably more time to electrical machines than to other subjects of comparable credits, because they are uncertain with regard to the selection and application of the theorems and principles involved. It is also often necessary for students to discover those "tricks" not revealed in their texts (or review books), that make it possible to solve problems easily. Students must usually resort to methods of trial-and-error to discover these "tricks," and as a result find that they may sometimes spend several hours to solve a single problem.

(e) When reviewing the exercises in classrooms, instructors usually request students to take turns in writing solutions on the boards and explaining them to the class. Students often find it difficult to explain in a manner that holds the interest of the class, and enables the remaining students to follow the material written on the boards. The remaining students seated in the class are, furthermore, too occupied with copying the material from the boards, to listen to the oral explanations and concentrate on the methods of solution.

This book is intended to aid students in electrical machines to overcome the difficulties described, by supplying detailed illustrations of the solution methods which are usually not apparent to students. The solution methods are illustrated by problems selected from those that are most often assigned for class work and given on examinations. The problems are arranged in order of complexity to enable students to learn and understand a particular topic by reviewing the problems in sequence. The problems are illustrated with detailed step-by-step explanations, to save the students the large amount of time that is often needed to fill in the gaps that are usually found between steps of illustrations in textbooks or review/outline books.

The staff of REA considers electrical machines a subject that is best learned by allowing students to view the methods of analysis. and solution techniques themselves. This approach to learning the subject matter is similar to that practiced in various scientific laboratories, particularly in the medical fields.

In using this book, students may review and study the illustrated problems at their own pace; they are not limited to the time allowed for explaining problems on the board in class.

When students want to look up a particular type of problem and solution, they can readily locate it in the book by referring to the index which has been extensively prepared. It is also possible to locate a particular type of problem by glancing at just the material within the boxed portions. To facilitate rapid scanning of the problems, each problem has a heavy border around it. Furthermore, each problem is identified with a number immediately above the problem at the right-hand margin.

To obtain maximum benefit from the book, students should familiarize themselves with the section, "How To Use This Book," located in the front pages.

To meet the objectives of this book, staff members of REA have selected problems usually encountered in assignments and examinations, and have solved each problem meticulously to illustrate the steps which are usually difficult for students to comprehend. Special gratitude is expressed to them for their efforts in this area, as well as to the numerous contributors who devoted brief periods of time to this work.

Gratitude is also expressed to the many persons involved in the difficult task of typing the manuscript with its endless changes, and to the REA art staff who prepared the numerous detailed illustrations together with the layout and physical features of the book.

The difficult task of coordinating the efforts of all persons was carried out by Carl Fuchs. His conscientious work deserves much appreciation. He also trained and supervised art and production personnel in the preparation of the book for printing.

Finally, special thanks are due to Helen Kaufmann for her unique talents to render those difficult border-line decisions and constructive suggestions related to the design and organization of the book.

Max Fogiel, PH. D.
Program Director

HOW TO USE THIS BOOK

This book can be an invaluable aid to students in electrical machines as a supplement to their textbooks. The book is subdivided into 12 chapters, each dealing with a separate topic. The subject matter is developed beginning with electromagnetism, magnetic and electric circuits, D.C. generators and motors, transformers and extending through synchronous motors and generators. Also included are induction machines, special machines and transmission and distribution of power. An extensive number of applications have been included, since these appear to be most troublesome to students. A special section summarizing electrical machinery has been included at the end of the book.

TO LEARN AND UNDERSTAND A TOPIC THOROUGHLY

1. Refer to your class text and read the section pertaining to the topic. You should become acquainted with the principles discussed there. These principles, however, may not be clear to you at that time.

2. Then locate the topic you are looking for by referring to the "Table of Contents" in front of this book, "The Electrical Machines Problem Solver."

3. Turn to the page where the topic begins and review the problems under each topic, in the order given. For each topic, the problems are arranged in order of complexity, from the simplest to the more difficult. Some problems may appear similar to others, but each problem has been selected to illustrate a different point or solution method.

To learn and understand a topic thoroughly and retain its contents, it will be generally necessary for students to review the problems several times. Repeated review is essential in order to gain experience in recognizing the principles that should be applied, and to select the best solution technique.

TO FIND A PARTICULAR PROBLEM

To locate one or more problems related to a particular subject matter, refer to the index. In using the index, be certain to note that the numbers given there refer to problem numbers, not to page numbers. This arrangement of the index is intended to facilitate finding a problem more rapidly, since two or more problems may appear on a page.

If a particular type of problem cannot be found readily, it is recommended that the student refer to the "Table of Contents" in the front pages, and then turn to the chapter which is applicable to the problem being sought. By scanning or glancing at the material that is boxed, it will generally be possible to find problems related to the one being sought, without consuming considerable time. After the problems have been located, the solutions can be reviewed and studied in detail. For this purpose of locating problems rapidly, students should acquaint themselves with the organization of the book as found in the "Table of Contents."

In preparing for an exam, it is useful to find the topics to be covered in the exam from the "Table of Contents," and then review the problems under those topics several times. This should equip the student with what might be needed for the exam.

CONTENTS

8 SYNCHRONOUS GENERATORS 447

9 SYNCHRONOUS MOTORS 511

10 INDUCTION MACHINES 553

11 SPECIAL MACHINES 669

12 TRANSMISSION AND DISTRIBUTION OF POWER 698

SECTION II

SUMMARY OF ELECTRICAL MACHINERY

CHAPTER 1

ELECTROMAGNETISM

MAGNETIC FIELDS AND FORCES

• PROBLEM 1-1

The figure shows a field which is in the x direction and is constant at the value K. Evaluate the line integral $\int \bar{F} \cdot d\bar{s}$ from a to c along two paths, abc and adc and around the closed path abcda.

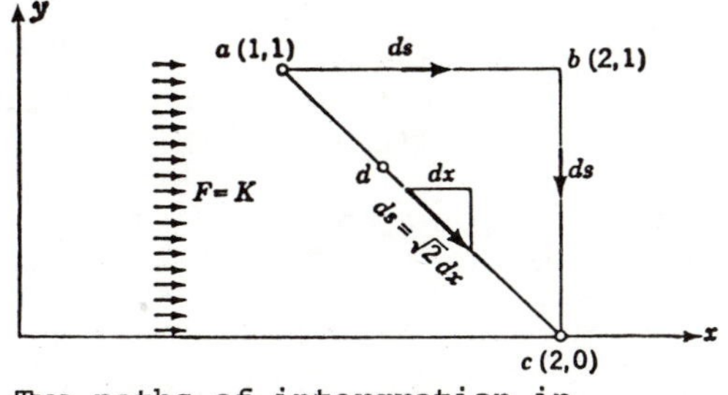

Two paths of intergration in a field of constant value.

Solution: First, for the path abc,

$$\int_{abc} \bar{F} \cdot d\bar{s} = \int_{ab} \bar{F} \cdot d\bar{s} + \int_{bc} \bar{F} \cdot d\bar{s}$$

In the integral from a to b, F = K, ds = dx, and the angle θ between $\bar{F}$ and $d\bar{s}$ is zero, so that $\cos\theta = 1$. In the integral from b to c, F = K, ds = dy, and the angle θ between $\bar{F}$ and $d\bar{s}$ is $90°$, so that $\cos\theta = 0$. Then,

$$\int_{abc} \bar{F} \cdot d\bar{s} = \int_{x=1}^{x=2} (K)(1)(dx) + \int_{y=1}^{y=0} (K)(0)(dy)$$

$$= Kx\Big]_1^2 = K$$

Next, on the path adc, F = K, the angle θ between $\bar{F}$ and $d\bar{s}$ is $45°$, and so $\cos\theta = 1\sqrt{2}$ and $ds = dx/\cos\theta = \sqrt{2}\,dx$.

Then,

$$\int_{adc} \bar{F} \cdot d\bar{s} = \int_{x=1}^{x=2} (K)\left(\frac{1}{\sqrt{2}}\right)(\sqrt{2}\ dx) = Kx\Big]_1^2 = K$$

The integrals along the two paths are equal, and around the closed path,

$$\oint \bar{F} \cdot d\bar{s} = \int_{abcda} \bar{F} \cdot d\bar{s} = 0.$$

• PROBLEM 1-2

The figure shows a field which is in the x direction and which varies linearly with y, such that F = 10y. Evaluate the line integral of $\bar{F} \cdot d\bar{s}$ from a to c along the two paths abc and adc, and along the closed curve abcda.

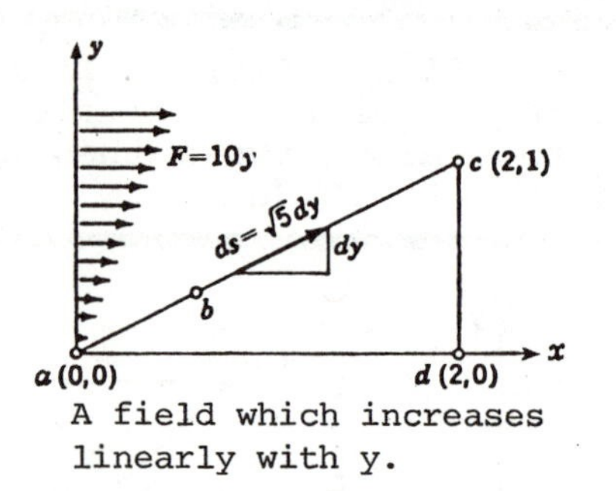

A field which increases linearly with y.

Solution: First, along the path abc, F = 10y, the cosine of the angle betweeen $\bar{F}$ and $d\bar{s}$ is $2/\sqrt{5}$, and $ds = dy/\sin\theta = \sqrt{5}\ dy$. Then

$$\int_{abc} \bar{F} \cdot d\bar{s} = \int_{y=0}^{y=1} (10y)\left(\frac{2}{\sqrt{5}}\right)(\sqrt{5}\ dy)$$

$$= 20 \int_0^1 y\ dy$$

$$= 10y^2\Big]_0^1$$

$$= 10$$

Along the path adc, F = 0 from a to d, thus contributing nothing to the integral. Also, from d to c, the path of integration is at right angles to $\bar{F}$, the cosine of the angle between $\bar{F}$ and $d\bar{s}$ is zero, and again nothing is contributed to the integral. Thus,

$$\int_{adc} \bar{F} \cdot d\bar{s} = 0$$

The line integral around the closed curve adcda is equal to 10.

• PROBLEM 1-3

The figure shows a long, straight, current-carrying wire which is encircled symmetrically by a ring of magnetic material that has an absolute permeability μ. The surrounding medium is air, which will be assumed to have the free-space permeability μ_0. There are no other magnetic objects in the vicinity. Find the magnetic field $\bar{B}$ both inside and outside the ring.

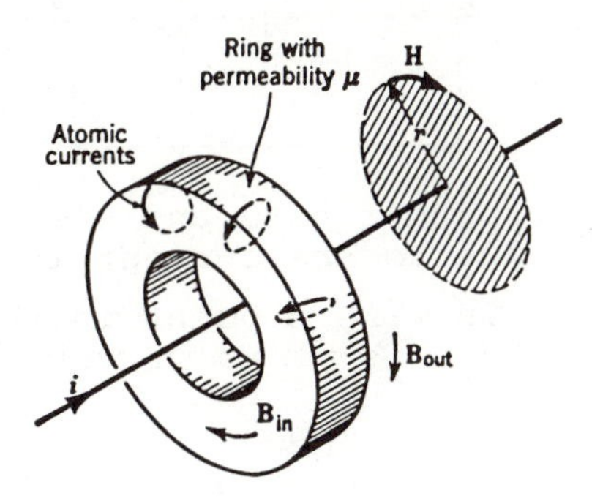

Solution: By symmetry, the magnetic field will be circular about the wire. Choose any circle of radius r as shown in the figure. This circle may be either inside or outside the ring. The magnetic field intensity $\bar{H}$ will be uniform around the periphery of such a circle. Choose a plane area spanning this circle, as shown in the figure. This surface is pierced by the current i. Then, the integral of H around the circle is

$$\oint \bar{H} \cdot d\bar{s} = H2\pi r = i$$

from which

$$H = \frac{i}{2\pi r}$$

Outside the ring,

$$B_{out} = \mu_0 H = \frac{\mu_0 i}{2\pi r} = \frac{4\pi \times 10^{-7} i}{2\pi r} = 2 \times 10^{-7} \frac{i}{r}.$$

Inside the ring,

$$B_{in} = \mu H = \frac{\mu i}{2\pi r},$$

which can also be expressed as

$$B_{in} = \frac{\mu}{\mu_0} \frac{\mu_0 i}{2\pi r} = 2 \times 10^{-7} \mu_{rel} \frac{i}{r}$$

where the fact that μ_{rel} is different from unity takes care of the effect of the atomic currents in the ring.

• PROBLEM 1-4

Two parallel wires a distance d apart carry equal currents i in opposite directions, as shown in the figure. Find the magnetic field $\overline{B}$ for points between the wires at a distance x from the midpoint.

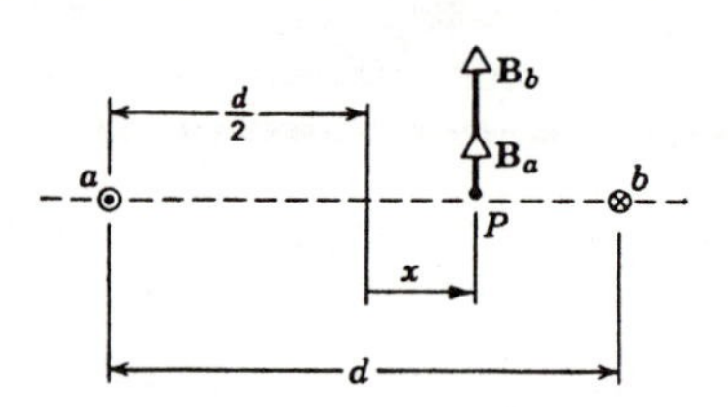

Solution: Study of the figure and use of the right-hand rule show that $\overline{B}_a$ due to the current in wire a and $\overline{B}_b$ due to the current in wire b point in the same direction (up) at point P. Each is given in magnitude by

$$B(r) = \frac{\mu_0 i}{2\pi r}$$

$$B(x) = B_a + B_b = \frac{\mu_0 i}{2\pi\left(\frac{d}{2} + x\right)} + \frac{\mu_0 i}{2\pi\left(\frac{d}{2} - x\right)}$$

$$= \frac{2\mu_0 id}{\pi(d^2 - 4x^2)}$$

• PROBLEM 1-5

A circular coil of diameter 20 mm is mounted with the plane of the coil perpendicular to the direction of a uniform magnetic flux density of 100 mT. Find the total flux threading the coil.

Solution: The flux is given by $\Phi = \int \overline{B} \cdot d\overline{s}$. As the flux density is uniform, the flux is given by

$$\Phi = \overline{B} \cdot \overline{A}$$

The vector area is normal to the plane of the area so that the vector flux density is parallel to the vector area and the required flux is the product of the flux density and the area. Therefore, $\Phi = BA$ or

$$\Phi = 10^{-1} \times \pi \times 10^{-4} = 3.14 \times 10^{-5} \text{ Wb} = 31.4 \ \mu\text{Wb}.$$

• **PROBLEM 1-6**

A semicircular loop of wire carrying current I is situated in a magnetic field $\overline{B}$ perpendicular to the plane defined by the loop. Deduce the force exerted on the loop.

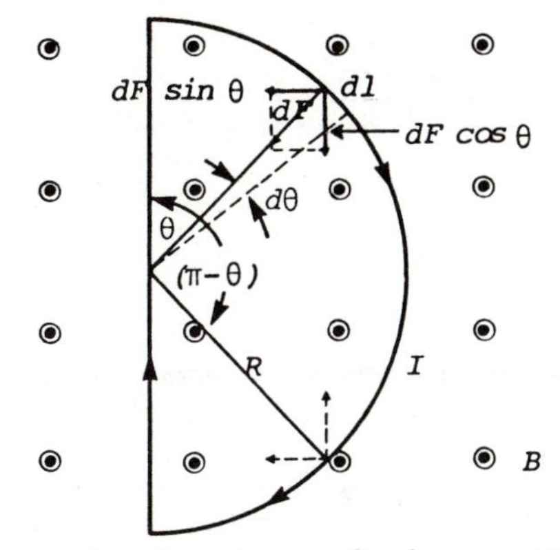

A semicircular loop of wire carrying a current *I* in a magnetic field *B*.

Solution: (see figure) A segment of wire of length $d\overline{\ell}$ has a force $d\overline{F}$ on it of magnitude

$$dF = IB\, d\ell = I\, BR\, d\theta$$

directed toward the center of the semicircle. The component $dF \cos\theta$ is canceled by the corresponding contribution from the arc segment located at $\pi - \theta$. Only the component $dF \sin\theta$ when summed over the complete semicircle gives a nonzero result. The resultant force is therefore

$$F = \int_0^{\pi} dF \sin\theta$$

$$= IBR \int_0^{\pi} \sin\theta \, d\theta$$

$$= 2IBR.$$

Note that the force experienced by the semicircle is the same as would be experienced by a straight wire of length 2R.

• **PROBLEM 1-7**

Determine the force in pounds upon each side of a coil of 40 turns carrying a current of 15 amp if the axial length of the coil is 10 in. and the flux density is 20,000 lines per square inch. The conductors are at right angles to the flux. There are 4.448×10^5 dynes in a pound.

Solution:

$$B = \frac{20,000}{(2.54)^2} \text{ lines per square centimeter}$$

$$\ell = 10 \times 2.54 = 25.4 \text{ cm}$$

$$F = \frac{B\ell I \sin\gamma}{10}$$

$$= \frac{20,000 \times 10 \times 2.54 \times 15 \times 40}{(2.54)^2 \times 10 \times 4.448 \times 10^5} = 10.6 \text{ lb}$$

on each side of the coil.

• PROBLEM 1-8

A rectangular flat coil of 20 turns lies with its plane parallel to a magnetic field (see the figure), the flux density in the field being 3,000 gauss. The axial length of the coil is 8 in. The current is 30 amp. Determine the force which acts on each side of the coil [see arrows in Fig. (a)] in pounds and in newtons.

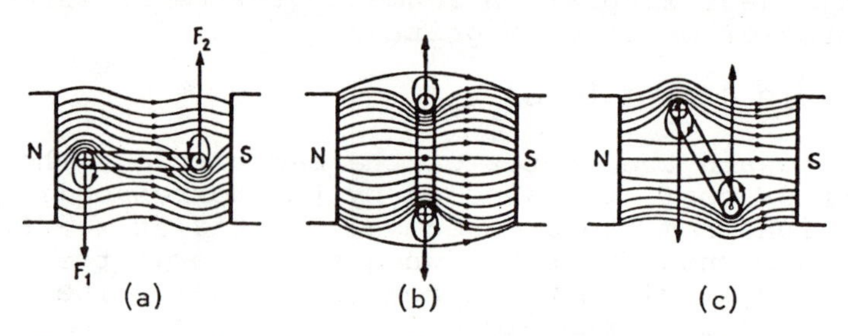

Torque developed at different positions of coil.

Solution: $F = \frac{B\ell I \sin\gamma}{10}$ dynes

$$B = 3,000 \text{ gauss}$$

$$\ell = 8 \times 2.54 = 20.32 \text{ cm.}$$

$$I = 30 \text{ amp.}$$

$$F = 3,000 \times 20.32 \times 30/10 = 182,900 \text{ dynes.}$$

As there are 20 turns,

$$F_1 = F_2 = 20 \times 182,900 = 3,658,000 \text{ dynes.}$$

$$\frac{3,658,000}{981} = 3,729 \text{ g.}$$

$$= 3.729 \text{ kg.}$$

$$3.729 \times 2.204 = 8.219 \text{ lb.}$$

In mks units,

$B = 0.3$ weber.

$\ell = 0.2032$ m.

$F = B\ell I$ newtons

$F = 0.3 \times 0.2032 \times 30 = 1{,}829$ newtons.

$F_1 = F_2 = 20 \times 1.829 = 36.58$ newtons.

36.58 newtons $= 36.58 \times 0.2248 = 8.223$ lb.

• PROBLEM 1-9

(a) What force will be required to move a short-circuited conductor across a magnetic field of 5,000 gausses if the conductor is 100 centimeters long, moves at a uniform speed of 100 centimeters per second and the circuit has a resistance of 0.04 ohm?

(b) What force will be required to force the conductor across at the rate of 200 centimeters per second?

Solution: (a) $F = \frac{IE}{V}$

$$I = \frac{E}{R}$$

$$E = B\ell V$$

$$E = 5{,}000 \times 100 \times 100$$

$= 50{,}000{,}000$ absolute units of pressure.

$$I = \frac{50{,}000{,}000}{0.04 \times 10^9}$$

$= 1.25$ absolute units of current

$$F = \frac{50{,}000{,}000 \times 1.25}{100}$$

$= 625{,}000$ dynes

$= \frac{625{,}000}{445{,}000} = 1.4$ pounds.

(b) $E = 5{,}000 \times 100 \times 200$

$I = \frac{5{,}000 \times 100 \times 200}{0.04 \times 10^9} = 2.5$ absolute units

$F = \frac{5{,}000 \times 100 \times 200 \times 2.5}{200} = 1{,}250{,}000$ dynes

$$= 2.8 \text{ pounds.}$$

The reaction is thus directly proportional to the speed. It must be noted that this is true only so long as the resistance of the circuit and the flux density remain constant.

BIOT-SAVART LAW

• PROBLEM 1-10

Derive an expression for $d\bar{H}$, the incremental magnetic field intensity produced at P(x,y,z) by a current element $Id\ell\ \hat{a}_z$ located at the origin in (a) Cartesian coordinates, (b) in cylindrical coordinates.

Solution: From Biot-Savart law,

$$d\bar{H} = \frac{I\ d\bar{\ell} \times \hat{a}_R}{4\pi R^2}$$

(a) In Cartesian coordinates, vector $\bar{R}$ from the origin to P(x,y,z) is

$$\bar{R} = x\ \hat{a}_x + y\ \hat{a}_y + z\ \hat{a}_z,$$

and

$$\hat{a}_R = \frac{x\ \hat{a}_x + y\ \hat{a}_y + z\ \hat{a}_z}{\sqrt{x^2 + y^2 + z^2}}$$

Since $d\bar{\ell} = d\ell\ \hat{a}_z$,

$$d\bar{H} = \frac{I\ d\ell[x\ \hat{a}_z \times \hat{a}_x + y\ \hat{a}_z \times \hat{a}_y + z\ \hat{a}_z \times \hat{a}_z\ ]}{4\pi[x^2 + y^2 + z^2]^{3/2}}$$

Now,

$$\hat{a}_z \times \hat{a}_x = \hat{a}_y, \quad \hat{a}_z \times \hat{a}_y = -\hat{a}_x, \quad \hat{a}_z \times \hat{a}_z = 0$$

Therefore,

$$d\bar{H} = \frac{I\ d\ell[x\ \hat{a}_y - y\ \hat{a}_x]}{4\pi[x^2 + y^2 + z^2]^{3/2}}$$

(b) In cylindrical coordinates,

$$\bar{R} = r\ \hat{a}_r + z\ \hat{a}_z$$

$$\hat{a}_R = \frac{r\ \hat{a}_r + z\ \hat{a}_z}{[r^2 + z^2]^{\frac{1}{2}}}$$

Therefore,

$$d\overline{H} = \frac{I\ d\ell\ \hat{a}_z \times [r\ \hat{a}_r + z\ \hat{a}_z]}{4\pi[r^2 + z^2]^{3/2}}$$

$$= \frac{I\ d\ell[r\ \hat{a}_z \times \hat{a}_r + z\ \hat{a}_z \times \hat{a}_z]}{4\pi[r^2 + z^2]^{3/2}}$$

Again, $\hat{a}_z \times \hat{a}_r = \hat{a}_\phi$, thus

$$d\overline{H} = \frac{I\ d\ell\ r}{4\pi[r^2 + z^2]^{3/2}}\ \hat{a}_\phi$$

• PROBLEM 1-11

An infinitely long, thin wire carrying current I_0 has a right angle bend as shown in Fig. 1. Using Biot-Savart law, find B along the positive x-axis.

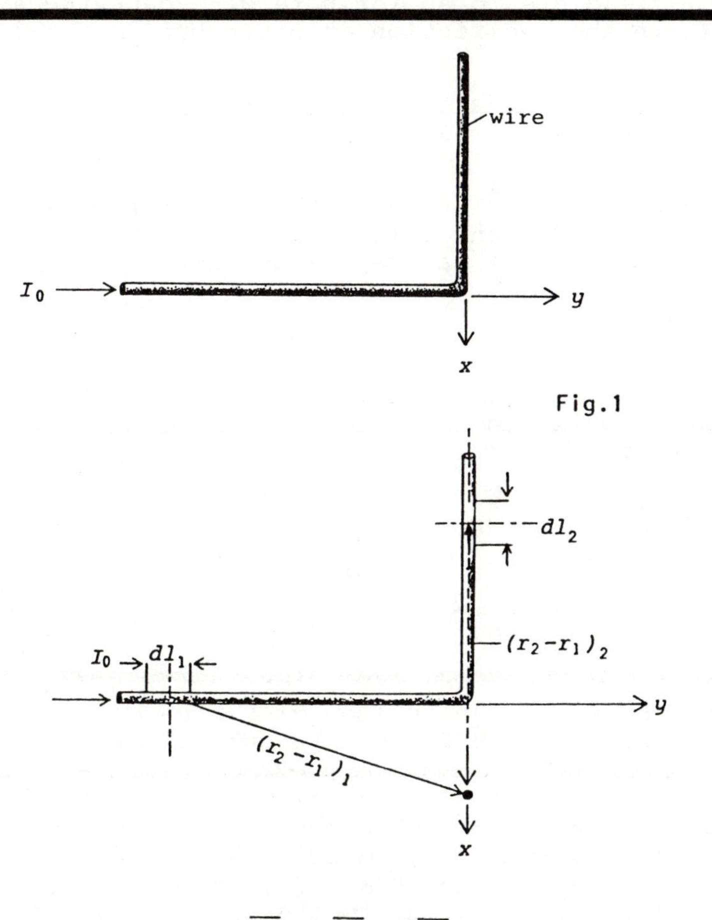

Solution: Using Biot-Savart law,

$$\overline{B} = \frac{\mu_0 I_0}{4\pi} \oint_C \frac{(\overline{r}_2 - \overline{r}_1) \times \overline{d\ell}}{|\overline{r}_2 - \overline{r}_1|^3}$$

where $\overline{r}_2 - \overline{r}_1$ and $\overline{d\ell}$ are shown in Fig. 2.

Note that $\overline{d\ell}_2$ does not contribute to the field since it is parallel to $(\overline{r}_2 - \overline{r}_1)_2$ and the cross-product is zero. Therefore, only the horizontal wire will contribute to the $\overline{B}$-field. Thus,

$$\overline{B} = \frac{\mu_0 I_0}{4\pi} \oint_C \frac{(\overline{r}_2 - \overline{r}_1)_1 \times \overline{d\ell}_1}{|(\overline{r}_2 - \overline{r}_1)_1|^3}$$

$$= \frac{\mu_0 I_0}{4\pi} \int_{-\infty}^{0} \frac{(\overline{r}_2 - \overline{r}_1)_1 \times \hat{a}_y\, dy}{|(\overline{r}_2 - \overline{r}_1)|^3}$$

$$= \frac{\mu_0 I_0}{8\pi} \int_{-\infty}^{\infty} \frac{(\overline{r}_2 - \overline{r}_1)_1 \times \hat{a}_y\, dy}{|(\overline{r}_2 - \overline{r}_1)_1|^3} \tag{1}$$

Now, the $\overline{B}$-field due to an infinite wire carrying a current I_0 in the y-direction is given by:

$$\overline{B}_{inf} = \frac{\mu_0 I_0}{4\pi} \int_{-\infty}^{\infty} \frac{(\overline{r}_2 - \overline{r}_1) \times \hat{a}_y\, dy}{|\overline{r}_2 - \overline{r}_1|^3} = \frac{\mu_0 I_0}{2\pi\rho}\hat{a}_\phi \tag{2}$$

where ρ is the perpendicular distance from the axis along the wire (y-axis) to the point under consideration. Using Eq. (2), Eq. (1) can be written as:

$$\overline{B} = \frac{\mu_0 I_0}{4\pi\rho}\hat{a}_\phi \tag{3}$$

where ρ is the distance along the x-axis = x. Noting that at the x-axis, the B-field is perpendicular to the plane of the paper, Eq. (3) becomes

$$\overline{B} = -\frac{\mu_0 I_0}{4\pi x}\hat{a}_z$$

• PROBLEM 1-12

The figure shows a circular loop of radius R carrying a current i. Calculate $\overline{B}$ for points on the axis.

Solution: The vector $d\overline{\ell}$ for a current element at the top of the loop points perpendicularly out of the page. The angle θ between $d\overline{\ell}$ and $\overline{r}$ is 90°, and the plane formed by $d\overline{\ell}$ and $\overline{r}$ is normal to the page. The vector $d\overline{B}$ for this element is at right angles to this plane and thus lies in the plane of the figure and at right angles to $\overline{r}$, as the figure shows.

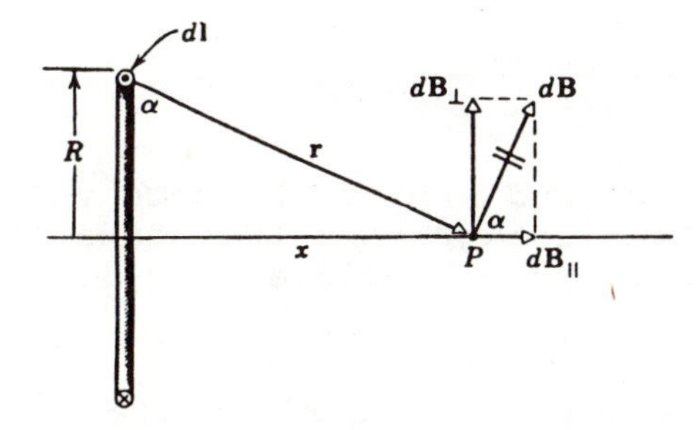

A ring of radius *R* carrying a current *i*.

Resolve $d\overline{B}$ into two components, one $d\overline{B}_{||}$, along the axis of the loop and another, $d\overline{B}_{\perp}$, at right angles to the axis. Only $d\overline{B}_{||}$ contributes to the total induction $\overline{B}$ at point P. This follows because the components $d\overline{B}_{||}$ for all current elements lie on the axis and add directly; however, the components $d\overline{B}_{\perp}$ point in different directions perpendicular to the axis, and their resultant for the complete loop is zero, from symmetry. Thus

$$B = \int dB_{||} ,$$

where the integral is a simple scalar integration over the current elements.

For the current element shown in the figure, we have, from the Biot-Savart law

$$dB = \frac{\mu_0 i}{4\pi} \frac{d\ell \sin 90°}{r^2}$$

Also

$$dB_{||} = dB \cos \alpha .$$

Combining gives

$$dB_{||} = \frac{\mu_0 i \cos \alpha \, d\ell}{4\pi r^2} .$$

The figure shows that r and α are not independent of each other. Express each in terms of a new variable x, the distance from the center of the loop to the point P. The relationships are

$$r = \sqrt{R^2 + x^2}$$

and

$$\cos \alpha = \frac{R}{r} = \frac{R}{\sqrt{R^2 + x^2}}$$

Substituting these values into the expression for $dB_{||}$ gives

$$dB_{||} = \frac{\mu_0 iR}{4\pi(R^2 + x^2)^{3/2}} d\ell.$$

Note that i, R, and x have the same values for all current elements. Integrating this equation, noting that $\int d\ell$ is simply the circumference of the loop ($= 2\pi R$), yields

$$B = \int dB_{||} = \frac{\mu_0 iR}{4\pi(R^2 + x^2)^{3/2}} \int dl$$

$$= \frac{\mu_0 iR^2}{2(R^2 + x^2)^{3/2}}$$

AMPERE'S LAW

• PROBLEM 1-13

Calculate the magnetic field $\overline{B}$ at a distance R from an infinitely long straight wire, using Ampere's Circuital law.

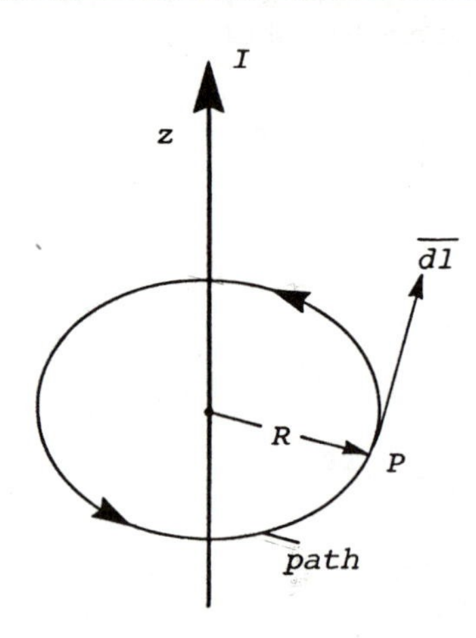

Solution: Ampere's Circuital law is

$$\oint \overline{B} \cdot d\overline{\ell} = \mu_0 I \qquad (1)$$

where μ_0 is the permeability of free space and I is the total current crossing any surface bounded by the line integral path. Using cylindrical coordinates,

$$\overline{B} = \hat{r} B_r + \hat{\theta} B_\theta + \hat{z} B_z$$

Take for the path a circle of radius R passing through

P and with its center at the wire, the plane of the circle being perpendicular to the axis of the wire (see figure).

Then,

$$d\bar{\ell} = \hat{\theta}\ R\ d\theta$$

and

$$\oint \bar{B}\cdot d\bar{\ell} = \oint [\hat{r}B_r + \hat{\theta}B_\theta + \hat{z}B_z] \cdot \hat{\theta}R d\theta$$

$$= RB_\theta \oint d\theta$$

$$= B_\theta R\ (2\pi).$$

From Eq. (1),

$$B_\theta R\ (2\pi) = \mu_0 I$$

or

$$B_\theta = \frac{\mu_0 I}{2\pi R}$$

Now, according to the Biot-Savart law,

$$d\bar{B} = \frac{\mu_0 I\ d\bar{z} \times \hat{r}}{4\pi r^2}$$

Therefore, $\bar{B}$ must be in the θ direction only, since if $\bar{C} = \bar{A} \times \bar{B}$, the vector $\bar{C}$ is perpendicular to the plane containing $\bar{A}$ and $\bar{B}$. Thus,

$$\bar{B} = \frac{\mu_0 I}{2\pi R}\hat{\theta}\ ,$$

that is, the B-field are circles whose centers lie on the axis of the wire.

• PROBLEM 1-14

The figure shown is a sectional view of two long parallel plates of width w. The plate on the left carries a current I toward the reader, and that on the right carries an equal current away from the reader. Find the field between the two parallel plates.

Solution: In the region between the plates and not too near the edges, the $\bar{B}$-field is uniform. The field outside the plates is small, and becomes smaller as the width w is increased.

Apply Ampere's law to the dotted rectangle, and assume the

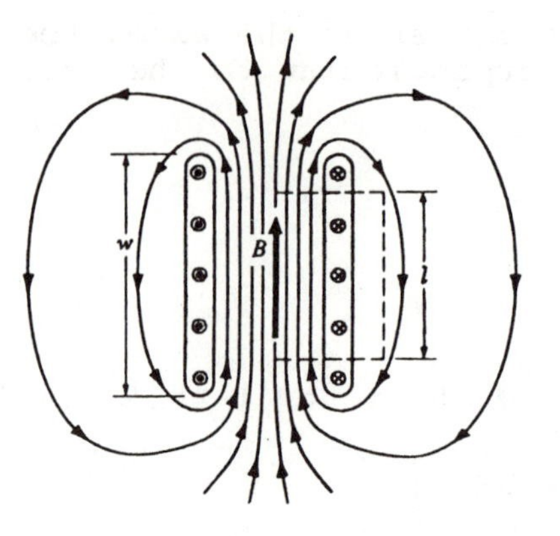

field outside the plates to be zero. Then $\overline{B} \cdot \overline{d\ell} = B\ell$. If I is the total current in either plate, the current per unit width is I/w and the current through the rectangle is $I\ell/w$. Hence

$$B\ell = \frac{\mu_0 I \ell}{w}, \quad B = \frac{\mu_0 I}{w}.$$

It is interesting to compare the nature of the magnetic field between two long, current-carrying plates, with the electric field between two large charged plates. Both fields are approximately uniform, the magnetic lines being parallel to the plates and the electric lines perpendicular. In the magnetic case, $B = \mu_0(I/w)$, whereas in the electric case $E = (1/\varepsilon_0)(Q/A)$.

• PROBLEM 1-15

A solenoid is constructed by winding wire in a helix around the surface of a cylindrical form, usually of circular cross section. The turns of the winding are ordinarily closely spaced and may consist of one or more layers. For simplicity, the solenoid is represented in the figure by a relatively small number of circular turns, each carrying a current I. Find the field at any point due to the solenoid.

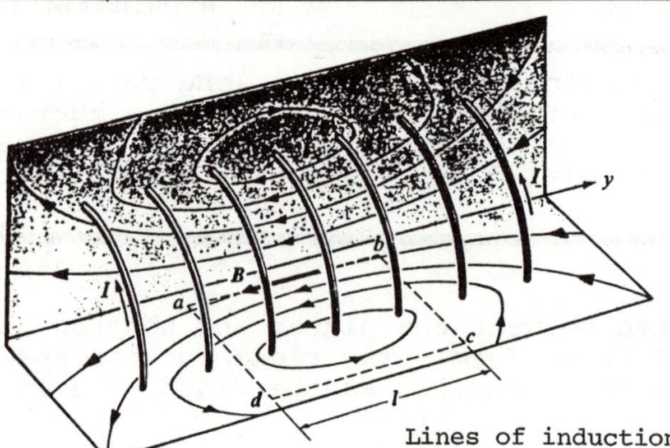

Lines of induction surrounding a solenoid. The dotted rectangle abcd is used to compute the flux density B in the solenoid from Ampère's law.

Solution: The resultant field at any point is the vector sum of the $\overline{B}$-vectors due to the individual turns. The diagram shows the lines of induction in the xy- and yz-planes. Exact calculations show that for a long, closely-wound solenoid, half of the lines passing through a cross section at the center emerge from the ends and half "leak out" through the windings between center and end.

If the length of the solenoid is large compared with its cross-sectional diameter, the internal field near its center is very nearly uniform and parallel to the axis, and the external field near the center is very small. The internal field at or near the center can then be found by use of Ampere's law.

Select as a closed path the dotted rectangle abcd as shown in the figure. Side ab, of length ℓ, is parallel to the axis of the solenoid. Sides bc and da are to be taken very long so that side cd is far from the solenoid and the field at this side is negligibly small.

By symmetry, the $\overline{B}$ field along side ab is parallel to this side and is constant, so that for this side $B_{\|} = B$ and

$$\oint \overline{B} \cdot d\ell = B\ell.$$

Along sides bc and da, $B_{\|} = 0$ since $\overline{B}$ is perpendicular to these sides; and along side cd, $B_{\|} = 0$ also since $B = 0$. The sum around the entire closed path therefore reduces to $B\ell$.

Let n be the number of turns per unit length in the windings. The number of turns in length ℓ is then $n\ell$. Each of these turns passes once through the rectangle abcd and carries a current I, where I is the current in the windings. The total current through the rectangle is then $n\ell I$, and from Amperes law,

$$B\ell = \mu_0 n\ell I,$$

$$B = \mu_0 nI \quad \text{(solenoid)}.$$

Since side ab need not lie on the axis of the solenoid, the field is uniform over the entire cross section.

FARADAY'S AND LENZ'S LAWS

• PROBLEM 1-16

Let B in Fig. 1 be increasing at the rate dB/dt. Let R be the effective radius of the cylindrical region in which the magnetic field is assumed to exist. What is the magnitude of the electric field E at any radius r? Assume that dB/dt = 0.10 T/s and R = 10 cm.

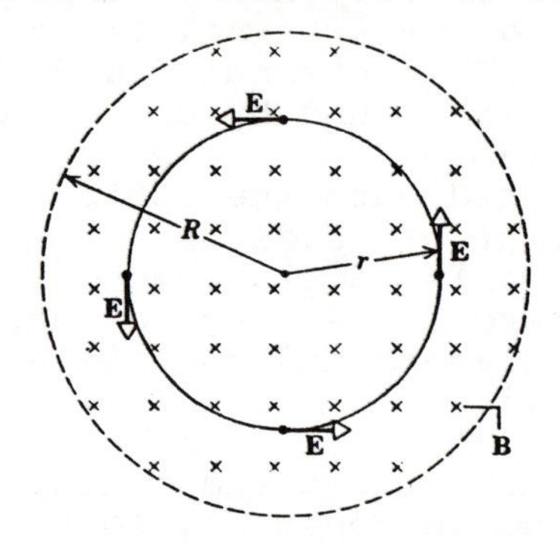

The induced electric field at four points produced by an increasing magnetic field.

Fig.1

Solution: (a) For $r < R$, the flux Φ_B through the loop is

$$\Phi_B = B(\pi r^2).$$

Substituting into Faraday's law

$$\oint \bar{E} \cdot d\bar{l} = -\frac{d\Phi_B}{dt},$$

yields

$$(E)(2\pi r) = -\frac{d\Phi_B}{dt} = -(\pi r^2)\frac{dB}{dt}.$$

Solving for E yields

$$E = -\tfrac{1}{2} r \frac{dB}{dt}.$$

The minus sign is retained to suggest that the induced electric field $\bar{E}$ acts to oppose the change of the magnetic field. Note that E(r) depends on dB/dt and not on B. Substituting numerical values, assuming r = 5.0 cm, yields, for the magnitude of E,

$$E = \tfrac{1}{2} r \frac{dB}{dt} = (\tfrac{1}{2})(50 \times 10^{-3} \text{m})(0.10 \text{ T/s}) = 2.5 \text{ mV/m}.$$

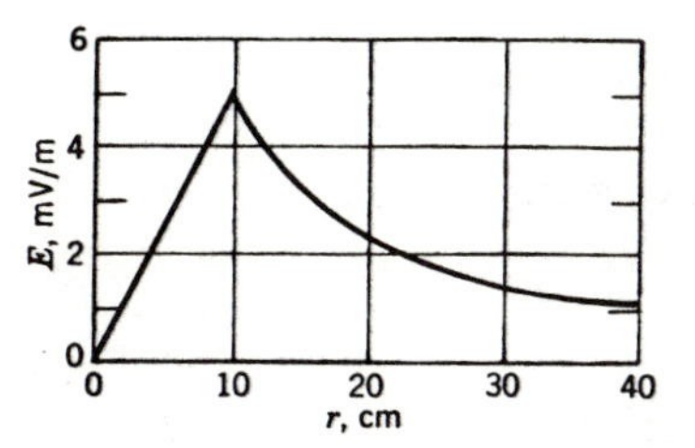

If the fringing of the field in Fig.1 were to be taken into account, the result would be a rounding of the sharp cusp at $r = R$ (= 10 cm).

Fig.2

(b) For r > R the flux through the loop is

$$\Phi_B = \int \overline{B} \cdot d\overline{S} = B(\pi R^2).$$

This equation is true because $\overline{B} \cdot d\overline{S}$ is zero for those points of the loop that lie outside the effective boundary of the magnetic field.

From Faraday's law:

$$(E)(2\pi r) = -\frac{d\Phi_B}{dt} = -(\pi R^2)\frac{dB}{dt}.$$

Solving for E yields

$$E = -\frac{1}{2}\frac{R^2}{r}\frac{dB}{dt}.$$

These two expressions for E(r) yield the same result, as they must, for r = R. Figure 2 is a plot of the magnitude of E(r) for the numerical values given.

• PROBLEM 1-17

(a) The dimensions of the pole faces in Fig. 1 are 20 by 40 centimeters. If the flux density B is 5,000 lines per square centimeter and the conductor moves across this flux in ½ second, what is the average value of the induced pressure?

(b) Suppose one side of a coil of 1,000 turns cuts the flux specified in part (a) in 0.1 second, what will be the pressure between the terminals of the coil?

Solution:

(a) $E = \dfrac{\Phi}{t \times 10^8}$ volts

$\Phi = B \times A$

$$= 5{,}000 \times 20 \times 40$$

$$= 4{,}000{,}000 \text{ maxwells,}$$

then

$$E = \frac{4{,}000{,}000}{\frac{1}{2} \times 10^8}$$

$$= 0.08 \text{ volt for one conductor.}$$

(b) $E = n \dfrac{\Phi}{t \times 10^8} \text{volts}$

$\Phi = 4{,}000{,}000$

$n = 1{,}000$

$t = 0.1$ sec.

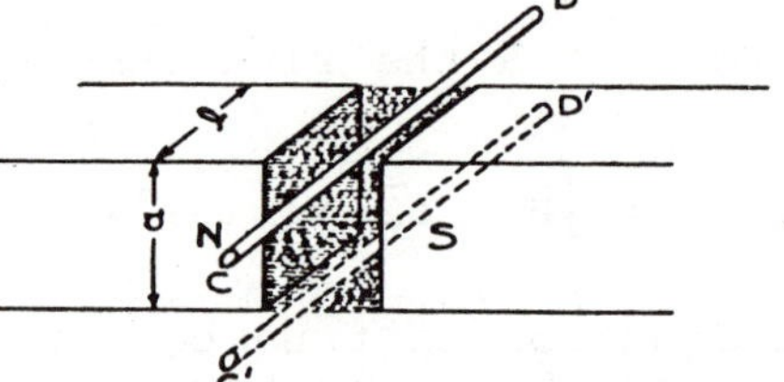

Fig. 1

Hence,

$$E = \frac{4{,}000{,}000 \times 1{,}000}{0.1 \times 10^8}$$

$$= 400 \text{ volts.}$$

• PROBLEM 1-18

A time-varying magnetic field is given by

$$\overline{B} = B_0 \cos \omega t \; \hat{i}_y$$

where B_0 is a constant. Find the induced emf around a rectangular loop in the xz plane as shown in Fig. 1.

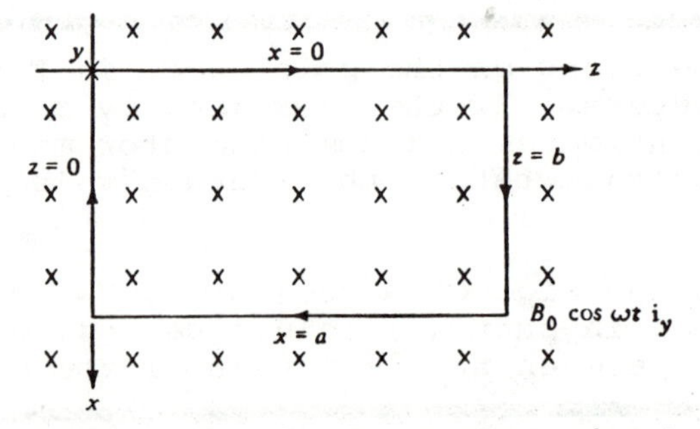

A rectangular loop in the *xz* plane situated in a time-varying magnetic field.

Fig.1

Solution: The magnetic flux enclosed by the loop and directed into the paper is given by

$$\Phi = \int_S \bar{B} \cdot d\bar{S} = \int_{z=0}^{b} \int_{x=0}^{a} B_0 \cos \omega t \, \hat{i}_y \cdot dx \, dz \, \hat{i}_y$$

$$= B_0 \cos \omega t \int_{z=0}^{b} \int_{x=0}^{a} dx \, dz = abB_0 \cos \omega t$$

The induced emf in the clockwise sense is then given by

$$\oint_C \bar{E} \cdot d\bar{l} = - \frac{d}{dt} \int_S \bar{B} \cdot d\bar{S}$$

$$= - \frac{d}{dt}[abB_0 \cos \omega t] = ab \, B_0\omega \sin \omega t$$

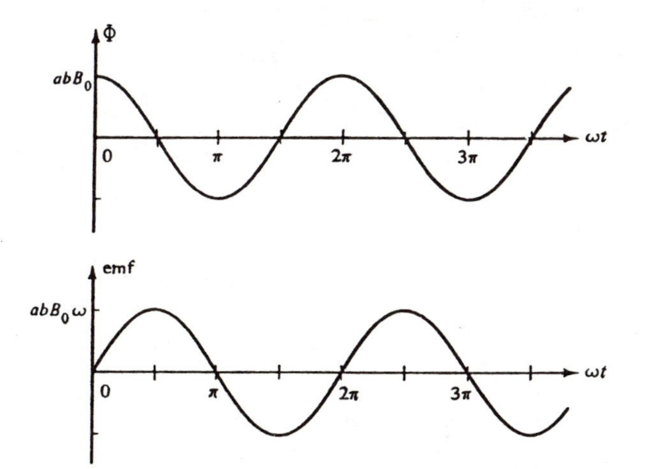

Time variations of magnetic flux Φ enclosed by the loop of Fig. 1, and the resulting induced emf around the loop.

Fig.2

The time variations of the magnetic flux enclosed by the loop and the induced emf around the loop are shown in Fig. 2. It can be seen that when the magnetic flux enclosed by the loop is decreasing with time, the induced emf is positive, thereby producing a clockwise current if the loop were a wire. This polarity of the current gives rise to a magnetic field directed into the paper inside the loop and hence acts to increase the magnetic flux enclosed by the loop. When the magnetic flux enclosed by the loop is increasing with time, the induced emf is negative, thereby producing a counterclockwise current around the loop. This polarity of the current gives rise to a magnetic field directed out of the paper inside the loop and hence acts to decrease the magnetic flux enclosed by the loop. These observations are consistent with Lenz's law.

• PROBLEM 1-19

Referring to Fig. (a), a circular loop described by the equation $x^2 + y^2 = 16$ is located in the x-y plane centered at the origin. The $\overline{B}$ field is described by

$$\overline{B} = \hat{z}2\sqrt{x^2 + y^2}\cos\omega t \quad (\text{Wb/m}^2)$$

Find the total emf induced in the loop.

<u>Solution</u>: From $\text{emf} = -\frac{d}{dt}\left(\int_S \overline{B} \cdot d\overline{s}\right) = -\int_S \frac{\partial \overline{B}}{\partial t} \cdot d\overline{s}$

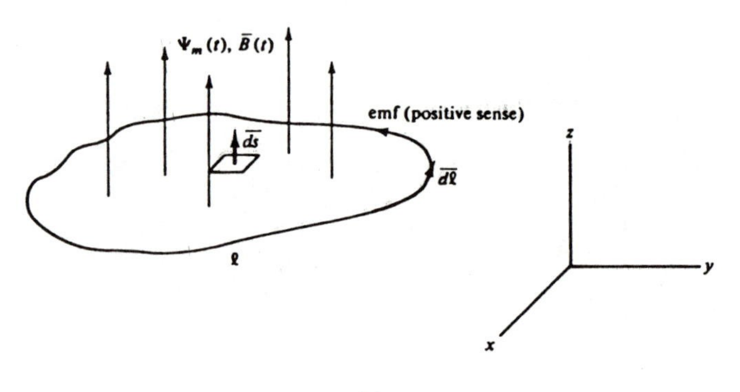

(a)

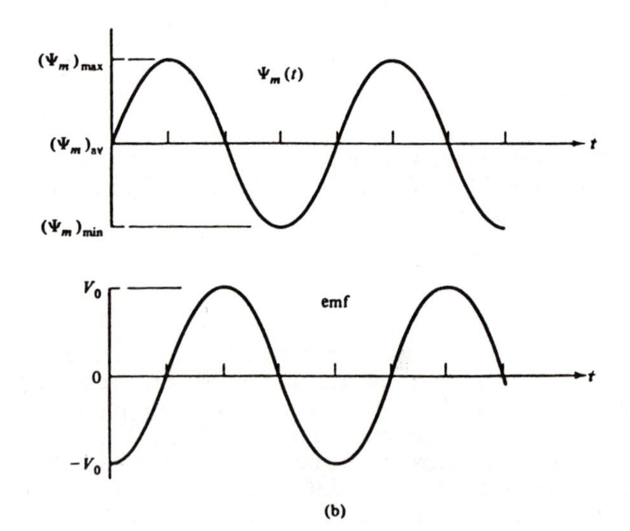

(b)

(a) Time-changing magnetic field linking a closed loop. (b) The time relationship between the time-changing magnetic field and the emf induced in the loop.

$$\text{emf} = -\int_y \int_x -2\omega \sin \omega t\sqrt{x^2 + y^2}\hat{z} \cdot \hat{z}\, dx\, dy$$

$$= 2\omega \quad \sin \omega t \iint \sqrt{x^2 + y^2}\, dx\, dy$$

Changing to cylindrical coordinates,

$$\text{emf} = 2\omega \sin \omega t \int_0^{2\pi}\int_0^4 r_c dr_c r_c d\phi$$

$$= 2 \sin \omega t (2\pi)\frac{4^3}{3}$$

$$= \frac{4\omega\pi \times 64}{3} \sin \omega t$$

$$= 268\omega \sin \omega t \quad (V)$$

• PROBLEM 1-20

Consider the circular wire loop of radius r shown in the figure. Assume that the wire loop is being heated in such a way that the radius r is a linear function of the time t, that is,

$$r = vt$$

where v is the constant radial velocity of a point on the loop. Also assume that the magnetic flux density is uniform and that the magnitude of $\overline{B}$ is given by

$$B = B_0(1 + kt)$$

where B_0 and k are constants. The vector $\overline{B}$ is normal to the page and directed toward the reader. Determine the induced emf in the wire loop.

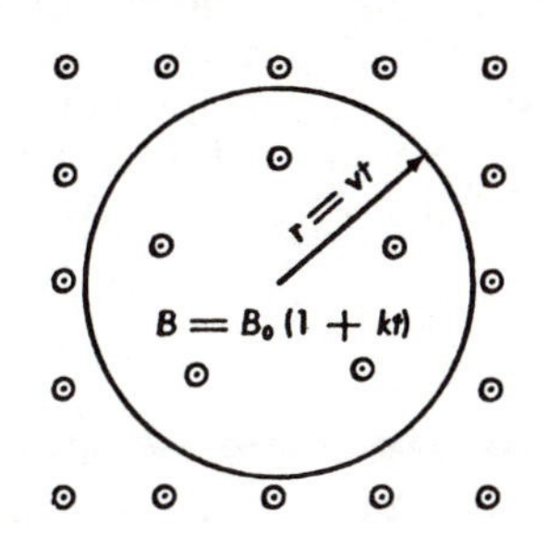

An expanding circular loop in a variable magnetic field.

Solution: The first step in the solution of this problem is to obtain an expression for the flux ϕ. Since

$$\phi = \iint \overline{B} \cdot d\overline{S}$$

and $\overline{B}$ is not a function of the coordinates,

$$\phi = \pi r^2 B$$

where the direction of dS was chosen to be parallel to that of B.

The second step is to find the value of the induced emf from the following equation:

$$E = -\frac{d\phi}{dt} = -\pi B(2r\frac{dr}{dt}) - \pi r^2 \frac{dB}{dt}$$

Since

$$\frac{dr}{dt} = v$$

and

$$\frac{dB}{dt} = B_0 k,$$

$$E = -vB(2\pi r) - B_0 k(\pi r^2)$$

Thus, the induced emf consists of two terms. The first depends on the speed v with which elements of the wire loop move through the magnetic field. For this reason, this part of the induced emf is called the motional emf. The second term depends on the time rate of change of the flux density $\overline{B}$. The value of the induced emf may be expressed in general terms as follows:

$$E = \oint \overline{v} \times \overline{B} \cdot d\overline{s} - \iint \frac{\partial \overline{B}}{\partial t} \cdot d\overline{S}$$

where $\oint \overline{v} \times \overline{B} \cdot d\overline{s}$ replaces $-vB\ (2\pi r)$ and $-\iint \partial\overline{B}/\partial t \cdot d\overline{S}$ replaces $-B_0 k(\pi r^2)$. In the case of the wire loop, since dS is directed toward the reader, the vector $d\overline{s}$ is oriented so as to yield counterclockwise circulation. The cross product, $\overline{v} \times \overline{B}$ is opposite in direction to $d\overline{s}$. Consequently, a minus sign automatically results when the term $\oint \overline{v} \times \overline{B} \cdot d\overline{s}$ is evaluated.

• PROBLEM 1-21

A coil of N turns and area A is rotated in a uniform magnetic field B, at an angular velocity $\omega_0$rad/sec as shown in the figure. Calculate the emf generated, using the Faraday induction law.

Solution: The flux linkage threading the coil at an angle θ between the plane of the coil and direction of B is given by

$$N\phi = NAB \sin \theta = NAB \sin \omega_0 t$$

where $\omega_0 t$ is the value of θ for a particular time t. Then

$$E = -N\frac{d\Phi}{dt} = -\omega_0 NAB \cos \omega_0 t$$

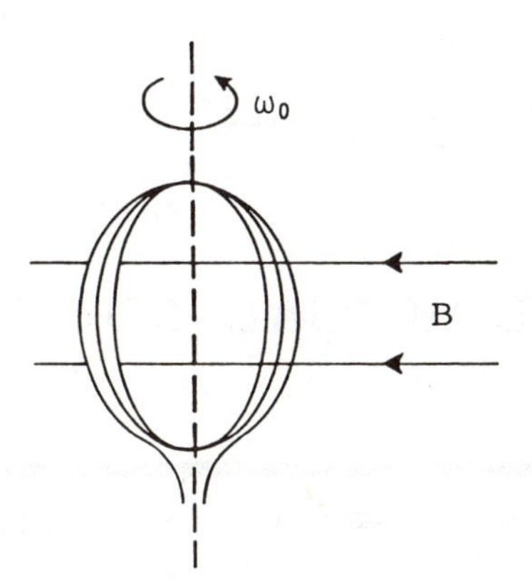

A simple generator: a rotating coil in an external magnetic field.

Thus this generator gives a sinusoidal emf of amplitude $\omega_0 NAB$. The source of this emf is the mechanical work done in rotating the coil. This is zero (in a frictionless system) if no current flows and increases linearly with increasing current.

• PROBLEM 1-22

A coil with five series-connected turns rotates at a speed of 1200 rpm. The flux per pole is $\Phi = 3 \times 10^6$ maxwells; the number of poles is p = 6. What is the average emf induced in the coil? What is the amplitude and the effective value of the emf induced in the coil if the flux is sinusoidally distributed?

Solution: From the following equation:

$$E_{av} = 4\Phi N \frac{p}{2}\frac{n}{60} 10^{-8} \text{ volt,}$$

with p = 6 and N = 5,

$$E_{av} = 4 \times 3 \times 10^6 \times 5 \times \frac{6}{2} \times \frac{1200}{60} \times 10^{-8} = 36 \text{ volts.}$$

$$f = \frac{p \times n}{120} \text{ cps}$$

$$= \frac{6 \times 1200}{120} = 60 \text{ cps.}$$

$$E_m = 2\pi f N\Phi 10^{-8} \text{ volt}$$

$$= 2\pi \times 60 \times 5 \times 3 \times 10^6 \times 10^{-8} = 56.6 \text{ volts}$$

and

$$E = \frac{56.5}{\sqrt{2}} = 40 \text{ volts}$$

For a sinusoidally distributed flux, E_{av} must be $\frac{2}{\pi}E_m$.

SELF AND MUTUAL INDUCTANCE

• PROBLEM 1-23

Find the self-inductance of a long solenoid shown in the figure.

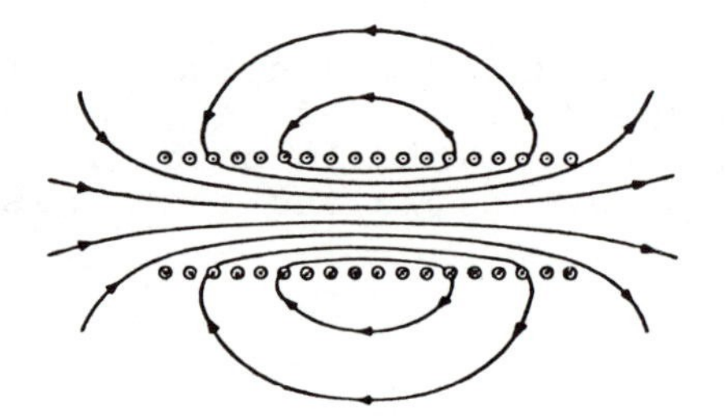

A cross section of a solenoid
and its magnetic field.

Solution: The magnetic induction inside a long solenoid, neglecting end effects, is constant,

$$B = \mu_0 N'I,$$

where N' is the number of turns/meter. Thus

$$\Phi = \frac{\mu_0 \; I}{\ell}\pi R^2,$$

where N is the total number of turns, ℓ is the length of the solenoid, and R is its radius. Then

$$L = \frac{N\Phi}{\ell} = \frac{\mu_0 N^2}{\ell}\pi R^2,$$

$$= \mu_0 N'^2 \; \ell \; \pi R^2.$$

• PROBLEM 1-24

Figure 1 shows a toroidal coil wound on a plastic ring of rectangular cross section. The coil has 200 turns of round copper wire which is 3 mm in diameter.

a) For a coil current of 50A, find the magnetic flux density at the mean diameter of the coil.

b) Find the inductance of the coil, assuming that the flux density within it is uniform and equal to that at the mean diameter.

c) Determine the percentage error incurred in assuming uniform flux density in the coil.

d) Given that the volume resistivity of copper is 17.2 x 10^{-9} ohm-meter, determine the parameters of the approximate electric circuit of Fig. 2.

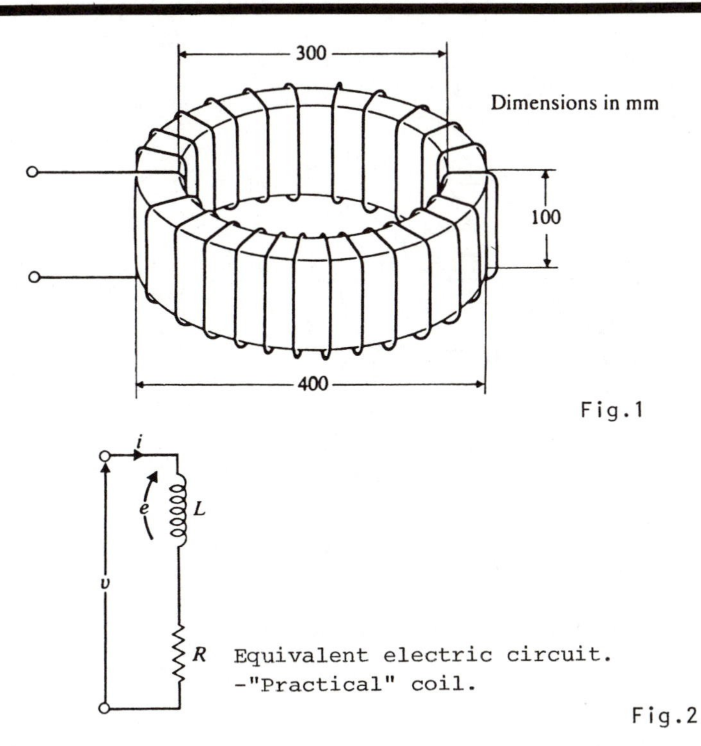

Fig.1

Equivalent electric circuit. -"Practical" coil.

Fig.2

Solution:

a) At the mean diameter:

$$H = \frac{Ni}{2\pi r} = \frac{200 \times 50}{0.35\pi} = 90.95 \text{ A/m}$$

$$B = \mu_0 H = 4\pi \times 10^{-7} \times 9095 = 11.43 \times 10^{-3} \text{ Wb/m}^2$$

b) Assume $B_{av} = 11.43 \times 10^{-3}$ Wb/m^2

$$\Phi = BA = 11.43 \times 10^{-3} \times 0.1 \times 0.05$$

$$= 57.15 \times 10^{-6} \text{ Wb}$$

$$\lambda = N\Phi = 200 \times 57.15 \times 10^{-6}$$

$$= 11.43 \times 10^{-3} \text{ Wb}$$

$$L = \frac{\lambda}{i} = \frac{11.43 \times 10^{-3}}{50} = 0.2286 \times 10^{-3} \text{ H}$$

Alternatively,

$$R = \frac{\ell}{\mu_0 A} = \frac{0.35\pi}{4\pi \times 10^{-7} \times 0.1 \times 0.05} = 175.0 \times 10^6 \text{ A/Wb}$$

$$L = \frac{N^2}{R} = \frac{200^2}{175.0 \times 10^6} = 0.2286 \times 10^{-3} \text{ H.}$$

c) At radius r, where $0.15 < r < 0.20$ m,

$$B = \frac{\mu_0 Ni}{2\pi r} = \frac{4\pi \times 10^{-7} \times 200i}{2\pi r} \text{ T.}$$

$$\Phi = \int_{0.15}^{0.20} B \times 0.1 \, dr \quad \text{Wb}$$

$$\lambda = N\Phi = \frac{0.1\mu_0 N^2 i}{2\pi} \int_{0.15}^{0.20} \frac{dr}{r} \text{ Wb}$$

$$L = \frac{\lambda}{i} = \frac{0.1 \times 4\pi \times 10^{-7} \times 200^2}{2\pi} \ln\left(\frac{0.2}{0.15}\right) = 0.2301 \times 10^{-3} \text{ H}$$

$$\text{Error} = \frac{0.2301 - 0.2286}{0.2301} \times 100\% = 0.651\%$$

d) $$R = \frac{\rho\ell_{wire}}{A_{wire}} = \frac{17.2 \times 10^{-9} \times 200 \times 0.3}{\pi \times 3^2 \times 10^{-6}/4} = 0.1460 \ \Omega$$

The parameters of an approximate equivalent circuit are therefore

$R = 0.1460 \ \Omega$; $L = 0.2286$ mH.

• PROBLEM 1-25

The figure shows a tightly wound toroidal coil with N turns and current I_1 and a square coil with current I_2. Calculate the mutual inductance of the coils.

Solution: Coil (2) intercepts all of the flux of coil (1) and thus:

$$L_{12} = \Phi_{21}/I_1$$

$$= \frac{1}{I_1} \iint_{S_2} \bar{B}_1 \cdot \hat{n}\, ds = \frac{1}{I_1} \iint_{S_1} \bar{B}_1 \cdot \hat{n}\, ds$$

From Ampere's law, B for a toroid is found to be

$$B = \frac{\mu_0 NI}{2\pi r} \qquad (r < b)$$

$$= 0 \qquad (r > b)$$

Therefore,

$$\frac{1}{I_1} \iint_{S_1} \bar{B}_1 \cdot \hat{n}\, ds = \frac{\mu_0 NI_1}{I_1 2\pi} \int_a^b \frac{Nd}{r} dr$$

$$= \frac{\mu_0 N^2 d}{2\pi} \ln\left(\frac{b}{a}\right)$$

Thus,

$$L_{12} = \frac{\mu_0 N^2 d}{2\pi} \ln\left(\frac{b}{a}\right)$$

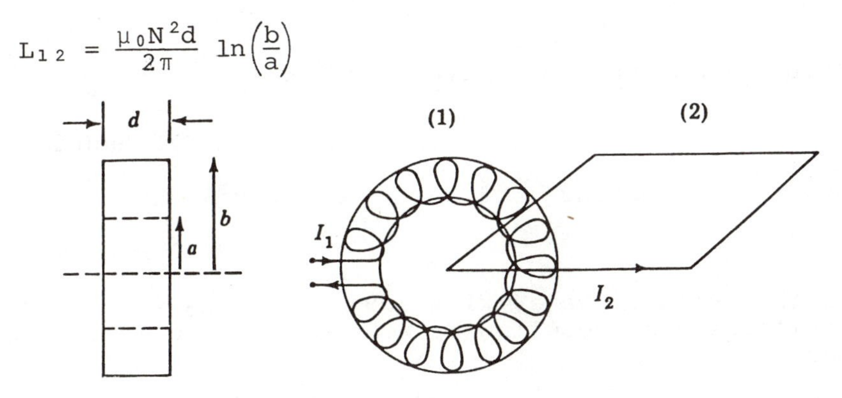

A toroid and a square loop.

which is the same as the self inductance of coil (1). In this case, Φ_{21}/I_1 is much easier to evaluate than Φ_{12}/I_2. Note that any coil encircling the toroid as shown in the figure would have the same mutual inductance.

• PROBLEM 1-26

Four coils of 500 turns each are in series addition on the same magnetic circuit, as indicated in the figure shown. Each coil procuces 0.03 weber of flux while carrying 10 amp, and 85 percent of the flux of any coil will link each of the other coils. Calculate the self-inductance of each coil, the mutual inductance between any two coils, and the over-all inductance of the entire circuit.

Solution:

Self-induced flux = 0.03 weber per coil.

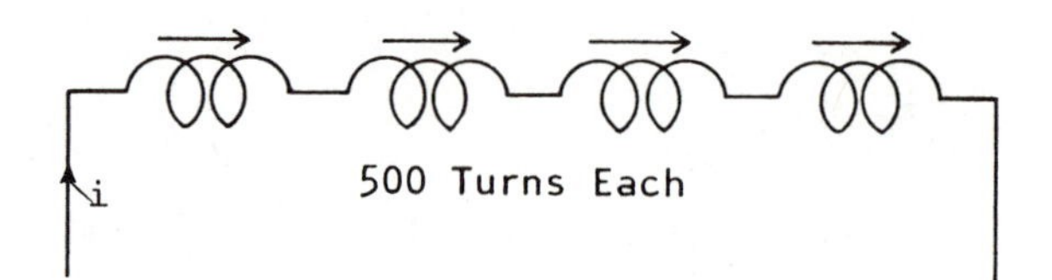

Calculations of components of inductance, and overall inductance, of coils in series.

Mutual flux per pair of coils = 0.85 x 0.03 = 0.0255 weber.

Leakage flux per coil = 0.15 x 0.03 = 0.0045 weber.

Total flux per coil = 0.03 + 3 x 0.85 x 0.03 = 0.1065 weber.

Flux common to all coils = 4 x 0.85 x 0.03 = 0.102 weber.

Self-inductance per coil is $L = \frac{n\phi}{i}$ henries

$$= \frac{500 \times 0.03}{10} = 1.5 \text{ henries.}$$

$$\text{Mutual inductance per pair of coils} = \frac{500 \times 0.85 \times 0.03}{10}$$

$$= 1.275 \text{ henries.}$$

Over-all inductance = 4 self + 12 mutual = 6 + 15.3

= 21.3 henries.

In this special case, $k_1 = 0.85 = k_2 = k_3 = k_4$, and the coefficient of coupling between any two coils is also 85 percent.

• PROBLEM 1-27

Find the apparent inductance of an iron-core choke consisting of laminations stacked ½-inch high and assembled as shown in fig. 1. The material used is silicon steel with about 4 percent silicon, for which the normal magnetization curve is given by fig. 2 and the incremental permeability curves are given by fig. 3. The stacking factor for the assembly as indicated is assumed to equal 0.94. Assume that the coil has 5,000 turns and is wound on the center leg of the core.

The coil carries 0.020 ampere direct current, and a voltage of 50 volts rms at 120 cycles per second is applied to the coil. There are no windings on the outer legs.

Solution: First determine the direct magnetizing force H_{dc} in the steel. The alternating maximum flux density corresponding to the alternating component of voltage is then computed. From these data the apparent incremental permeability of the steel is determined from Fig. 3. The apparent inductance is then computed on the assumption that the magnetic circuit is linear within the range of the superposed alternating magnetomotive force.

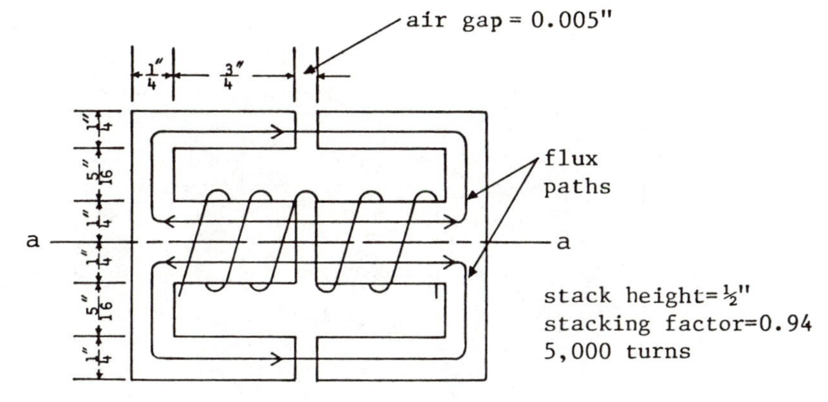

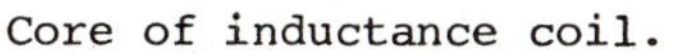
Core of inductance coil.

Fig.1

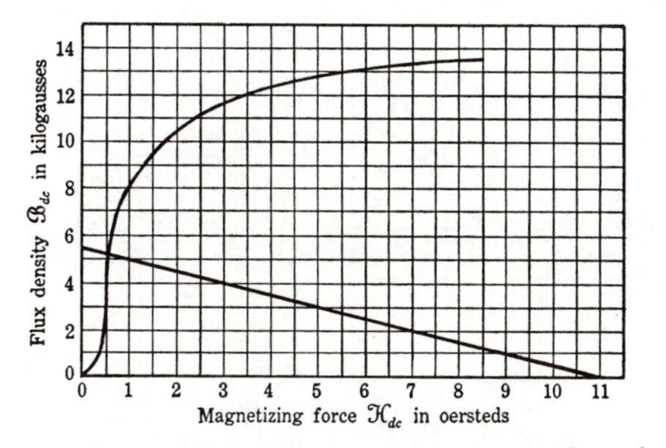

Normal d-c magnetization curve for the steel of Fig.3, and graphical construction for the determination of $\mathcal{H}_{dc}$.

Fig.2

To determine the value of H_{dc}, the total direct magnetizing force must be divided between the air gaps and the steel for the condition that the flux across the air gap equals the flux in the iron. For this calculation the core may be divided into two parts in parallel magnetically, as shown by the plane aa perpendicular to the paper in Fig. 1. The magnetic dimensions for one half can be computed and dealt with alone, except that the total flux linking the coil is then twice that in either half of the core.

The direct magnetizing force H_{dc} is determined graphically by finding the point of intersection of the magnetization curve of the steel shown in Fig. 2, and the negative air-gap line. The parameters required to determine the location of the negative air-gap line on Fig. 2 are obtained as follows:

The direct magnetomotive force F_{dc} is

$$F_{dc} = 5{,}000 \times 0.020$$

$$= 100 \text{ amp-turns or } 126 \text{ gilberts.}$$

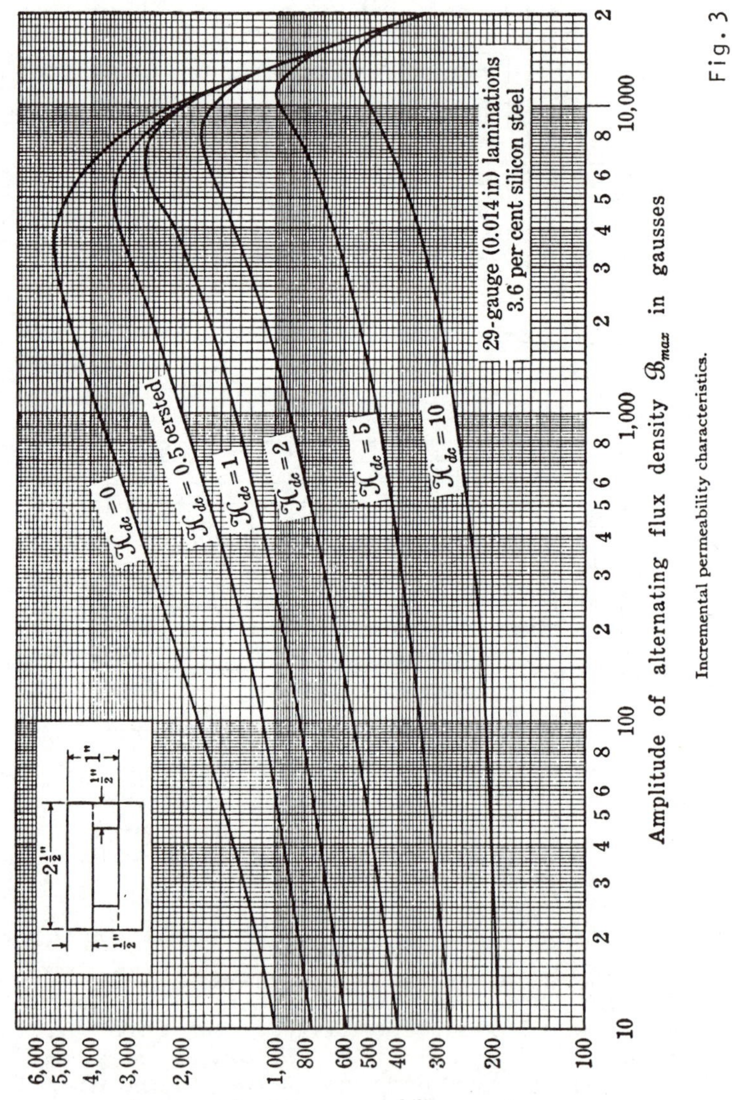

Incremental permeability characteristics.

Fig. 3

To compute the length ℓ_s of the flux path in the steel core, assume that the flux rounds the corners in the steel on a mean radius of 1/8 inch. Then

$$\ell_s = 4\left(\frac{3}{4}\right) + 2\left(\frac{5}{16}\right) + \frac{\pi}{4}$$

$$= 4.4 \text{ in., or } 11.2 \text{ cm.}$$

The cross-sectional area A_s of the steel is

$$A_s = (0.94 \times 0.25 \times 0.50)$$

$$= 0.117 \text{ sq. in., or } 0.76 \text{ sq. cm.}$$

The total length ℓ_a of the two air gaps in series with the flux path is

$$\ell_a = 0.010 \text{ in., or } 0.0254 \text{ cm.} \qquad (1)$$

The equivalent cross-sectional area A_a of the gap is determined by the effect of fringing. To allow for fringing, each gap may be assumed to have effective cross-sectional dimensions greater than the steel dimensions by the gap length. Thus the equivalent area of the outer-leg gap is

$$A_a = (0.25 + 0.005)(0.5 + 0.005)$$

$$= 0.129 \text{ sq. in., or } 0.83 \text{ sq. cm.} \qquad (2)$$

The equivalent area of the center-leg gap is

$$(0.50 + 0.005)(0.50 + 0.005) = 0.255 \text{ sq. in.,}$$

and therefore the equivalent area on each side of the central plane aa, Fig. 1 is

$$\frac{0.255}{2} = 0.1275 \text{ sq. in., or } 0.82 \text{ sq. cm.}$$

Although this equivalent area is slightly less than the equivalent area of the outer-leg gap, subsequent calculations are based on the assumption that the combined effect of both gaps is the same as the effect of a single gap whose area A_a is 0.83 sq. cm., Eq. (1) and whose length ℓ_a is 0.0254 cm, Eq. (2).

The intersection of the negative air-gap line with the H-axis is then

$$\frac{F_{dc}}{\ell_s} = \frac{126}{11.2} = 11.2 \text{ oersteds,}$$

and its intersection with the B axis is

$$\frac{\mu_0 A_a F_{dc}}{A_s \ell_a} = \frac{1.00 \times 0.83 \times 126}{0.76 \times 0.0254} = 5{,}420 \text{ gausses,}$$

where μ_0 is the permeability of free space.

The co-ordinates of the point of intersection of the magnetization curve of the steel and the negative air-gap line give the values of direct magnetizing force

H_{dc} and the flux density in the steel, in the absence of any alternating magnetizing force. From Fig. 2, the value of H_{dc} is obtained as 0.55 oersted approximately.

Before the incremental permeability μ_{ac} can be obtained, the maximum value B_{max} of the alternating component of flux density in the steel must be calculated. To determine B_{max}, the maximum value ϕ_{max} of the alternating component of flux required to generate 50 v rms at 120 ~ must be computed.

$$\phi_{max} = \frac{E}{4.44fN} = \frac{50}{4.44 \times 120 \times 5{,}000}$$

$$= 1.88 \times 10^{-5} \text{ weber or } 1{,}880 \text{ maxwells.}$$

For a stacking factor of 0.94, the net area of steel in the center leg, which carries the total flux, is

$$0.94 \times 0.500 \times 0.500 = 0.235 \text{ sq in.} = 1.52 \text{ sq cm.}$$

Hence the alternating component of flux density in the steel is

$$B_{max} = \frac{1{,}880}{1.52} = 1{,}240 \text{ gausses.}$$

From Fig. 3, the incremental relative permeability μ_{ac}/μ_0 corresponding to H_{dc} of 0.55 oersted and B_{max} of 1,240 gausses is approximately 2,000. Therefore, in mks units,

$$\mu_{ac} = 2{,}000 \times 10^{-7}.$$

To compute the apparent inductance, the incremental flux linkages per ampere in the coil are determined, the reluctance of the magnetic circuit being calculated first. Since the plane aa, Fig. 1, divides the magnetic circuit into two identical parallel paths, the reluctance is one-half the value for either of these paths. If all quantities are expressed in mks units, the desired inductance is given in henries.

The reluctance R_s of the steel part of the path in each half of the reactor is

$$R_s = \frac{\ell_s}{\mu_{ac} A_s},$$

and on substitution of the values of ℓ_s and A_s converted to mks units,

$$R_s = \frac{0.112}{2{,}000 \times 10^{-7} \times 0.76 \times 10^{-4}}$$

$$= 0.74 \times 10^7 \text{ mks units, or pragilberts/weber.}$$

The air gaps in each half of the magnetic circuit are considered to be equivalent to a single gap whose length ℓ_a is given by Eq. (1) and whose area A_a is given by Eq. (2). Therefore the reluctance R_a of the air gaps in each half of the magnetic circuit is

$$R_a = \frac{\ell a}{\mu_0 A_a} = \frac{0.0254 \times 10^{-2}}{10^{-7} \times 0.83 \times 10^{-4}}$$

$$= 3.06 \times 10^7 \text{ pragilberts/weber.}$$

The effective reluctance of each half of the magnetic circuit is the sum of R_s and R_a, and the combined reluctance R of the two halves in parallel is

$$\therefore R = \frac{0.74 + 3.06}{2} \times 10^7$$

$$= 1.90 \times 10^7 \text{ pragilberts/weber.}$$

The apparent inductance L_a in henries is the flux linkages produced per ampere in the coil. One ampere in the coil produces a magnetomotive force of $4\pi(5{,}000)$ pragilberts, resulting in a flux Φ of

$$\Phi = \frac{4\pi \times 5{,}000}{1.90 \times 10^7} = 0.00330 \text{ weber.}$$

This flux links 5,000 turns; whence the inductance is

$$L_a = \frac{N\Phi}{I}$$

$$= 5{,}000 \times 0.00330 = 16.5 \text{ h.}$$

• PROBLEM 1-28

The rotary electromagnet in the figure shown has windings of $N_1 = 400$ turns and $N_2 = 200$ turns on the stator and rotor, respectively. The resistance of the windings is negligible. Tests were made at a frequency of 60 Hz with the rotor stationary and with its magnetic axis displaced from that of the stator at an angle of 25°. The test data follow:

Excitation applied to the stator (400-turn) winding, rotor open-circuited:

Stator	Rotor

Volts	Amperes	Volts
80.0	0.93	26.0

Excitation applied to the rotor (200-turn) winding, stator open-circuited:

Rotor		Stator
Volts	Amperes	Volts
40.0	2.50	70.0

Calculate (a) the self-inductance L_{ss} of the stator winding, (b) the self-inductance L_{rr} of the rotor winding, (c) the mutual inductance M between the stator and rotor windings, and (d) the coefficients k_1 and k_2 for the stator and rotor windings and the coefficient of coupling k.

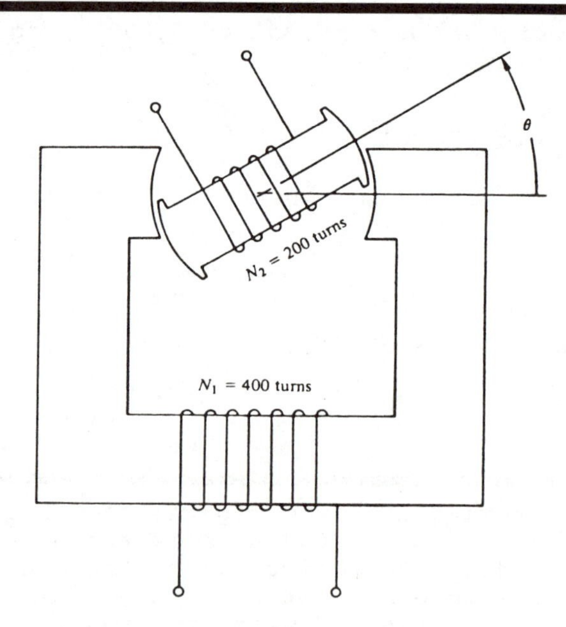

Rotary electromagnet with two inductively coupled windings.

Solution:

a. The self-inductance of the stator may be calculated from the following equation:

$$L_{ss} = \frac{\lambda_{ss}}{\ell_s}$$

The maximum value of λ_{ss} can be determined from

$$E = \frac{E_m}{\sqrt{2}} = \frac{N\omega AB_m}{\sqrt{2}} = 4.44f \quad NA\ B_m,$$

in which the maximum instantaneous flux is $\Phi_m = AB_m$ and the maximum instantaneous flux linkage $N\Phi_m$, so that

$$\lambda_{ss_m} = \frac{E}{4.44f} = \frac{80}{4.44 \times 60} = 0.300 \text{ weber turn.}$$

The maximum instantaneous current is

$$\sqrt{2}\, I_s = \sqrt{2} \times 0.93 = 1.32 \text{ A.}$$

Hence,

$$L_{ss} = 0.300 \div 1.32 = 0.227 \text{ H}$$

b. The self-inductance of the rotor is

$$L_{rr} = \frac{40}{\sqrt{2} \times 2.5 \times 4.44 \times 60} = 0.0425 \text{ H}$$

c. The mutual inductance on the basis of current in the stator winding with the rotor open is

$$M = \frac{26.0}{\sqrt{2} \times 0.93 \times 4.44 \times 60} = 0.0742 \text{ H.}$$

and on the basis of rotor current with the stator open

$$M = \frac{70.0}{\sqrt{2} \times 2.50 \times 4.44 \times 60} = 0.0742 \text{ H.}$$

The two values of M check.

d.

$$k_1 = \frac{\phi_{21}}{\phi_{11}}$$

$$\Phi_{11} \equiv \frac{\lambda_{11}}{N_1}, \quad \Phi_{21} \equiv \frac{\lambda_{21}}{N_2}, \text{ hence,}$$

$$k_1 = \frac{N_1}{N_2}\frac{\lambda_{21}}{\lambda_{11}}$$

$$= \frac{N_1}{N_2}\frac{E_2}{E_1}$$

$$= \frac{400}{200} \times \frac{26}{80} = 0.65.$$

$$k_2 = \frac{N_2}{N_1}\frac{\lambda_{12}}{\lambda_{22}}$$

$$= \frac{200}{400} \times \frac{70}{40} = 0.875.$$

and

$$k = \sqrt{k_1 k_2} = \sqrt{0.65 \times 0.875} = 0.754$$

As a check,

$$M = k\sqrt{L_{ss} L_{rr}}$$

$$= 0.754\sqrt{0.227 \times 0.0425} = 0.0741 \text{ H}.$$

RELUCTANCE

• PROBLEM 1-29

The figure shows a toroidal core consisting of two different magnetic materials. Find the reluctance of this circuit.

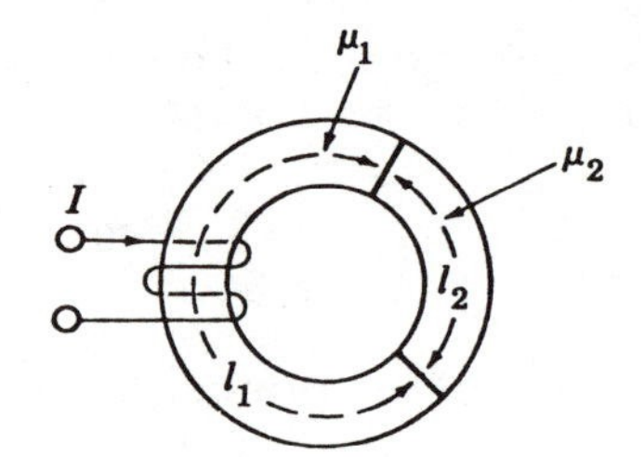

An inhomogeneous magnetic circuit.

Solution: Here use the continuity of normal $\bar{B}$, that is, $\bar{B}$ and the flux are continuous at the boundary between the two materials, but $\bar{H}$ has a discontinuity.

$$NI = H_1\ell_1 + H_2\ell_2 = \frac{B\ell_1}{\mu_1} + \frac{B\ell_2}{\mu_2}$$

$$R = \frac{NI}{\Phi_S} = \frac{\ell_1}{\mu_1 A} + \frac{\ell_2}{\mu_2 A} = R_1 + R_2$$

where $R_1 = \frac{\ell_1}{\mu_1 A}$ and $R_2 = \frac{\ell_2}{\mu_2 A}$

• PROBLEM 1-30

The electromagnet of the figure has an effective closed magnetic path length ℓ, a gap width of g, pole faces of area S, and is wound with n turns. Find the total reluctance and the inductance of the electromagnet.

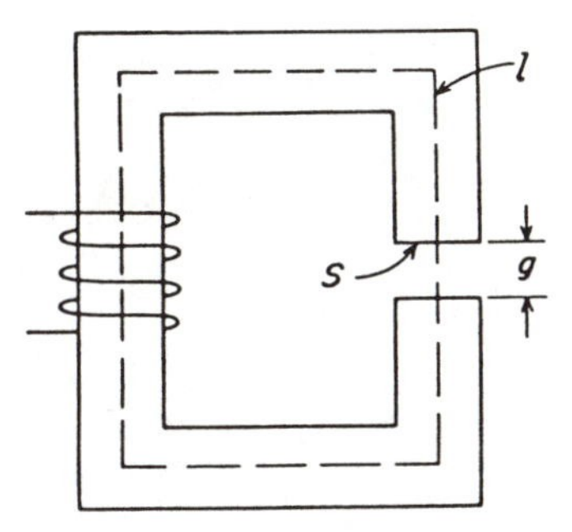

Solution: The reluctance of the core is by definition,

$$R_C = \frac{\ell - g}{K_M \mu_0 S}$$

and of the gap is

$$R_G = \frac{g}{\mu_0 S}$$

so that the total reluctance is

$$R = \frac{\ell - g}{K_M \mu_0 S} + \frac{g}{\mu_0 S}$$

Assume that the gap width is small compared to the core dimensions, so that the lines of force do not spread out. In addition, assume that the permeability of iron is high, so that $K_M >> 1$, then

$$R = \frac{\ell / K_M + g}{\mu_0 S} = \frac{g}{\mu_0 S}$$

The flux is then

$$\Phi = \frac{nI}{R} = \frac{nI\mu_0 A}{g}$$

and the inductance of the magnet is

$$L = \frac{d\Phi}{dI} = \frac{n\mu_0 A}{g}$$

• PROBLEM 1-31

A toroidal iron core of square cross-section, with a 2 mm air gap and wound with 100 turns, has the dimensions shown. Assume the iron has the constant $\mu = 1000\mu_0$. Find (a) the reluctances of the iron path and the air gap and (b) the total flux in the circuit if I = 100 mA.

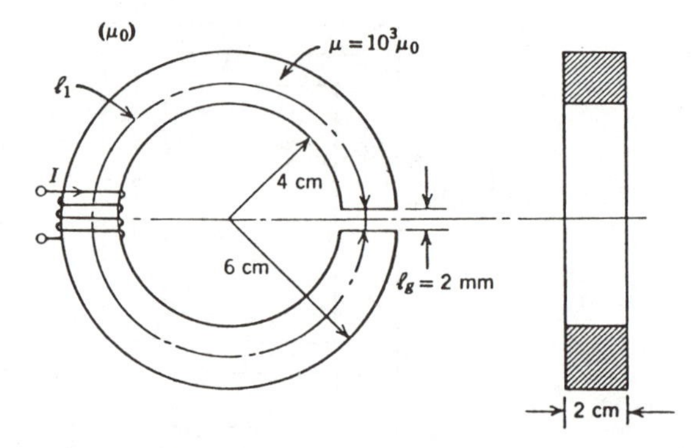

Solution: (a) The reluctance of the iron path, having a median length $\ell_1 \cong 2\pi(0.05) = 0.314$ m and cross-sectional area $A_1 = 4 \times 10^{-4}$ m^2, is

$$R_1 \cong \frac{\ell_1}{\mu_1 A_1} = \frac{0.314 - 0.002}{10^3(4\pi \times 10^{-7})4 \times 10^{-4}} = 0.621 \times 10^6 H^{-1}$$

The air gap reluctance, assuming no fringing, becomes

$$R_g = \frac{\ell_g}{\mu_0 A_1} = \frac{0.002}{4\pi \times 10^{-7}(4 \times 10^{-4})} = 3.98 \times 10^6 H^{-1}$$

(b) The magnetic flux is given by

$$\Phi_m = \frac{nI}{R}$$

i.e., the magnetomotive force nI of the coil divided by the reluctance of the series circuit

$$\Phi_m = \frac{nI}{R_1 + R_g} = \frac{10^2(0.1)}{4.6 \times 10^6} = 2.18 \times 10^{-6} \text{ Wb.}$$

With the air gap absent, Φ_m is limited only by the reluctance R_1 of the iron path, becoming $\Phi_m = 15.97 \times 10^{-6}$ Wb.

HYSTERESIS

• PROBLEM 1-32

A B-H loop for a type of electric steel sheet is shown in the figure. Determine approximately the hysteresis loss per cycle in a torus of 300 mm mean diameter and a square cross section of 50 x 50 mm.

Solution: The area of each square in the figure represents

(0.1 tesla) x (25 amperes/meter)

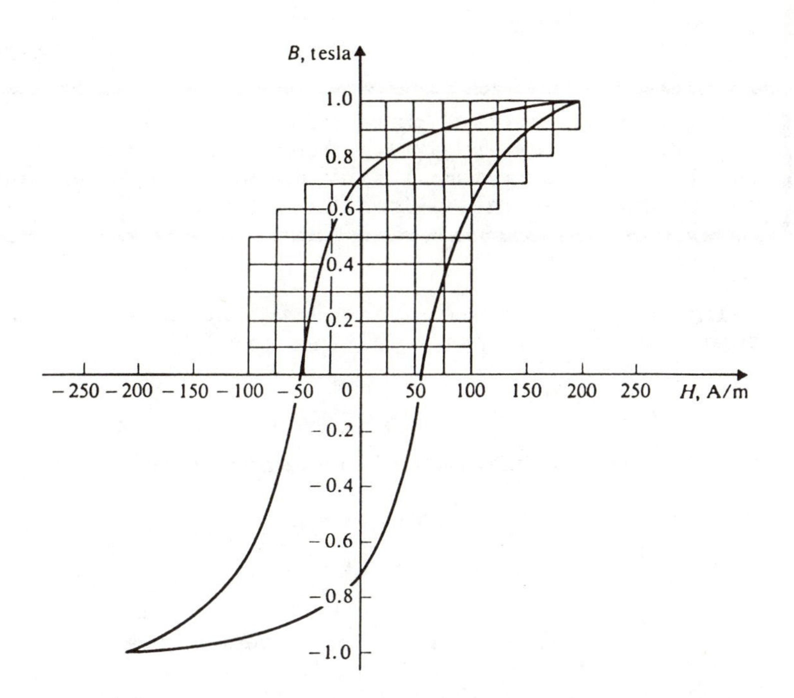

B-H loop for M-36 electric steel.

$$= 2.5 \frac{\text{weber}}{\text{meter}^2} \times \frac{\text{ampere}}{\text{meter}}$$

$$= 2.5 \frac{\text{volt x second x ampere}}{\text{meter}^3} = 2.5 \frac{\text{joule}}{\text{meter}^3}$$

If a square that is more than half within the loop is regarded as totally enclosed, and one that is more than half outside the loop is disregarded, then the area of the loop is

$$2 \times 43 \times 2.5 = 215 \text{ J/m}^3$$

The volume of the torus is

$$0.05^2 \times 0.3\pi = 2.36 \times 10^{-3} \text{ m}^3.$$

Energy loss in the torus per cycle is thus

$$2.36 \times 10^{-3} \times 215 = 0.507 \text{ J.}$$

• PROBLEM 1-33

Using Steinmetz law, determine the ergs loss per cycle in a core of sheet iron having a net volume of 40 cu cm, in which the maximum flux density is 8,000 gauss. The value of η for sheet iron is 0.004.

Solution: Using Steinmetz law, the hysteresis loss per cubic cm in ergs per cycle is given by

$$w_h = \eta B^{1.6}$$

where B is the maximum flux density in gauss

$$\eta = 0.004 \text{ for sheet iron}$$

$$\therefore w_h = 0.004 \times (8000)^{1.6}$$

$$= 7{,}028 \text{ ergs per cu.cm.per cycle}$$

Total loss W = 7,028 x 40 = 281,000 ergs per cycle, or 281,000 x 10^{-7} = 0.0281 joules per cycle.

CHAPTER 2

MAGNETIC CIRCUITS

DETERMINATION OF AMPERE-TURNS

• PROBLEM 2-1

A cast-iron core of the shape indicated in figure 1 has the following pertinent dimensions: $A_1 = 2\ in^2$, $A_2 = 3\ in^2$, $\ell_1 = 10$ in., and $\ell_2 = 25$ in. Find the mmf required to produce a flux of 90 KL in this core.

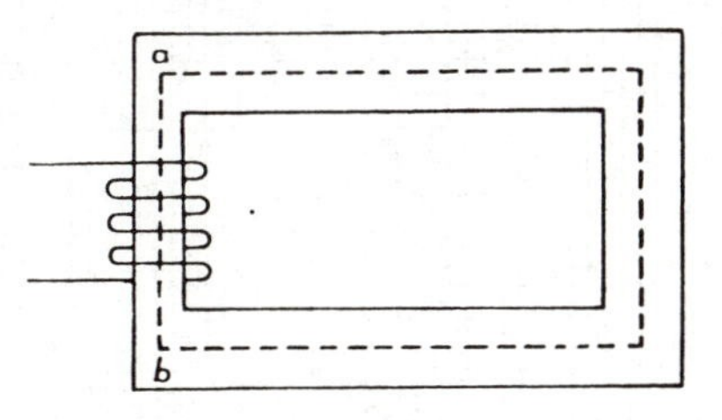

Magnetic series circuit. Fig. 1

Solution: The flux densities are, from the equation

$$B = \Phi/A,$$

$B_1 = 90/2 = 45\ KL/in.^2$ and $B_2 = 90/3 = 30\ KL/in.^2$; the corresponding field intensities, taken from the curve of Fig. 2 for cast iron, are $H_1 = 100$ A-T/in. and $H_2 = 46$ A-T/in.; the mmf is given by the following equation:

$$NI = \oint H\, d\ell = H_1 \ell_1 + H_2 \ell_2;$$

thus,

$$NI = 100 \times 10 + 46 \times 25 = 2{,}150\ A\text{-}T.$$

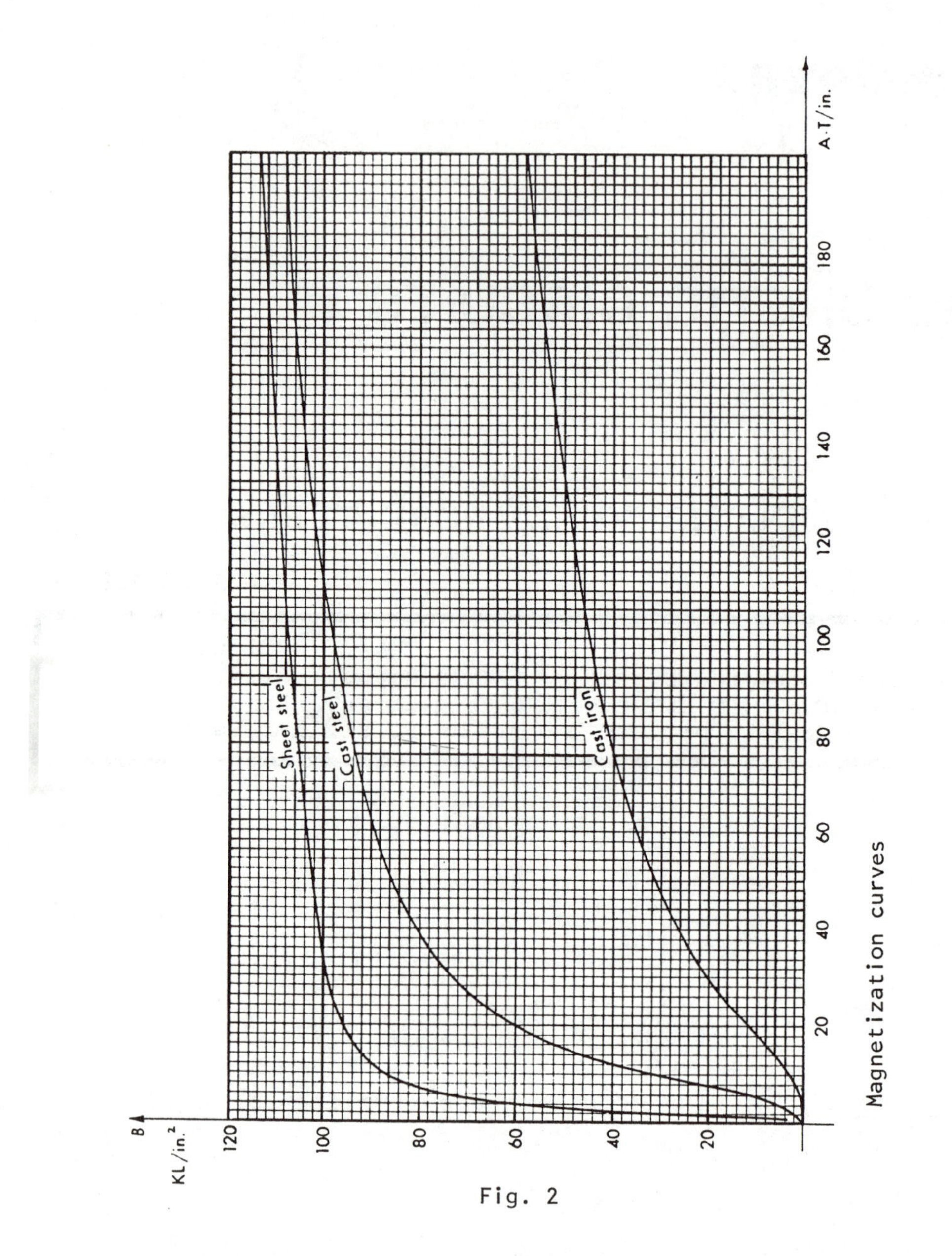

Fig. 2

• PROBLEM 2-2

> A ring has a circular cross-section of 8 sq. cm. (Fig. 1). The ring has a mean radius of 10 cm, and is composed half of cast steel and half of wrought iron. Determine the ampere-turns required to set up a total flux of 80,000 lines.

Solution: Flux density = 10,000 lines per square centimeter

Ampere-turns per centimeter for cast steel

$= 6.3$

Ampere-turns per centimeter for wrought iron

$= 12.2$

Total ampere-turns, cast steel

$= \pi \times 10 \times 6.3 = 198$

Total ampere-turns, wrought iron

$= \pi \times 10 \times 12.2 = \underline{383}$

Total ampere-turns 581.

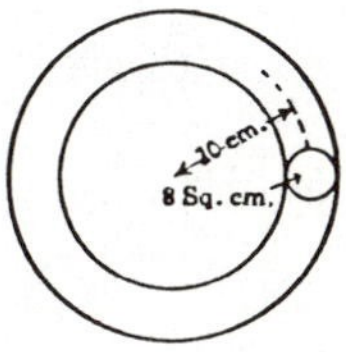

Fig. 1

The simplest type of magnetic circuit.

If the ring were cut with a hack saw, making an air gap 0.1 cm. long, additional ampere-turns would be required to maintain the same flux density. From the expression

$\ell H = 0.4\pi NI,$

when

$\mu = 1,\ B = H,$

we require for $B = 10{,}000$,

$$\frac{10,000}{0.4\pi} = 7958 \text{ ampere-turns per centimeter}$$

or 796 ampere-turns for an air gap of 0.1 cm.

Thus the total magnetic circuit would require 581 + 796 = 1377 ampere turns.

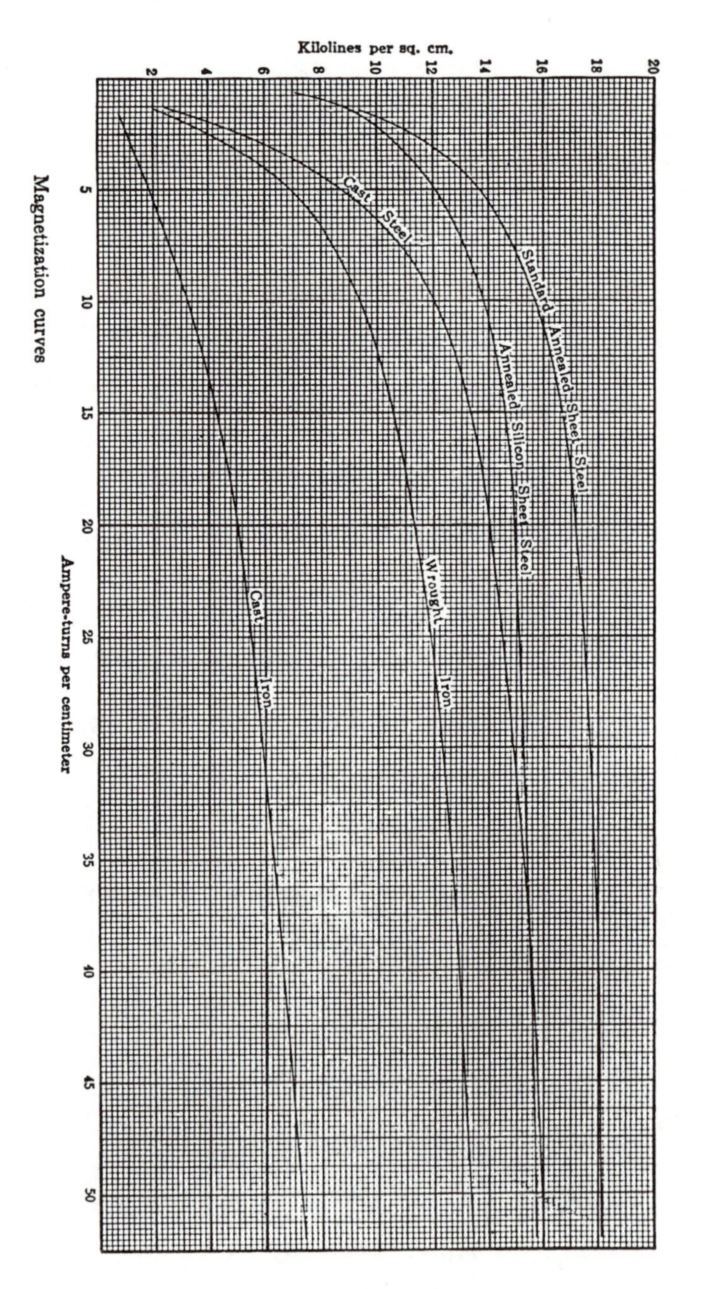

Fig. 2

The magnetic structure shown in Fig. 2 is made of 4.25 percent silicon transformer sheet-steel laminations 0.014 inch thick stacked into a pile 2 inches thick (dimension a). The stacking factor is 0.90 for this structure. Dimension b is 2.5 inches. The air-gap length δ is 0.10 inch. The mean length of the steel part of the magnetic circuit is 30 inches. Find the magnetomotive force necessary to establish a total flux of 250 kilolines.

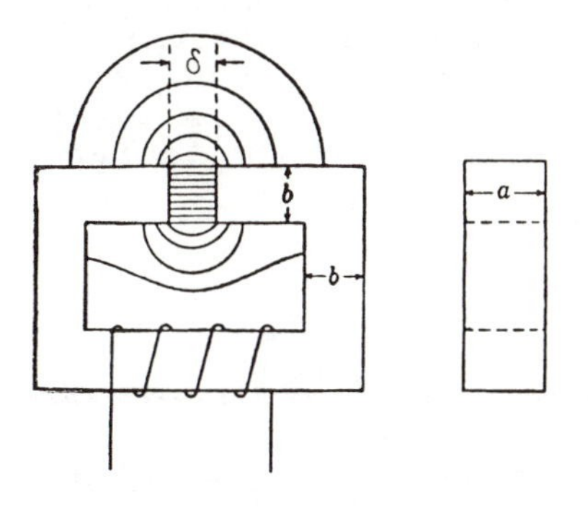

Fig. 2

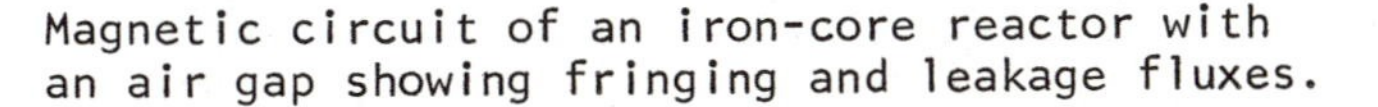
Magnetic circuit of an iron-core reactor with an air gap showing fringing and leakage fluxes.

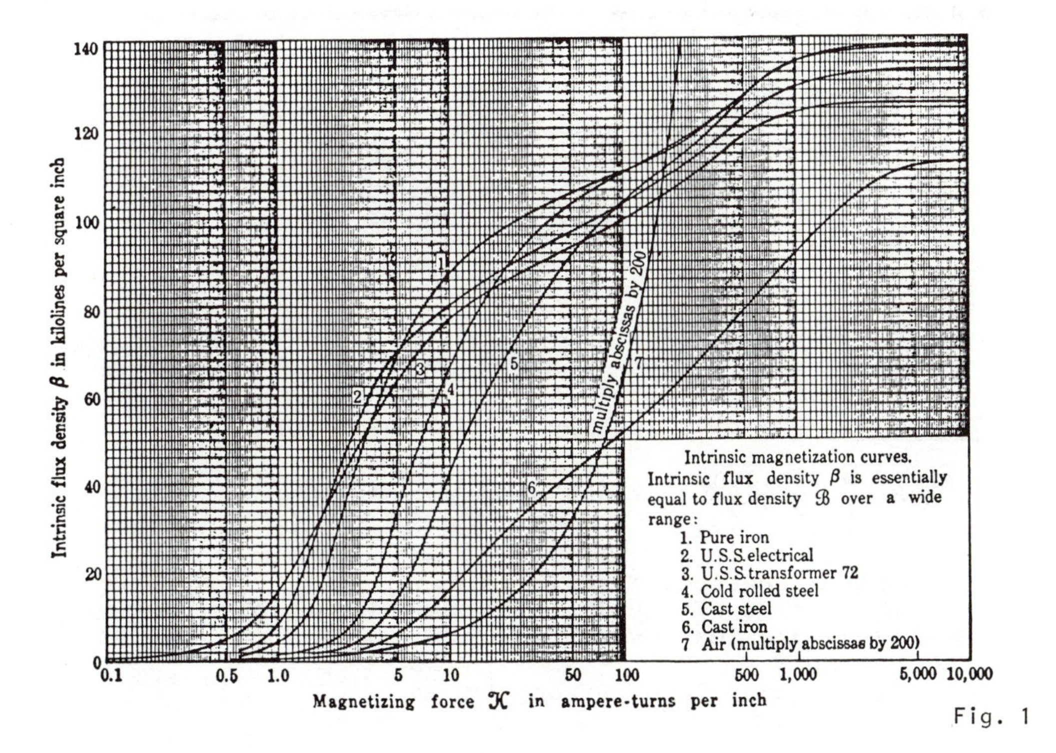

Fig. 1

Solution: For a short air-gap, the equivalent cross-sectional area of the gap is $A = (a + \delta)\ (b + \delta) =$

(2 + 0.1) (2.5 + 0.1), or 5.45 sq. in. The average flux density in the gap therefore equals 250/5.45, or 45.9 kilolines/sq in. From the curve for air in Fig. 1, the magnetizing force for the gap with this average flux density is 72 x 200, or 14,400 amp-turns/in. Since the length of the gap is 0.1 in., the ampere-turns required for the gap are 1,440 approximately.

The net cross-sectional area of the steel is 5 x 0.9, or 4.5 sq. in. The flux density in the steel is then 250/4.5, or 55.5 kilolines/sq in. If curve 3, Fig. 1, is taken to represent the properties of this steel the magnetizing force in the laminations is about 3.7 amp-turns/in. The calculated magnetic potential drop required for the laminations is therefore 3.7 x 30 or 111 amp-turns. The magnetomotive force required for the complete magnetic structure is therefore 1,550 amp-turns.

• PROBLEM 2-4

Determine the ampere-turns necessary to produce an air-gap flux of 750,000 maxwells in the electromagnet of Fig. 1. The cores are cast iron, and the yoke and pole pieces are cast steel. Neglect fringing and leakage.

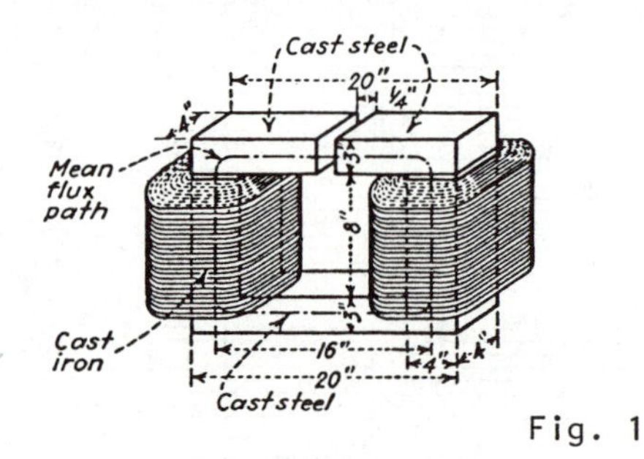

Fig. 1

Electromagnet

Solution: The flux density in the lower yoke is

$$B_1 = \frac{750,000}{3 \times 4} = 62,500 \text{ maxwells per sq in.}$$

The ampere-turns per inch for a density of 62,500 from Fig. 2 (cast steel) is 18. The mean length of flux-path is (approximately) 16 in.

$$I_1N_1 = 16 \times 18 = 288 \text{ amp-turns,}$$

or 288 amp-turns is required to produce a flux of 750,000 maxwells in the lower yoke. The density in the cores is

$$B_2 = \frac{750,000}{4 \times 4} = 46,900 \text{ maxwells per sq in.}$$

From the curve (cast iron) the ampere-turns per inch = 86.

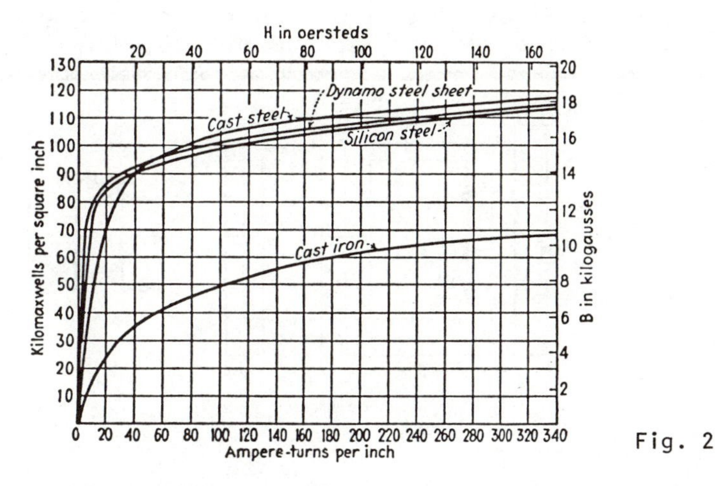

Fig. 2

Typical magnetization curves for iron and steel.

As there are two cores, the total length is 16 in., neglecting the quarter turn at each corner of the magnetic circuit.

$$I_2N_2 = 16 \times 86 = 1,376 \text{ amp-turns.}$$

The pole pieces are in every way identical with the yoke, except that the path is 0.25 in. shorter. This small difference if neglected will not introduce an appreciable error, and so for the two pole pieces

$$I_3N_3 = 288 \text{ amp-turns.}$$

For the air-gap, $I_4N_4 = 0.313\ B_1\ell_1$.

$$I_4N_4 = 0.313 \times 62,500 \times 0.25 = 4,890 \text{ amp-turns.}$$

As all the various parts are in series,

The total ampere-turns = 288 + 1,376 + 288 + 4,890

= 6,842.

Determine the ampere-turns necessary to produce 18 x 10^4 lines of flux in the air gap of the magnetic circuit shown in Fig. 1 (the core of silicon steel). To take care of the fringing in this small gap, the effective area is obtained by adding the length of the gap to the dimensions of the section of the steel path.

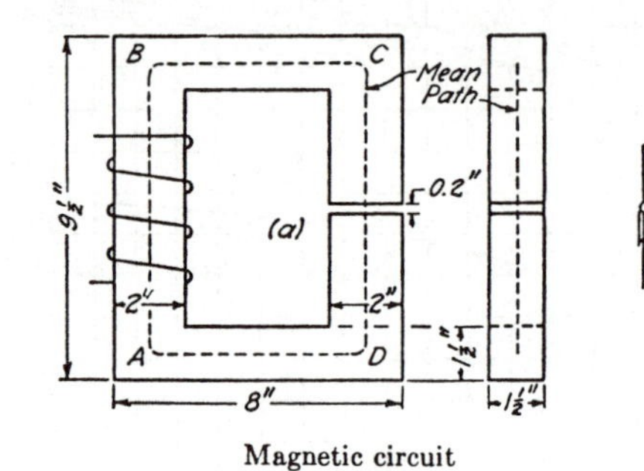

Magnetic circuit

Fig. 1

Solution:

$$B_{AB} = B_{CD} = \frac{\Phi}{A} = \frac{18 \times 10^4}{2 \times 1.5} = 60{,}000 \text{ lines per square inch}$$

$$\ell_{AB} + \ell_{CD} = 8 + 7.8 = 15.8 \text{ in.}$$

From Fig. 2, 1.5 ampere-turns per inch are required and

$$NI = 1.5 \times 15.8 = 23.7 \text{ ampere-turns.}$$

$$B_{BC} = B_{AD} = \frac{18 \times 10^4}{1.5 \times 1.5} = 80{,}000 \text{ lines per square inch}$$

$$\ell_{BC} + \ell_{AD} = 6 \text{ in.} + 6 \text{ in.} = 12 \text{ in.}$$

From Fig. 2, 3.5 ampere-turns per inch are required and

$$NI = 3.5 \times 12 = 42 \text{ ampere-turns}$$

$$\text{Air-gap } NI = 0.313B\ell = 0.313 \times \frac{18 \times 10^4}{1.7 \times 2.2} \times 0.2$$

$$= 3013 \text{ ampere-turns.}$$

$$\text{Total } NI = 24 + 42 + 3013 = 3079 \text{ ampere-turns.}$$

Note that 98 percent of the total ampere-turns are needed for the air gap alone.

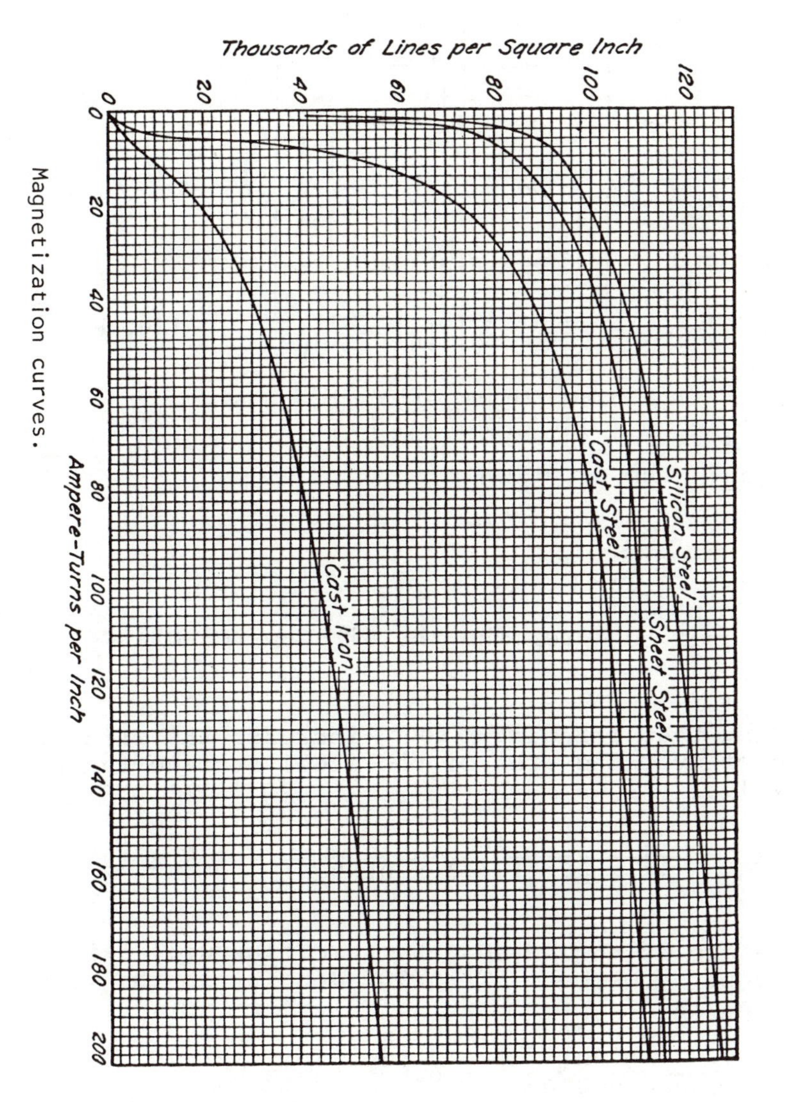

Magnetization curves.

Fig. 2

• PROBLEM 2-6

Using Fig. 1, determine the number of ampere-turns required to establish a flux of 0.001 Wb in the air gap. Assume I_2 and I_3 are zero. The air-gap length ℓ_g, is 0.1 mm; the magnetic member is constructed of laminated M-19 steel with a stacking factor of 0.9 and of length ℓ_m, equal to 100 mm.

Solve this problem using reluctances, neglecting leakage but not fringing effects. The cross-sectional area of the magnetic member is $A_m = 16\ cm^2$ (gross).

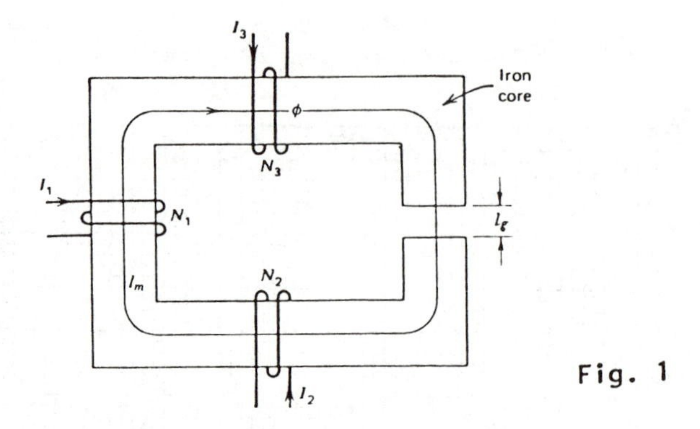

Fig. 1

A composite magnetic circuit, with multiple excitation (mmf).

Solution: The equation,

$$R_g = \frac{\ell_g}{\mu_0 A_g},$$

can be used to determine the air-gap reluctance.

Assuming that fringing effects increase the effective gap area over the area of the steel surface facing it by 10%,

$$R_g = \frac{10^{-4}}{(4\pi \times 10^{-7})(1.1 \times 0.0016)} = 4.5 \times 10^4 \text{ A/wb.}$$

If leakage is neglected, the same flux will exist in the magnetic member. The flux density in the magnetic material is

$$B_m = \frac{0.001}{0.9 \times 0.0016} = 0.695 \text{ tesla}$$

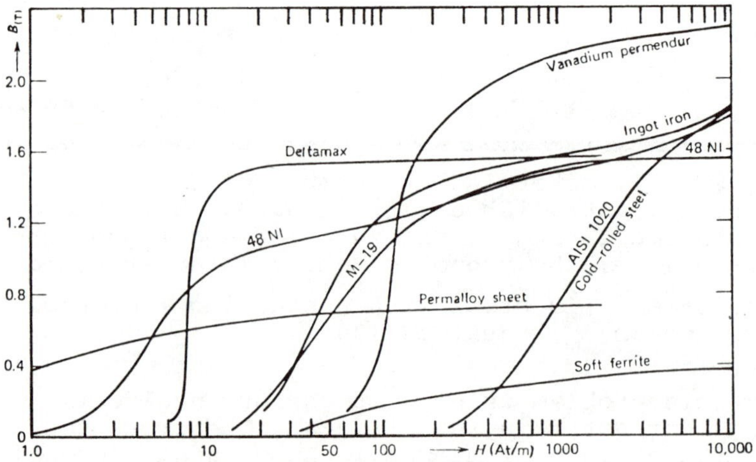

B-H curves of selected soft magnetic materials.

Fig. 2

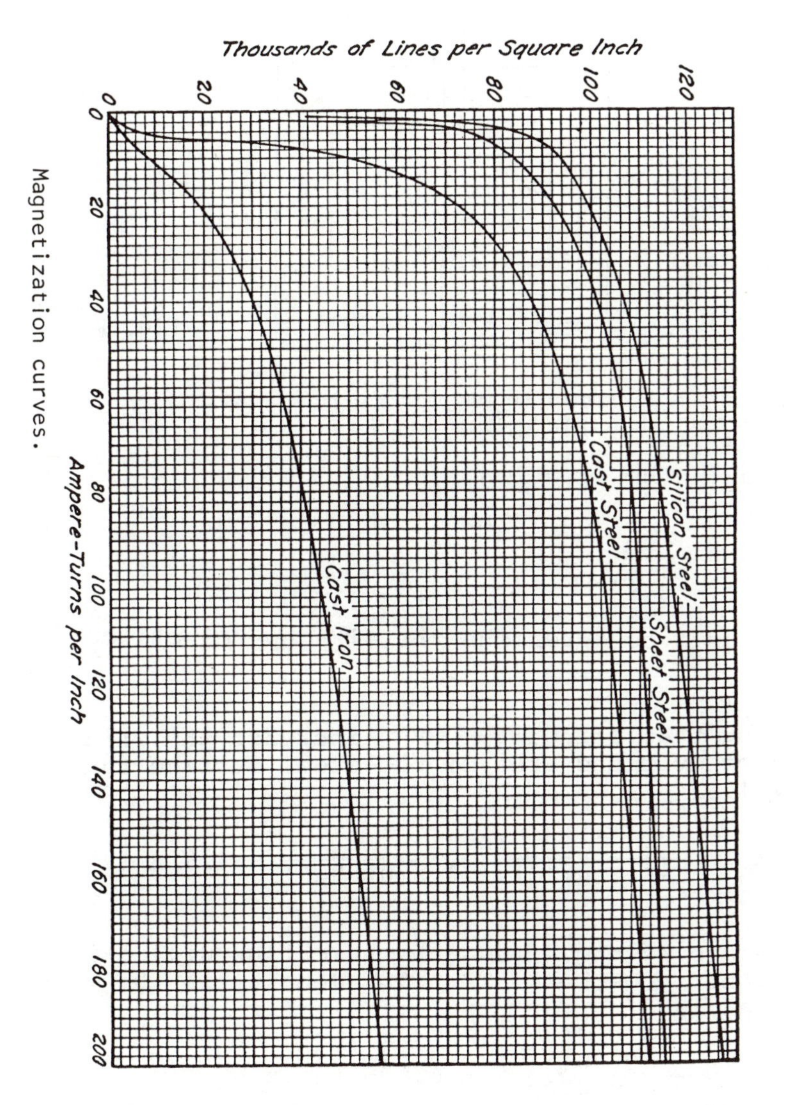

Magnetization curves.

Fig. 2

• PROBLEM 2-6

Using Fig. 1, determine the number of ampere-turns required to establish a flux of 0.001 Wb in the air gap. Assume I_2 and I_3 are zero. The air-gap length ℓ_g, is 0.1 mm; the magnetic member is constructed of laminated M-19 steel with a stacking factor of 0.9 and of length ℓ_m, equal to 100 mm.

Solve this problem using reluctances, neglecting leakage but not fringing effects. The cross-sectional area of the magnetic member is $A_m = 16\ \text{cm}^2$ (gross).

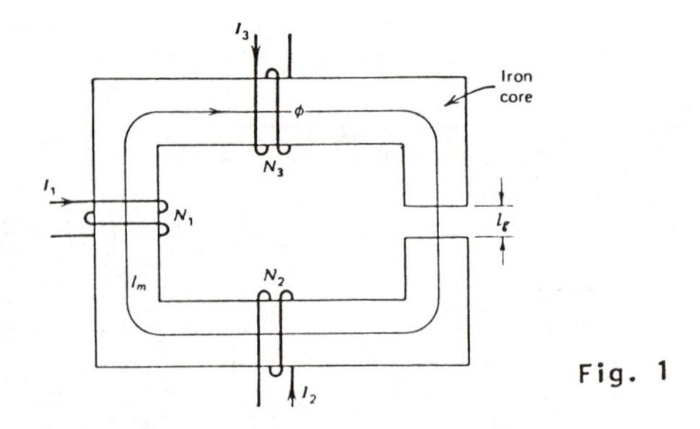

Fig. 1

A composite magnetic circuit, with multiple excitation (mmf).

Solution: The equation,

$$R_g = \frac{\ell_g}{\mu_0 A_g},$$

can be used to determine the air-gap reluctance.

Assuming that fringing effects increase the effective gap area over the area of the steel surface facing it by 10%,

$$R_g = \frac{10^{-4}}{(4\pi \times 10^{-7})(1.1 \times 0.0016)} = 4.5 \times 10^4 \text{ A/wb.}$$

If leakage is neglected, the same flux will exist in the magnetic member. The flux density in the magnetic material is

$$B_m = \frac{0.001}{0.9 \times 0.0016} = 0.695 \text{ tesla}$$

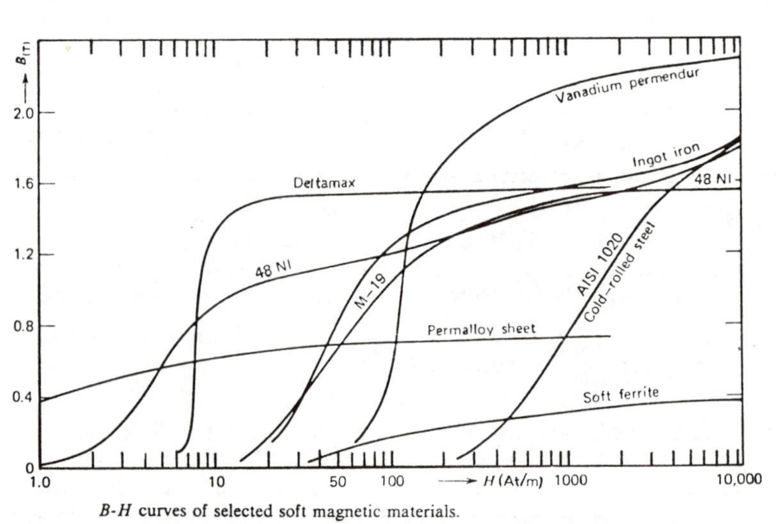

B-H curves of selected soft magnetic materials.

Fig. 2

From the M-19 curve in Figure 2, the amplitude perm-eability is

$$\mu_a = \frac{B_m}{\mu_0 H_m} = \frac{0.695}{(4\pi \times 10^{-7})(54)} = 10.240 = \mu_R$$

The reluctance of the magnetic member is

$$R_m = \frac{\ell_m}{\mu_R \mu_0 A_m} = \frac{0.1}{10240(4\pi \times 10^{-7})(0.9 \times 0.0016)}$$

$$= 0.54 \times 10^4 \quad \text{A/wb.}$$

The required exciting ampere-turns are

$$N_1 I_1 = \phi(R_g + R_m) = 0.001(5.04 \times 10^4) = 50.4 \text{ A.}$$

• PROBLEM 2-7

The magnetic circuit in Fig. 2 is comprised of a laminated core of steel with an air gap g = 0.25 cm and has an exciting winding of 350 turns. The magnetization curve for the iron is shown in Fig. 1. The presence of nonmagnetic material between laminations is taken into account by a stacking factor of 0.93 for this core. Neglect leakage but correct for fringing and calculate the current in the exciting winding to produce a flux of 5.0×10^{-4} Wb in the core.

Solution: Since magnetic leakage is neglected, there is no parallel flux path and the flux is confined to a path through the iron and air gap in series.

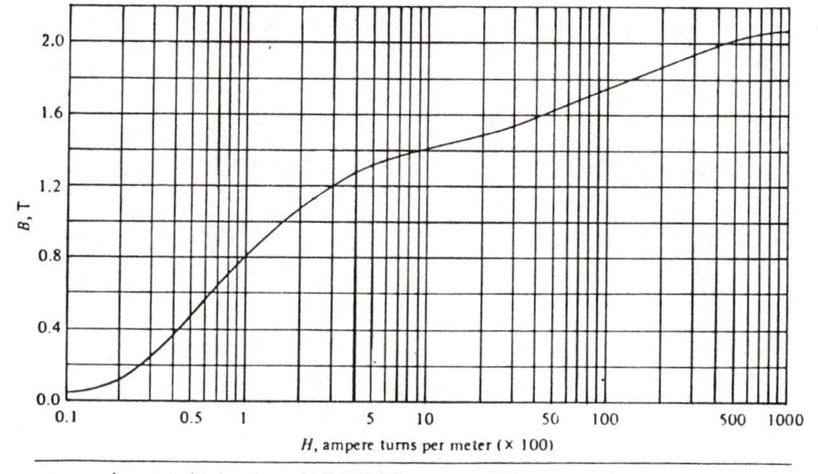

dc magnetization curve for M-19 fully processed 29-gauge steel.

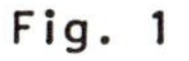
Fig. 1

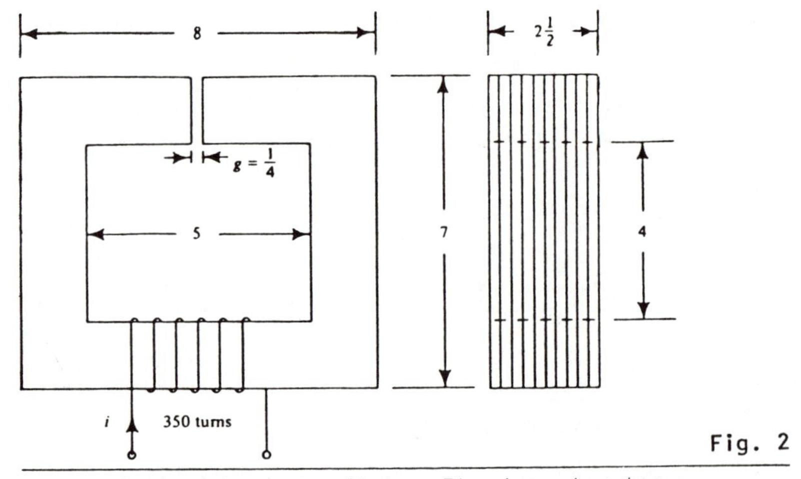

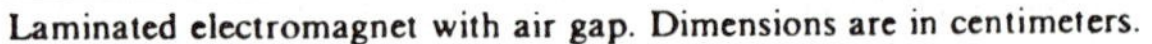

Laminated electromagnet with air gap. Dimensions are in centimeters.

Net area of core = $(1.5 \times 2.5 \times 0.93) \times 10^{-4}$

$= 3.49 \times 10^{-4}\ m^2$

Mean length of flux path in iron = $2(6.5 + 5.5) \times 10^{-2}$

$= 0.24$ m

Flux density in iron $B_{iron} = \dfrac{\Phi}{A_{iron}}$

$= (5.0 \times 10^{-4})/(3.49 \times 10^{-4})$

$= 1.43$ T

H for iron from Fig. 1 = 1000 ampere turns/m

Mmf for iron $F_{iron} = H_{iron}\ell_{iron} = 1000 \times 0.24 = 240$ ampere turns.

Corrected area of air gap = $(1.5 + 0.25)(2.5 + 0.25) \times 10^{-4}$

$= 4.81 \times 10^{-4}\ m^2$

Flux density in the air gap, $B_{air} = \dfrac{\Phi}{A_{air}}$

$= \dfrac{5.0 \times 10^{-4}}{4.81 \times 10^{-4}} = 1.04$ T

It should be noted that the stacking factor applies only to laminated iron and not to the air gap:

$$H_{air} = \frac{B_{air}}{\mu_0} = \frac{1.04}{4\pi \times 10^{-7}} = 8.27 \times 10^5 \text{ ampere turns/m}$$

Mmf for air gap $F_{air} = H_{air}g = 8.27 \times 10^5 \times 0.25 \times 10^{-2}$

$$= 2070 \text{ ampere turns}$$

Total mmf for the iron and air gap in series:

$$F_t = F_{iron} + F_{air} = 240 + 2070 = 2310 \text{ ampere turns}$$

and the current is

$$I = \frac{F_t}{N} = \frac{2310}{350} = 6.6\text{A}$$

• PROBLEM 2-8

(i) Refer to Figure 1. $A_{Fe} = A_g = 0.01\text{m}^2$, $g = 0.005$ m, $N = 1000$ turns. What current would be required to produce a flux of 0.01 Wb in the air gap, neglecting iron reluctance?

The total iron cross section is $A_{Fe} = 0.01 \text{ m}^2$ and the length of the iron flux path is $\ell_{Fe} = 0.25$ m.

(ii) What current is required to produce a flux density in the air gap of 1.5T?

(iii) Repeat Part (ii) taking into account the iron: (a) assuming a soft steel casting, and (b) assuming M-19 sheets with a stacking factor of 0.90.

Solution:

(i) $$R = \frac{g}{\mu_0 A_g} = \frac{0.005}{4\pi \times 10^{-7} \times 0.01} = 3.98 \times 10^5 \text{ A/wb.}$$

$$F = NI = 1000I \text{ A.}$$

$$\Phi = \frac{F}{R} = \frac{1000I}{3.98 \times 10^5} = 0.01 \text{ Wb}$$

$$I = \frac{3.98 \times 10^3}{1000} = 3.98 \text{ A.}$$

$F = NI = 3980$ A.

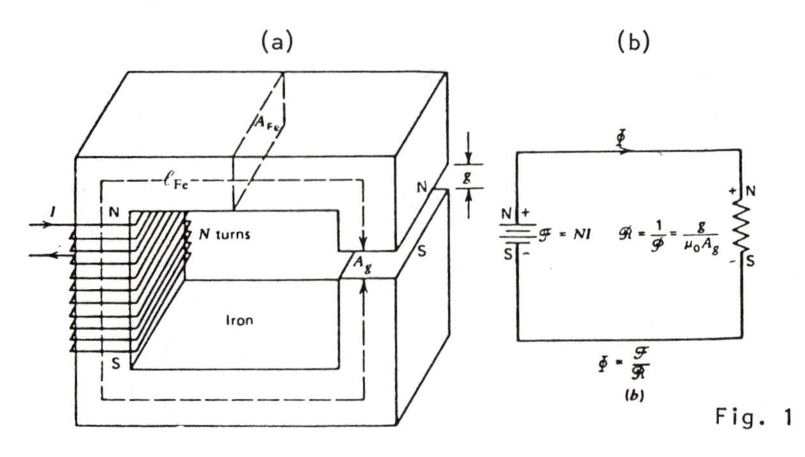

A magnetic circuit and its electrical analog.(a) A simple magnetic circuit.(b)Electric analog.

(ii) $\Phi = BA = 1.5 \times 0.01 = 0.015$ Wb, or 1.5 times the flux of part (i) which implies $I = 1.5 \times 3.98 = 5.97$ A

$$H = B/\mu_0 = \frac{1.5}{4\pi \times 10^{-7}} = 1.194 \times 10^6 \text{ A/m}$$

$$F = Hg = 1.194 \times 10^6 \times 5 \times 10^{-3} = 5.97 \times 10^3 \text{ A}$$

$$F = NI = 10^3 I = 5.97 \times 10^3 \text{ A}$$

$$I = 5.97 \text{ A}$$

(iii) (a) $B_{Fe} = B_g = 1.5$ T

From the curve, $H_{Fe} = 3000$ A/m

$$F = F_g + H_{Fe}\ell_{Fe} = 5970 + 3000 \times 0.25 = 6720 \text{ A}$$

$I = \frac{F}{N} = 6.72$ A, an increase of 13 percent

above the value obtained with iron neglected.

(b) $B_{Fe} = \dfrac{B_{gross}}{\text{stacking factor}} = \dfrac{1.5}{0.90} = 1.67$ T

From the curve, H_{Fe} is very indefinite, say 5500 A/m.

$F_{Fe} = H_{Fe} \ell_{Fe} = 5500 \times 0.25 = 1375$ A.

I = 1.375 + 5.97 = 7.34 A, an increase of 23 percent over that required for the air gap alone.

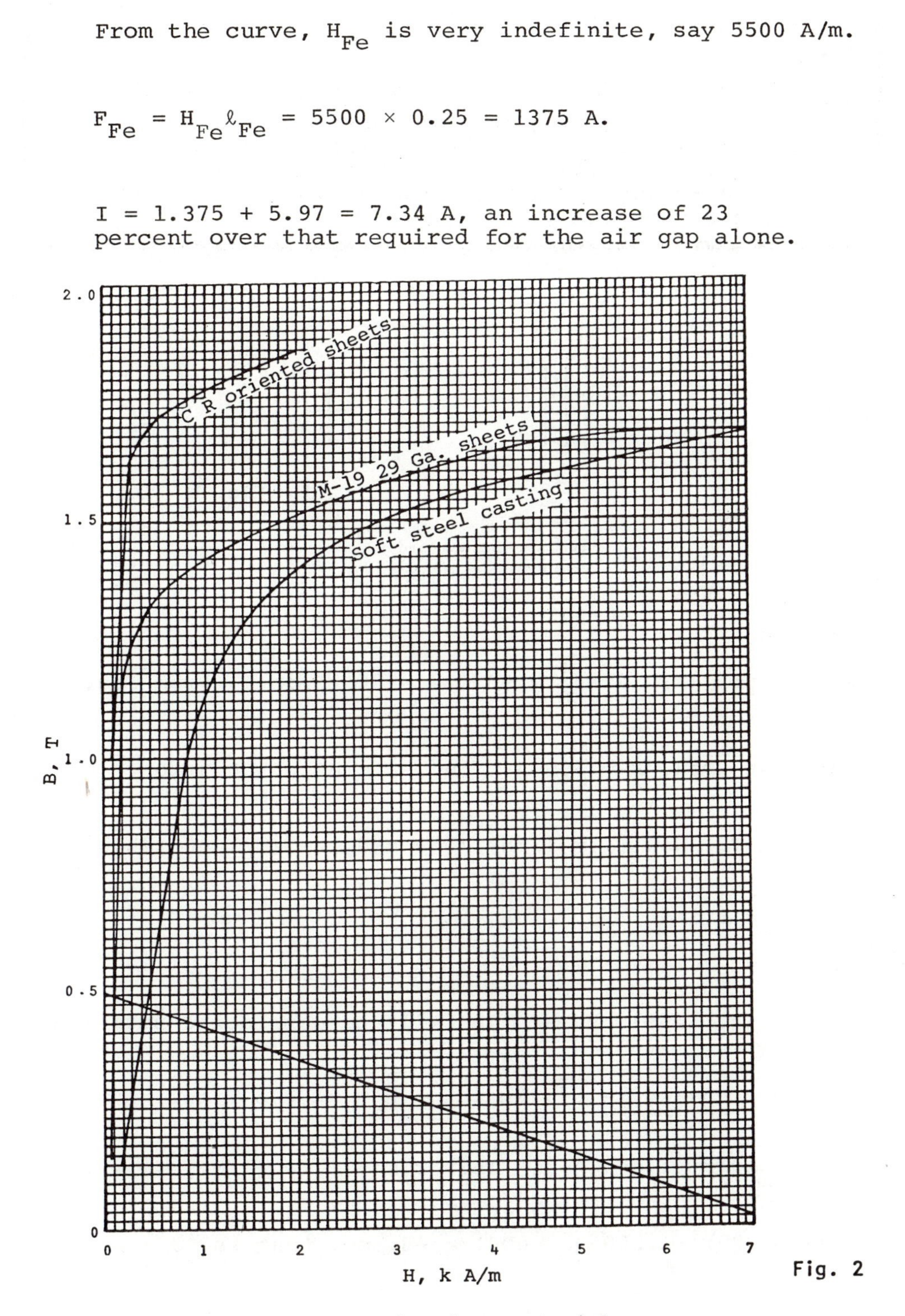

Fig. 2

B-H curves for three materials.

• PROBLEM 2-9

The core shown in Fig. 1 is made of cast iron. Each leg has a mean length of 2.5 in., and the cross-sectional area is uniformly 1.2 in.^{2} Each of the two coils has 1,000 turns.

Of the two currents, $i_1 = 2$ amp is given. (This is a positive value, meaning that the actual direction of i_1 is in the direction of the arrow.) Find the current i_2 if the flux in the core is to be: (a) 90 KL in the counter-clockwise direction, and (b) 54 KL in the clockwise direction.

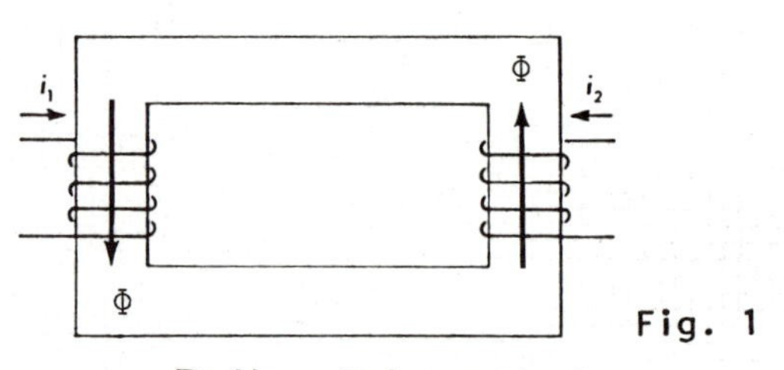

Fig. 1

Doubly excited magnetic circuit, showing assumed positive current directions.

Solution: (a) $B = \frac{\Phi}{A} = 90/1.2 = 75$ KL/in.^{2} From Fig. 2 for cast iron, $H = 500$ A-T/in., $F = H\ell = 500 \times 10 = 5{,}000$ A-T. Now $i_2 = (1/N_2)(F - i_1N_1) = (1/1{,}000)(5{,}000 - 2{,}000) = 3$ amp.

(b) A clockwise flux will be considered negative. Thus $B = -54/1.2 = -45$ KL/in.^{2}, $H = -100$ A-T/in., $F = -100 \times 10 = -1{,}000$ A-T. This time, $i_2 = (1/N_2)\ (F - i_1N_1) = (1/1{,}000)(-1{,}000 - 2{,}000) = -3$ amp. that is, 3 amp opposite to the arrow for i_2.

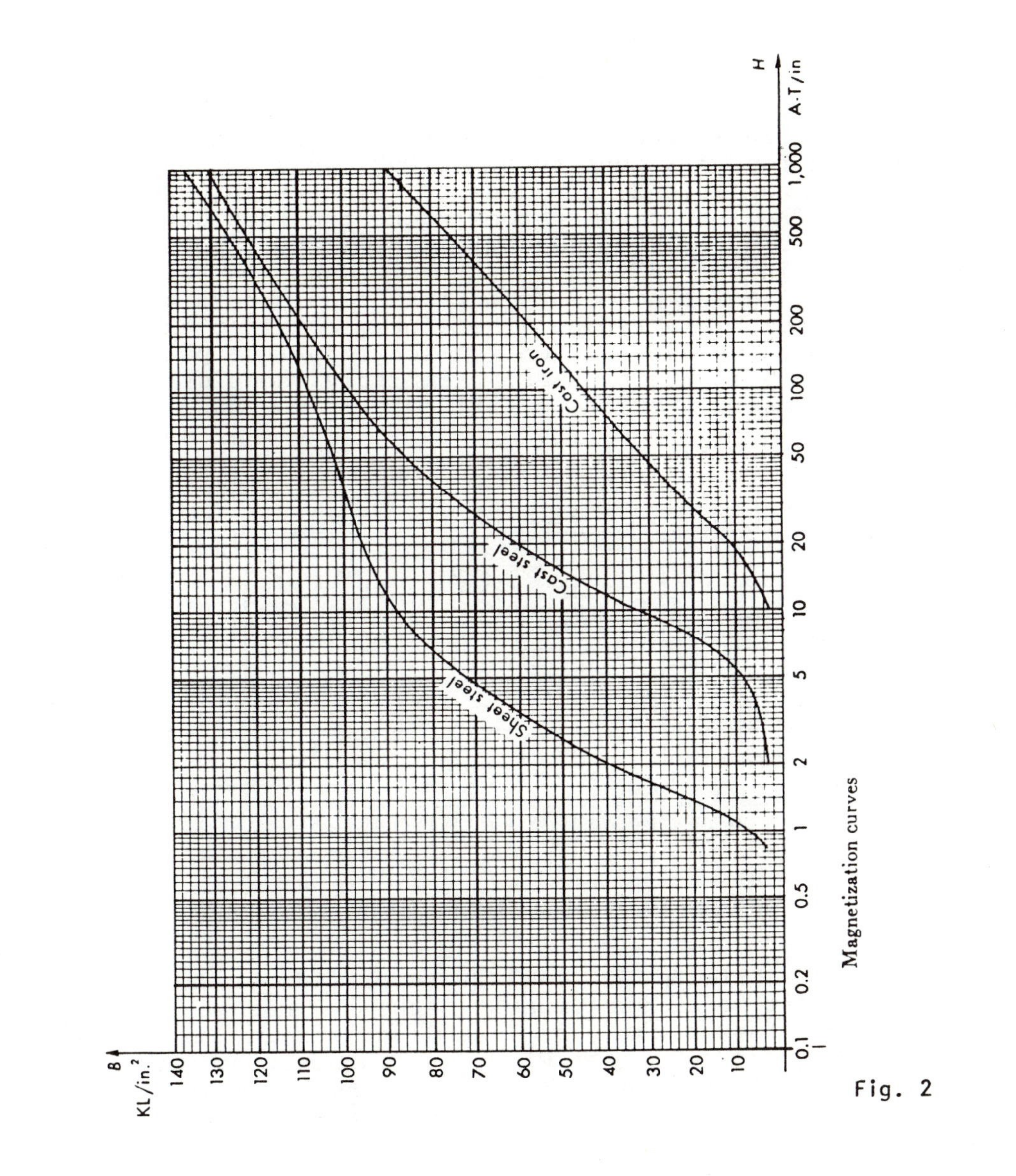

Fig. 2

• PROBLEM 2-10

In the magnetic circuit shown in Fig. 1 a total flux of 1,200,000 or 12×10^5 lines is to be set up in the poles. As noted in the figure, the poles are of wrought iron, the top of silicon-steel sheets, the base of cast iron. There is an air gap, 0.95 cm. long, between each pole and the base. Determine the total ampere-turns per pole required.

Solution: The flux density in the poles and in the

air gaps is $B = \frac{\Phi}{A} = \frac{12 \times 10^5}{10 \times 12} = 10{,}000$ gausses, the total flux at points a and c of the top and at points b and d of the base is 6×10^5 maxwells. The flux density at a and c of the top is $\frac{6 \times 10^5}{4 \times 10} = 15{,}000$ gausses. At the joint between the poles and the top, z, the flux enters the top at a density of 10,000 gausses; and as the flux bends around, the density rises to 15,000 gausses. These short distances where the flux density in the top and bases changes might be subject to a refinement in the calculation, but it would be rather complicated, and unnecessary in this case, where there is an appreciable air gap. Therefore, consider that the flux density over the entire length of the path in the top is 15,000 gausses. Similarly, consider the flux density over the entire path in the base to be $\frac{6 \times 10^5}{10 \times 10} = 6000$ gausses.

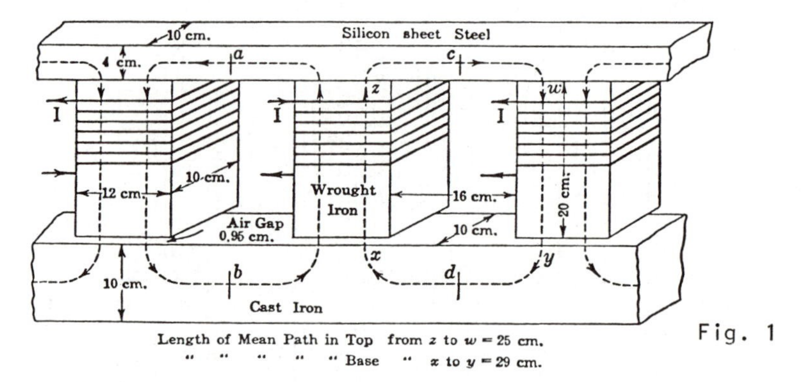

Fig. 1

A parallel magnetic circuit involving an air gap and parts of different materials.

Consider that the middle coil is to provide the m.m.f. for that portion of the circuit between lines ab and cd. To set up a flux density of 10,000 gausses in the wrought-iron poles requires, from Fig. 2, a magnetizing force of 12.3 ampere-turns per centimeter. For a length of 20 cm., $12.3 \times 20 = 246$ ampere-turns are needed.

The magnetizing force needed in the silicon-steel sheets of the top, to set up 15,000 gausses, is 22.5 ampere-turns per centimeter. The distance from z to c being 12.5 cm., $22.5 \times 12.5 = 281$ ampere-turns are needed.

For the cast-iron base, with an average flux density of 6000 gausses, a magnetizing force of 32 ampere-turns per centimeter and a total m.m.f. of $32 \times 14.5 = 464$ ampere-turns are needed.

For an air-gap density of 10,000 gausses, the magnetizing force is given by the following equation:

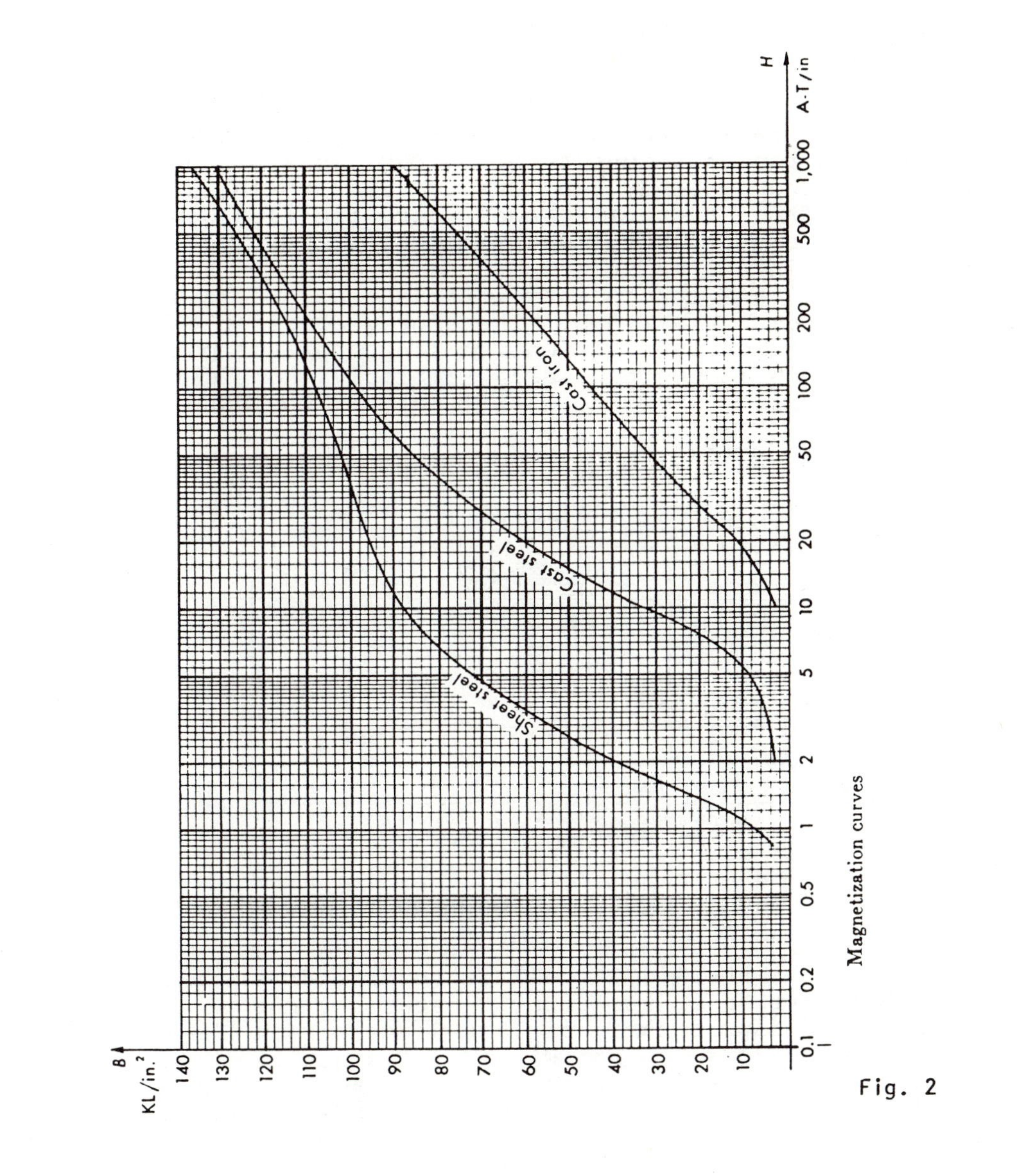

Fig. 2

• PROBLEM 2-10

In the magnetic circuit shown in Fig. 1 a total flux of 1,200,000 or 12×10^5 lines is to be set up in the poles. As noted in the figure, the poles are of wrought iron, the top of silicon-steel sheets, the base of cast iron. There is an air gap, 0.95 cm. long, between each pole and the base. Determine the total ampere-turns per pole required.

Solution: The flux density in the poles and in the

air gaps is $B = \frac{\Phi}{A} = \frac{12 \times 10^5}{10 \times 12} = 10{,}000$ gausses, the total flux at points a and c of the top and at points b and d of the base is 6×10^5 maxwells. The flux density at a and c of the top is $\frac{6 \times 10^5}{4 \times 10} = 15{,}000$ gausses. At the joint between the poles and the top, z, the flux enters the top at a density of 10,000 gausses; and as the flux bends around, the density rises to 15,000 gausses. These short distances where the flux density in the top and bases changes might be subject to a refinement in the calculation, but it would be rather complicated, and unnecessary in this case, where there is an appreciable air gap. Therefore, consider that the flux density over the entire length of the path in the top is 15,000 gausses. Similarly, consider the flux density over the entire path in the base to be $\frac{6 \times 10^5}{10 \times 10} = 6000$ gausses.

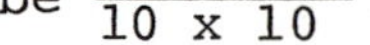

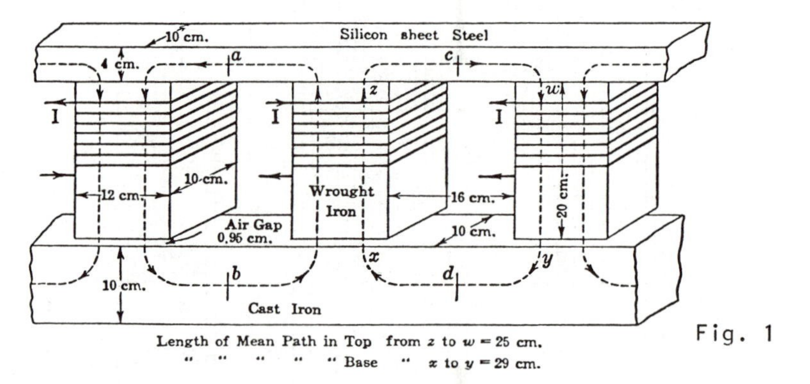

Fig. 1

A parallel magnetic circuit involving an air gap and parts of different materials.

Consider that the middle coil is to provide the m.m.f. for that portion of the circuit between lines ab and cd. To set up a flux density of 10,000 gausses in the wrought-iron poles requires, from Fig. 2, a magnetizing force of 12.3 ampere-turns per centimeter. For a length of 20 cm., $12.3 \times 20 = 246$ ampere-turns are needed.

The magnetizing force needed in the silicon-steel sheets of the top, to set up 15,000 gausses, is 22.5 ampere-turns per centimeter. The distance from z to c being 12.5 cm., $22.5 \times 12.5 = 281$ ampere-turns are needed.

For the cast-iron base, with an average flux density of 6000 gausses, a magnetizing force of 32 ampere-turns per centimeter and a total m.m.f. of $32 \times 14.5 = 464$ ampere-turns are needed.

For an air-gap density of 10,000 gausses, the magnetizing force is given by the following equation:

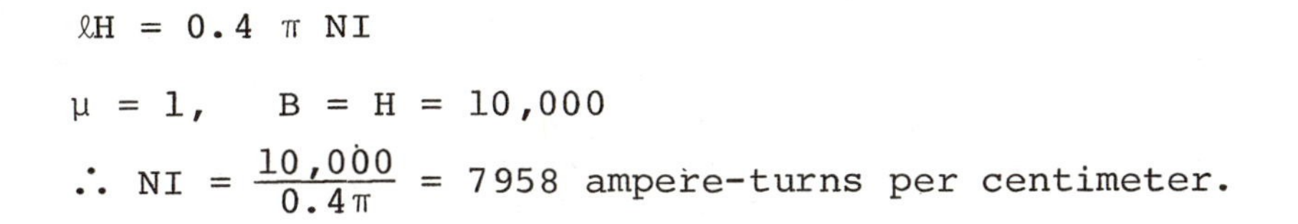

$$\ell H = 0.4\ \pi\ NI$$

$$\mu = 1, \quad B = H = 10{,}000$$

$$\therefore\ NI = \frac{10{,}000}{0.4\pi} = 7958 \text{ ampere-turns per centimeter.}$$

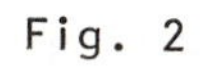

Fig. 2

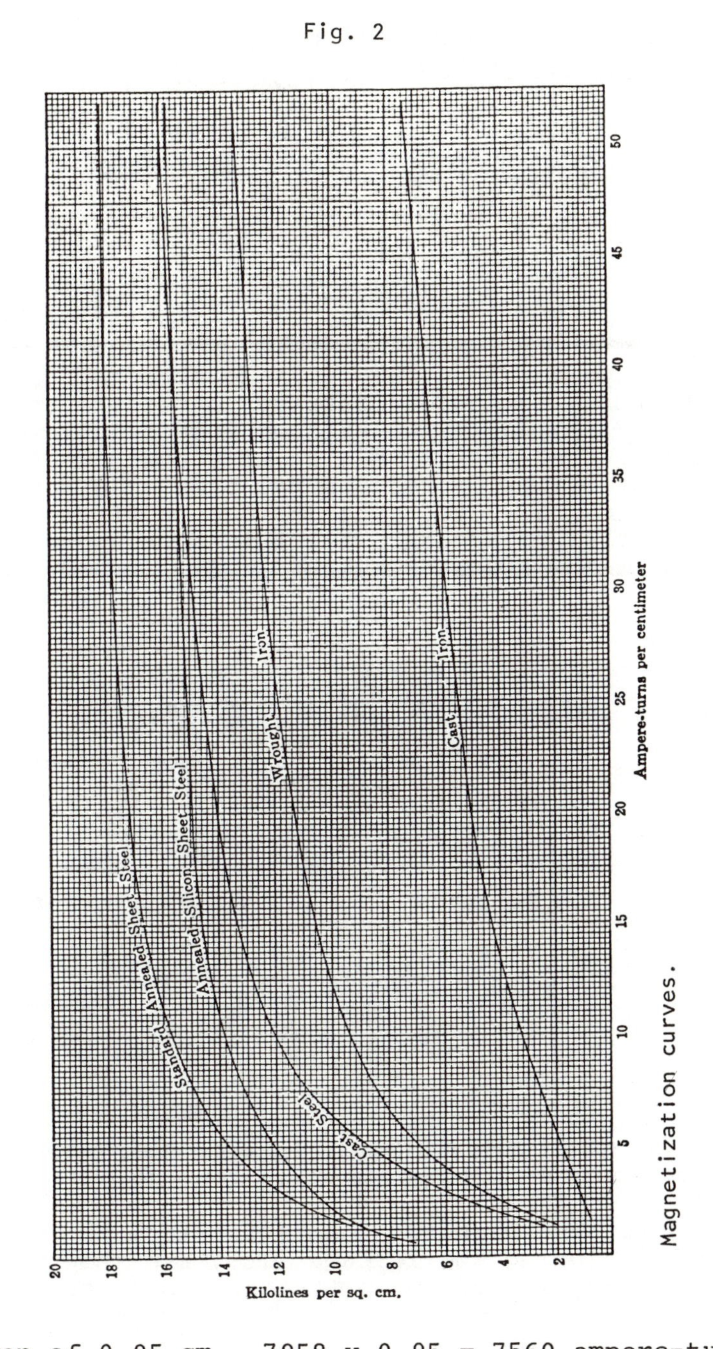

Magnetization curves.

For a gap of 0.95 cm., 7958 x 0.95 = 7560 ampere-turns are needed. The above figures are grouped in Table 1.

Table 1

Part	Material	B in Gausses	H in Amp.-turns per Cm.	Length in Cm.	M.m.f. in Amp-turns
Poles.....	Wrought iron.......	10,000	12.3	20	246
Top......	Silicon-steel sheets...	15,000	22.5	12.5	281
Base......	Cast iron...........	6,000	32	14.5	464
Gaps......	Air.................	10,000	7958	0.95	7560
Total ampere-turns per pole.................................					8551

• **PROBLEM 2-11**

For the dynamo shown in Fig. 1 compute the number of ampere-turns per pole necessary to produce an air-gap flux of 7,500,000 maxwells from each pole into the armature. The air-gap has an effective length of 0.235 in., after correction has been made for armature teeth, fringing, etc. The leakage coefficient (ratio of core flux to armature flux) is assumed to be 1.15.

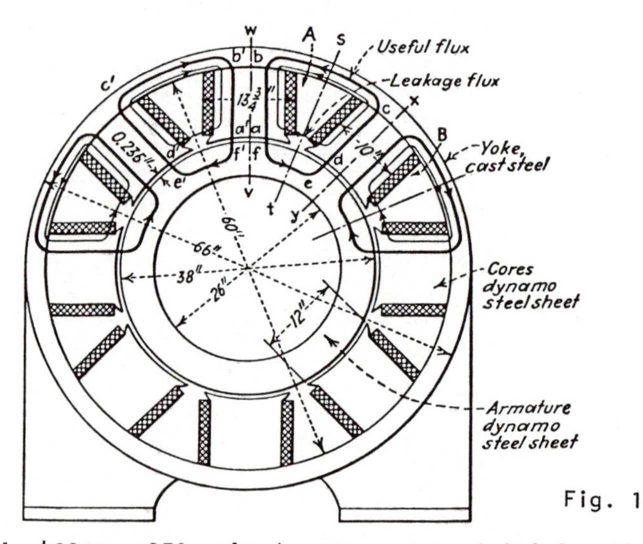

Eight-pole 400-rpm 250-volt d-c generator. Axial length of armature stampings and pole faces = 16 in.

Solution: The paths of the fluxes from the various poles, including the leakage flux, are shown in the figure. The lengths of paths are easily determined.

Since the ampere-turns required by the iron are small compared with those required for the air-gap, a high degree of accuracy in determining the dimensions of the iron part of the magnetic circuit is not necessary.

Consider the flux path abcdef.

The length ab = $\frac{60 - 38}{2}$ - 0.235 = 10.8 in. (approximately); bc (approximately one-eighth the mean circumference of the yoke, less 5 in.) = $\frac{\pi 63''}{8} - 5 =$ 24.7 - 5 = 19.7 in.

$$\text{ad (excluding air-gaps)} = \frac{\pi 32}{8}$$

$$= 12.6 \text{ in. (approximately).}$$

(For example, the distance from the surface of the armature at a, to f is approximately equal to the distance from the line wv to f.)

The flux densities are as follows:

Flux in cores = 7,500,000 x 1.15 = 8,630,000 maxwells, as the flux in the core is equal to the armature flux plus the leakage flux.

Flux density in cores = $\frac{8,630,000}{16 \times 10}$ = 53,900 maxwells per sq. in.

This should be increased by about 10 percent to allow for the thickness of the oxide on the surface of the laminations:

53,900 x 1.10 = 59,300.

Flux density in yoke = $\frac{8,630,000}{2(16 \times 3)}$ = 90,000 maxwells per sq in., as the pole flux divides, one-half going each way in the yoke.

Flux density in armature = $\frac{7,500,000}{2(6 \times 16)}$ = 39,100 maxwells per sq in.

This must be increased by about 25 percent to allow for the air-duct space and the oxides between laminations.

Therefore, the density in the armature is

39,100 x 1.25 = 49,000 maxwells per sq in.

The corresponding net area of iron thus becomes

0.80 x 2(6 x 16) = 76.8 sq in.

Hence B = $\frac{3,750,000}{76.8}$ = 48,900 maxwells per sq in. (check).

The air-gap density = $\frac{7,500,000}{16 \times 12}$ = 39,100 maxwells per sq in.

Knowing the above quantities, and using the magnetization curves of Fig. 2, it is a comparatively simple matter to determine the total ampere-turns per pole.

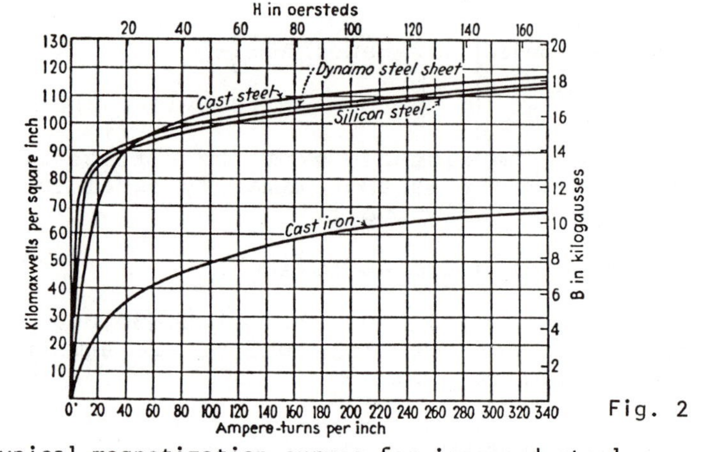

Fig. 2

Typical magnetization curves for iron and steel.

For example, with dynamo steel sheet 4 amp-turns per in. are necessary with 59,300 maxwells per sq in. Hence, for the core ab,

$$I_1N_1 = 4 \times 10.8 = 43 \text{ amp-turns.}$$

The ampere-turns are then found as follows:

Part	Material	Flux, maxwells	Area, square inches	Flux density, maxwells per square inch	Ampere-turns per inch	Length, inches	Ampere-turns
Core *ab*......	Dynamo steel sheet	8,630,000	160	59,300	4.0	10.8	43
Yoke *bc*......	Cast steel	4,315,000	48	90,000	40.0	19.7	788
Core *cd*......	Dynamo steel sheet	8,630,000	160	59,300	4.0	10.8	43
Gap *de*......	Air	7,500,000	192	39,100	12,240	0.235	2,876
Armature *ef*..	Dynamo steel sheet	3,750,000	76.8	49,000	3.0	12.6	38
Gap *fa*.......	Air	7,500,000	192	39,100	12,240	0.235	2,876
			Total ampere-turns for two poles				6,664
			Total ampere-turns per pole				3,332

As the machine is symmetrical, each complete magnetic circuit requires the same number of ampere-turns per pole.

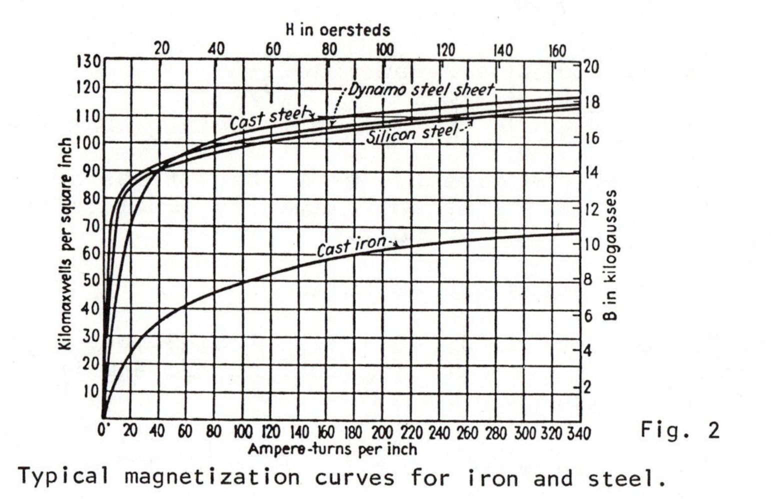

• PROBLEM 2-12

As an example, a 250-kw, 6-pole, 240-volt, 1200-rpm generator will be considered. The dimensions of this generator, in inches, are listed below: (also see Fig. 1).

Outside diameter of armature core . .	D	= 22.5
Inside diameter of armature core . .	d	= 12.5
Gross core length	L	= 7.0
Number of radial vents	n_v	= 2
Width of each vent	b_v	= $\frac{3}{8}$
Net length of core $\ell = 7 - 2 \times \frac{3}{8}$ . .	ℓ	= 6.25
Single air gap	g	= 0.218
Pole embrace (arc)	b_p	= 7.65
Pole length	l_p	= 7.0
Pole width		= 6.0
Frame ID		= $37\frac{11}{16}$
Frame OD		= $42\frac{7}{16}$
Frame width		= 10
No. of arm. slots		= 72
Slot depth		= 2.0
Slot width		= 0.36

Electrical steel is used for the armature. USS hot-rolled steel is used for the poles and frame. Find (a) flux per pole, (b) ampere-turns necessary to drive the pole flux through the two gaps, (c) ampere-turns for the teeth, (d) ampere-turns for the armature core, (e) ampere-turns for the poles, (f) ampere-turns for the yoke (frame), (g) total ampere-turns.

Solution: (a) Flux per pole. The armature has a lap winding and, therefore, a = p = 6. Each slot has six conductors, three per layer. The total number of conductors is, then, Z = 6 x 72 = 432. From the following equation:

$$E = Z \frac{p}{a} \frac{n}{60} \Phi 10^{-8} \text{ volt.,}$$

the flux necessary to induce an emf of 240 volts at no-load in the armature winding is

$$\Phi = \frac{240 \times 10^8}{432} \times \frac{60}{1200} \times \frac{6}{6} = 2.78 \times 10^6 \text{ maxwells}$$

(b) Ampere-turns necessary to drive the pole flux through the two gaps.

The ampere-turns necessary to drive the flux Φ through the air-gap is the largest of the 5 components. Therefore, the area of the gap A_g must be determined with satisfactory accuracy.

Setting

$$A_g = \ell_e b_e$$

and

$$B_g = \frac{\Phi}{\ell_e b_e}$$

the effective armature length ℓ_e and the effective pole arc b_e are obtained from the following considerations.

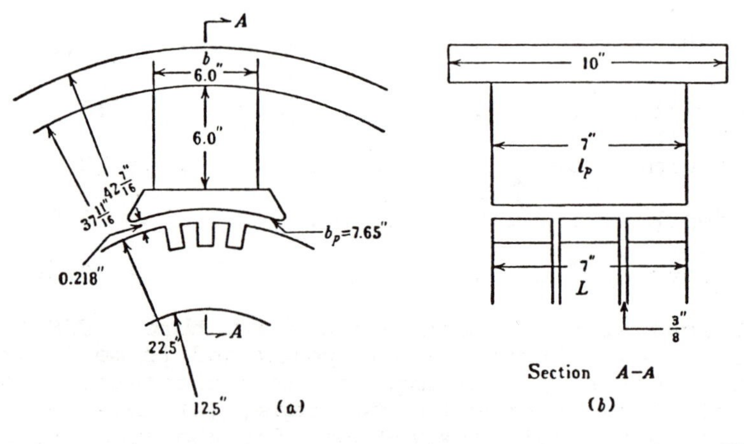

Machine dimensions.

Fig. 1

The laminated core of the armature of medium size and larger machines is divided into several stacks by radial vents. The flux density in the vents is much smaller than that in the gap between pole and armature iron (Fig. 2a). The effective armature length ℓ_e is, therefore, smaller than the gross core length L and larger than the net core length $\ell = L - n_v b_v$. It will be

assumed here that $\ell_e = (L + \ell)/2$.

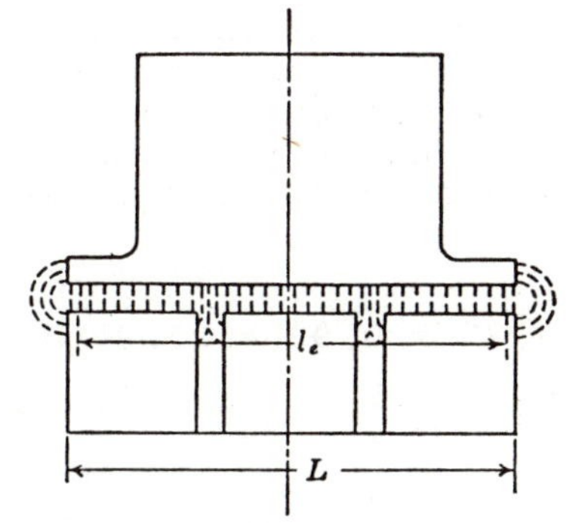

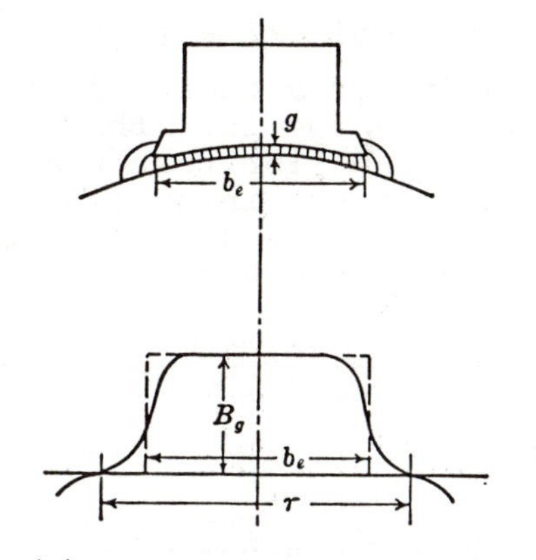

Fig. 2

(a) Determination of the equivalent armature length.

(b) Field distribution curve of a d-c machine.

Because of fringing on both sides of the pole shoe (Fig. 2b) the effective pole arc b_e is larger than the actual pole arc b_p. It can be assumed that $b_e = b_p + 2g$. Thus

$$\ell_e = \frac{7 + 6.25}{2} = 6.62 \text{ in.}$$

$$b_e = 7.65 + (2 \times 0.218) = 8.1 \text{ in.}$$

and

$$B_g = \frac{\Phi}{A_g} = \frac{2.78 \times 10^6}{6.62 \times 8.1} = 51{,}800 \text{ lines/in.}^2$$

The ampere-turns for two gaps are given by the following equation:

$$NI = 2\frac{B_g}{0.4\pi}g \ .$$

$$\therefore \quad AT_g = \frac{51{,}800}{0.4\pi \times (2.54)^2} \times 2 \times 0.218 \times 2.54 \times 1.10$$

$$= 7780 \text{ AT}$$

2.54 is the conversion factor from inches to centimeters. The last factor 1.10 takes into account the fact that the gap appears increased at the points opposite the slots, so that the effective air gap is larger than 0.218 in.

(c) Ampere-turns for the teeth. The pole flux Φ goes into the armature through the teeth lying within the pole arc $b_e = 8.1$ in. The pole pitch is

$$\tau = \frac{\pi D}{6} = \frac{\pi \times 22.5}{6} = 11.8 \text{ in.}$$

Thus the pole flux Φ goes through $\frac{72}{6} \times \frac{8.1}{11.8}$ teeth. The width of a tooth in the middle of the tooth is equal to 0.534 in. The iron length of the core is somewhat smaller than the net core length $\ell = 6.25$ in., because of the insulation between the laminations. Assuming a loss due to insulation of 8°/₀, the cross-section of the teeth per pole is

$$A_t = \frac{72}{6} \times \frac{8.1}{11.8} \times 0.534 \times 6.25 \times 0.92 = 25.3 \text{ in.}^2$$

and

$$B_t = \frac{\Phi}{A_t} = \frac{2.78 \times 10^6}{25.3} = 111{,}000 \text{ lines/in.}^2$$

The B-H curve for electrical steel yields for B = 110,000 lines/in.^{2}, 180 AT per in. The length of the magnetic path in a tooth is equal to 2 in. (=slot depth). Thus the AT necessary to drive the flux through the teeth twice is

$$AT_t = 2 \times 2 \times 180 = 720 \text{ AT}$$

(d) Ampere-turns for the armature core. The height of the armature core below the teeth is equal to $\frac{22.5 - 12.5}{2} - 2 = 3.0$ in. The iron length of the core is the same as that of the teeth, i.e., 6.25 x 0.92 in. The cross-section of the core is thus

$$(3 \times 6.25 \times 0.92) \text{ sq. in.}$$

Since the pole flux Φ divides into two parts within the core

$$B_e = \frac{2.78 \times 10^6}{2 \times 3 \times 6.25 \times 0.92} = 80{,}600 \text{ lines/in.}^2$$

The B-H curve for electrical steel yields, for B = 80,600 lines, 10 AT per in. The length of the path in the core is equal to the pole pitch in the middle of the core, i.e., to

$$\frac{\pi(d_i + h_c)}{p} = \frac{\pi(12.5 + 3.0)}{6} = 8.1 \text{ in.}$$

Thus the ampere-turns necessary to drive the flux through the core are

$$AT_c = 10 \times 8.1 = 81 \text{ AT}$$

(e) Ampere-turns for the poles. It will be assumed

that the leakage flux between the pole shoes and pole bodies is equal to 20°/₀ of the armature flux. Therefore, the flux through the pole and the yoke is

$$\Phi_p = 1.2 \times 2.78 \times 10^6 = 3.34 \times 10^6 \text{ maxwells}$$

The pole laminations are not insulated. The loss of iron thickness due to air layers between the laminations can be assumed 5%. The cross-section of a pole is, therefore, equal to (0.95 x 7 x 6) sq. in. (see Fig. 1), and the flux density in the pole is

$$B_p = \frac{3.34 \times 10^6}{0.95 \times 7 \times 6} = 83{,}600 \text{ lines/in.}^2$$

The B-H curve for USS hot-rolled steel yields, for B = 83,600 lines, 18 AT per in. The length of the path in the pole is assumed equal to the height of the pole without pole shoe (the flux density is low in the pole shoe). This length is equal to 6.0 in., and thus the ampere-turns necessary for two poles are

$$AT_p = 2 \times 18 \times 6.0 = 216 \text{ AT}$$

(f) Ampere-turns for the yoke (frame). The height of the yoke is

$$\frac{(42\frac{7}{16} - 37\frac{11}{16})}{2} = 2\frac{3}{8} \text{ in.}$$

The yoke is of solid steel. Thus the flux density in the yoke is

$$B_y = \frac{3.34 \times 10^6}{2 \times 2.375 \times 10} = 70{,}000 \text{ lines/in.}^2$$

The B-H curve for USS hot-rolled steel yields, for B = 70,000 lines, 13.0 AT per in. The length of the magnetic path in the yoke is $\frac{\pi(42\frac{7}{16} - 2\frac{3}{8})}{6} = 20.9$ in., and the ampere-turns necessary to drive the flux through the yoke are

$$AT_y = 13 \times 20.9 = 270 \text{ AT}$$

(g) Total ampere-turns. The ampere-turns for two poles (for one magnetic circuit) are

$$AT_{cir} = AT_y + AT_p + AT_g + AT_t + AT_c$$

$$= 270 + 216 + 7780 + 720 + 81 = 9067 \text{ AT}$$

and for all six poles (all three magnetic circuits)

$$AT_{total} = 3 \times 9067 = 27{,}200 \text{ AT}$$

It does not matter whether these AT are produced by a large number of turns on the poles and a small current in the conductors or by a small number of turns and a large current in the conductors. It is necessary, however, that the product of the number of turns and the current in the conductors is equal to the required number of ampere-turns. The machine considered has 1064 turns per pole. Therefore, the field current necessary to drive the flux $\Phi = 2.78 \times 10^6$ maxwells through the magnetic path is

$$I_f = \frac{9067}{2 \times 1064} = \frac{27,200}{6 \times 1064} = 4.25 \text{ amp}$$

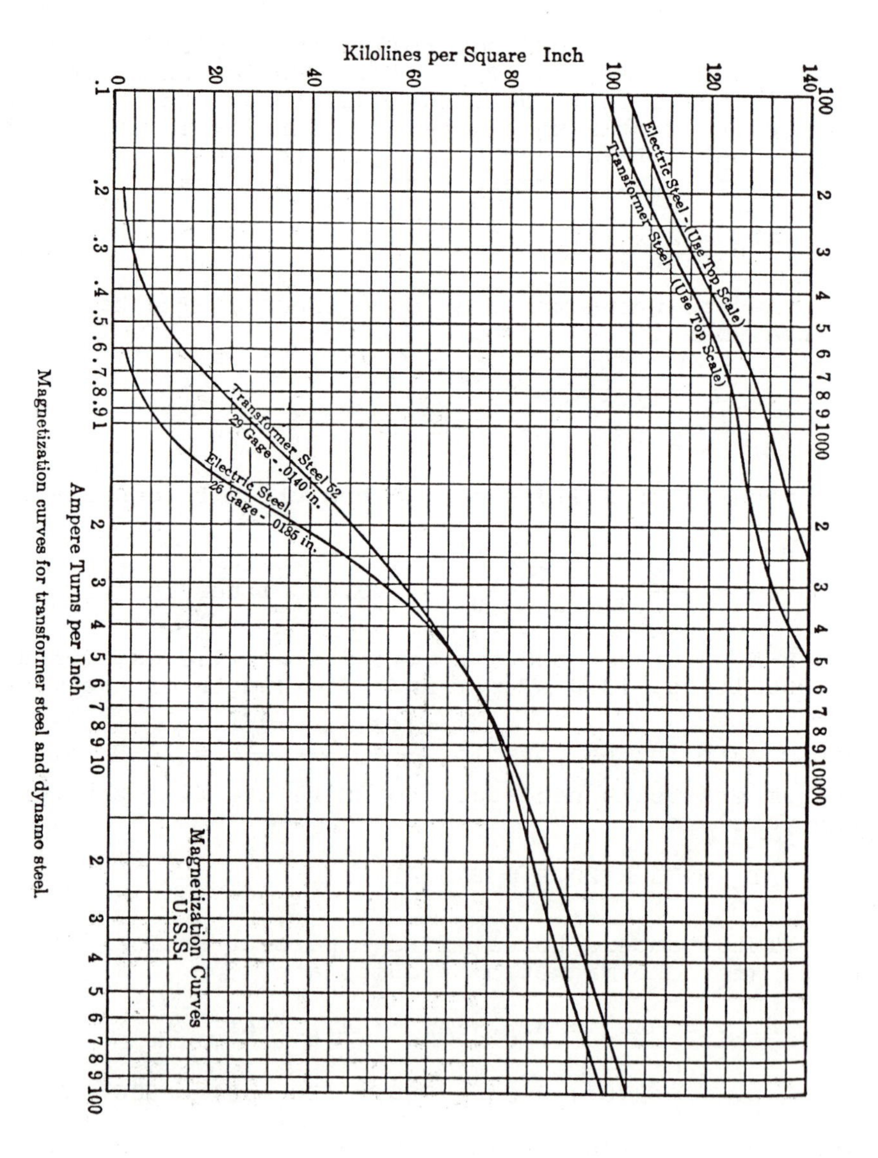

Magnetization curves for transformer steel and dynamo steel.

Fig. 3

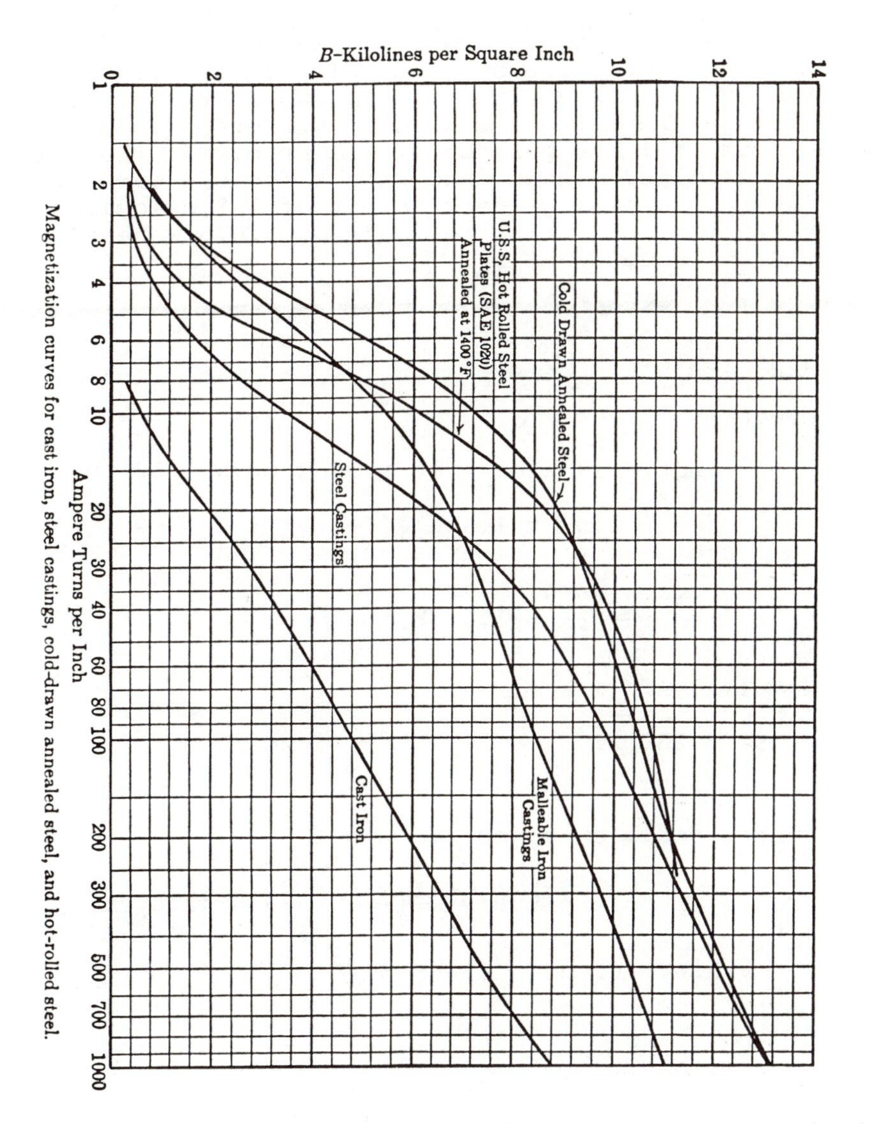

Magnetization curves for cast iron, steel castings, cold-drawn annealed steel, and hot-rolled steel.

Fig. 4

• PROBLEM 2-13

The various dimensions of a magnetic circuit are shown in Fig. 1, some of them having been calculated from the given inside and outside diameters of yoke and armature. There are 66 teeth and slots, only indicated schematically, the full dimensions being given on Fig. 2. The axial length of pole and armature is 35 cm and the corresponding dimension for the yoke is 45 cm. The effective air-gap length corrected for slotting effects may be taken as 0.5 cm. The ratio of B_{av}/B_{max} may be based on a rectangular field form as in Fig. 3, i.e. equal to the pole arc/pole pitch ratio. Pole and armature laminations are of armature iron and a 95% building factor can be used. In practice, an armature of this length would have two or three radial ventilating ducts along the core length but these will be neglected for the purpose of the problem. The yoke is made of rolled steel. See Fig. 4 for B/H curves. Assuming a pole leakage coefficient of 1.2, calculate the excitation for a flux per pole of 0.075 webers. Assume $B_{real} = 0.95 \times B_{app}$.

Solution: Pole

Length = 20 cm

Cross-sectional area = $18 \times 35 \times 0.95 = 600\ \text{cm}^2$

$$B_p = \frac{\text{Useful flux x leakage factor}}{\text{Cross-sectional area of pole body}}$$

$$= \frac{0.075 \times 1.2}{0.06} = 1.5\ \text{Wb/m}^2$$

H = 11.5 AT/cm

$\therefore\ H\ell = 230$ AT

Air Gap

Length = 0.5 cm effective

$$\text{Area over pole pitch} = \frac{45 \times \pi}{4} \times 35 = 1240\ \text{cm}^2.$$

$B_{av} = 0.075/0.124 = 0.605\ \text{Wb/m}^2$.

$$\therefore\ \frac{\text{pole arc}}{\text{pole pitch}} = \frac{25}{45\pi/4} = 0.705$$

hence $B_{max} = 0.605/0.705 = 0.86\ \text{Wb/m}^2$

$H\ell = (B_{max}/\mu_0) \times$ effective gap length

$$= \frac{0.86}{4\pi \times 10^{-7}} \times \frac{0.5}{10^2} = 3420\ \text{AT}$$

Teeth

Length = 3 cm

The area is calculated 2/3 the way down the slot. The gap flux over a tooth pitch is apparently carried by this reduced area so that the apparent tooth density is obtained on increasing B_{max} by the area ratio:

$$\frac{\text{area over tooth pitch}}{\text{area of one tooth}} = \frac{2.14 \times 35}{[(\pi \times 41)/66 - 1] \times 35 \times 0.95}$$

$$= 2.36$$

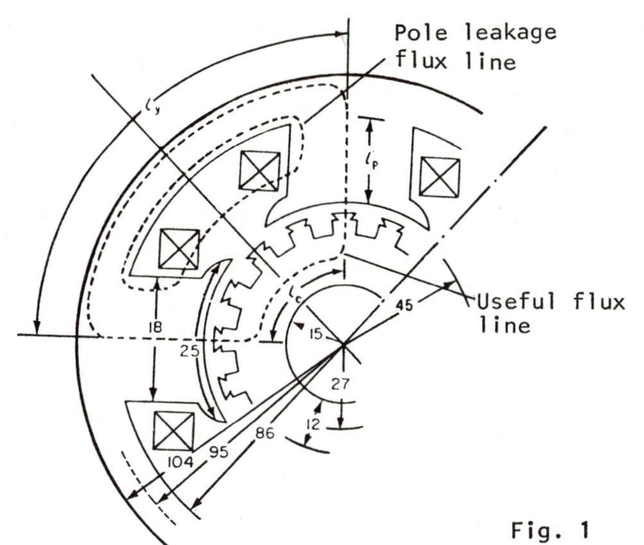

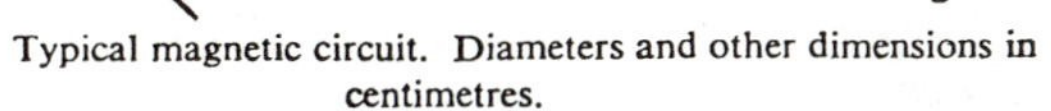
Typical magnetic circuit. Diameters and other dimensions in centimetres.

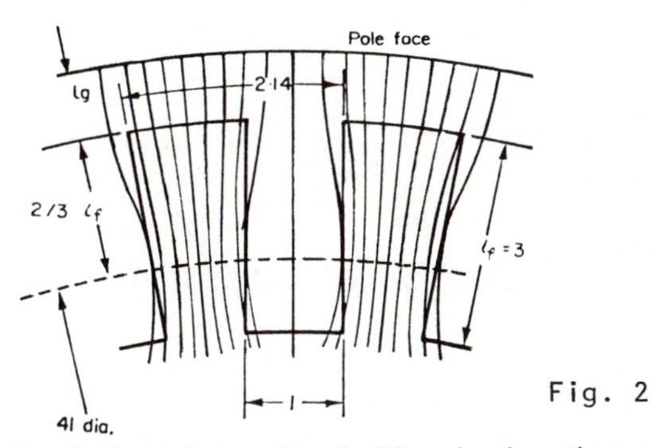

Flux distribution in tapered tooth. Dimensions in centimetres.

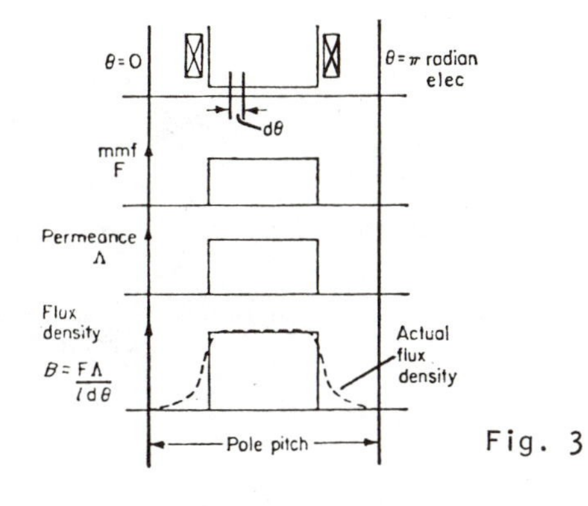

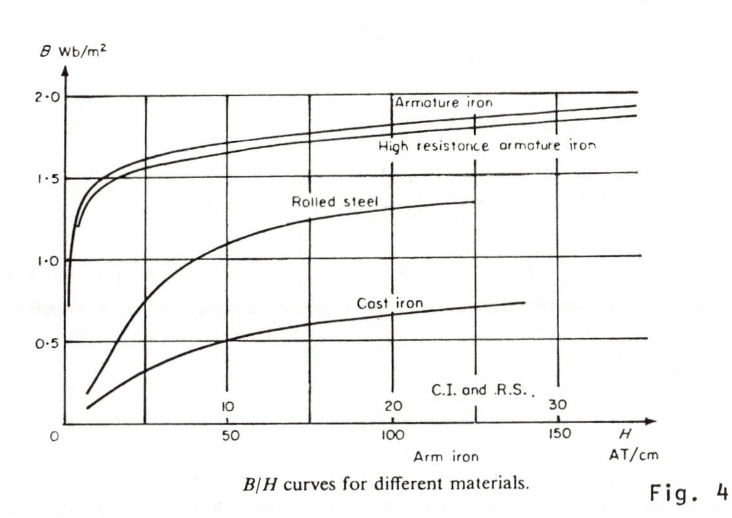

B/H curves for different materials.

Apparent tooth density = 2.36 x 0.86 = 2.03 Wb/m^2

Real tooth density = 2.03 x 0.95 = 1.93 Wb/m^2

for which H = 173 AT/cm

Tooth ampere-turns = 3 x 173 = 520 AT.

Yoke

$$\text{Length} = \ell_y/2 = \frac{95 \times \pi}{4 \times 2} = 37.4 \text{ cm.}$$

Area = 9 x 45 = 405 cm^2

$$B_y = \frac{\text{useful flux x leakage factor}}{\text{2 x yoke cross-sectional area}}$$

$$= \frac{0.075 \times 1.2}{0.0405 \times 2} = 1.11 \text{ Wb/m}^2$$

H = 10.5 AT/cm

$H\ell$ = 390 AT

Armature Core

$$\text{Length} = \ell_c/2 = \frac{27 \times \pi}{4 \times 2} = 10.6 \text{ cm.}$$

Area = 12 x 35 x 0.95 = 398 m^2

$$B_c = \frac{\text{useful flux}}{\text{2 x cross-sectional area of armature core}}$$

$$= \frac{0.075}{0.0398 \times 2} = 0.94 \text{ Wb/m}^2$$

H = 2 AT/cm

$H\ell$ = 20 AT

Summing the ampere-turns for the five parts, the total excitation is 4580 AT per pole.

FLUX PRODUCED BY A CURRENT

• PROBLEM 2-14

Figure 1 shows an electromagnet with two air gaps in parallel. Neglect leakage and the reluctance of the iron but correct for fringing by adding the length of the air gap to each of the other two dimensions and determine the flux in each air gap and the total flux when the current in the 900-turn winding is 0.2 A.

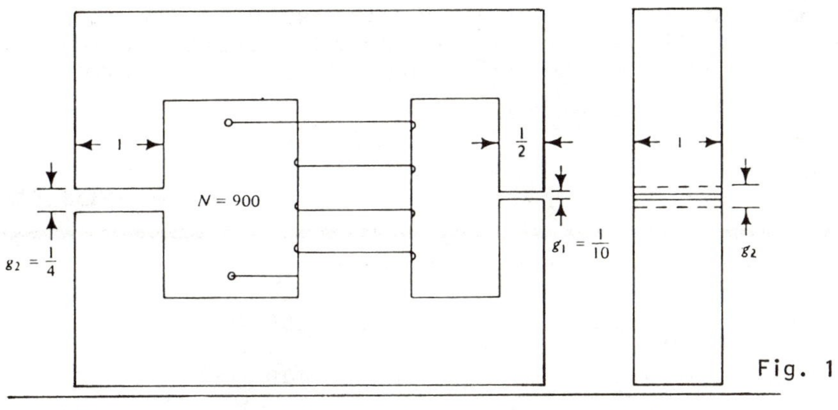

Electromagnet with two air gaps in parallel.
Dimensions are in centimeters.

Solution: Since there are two air gaps in parallel, their combined permeance is the sum of their permeances. The permeance of the 0.10-cm air gap g_1 is

$$\tau_1 = \frac{\mu_0 A_1}{\ell_1} = \frac{4\pi \times 10^{-7}(0.5 + 0.1)(1.0 + 0.1) \times 10^{-2}}{0.1}$$

$$= 8.31 \times 10^{-8} \text{ H}$$

and the flux in the 0.1-cm air gap is

$$\Phi_1 = F\tau_1 = Ni\tau_1 = 900 \times 0.2 \times 8.31 \times 10^{-8}$$

$$= 1.50 \times 10^{-5} \text{ Wb}$$

The permeance of the 0.25-cm air gap g_2 is

$$\tau_2 = \frac{\mu_0 A_2}{\ell_2} = \frac{4\pi \times 10^{-7}(1.0 + 0.25)(1.0 + 0.25) \times 10^{-2}}{0.25}$$

$$= 7.85 \times 10^{-8} \text{ H}$$

and the flux in the 0.25-cm air gap is

$$\Phi_2 = F\tau_2 = 900 \times 0.2 \times 7.85 \times 10^{-8} = 1.41 \times 10^{-5}\text{Wb}$$

The total flux which is that in the middle leg is

$$\Phi_T = \Phi_1 + \Phi_2 = (1.50 + 1.41) \times 10^{-5} = 2.91 \times 10^{-5}\text{Wb.}$$

The total flux can also be found from the total permeance and the mmf as follows:

$$\tau_T = \tau_1 + \tau_2 = (8.31 + 7.85) \times 10^{-8} = 16.16 \times 10^{-8}\text{H}$$

$$\Phi_T = 900 \times 0.2 \times 16.16 \times 10^{-2} = 2.91 \times 10^{-5} \text{ Wb.}$$

The leakage flux in the magnetic circuit of Fig. 1 is appreciable, particularly that through the air path in parallel with the 0.25-cm air gap g_2. The flux that links the exciting winding is therefore appreciably greater than the calculated value if the iron is unsaturated.

• PROBLEM 2-15

In the magnetic system shown in Fig. 1,

$$\ell_1 = \ell_3 = 300 \text{ mm} \qquad \ell_2 = 100 \text{ mm.}$$

$$A_1 = A_3 = 200 \text{ mm}^2 \qquad A_2 = 400 \text{ mm}^2$$

$$\mu_{r1} = \mu_{r3} = 2250 \qquad \mu_{r2} = 1350$$

$$N = 25$$

Determine the flux densities B_1, B_2, and B_3 in the three branches of the circuit when the coil current is 0.5 A.

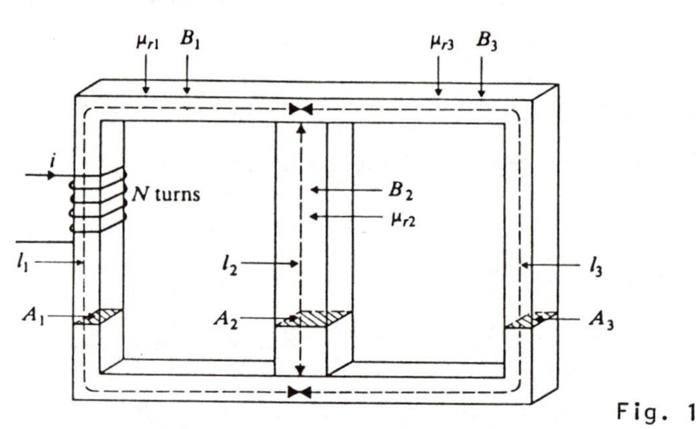

Fig. 1

Magnetic circuit.

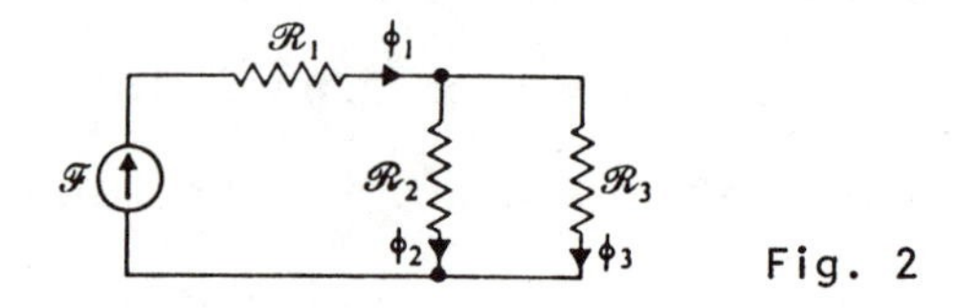

Fig. 2

Equivalent magnetic circuit for the system of Fig. 1.

Solution:

$$R_1 = R_3 = \frac{300 \times 10^{-3}}{2250 \times 4\pi \times 10^{-7} \times 200 \times 10^{-6}}$$

$$= 0.531 \times 10^6 \quad \text{A/Wb}$$

$$R_2 = \frac{100 \times 10^{-3}}{1350 \times 4\pi \times 10^{-7} \times 400 \times 10^{-6}}$$

$$= 0.148 \times 10^6 \quad A/Wb$$

The equivalent magnetic circuit for this system is shown in Fig. 2, and the problem may be solved by writing mmf equations for the two loops employing branch fluxes. Thus

$$F = R_1\Phi_1 + R_2\Phi_2$$

$$0 = R_3\Phi_3 - R_2\Phi_2$$

These are analogous to equations of potential difference for a dc circuit. Also,

$$\Phi_1 = \Phi_2 + \Phi_3$$

This is analogous to a current equation at a node. Substitution of the values of mmf and reluctances in these equations gives

$$12.5 \times 10^{-6} = 0.531\ \Phi_1 + 0.148\ \Phi_2$$

$$0 = -0.148\Phi_2 + 0.531\ \Phi_3$$

$$0 = -\Phi_1 + \Phi_2 + \Phi_3$$

Solution of these equations yields

$$\Phi_1 = 19.3 \times 10^{-6}\ Wb$$

$$\Phi_2 = 15.1 \times 10^{-6}\ Wb$$

$$\Phi_3 = 4.21 \times 10^{-6}\ Wb$$

from which the following values are obtained:

$$B_1 = \frac{\Phi_1}{A_1} = \frac{19.3 \times 10^{-6}}{200 \times 10^{-6}} = 0.0965\ T$$

$$B_2 = \frac{\Phi_2}{A_2} = \frac{15.1 \times 10^{-6}}{400 \times 10^{-6}} = 0.0377\ T$$

$$B_3 = \frac{\Phi_3}{A_3} = \frac{4.21 \times 10^{-6}}{200 \times 10^{-6}} = 0.0210\ T$$

• PROBLEM 2-16

Refer to Figure 1. $A_{Fe} = A_g = 0.01\ m^2$, $g = 0.005$ m, $N = 1000$ turns.

The length of the iron flux path is $\ell_{Fe} = 0.25$ m.

How much flux would be produced in the air gap for a coil current of 1.8 A? The iron is M-19, 29-gage steel laminations, stacking factor = 0.9.

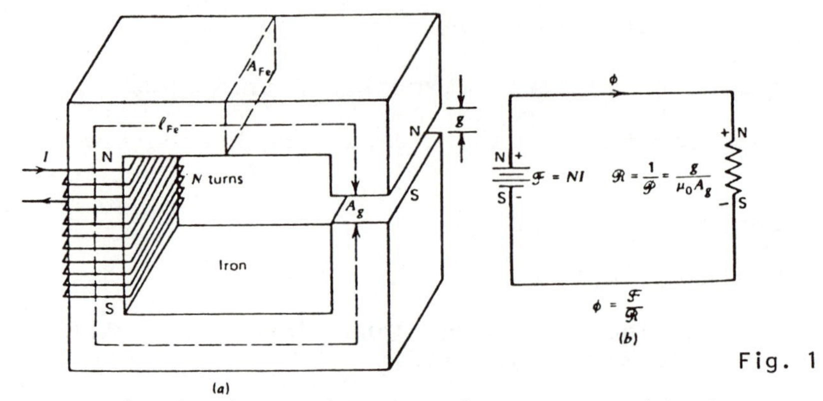

Fig. 1

A magnetic circuit and its electrical analog. (a)Asimple magnetic circuit. (b)Electric analog.

Solution:

$$B_{Fe} = \frac{\phi}{A_{Fe}}, \text{ where } \phi \text{ is the desired flux.}$$

$$H_{Fe} - \frac{F_{Fe}}{\ell_{Fe}} = \frac{NI - F_g}{\ell_{Fe}} = \frac{F_{coil}}{\ell_{Fe}} - \frac{\phi_g R_g}{\ell_{Fe}}$$

$$\therefore H_{Fe} = \frac{F_{coil}}{\ell_{Fe}} - \frac{B_{Fe} A_{Fe} R_g}{\ell_{Fe}}$$

(because $\Phi_g = B_{Fe} \cdot A_{Fe}$)

or

$$H_{Fe} = \frac{NI}{\ell_{Fe}} - B_{Fe} \frac{1}{\mu_0} \frac{A_{Fe}}{A_g} \frac{g}{\ell_{Fe}} \tag{1}$$

Note that Equation (1) is that of a straight line with an intercept F_{coil}/ℓ_{Fe} and a slope $(-A_{Fe}g/\mu_0 A_g \ell_{Fe})$. This is the air-gap characteristic in terms of B_{Fe} and H_{Fe}. For this example,

$$NI/\ell_{Fe} = 1000 \times 1.8/.25 = 7200 \text{ A/m}$$

The gross iron area is equal to the gap area. The stacking factor makes $(A_{Fe}/A_g) = 0.9$.

$$A_{Fe}g/\mu_0 A_g \ell_{Fe} = (0.9 \times 0.005)/(4\pi \times 10^{-7} \times 0.25) =$$

$$1.432 \times 10^4$$

and Equation (1) becomes

$$H_{Fe} = 7200 - 1.432 \times 10^4 \ B_{Fe}$$

To find the intercept on the B axis, let $H_{Fe} = 0$

$$B_{Fe} = \frac{7200}{1.432 \times 10^4} = 0.503 \text{ T}$$

This line is plotted on Figure 2. The intersection of the line with M-19 curve is the simultaneous solution of the nonlinear equation for the iron and the linear equation for the air gap. The intercept shows that for 1.8 coil amperes, B_{Fe} is 0.49 T. Then

$$\phi = B_{Fe}A_{Fe} = 0.50 \times (0.9 \times 0.01) = 4.5 \times 10^{-3}\text{Wb}.$$

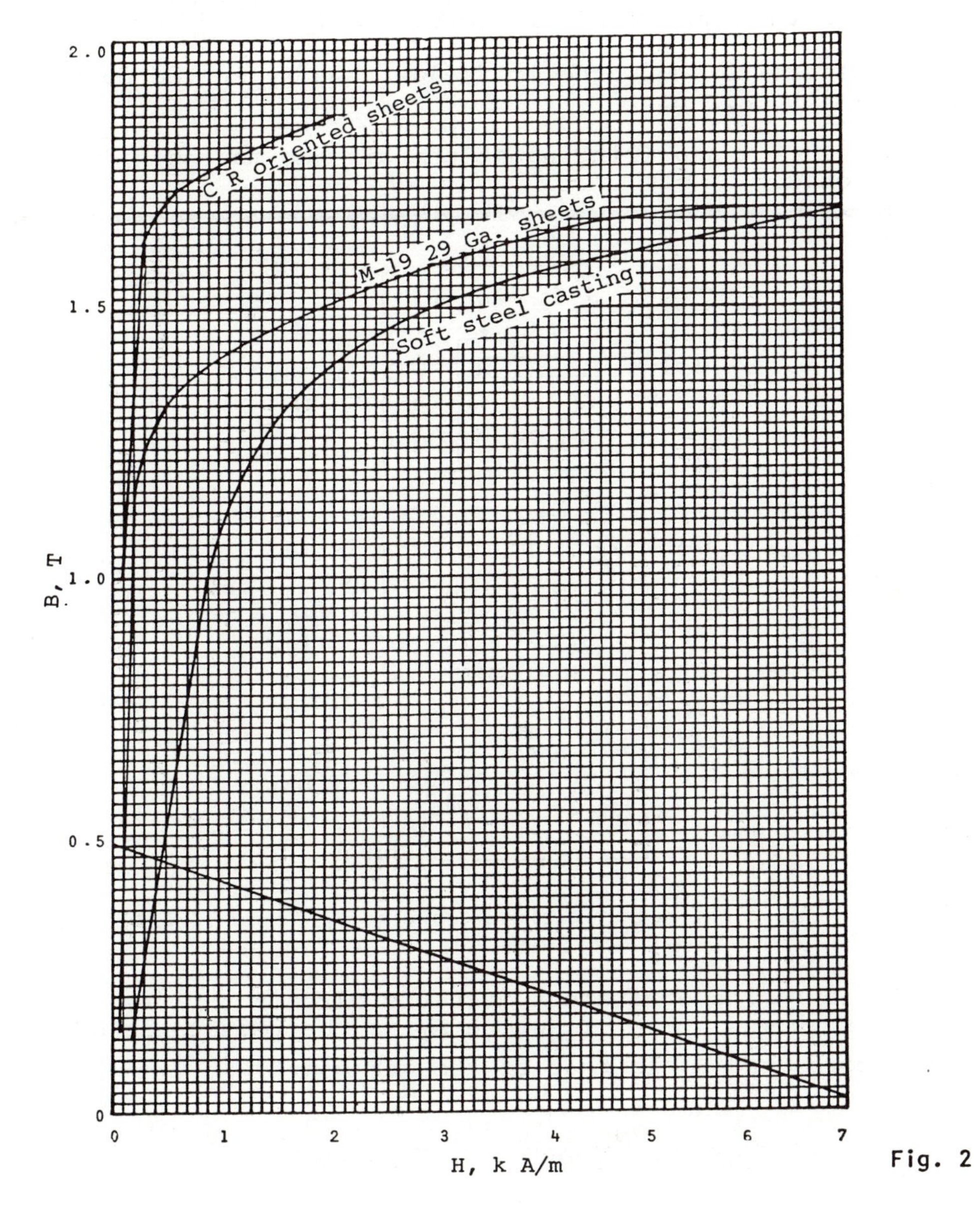

Fig. 2

B-H curves for three materials.

• PROBLEM 2-17

A 1000-turn coil is located on a ferromagnetic code that has an air gap, as illustrated in Fig. 1. If the coil current is 1.5 amperes and the relative permeability of the core is 1450,

a) Determine the proportion of the total mmf required to overcome the reluctance of the air gap.

b) Determine the flux density produced in the air gap.

c) Determine the ratio of the flux density in the air gap to the flux density produced at the center of the coil in the absence of the core.

d) Determine the magnetic field intesity in the core and in the air gap.

Leakage flux and fringing at the air gap may be neglected.

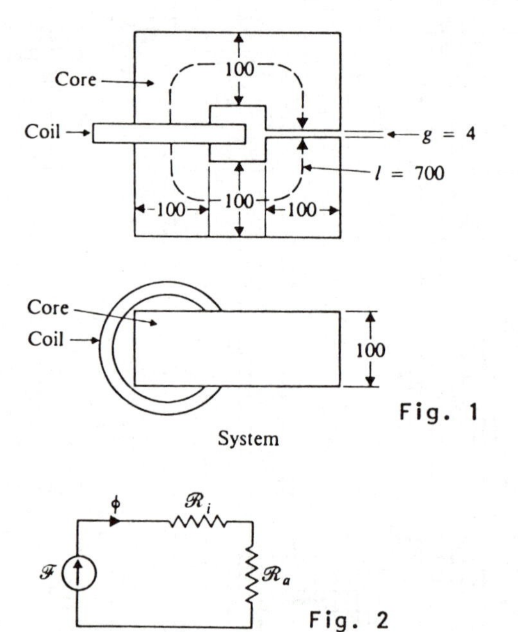

Equivalent magnetic circuit of the system of Fig. 1.

Solution: The equivalent magnetic circuit for the system is shown in Fig. 2, where R_i is the reluctance of the core, and R_a is that of the air gap.

a) $$R_i = \frac{700 \times 10^{-3}}{1450 \times 4\pi \times 10^{-7} \times 100^2 \times 10^{-6}}$$

$$= 38.42 \times 10^3 \text{ A/Wb}$$

$$R_a = \frac{4 \times 10^{-3}}{4\pi \times 10^{-7} \times 100^2 \times 10^{-6}} = 318.3 \times 10^3 \quad \text{A/Wb}$$

The total reluctance of the flux path is

$$R = R_i + R_a = (38.42 + 318.3)10^3 = 356.7 \times 10^3 \text{ A/Wb}$$

The proportion of the total mmf required to overcome the air-gap reluctance is

$$\frac{R_a}{R} = \frac{318.3}{356.7} = 0.892.$$

b) $$\Phi = \frac{F}{R} = \frac{10^3 \times 1.5}{356.7 \times 10^3} = 4.205 \times 10^{-3} \text{ Wb.}$$

$$B = \frac{\phi}{A} = \frac{4.205 \times 10^{-3}}{100^2 \times 10^{-6}} = 0.4205 \text{ T.}$$

c) $$\text{Ratio} = \frac{0.4205}{10.77 \times 10^{-3}} = 39.0.$$

d) In the air gap,

$$H_a = \frac{B}{\mu_0} = \frac{0.4205}{4\pi \times 10^{-7}} = 0.3346 \times 10^6 \quad \text{A/m.}$$

In the core,

$$H_i = \frac{B}{\mu_r \mu_0} = \frac{0.4205}{1450 \times 4\pi \times 10^{-7}} = 230.8 \text{ A/m}$$

• PROBLEM 2-18

A magnetic circuit consists of two parts in series. The dimensions are $A_1 = 4.8 \text{ in.}^2$, $A_2 = 3.8 \text{ in.}^2$, $\ell_1 = 12$ in., and $\ell_2 = 30$ in. There is a coil with 1,000 A-T wound around the core, which is made of cast steel. Find the flux.

Solution: There is nothing to indicate the magnitude of the flux. But since the given 1000 A-T are to be split into two parts , one for each part of the magnetic circuit, they can be split into halves as a first guess, giving $H_1\ell_1 = 500$ A-T. The corresponding flux (ϕ'), and the total mmf (F') required for producing this flux can, then, be determined. In terms of the tabulated arrangement, this means working from right to left to fill the empty boxes in the first row, and then from left to right for the second row.

For part-1

$$H\ell = 500$$

$$H = \frac{500}{\ell} = \frac{500}{12} = 41.5 \text{ A-T/in.}$$

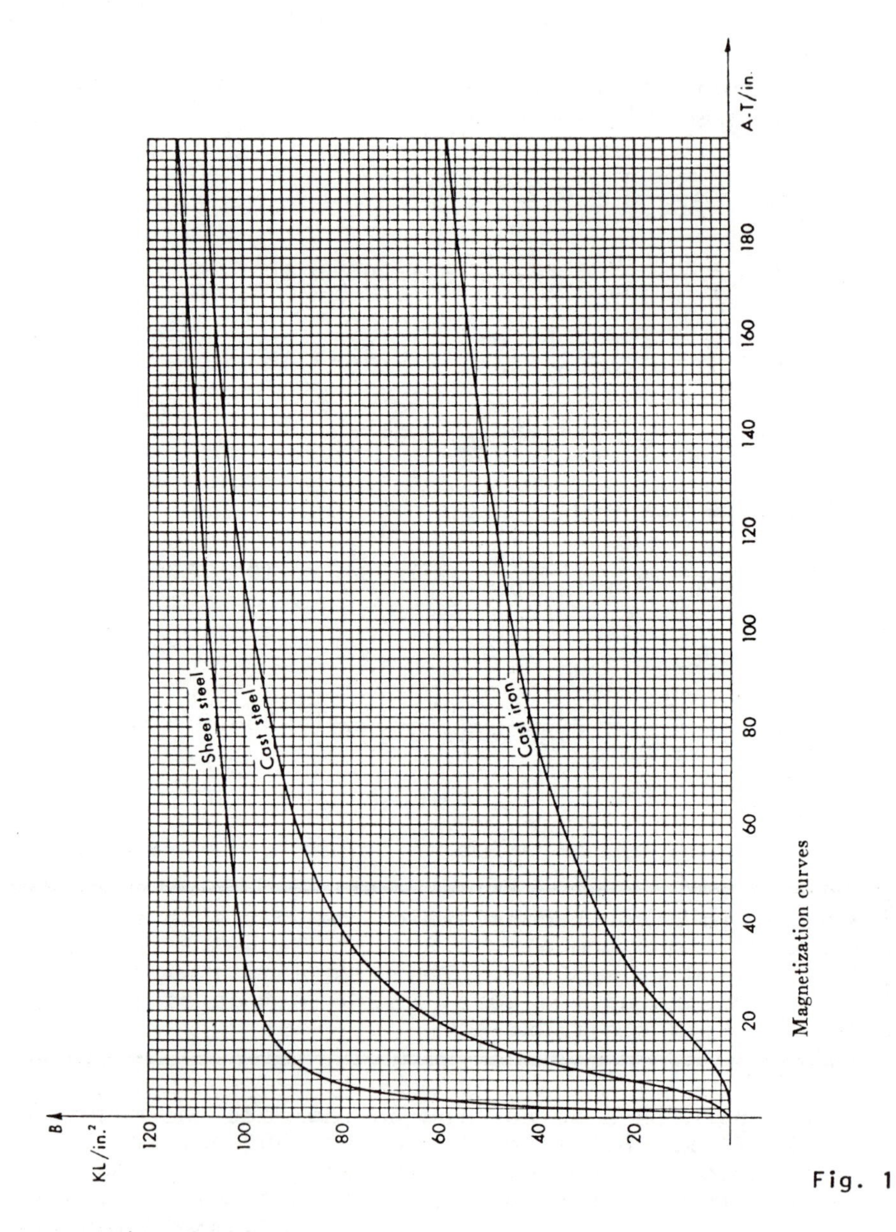

Magnetization curves

Fig. 1

From Fig. 1, for cast steel,

$$B = 81.5 \quad \text{KL/in.}^2$$

$$\Phi = BA$$

$$= 81.5 \times 4.8$$

$$= 390 \text{ KL.}$$

For part-2

$$B = \frac{\phi}{A} = \frac{390}{3.8} = 103 \text{ KL/in.}^2$$

From Fig. 1 for cast steel,

$$H = 130 \text{ A-T/in.}$$

$$H\ell = 130 \times 30 = 3{,}900 \text{ A-T.}$$

Part	Φ	A	B	H	ℓ	$H\ell$
1	390	4.8	81.5	41.5	12	500
2	390	3.8	103	130	30	3,900
						4,400 A-T

Since it takes so much more than the given 1,000 A-T to produce a flux of 390 KL, a lower value of 200 KL may be chosen as a second guess.

Part	Φ	A	B	H	ℓ	$H\ell$
1	200	4.8	41.5	12.5	12	150
2	200	3.8	52.5	16	30	480
						630 A-T

200 KL is too low. A third try may be made with 250 KL.

Part	Φ	A	B	H	ℓ	$H\ell$
1	250	4.8	52	16	12	192
2	250	3.8	66	24	30	720
						912 A-T

This is closer, but still a little too low. Try 260 KL next.

Part	Φ	A	B	H	ℓ	$H\ell$
1	260	4.8	54	17	12	204
2	260	3.8	68.5	26	30	780
						984 A-T

In view of the limited accuracy expected of any magnetic-circuit calculation (considering the various approximations inherent in the application of the circuit concept to ferromagnetic cores), there is no necessity for going further. 260 (or possibly 261) KL can be taken to be the answer to this problem.

The purpose of this example is to demonstrate the trial-and-error procedure, not to solve a numerical problem as fast as possible. Actually, in this example, the first guess could be considerably improved. Since part 1 of the magnetic circuit has the larger cross-sectional area and the smaller length, its reluctance must be quite a bit less than that of part 2. Thus we know that $H_1\ell_1 < H_2\ell_2$, and we might try $H_1\ell_1 = 250$ A-T (or, better still, $H_2\ell_2 = 750$ A-T) as our first guess.

DESIGN OF PERMANENT MAGNETS

• PROBLEM 2-19

Design a permanent magnet of Alnico 6, in which the air-gap is 0.6 cm long and has an area of 2 sq cm and the flux in the gap is 15,000 maxwells. The air-gap area is produced by the soft-steel pole pieces, each 2 sq cm area, Fig. 1, in which the mmf drop is negligible. Neglect fringing and leakage flux. Determine: (a) cross section of magnet; (b) length of magnet; (c) slope of shear line.

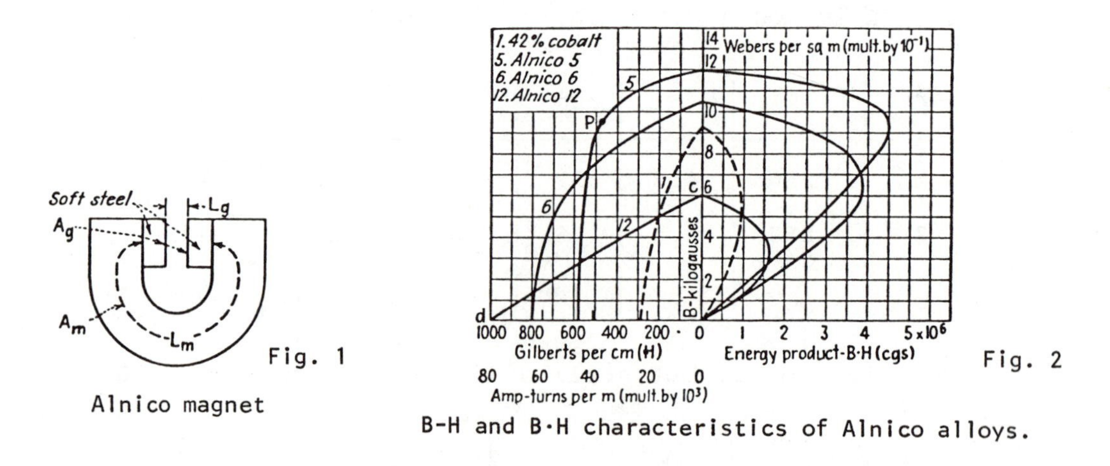

Fig. 1

Alnico magnet

Fig. 2

B-H and B·H characteristics of Alnico alloys.

Solution: (a) $B_g = 15,000/2 = 7,500$ gauss. Figure 2 shows that the maximum value of B·H for Alnico 6 occurs when $B = B_1 = 6,400$ gauss.

$$\frac{B_g}{B_m} = \frac{A_m}{A_g}$$

For part-2

$$B = \frac{\phi}{A} = \frac{390}{3.8} = 103 \text{ KL/in.}^2$$

From Fig. 1 for cast steel,

$$H = 130 \text{ A-T/in.}$$

$$H\ell = 130 \times 30 = 3{,}900 \text{ A-T.}$$

Part	Φ	A	B	H	ℓ	$H\ell$
1	390	4.8	81.5	41.5	12	500
2	390	3.8	103	130	30	3,900
						4,400 A-T

Since it takes so much more than the given 1,000 A-T to produce a flux of 390 KL, a lower value of 200 KL may be chosen as a second guess.

Part	Φ	A	B	H	ℓ	$H\ell$
1	200	4.8	41.5	12.5	12	150
2	200	3.8	52.5	16	30	480
						630 A-T

200 KL is too low. A third try may be made with 250 KL.

Part	Φ	A	B	H	ℓ	$H\ell$
1	250	4.8	52	16	12	192
2	250	3.8	66	24	30	720
						912 A-T

This is closer, but still a little too low. Try 260 KL next.

Part	Φ	A	B	H	ℓ	$H\ell$
1	260	4.8	54	17	12	204
2	260	3.8	68.5	26	30	780
						984 A-T

In view of the limited accuracy expected of any magnetic-circuit calculation (considering the various approximations inherent in the application of the circuit concept to ferromagnetic cores), there is no necessity for going further. 260 (or possibly 261) KL can be taken to be the answer to this problem.

The purpose of this example is to demonstrate the trial-and-error procedure, not to solve a numerical problem as fast as possible. Actually, in this example, the first guess could be considerably improved. Since part 1 of the magnetic circuit has the larger cross-sectional area and the smaller length, its reluctance must be quite a bit less than that of part 2. Thus we know that $H_1\ell_1 < H_2\ell_2$, and we might try $H_1\ell_1 = 250$ A-T (or, better still, $H_2\ell_2 = 750$ A-T) as our first guess.

DESIGN OF PERMANENT MAGNETS

• PROBLEM 2-19

Design a permanent magnet of Alnico 6, in which the air-gap is 0.6 cm long and has an area of 2 sq cm and the flux in the gap is 15,000 maxwells. The air-gap area is produced by the soft-steel pole pieces, each 2 sq cm area, Fig. 1, in which the mmf drop is negligible. Neglect fringing and leakage flux. Determine: (a) cross section of magnet; (b) length of magnet; (c) slope of shear line.

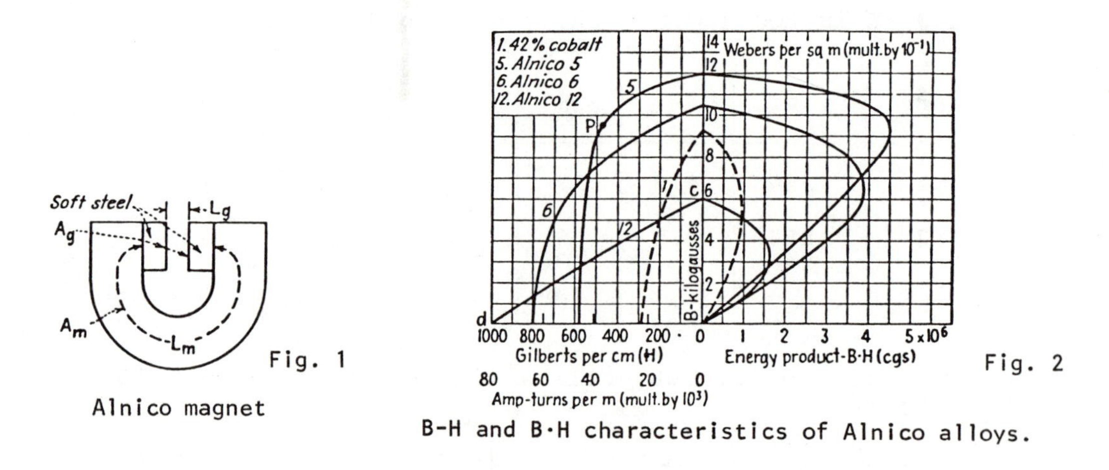

Fig. 1

Alnico magnet

Fig. 2

B-H and B·H characteristics of Alnico alloys.

Solution: (a) $B_g = 15{,}000/2 = 7{,}500$ gauss. Figure 2 shows that the maximum value of B·H for Alnico 6 occurs when $B = B_1 = 6{,}400$ gauss.

$$\frac{B_g}{B_m} = \frac{A_m}{A_g}$$

$\therefore \quad A_m = 2 \times \frac{7{,}500}{6{,}400} = 2{,}344$ sq cm.

(b) Since the gap is air, $H_g = B_g$ numerically. B_g = 7,500 gauss. Hence, from $H_m L_m - H_g L_g = 0$ and Fig. 3,

$$L_m = \frac{7{,}500 \times 0.6}{600} = 7.5 \text{ cm.}$$

(c) $\frac{B_m}{H_m} = \frac{B_g A_g / A_m}{B_g L_g / L_m} = \frac{A_g L_m}{A_m L_g} = -\tan\alpha$

$$\frac{B_m}{H_m} = \frac{A_g L_m}{A_m L_g} = \frac{2 \times 7.5}{2.344 \times 0.6} = 10.67 = -\tan\alpha.$$

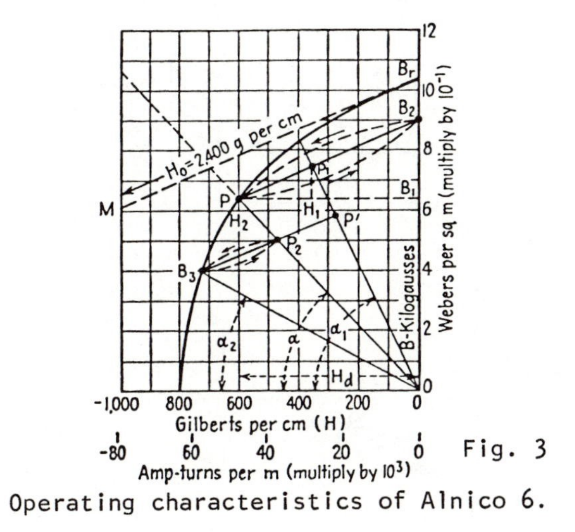

Fig. 3
Operating characteristics of Alnico 6.

Draw a shearing line from the origin which meets the -H abscissa at -1,000 and the ordinate B at 10,670. This gives tan α = -10.67. The shearing line intersects the curve at P, where B = 6,400 gauss and H = -600 oersteds.

• PROBLEM 2-20

In the permanent magnet system of Fig. 1(a), the magnet is made of Alnico 5, whose demagnetization curve is shown in Fig.2. The dimensions of the magnet are ℓ_m = 5 cm, A_m = 9 cm^2. The air gap dimensions are ℓ_g = 1 mm, A_g = 8 cm^2. The coil has 500 turns. The magnet is stabilized by a pulse of current giving H_n = -40 x 10^3 A/m in the magnet. The permeability of the soft magnetic material is so high that the mmf required to produce flux in it may be neglected. Also, fringing around the air gap and leakage flux may be ignored.

a) Determine the flux densities in the air gap and the magnet.

b) Determine the maximum negative coil current that may be permitted without loss of permanent magnetization.

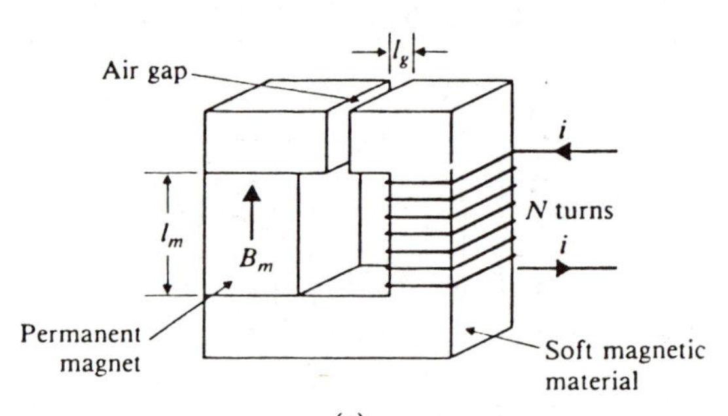

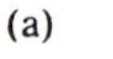

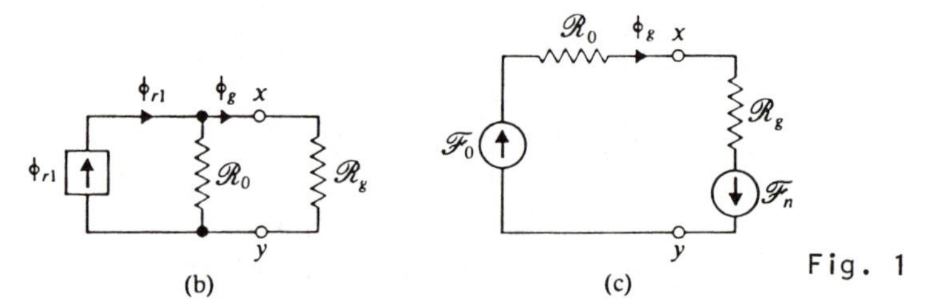

Linear model for a permanent-magnet with air gap. (a)Magnet system. (b)Equivalent magnetic circuit , i=0. (c)Equivalent magnetic circuit, i≠0.

Solution: a) Assuming that the slope of the recoil line is equal to that of the demagnetization curve at H_n = 0, the recoil permeability is

$$\mu_r\mu_0 = \frac{1.230 - 1.125}{24 \times 10^3} = 4.375 \times 10^{-6}$$

The recoil line terminates at point (-40 x 10^3, 0.99)

so that the minimum flux density permissible without danger of further demagnetization is 0.99 T. From the coordinates of the end point and the value of $\mu_r\mu_0$, the equation of the recoil line is

$$B = 1.165 + 4.375 \times 10^{-6} H_n \quad T,$$

$$-40 \times 10^3 < H_n < 0 \quad A$$

$$R_0 = \frac{\ell_m}{\mu_r\mu_0 A_m} \quad A/wb$$

$$= \frac{5 \times 10^{-2}}{4.375 \times 10^{-6} \times 9 \times 10^{-4}} = 12.70 \times 10^6 \quad A/Wb.$$

$$R_g = \frac{\ell_g}{\mu_0 A_g} \quad A/wb$$

$$= \frac{1.0 \times 10^{-3}}{4\pi \times 10^{-7} \times 8 \times 10^{-4}} = 0.995 \times 10^6 \quad A/Wb.$$

From the equation to the recoil line, the remanence for this degree of stabilization is $B_{r1} = 1.165$ T, so that from $\phi_m = B_m A_m$ Wb

$$\phi_{r1} = 9 \times 10^{-4} \times 1.165 = 1.049 \times 10^{-3} \quad Wb.$$

From the equivalent magnetic circuit of Fig. 1 (b), or from

$$\phi_g = \frac{R_0}{R_0 + R_g} \phi_{r1} \quad Wb$$

$$\phi_g = \frac{12.70 \times 10^6}{(12.70 + 0.995) 10^6} \times 1.049 \times 10^{-3}$$

$$= 0.9728 \times 10^{-3} \text{ Wb.}$$

$$B_g = \frac{\phi_g}{A_g} = \frac{0.9728 \times 10^{-3}}{8 \times 10^{-4}} = 1.216 \text{ T.}$$

$$B_m = \frac{\phi_g}{A_m} = \frac{0.9728 \times 10^{-3}}{9 \times 10^{-4}} = 1.081 \text{ T.}$$

b) To prevent loss of permanent magnetization, the magnet flux must not be reduced below the value

$$\phi_g = \phi_a = 0.99 \times 9 \times 10^{-4} = 0.891 \times 10^{-3} \text{ Wb.}$$

The source mmf in Fig. 1 (c) is

$$F_0 = \phi_{r1} R_0 = 1.049 \times 10^{-3} \times 12.7 \times 10^6$$

$$= 13.32 \times 10^3 \text{ A.}$$

Then, from Fig. 1(c),

$$F_n = F_0 - (R_0 + R_g)\phi_a$$

$$= 13.32 \times 10^3 - (12.7 + .995)10^6 \times 0.891 \times 10^{-3}$$

$$= 1119 \text{ A.}$$

The coil current is

$$i = F_n/N = \frac{1119}{500} = 2.24 \text{ A.}$$

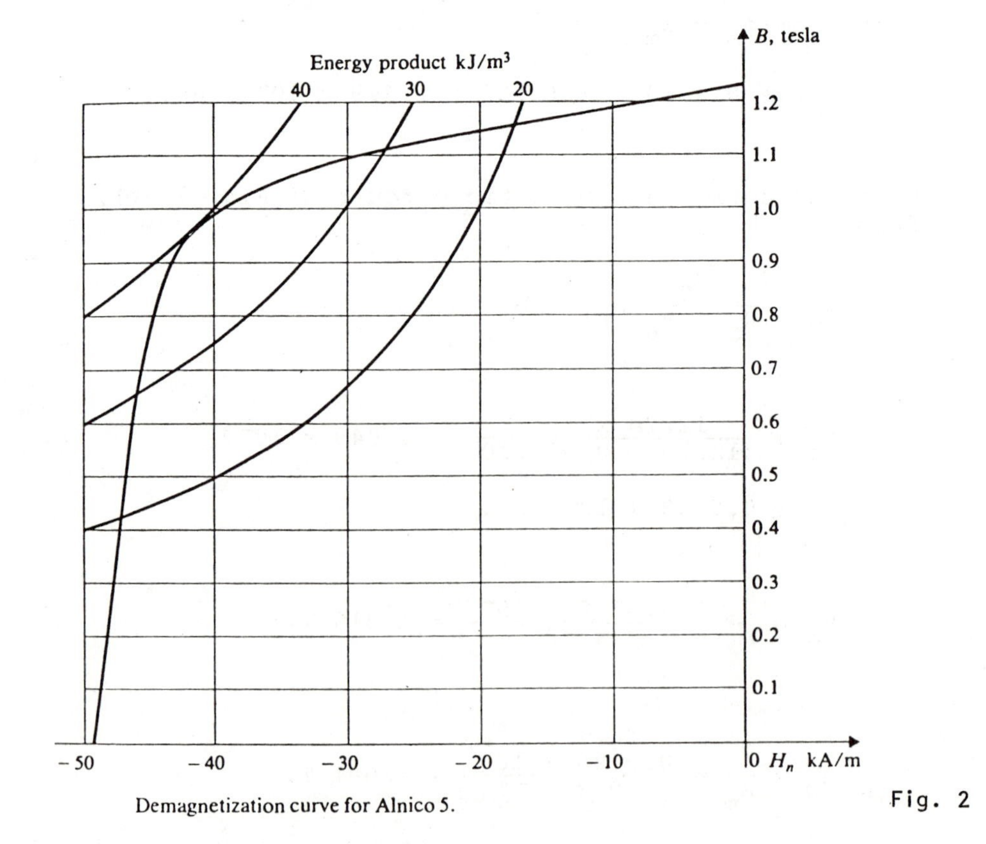

Demagnetization curve for Alnico 5.

Fig. 2

HYSTERESIS AND EDDY CURRENT LOSSES

• PROBLEM 2-21

The total core losses (hysteresis plus eddy current) for a sheet steel core are found to be 500 w at 25 Hz. When the frequency is increased to 50 Hz and the maximum flux density is kept constant, the total core loss becomes 1400 w. Find the hysteresis and eddy-current losses for both frequencies.

Solution: Since B_{max} is constant, the equation,

$$P_c = k'_h \ (B_{max})^n f + k'_e \ (B_{max})^2 f^2,$$

can have the following form:

$$P_c = Af + Bf^2 \text{ where } A = k'_h \ (B_{max})^n \text{ and } B = k'_e (B_{max})^2$$

For a frequency of 25 Hz, the core loss is

$$P_{c1} = 500 = A(25) + B(25)^2$$

For a frequency of 50 Hz, the core loss is

$$P_{c2} = 1400 = A(50) + B(50)^2$$

Solve the two equations to find A = 12 and B = 0.32. Now, we can find the individual losses.

$$P_{h1} = Af_1 = 300\text{w} \qquad P_{e1} = Bf_1^2 = 200 \text{ w}$$

$$P_{h2} = Af_2 = 600 \text{ w} \qquad P_{e2} = Bf_2^2 = 800 \text{ w.}$$

• PROBLEM 2-22

(i) A sheet steel core made from good sheet steel measures 3 in. by 3 in. and is 14 in. long. What will be the hysteresis loss per cycle in watts if the maximum flux density is 95 x 10^3 lines per square inch?

(ii) What will be the effect on the eddy current losses in a transformer (a) if the frequency is doubled, (b) if the flux density is doubled, (c) if both the frequency and the flux density are doubled at the same time?

Solution: (i) The hysteresis loss is given by

$$W_h = n\, B_m^{1.6}\, V\, 10^{-7}$$

joules per cycle.

$$= 0.002 \times \left[\frac{95 \times 10^3}{(2.54)^2}\right]^{1.6} \times (3 \times 3 \times 14)(2.54)^3 \times 10^{-7}$$

$$= 1.93 \text{ joules per cycle}$$

(ii) (a) $W_e = K_e(2f)^2 B_m^{\,2} = 4K_e f^2 B_m^{\,2}$.

(b) $W_e = K_e f^2 (2B_m)^2 = 4K_e f^2 B_m^{\,2}$.

(c) $W_e = K_e (2f)^2 (2B_m)^2 = 16K_e f^2 B_m^{\,2}$.

in terms of the original frequency and flux density.

ENERGY AND TORQUE IN MAGNETIC CIRCUITS

• PROBLEM 2-23

The distance between two parallel pole faces, similar to those in the figure is 10 cm, and the cross section in which the field is uniform is 8 by 5 cm, perpendicular to the paper. The mmf between surfaces AB and CD is 4,800 gilberts. Determine: (a) work done in carrying a unit N pole between surfaces CD and AB; (b) reluctance of flux path; (c) total flux; (d) flux density in gauss; (e) mmf per centimeter length of flux path; (f) flux in cross section of a centimeter cube perpendicular to field, surfaces of cube being either parallel with or perpendicular to field; (g) field intensity in oersteds or dynes per unit pole.

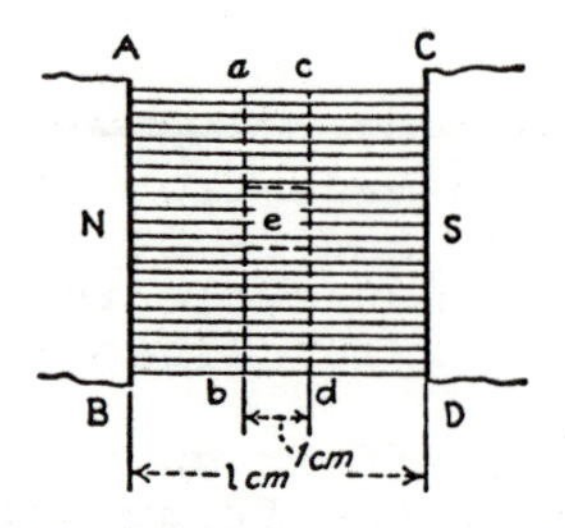

Field intensity and mmf.

Solution: (a) Since mmf between two points is defined

as work performed in carrying a unit pole between those points, the answer to (a) must be 4,800 ergs.

(b) $R = 10/(8 \times 5) = 0.250$ cgs reluctance units.

(c) $\phi = 4{,}800/0.250 = 19{,}200$ maxwells.

(d) $B = 19{,}200/(8 \times 5) = 480$ gauss.

(e) Mmf per centimeter is $4{,}800/10 = 480$ gilberts.

(f) Since the reluctance of a centimeter cube is unity, the flux in the cross section of the cube must be

$\phi = 480/1 = 480$ maxwells, which is also the flux density B.

(g) Since the mmf per centimeter is 480 gilberts, 480 ergs is involved in carrying a unit N pole 1 cm in the direction of the field. Since the field is uniform, the force is constant and must be $480/1 = 480$ dynes (force = work/distance). Field intensity is defined as force in dynes per unit pole; hence $H = 480$ oersteds.

It is to be noted that the gauss (d), the mmf per centimeter in gilberts (e), and the field intensity in oersteds (g) are all equal numerically.

• PROBLEM 2-24

The circuit shown carries the information given in the figure. For a coil of 800 turns carrying 1.5 amp, determine the flux densities in each material, the inductance of the coil, the energy storage of the circuit, and the pull between the faces of the air gap. Leakage flux and fringing at the air gap will be neglected.

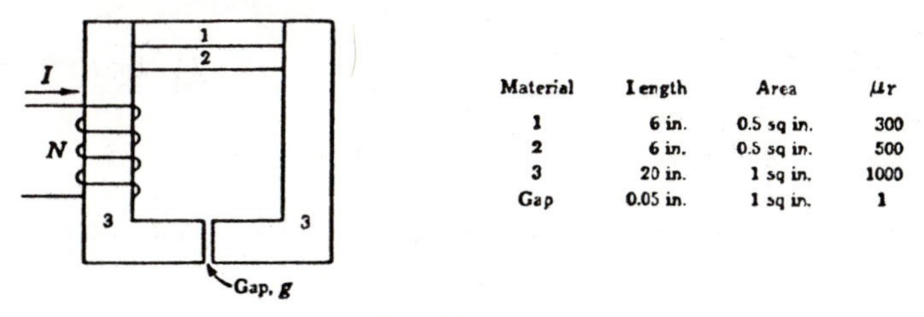

Material	Length	Area	μr
1	6 in.	0.5 sq in.	300
2	6 in.	0.5 sq in.	500
3	20 in.	1 sq in.	1000
Gap	0.05 in.	1 sq in.	1

Magnetic circuit with air gap.

Solution: The reluctances are calculated first, using conversion factors.

$$R_1 = \frac{\ell}{A\mu} = \frac{6 \times 0.0254}{0.5 \times 6.45 \times 10^{-4} \times 300 \times 4\pi \times 10^{-7}}$$

$$= 1.252 \times 10^6 \text{ amp/weber}$$

Similarly, $R_2 = 0.752 \times 10^6$, $R_3 = 0.626 \times 10^6$, and $R_g = 1.565 \times 10^6$ amp/weber. From these,

$$R_t = \frac{R_1 R_2}{R_1 + R_2} + R_3 + R_g = 2.66 \times 10^6 \text{ amp/weber}$$

Hence

$$\phi_t = \phi_3 = \phi_g = \phi_1 + \phi_2 = \frac{NI}{R_t}$$

$$= \frac{800 \times 1.5}{2.66 \times 10^6} = 4.51 \times 10^{-4} \text{ webers}$$

$$= 45{,}100 \text{ lines (maxwells)}$$

From $\frac{\phi_1}{\phi_2} = \frac{R_2}{R_1}$, we obtain $\phi_1 = 16{,}900$ lines and $\phi_2 = 28{,}200$ lines. From these values for the flux,

$$\beta_1 = 33{,}800 \text{ lines/sq in.} = 0.523 \text{ weber/sq meter,}$$

$$\beta_2 = 56{,}400 \text{ lines/sq in., and}$$

$$\beta_3 = \beta_g = 45{,}100 \text{ lines/sq in.}$$

At this point, the distribution of the magnetic potential drops follows readily: $H_1\ell_1 = \phi_1 R_1 = H_2\ell_2 = \phi_2 R_2 = 212$; $H_3\ell_3 = \phi_3 R_3 = 282$; $H_g\ell_g = \phi_g R_g = 706$; and $(H_1\ell_1 \text{ or } H_2\ell_2) + H_3\ell_3 + H_g\ell_g = 1{,}200 = NI = 800 \times 1.5$ amp turns.

The inductance of the coil is

$$L = \frac{N\phi_3}{I} = \frac{800 \times 4.51 \times 10^{-4}}{1.5} = 0.24 \text{ henry.}$$

The energy stored in the magnetic field is

$$W = \tfrac{1}{2}LI^2 = \tfrac{1}{2} \times 0.24 \times (1.5)^2 = 0.27 \text{ watt-second}$$

The pull between the faces of the gap is

$$f = \frac{\beta_g{}^2 A}{2\mu_0} = \frac{(45{,}100 \times 1.55 \times 10^{-5})^2 \times (6.45 \times 10^{-4})}{2 \times 4\pi \times 10^{-7}}$$

$$= 125 \text{ newtons}$$

$$= 28.2 \text{ lb.}$$

• PROBLEM 2-25

Determine the potential magnetic energy in the air-gap and magnetic material of the magnetic circuit of Fig. 1.

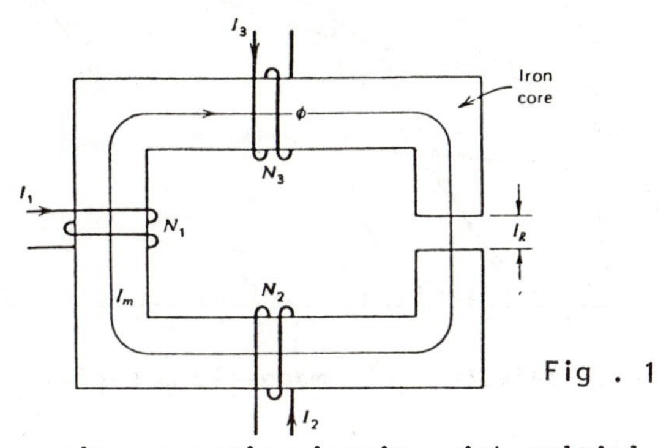

A composite magnetic circuit, with multiple excitation (mmf's).

Solution: The field distribution is uniform in both the gap and the magnetic material. In the gap from the equation, $B = \Phi/A$, $B_g = 0.001/(1.1 \times 0.0016) = 0.57$ tesla.

$$W = \frac{1}{2}\left(\frac{B_g^2}{\mu_0}\right)(\text{vol}) = \frac{1}{2}\left(\frac{0.57^2}{4\pi \times 10^{-7}}\right)(1.1 \times 0.0016 \times 10^{-4})$$

$$= 0.0228 \text{ J}$$

In the magnetic material, $B_m = 0.695$ tesla.

$$H_m = 54 \text{ A/m (from Figure 2)}$$

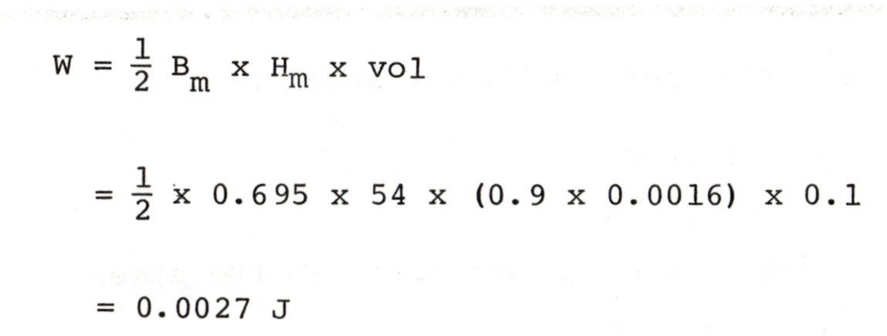

$$W = \frac{1}{2} B_m \times H_m \times \text{vol}$$

$$= \frac{1}{2} \times 0.695 \times 54 \times (0.9 \times 0.0016) \times 0.1$$

$$= 0.0027 \text{ J}$$

It is evident that most of the energy is required to establish the flux in the air-gap.

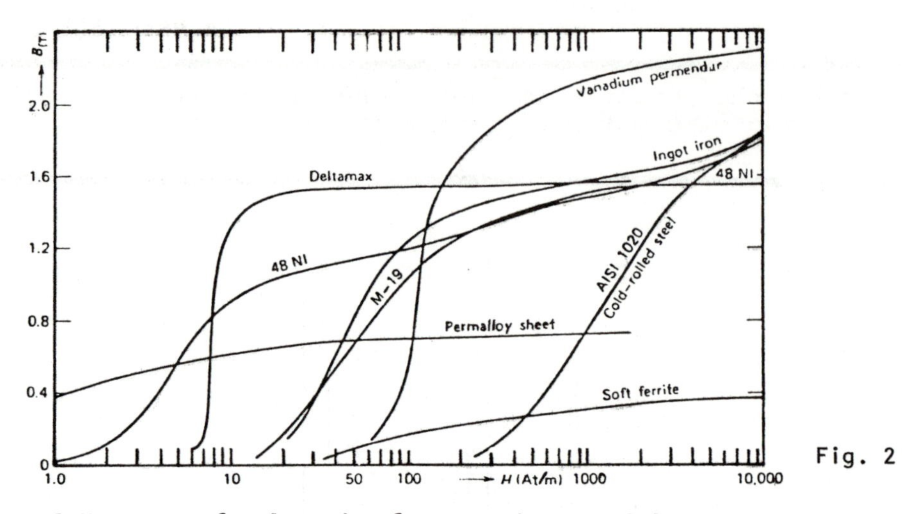

Fig. 2

B-H curves of selected soft magnetic materials.

• PROBLEM 2-26

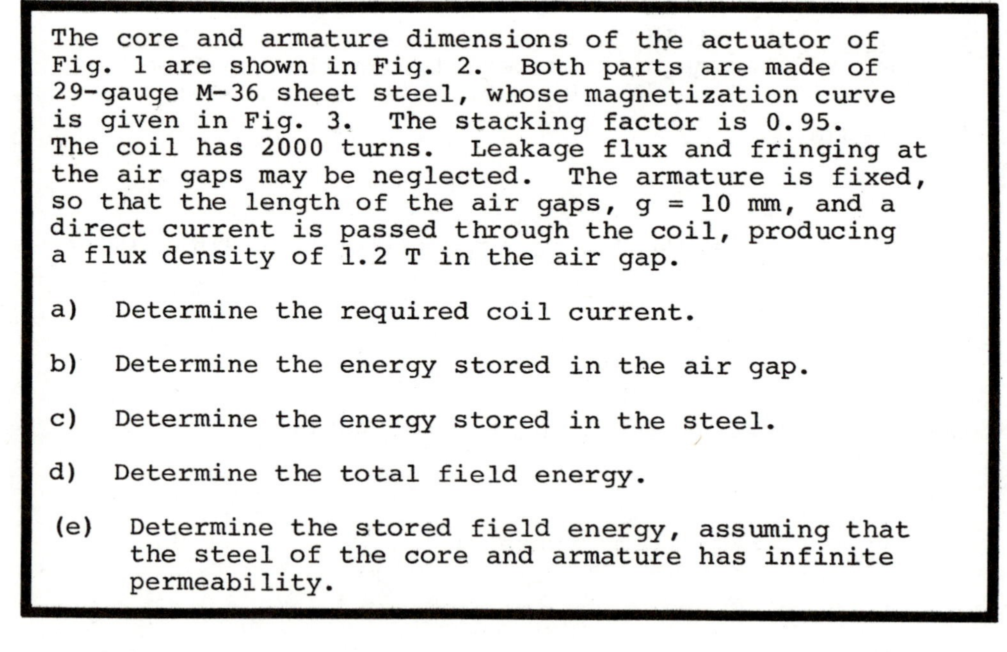

The core and armature dimensions of the actuator of Fig. 1 are shown in Fig. 2. Both parts are made of 29-gauge M-36 sheet steel, whose magnetization curve is given in Fig. 3. The stacking factor is 0.95. The coil has 2000 turns. Leakage flux and fringing at the air gaps may be neglected. The armature is fixed, so that the length of the air gaps, g = 10 mm, and a direct current is passed through the coil, producing a flux density of 1.2 T in the air gap.

a) Determine the required coil current.

b) Determine the energy stored in the air gap.

c) Determine the energy stored in the steel.

d) Determine the total field energy.

(e) Determine the stored field energy, assuming that the steel of the core and armature has infinite permeability.

Solution: a) Flux density in the steel is

$$B_s = \frac{1.2}{0.95} = 1.26 \text{ T}$$

From Fig. 3, magnetic field intensity in the steel is

$$H_s = 320 \text{ A/m}$$

Length of flux path in the steel is

$$\ell_s \simeq 2 \times 80 + 2(70 - 17.5/2) = 282.5 \text{ mm}$$

The mmf required by the steel is

$$F_s = 320 \times 0.2825 = 90.4 \text{ A}$$

Total flux in the magnetic system is

$$\phi = 35^2 \times 10^{-6} \times 1.2 = 1.470 \times 10^{-3} \text{ Wb}$$

Reluctance of air gaps in the magnetic path is

$$R_a = \frac{2 \times 10 \times 10^{-3}}{4\pi \times 10^{-7} \times 35^2 \times 10^{-6}} = 12.99 \times 10^6 \text{ A/Wb}$$

The mmf required by air gaps is

$$F_a = 1.470 \times 10^{-3} \times 12.99 \times 10^6 = 19.09 \times 10^3 \text{ A}$$

Total mmf required is

$$F = F_a + F_s = 19.09 \times 10^3 + 90.4 = 19.18 \times 10^3 \text{ A}$$

Coil current is

$$I = \frac{19.18 \times 10^3}{2000} = 9.59 \text{ A}$$

b) Energy density in the air gaps is

$$W_a = \frac{1}{2}\frac{B}{\mu_0} \text{ J/m}^3 = \frac{1}{2} \times \frac{1.2^2}{4\pi \times 10^{-7}} = 0.573 \times 10^6 \text{ J/m}^3.$$

Volume of air gaps is

$$V_a = 2 \times 10 \times 35^2 \times 10^{-9} = 24.5 \times 10^{-6} \text{ m}^3$$

Energy stored in the air gap is

$$W_a = 24.5 \times 0.573 = 14.0 \text{ J}$$

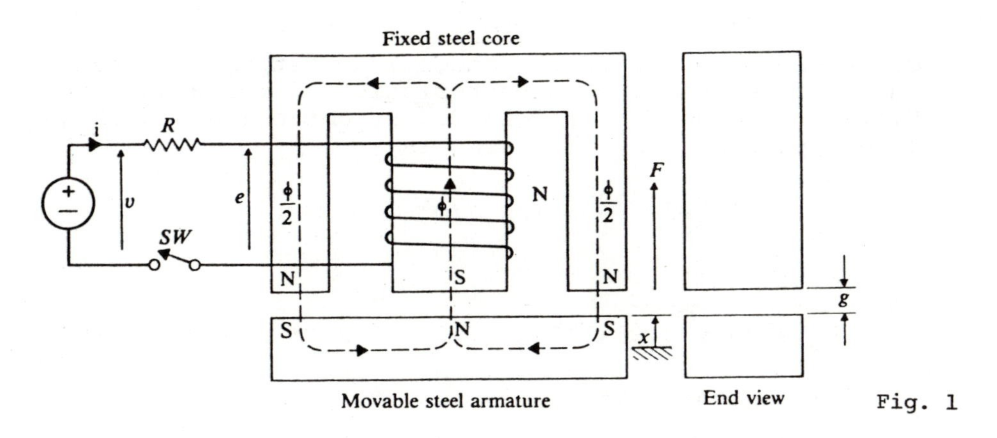

Fig. 1

Actuator of vertical-lift contactor (x=0 is "DOWN" position).

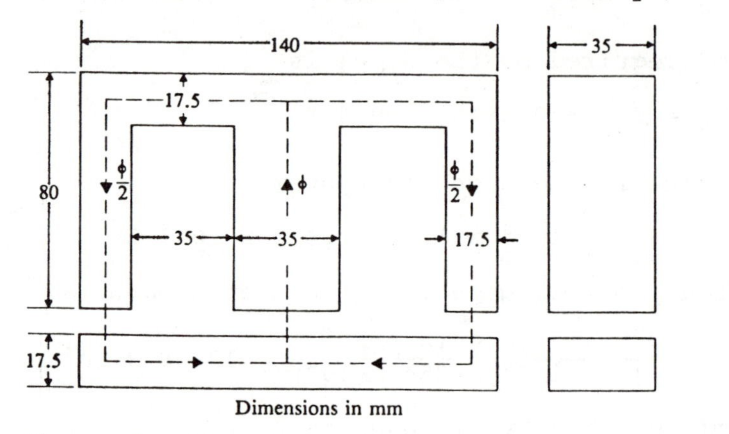

Fig. 2

Diagram for dimensions of the actuator.

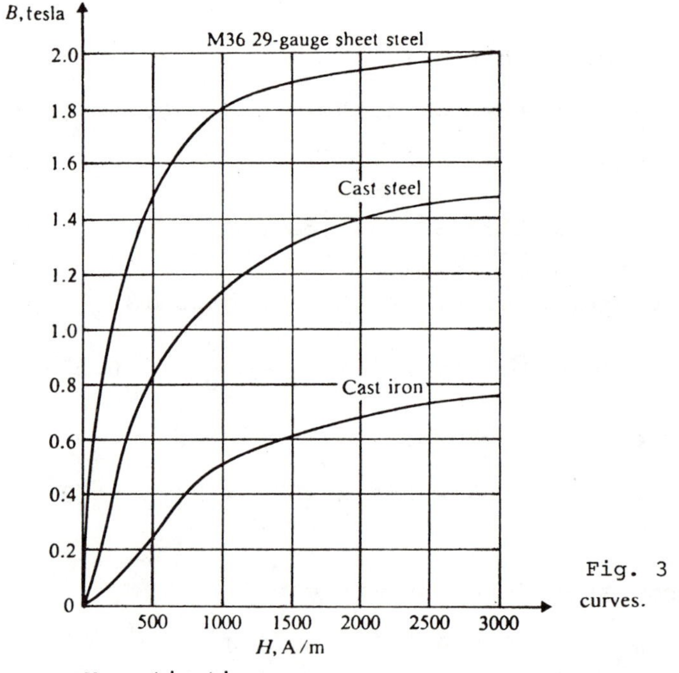

Fig. 3 curves.

Magnetization curves.

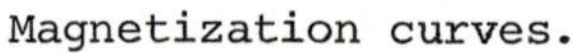

c) Energy density in the steel is given by the area enclosed between the characteristic and the B axis in Fig. 3 up to a value of 1.26 T. By employing a straight-line approximation, this area is

$$w_s = \frac{1.26 \times 300}{2} = 189 \text{ J/m}^3$$

Volume of steel is

$$V_s = 35 \times 0.95(140 \times 97.5 - 2 \times 35 \times 62.5) \times 10^{-9}$$

$$= 0.308 \times 10^{-3} \text{ m}^3$$

Energy stored in the steel is

$$W_s = 0.308 \times 10^{-3} \times 189 = 0.0582 \quad J$$

d) Total field energy is

$$W_f = W_a + W_s = 14.0 + 0.0582 \simeq 14.06 \text{ J}$$

The proportion of field energy stored in the steel is, therefore, seen to be negligibly small.

(e) From (a),

$$R_a = 12.99 \times 10^6$$

$$= 1.470 \times 10^{-3}$$

$$W_f = \frac{R\phi^2}{2}$$

$$\phi = \frac{12.99 \times 10^6 \times 1.470^2 \times 10^{-6}}{2} = 14.0 \text{ J.}$$

• PROBLEM 2-27

For the doubly excited system in Fig. 1, the inductances are approximated as follows: $L_1 = 11 + 3 \cos 2\theta$, $L_2 = 7 + 2 \cos 2\theta$, $M = 11 \cos \theta$ henrys. The coils are energized with direct currents. $I_1 = 0.7$ amp. $I_2 = 0.8$ amp. (a) Find the torque as a function of θ. (b) Find the energy stored in the system as a function of θ.

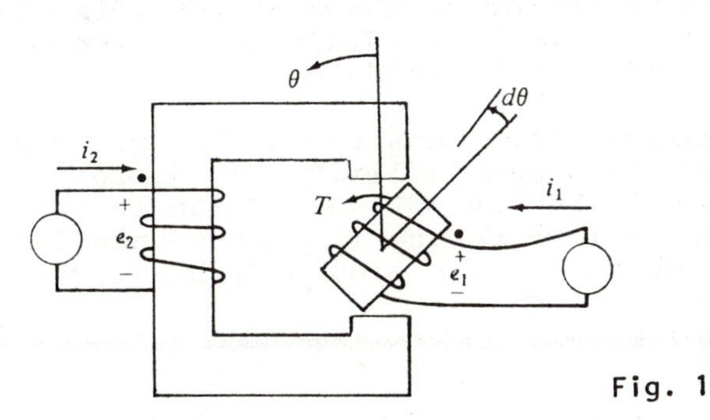

Fig. 1

Solution: (a) Find the derivatives of the inductances with respect to θ.

$$dL_1/d\theta = -6 \sin 2\theta, \; dL_2/d\theta = -4 \sin 2\theta,$$

$$dM/d\theta = -11 \sin \theta$$

$$T = \tfrac{1}{2} i_1^2 \, dL_1/d\theta + \tfrac{1}{2} i_2^2 \, dL_2/d\theta + i_1 i_2 \, dM/d\theta$$

$$= \tfrac{1}{2}(0.7)^2(-6 \sin 2\theta) + \tfrac{1}{2}(0.8)^2(-4 \sin 2\theta) + (0.7)(0.8)(-11 \sin \theta)$$

$$= -2.75 \sin 2\theta - 6.16 \sin \theta \text{ newton-meters.}$$

For the position shown in Fig. 1, $\theta = -50°$. The value of torque is $T = +7.43$ nm. This torque acts counterclockwise on the rotor. If this rotor is allowed to turn, it will move to the position where $\theta = 0°$ and where the torque is zero. (The torque is also zero at $\theta = 180°$, but if the position is away from 180°, the direction of the torque will tend to move the rotor toward 0°.) The sign of the mutual term is sensitive to the polarity of the currents. The reluctance torque terms are independent of the polarity of the currents.

(b) The stored energy is given by

$$W_m = \tfrac{1}{2}L_1 i_1^2 + \tfrac{1}{2}L_2 i_2^2 + M i_1 i_2$$

$$= \tfrac{1}{2}(11 + 3 \cos 2\theta)(0.7)^2 + \tfrac{1}{2}(7 + 2 \cos 2\theta)(0.8)^2 + (11 \cos \theta)(0.7)(0.8)$$

$$= 4.935 + 1.375 \cos 2\theta + 6.16 \cos \theta$$

• PROBLEM 2-28

The magnetic circuit shown in Fig. 1 is made of cast steel. The rotor is free to turn about a vertical axis. The dimensions are shown in the figure.

(a) Derive an expression in mks rationalized units for the torque acting on the rotor in terms of the dimensions and the magnetic field in the two air gaps. Neglect the effects of fringing.

(b) The maximum flux density in the overlapping portions of the air gaps is limited to approximately 130 kilolines/in.^{2}, because of saturation in the steel. Compute the maximum torque in inch-pounds for the following dimensions: $r_1 = 1.00$ in.; $h = 1.00$ in.; $g = 0.10$ in.

Solution: The torque can be derived from the derivative of air-gap reluctance, of air-gap permeance, or of field energy.

(a) The field energy density is $\mu_0 H_{ag}^2/2$ and the volume of the two overlapping air gaps is $2gh(r_1 + 0.5g)\theta$. Consequently the field energy is

$$W_{ag} = \mu_0 H_{ag}^2 gh(r_1 + 0.5g)\theta \qquad (1)$$

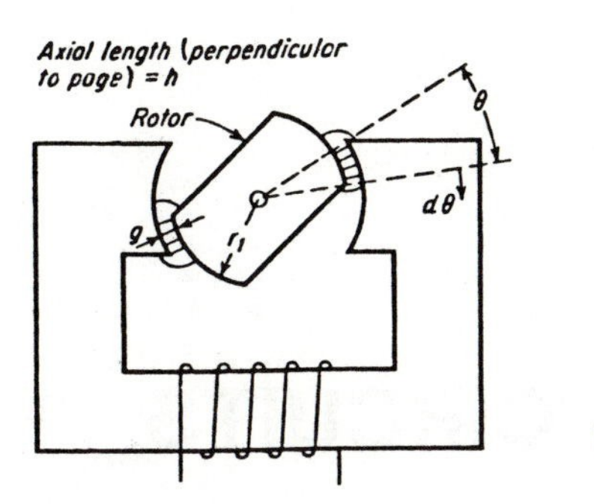

Fig. 1

Magnetic system.

At constant mmf H_{ag} is constant, and therefore differentiation of Eq. (1) with respect to θ at constant mmf in accordance with the following equation:

$$f = \frac{\partial W_{fld}}{\partial x}(F,x) = + \frac{\partial W_{fld}}{\partial x}(i,x), \text{ yields}$$

$$\text{Torque, } T = \mu_0 H_{ag}^2 gh(r_1 + 0.5g) = \frac{B_{ag}^2 gh(r_1 + 0.5g)}{\mu_0}$$

The torque acts in a direction to align the rotor with the stator pole faces.

(b) Convert the flux density and dimensions to mks units.

$$B_{ag} = \frac{130,000}{6.45} \times 10^4 \times 10^{-8} = 2.02 \text{ webers/m}^2$$

$$g = 0.1 \times 2.54 \times 10^{-2} = 0.00254 \text{ m}$$

$$h = r_1 = 1.00 \times 2.54 \times 10^{-2} = 0.0254 \text{ m}$$

$$\mu_0 = 4\pi \times 10^{-7}$$

Substitution of these numerical values in Eq. 2 gives

$$T = 5.56 \text{ newton-m}$$

$$= 5.56 \times 0.738 \times 12 = 49.3 \text{ in.-lb.}$$

CHAPTER 3

ELECTRIC CIRCUITS

RESISTANCE AND OHM'S LAW

• PROBLEM 3-1

(i) What is the circular-mil area of a wire (a) 0.1 in., (b) 0.2 in., (c) 0.325 in. in diameter?

(ii) The resistance of a length of copper wire is 3.60 ohms at 20°C. What is its resistance at 80°C?

Solution:

(i) (a) Diameter in mils = 0.1 x 1,000 = 100 mils

Area = diameter in mils, squared

= 100 x 100 = 10,000 cir mils

(b) Diameter = 0.2 x 1,000 = 200 mils

Area = 200 x 200 = 40,000 cir mils

(c) Diameter = 0.325 x 1,000 = 325 mils

Area = 325 x 325 = 105,625 cir mils

(ii) $R_2 = R[1 + a(t_2 - t)]$

where a = temperature coefficient of resistance at 20°C = 0.00393 (for copper).

$= 3.60[1 + 0.00393(80 - 20)]$

$= 3.60 \times 1.236 = 4.45$ ohms

• **PROBLEM 3-2**

Five resistances of 10, 15, 20, 25, and 30 ohms are connected in parallel. Calculate the joint resistance.

Solution: Given

$R_1 = 10$ ohms,

$R_2 = 15$ ohms,

$R_3 = 20$ ohms,

$R_4 = 25$ ohms,

$R_5 = 30$ ohms,

$$R = \frac{R_1R_2R_3R_4R_5}{R_1R_2R_3R_4 + R_1R_2R_3R_5 + R_1R_2R_4R_5 + R_1R_3R_4R_5 + R_2R_3R_4R_5}.$$

$R_1R_2R_3R_4R_5 = 10 \times 15 \times 20 \times 25 \times 30 = 2{,}250{,}000$

$R_1R_2R_3R_4 = 10 \times 15 \times 20 \times 25 = 75{,}000$

$R_1R_2R_3R_5 = 10 \times 15 \times 20 \times 30 = 90{,}000$

$R_1R_2R_4R_5 = 10 \times 15 \times 25 \times 30 = 112{,}500$

$R_1R_3R_4R_5 = 10 \times 20 \times 25 \times 30 = 150{,}000$

$R_2R_3R_4R_5 = 15 \times 20 \times 25 \times 30 = 225{,}000$

Sum of partial products $= 652{,}500$

$$\therefore R = \frac{2{,}250{,}000}{652{,}500} = 3.46 \text{ ohms.}$$

• **PROBLEM 3-3**

A coil that has a resistance of 0.05 ohms is connected to a cell that develops an emf of 1.5 volts. Find the current flowing if the internal resistance of the cell is (a) 0.1 ohm, and (b) 0.01 ohm.

Solution: (a) $I = \frac{E}{R_i + R_\ell} = \frac{1.5}{0.1 + 0.05} = 10$ amp

(b) $I = \frac{E}{R_i + R_\ell} = \frac{1.5}{0.01 + 0.05} = 25$ amp.

• PROBLEM 3-4

Determine the range of resistance values required in a rheostat for maintaining constant voltage at the terminals of a 2-kw 115-volt load, the battery consisting of 64 cells of the lead-acid type, with terminal voltage varying from 2.1 volts at full charge to 1.8 volts at the lower discharge limit.

Solution:
Voltage at the beginning of the discharge = 64 x 2.1 = 134.5 volts
Voltage at the end of discharge = 64 x 1.8 = 115 volts
$\therefore$ Max. drop across rheostat = 134.5 - 115 = 19.5 volts

$$\therefore \text{ Load current} = \frac{2 \times 1000}{115} = 17.4 \text{ amp}$$

$$\text{Rheostat Resistance} = \frac{19.5}{17.4} = 1.12 \text{ ohms.}$$

A suitable rheostat must therefore have a resistance ranging from 1.12 ohms at beginning of discharge to 0 at end of discharge, with a current capacity of 17.4 amp.

• PROBLEM 3-5

A 115-volt dc motor draws a current of 200 amperes and is located 1000 ft. from the supply source. If the copper transmission wire has a diameter of 0.45 inch, what must be the voltage of the supply source? See the figure.

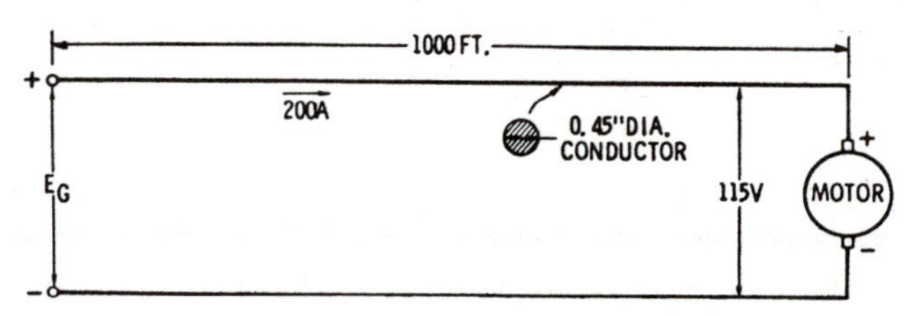

Voltage-drop calculation in a motor feeder line.

Solution: The resistance of any conductor varies directly with its length and inversely as its cross-sectional area. Therefore, the formula is

$$R = \frac{P \times L}{A},$$

where,

R is the resistance in ohms,

• PROBLEM 3-2

Five resistances of 10, 15, 20, 25, and 30 ohms are connected in parallel. Calculate the joint resistance.

Solution: Given

$R_1 = 10$ ohms,

$R_2 = 15$ ohms,

$R_3 = 20$ ohms,

$R_4 = 25$ ohms,

$R_5 = 30$ ohms,

$$R = \frac{R_1R_2R_3R_4R_5}{R_1R_2R_3R_4 + R_1R_2R_3R_5 + R_1R_2R_4R_5 + R_1R_3R_4R_5 + R_2R_3R_4R_5}.$$

$R_1R_2R_3R_4R_5 = 10 \times 15 \times 20 \times 25 \times 30 = 2,250,000$

$R_1R_2R_3R_4 = 10 \times 15 \times 20 \times 25 = 75,000$

$R_1R_2R_3R_5 = 10 \times 15 \times 20 \times 30 = 90,000$

$R_1R_2R_4R_5 = 10 \times 15 \times 25 \times 30 = 112,500$

$R_1R_3R_4R_5 = 10 \times 20 \times 25 \times 30 = 150,000$

$R_2R_3R_4R_5 = 15 \times 20 \times 25 \times 30 = 225,000$

Sum of partial products $= 652,500$

$$\therefore R = \frac{2,250,000}{652,500} = 3.46 \text{ ohms.}$$

• PROBLEM 3-3

A coil that has a resistance of 0.05 ohms is connected to a cell that develops an emf of 1.5 volts. Find the current flowing if the internal resistance of the cell is (a) 0.1 ohm, and (b) 0.01 ohm.

Solution: (a) $$I = \frac{E}{R_i + R_\ell} = \frac{1.5}{0.1 + 0.05} = 10 \text{ amp}$$

(b) $$I = \frac{E}{R_i + R_\ell} = \frac{1.5}{0.01 + 0.05} = 25 \text{ amp.}$$

• PROBLEM 3-4

Determine the range of resistance values required in a rheostat for maintaining constant voltage at the terminals of a 2-kw 115-volt load, the battery consisting of 64 cells of the lead-acid type, with terminal voltage varying from 2.1 volts at full charge to 1.8 volts at the lower discharge limit.

Solution:
Voltage at the beginning of the discharge = 64 x 2.1 = 134.5 volts
Voltage at the end of discharge = 64 x 1.8 = 115 volts
$\therefore$ Max. drop across rheostat = 134.5 - 115 = 19.5 volts

$$\therefore \text{ Load current } = \frac{2 \times 1000}{115} = 17.4 \text{ amp}$$

$$\text{Rheostat Resistance } = \frac{19.5}{17.4} = 1.12 \text{ ohms.}$$

A suitable rheostat must therefore have a resistance ranging from 1.12 ohms at beginning of discharge to 0 at end of discharge, with a current capacity of 17.4 amp.

• PROBLEM 3-5

A 115-volt dc motor draws a current of 200 amperes and is located 1000 ft. from the supply source. If the copper transmission wire has a diameter of 0.45 inch, what must be the voltage of the supply source? See the figure.

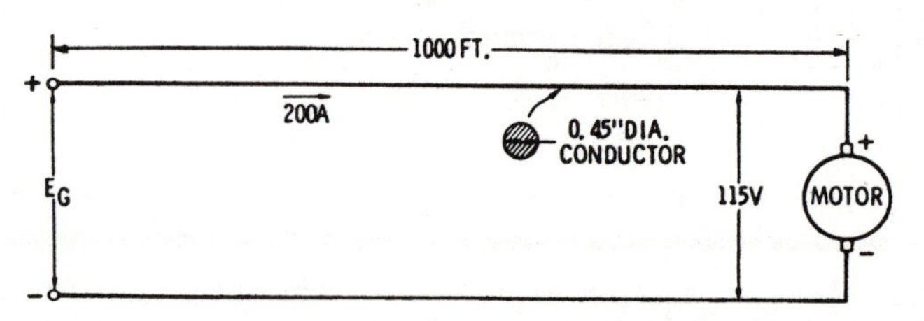

Voltage-drop calculation in a motor feeder line.

Solution: The resistance of any conductor varies directly with its length and inversely as its cross-sectional area. Therefore, the formula is

$$R = \frac{P \times L}{A},$$

where,

R is the resistance in ohms,

P is the specific resistance of one mil. ft of copper wire (10.4),

L is the length in feet,

A is the area in circular mils (diameter of wire in thousandths squared).

The cross-sectional area of a circular conductor can be expressed in circular mils by squaring the diameter of the conductor expressed in thousandths.

$$\therefore A = (0.45 \times 1000)^2 = 450^2$$

Thus, the resistance of the feed lines is

$$R = \frac{10.4 \times 2000}{450^2} = 0.103 \text{ ohm}$$

and the voltage of the supply source is

$$E_G = E_R + I_L R = 115 + (200 \times 0.103) = 135.6 \text{ volts.}$$

• PROBLEM 3-6

A 10-hp, 230-volt dc motor of 84% full-load efficiency is located 500 ft. from the supply mains. If the motor-starting current is 1.5 times the full-load current, what is the smallest cross-sectional area of copper wire required when the allowable voltage drop in the feeder at starting is 24 volts? See the figure.

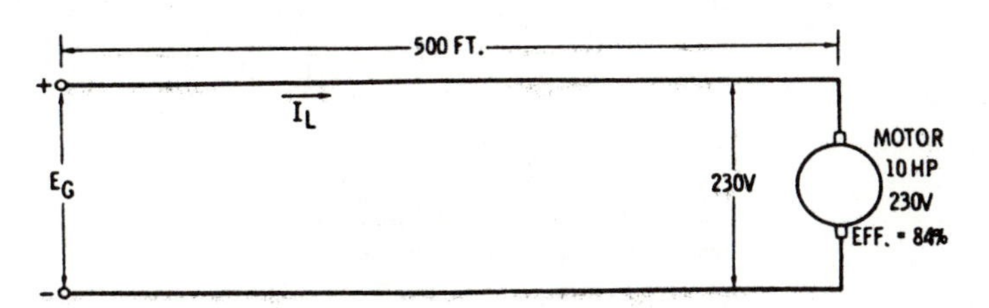

Cross-sectional area calculation for the feeder line to a 10-hp motor.

Solution: The motor full-load current is

$$I_L = \frac{\text{hp} \times 746}{E \times \text{efficiency}} = \frac{10 \times 746}{230 \times 0.84} = 38.6 \text{ amperes.}$$

Motor-starting current is

$$I_S = 38.6 \times 1.5 = 57.9 \text{ amperes}$$

Since the voltage drop in the feeder at starting is 24

volts, then, according to Ohm's law $R = \frac{24}{57.9} = 0.415$ ohm.

The minimum cross-sectional area is therefore

$$A = \frac{P \times L}{R}$$

where P is the specific resistance of one mil. ft. of copper wire (10.4).

$$A = \frac{10.4 \times 2 \times 500}{0.415} = 26,000 \text{ circular mils (approximately)}$$

• PROBLEM 3-7

> It is desired to measure the resistance of the insulation between a motor winding and the motor frame. A 300-volt 30,000-ohm voltmeter is used in making the measurements. The voltmeter when connected across the source reads 230 volts and when connected in series with the insulation reads 5 volts. Find the insulation resistance.

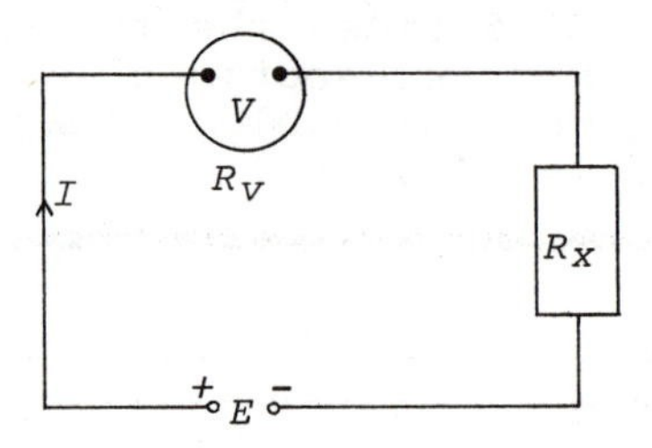

The voltmeter method of measuring resistance.

Solution: From the figure,

$$R_x = R_V\left(\frac{E}{V}\right) - 1 \quad \text{where } R_x \text{ is the resistance of insulation.}$$

$$= 30,000\left(\frac{230}{5} - 1\right) = 30,000 \times 45$$

$$= 1,350,000 \text{ ohms, or } 1.35 \text{ megohms.}$$

KIRCHHOFF'S LAW

• PROBLEM 3-8

A circuit contains six nodes lettered A, B, C, D, E, and F. Let v_{AB} be the voltage between nodes A and B with its positive reference at the first-named node, here A. Find v_{AC}, v_{AD}, v_{AE}, and v_{AF} if $v_{AB} = 6$ V, $v_{BD} = -3$ V, $v_{CF} = -8$ V, $v_{EC} = 4$ V, and: (a) $v_{DE} = 1$ V; (b) $v_{CD} = 1$ V; (c) $v_{FE} = 4$ V.

Solution: (a) Position the nodes in any arbitrarily chosen order and indicate the voltages between the nodes. In order to find the voltages in question, write a KVL equation around a loop which contains the unknown voltage.

To find v_{AC} write,

$$v_{AC} = v_{AB} + v_{BD} + v_{DE} + v_{EC}$$

$$v_{AC} = 6 + (-3) + 1 + 4$$

yielding $v_{AC} = 8V$.

Fig.1

To solve for v_{AD} write,

$$v_{AD} = v_{AB} + v_{BD}$$

$$v_{AD} = 6 + (-3)$$

yielding $v_{AD} = 3V$.

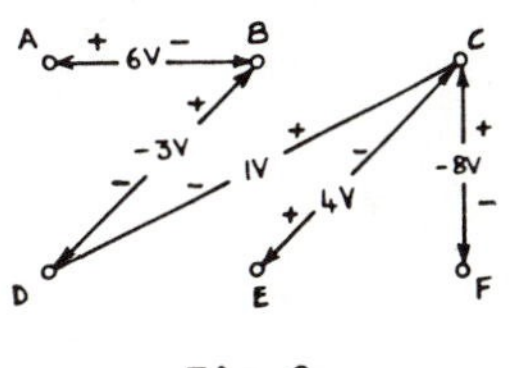

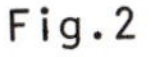
Fig.2

To find v_{AE} write,

$$v_{AE} = v_{AB} + v_{BD} + v_{DE}$$

$$v_{AE} = 6 + (-3) + 1$$

yielding $v_{AE} = 4V$.

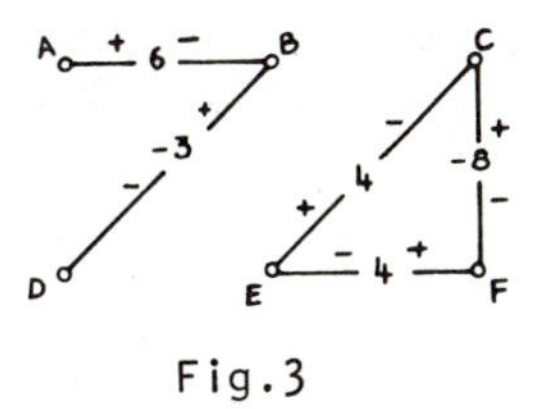

Fig.3

To find v_{AF}:

$$v_{AF} = v_{AB} + v_{BD} + v_{DE} + v_{EC} + v_{CF}$$

$$v_{AF} = 6 + (-3) + 1 + 4 + (-8)$$

yielding $v_{AF} = 0$.

(b) $v_{AC} = 6 + (-3) - 1 = 2V$

$v_{AD} = 6 + (-3) = 3V$

$v_{AE} = 6 + (-3) - 1 - 4 = -2V$

$v_{AF} = 6 + (-3) - 1 - 8 = -6V.$

(c) Only one loop can be formed to include the unknown voltages and that loop is

$$v_{AD} = v_{AB} + v_{BD} = 6 + (-3) = 3V$$

• PROBLEM 3-9

Determine v_x in each of the circuits of Fig. 1.

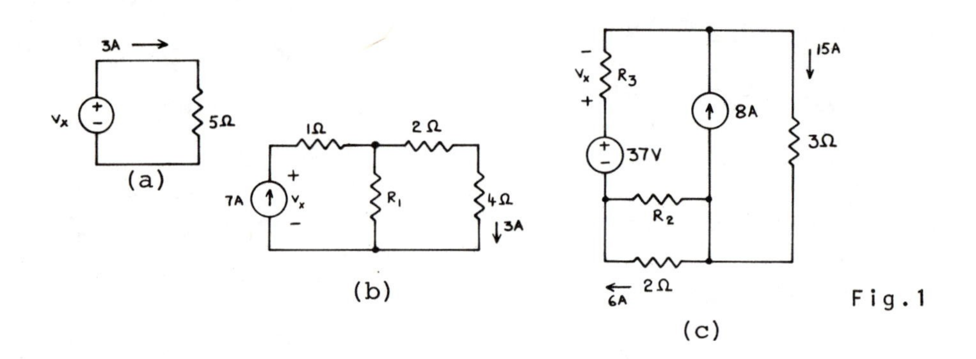

Fig. 1

Solution: (a) Apply Kirchhoff's Voltage Law to establish the sum of the voltage drops around a loop. The sum of the voltage drops around a loop must be equal to the sum of the sources around that loop.

In Fig. 1(a) the voltage drop across the 5Ω resistor is 3 × 5 = 15V. Therefore, v_x must be 15V.

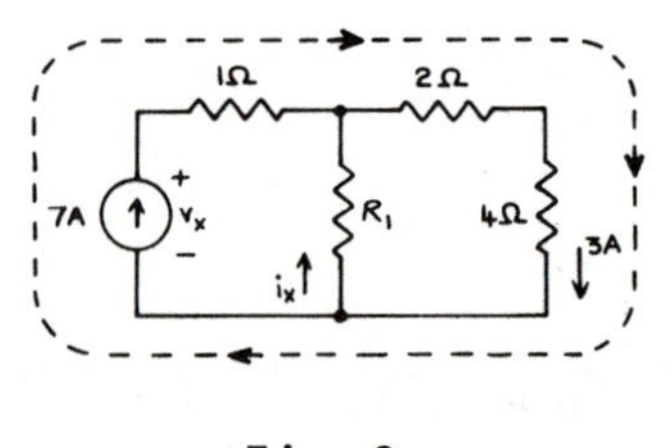

Fig. 2

(b) Fig. 2 shows the loop path taken for this problem.

The sum of the voltage drops $= 7 \times 1 + 3(2 + 4)$

$= 25$

The source voltage v_x $= 25V.$

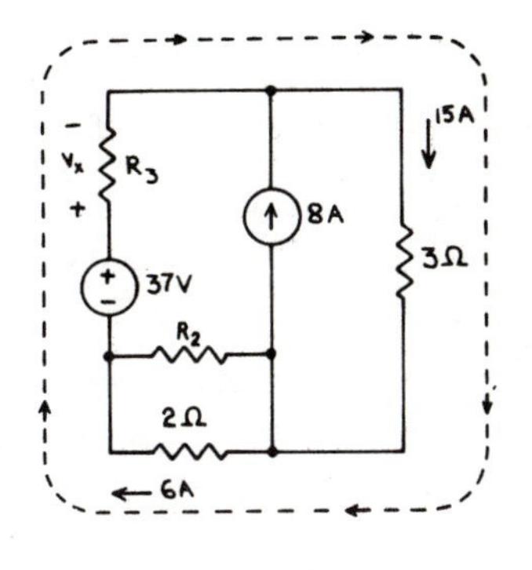

Fig.3

(c) Fig. 3 illustrates the loop path taken to solve this problem.

The sum of the voltage drops $= v_x + 15 \times 3 + 2 \times 6$

$= v_x + 57V$

The voltage source $= 37V.$

Therefore $37V = v_x + 57V$

$v_x = -20V.$

Note: There are several methods one can use to apply Kirchhoff's laws. In all methods, one must pay careful attention to the signs of voltage sources and current directions.

• PROBLEM 3-10

A circuit contains four nodes lettered A, B, C, and D. There are six branches, one between each pair of nodes. Let i_{AB} be the current in branch AB directed from node A to node B through the element. Then given $i_{AB} = 16$ mA, and $i_{DA} = 39$ mA, find i_{AC}, i_{BD} if $i_{CD} =$ (a) 23 mA; (b) -23 mA.

Solution: Fig. 1 represents the circuit described in this problem.

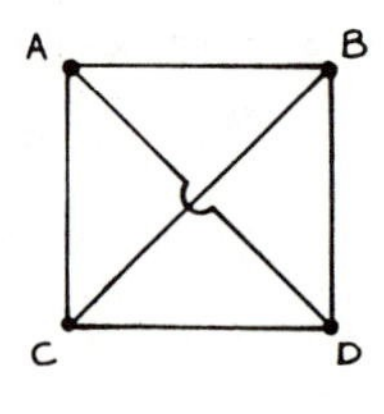

Fig.1

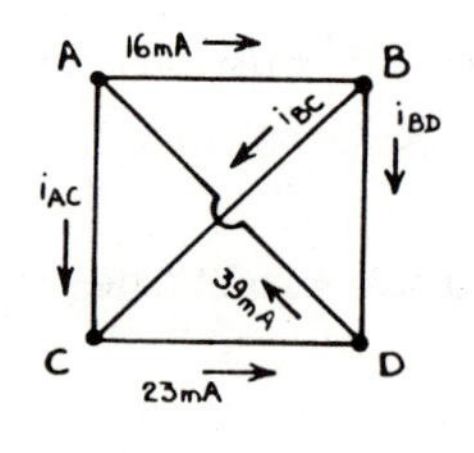

Fig.2

All possible interconnections of 6 different branches between the 4 nodes will give the same results.

(a) Fig. 2 shows the current values given and those to be found.

We can proceed to write a KCL equation for either node A or D.

Choosing node A,

$$39 \text{ mA} = i_{AC} + 16 \text{ mA}$$

$$i_{AC} = 23 \text{ mA}.$$

Now an equation for either node C or D can be written.

Choosing node C:

$$i_{AC} + i_{BC} = 23 \text{ mA}$$

$$23 \text{ mA} + i_{BC} = 23 \text{ mA}$$

$$i_{BC} = 0$$

At node B, since $i_{BC} = 0$,

$$i_{BD} = i_{AB} = 16 \text{ mA}$$

(b) If $i_{CD} = -23$ mA, then $i_{DC} = 23$ mA.

The current i_{AC} remains unchanged $i_{AC} = 23$ mA.

At node C we have,

$$i_{AC} + 23 \text{ mA} + i_{BC} = 0$$

$$i_{BC} = -23 - 23 = -46 \text{ mA}$$

At node D we have,

$$i_{BD} = 39 \text{ mA} + 23 \text{ mA}$$

$$i_{BD} = 62 \text{ mA}.$$

Determine i_x in each of the circuits of Fig. 1.

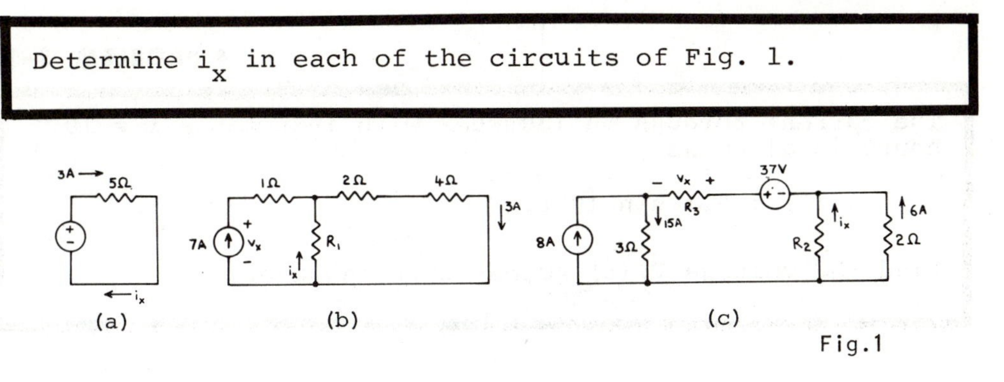

Fig.1

Solution: (a) The circuit in Fig. 1(a) is a single loop circuit. Hence, the current flowing through the loop is the same everywhere on the loop. Thus we can conclude

$$i_x = 3A.$$

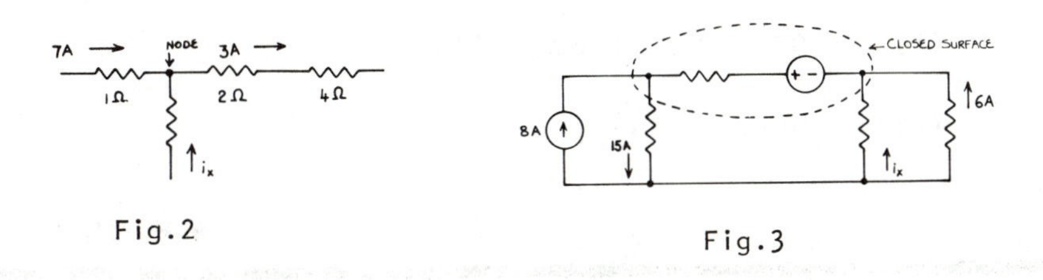

Fig.2 Fig.3

(b) Use Kirchhoff's Current Law to sum the currents at the node in Fig. 2.

The sum of the currents entering the node must be equal to the sum of the current(s) leaving the node. Thus,

$$7A + i_x = 3A.$$

Solving for i_x we obtain

$$i_x = 3A - 7A = -4A.$$

(c) Kirchhoff's current law can be extended by summing the currents entering and leaving a closed surface. Fig. 3 shows the closed surface used in this problem.

Thus,

$$8A + 6A + i_x = 15A;$$

solving for i_x,

$$i_x = 15A - 8A - 6A = 1A.$$

R L C CIRCUITS

• PROBLEM 3-12

The current through an inductor with inductance $L = 10^{-3}$ henry is given as

$$i_L(t) = 0.1 \sin 10^6 t.$$

Find the voltage $V_L(t)$ across this inductor.

Solution: We apply the definition for voltage across an inductor

$$V_L(t) = L \frac{di_L(t)}{dt}$$

$$V_L(t) = L \frac{d}{dt} (0.1 \sin 10^6 t)$$

$$V_L(t) = 10^{-3} 10^6 (.1) \cos 10^6 t$$

$$V_L(t) = 100 \cos 10^6 t.$$

• PROBLEM 3-13

The figure shows a series circuit made up of a voltage source and two linear circuit elements. Write an equation which describes the behavior of the circuit.

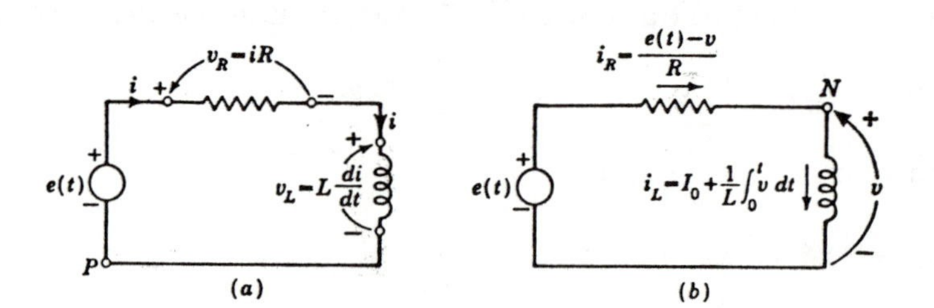

A series *RL* circuit: (a) current *i* as the unknown; (b) voltage across the inductance element as the unknown.

Solution: By Kirchhoff's current law, the same current flows through all elements; it is designated in the sketch by the symbol i, with the positive direction clockwise around the circuit. The current will be the dependent variable in the equation that is to be written.

As indicated in Fig. a, the voltage across the resistance is $v_R = Ri$, and that across the inductance is $v_L =$

L di/dt. The positive directions of these are taken in accordance with the positive direction of i. Starting at the point P and going once clockwise around the circuit, from Kirchhoff's voltage law:

$$\sum v = e(t) - Ri - L\frac{di}{dt} = 0.$$

This can be rearranged into a more conventional form with the terms involving i on the left side and the driving function on the right:

$$L\frac{di}{dt} + Ri = e(t)$$

This is a differential equation. If e(t) is a prescribed function of time, the equation can be solved for i as a function of time. On the other hand, if i is given as a function of time, one can solve for the e(t) which would produce this current.

• PROBLEM 3-14

For the circuit shown in Fig. 1, find i and v as functions of time for t > 0.

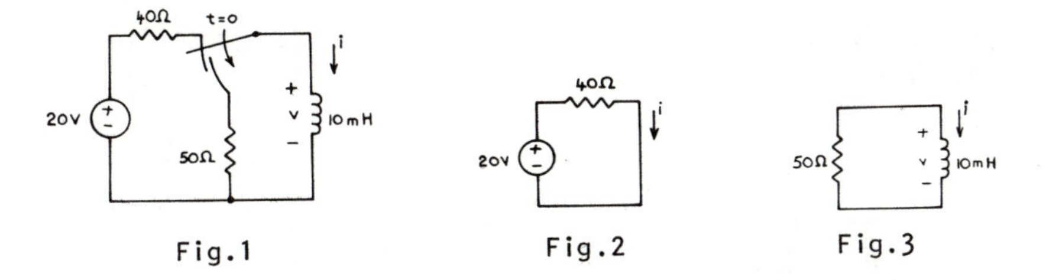

Solution: First, find the current through the inductor i, just before the switch is thrown. Fig. 2 shows the circuit at $t = 0^-$. In Fig. 2, $i = \frac{20v}{40\Omega} = 0.5A$. Fig. 3 shows the circuit in Fig. 1 at $t = 0^+$.

In Fig. 3, in order for the voltage drops to sum to zero around loop, the voltage v must be $-(50\Omega)(0.5A) = -25V$. The time constant is found to be

$$\frac{L}{R} = \frac{10mH}{50\Omega} = \frac{1}{5000}.$$

Write the response,

$$i(t) = 0.5e^{-5000t}A;\ t > 0$$

$$v(t) = -25e^{-5000t}V;\ t > 0 \quad .$$

Given two circuits (see figure) in parallel, one branch consisting of a resistance of 15 ohms and the other of an inductive reactance of 10 ohms. When the impressed voltage is 110, find the

(a) Current through the ohmic resistance.

(b) Current through the inductive reactance.

(c) Line current.

(d) Power factor.

(e) Angle of lag of the line current.

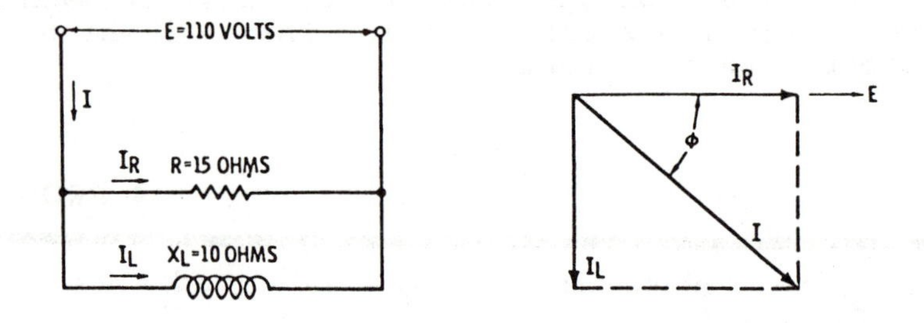

Resistance and inductance in parallel.

Solution: With reference to the vector diagram in the figure, the

(a) Current through the ohmic resistance is

$$I_R = \frac{E}{R} = \frac{110}{15} = 7.34 \text{ amperes.}$$

(b) Current through the inductive reactance is

$$I_L = \frac{E}{X_L} = \frac{110}{10} = 11 \text{ amperes,}$$

(c) Total line current (I_T) is

$$I = \sqrt{I_R^2 + I_L^2} = \sqrt{7.34^2 + 11^2} = 13.2 \text{ amperes,}$$

(d) Power factor is

$$\cos\phi = \frac{I_R}{I_T} = \frac{7.34}{13.2} = 0.556.$$

(e) Angle of lag of the line current is

$$\phi = 56.2°.$$

• PROBLEM 3-16

Consider the capacitor shown in the figure. The capacitance C(t) is given by

$$C(t) = C_0(1 + 0.5 \sin t)$$

The voltage across this capacitor is given by

$$v(t) = 2 \sin \omega t$$

Find the current through the capacitor.

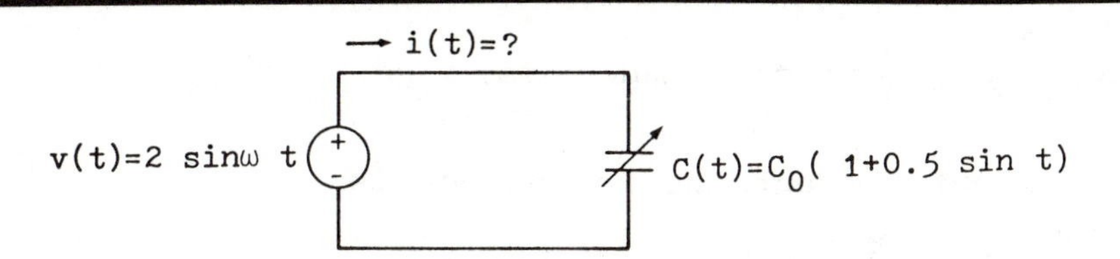

Solution: We can find the charge on the capacitor q(t) by using the definition q(t) = C v(t). In this problem C is a time varying function C (t).

$$q(t) = C(t)\ v(t)$$

$$q(t) = C_0\ (1 + 0.5 \sin t)(2 \sin \omega t).$$

Since $i(t) = \frac{dq}{dt}$, we have

$$i(t) = \frac{d}{dt}\left[C_0\ (1 + 0.5 \sin t)(2 \sin \omega t)\right]$$

$$= (2 \sin \omega t)(0.5\ C_0 \cos t) +$$

$$C_0\ (1 + 0.5 \sin t)(2\omega \cos \omega t).$$

$$i(t) = C_0 \sin \omega t \cos t + 2\omega\ C_0 \cos \omega t\ (1 + 0.5 \sin t).$$

• PROBLEM 3-17

As shown in the figure, a resistance of 130 ohms and a capacitance of 30 microfarads are connected in parallel across a 230-volt; 50 hertz supply. Find the following:

(a) Current in each circuit.

(b) Total current.

(c) Phase difference between the total current and the applied voltage.

(d) Power consumed.

(e) Power factor.

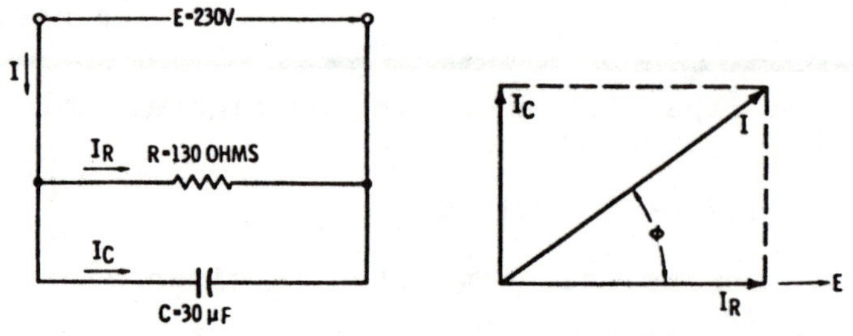

Resistance and capacitance in parallel.

Solution: The capacitive reactance of the circuit is

$$X_C = \frac{10^6}{2\pi \times 50 \times 30} = 106 \text{ ohms.}$$

(a) Current through the resistance is

$$I_R = \frac{E}{R} = \frac{230}{130} = 1.77 \text{ amperes.}$$

Current through the capacitance is

$$I_C = \frac{E}{X_C} = \frac{230}{106} = 2.17 \text{ amperes.}$$

(b) Total current is

$$I = \sqrt{I_R{}^2 + I_C{}^2} = \sqrt{1.77^2 + 2.17^2} = 2.8 \text{ amperes.}$$

(c) Phase difference is

$$\cos\phi = \frac{I_R}{I} = \frac{1.77}{2.8} = 0.632.$$

$$\therefore \quad \phi = 50.8° \quad \text{(angle of lead).}$$

(d) Power consumed is

$$P = I_R{}^2 R = 1.77^2 \times 130 = 407 \text{ watts.}$$

(e) Power factor according to (c) is 0.632, or 63.2%.

• PROBLEM 3-18

In the circuit shown in the figure, the voltage across the capacitance element at a certain instant of time is measured to be 40 volts as shown. At this instant, what is the current in the circuit, and at what rate is v_C changing?

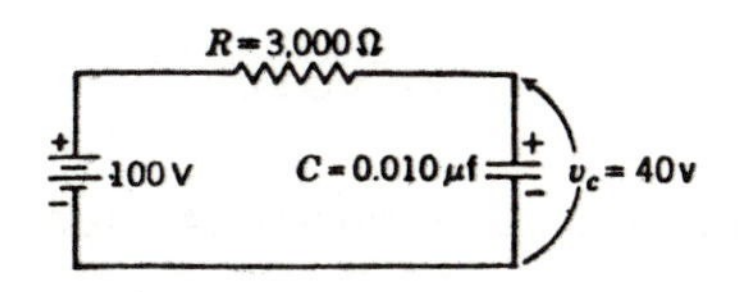

Solution: By Kirchhoff's voltage law,

$$E_S = v_R + v_C$$

$$\therefore v_R = E_S - v_C$$

$$= 100 - 40 \text{ V.}$$

∴ Voltage across the resistance is 60 volts; hence, the current is $i = v_R/R = \frac{60}{3000} = 0.020$ amp.

The current causes the voltage across the capacitance to change at a rate determined by $C\, dv_C/dt = i$, or $0.010 \times 10^{-6}\, dv_C/dt = 0.020$ amp, from which $dv_C/dt = 2 \times 10^6$ volts/sec.

• PROBLEM 3-19

A series circuit consists of a 30-microfarad capacitance and a resistance of 50 ohms connected across a 110-volt, 60-hertz supply. Calculate the

(a) Impedance of the circuit.

(b) Current in the circuit.

(c) Voltage drop across the resistance.

(d) Voltage drop across the capacitance.

(e) Angle between the voltage and the current.

(f) Power loss.

(g) Power factor of the circuit.

Solution: (a) Impedance of the circuit is

$$X_C = \frac{1}{2\pi fC} = \frac{1}{2\pi \times 60 \times 0.000030} = 88.4 \text{ ohms.}$$

$$\therefore Z = \sqrt{R^2 + X_C^2} = \sqrt{50^2 + 88.4^2} = 101.6 \text{ ohms.}$$

(b) Current in the circuit is

$$I = \frac{E}{Z} = \frac{110}{101.6} = 1.08 \text{ amperes.}$$

(c) Voltage drop across the resistance is

$E_R = IR = 1.08 \times 50 = 54$ volts.

(d) Voltage drop across the capacitance is

$E_C = IX_C = 1.08 \times 88.5 = 95.6$ volts.

(e) Angle between voltage and current is

$$\cos\phi = \frac{R}{Z} = \frac{50}{101.6} = 0.492$$

$$\therefore \phi = 60.5°$$

(f) Power loss is

$P = I^2R = 1.08^2 \times 50 = 58.3$ watts.

(g) Power factor is

$\cos\phi = 0.492$, or 49.2 percent.

• PROBLEM 3-20

In the network shown in the figure, the voltage across the capacitance element at a certain instant of time is measured to be 40 volts as shown. Determine all currents and the rate of change of voltage across the capacitance at this instant.

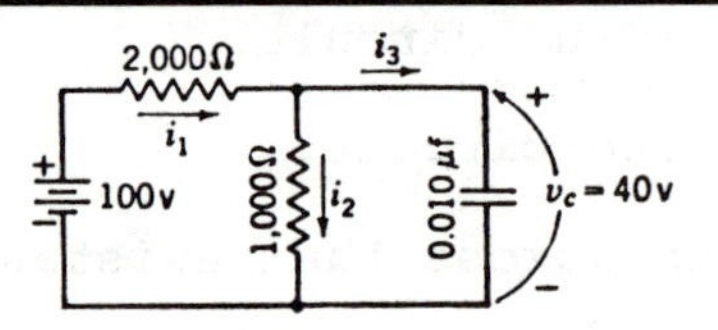

Solution: By Kirchhoff's voltage law, the voltage across the 2000-ohm resistance is 60 volts; hence, $i_1 = \frac{60}{2000} =$ 0.030 amp. The voltage across the 1000-ohm resistance is the same as the voltage across the capacitance; hence, $i_2 = \frac{40}{1000} = 0.040$ amp. By Kirchhoff's current law,

$$i_3 = 0.030 - 0.040 = -0.010 \text{ amp}$$

that is, the current is actually flowing opposite to the arrow. The rate of change of v_c is determined from $C\, dv_c/dt = i_3$, or

$$0.010 \times 10^{-6} \frac{dv_c}{dt} = -0.010,$$

from which $dv_c/dt = -10^6$ volts/sec; thus v_c is decreasing at this instant.

• PROBLEM 3-21

Find the homogeneous solution to a differential equation for the voltage $v_C(t)$ across a $\frac{1}{2}$ F capacitor in a series RLC circuit where R = 4Ω, L = 4H, $i_L(0) = \frac{1}{4}$ A, and $v_C(0) = 0$.

Solution: Since we are interested in the homogeneous solution of an RLC circuit, start the formulation of the differential equation by writing

$$v_L + v_R + v_C = 0 \quad \text{where}$$

$v_L = L\frac{di_L}{dt}$ and, $v_R = i_L R$. Substituting $C\frac{dv_C}{dt}$ for i_L gives the homogeneous second order differential equation in terms of v_C.

$$LC\frac{d^2v_C}{dt^2} + RC\frac{dv_C}{dt} - v_C = 0$$

It is known that the solution must be of the form $e^{-\alpha t}$ $(A\cos\omega_d t + B\sin\omega_d t)$ when $\alpha^2 = \left(\frac{R}{2L}\right)^2 < \omega_0^2 = \frac{1}{LC}$. Since $\alpha^2 = \frac{1}{4}$ and $\omega_0^2 = \frac{1}{2}$ this is the proper solution form for our circuit. Find ω_d by solving

$$\omega_d = \sqrt{\omega_0^2 - \alpha^2}$$

$$\omega_d = \sqrt{\frac{1}{2} - \frac{1}{4}}$$

$$\omega_d = \sqrt{\frac{1}{4}} = \frac{1}{2}$$

Hence, $v_C(t) = e^{-\frac{1}{2}t}\left[A\cos\frac{1}{2}t + B\sin\frac{1}{2}t\right]$

At t = 0 $\quad v_C(0) = A = 0$ and, since

$$\frac{i_L}{C} = \frac{dv_C}{dt} = -\frac{1}{2}e^{-\frac{1}{2}t}A\cos\frac{1}{2}t - \frac{1}{2}e^{-\frac{1}{2}t}A\sin\frac{1}{2}t$$

$$-\frac{1}{2}e^{-\frac{1}{2}t}B\sin\frac{1}{2}t + \frac{1}{2}e^{-\frac{1}{2}t}B\cos\frac{1}{2}t,$$

$$\frac{i_L(0)}{C} = \frac{dv_C(0)}{dt} = -\frac{1}{2}A + \frac{1}{2}B$$

$$\frac{i_L(0)}{C} = \frac{\frac{1}{4}}{\frac{1}{2}} = \frac{1}{2} = -\frac{1}{2} A + \frac{1}{2} B; \text{ since } A = 0 \quad B = 1.$$

Therefore, the homogeneous solution is $e^{-\frac{1}{2}t} \sin \frac{1}{2} t$ V.

• PROBLEM 3-22

Note that the current source in the circuit of Fig. 1 goes to zero at $t = 0$, and find $i(0^+)$ and $i'(0^+)$ if $R =$: (a) 500 Ω; (b) 400 Ω; (c) 320 Ω.

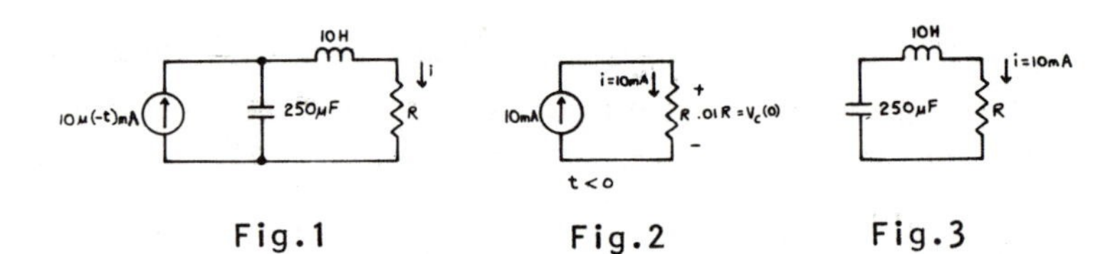

Fig.1 Fig.2 Fig.3

Solution: For $t < 0$ assuming that the circuit has been operating for a long time, the circuit shown in Fig. 2 applies. Note that the capacitor is an open circuit while the inductor is a short circuit.

At $t = 0^+$, as shown in Fig. 3, i must remain at 10 mA no matter what value of R is placed in the circuit, since the inductor current cannot change instantaneously for a finite voltage change. Since the voltage across a capacitor is subject to the same restrictions,

$$v_L(0^-) = v_L(0^+) = 0 \quad \text{thus} \quad \frac{di(0^+)}{dt} = \frac{v_L(0^+)}{L} = \frac{0}{10} = 0$$

for all values of R.

• PROBLEM 3-23

A coil of inductance of 0.1 henry and resistance of 2 ohms is connected in series with a condenser of 70.4-microfarad capacitance, across a 60-cycle, 110-volt line. How much current will flow and what will be the voltage across each part of the circuit?

Solution:
$$I = \frac{E}{\sqrt{R^2 + (2\pi fL - \frac{1}{2\pi fC})^2}}$$

$$= \frac{110}{\sqrt{2^2 + (37.7 - 37.7)^2}} = 55 \text{ amperes.}$$

The inductance drop of the coil is

$$2\pi fLI = 377 \times 0.1 \times 55 = 2073 \text{ volts.}$$

The resistance drop of the coil is

$$55 \times 2 = 110 \text{ volts.}$$

The total drop across the coil is

$$\sqrt{110^2 + 2073^2} = 2076 \text{ volts.}$$

The drop across the condenser is

$$\frac{I}{2\pi fC} = \frac{55}{377 \times 70.4 \times 10^{-6}} = 2073 \text{ volts.}$$

• PROBLEM 3-24

(i) Figure (a) shows a circuit containing a resistance of 20 ohms, an inductance of 0.10 henry, and a 100-μf capacitor all connected in series. The applied voltage is 240 volts at 60 cycles. It is desired to find (a) the circuit impedance, (b) the current flow, (c) the voltage across each part, (d) the power factor, (e) the volt-amperes, and (f) the power.

(ii) Find the total impedance of two coils and a capacitor in series. Coil 1 has a resistance of 15 ohms and an inductive reactance of 35 ohms. Coil 2 has a resistance of 20 ohms and an inductive reactance of 45 ohms. The capacitor has a capacitive reactance of 60 ohms and has negligible resistance.

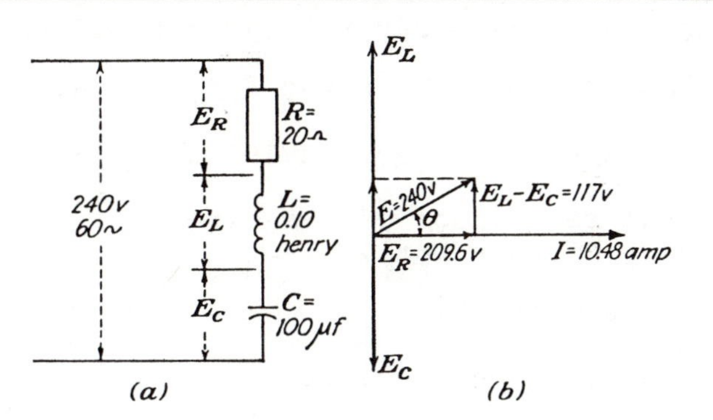

Circuit and vector diagram.

Solution:

(i) (a) $X_L = 2\pi fL = 2 \times 3.14 \times 60 \times 0.10 = 37.7$ ohms.

$$X_C = \frac{1,000,000}{2\pi fC_{\mu f}} = \frac{1,000,000}{2 \times 3.14 \times 60 \times 100} = 26.5 \text{ ohms.}$$

$$X = X_L - X_C = 37.7 - 26.5 = 11.2 \text{ ohms.}$$

$$Z = \sqrt{R^2 + X^2} = \sqrt{20^2 + 11.2^2} = 22.9 \text{ ohms.}$$

(b) $I = \frac{E}{Z} = \frac{240}{22.9} = 10.48$ amp.

(c) $E_R = IR = 10.48 \times 20 = 209.6$ volts

$$E_L = IX_L = 10.48 \times 37.7 = 395 \text{ volts.}$$

$$E_C = IX_C = 10.48 \times 26.5 = 278 \text{ volts.}$$

The vector diagram is drawn in Fig. (b) showing the position of the three voltages relative to the current. As a check on the work, the three voltages may be added vectorially by first adding E_L and E_C and then adding the result to E_R. Since E_L and E_C are exactly opposed, their resultant is $E_L - E_C$ as shown. The resultant is added to E_R by the triangle method as shown.

Mathematically

$$E = \sqrt{E_R{}^2 + (E_L - E_C)^2}$$

$$= \sqrt{(209.6)^2 + (395 - 278)^2} = 240 \text{ volts (check)}$$

(d) Power factor $= \cos\theta = \frac{R}{Z} = \frac{20}{22.9} = 0.873.$

The current lags the voltage since X_L is greater than X_C.

(e) Volt-amperes = EI = 240 x 10.48 = 2,515 VA.

(f) $P = EI \cos\theta = 2{,}515 \times 0.873 = 2{,}195$ watts

or $P = I^2R = (10.48)^2 \times 20 = 2{,}195$ watts (check).

(ii) The total resistance of the circuit is the sum of the resistances of each part, or

$$R = 15 + 20 = 35 \text{ ohms}$$

The total inductive reactance is

$$X_L = 35 + 45 = 80 \text{ ohms}$$

The net reactance of the circuit is

$$X_L - X_C = 80 - 60 = 20 \text{ ohms (inductive).}$$

The combined impedance is

$$Z = \sqrt{R^2 + X^2} = \sqrt{35^2 + 20^2} = 40.3 \text{ ohms.}$$

• PROBLEM 3-25

A circuit connected as shown in the figure contains a 10-ohm resistance and 0.5-henry inductance in parallel with a capacitor of 20 microfarads. The voltage and frequency of the source are 1000 and 60, respectively. Find the

(a) Current through the coil.

(b) Phase angle between the current through the coil and the potential across it.

(c) Current through the capacitor.

(d) Total current.

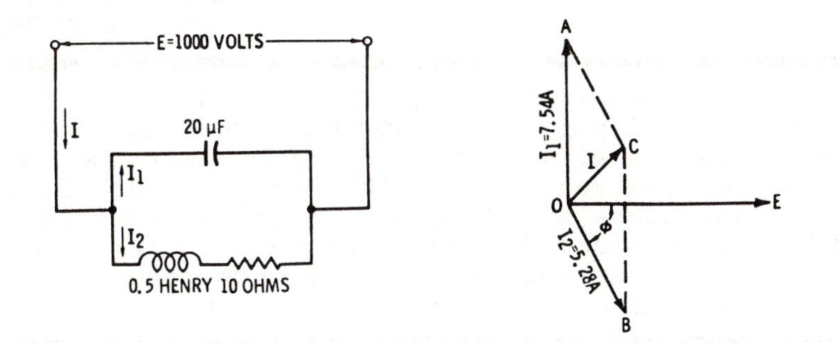

Impedance and capacitance in parallel.

Solution: (a) The current through the coil is

$$I_2 = \frac{E}{\sqrt{(R)^2 + (x_L)^2}} \quad \text{where } x_L = 2\pi fL.$$

$$= \frac{1000}{\sqrt{(10)^2 + (2\pi \times 60 \times 0.5)^2}} = 5.30 \text{ amperes.}$$

(b) Phase angle is $\tan \phi = \frac{x_L}{R}$

$$= \frac{2\pi 60 \times 0.5}{10} = 18.85$$

$$\therefore \phi = 86.9°.$$

(c) Current through the capacitor is

$$I_1 = \frac{1000}{X_C} = 20 \times 10^{-6} \times 2\pi \times 60 \times 1000 = 7.54 \text{ amps.}$$

(d) With reference to the vector diagram

$|\overrightarrow{OA}| = |\overrightarrow{I_1}| = 7.54$ amps, and $|\overrightarrow{OB}| = |\overrightarrow{I_2}| = 5.30$ amps.

As the current, I, is the resultant of these two vectors, it is now possible to construct the parallelogram as indicated by the dotted lines. It follows from the construction that β OBC = 90°, and from the law of cosines

$$|\vec{I}| = \sqrt{(OB^2 + OA^2) - (2 \times OB \times OA \sin \phi)}$$

$$= \sqrt{(5.30^2 + 7.54^2) - (2 \times 5.30 \times 7.54 \sin\phi)}$$

$$|\vec{I}| = 2.26 \text{ amperes (approximately)}.$$

ANALYSIS OF CIRCUITS USING COMPLEX NUMBERS

• PROBLEM 3-26

A voltage of 104 + j60 volts is applied to a circuit consisting of 2 parallel branches. One branch contains 6 ohms resistance, 10 ohms inductive reactance, and 5 ohms capacitive reactance. The second branch contains 8 ohms resistance and 3 ohms capacitive reactance. Using complex quantities, determine the branch currents, the line current, the equivalent impedance, the power input to each branch, and the total power input.

Solution: Solving for the current in the first branch,

$$\vec{Z}_1 = 6 + j10 - j5 = 6 + j5 \ \Omega.$$

and

$$\vec{I}_1 = \frac{E}{Z_1} = \frac{104 + j60}{6 + j5} = \frac{104 + j60}{6 + j5} \times \frac{6 - j5}{6 - j5}$$

$$= \frac{924 - j160}{36 + 25}$$

$$= 15.1 - j2.62$$

$$= 15.3 \text{ amperes in magnitude.}$$

$$\vec{I}_1\vec{E} = (15.1 - j2.62)(104 - j60)$$

$$= (1{,}570 - 157) - j(272 + 906)$$

$$= 1{,}413 \text{ watts} - j1{,}178 \text{ volt-amperes}$$

Therefore, the power input to $\vec{Z}_1$ is 1,413 watts, which is also equal to $(15.3)^2 R_1$, and the reactive power taken by $\vec{Z}_1$ is -1,178 volt-amperes. The second branch can be solved in a similar manner, giving

$$\vec{Z}_2 = 8 - j3$$

and $\vec{I}_2 = \frac{E}{Z_2} = \frac{104 + j60}{8 - j3}$

$= 8.93 + j10.85$

$= 14.05$ amperes in magnitude

$\vec{I}_2\overset{*}{\vec{E}} = (8.93 + j10.85)(104 - j60)$

$= 1{,}580$ watts $+ j592$ volt-amperes

The input to $\vec{Z}_2$ is 1,580 watts power and +592 volt-amperes reactive power.

The line current is the vector sum of the two branch currents, that is,

$$\vec{I} = \vec{I}_1 + \vec{I}_2$$
$$= (15.1 - j2.62) + (8.93 + j10.85)$$
$$= 24.03 + j8.23$$

$= 25.4$ amperes in magnitude

The power input and reactive power input are equal to the respective sums of the power inputs and the reactive power inputs to the 2 branches. Therefore,

power input $= 1{,}413 + 1{,}580$

$= 2{,}993$ watts

and reactive power $= -1{,}178 + 592$

$= -586$ volt-amperes

The values of power and reactive power are also given by

$\vec{I}\overset{*}{\vec{E}} = (24.03 + j8.23)(104 - j60)$

$= 2{,}993$ watts $- j586$ volt-amperes

The input impedance is equal to

$$\vec{Z} = \frac{\vec{E}}{\vec{I}} = \frac{104 + j60}{24.03 + j8.23}$$

$= 4.63 + j0.91$ ohms.

• PROBLEM 3-27

A branch consisting of a resistance of 3 ohms in series with a condenser of 7 ohms reactance is placed in parallel with an inductive reactance, 5 + j6. What current flows through each branch and in the line when the circuit is supplied from a 110-volt source? (Fig. 1 and 2) Also, find the equivalent resistance and reactance of the combined circuit and total power.

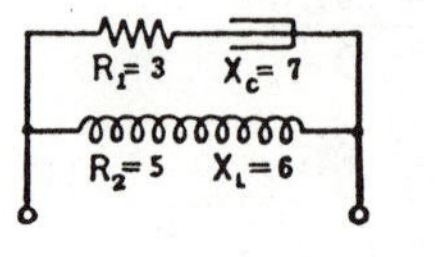

Fig.1

Solution:

$$Z_1^2 = 3^2 + 7^2 = 58. \qquad Z_2^2 = 5^2 + 6^2 = 61,$$

$$\vec{Y}_1 = \frac{R_1}{Z_1^2} + j\frac{X_C}{Z_1^2} = \frac{3}{58} + j\frac{7}{58} = 0.0517 + j0.1207 \text{ mho.}$$

$$\vec{Y}_2 = \frac{R_2}{Z_2^2} - j\frac{X_L}{Z_2^2} = \frac{5}{61} - j\frac{6}{61} = 0.0820 - j0.0984 \text{ mho.}$$

$$\vec{Y} = (G_1 + G_2) + j(B_1 + B_2) = 0.1337 + j0.0223 \text{ mho.}$$

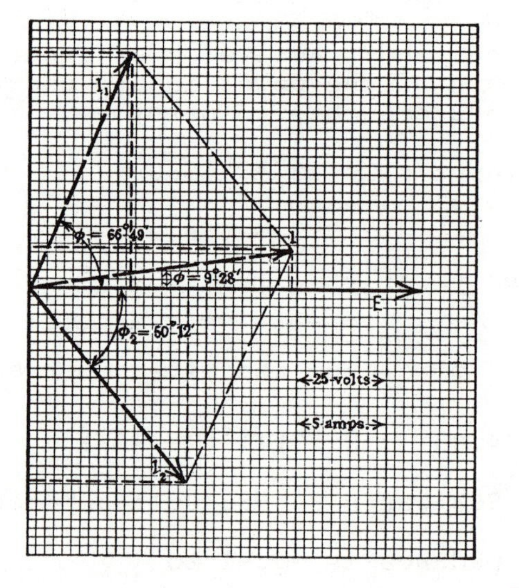

Fig.2

$$Y = \sqrt{0.1337^2 + 0.0223^2} = 0.1335 \text{ mho,}$$

$$Z = \frac{1}{0.1335} = 7.3801 \text{ ohms,}$$

$$\tan\phi = \frac{0.0223}{0.1337} = 0.1668, \ \phi = 9°\ 28', \quad \cos\phi = 0.9864.$$

$$Y_1 = \sqrt{0.0517^2 + 0.1207^2} = 0.1313 \text{ mho,}$$

$$Y_2 = \sqrt{0.0820^2 + 0.0984^2} = 0.1281 \text{ mho.}$$

$$\vec{I} = E\vec{Y} = 110(0.1337 + j0.0223) = 14.707 + j2.453,$$

$$I = EY = 110 \times 0.1335 = 14.905 \text{ amperes,}$$

and $\vec{I}_2 = \frac{E}{Z_2} = \frac{104 + j60}{8 - j3}$

$= 8.93 + j10.85$

$= 14.05$ amperes in magnitude

$\vec{I}_2 \overset{*}{\vec{E}} = (8.93 + j10.85)(104 - j60)$

$= 1,580$ watts $+ j592$ volt-amperes

The input to $\vec{Z}_2$ is 1,580 watts power and +592 volt-amperes reactive power.

The line current is the vector sum of the two branch currents, that is,

$$\vec{I} = \vec{I}_1 + \vec{I}_2$$
$$= (15.1 - j2.62) + (8.93 + j10.85)$$
$$= 24.03 + j8.23$$
$$= 25.4 \text{ amperes in magnitude}$$

The power input and reactive power input are equal to the respective sums of the power inputs and the reactive power inputs to the 2 branches. Therefore,

$$\text{power input} = 1,413 + 1,580$$

$$= 2,993 \text{ watts}$$

and $\text{reactive power} = -1,178 + 592$
$= -586$ volt-amperes

The values of power and reactive power are also given by

$$\vec{I}\overset{*}{\vec{E}} = (24.03 + j8.23)(104 - j60)$$

$$= 2,993 \text{ watts} - j586 \text{ volt-amperes}$$

The input impedance is equal to

$$\vec{Z} = \frac{\vec{E}}{\vec{I}} = \frac{104 + j60}{24.03 + j8.23}$$
$$= 4.63 + j0.91 \text{ ohms.}$$

• PROBLEM 3-27

A branch consisting of a resistance of 3 ohms in series with a condenser of 7 ohms reactance is placed in parallel with an inductive reactance, 5 + j6. What current flows through each branch and in the line when the circuit is supplied from a 110-volt source? (Fig. 1 and 2) Also, find the equivalent resistance and reactance of the combined circuit and total power.

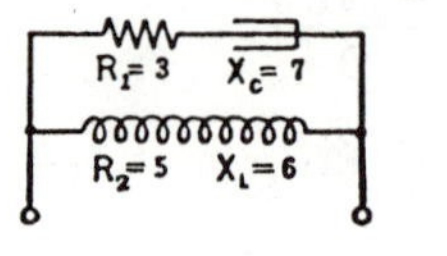

Fig.1

Solution:

$$Z_1^{\,2} = 3^2 + 7^2 = 58. \qquad Z_2^{\,2} = 5^2 + 6^2 = 61,$$

$$\vec{Y}_1 = \frac{R_1}{Z_1^{\,2}} + j\,\frac{X_C}{Z_1^{\,2}} = \frac{3}{58} + j\,\frac{7}{58} = 0.0517 + j0.1207 \text{ mho.}$$

$$\vec{Y}_2 = \frac{R_2}{Z_2^{\,2}} - j\,\frac{X_L}{Z_2^{\,2}} = \frac{5}{61} - j\,\frac{6}{61} = 0.0820 - j0.0984 \text{ mho.}$$

$$\vec{Y} = (G_1 + G_2) + j(B_1 + B_2) = 0.1337 + j0.0223 \text{ mho.}$$

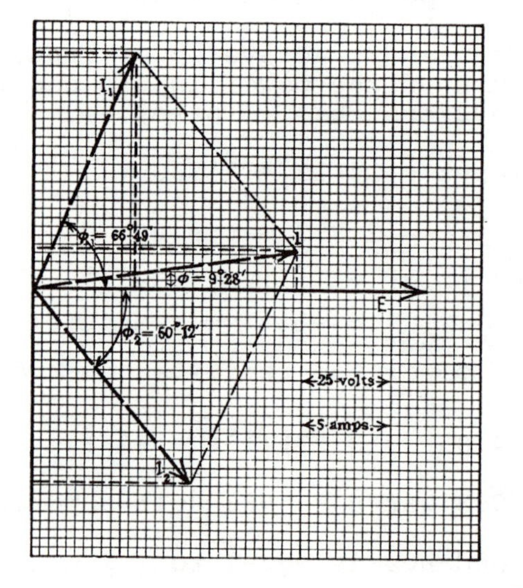

Fig.2

$$Y = \sqrt{0.1337^2 + 0.0223^2} = 0.1335 \text{ mho,}$$

$$Z = \frac{1}{0.1335} = 7.3801 \text{ ohms,}$$

$$\tan\phi = \frac{0.0223}{0.1337} = 0.1668, \ \phi = 9°\ 28', \quad \cos\phi = 0.9864.$$

$$Y_1 = \sqrt{0.0517^2 + 0.1207^2} = 0.1313 \text{ mho,}$$

$$Y_2 = \sqrt{0.0820^2 + 0.0984^2} = 0.1281 \text{ mho.}$$

$$\vec{I} = E\vec{Y} = 110(0.1337 + j0.0223) = 14.707 + j2.453,$$

$$I = EY = 110 \times 0.1335 = 14.905 \text{ amperes,}$$

$$\vec{I}_1 = E\vec{Y}_1 = 110\left(\frac{3}{58} + j\,\frac{7}{58}\right) = 5.6897 + j13.2759,$$

$$\vec{I}_2 = E\vec{Y}_2 = 110\left(\frac{5}{61} - j\,\frac{6}{61}\right) = 9.0164 - j10.8197,$$

$$\vec{I} = E(\vec{Y}_1 + \vec{Y}_2) \qquad = 14.7071 + j2.4562 \text{ (check)},$$

(Closer agreement could be obtained by carrying out some of the previous arithmetic to another significant figure.)

$$I_1 = EY_1 = 110 \times 0.1313 = 14.443 \text{ amperes},$$

$$I_2 = EY_2 = 110 \times 0.1281 = 14.091 \text{ amperes},$$

$$\tan\phi_1 = \frac{0.1207}{0.0517} = 2.3346, \qquad \phi_1 = 66°\ 49',$$

$$\cos\phi_1 = 0.3937, \qquad \sin\phi_1 = 0.9142,$$

$$\tan\phi_2 = \frac{-0.0984}{0.0820} = -1.2000\ , \quad \phi_2 = -50°\ 12',$$

$$\cos\phi_2 = 0.6401, \qquad \sin\phi_2 = -0.7683.$$

$$P_1 = EI_1\cos\phi_1 = 110 \times 14.443 \times 0.3937 = 625.5 \text{ watts}$$

$$= I_1{}^2R_1 = 14.443^2 \times 3 = 625.8 \text{ watts}.$$

$$P_2 = EI_2\cos\phi_2 = 110 \times 14.091 \times 0.6401 = 992.2 \text{ watts}$$

$$= I_1{}^2R_1 = 14.091^2 \times 5 = 992.7 \text{ watts}.$$

The equivalent resistance and reactance of the combined circuit are obtained from

$$G = \frac{R}{Z^2} \quad \text{and} \quad B = \frac{X}{Z^2}\ .$$

$$R = GZ^2 = 0.1337 \times 7.3801^2 = 7.282 \text{ ohms},$$

$$X = BZ^2 = 0.0223 \times 7.3801^2 = 1.215 \text{ ohms}.$$

Then

$$P = EI\cos\phi = 110 \times 14.905 \times 0.9864 = 1617.3 \text{ watts}.$$

$$= I^2R = 14.905^2 \times 7.282 = 1617.7 \text{ watts}$$

$$= P_1 + P_2 = 625.5 + 992.2 = 1617.7 \text{ watts (check)}.$$

• PROBLEM 3-28

In the circuit of Fig. 1 find the steady-state current i flowing through the voltage generator. Draw the admittance diagram.

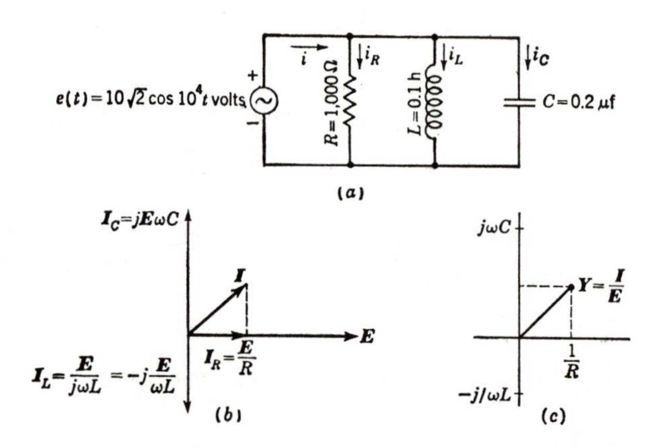

Fig.1: *(a)* the circuit; *(b)* phasor diagram; *(c)* admittance.

Solution: Since the voltage $\vec{E}$ is common to all branches, draw it as the first phasor of the diagram. Draw the phasor currents $\vec{I}_R$, $\vec{I}_L$, and I_C. The phasor $\vec{I}$ representing our unknown current i in the left branch may then be found by adding the phasor currents in the other three branches. Thus

$$\vec{I} = \vec{I}_R + \vec{I}_L + \vec{I}_C \tag{1}$$

or $$\vec{I} = \frac{\vec{E}}{R} - \frac{j\vec{E}}{\omega L} + j\vec{E}\omega C = \left(\frac{1}{R} - \frac{j}{\omega L} + j\omega C\right)\vec{E} = \vec{Y}\vec{E} \tag{2}$$

Substituting numerical values into Eq. 2 (note rms E = 10 volts),

$$\vec{I} = 10^{-2} - j10^{-2} + j2 \times 10^{-2}$$

$$= 10^{-2} + j10^{-2} = 10^{-2}\sqrt{2}\ \underline{/45°}\ \text{amp (rms)} \tag{3}$$

From Eq. 3 and the information that $\omega = 10{,}000$, the expression for i may be written at once (use $\sqrt{2}$ to obtain maximum value):

$$i = I_{max} \cos(\omega t + \phi) = \sqrt{2}\, I \cos(\omega t + \phi)$$

$$= 0.020 \cos(10{,}000t + 45°) \quad \text{amp.} \tag{4}$$

Although the solution could be found by considering Eq. 1 without resorting to the phasor diagram, the diagram serves as a useful visualization and check on the phase angles of the component parts of the solution. The magnitudes of $\vec{I}_R$, $\vec{I}_L$, and $\vec{I}_C$ cannot be merely added algebraically to obtain the magnitude of I. The phasor diagram emphasizes the necessary vector addition.

The admittance diagram for $\vec{Y}$, as defined in Eq. 2, is shown in Fig. 1c. Because the elements of the circuit are in parallel, the total complex admittance is the sum of the individual complex admittances of the elements.

Thus, on the complex plane, the admittance Y is the vector sum of 1/R, jωC, and -j/ωL. The diagram of complex admittances shown in sketch c contains essentially the same information as the phasor diagram of sketch b.

REDUCTION OF COMPLEX NETWORKS

• PROBLEM 3-29

The circuit of Fig.(a) is in the d-c steady state. Find all currents and voltages.

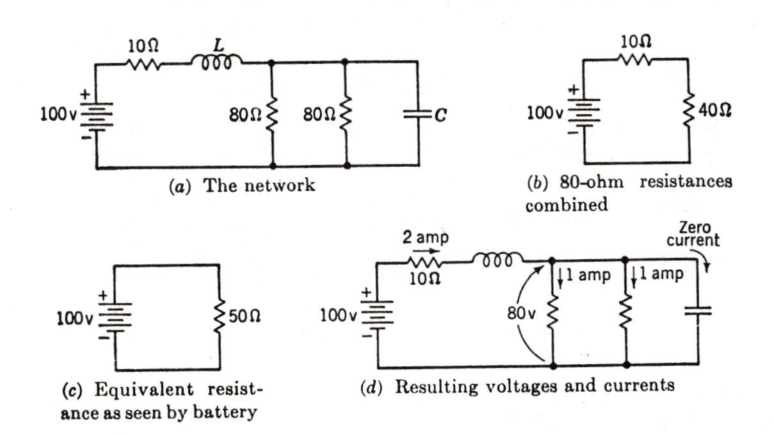

(a) The network

(b) 80-ohm resistances combined

(c) Equivalent resistance as seen by battery

(d) Resulting voltages and currents

Solution: Consider the capacitance to be an open circuit and the inductance to be a short circuit. Combine the two 80-ohm resistances in parallel, giving an equivalent resistance of 40 ohms, then combine this in series with the 10-ohm resistance, and obtain the equivalent resistance, as seen by the battery, of 50 ohms. The battery current will be $I = \frac{100}{50} = 2$ amp. Now go back to the original circuit as shown in Fig. (d) The voltage drop in the 10-ohm resistance is 2 x 10 = 20 volts, leaving 80 volts across the 80-ohm resistances. This produces a current of 1 amp in each, as shown.

• PROBLEM 3-30

In the circuit shown in Fig. 1, the driving force is an emf of 10 volts rms. Determine (a) the rms value of the input current from the generator, (b) the voltage across the capacitor and (c) the instantaneous values of the input current and the voltage across the capacitor.

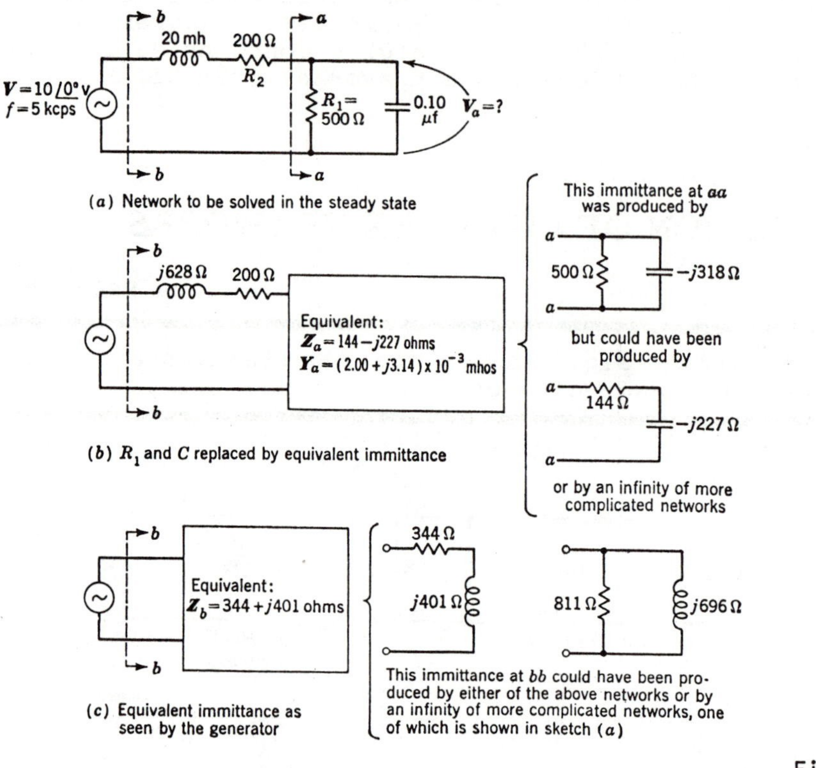

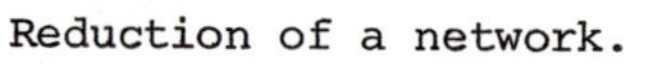

Reduction of a network.

Fig. 1

Solution: First compute the admittance of the capacitor at the given frequency:

$$\vec{Y}_C = j\omega C = j2\pi \times 5000 \times 0.100 \times 10^{-6}$$

$$= j3.14 \times 10^{-3} \text{ mho}$$

Next, compute the admittance of the shunt resistor:

$$\vec{Y}_R = \frac{1}{R} = \frac{1}{500} = 2.00 \times 10^{-3} \text{ mho.}$$

Now compute the admittance of the parallel combination by adding the complex admittances. This gives the equivalent admittance of everything to the right of the cut made in Fig. 1a along the line aa. This equivalent admittance $\vec{Y}_a$ is given by

$$\vec{Y}_a = \vec{Y}_R + \vec{Y}_C = (2.00 + j3.14) \times 10^{-3} \text{ mho.}$$

Proceeding to the left in the network, there are elements in series with the foregoing equivalent. Hence, compute the impedance that corresponds to $\vec{Y}_a$:

$$\vec{Z}_a = \frac{1}{\vec{Y}_a} = \frac{1}{(2.00 + j3.14) \times 10^{-3}} = 144 - j227 \text{ ohms}$$

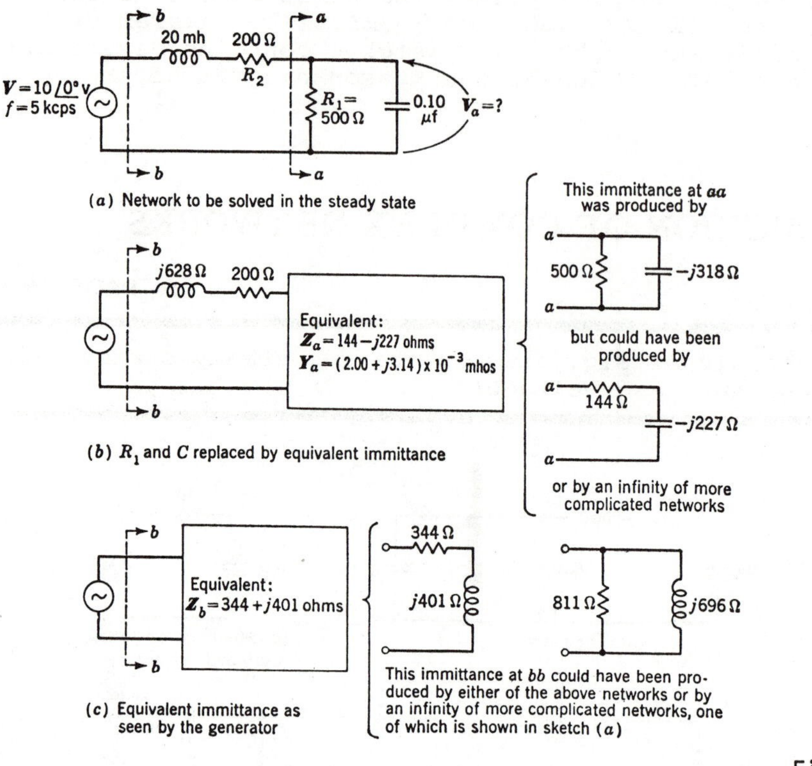

This is shown in Fig. 1b. The impedance seen by the generator (at cut bb) is

$$\vec{Z}_b = \vec{Z}_a + R_2 + j\omega L = 133 - j227 + 200 + j2\pi \times 5000 \times 0.020$$

$$= 344 + j401 \text{ ohms.}$$

or, in polar form,

$$\vec{Z}_b = \sqrt{(344)^2 + (401)^2} \; \underline{/\tan^{-1} \frac{401}{344}} = 528 \; \underline{/49.4°} \text{ ohms.}$$

(a) The input current from the generator is

$$\vec{I} = \frac{\vec{V}}{\vec{Z}_b} = \frac{10.0\underline{/0°}}{528\underline{/49.4°}} = 18.9 \times 10^{-3} \; \underline{/-49.4°} \text{ amp.}$$

(b) The voltage $\vec{V}_a$ (see Fig. 1) can be obtained from the current $\vec{I}$ and the equivalent impedance $\vec{Z}_a$:

$$\vec{V}_a = \vec{I}\vec{Z}_a$$

$$= (18.9 \times 10^{-3}\underline{/-49.4°})(144 - j227)$$

$$= (18.9 \times 10^{-3}\underline{/-49.4°})(269\underline{/-57.6°})$$

$$= 5.10\underline{/-107.0°} \text{ volts.}$$

(c) The foregoing current and voltages are expressed in terms of their rms values. Expressed in terms of instantaneous quantities, we have

Input voltage: $v = 10\sqrt{2} \cos \omega t$

$= 14.4 \cos \omega t$ volts

Input current: $i = \sqrt{2}\; 18.9 \times 10^{-3} \times \cos(\omega t - 49.4°)$

$= 0.0267 \cos(\omega t - 49.4°)$ amp

Voltage across aa: $v_a = 5.10\sqrt{2} \cos(\omega t - 107.0°)$

$= 7.21 \cos(\omega t - 107.0°)$ volts.

• PROBLEM 3-31

In the figure shown, find $\vec{V}_1$ and $\vec{V}_2$ in response to a sinusoidal driving current source of 2.0 amp rms, by reducing the number of node equations using series parallel combinations. The impedances shown are those evaluated at the frequency of the driving source.

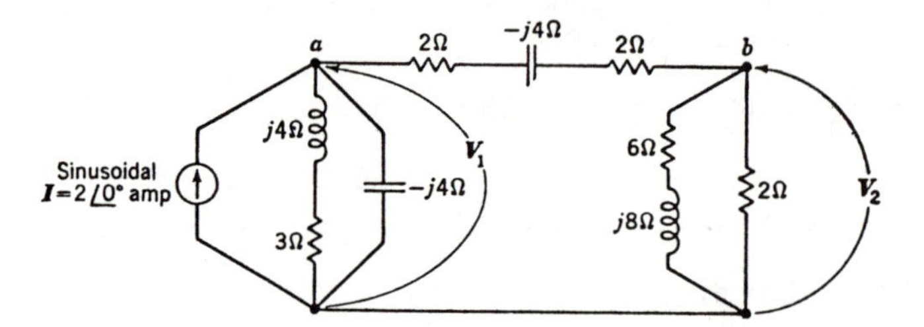

A network to be solved for the steady-state voltages V_1 and V_2. The values of impedance shown are evaluated at the frequency of the driving source.

Solution: The total number of nodes is seven, and the number of unknown node voltages is six. There are four meshes, and three unknown mesh currents. By using series parallel combinations, the number of node voltage equations are reduced to two. These equations are

At node a: $$\sum\vec{I} = \frac{\vec{V}_1}{3 + j4} + \frac{\vec{V}_1}{-j4} + \frac{\vec{V}_1 - \vec{V}_2}{4 - j4} - 2\underline{/0^\circ} = 0$$

At node b: $$\sum\vec{I} = \frac{\vec{V}_2 - \vec{V}_1}{4 - j4} + \frac{\vec{V}_2}{2(6 + j8)/(8 + j8)} = 0$$

The currents through the two right-hand branches could have been written as $\vec{V}_2/2 + \vec{V}_2/(6 + j8)$ instead of using the equivalent impedance $2(6 + j8)/(8 + j8)$.

Algebraic simplification of these two equations produces

$$(0.245 + j0.215)\vec{V}_1 - (0.125 + j0.125)\vec{V}_2 = 2\underline{/0^\circ}$$

$$-(0.125 + j0.125)\vec{V}_1 + (0.685 + j0.045)\vec{V}_2 = 0$$

The coefficient of $\vec{V}_1$ in the first equation is the total admittance that terminates on node a. The coefficient of $\vec{V}_2$ in the second equation is the total admittance that terminates on node b. The coefficient of $\vec{V}_2$ in the first equation and the coefficient of $\vec{V}_1$ in the second equation represent the admittance between nodes a and b.

Solving for $\vec{V}_1$ and $\vec{V}_2$,

$$\vec{V}_1 = 6.79\underline{/-35.0^\circ} \text{ volts}$$

$$\vec{V}_2 = 1.74\underline{/6.2^\circ} \text{ volts}$$

Since the driving force is expressed in terms of its rms value, the solution is expressed in rms volts, as is implied by the omission of the subscript m.

Any current or voltage may now be found by a simple single equation involving $\vec{V}_1$ or $\vec{V}_2$ or both. For example, the voltage across the 8-ohm inductance is

$$\vec{V} = \vec{V}_2 \frac{j8}{6 + j8} = \frac{(1.74\underline{/6.2^\circ})(8\underline{/90^\circ})}{10\underline{/53.1^\circ}} = 1.40\underline{/43.1^\circ} \text{ volts}$$

The power lost in the 6-ohm resistor is, for example,

$$P = I^2R = \left|\frac{\vec{V}_2}{6 + j8}\right|^2 \times 6 = \left(\frac{1.74}{10}\right)^2 \times 6 = 0.182 \text{ watt.}$$

• **PROBLEM 3-32**

Consider the bridged-T network of Fig. (a). Find the output voltage V_0.

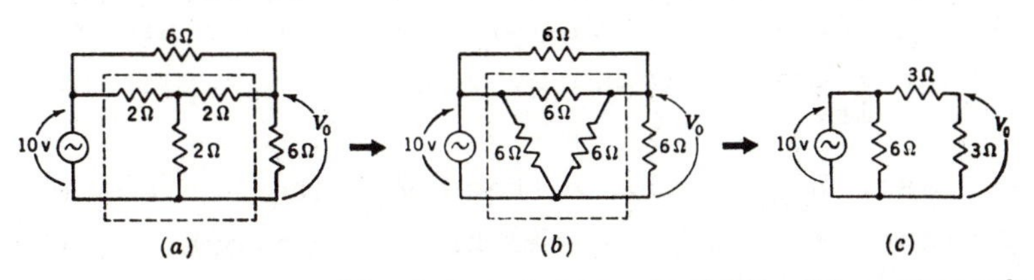

A network with a bridged T and its simplification through a wye-delta transformation: *(a)* original network; *(b)* trans-formed network; *(c)* simplified network.

Solution: First transform the wye-connected network inside the dotted square in Fig. (a) to a delta connected network shown in Fig. (b). If z_1, z_2 and z_3 are the impedances in the wye-connected network, the impedances in the delta-connected network are given by

$$z_a = \frac{z_1z_2 + z_2z_3 + z_3z_1}{z_1} \ \Omega$$

$$z_b = \frac{z_1z_2 + z_2z_3 + z_3z_1}{z_2} \ \Omega$$

$$z_c = \frac{z_1z_2 + z_2z_3 + z_3z_1}{z_3} \ \Omega$$

The impedances in the transformed network are then

$$z_a = z_b = z_c = \frac{2 \times 2 + 2 \times 2 + 2 \times 2}{2} = 6 \text{ ohms}$$

The transformed network can now be simplified to the network shown in Fig. (c). The voltage V_0 is then given by

$$V_0 = 10 \times \left(\frac{3}{3 + 3}\right) = 5 \text{ volts.}$$

• **PROBLEM 3-33**

Find the phasor current $\vec{I}$ in the network shown in Fig. (a), using Thévenin's theorem.

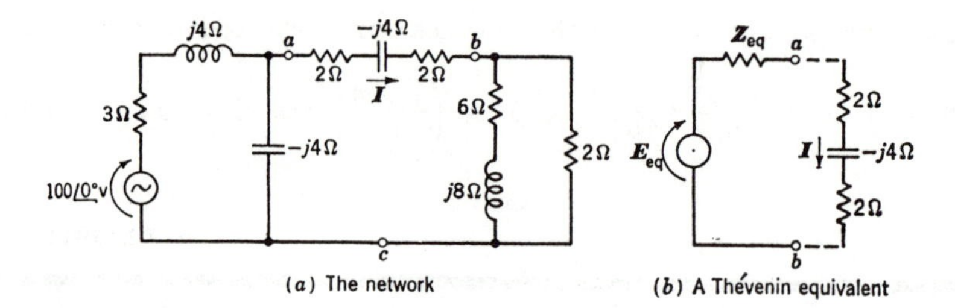

A network to be solved for *I*.

Solution: First break the network at points a and b. Find the equivalent Thévenin voltage generator as specified by $\vec{E}_{eq}$ and $\vec{Z}_{eq}$ in Fig. (b). Once these are known, the current $\vec{I}$ may be found by solving this simple series circuit.

In this case, $\vec{E}_{eq}$ is the voltage $\vec{V}_{ac}$ (no current can get to the right-hand side of the broken network and therefore $\vec{V}_{cb} = 0$). Thus

$$\vec{E}_{eq} = \vec{V}_{ac}$$

$$= \vec{I}_{ac} \times \vec{X}_c$$

$$= \frac{\vec{V}_s}{z} \times \vec{X}_c$$

$$= \frac{100\angle 0°}{3 + j4 - j4}(-j4) = -j133 \text{ volts.}$$

Now, for $\vec{Z}_{eq}$, one sees (from terminals ab) two sets of parallel branches which are in series with each other. Thus,

$$\vec{Z}_{eq} = \frac{2(6 + j8)}{2 + 6 + j8} + \frac{-j4(3 + j4)}{-j4 + 3 + j4} = 7.1 - j3.8 \text{ ohms.}$$

Hence

$$\vec{I} = \frac{\vec{E}_{eq}}{\vec{Z}_{eq} + [(2 + 2) - j4]}$$

$$= \frac{-j133}{(7.1 - j3.8) + (4 - j4)} = 9.8\angle -55° \text{ amp.}$$

POLYPHASE SYSTEMS, WYE-DELTA CONNECTIONS

• PROBLEM 3-34

Calculate the magnitude of a line current in the circuit shown in Fig. 1.

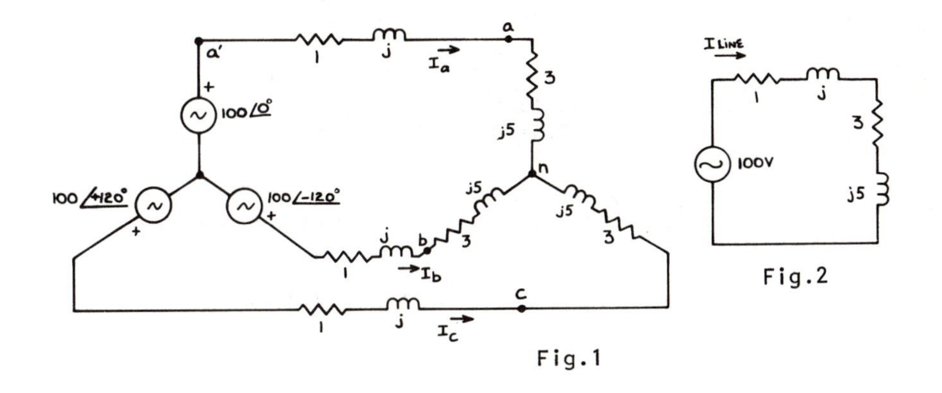

Fig.1

Fig.2

Solution: The circuit is balanced because the load for each phase is the same, the magnitude of the source for each phase is the same, and the angle for each phase is displaced by 120°. Since the circuit is balanced, the magnitude of the line current in each phase is the same. To find the line current, Fig. 1 is redrawn in Fig. 2 to include only one phase.

The total impedance of one phase is 1 + j + 3 + j5 = 4 + j6Ω. The magnitude of the line current is

$$\left|\vec{I}_{line}\right| = \left|\frac{100}{4 + j6}\right| = \frac{100}{\sqrt{4^2 + 6^2}} = \frac{100}{7.21} = 13.85 \text{ A.}$$

• PROBLEM 3-35

A three-phase, Y-connected system with 220V per phase is connected to three loads: 10Ω, $20\underline{/20°}$ Ω, and $12\underline{/-35°}$ Ω to phases 1, 2 and 3, respectively. Find the current in each line and in the neutral line.

Solution: This three-phase, Y-connected system has a conductor between the generator neutral and the load neutral. Because of this the current in each line can be calculated as if we had single phase circuits.

The generator terminals will be unprimed and the load terminals primed. The line currents are (assume the phasing is 1-3-2)

$$\vec{I}_{11'} = \frac{\vec{V}_{1'n'}}{\vec{Z}_1} = \frac{220\underline{/0°}}{10\underline{/0°}} = 22.0\underline{/0°} \text{ A}$$

$$\vec{I}_{22'} = \frac{\vec{V}_{2'n'}}{\vec{Z}_2} = \frac{220\angle 120°}{20\angle 20°} = 11.0\angle 100° \text{ A}$$

$$\vec{I}_{33'} = \frac{\vec{V}_{3'n'}}{\vec{Z}_3} = \frac{220\angle -120°}{12\angle -35°} = 18.3\angle -85° \text{ A}$$

The current in the neutral line is found by summing the above three currents.

$$\vec{I}_{nn'} = \vec{I}_{11'} + \vec{I}_{22'} + \vec{I}_{33'} = 22.9\angle -18.8° \text{ A}$$

• PROBLEM 3-36

A balanced set of three-phase voltages is connected to an unbalanced set of Y-connected impedances as shown in Fig. 1. The following values are known:

$$\vec{V}_{ab} = 212\angle 90° \text{ V} \qquad \vec{Z}_{an} = 10 + j0\ \Omega$$

$$\vec{V}_{bc} = 212\angle -150° \text{ V} \qquad \vec{Z}_{bn} = 10 + j10\Omega$$

$$\vec{V}_{ca} = 212\angle -30° \text{ V} \qquad \vec{Z}_{cn} = 0 - j20\Omega$$

Find the line currents $\vec{I}_{a'a}$, $\vec{I}_{b'b}$ and $\vec{I}_{c'c}$.

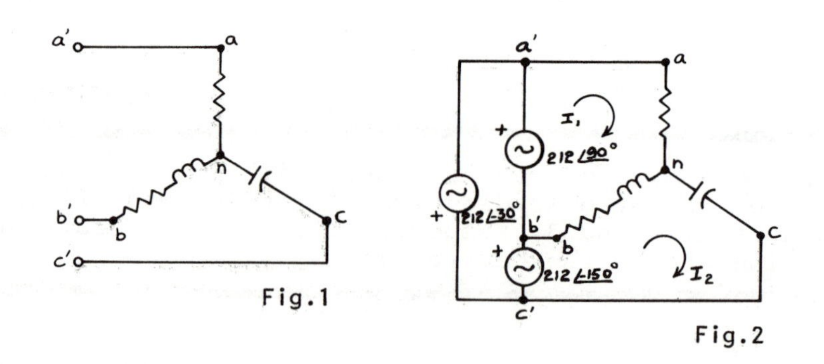

Fig.1

Fig.2

Solution: In the unbalanced case, any of the usual methods of analysis can be applied. If we draw the circuit with the sources attached, as in Fig. 2, we can see that two mesh current equations can be solved and, thus, all three line currents are determined.

The equations are:

$$(10 + j0)\vec{I}_1 + (10 + j10)\vec{I}_1 - (10 + j10)\vec{I}_2 = 212\angle 90°$$

$$-(10 + j10)\vec{I}_1 + (10 + j10)\vec{I}_2 + (0 - j20)\vec{I}_2 = 212\angle -150°$$

Solving by determinants,

$$\vec{I}_1 = \frac{\begin{vmatrix} 0+j212 & -10-j10 \\ -183.6-j106 & 10-j10 \end{vmatrix}}{\begin{vmatrix} 20+j10 & -10-j10 \\ -10-j10 & 10-j10 \end{vmatrix}} = 3.66\underline{/15°}\ \text{A}$$

$$\vec{I}_2 = \frac{\begin{vmatrix} 20+j10 & 0+j212 \\ -10-j10 & -183.6+j100 \end{vmatrix}}{\text{same}} = 11.96\underline{/-113.8°}\ \text{A}$$

Now, $\vec{I}_{aa'} = \vec{I}_1$ and $\vec{I}_{cc'} = -\vec{I}_2 = 11.96\underline{/66°}$.

The third line current is found from KCL applied at n:

$$\vec{I}_{bb'} = -\vec{I}_{aa'} + \vec{I}_{c'c} = -\vec{I}_1 + \vec{I}_2$$

$$\vec{I}_{bb'} = -\ (3.54 + j.95) + -4.83 - j10.94$$

$$= -8.37 - j11.89 = 14.54\underline{/-125.1°}\ \text{A.}$$

• PROBLEM 3-37

Figure 1a shows an unbalanced set of impedances connected in Y, to which are applied balanced line voltages. The impedances are given by their ohmic values and all the figures have been chosen as numbers easily handled in order to show the method of solution with a minimum of numerical computation. The same analysis will hold for other impedances and for unbalanced voltages applied to them. Find the line currents and line voltages.

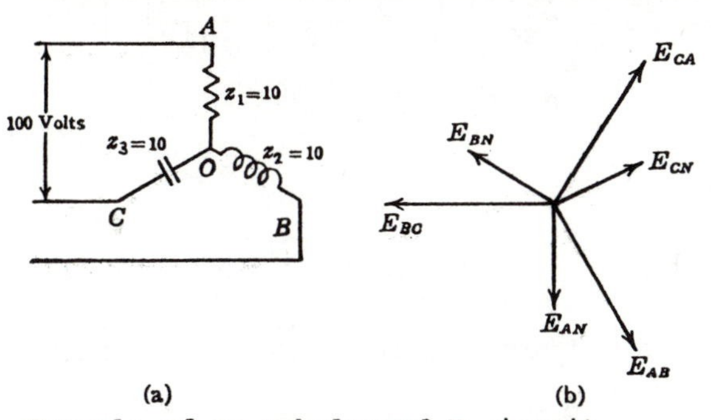

Example of an unbalanced Y-circuit.

Fig.1

Solution: Solving the problem first with a neutral line, ON, assumed present, from Fig. 2

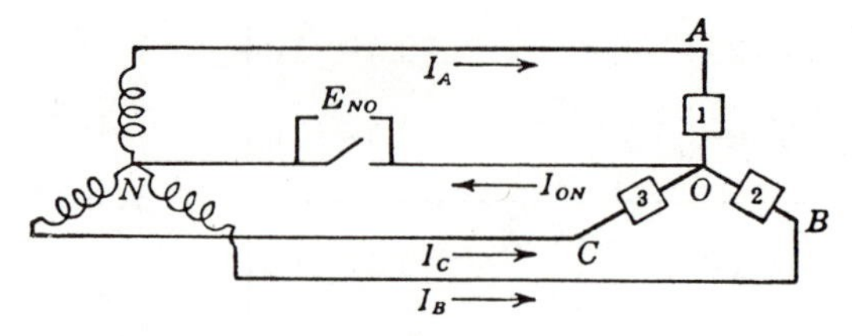

Unbalanced Y-connected load.

Fig.2

$$\vec{I}_{ON} = \frac{\vec{E}_{AN}}{\vec{Z}_1} + \frac{\vec{E}_{BN}}{\vec{Z}_2} + \frac{\vec{E}_{CN}}{\vec{Z}_3}$$

$$= \frac{-j57.7}{10} + \frac{-50 + j28.85}{j10} + \frac{50 + j28.85}{-j10}$$

$$= j4.23 \text{ amperes,}$$

where the complex magnitudes of the leg voltages have been determined from the voltage diagram of Figure 1b.

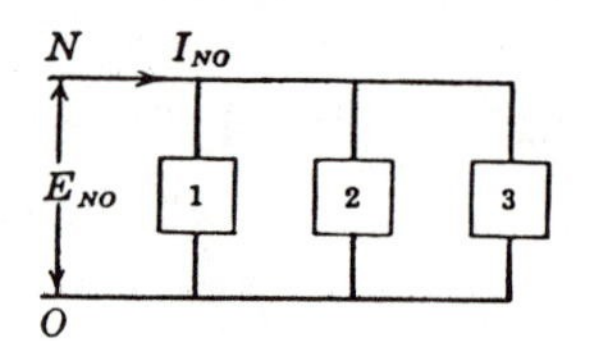

Determination of the voltage E_{on} of Figure 2.

Fig.3

From Fig. 3,

$$\vec{I}_{NO} = -\vec{I}_{ON} = \vec{E}_{NO}\left[\frac{1}{\vec{Z}_1} + \frac{1}{\vec{Z}_2} + \frac{1}{\vec{Z}_3}\right]$$

$$-j4.23 = \vec{E}_{NO}\left[\frac{1}{10} + \frac{1}{j10} + \frac{1}{-j10}\right]$$

$$= 0.1\vec{E}_{NO}$$

whence

$$\vec{E}_{NO} = -j42.3 \text{ volts}$$

The leg voltages thus become, with the neutral line open,

$$\vec{E}_{AO} = \vec{E}_{AN} + \vec{E}_{NO}$$

$$= -j57.7 - j42.3$$

$$= -j100 \text{ volts.}$$

$$\vec{E}_{BO} = -50 - j13.45 \text{ volts}$$

$$\vec{E}_{CO} = 50 - j13.45 \text{ volts}$$

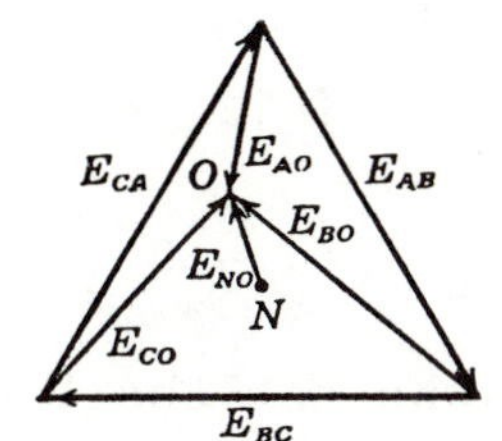

Location of neutral in unbalanced Y-circuit.

Fig.4

These quantities are shown in the vector diagram of Fig. 4. The leg (and line) currents are

$$\vec{I}_A = \frac{\vec{E}_{AO}}{\vec{Z}_1} = \frac{-j100}{10} = -j10 \text{ amp}$$

$$\vec{I}_B = \frac{\vec{E}_{BO}}{\vec{Z}_2} = -1.34 + j5 \text{ amp}$$

$$\vec{I}_C = \frac{\vec{E}_{CO}}{\vec{Z}_3} = 1.34 + j5 \text{ amperes}$$

The total power will be measured by wattmeters placed in any 2 of the 3 lines, and it can be calculated by the graphical method, by means of complex numbers, or as the sum of I^2R for the 3 legs. The latter usually is the most direct and for this problem shows 1,000 watts immediately, all the power being consumed in the 10-ohm resistance of the one leg. It can be shown that the other methods give the same result.

Note that, although the solution to this problem is shown by way of complex numbers, each step and the entire analysis can be carried out graphically. The second form of solution, following directly, in which impedance drops of adjacent legs are added to give line voltages, permits of analysis only in terms of complex numbers.

The second solution is carried through as follows: Referring to Figure 1a,

$$\vec{E}_{BC} = \vec{I}_B\vec{Z}_2 - \vec{I}_C\vec{Z}_3 .$$

$$= -100 + j0 = \vec{I}_B(j10) - \vec{I}_C(-j10).$$

$$\vec{E}_{AB} = \vec{I}_A\vec{Z}_1 - \vec{I}_B\vec{Z}_2$$

$$= 50 - j86.6 = \vec{I}_A(10) - \vec{I}_B(j10).$$

$$\vec{E}_{CA} = \vec{I}_C\vec{Z}_3 - \vec{I}_A\vec{Z}_1$$

$$= 50 + j86.6 = \vec{I}_C(-j10) - \vec{I}_A(10).$$

and $$\vec{I}_A + \vec{I}_B + \vec{I}_C = 0$$

Solving these equations for the leg currents,

$$\vec{I}_A = -j10 \text{ amperes}$$

$$\vec{I}_B = -1.34 + j5 \text{ amperes}$$

$$\vec{I}_C = 1.34 + j5 \text{ amperes}$$

from which the leg voltages are found to be

$$\vec{E}_{AO} = \vec{I}_{AO}\vec{Z}_{AO} = -j100 \text{ volts}$$

$$\vec{E}_{BO} = -50 - j13.4 \text{ volts}$$

$$E_{CO} = 50 - j13.4 \text{ volts.}$$

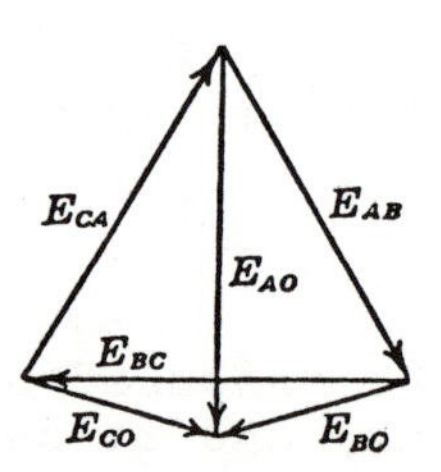

Vector diagram for the unbalanced Y-circuit of Figure 1.

Fig.5

The voltage diagram for the circuit is shown in Figure 5. The floating neutral O now is outside the triangle of line voltages.

• PROBLEM 3-38

A balanced three-phase three-wire system has a Δ-connected load with a 50Ω resistor, a 5 μF capacitor, and a 0.56 H inductor in series in each phase. Using positive phase sequence with $\vec{V}_{an} = 390\underline{/30°}$ V rms and $\omega = 500$ rad/s, find the magnitude and phases of all line currents.

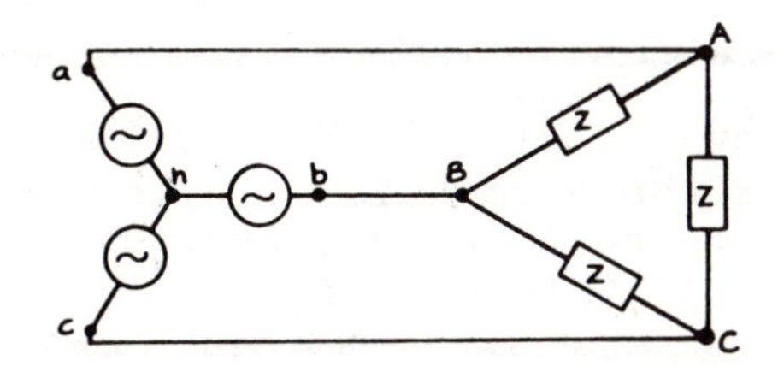

Solution: The circuit described is as shown. Since $\vec{V}_{an} = 390\underline{/30°}$ V rms and a positive phase sequence is specified, $\vec{V}_{bn} = 390\underline{/-90°}$ V rms and $\vec{V}_{cn} = 390\underline{/+150°}$ V rms.

The line voltages are:

$$\vec{V}_{AB} = \vec{V}_{an} - \vec{V}_{bn} = 390\underline{/30°} - 390\underline{/-90°} = 390\sqrt{3}\underline{/60°} \text{ V rms}$$

$$\vec{V}_{BC} = \vec{V}_{bn} - \vec{V}_{cn} = 390\underline{/-90°} - 390\underline{/+150°} = 390\sqrt{3}\underline{/-60°} \text{ V rms}$$

$$\vec{V}_{CA} = \vec{V}_{cn} - \vec{V}_{an} = 390\underline{/+150°} - 390\underline{/30°} = 390\sqrt{3}\underline{/180°} \text{ V rms.}$$

For a frequency $\omega = 500$ rad/sec the load impedance in each phase is

$$\vec{Z} = 50 + j\left[(500)(0.56) - \frac{1}{(500)(5 \times 10^{-6})}\right]$$

$$= 50 - j120 = 130\underline{/-67.4°}\ \Omega$$

The load phase currents are then

$$\vec{I}_{AB} = \frac{\vec{V}_{AB}}{\vec{Z}} = \frac{390\sqrt{3}\underline{/60°}}{130\underline{/-67.4°}} = 3\sqrt{3}\underline{/127.4°} \text{ A rms}$$

$$\vec{I}_{BC} = \frac{\vec{V}_{BC}}{\vec{Z}} = \frac{390\sqrt{3}\underline{/-60°}}{130\underline{/-67.4°}} = 3\sqrt{3}\underline{/7.4°} \text{ A rms}$$

$$\vec{I}_{CA} = \frac{\vec{V}_{CA}}{\vec{Z}} = \frac{390\sqrt{3}\underline{/180°}}{130\underline{/-67.4°}} = 3\sqrt{3}\underline{/-112.6°} \text{ A rms}$$

Finally the line currents are

$$\vec{I}_{aA} = \vec{I}_{AB} - \vec{I}_{CA} = 3\sqrt{3}\underline{/127.4°} - 3\sqrt{3}\underline{/-112.6°} = 9\underline{/97.4°} \text{ A rms}$$

$$\vec{I}_{bB} = \vec{I}_{BC} - \vec{I}_{AB} \quad 3\sqrt{3}\underline{/7.4°} - 3\sqrt{3}\underline{/127.4°} = 9\underline{/-22.6°} \text{ A rms}$$

$$\vec{I}_{cC} = \vec{I}_{CA} - \vec{I}_{BC} = 3\sqrt{3}\underline{/-112.6°} - 3\sqrt{3}\underline{/7.4°} = 9\underline{/-142.6°} \text{ A rms}$$

• PROBLEM 3-39

In the circuit of Fig. 1, $\vec{Z}_1 = 1\angle 60°$ and $\vec{Z}_2 = 5\angle 36.9°$. The line voltages at the terminals a, b, c, are 230 volts. Find the magnitude of the line currents, and that of the line voltages at the load terminals.

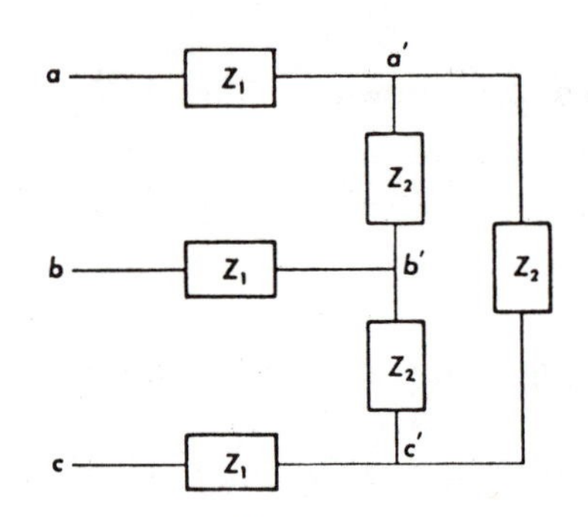

Circuit with line impedances.

Fig.1

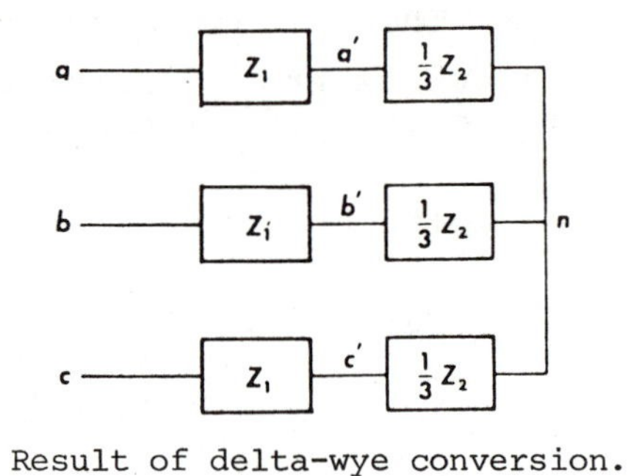

Result of delta-wye conversion.

Fig.2

Solution: After conversion of the delta into a wye, the circuit is shown in Fig. 2. It contains only three impedances $(Z_1 + Z_2/3)$ in wye-connection. The phase voltage is $230/\sqrt{3} = 133$ volts. Dividing this voltage by the total phase impedance, the current I is given by

$$\vec{I} = \frac{133\ \angle 0°}{1\ \angle 60° + 5\ \angle 36.9°}$$

$$= 22.4\ \angle -40.7° \quad \text{A.}$$

The voltage to neutral at the load terminals is $5\ \angle 36.9° \times 22.4\ \angle -40.7° = 112\ \angle -3.8°$ V. Finally, the magnitude of the line voltage at the load terminals

$$= \sqrt{3} \times 112 = 194 \text{ volts.}$$

• PROBLEM 3-40

(i) Three impedances of $3 + j4\Omega$ are connected in Δ to a 220-V, three-phase line. Find the line current, the power factor, and the complex power.

(ii) Three impedances, each $4-j3\Omega$, are connected in Y to a 208-V, three-phase line. Find the line current, power factor, and the complex power.

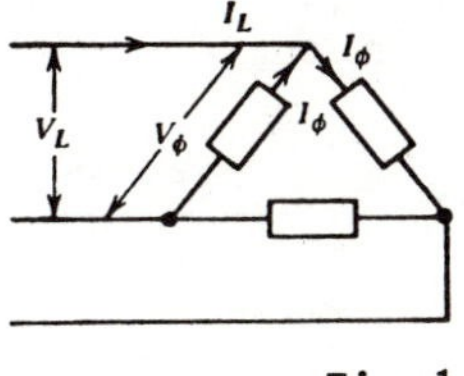

Fig.1

Solution: (i) From Fig. 1,

$$V_L = V_\phi$$

$$I_L > I_\phi$$

since,

$$I_L = \sqrt{3}I_\phi \text{ and } \sqrt{3} > 1.$$

$$\vec{Z}_L = 3 + j4 = 5\angle 53.1^\circ\ \Omega.$$

$$I_\phi = \frac{V_\phi}{|\vec{Z}_\phi|} = \frac{220}{5} = 44 \text{ A.}$$

$$I_L = \sqrt{3} \times 44 = 76.2 \text{ A.}$$

$$\text{p.f.} = \cos\theta = \cos 53.1 = 0.600.$$

$$\sin\theta = 0.800$$

$$|\vec{S}| = \sqrt{3}V_L I_L = \sqrt{3} \times 220 \times 76.2 = 29{,}000 \text{ VA}$$

$$\vec{S} = |\vec{S}|\ (\cos\theta + j\sin\theta) =$$

$$= 17{,}400 + j23{,}200 \text{ VA.}$$

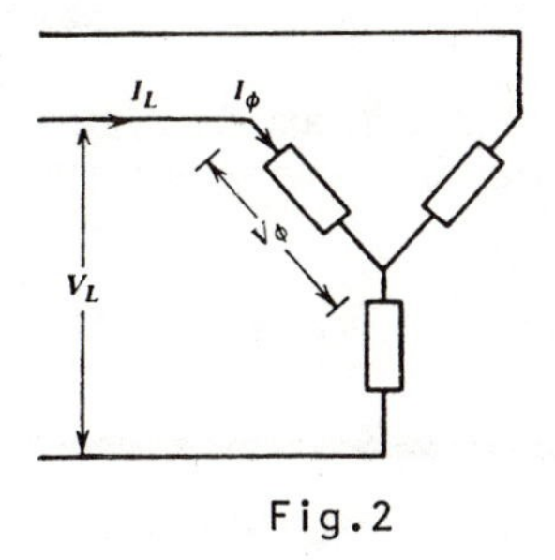

Fig.2

(ii) From Fig. 2,

$$I_L = I_\phi$$

$$V_\phi < V_L,$$

since,

$$V_\phi = V_L/\sqrt{3} \quad \text{and} \quad \sqrt{3} > 1.$$

$$\vec{Z}_L = 4 - j3 = 5\underline{/-36.9°} \quad \Omega.$$

$$V_\phi = \frac{208}{\sqrt{3}} = 120 \text{ V}.$$

$$I_L = I_\phi = \frac{V_\phi}{|\vec{Z}|} = \frac{120}{5} = 24 \text{ A}.$$

Power factor = cos(-36.9°) = 0.800 leading.

$$|\vec{S}| = \sqrt{3}\, V_L I_L = \sqrt{3} \times 208 \times 24 = 8.65 \text{ kVA}.$$

$$P = |\vec{S}| \cos(-36.9°) = 6.92 \text{ kW}.$$

$$Q = |\vec{S}| \sin(-36.9°) = -5.19 \text{ kVAR}.$$

$$\vec{S} = P + jQ = 6.92 - j5.19 \text{ kVA}.$$

• PROBLEM 3-41

A balanced 3-phase load consumes a total power of 100 kW at a power factor of 0.6 lagging; the line voltage being 1000 V. Determine the values of the impedance components per phase assuming:

(a) Delta connection (i) resistance and reactance in series.
(ii) resistance and reactance in parallel.

(b) Star connection (i) resistance and reactance in series.
(ii) resistance and reactance in parallel.

Sketch a vector diagram for cases (a)(i) and (b)(i) showing at least one line and phase vector for current and voltage.

Reference should be made to the diagrams to follow the solution.

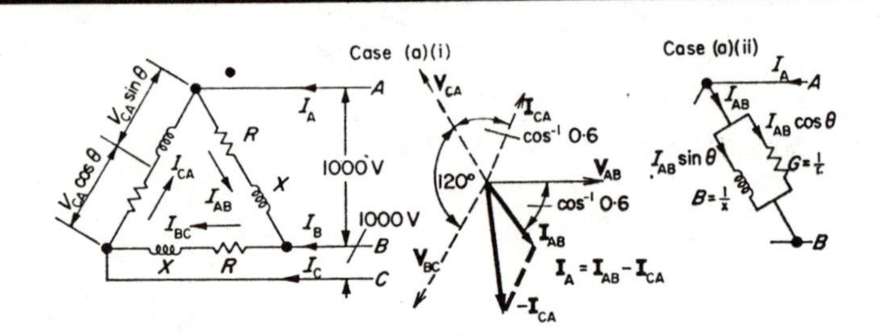

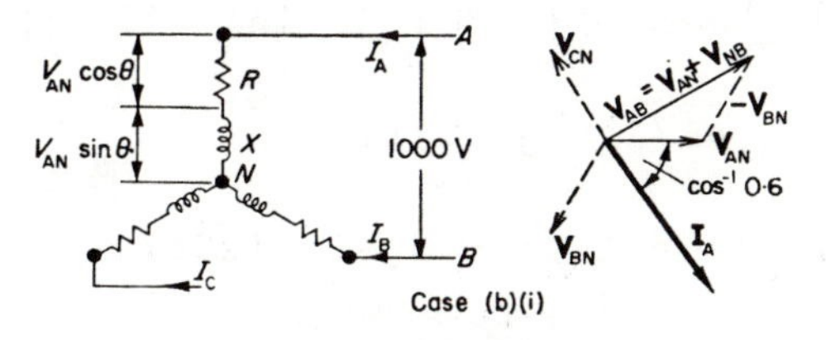

Case (b)(i)

<u>Solution</u>: Line current $= \dfrac{100{,}000}{1000 \times \sqrt{3} \times 0.6} = 96$ A.

Case (a)(i)

Phase current $= 96/\sqrt{3} = 55.5$ A.

$Z = V/I = 1000/55.5 = 18\Omega.$

$\vec{Z} = (V \cos \theta)/I + j(V \sin \theta)/I = 18(0.6 + j0.8)$

$= 10.8 + j14.4\Omega = R + jX$

Case (a)(ii)

$I \cos \theta = 55.5 \times 0.6 = 33.3$ A

$I \sin \theta = 55.5 \times 0.8 = 44.4$ A

$r = 1000/33.3 = 30\Omega$

$x = 1000/44.4 = 22.5\Omega$

$\vec{Y} = 1/30 - j/22.5 = 0.0333 - j0.0444\mho$

alternatively

$$\vec{Y} = \frac{1}{R + jX} = \frac{1}{10.8 + j14.4} \cdot \frac{10.8 - j14.4}{10.8 - j14.4}$$

$$= 0.0333 - j0.0444\mho = G - jB$$

Case (b)(i)

Phase voltage $= 1000/\sqrt{3} = 577$ V

$\vec{Z} = (577/96)(0.6 + j0.8) = 3.6 + j4.8\Omega$

Case (b)(ii)

$I \cos \theta = 96 \times 0.6 = 57.6$ A

$I \sin \theta = 96 \times 0.8 = 76.8$ A

$r = 577/57.6 = 10\Omega \qquad x = 577/76.8 = 7.5\Omega$

$\vec{Y} = 1/10 - j/7.5$; alternatively $= 1/(3.6 + j4.8)\mho$.

Note, in the vector diagrams that if $\vec{V}_{AB}$ is taken as reference $= V_{AB} + j0$ then

$$\vec{V}_{BC} = \vec{V}_{AB}\left[\cos -\left(120°\right) + j \sin\left(- 120°\right)\right]$$

$$= \vec{V}_{AB}[-1/2. - j\surd(3)/2]$$

$$\vec{V}_{CD} = \vec{V}_{AB}(\cos 120° + j \sin 120°)$$

$$= \vec{V}_{AB}[-1/2 + j\surd(3)/2]$$

For case (b) it is more convenient to take $\vec{V}_{AN}$, the voltage to neutral, as reference.

• PROBLEM 3-42

The figure shows a star-connected set of impedances with line impedance 1 + j3 between the supply terminals and the load. At the load, the voltage is 220 volts, with sequence A-B-C. W_B = 866 watts, and W_C = 1,732 watts. Calculate the voltage, power, and power factor at the supply terminals.

(b) Three impedances of 15 + j9 ohms each are in delta on a 440-volt, 3-phase power supply. Find line current and load power.

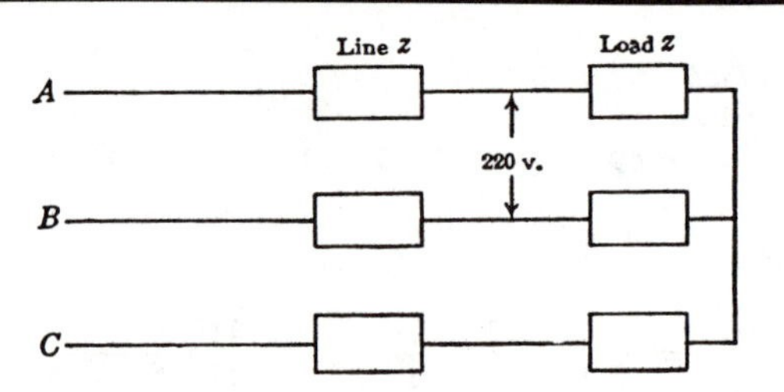

Star-connected load with line impedances between supply and load.

Solution: (a) $\tan\theta = \sqrt{3}\ \dfrac{W_B - W_C}{W_B + W_C} = -0.577$, whence $\theta = 30°$, leading power factor,

From the following equation:

$$\text{Power} = \sqrt{3}\ E_{line}I_{line} \cos\theta \text{ watts,}$$

$$I_{line} = \frac{\text{Power}}{\sqrt{3}\ E_{line} \cos\theta}$$

$$= \frac{866 + 1{,}732}{\sqrt{3} \times 220 \times 0.866}$$

$$= 7.87 \text{ amperes.}$$

Leg impedance is $Z = \frac{220/\sqrt{3}}{7.87}$

$= 16.14$ ohms mag. $= 14.0 - j8.1$

Over-all leg impedance, including one line and one leg $= 15.0 - j5.1$

$= 15.9$ ohms mag.

Line voltage at input terminals $= \sqrt{3} \times 7.87 \times 15.9 = 216$ volts, which also can be obtained by adding the impedance drop of one line to the leg voltage of 127 volts and multiplying the result by $\sqrt{3}$.

Power factor at input terminals $= \frac{15}{15.9} = 0.943$ leading

Input power $= \sqrt{3}\, E_{line}\, I_{line} \cos\theta$

$= \sqrt{3} \times 216 \times 7.87 \times 0.943 = 2{,}785$ watts

$= 3 \times (7.87)^2 \times 15$

$= (866 + 1{,}732) + 3 \times (7.87)^2 \times 1$, the sum of power taken by the load and the line losses

(b) Leg impedance is 17.49 ohms mag., whence leg current $= 440/17.49 = 25.2$ amp, and line current $= \sqrt{3} \times 25.2 = 43.6$ amp. Power factor $= 15/17.49 = 0.86$ approx. The phase angle for the leg is 31° approx.

3-phase power $= 3\, I_{leg}{}^2 R_{leg}$

$= 3 \times (25.2)^2 \times 15 = 28{,}500$ watts

$= \sqrt{3} \times 440 \times 43.6 \times \frac{15}{17.49}$

Leg voltage for an equivalent Y system is 254 volts, leg current is 43.6 amp, and leg impedance then is $254/43.6 = 5.83$ ohms mag. $= 5 + j3$. The equivalent Y system has the same leg power factor and takes the same total power. Note that for the equivalent systems the impedance ratio between the legs of the delta and the Y is 3. Once the equivalent Y circuit is obtained, line impedances can be taken into account easily.

With wattmeters in lines A and C

$W_A = E_{AC}\, I_A \cos(E_{AC}, I_A)$

$= 43.6 \times 440 \times \cos 61° = 9{,}320$ watts

$W_C = 43.6 \times 440 \times \cos 1° = 19{,}180$ watts

• PROBLEM 3-43

A 120 V per phase, three-phase, Y connected source delivers power to the following delta-connected load:

Phase 1: $40\angle 0° \ \Omega$

Phase 2: $20\angle -60° \ \Omega$

Phase 3: $15\angle 45° \ \Omega$.

Determine the phase currents, line currents, and line voltages. Show that the line currents add up to zero.

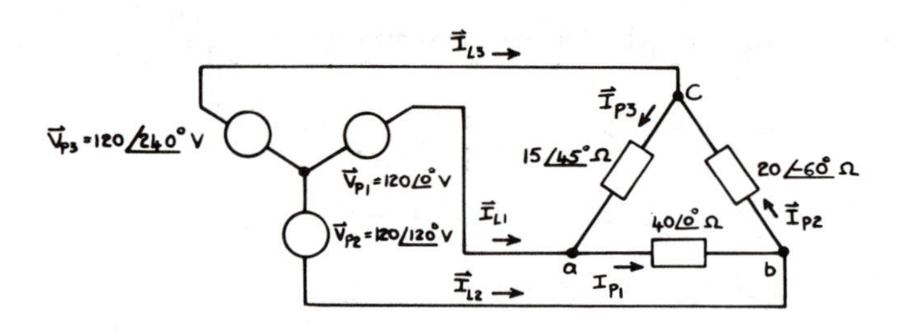

Solution: The circuit diagram for this problem is indicated in the figure. The line voltages are found from the formula:

$$|\vec{V}_L| = \sqrt{3}|\vec{V}_P|$$

Once the line voltages are found, the phase currents can be found by Ohm's Law. The line currents are then found by summing the proper phase currents. Notice that the line current cannot be found from the formula

$$I_L = \sqrt{3}\ I_P$$

because the delta-connected load is not balanced.

Since the Y-connected source is balanced, the line voltage is

$$V_L = \sqrt{3}\ 120 = 208\ \text{V}.$$

This is the magnitude of the voltage appearing across each load. $\vec{V}_{ab}$ will be adopted as the reference voltage, hence

$$\vec{V}_{L1} = \vec{V}_{ab} = 208\angle 0°\ \text{V}$$

$$\vec{V}_{L2} = \vec{V}_{bc} = 208\angle 120°\ \text{V}$$

$$\vec{V}_{L3} = \vec{V}_{ca} = 208\angle 240°\ \text{V}.$$

Now the phase current can be found by Ohm's Law.

$$\vec{I}_{P1} = \frac{\vec{V}_{L1}}{40\underline{/0°}} = \frac{208\underline{/0°}}{40\underline{/0°}} = 5.2\underline{/0°} \text{ A}$$

$$\vec{I}_{P2} = \frac{\vec{V}_{L2}}{20\underline{/-60°}} = \frac{208\underline{/120°}}{20\underline{/-60°}} = 10.4\underline{/180°} \text{ A}$$

$$\vec{I}_{P3} = \frac{\vec{V}_{L3}}{15\underline{/45°}} = \frac{208\underline{/240°}}{15\underline{/45°}} = 13.8\underline{/195°} \text{ A.}$$

From the figure, it is seen that

$$\vec{I}_{L1} = \vec{I}_{P1} - \vec{I}_{P3}.$$

Converting the phase currents to rectangular form we have

$$\vec{I}_{L1} = 5.2 + j0 - (-13.33 - j3.57) = 18.5 + j3.57\text{A}.$$

Similarly,

$$\vec{I}_{L2} = \vec{I}_{P2} - \vec{I}_{P1}$$

$$= -10.4 + j0 - (5.2 + j0) = -15.6 + j0 \text{ A}$$

$$\vec{I}_{L3} = \vec{I}_{P3} - \vec{I}_{P2}$$

$$= -13.33 - j3.57 - (-10.4 + j0) = -2.9 - j3.57\text{A}.$$

When the three line currents are added it is seen that

$$\vec{I}_{L1} + \vec{I}_{L2} + \vec{I}_{L3} = 0 + j0 \text{ A}.$$

This must be true because there is no neutral line in a delta-connected load.

• PROBLEM 3-44

(a) In the balanced three-phase circuit of Fig. 1, load 1 draws 80 kw and 60 leading kvar; load 2 draws 120 kw and 160 lagging kvar. The line voltage of the system is 1,000 volts. Find the line current at the generator terminals by adding the line currents of the two loads.

(b) In the circuit of Fig. 2, the two loads are the same as in part a, but the transmission-line impedances are now also to be taken into consideration. Load 2 is at the far end of the transmission line, and load 1 is somewhere in between the generator and load 2. The transmission-line impedances are $\vec{Z}_1 = 0.3 \underline{/60°}$ and $\vec{Z}_2 = 0.5 \underline{/70°}$ ohms. The line voltage has a magnitude of 1,000 volts at the terminals of load 2. Find the magnitudes of the line voltage and line current at the generator terminals.

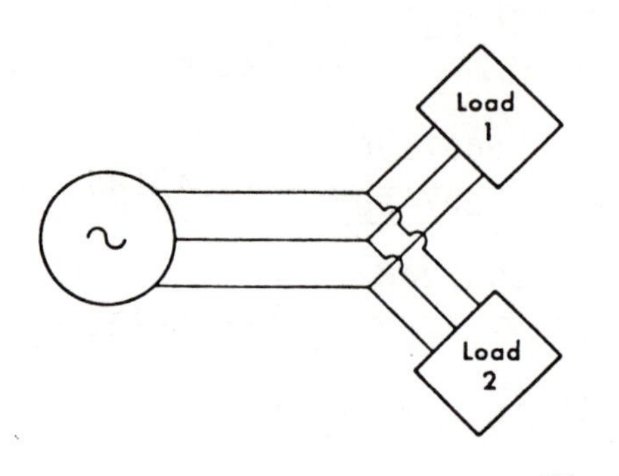

Fig.1

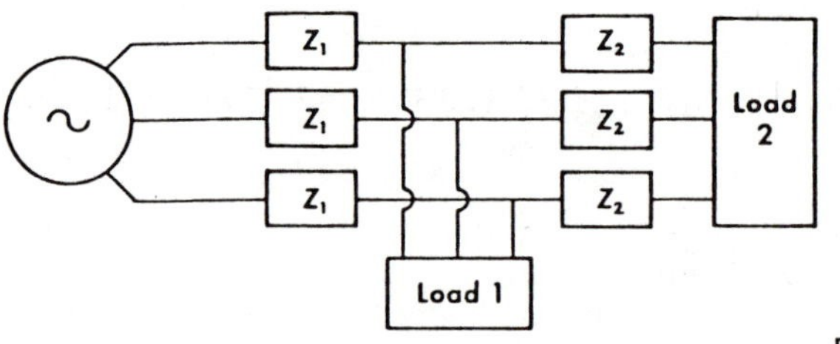

Fig.2

Solution: (a) For load 1, $\tan\theta_1 = -Q_1/P_1 = -60/80 = -0.75$, $\theta_1 = -36.9°$, $\cos\theta_1 = 0.8$. The magnitude of the line current of load 1 is

$$|\vec{I}_1| = \frac{P_1}{\sqrt{3}|\vec{V}_L|\cos\theta_1} = \frac{80{,}000}{\sqrt{3} \times 1{,}000 \times 0.8} = 57.8 \text{ amp.}$$

For load 2, $\tan\theta_2 = 160/120 = 1{,}333$, $\theta_2 = 53.1°$, $\cos\theta_2 = 0.6$,

$$|\vec{I}_2| = \frac{P_2}{\sqrt{3}|\vec{V}_L|\cos\theta_2} = \frac{120{,}000}{\sqrt{3} \times 1{,}000 \times 0.6} = 115.5 \text{ amp.}$$

The phasors representing these two line currents must be added. Since we are dealing with line currents as part of a per-phase calculation, we choose the voltage to neutral as our axis of reference. Thus $\vec{V} = \frac{1{,}000}{\sqrt{3}} \underline{/0°}$ $= 578 \underline{/0°}$ V, $\vec{I}_{L1} = |\vec{I}_{L1}|\underline{/-\theta_1} = 57.8 \underline{/36.9°}$ A, and $\vec{I}_{L2} = |\vec{I}_{L2}|\underline{/-\theta_2} = 115.5 \underline{/-53.1°}$ A. Adding the current phasors, $\vec{I}_L = \vec{I}_{L1} + \vec{I}_{L2} = 57.8 \underline{/36.9°} + 115.5 \underline{/-53.1°}$ $= 129 \underline{/-26.6°}$ A. The cosine of this angle of 26.6° is the power factor at the generator terminals.

(b) Use the voltage to neutral at the terminals of load 2 as the axis of reference. Thus $\vec{V}_2 = 578 \underline{/0°}$ volts and $\vec{I}_{L2} = 115.5 \underline{/-53.1°}$ amp, as in the preceding example. Next, obtain the voltage to neutral at the terminals of load 1: $\vec{V}_1 = \vec{V}_2 + \vec{Z}_2\vec{I}_{L2} = 578 \underline{/0°} +$

$0.5 \angle 70° \times 115.5 \angle -53.1°$ V, which comes out $\vec{V}_1 = 633 \angle 1.5°$ V. This changes the line current of load 1 to $|\vec{I}_{L1}| = \frac{P_1}{3|\vec{V}_1|\cos\theta_1} = \frac{80,000}{3 \times 633 \times 0.8} = 52.6$ amp, and its phase angle to $36.9° + 1.5° = 38.4°$. The total line current is $\vec{I}_L = \vec{I}_{L1} + \vec{I}_{L2} = 52.6 \angle 38.4° + 115.5 \angle -53.1° = 126 \angle -28.4°$ amp. The voltage to neutral at the generator terminals is $\vec{V}_1 + \vec{Z}_1\vec{I}_L = 633 \angle 1.5° + 0.3 \angle 60° \times 126 \angle -28.4° = 665 \angle 3.2°$ volts. The line voltage at the generator terminals is $\sqrt{3} \times 665 = 1,150$ volts. The power factor at the generator terminals is the cosine of the angle between the voltage to neutral and the line current i.e. $\cos(28.4° + 3.2°) = 0.85$ (lagging).

ENERGY, POWER AND POWER FACTOR

• PROBLEM 3-45

(a) A transmission line consists of two No. 0 copper wires and is 5 miles long. How much power is wasted in the line when 75 amperes are flowing?

(b) How many calories of heat per hour are developed in an electric heater which takes 7.5 amperes at 110 volts pressure?

Solution: (a) $P = I^2R$

$I = 75$ amperes

R = resistance of 10 miles of No. 0 wire

$= 10 \times 0.528 = 5.28$ ohms (resistance of No. 0 wire = 0.528 ohm per mile).

Then $P = 75^2 \times 5.28$

$= 29,700$ watts

$= 29.7$ kilowatts.

(b) Heat in calories $= 0.24I^2Rt$

$= 0.24EIt$

$I = 7.5$ amperes

$E = 110$ volts

$t = 3,600$ seconds

Hence $\text{Heat} = 0.24 \times 7.5 \times 110 \times 3,600$

$= 712,800$ calories.

• PROBLEM 3-46

(i) How much energy is used by a 1,500-watt heater in 8 hr?

(ii) A 12-kw load is supplied from a line that has a resistance of 0.1 ohm in each conductor. Find the line loss in watts when the load is supplied at (a) 120 volts and (b) 240 volts.

Solution: Energy = power x time

$= 1,500 \times 8 = 12,000$ whr

$= \frac{12,000}{1,000} = 12$ kwhr.

(a) $I = \frac{P}{E} = \frac{12,000}{120} = 100$ amp.

Line loss (both lines) $P = I^2R$

$P = (100)^2 \times 0.2 = 2,000$ watts.

(b) $I = \frac{P}{E} = \frac{12,000}{240} = 50$ amp.

Line loss (both lines) $P = I^2R$

$P = (50)^2 \times 0.2 = 500$ watts.

Note in the above example that the same power may be transmitted at 240 volts with one-fourth the line loss as at 120 volts. Or the power could be transmitted at the same loss with conductors one-fourth as large at 240 volts as at 120 volts.

• PROBLEM 3-47

(a) Find the value of an inductor which carries 2 A of current and in which 20 J of energy is stored.

(b) Find the value of a capacitor with 500 volts across it, in which 20 J of energy is stored.

Solution: (a) Since $E_L = 1/2\ Li^2$ we can solve for the unknown L.

$$L = \frac{2\ E_L}{i^2} = \frac{40}{(2)^2} = 10\ H$$

(b) Similarly,

$$E_C = 1/2\ C\ v^2$$

$$C = \frac{2\ E_C}{v^2} = \frac{40}{(500)^2} = 1.6 \times 10^{-4}\ F. = 160\ \mu F.$$

• PROBLEM 3-48

> Let an alternating voltage of 110 volts at a frequency of 60 cycles per second be impressed on a series circuit of 8.66 ohms resistance, 0.106 henry inductance, and 75.8 μf capacitance (Figure 1). Determine (a) the current and its phase angle with the impressed voltage; (b) the voltage drop across each element of the circuit; and (c) the power input to the circuit.

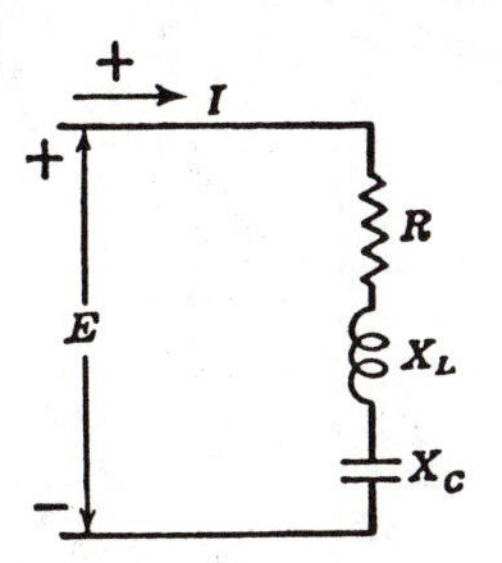

Series circuit of *R*, *L*, and *C*. Fig. 1

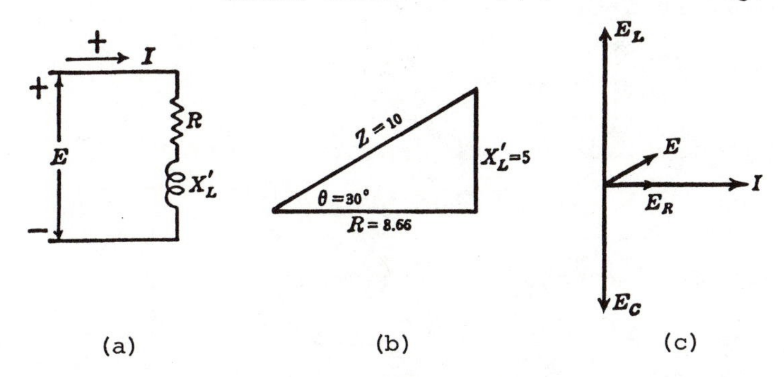

Equivalent circuit, impedance triangle, and vector diagram of circuit of Figure 1. Fig. 2

<u>Solution</u>: The inductive reactance is

$$X_L = 2\pi fL$$

$$= 377(0.106)$$

$$= 40\ \text{ohms}.$$

The capacitive reactance is

$$X_C = \frac{1}{2\pi fC} = \frac{1}{377(75.8)10^{-6}}$$

$$= 35 \text{ ohms.}$$

The net reactance of the circuit is

$$X_L' = (X_L - X_C) = 5 \text{ ohms}$$

and is inductive. The actual circuit, therefore, can be reduced to that shown in Figure 2a, which has the impedance triangle of Figure 2b. The complete vector diagram for the actual circuit is given by Figure 2c in which $\vec{E}_L - \vec{E}_C$ is the voltage drop across the net reactance X_L'.

The impedance of the circuit is

$$Z = \sqrt{R^2 + (X_L - X_C)^2} = \sqrt{(8.66)^2 + (5)^2}$$

$$= 10 \text{ ohms}$$

and the current is

$$I = \frac{E}{Z} = \frac{110}{10} = 11 \text{ amperes.}$$

The phase angle is

$$\theta = \tan^{-1} \frac{5}{8.66} = 30°$$

whence the power factor is

$$\cos \theta = \frac{8.66}{10} = 0.866.$$

The voltage drop across the resistance is

$$E_R = IR = 11 \times 8.66 = 95.2 \text{ volts,}$$

the voltage drop across the inductance is

$$E_L = IX_L = 11 \times 40 = 440 \text{ volts}$$

and the voltage drop across the capacitance is

$$E_C = IX_C = 11 \times 35 = 385 \text{ volts}$$

It can readily be shown that the vector sum--not the algebraic sum-- of these three component voltages is equal to the applied voltage of 110 volts.

If a voltmeter were connected so as to read the voltage drop across the series combination of L and C, it would indicate (440 - 385) = 55 volts, since the voltage drop across the inductance is directly opposite in polarity to that across the capacitance. Note that voltage components greater in magnitude than the applied voltage exist in the circuit. These voltages actually are present in the circuit and can be measured with an ordinary voltmeter. Analogous physical conditions often are met in mechanical systems.

The power input to the circuit is

$$P = EI \cos \theta$$

$$= 110 \times 11 \times 0.866$$

$$= 1{,}050 \text{ watts}$$

This must also equal

$$P = I^2R$$

$$= (11)^2 \times 8.66 = 1{,}050 \text{ watts}$$

• PROBLEM 3-49

Determine (a) the current I, (b) the power-factor and (c) the total power in the circuit shown in the figure by calculating admittance and susceptance.

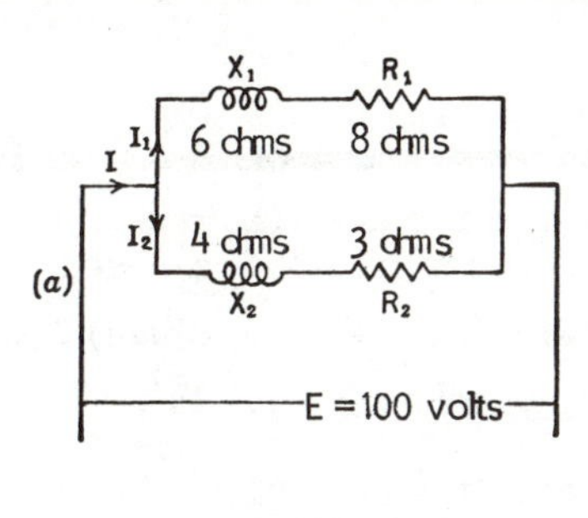

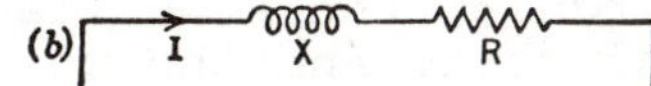

The general solution of circuits having branched sections requires that the parallel portions be replaced by equivalent series circuits; the circuit in (a) can be reduced to that in (b).

<u>Solution</u>:

(a) $$G_1 = \frac{R_1}{R_1^2 + X_1^2} = \frac{8}{8^2 + 6^2} = 0.08 \text{ mho.}$$

$$G_2 = \frac{R_2}{R_2^2 + X_2^2} = \frac{3}{3^2 + 4^2} = 0.12 \text{ mho.}$$

$$\therefore\ G = G_1 + G_2 = 0.20 \text{ mho.}$$

$$B_1 = \frac{x_1}{R_1^2 + x_1^2} = \frac{6}{8^2 + 6^2} = 0.06 \text{ mho.}$$

$$B_2 = \frac{x_2}{R_2^2 + x_2^2} = \frac{4}{3^2 + 4^2} = 0.16 \text{ mho.}$$

$$\therefore\ B = B_1 + B_2 = 0.22 \text{ mho.}$$

$$Y = \sqrt{G^2 + B^2} = \sqrt{0.20^2 + 0.22^2} = 0.2973 \text{ mho.}$$

$$I = EY = 100 \times 0.2973 = 29.73 \text{ amperes.}$$

(b)

$$\tan\phi = \frac{B}{G} = \frac{0.22}{0.20} = 1.1,$$

$$\sin\phi = \frac{B}{Y} = \frac{0.22}{0.2973} = 0.7400,$$

$$\cos\phi = \frac{G}{Y} = \frac{0.20}{0.2973} = 0.6727.$$

(c) The power may be calculated using the equation, $P = E^2G$.

Thus $P_1 = 100^2 \times 0.08 = 800$ watts.

$$P_2 = 100^2 \times 0.12 = 1200 \text{ watts.}$$

$$\therefore\ P = P_1 + P_2 = 100^2 \times 0.20 = 2000 \text{ watts.}$$

• PROBLEM 3-50

An air-core inductance coil ($L = 0.03$ henry, and resistance $R_L = 2.0$ ohms), is placed in series with a non-inductive resistance ($R_R = 5.0$ ohms), across a 110-volt, 60-cycle source of power (Fig. 1).

Find the current which will flow, the voltage across each part of the circuit, the power factor of the coil and of the entire circuit, and the power dissipated in the entire circuit and in each part of it.

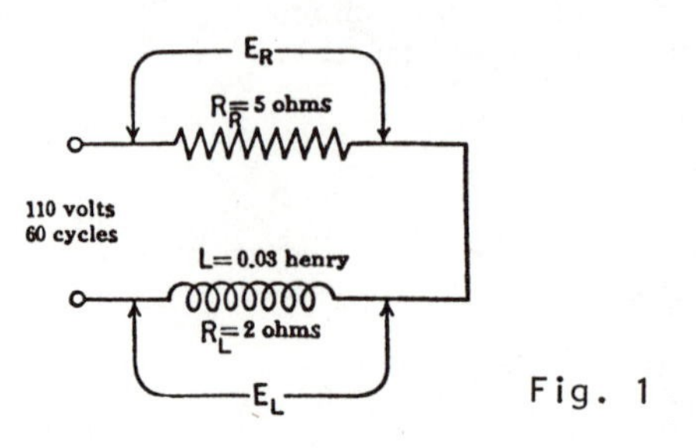

Fig. 1

A resistance of 5 ohms in series with an inductance of 0.03 henry and 2 ohms resistance. Whereas this 2 ohms will be considered with the 5 ohms, it is to be remembered that actually this 2 ohms resistance cannot be dissociated from the coil.

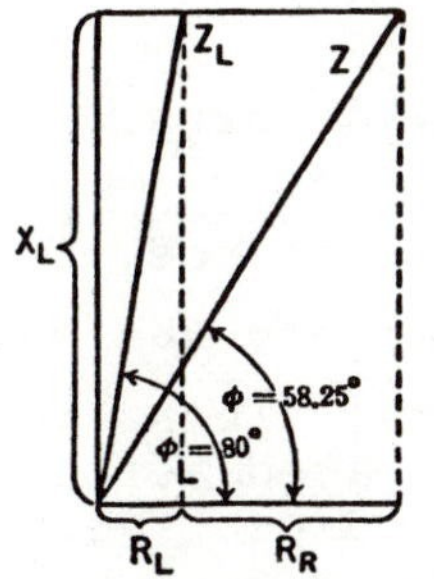

The impedance diagram of the circuit
circuit given in Figure 1.

Fig.2

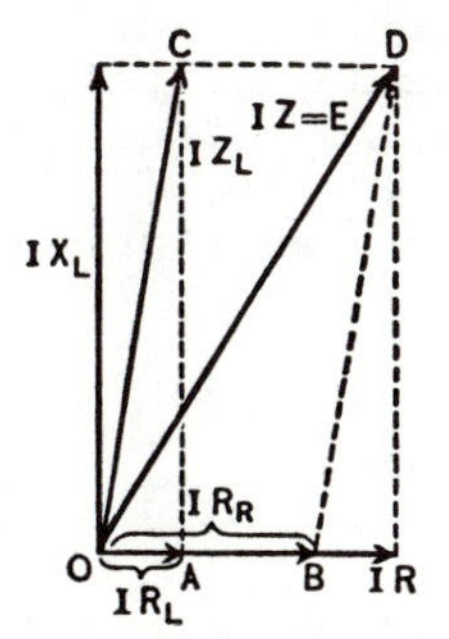

The voltage diagram for the
he circuit of Figure 1.

Fig.3

Solution: The resistance, R, of the entire circuit is the sum of the resistances of the two parts of the circuit, i.e.,

$$R = R_R + R_L = 2.0 + 5.0 = 7.0 \text{ ohms.}$$

The reactance, X_L, of the entire circuit, is that of the coil,

$$X_L = 2\pi fL = 2\pi \times 60 \times 0.03 = 377 \times 0.03 = 11.31 \text{ ohms}$$

The impedance of the entire circuit is then (Fig. 2).

$$Z = \sqrt{(R_R + R_L)^2 + (X_L)^2} = \sqrt{7^2 + 11.31^2}$$

$$= \sqrt{49 + 127.9} = 13.30 \text{ ohms.}$$

The impedance, Z_L, of the coil alone is

$$Z_L = \sqrt{R_L{}^2 + X_L{}^2} = \sqrt{2^2 + 11.31^2} = \sqrt{4 + 127.9}$$

$$= 11.48 \text{ ohms.}$$

When the circuit is placed across a 110-volt, 60-cycle source of power, the current,

$$I = \frac{E}{Z} = \frac{110}{13.30} = 8.27 \text{ amperes.}$$

The voltage across the resistance, $E_R = IR_R = 8.27 \times 5 = 41.35$ volts, and that across the coil, $E_L = IZ_L = 8.27 \times 11.48 = 94.94$ volts. These voltages will be measured by a voltmeter placed across the resistance and the inductance coil, respectively.

The total resistance drop of the circuit is $IR = IR_R + IR_L = 8.27 \times 2 + 8.27 \times 5 = 16.54 + 41.35 = 57.89$ volts. The reactance drop of the circuit is $IX_L = 8.27 \times 11.31 = 93.53$ volts. These voltages cannot be obtained by any voltmeter measurements, since, the resistance and inductance of a coil cannot be dissociated one from the other. When combined at right angles, the total IR drop and the IX drop yield the impressed voltage (Fig. 3).

$$E = \sqrt{(IR)^2 + (IX_L)^2} = \sqrt{57.89^2 + 93.53^2}$$

$$= \sqrt{3351 + 8748} = \sqrt{12{,}099} = 110 \text{ volts.}$$

The power factor of the entire circuit is, from the impedance diagram (Fig. 2),

$$\cos\phi = \frac{R_L + R_R}{Z} = \frac{7}{13.30} = 0.5263,$$

and from the voltage diagram (Fig. 3),

$$\cos\phi = \frac{IR_R + IR_L}{E} = \frac{16.54 + 41.35}{110} = \frac{57.89}{110} = 0.5263,$$

from which

$$\phi = 58.25°.$$

The power factor of the coil alone is

$$\cos\phi_L = \frac{R_L}{Z_L} = \frac{2.0}{11.48} = 0.1742,$$

or

$$\cos\phi_L = \frac{IR_L}{E_L} = \frac{16.54}{94.94} = 0.1742,$$

from which

$$\phi_L = 80°.$$

The power dissipated by the entire circuit is then

$$P = EI \cos\phi = 110 \times 8.27 \times 0.5263 = 478.8 \text{ watts},$$

or,

$$P = I^2(R_R + R_L) = 8.27^2 \times 7.0 = 478.8 \text{ watts}.$$

The power dissipated in the inductance coil is

$$P_L = E_L I \cos\phi_L = 94.94 \times 8.27 \times 0.1742 = 136.8 \text{ watts},$$

or

$$P_L = I^2 R_L = 8.27^2 \times 2.0 = 136.8 \text{ watts}.$$

The power dissipated in the resistance, R_R, is

$$P_R = E_R I = 41.35 \times 8.27 = 342.0 \text{ watts},$$

or

$$P_R = I^2 R_R = 8.27^2 \times 5.0 = 342.0 \text{ watts}.$$

Evidently

$$P = P_R + P_L = 342.0 + 136.8 = 478.8 \text{ watts}.$$

• PROBLEM 3-51

In a balanced three-phase, 208-volt circuit, the line current is 100 amperes. The power is measured by the two-wattmeter method; one meter reads 18 kW and the other zero. What is the power factor of the load? If the power factor were unity and the line current the same, what would each wattmeter read?

Solution: The expression for power is

$$P = EI \cos\phi \sqrt{3},$$

Since one wattmeter reads zero, then

$$P = 18{,}000 = 208 \times 100 \times \cos\phi \times \sqrt{3}$$

so

$$\cos\phi = \frac{18{,}000}{208 \times 100 \times \sqrt{3}} = 0.5.$$

With the power factor unity and with the same line current,

$$\tan \phi = 3 \left[\frac{W_1 - W_2}{W_1 + W_2}\right] = 0, \text{ or } W_1 = W_2$$

Also

$$W_1 + W_2 = 208 \times 100 \times \sqrt{3} = 36 \text{ kW.}$$

That is, each wattmeter reads 36/2, or 18 kW.

• PROBLEM 3-52

(a) Each load in Fig. 1 has 6 ohms resistance and 3 ohms inductive reactance. The alternator generates 110 volts per phase at 60 cycles. Determine the current in each line, the voltage between lines, the reading of each wattmeter, and the power factor of the system.

(b) Each load in Fig. 4 has 6 ohms resistance and 3 ohms inductive reactance. The alternator generates 110 volts per phase. Determine the line currents, the reading of each wattmeter, and the power factor of the system.

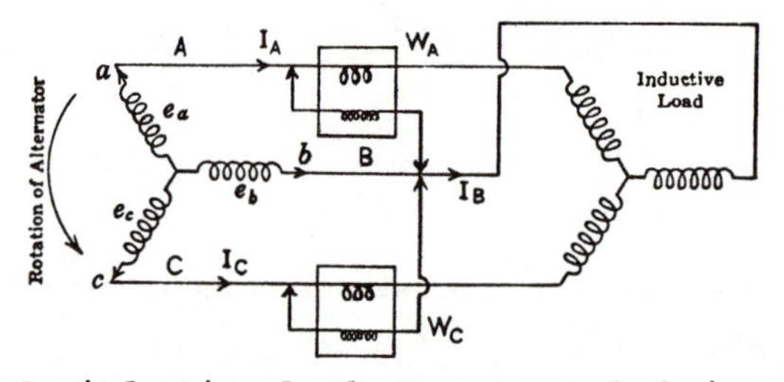

An inductive load, Y-connected, being supplied from a three phase generator, with wattmeters properly connected for power measurement. Fig. 1

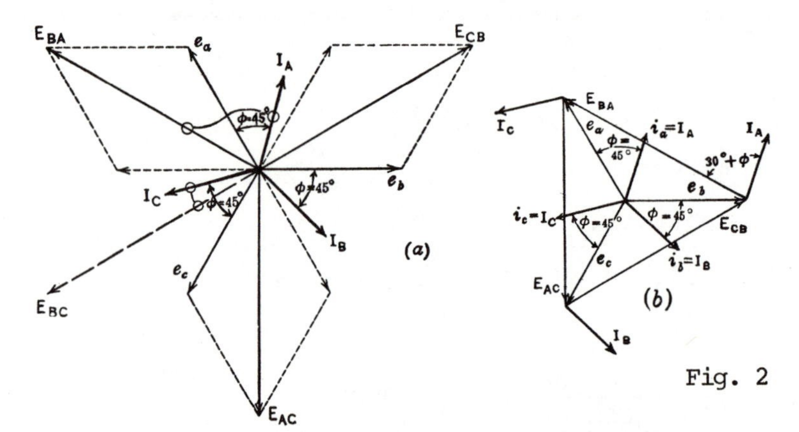

Fig. 2

Voltage and current relations for the circuit of Fig.1.

Solution: (a) The current per phase and the line current $= \frac{e}{z} = \frac{110}{\sqrt{6^2 + 3^2}} = \frac{110}{6.708} = 16.4$ amperes.

The power factor per phase, and of the system, is

$$\cos \phi = \frac{6}{6.708} = 0.8944 \quad \text{and} \quad \phi = 26.6°.$$

Upon constructing a vector diagram similar to Fig. 2, we find the line voltages to be $E = 110 \times \sqrt{3} = 190.5$ volts.

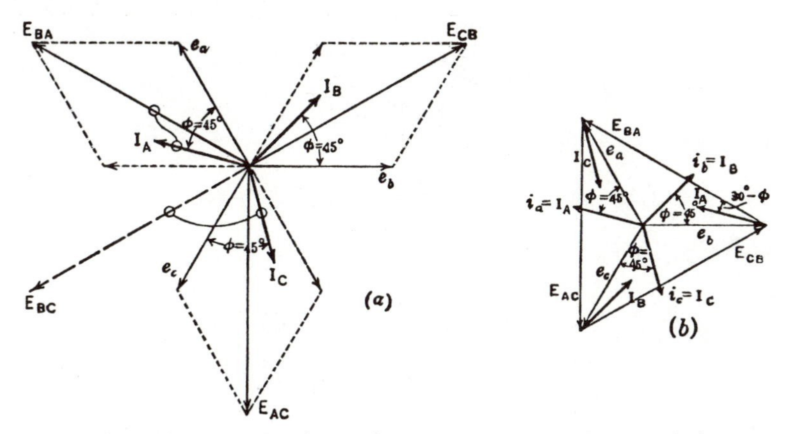

Voltage and current relations in a Y-connected capacitive load.

Fig.3

The readings of the two wattmeters are

$$W_A = E_{BA} \times I_A \times \cos (30° + \phi)$$

$$\therefore W_A = 190.5 \times 16.4 \times \cos(30° + 26.6°)$$

$$= 190.5 \times 16.4 \times 0.5505 = 1720 \text{ watts}$$

$$W_C = E_{BC} \times I_C \times \cos (30° - \phi)$$

$$W_C = 190.5 \times 16.4 \times \cos (30° - 26.6°)$$

$$= 190.5 \times 16.4 \times 0.9982 = 3119 \text{ watts.}$$

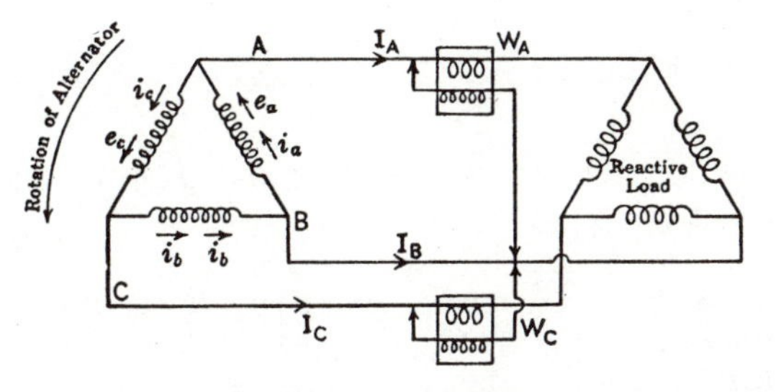

A reactive, Δ-connected load, with wattmeters for power measurement.

Fig.4

The total power is

$$W = W_A + W_C = 1720 + 3119 = 4839 \text{ watts}$$

$$= \sqrt{3}EI \cos\phi = \sqrt{3} \times 190.5 \times 16.4 \times 0.8944 = 4839 \text{ watts.}$$

$$= 3ei \cos\phi = 3 \times 110 \times 16.4 \times 0.8944 = 4840 \text{ watts.}$$

$$= 3i^2r = 3 \times 16.4^2 \times 6 = 4841 \text{ watts.}$$

The current per phase is

$$i = \frac{e}{z} = \frac{110}{\sqrt{6^2 + 3^2}} = \frac{110}{6.708} = 16.4 \text{ amperes.}$$

The current per line is

$$I = \sqrt{3}i = \sqrt{3} \times 16.4 = 28.4 \text{ amperes.}$$

The power factor is

$$\cos\phi = \frac{6}{6.708} = 0.8944 \quad \text{and} \quad \phi = 26.6°.$$

The readings of the two wattmeters are

$$W_A = E_{BA} \times I_A \times \cos(30° + \phi)$$

$$= 110 \times 28.4 \times \cos(30° + 26.6°)$$

$$= 110 \times 28.4 \times 0.5505 = 1720 \text{ watts;}$$

$$W_C = E_{BC} \times I_C \times \cos(30° - \phi)$$

$$= 110 \times 28.4 \times \cos(30° - 26.6°)$$

$$= 110 \times 28.4 \times 0.9982 = 3119 \text{ watts.}$$

The total power is

$$W = W_A + W_C = 1720 + 3119 = 4839 \text{ watts.}$$

$$= \sqrt{3}EI \cos\phi = \sqrt{3} \times 110 \times 28.4 \times 0.8944 = 4839 \text{ watts.}$$

The loads used Δ-connected in this problem are the same as those used Y-connected in part (a) with the same voltage impressed per phase. It will be seen that the wattmeter readings are identical.

• PROBLEM 3-53

Find the total power in the delta-connected load shown.

$$\vec{V}_{ba} = 220 \angle 0° \text{ V}$$

$$\vec{V}_{cb} = 220 \angle -120° \text{ V}$$

$$\vec{V}_{ac} = 220 \angle 120° \text{ V}$$

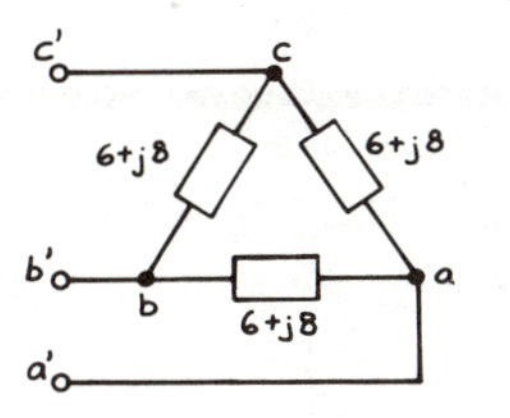

Solution: For a delta-connected load, the phase voltage $\vec{V}_P$ equals the line voltage $\vec{V}_L$ in each case. Therefore, each phase current can be found by dividing the phase voltage by phase impedance.

$$\vec{I}_P = \frac{\vec{V}_P}{\vec{Z}_P} = \frac{220\angle\phi}{10\angle 53°}$$

Phase impedance $\vec{Z}_P$ is converted from rectangular form of $x + jy$ to polar form $Z\angle\phi$ by using the following equations.

1) $x + jy$ - Rectangular form

2) $Z\angle\phi$ - Polar form

where $Z = \sqrt{x^2 + y^2}$ and $\phi = \tan^{-1} \frac{y}{x}$.

Therefore, using $x + jy = 6 + j8$,

$$Z = \sqrt{6^2 + 8^2} = 10$$

$$\phi = \tan^{-1} \frac{y}{x} = \tan^{-1} \frac{8}{6} = 53°$$

Then, using complex division, $\vec{I}_P = 22\angle\theta - 53°$

The power per phase is given by $P_P = |\vec{V}_P||\vec{I}_P| \cos \phi$ where ϕ = angle of phase impedance.

Therefore, $P_P = (220)(22)(\cos 53°) = 2912$, and the total power is $P_T = 3|\vec{V}_P||\vec{I}_P|\cos \phi = 3(P_P)$

$$= (3)(2912) = 8736\text{W}.$$

A certain load takes 40 kVA at 50% lagging power factor, while another load connected to the same source takes 80 kVA at 86.7% lagging power factor. Find the

(a) Total effective power.

(b) Reactive power.

(c) Power factor.

(d) Apparent power.

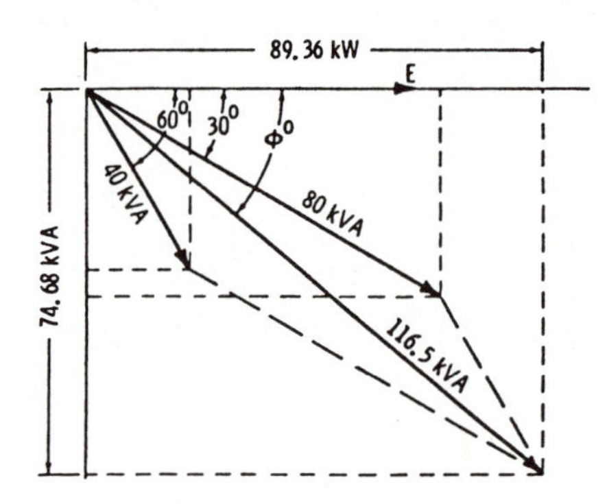

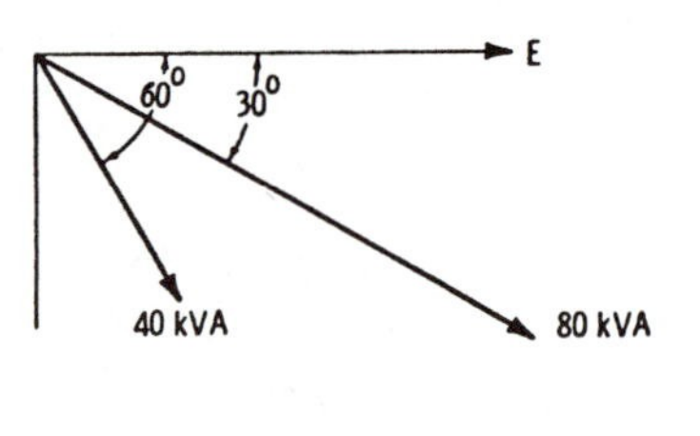

Vector diagrams of true, reactive, and apparent power in an ac circuit.

Solution: The apparent power taken by each load, expressed with reference to the voltage, is given in the figure. From a trigonometric table, the angle corresponding to a 50% power factor is 60°, and the angle corresponding to an 86.7% power factor is 30°. To find the resultant, it is necessary to complete the parallelogram. As shown in the figure, the sum of the effective power is indicated along the horizontal, and is as follows:

(a) $(40 \times 0.5) + (80 \times 0.867) = 89.36$ kW

(b) The total reactive power is indicated along the vertical, and is

$$(40 \times 0.867) + (80 \times 0.5) = 74.68 \text{ kVA}$$

(c) $$\tan\phi = \frac{74.68}{89.36} = 0.836$$

$$\cos\phi = 0.767$$

(d) The apparent power is

$$\frac{89.36}{\cos\phi} = \frac{89.36}{0.767} = 116.5 \text{ kVA.}$$

CHAPTER 4

D.C. GENERATORS

GENERATED EMF

• PROBLEM 4-1

When a generator is being driven at 1,200 rpm, the generated emf is 125 volts. Determine the generated emf (a) if the field flux is decreased by 10 percent with the speed remaining unchanged, and (b) if the speed is reduced to 1,100 rpm, the field flux remaining unchanged.

Solution: (a) $E_{g2} = 125 \times 0.90 = 112.5$ volts,

(b) $$\frac{E_{g1}}{E_{g2}} = \frac{n_1}{n_2} \quad \text{or} \quad \frac{125}{E_{g2}} = \frac{1,200}{1,100}$$

$$\therefore E_{g2} = \frac{1,100 \times 125}{1,200} = 114.6 \text{ volts.}$$

• PROBLEM 4-2

A four-pole generator has 500 conductors on the armature. If the generator is running at 1200 rpm, find the average voltage generated between brushes for (a) a lap winding, (b) a wave winding. The total flux per pole is 10^6 lines.

Solution: (a) For a simplex lap winding, there are as many paths through the armature as there are poles. Therefore, P = 4.

$$E = \frac{z}{P}\frac{\phi}{t} \times 10^{-8} \text{ volts}$$

$$= \frac{500 \times 10^6 \times 10^{-8}}{4 \times \frac{60}{4800}} = 100 \text{ volcs.}$$

(b) For the simplex wave winding, there are only two paths, regardless of the number of poles. Therefore, P = 2.

$$E = \frac{500 \times 10^6 \times 10^{-8}}{2 \times \frac{60}{4800}} = 200 \text{ volts.}$$

• PROBLEM 4-3

A four-pole d-c generator has an armature winding containing a total of 648 conductors connected in two parallel paths. If the flux per pole is 0.321 x 10^6 maxwells and the speed of rotation of the armature is 1,800 rpm, (a) calculate the average generated voltage, (b) calculate the rated current in each conductor (per path) if the power delivered by the armature is 5 kw.

Solution:

(a) Number of conductors in series per path = 648/2 = 324.

$$\text{Flux cut per revolution} = 4 \times 0.321 \times 10^6$$

$$= 1.284 \times 10^6 \text{ maxwells.}$$

$$\text{Revolutions per second of armature} = \frac{1,800}{60} = 30 \text{ rps.}$$

$$\text{Seconds per revolution of armature} = 1/30 = 0.0333 \text{ sec}$$

$$E_{av} = \frac{\phi}{t \times 10^8} \text{ volts.}$$

$$E_{av} \text{ (per conductor)} = \frac{1.284 \times 10^6}{0.0333} \times 10^{-8} = 0.386 \text{ volt}$$

$$E_g \text{ (total generated voltage)} = 0.386 \times 324 = 125 \text{ volts}$$

(b) $\text{Total armature current} = \frac{\text{watts}}{\text{volts}} = \frac{5,000}{125} = 40 \text{ amp}$

$$\text{Current per armature circuit (per conductor)} = 40/2$$

$$= 20 \text{ amp.}$$

• PROBLEM 4-4

A 4-pole commutator machine has 124 lap coils each having two turns. The flux per pole is 0.015 weber.

Calculate:

(a) the d.c. voltage appearing across quadrature brushes when running at 1500 rev/min in a steady field.

(b) The r.m.s. voltage with three sets of brushes per pole pair for 3-phase working and with relative field/ conductor speed = 1400 rev/min.

Solution: (a) No. of conductors in series = $(124 \times 2 \times 2)/4 = 124 = z_s$

d.c. voltage = $2p\phi n z_s = 4 \times 0.015 \times (1500/60) \times 124 = 186$ V.

(b) Brush separation $\theta = 360°/3 = 120°$ electrical.

r.m.s. diametral voltage = $1.11 \times (2p\phi n z_s)(2/\pi)$

$= 1.11 \times 186 \times 1400/1500 \times 2/\pi = 123$ V

Between brushes spaced at 120° electrical,

a.c. voltage = r.m.s. diametral voltage $\times \sin(\theta/2)$.

$= 123 \sin(120°/2) = 106.5$ V. r.m.s.

• PROBLEM 4-5

Consider a bipolar machine wound with 44 coils of 8 turns each of No. 13 wire. The total length per turn is 2 feet, of which 1 foot is active. Assume that the peripheral speed of the armature is 3000 feet per minute, that 60 percent of the inductors are active (i.e., lie under the pole face at the same time), and that the flux density in the air gap is 10,000 lines per square centimeter, the pole fringe not considered. Find (1) the generated e.m.f.; (2) the safe current capacity; (3) the armature resistance; (4) the full-load terminal voltage.

Solution: First determine the active length of inductor per path. This must be a two-path winding, hence,

Coils per path = 22

Turns per path = 176;

Length of inductor per path = 176 × 1 = 176 ft.

The flux density is given in lines per square centimeter.

Hence, using the metric system,

Length of inductor per path = 5370 cm.

Active length of inductor per path = 5370 × 60 percent

= 3220 cm.

The velocity of the inductors is 3000 feet per minute

= 1525 cm. per sec.

Flux density = 10,000 lines per sq. cm.

Flux cut per second = ℓvB

= 3220 × 1525 × 10,000

= 491×10^8 lines.

Hence the generated voltage per path = 491 volts, and this is the generated e.m.f. of the machine.

• PROBLEM 4-6

A two-pole dc generator has an armature containing a total of 40 conductors connected in two parallel paths. The flux per pole is 6.48×10^8 lines, and the speed of the prime mover is 30 rpm. The resistance of each conductor is 0.01 ohm, and the current-carrying capacity of each conductor is 10 A. Calculate:

(a) The average generated voltage per path and the generated armature voltage

(b) The armature current delivered to an external load

(c) The armature resistance

(d) The terminal voltage of the generator.

Solution: (a) Total ϕ linked in one revolution = P (ϕ/pole) = 2 poles × 6.48×10^8 lines/pole.

Time for one revolution, $t/\text{rev} = \frac{1}{30}$ min/rev

$= (60 \text{ s/min}) \times \frac{1}{30}$ min/rev

= 2 s/rev.

$$e_{av}/\text{cond} = \frac{\phi}{t} \times 10^{-8} \text{ V}$$

$$= \frac{2 \times 6.48 \times 10^8 \text{ lines}}{2 \text{ s/rev}} \times 10^{-8} \text{ V}$$

$$= 6.48 \text{ V/conductor.}$$

Generated voltage per path,

$$E_g = (\text{voltage/cond}) \times \text{no. of cond/path}$$

$$= (6.48 \text{ V/cond}) \times 40 \text{ cond/2 paths}$$

$$= 129.6 \text{ V/path}$$

Generated armature voltage,

$$E_g = \text{generated voltage/path}$$

$$= 129.6 \text{ V.}$$

(b) $I_a = (\text{I/path}) \times 2 \text{ paths} = (10 \text{ A/path}) \times 2 \text{ paths} = 20 \text{ A.}$

(c) $R_a = \dfrac{\text{R per path}}{\text{no. of paths}} = \dfrac{0.01 \text{ ohm/cond}}{2 \text{ paths}} \times 20 \text{ cond} = 0.1\ \Omega$

(d) $V_t = E_g - I_aR_a = 129.6 \text{ V} - [20 \text{ A} \times 0.1\ \Omega] = 127.6 \text{ V}$

ARMATURE WINDINGS

• PROBLEM 4-7

A six-pole dynamo, with interpoles, requires 360 inductors on its armature, with three turns per coil and two coil sides per slot. (a) Determine the number of coils, slots, and commutator bars. (b) Select suitable pitches for a simplex lap winding.

Solution:

(a) 3 turns = 6 inductors per coil

360 ÷ 6 = 60 coils = 120 coil sides

60 coils requires 60 commutator bars and

2 coil sides per slot requires 60 slots

Therefore coils = bars = slots = 60

(b) 60 slots ÷ 6 poles = 10 slots, back pitch

Therefore front pitch = 9 slots for progressive and 11 slots for retrogressive winding and

Commutator pitch = 1 bar

• **PROBLEM 4-8**

(a) A 4-pole machine has lap winding with 36 winding elements. What are the back and front pitches of this winding, expressed in commutator bars?

(b) A 4-pole machine has wave winding with 37 winding elements. What are the three pitches of this winding expressed in commutator bars?

Solution: (a) Since there are 36 winding elements, the number of commutator bars is 36. The number of commutator bars per pole is then 36/4 = 9, making the back pitch $y_b = 9$; the front pitch is given by the following equation:

$$y = y_b - y_f$$

Hence, $y_f = y_b - 1 = 8$ commutator pitches

(b) Since there are 37 winding elements. the number of commutator bars is equal to 37. Winding pitch, y is given by

$$y = \frac{K \mp 1}{P/2}$$

where K = no. of commutator bars

P = no. of poles

$$\therefore y = \frac{37 \mp 1}{2} = 18 \text{ or } 19.$$

The number of commutator bars per pole is 37/4 = 9.25.

The back pitch is made $y_b = 9$.

Therefore, from the equation: $y = y_b \pm y_f$,

$$y_f = y - y_b,$$

the front pitch is either $y_f = 18 - 9 = 9$ or $19 - 9 = 10$, commutator pitches.

• **PROBLEM 4-9**

On a certain 8-pole d.c. machine rated at 1200-kW, 600-V, 500-rev/min, the average value of the reactance voltage per coil is 4.4 V. The machine has a single-turn lap winding with 624 conductors. The interpole length is 30 cm, and its air gap is 0.85 cm long. The armature diameter is 1.3 m. The ratio of B_{max} to B_{av} in the commutating zone is 1.4.

Find the total number of turns required per pole on the quadrature axis, allowing 10% extra magnetising ampere-turns for the iron and slots.

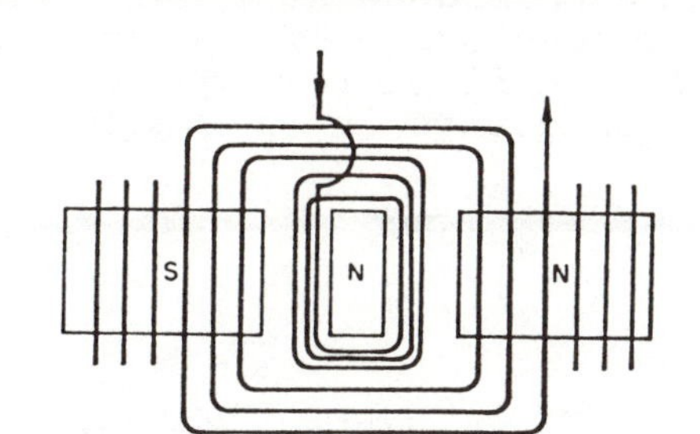

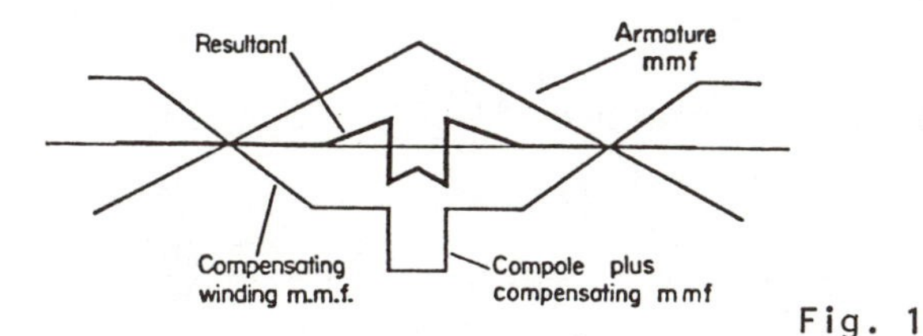

Fig. 1

Compensating winding.

Solution: Armature current per conductor $I_c = I/2p =$

$$\frac{1200 \times 10^3}{600} \times \frac{1}{8} = 250A$$

$$\text{Armature AT/pole} = \frac{I_c}{2p} \cdot \frac{Z}{2} = \frac{250}{8} \times \frac{624}{2} = 9750$$

$$e_r = 2B\ell v$$

$$4.4 = 2B \times \frac{30}{10^2} \times \pi \times 1.3 \times \frac{500}{60} \text{ from which } B = 0.216 \text{ Wb/m}^2.$$

This is the average value of the compole air-gap flux density. The maximum value is $1.4 \times 0.216 = 0.302$ Wb/m² and the magnetising ampere-turns allowing an extra 10% are

$$H_g \ell_g = \frac{0.302}{4\pi/10^7} \times \frac{0.85}{10^2} \times 1.1 = 2250 \text{ AT}$$

The total AT required on the quadrature axis 9750 + 2250

= 12,000. With full armature current through each turn this would require 6 turns. For a machine without a compensating winding (see Fig. 1), these turns would all be wound on the interpole. Otherwise there would usually be three on each interpole and three turns per pole on the compensating winding. If the number of turns had not been an integer the winding could have been split into parallel circuits and small adjustments made to the interpole air-gap length.

• PROBLEM 4-10

A 2,500-kw 600-volt 16-pole generator has a lap-wound armature with 2,360 conductors. If the pole faces cover 65 per cent of the entire circumference, calculate the number of pole-face conductors in each pole of a compensating winding.

Solution:

$$\text{Total armature current} = \frac{\text{watts}}{\text{volts}} = \frac{2,500,000}{600} = 4,170 \text{ amp.}$$

$$\text{Armature current per path} = \frac{4,170}{16} = 261 \text{ amp.}$$

$$\text{Armature conductors under each pole} = \frac{2,360}{16} \times 0.65 = 96$$

$$\text{Armature ampere-conductors under each pole face} = 96 \times 261 = 25,000$$

Compensating winding ampere-conductors per pole face

$$= 4,170 \times C$$

Equating $4,170C = 25,000$

$$C = \frac{25,000}{4,170} = 6 \text{ conductors in each pole face.}$$

• PROBLEM 4-11

A simplex lap wound dc dynamo has 800 conductors on its armature, a rated armature current of 1000 A and 10 poles. Calculate the number of pole face conductors per pole to give full armature reaction compensation, if the pole face covers 70 per cent of the pitch.

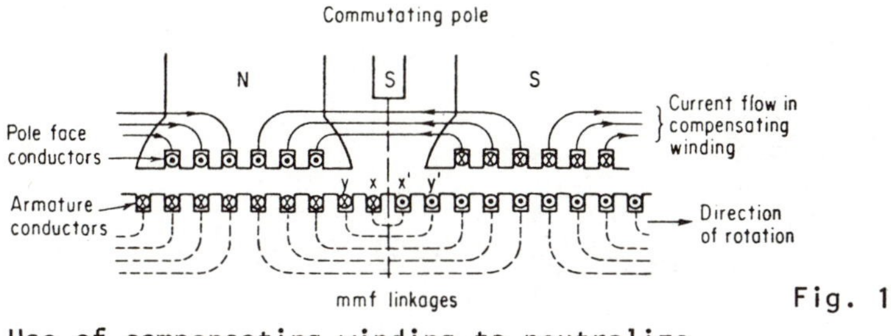

Fig. 1

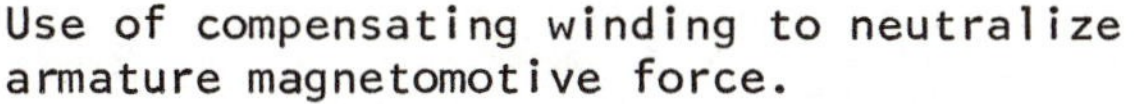
Use of compensating winding to neutralize armature magnetomotive force.

Solution:

$$Z = \frac{800}{10} = 80 \text{ cond/path under each pole}$$

Active conductors/pole,

$$Z_a = 80 \text{ cond/path} \times 0.7 = 56 \text{ conductors/pole}$$

Solving for Z_p in the following equation:

$$Z_p I_\ell = Z_a I_a \text{ ,}$$

$$Z_p \times 1000 \text{ A} = 56 \text{ cond} \times 1000 \text{ A/10 paths}$$

$$Z_p = 5.6 \text{ or } 6 \text{ conductors/pole}$$

Using the equation: $Zp = Z_a/a$,

$$Z_p = \frac{56}{10} \cong 6 \text{ conductors/pole}$$

as shown in Fig. 1.

NO-LOAD AND LOAD CHARACTERISTICS

• PROBLEM 4-12

Assuming constant field excitation, calculate the no-load voltage of a separately excited generator whose armature voltage is 150 V at a speed of 1800 rpm, when

(a) The speed is increased to 2000 rpm.

(b) The speed is reduced to 1600 rpm.

Solution: $E_g = K''S$ at constant field excitation,

and therefore

$$\frac{E_{final}}{E_{orig}} = \frac{S_{final}}{S_{orig}}$$

(a) $E_{final} = (E_{orig}) \frac{S_{final}}{S_{orig}} = (150 \text{ V}) \frac{2000}{1800} = 166.7 \text{ V}.$

(b) $E_{final} = (150 \text{ V}) \frac{1600}{1800} = 133.3 \text{ V}.$

• PROBLEM 4-13

A d-c machine has the following constants and ratings:

voltage = 250 $J = 1.2$ n-m-sec^2

line current = 95 amperes

speed = 863 rpm

armature resistance = 0.10 ohm

field resistance = 225 ohms

The d-c machine is to be used as a separately excited generator. It is to deliver a full load current of 95 amperes at a voltage of 250 when driven at a constant speed of 900 rpm. The field is supplied from a constant voltage source of 250 volts. Determine, (a) the gross field circuit resistance and field current for the stated full load conditions, (b) the electromagnetic torque and power, (c) the no load terminal voltage and the voltage regulation in percent.

Solution:

(a) The rated load armature generated voltage is

$$E_A = V_A + R_A I_A = 250 + (0.10)(95) = 259.5 \text{ volts}$$

The generator speed

$$\omega = \frac{2\pi N}{f}$$

$$= \left(\frac{900}{60}\right)(2\pi) = 30\ \pi \text{ rad/sec}$$

$$I_1 = \frac{E_A}{M_{A1}\omega} = \frac{259.5}{(2.4)(30\pi)} = 1.144 \text{ amp}$$

$$R_1 = \frac{250}{1.144} = 218.3 \text{ ohms}$$

(b) $T_g = M_{A1} I_1 I_A = (2.4)(1.144)(95) = 261$ newton-meters

$P_g = T_g \omega = (261)(30\pi) = 24{,}650$ watts

(c) At no load, $V_A = E_A = \omega M_{A1} I_1 = (30\pi)(2.4)(1.144) =$ 259.5 volts

$$\text{voltage regulation} = \frac{259.5 - 250}{250}(100) = 3.8\%.$$

• PROBLEM 4-14

The voltage of a 100-kw 250-volt shunt generator rises to 260 volts when the load is removed. (a) What full-load current does the machine deliver, and what is its percent regulation? A 25-kw 230-volt shunt generator has a regulation of 8.7 percent. (b) What will be the terminal voltage of the generator at no load? (c) If the change in voltage is assumed to be uniform between no-load and full-load kilowatts, calculate the kilowatt output of the generator when the terminal voltages are 240 and 235 volts.

Solution: $I_{FL} = \frac{100{,}000}{250} = 400$ amp.

$$\text{Percent regulation} = \frac{V_{NL} - V_{FL}}{V_{FL}} \times 100$$

$$= \frac{260 - 250}{250} \times 100 = 4 \text{ percent.}$$

(b)
$$8.7 = \frac{V_{NL} - 230}{230} \times 100$$

$$V_{NL} = \frac{8.7 \times 230}{100} + 230 = 250 \text{ volts}$$

(c)
$$P_{out} = \frac{V_{NL} - V_t}{V_{NL} - V} \times P_{in}$$

$$P_{240} = \frac{250 - 240}{250 - 230} \times 25 = 12.5 \text{ kw}$$

$$P_{235} = \frac{250 - 235}{250 - 230} \times 25 = 18.75 \text{ kw.}$$

• PROBLEM 4-15

A 100-kw 250-volt 400-amp long-shunt compound generator has armature resistance (including brushes) of 0.025 ohm, a series-field resistance of 0.005 ohm, and the magnetization curve of Fig. 1. There are 1,000 shunt-field turns per pole and 3 series-field turns per pole.

Compute the terminal voltage at rated current output when the shunt-field current is 4.7 amp and the speed is 1,150 rpm. Neglect armature reaction.

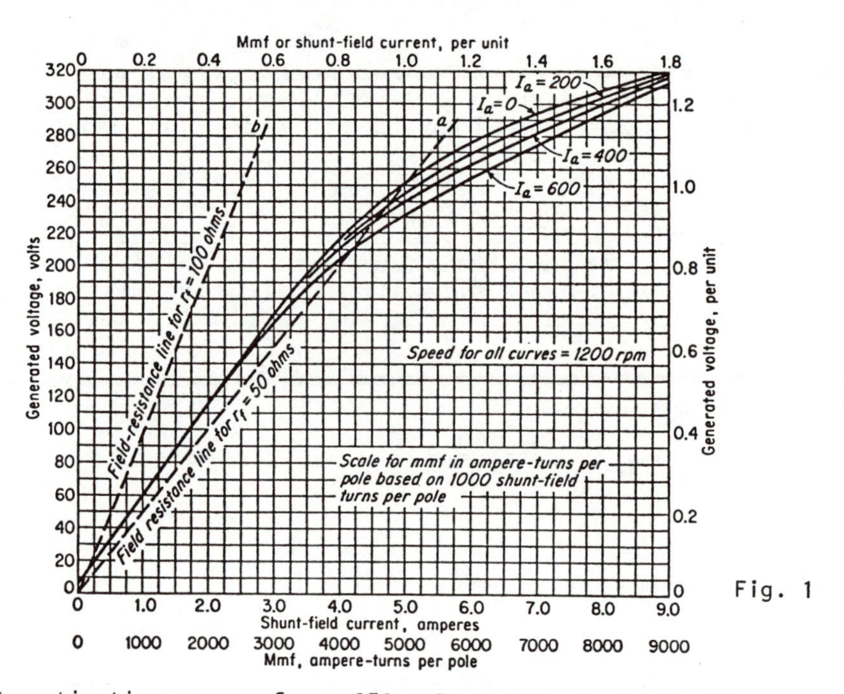

Fig. 1

Magnetization curves for a 250-volt 1,200-rpm d-c machine.

Solution: $I_s = I_a = I_L + I_f = 400 + 4.7 = 405$ amp. The mainfield mmf is given by:

$$\text{Main-field mmf} = I_f \pm \frac{N_s}{N_f} I_s$$

$$= 4.7 + \frac{3}{1{,}000} \times 405$$

$$= 5.9 \text{ equivalent shunt-field amp.}$$

By entering the $I_a = 0$ curve of Fig. 1 with this current, one reads 274 volts. Accordingly, the actual emf is

$$E_a = E_{ao} \frac{\omega_m}{\omega_{mo}}$$

$$= 274 \times \frac{1,150}{1,200} = 262 \text{ volts}$$

Then $V_t = E_a - I_s(r_a + r_s)$

$$= 262 - 405(0.025 + 0.005) = 250 \text{ volts.}$$

• PROBLEM 4-16

In a compound generator, connected short-shunt, the terminal voltage is 230 volts when the generator delivers 150 amp, Fig. 1. The shunt-field current is 2.5 amp, armature resistance 0.032 ohm, series-field resistance 0.015 ohm, and diverter resistance 0.030 ohm. Determine: (a) induced emf in armature; (b) total power generated in armature; (c) distribution of this power.

The combined series-field and diverter current is 150 amp.

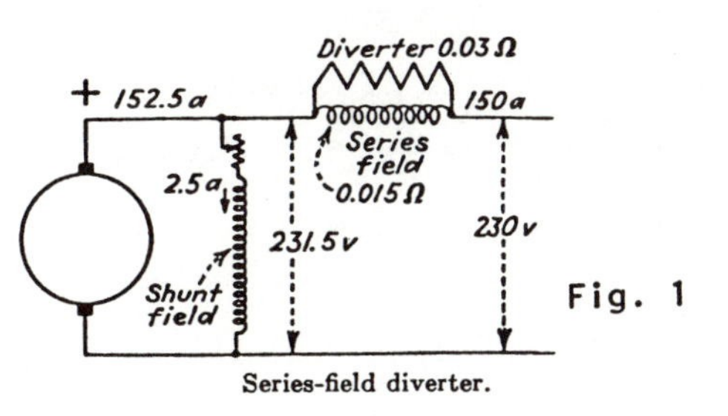

Series-field diverter.

Fig. 1

Solution: (a) Series-field current

$$I_s = 150 \frac{0.030}{0.015 + 0.030} = 100 \text{ amp.}$$

Diverter current

$$I_d = 150 \frac{0.015}{0.015 + 0.030} = 50 \text{ amp.}$$

Combined equivalent resistance of series field and diverter,

$$\frac{1}{R'} = \frac{1}{0.015} + \frac{1}{0.030}; \quad R' = 0.010 \text{ ohm.}$$

Voltage drop in series field and diverter

$$E' = 150 \times 0.010 = 1.50 \text{ volts.}$$

Armature current

$$I_a = 152.5 \text{ amp.}$$

Induced emf

$$E = 230 + 1.5 + 152.5 \times 0.032 = 236.4 \text{ volts.}$$

(b) Total power generated,

$$P_a = 236.4 \times 152.5 = 36{,}050 \text{ watts} = 36.05 \text{ kw.}$$

(c) Armature loss

$$P_a' = 152.5^2 \times 0.032 = 744 \text{ watts.}$$

Series-field loss

$$P_s = 100^2 \times 0.015 = \quad = 150 \text{ watts.}$$

Diverter loss

$$P_d = 50^2 \times 0.030 \quad = 75 \text{ watts.}$$

Shunt-field loss

$$P_{sh} = (230 + 1.5)2.5 \quad = 579 \text{ watts.}$$

Power delivered

$$P = 230 \times 150 \quad = \underline{34{,}500} \text{ watts.}$$

$$\text{Total} \quad = 36{,}048 \text{ watts.}$$

• PROBLEM 4-17

Consider the conditions shown in Fig. 1(a). A certain load is 4,000 ft distant from the generator. The load is supplied over a 500,000-cir-mil feeder. The no-load voltage of the generator is 500 volts. It is desired to maintain the load voltage at a substantially constant value of 500 volts from no load to the maximum demand of 300 amp. Determine characteristic of generator.

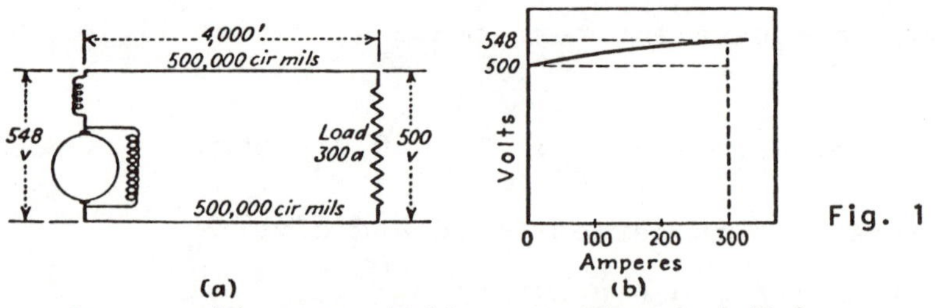

Fig. 1

Overcompounded generator maintaining constant voltage at end of feeder.

Solution: If the feeder is operated at "normal" density, the current would be 500 amp, or 0.001 amp per cir mil and the drop would be 0.01 volt per ft, making a total drop of 80 volts.

The actual drop is

$$(300 \div 500)80 = 48 \text{ volts.}$$

The generator terminal voltage should rise from a no-load value of 500 volts to 548 volts when 300 amp is being delivered to the load, Fig. 1(b).

• PROBLEM 4-18

A 10-kW 250-V self-excited generator, when delivering rated load, has an armature-circuit voltage drop that is 5% of the terminal voltage and a shunt-field current equal to 5% of rated load current. Calculate the resistance of the armature circuit and that of the field circuit.

Solution:

$V = 250$ V, rated value

$I = 10{,}000 \div 250 = 40$ A, rated load current.

$I_f = 0.05 \times 40 = 2$ A, field current.

From the equation, $I_a = I + I_f$,

$$I_a = 40 + 2 = 42 \text{ A}$$

$$r_a I_a = 0.05 \times 250 = 12.5 \text{ V}$$

$r_a = 12.5 \div 42 = 0.298\ \Omega$ the resistance of the armature circuit.

$$r_f I_f = 250$$

$r_f = 250 \div 2 = 125\ \Omega$, the resistance of the field circuit.

• PROBLEM 4-19

A six-pole generator requires 4.5 amp shunt-field excitation to give rated voltage at no load, and 7.0 amp to give the same voltage at rated load of 200 amp. There are 750 shunt-field turns per pole. (a) How many series turns must be added to give flat-compound operation, using the short-shunt connection? (b) If 12 turns per pole are added, having a resistance of 0.005 ohm per pole, what should be the resistance of a shunt across the series-field terminals to give the desired flat-compound effect?

Solution:

NI per pole required = (7.0 - 4.5) 750 = 1875

(a) Turns per pole = $\frac{1875}{200}$ = 9.4

(b) Current required for excitation = $\frac{1875}{12}$ = 156.25 amp

Current through diverter = 200 - 156.25 = 43.75 amp

Since for parallel circuits the voltage drops are equal across the parallel paths,

$$I_F R_F = I_D R_D$$

or

$$156.25\ (0.005 \times 6) = 43.75 \times R_D$$

and

$$R_D = \frac{156.25\ (0.005 \times 6)}{43.75} = 0.107 \text{ ohm.}$$

• PROBLEM 4-20

A long-shunt compound generator has a shunt field winding of 1000 turns per pole, and a series field winding of 4 turns per pole. In order to obtain the same (rated) voltage at full load as at no load, when operated as a shunt generator, it is necessary to increase the field current 0.2 A. The full-load armature current of the compound generator is 80 A and the series field resistance is 0.05 ohm. Calculate:

(a) The number of series field ampere-turns (At) required for flat-compound operation.

(b) The diverter resistance required for flat-compound operation.

Solution:

(a) $\delta I_f N_f$ = 0.2 A × 1000 turns = 200 At = $I_s\ N_s$ for flat-compound operation

(b) $I_s = \frac{I_s N_s}{N_s} = \frac{200 \text{ At}}{4 \text{ t}}$ = 50 A, required in series field winding for flat-compound operation

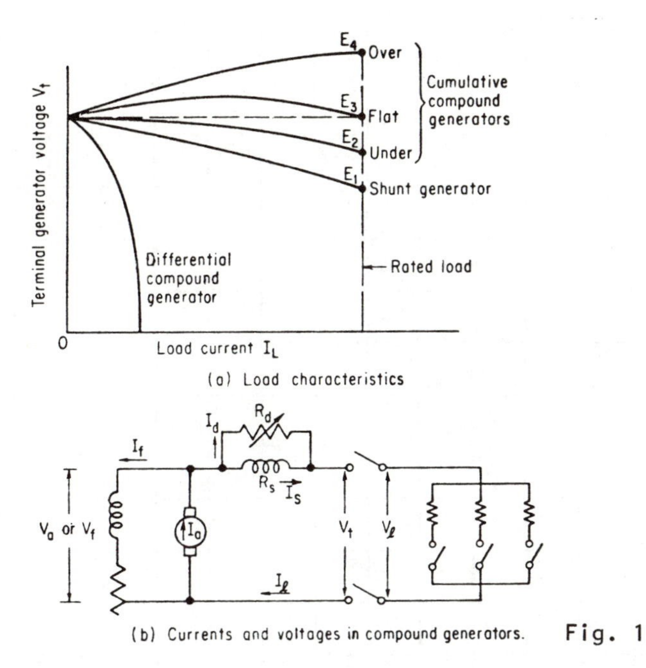

Fig. 1

External load voltage characteristic of cumulative and differential compound generators.

$$I_d = I_a - I_s = 80 \text{ A} - 50 \text{ A} = 30 \text{ A}$$

$$R_d = \frac{I_s R_s}{I_d} = \frac{50 \text{ A} \times 0.05 \ \Omega}{30 \text{ A}} = 0.0833 \ \Omega .$$

• PROBLEM 4-21

A 250-volt, 250-kw, 1000-ampere, long-shunt compound generator has an armature resistance of 0.011 ohm, including brushes and a series field resistance of 0.0022 ohm. One hundred per cent values on the magnetization curve of Fig. 1 are, 250 volts, 10 amp, and 900 rpm. The shunt field has 800 turns per pole, the series field 2 turns per pole. The effective demagnetizing ampere-turns per pole are 500.

(a) Determine the rated current terminal voltage when the shunt field current is 10.4 amperes, the speed remaining at 900 rpm.

(b) If the shunt field current at rated load is the same as in part a, but the rated speed decreased to 850 rpm, determine the terminal voltage at rated current.

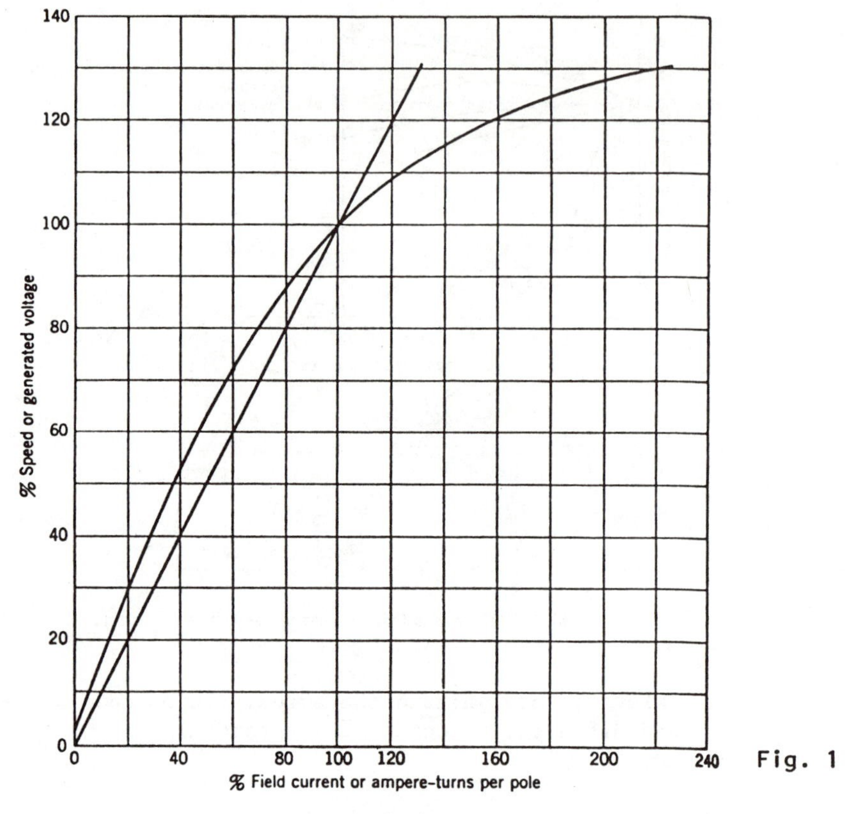

Magnetization curve.

Solution:

(a) Base ampere turns per pole = 800 x 10 = 8000

Effective ampere turns per pole at full load is

$$(NI)_{NET} = N_S I_S - (NI)_{DM}$$

where

$(NI)_{DM}$ = Demagnetizing ampere-turns

$$NI = 800 \times 10.4 + 2020.8 - 500 = 9840$$

Percent ampere-turns per pole = (9840/8000)100 = 123

From the magnetization curve entering at 123% excitation, the speed voltage is determined at 111% of 250 volts. Thus

$$E_a \text{ at rated load} = 1.11 \times 250 = 277.5 \text{ volts}$$

and

$$V_t = E_a - (I + I_f)(r_a + r_f)$$

$$= 277.5 - 1010.4(0.011 + 0.0022) = 264 \text{ volts.}$$

(b) If the speed had remained at 900 rpm, the armature generated voltage would remain at 277.5. Due to the decreased speed the induced voltage must be decreased a like amount. Thus, the actual emf is

$$E_a = 277.5(850/900) = 262 \text{ volts}$$

and

$$V_t = 262 - 1010.4(0.011 + 0.0022) = 249 \text{ volts.}$$

• PROBLEM 4-22

A 4½-kW, 125-V, 1150-r/min, separately excited dc generator has an armature-circuit resistance of 0.37 Ω. When the machine is drive at rated speed, the no-load saturation curve obtained is that shown in Fig. 1.

If the field rheostat is adjusted to give a field current of 2A, and the machine is driven at 1000 r/min, what will be the terminal potential difference when the load current is at the rated value? (The effects of armature reaction and brush contact resistance may be neglected.)

The above machine is driven as a shunt generator at rated speed, and the field rheostat is adjusted to give a no-load terminal potential difference of 135 V.

What will be the terminal potential difference of the machine when it is delivering its rated current of 36 A?

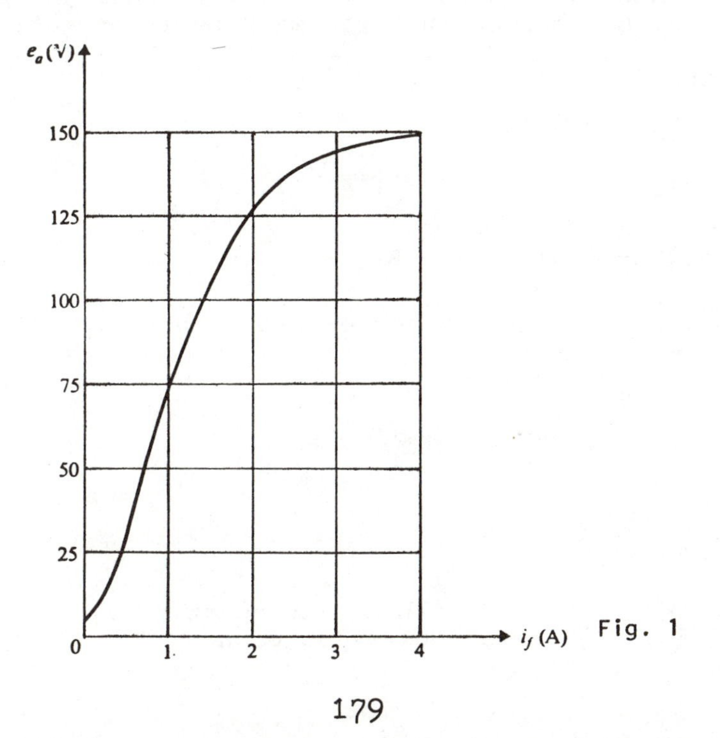

Fig. 1

Solution: To determine the steady-state condition of operation, it is necessary to know the armature current i_a. However, since the terminal potential difference and consequently the field current are not known, it is not possible to say what the armature current will be when the load current is 36 A. Since i_f is small, and

$$i_a = i_f + i_L \quad \text{A}$$

it is best to assume two values of i_a close to 36 A, determine the corresponding points on the external characteristic, and then determine the required point by interpolation.

From the magnetization characteristic of Fig. 1, when $e_a = 135$ V, $i_f = 2.42$ A. The slope of the field-resistance line shown in Fig. 2 is thus

$$R_f = \frac{135}{2.42} = 55.8 \ \Omega$$

1. Assume $i_a = 36$ A

$$e_a = R_a i_a + (R_e + R_f) i_f \quad \text{V}$$

$$e_{a1} = 0.37 \times 36 + 55.8\, i_{f1} = 13.3 + 55.8\, i_{f1} \quad \text{V.}$$

A line drawn in Fig. 2 with an intercept of 13.3 V and a slope of 55.8 Ω gives an operating point

$$e_{a1} = 127 \text{ V}, \quad i_{f1} = 2.05 \quad \text{A.}$$

This line also gives another much lower operating point, but it will be assumed that the machine is operating normally on the upper part of the external characteristic. Then using

$$v_t = e_a - R_a i_a \quad \text{V}$$

$$v_{t1} = 127 - 0.37 \times 36 = 114 \quad \text{V.}$$

$$i_{L1} = 36 - 2.05 = 34 \quad \text{A.}$$

which gives one point on the external characteristic.

2. Assume $i_a = 39$ A.

$$e_a = R_a i_a + (R_e + R_f) i_f \quad \text{V}$$

$$\therefore e_{a2} = 0.37 \times 39 + 55.8\, i_{f2} = 14.4 + 55.8\, i_{f2} \quad \text{V}$$

A line drawn in Fig. 2 with an intercept of 14.4 V and a slope of 55.8Ω gives an operating point

$$e_{a2} = 126 \quad V, \qquad i_{f2} = 2.0 \quad A$$

Thus

$$v_{t2} = 126 - 0.37 \times 39 = 112 \quad V$$

$$i_{L2} = 39 - 2.0 = 37 \quad A$$

From these two results, it may be seen that for $i_L = 36A$, v_t will be 113 V.

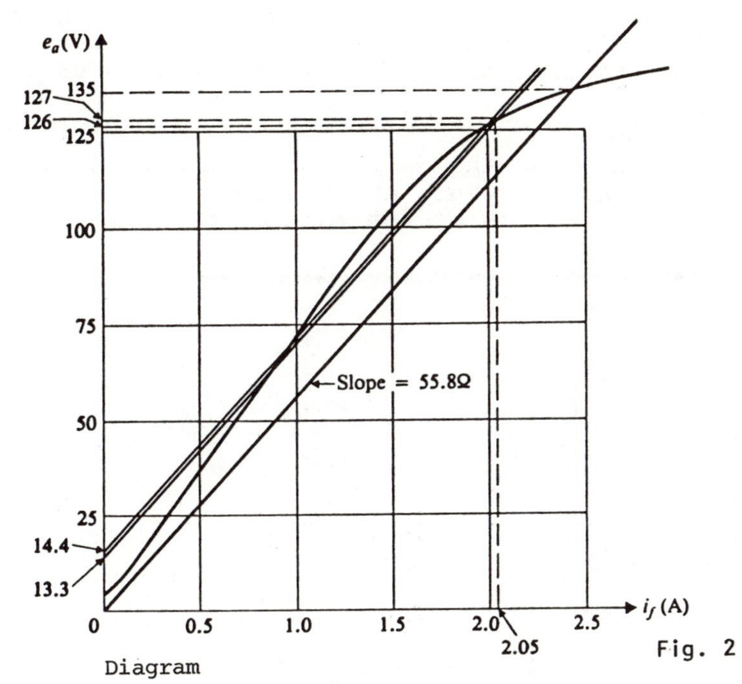

Diagram

Fig. 2

Another procedure, which eliminates the need for a graph, is to carry out calculations for a series of values of i_f below the no-load value, that yield the results shown in Table 1.

Without drawing the external characteristic, this table shows an operating condition of v_t = 112 V for i_L = 35.8 A, which closely approximates the result from the graph.

Table 1

Formula	Value of i_f				
	2.4	2.2	2.0	1.8	1.6
$v_t = R_f i_f$	134.0	123.0	112.0	100.0	89.0
e_a from Fig.1	135.0	131.0	126.0	120.0	110.0
$R_a i_a = e_a - v_{t2}$	1.0	8.0	14.0	20.0	21.0
$i_a = (e_a - v_t)/R_a$	2.7	21.6	37.8	54.1	56.8
$i_L = i_a - i_f$	0.3	19.4	35.8	52.3	55.2

A 600-V dc source is provided from a 440-V, 3-phase, ac system by means of a bank of rectifiers. The dc load is supplied via a feeder of resistance 0.41 Ω per conductor. In order to maintain approximately constant potential difference at the load end of the feeder, a series booster generator is fitted in one side of the feeder, as shown in Fig. 1.

The booster generator is rated at 1 kW, 25 V. Its no-load saturation curve, obtained at rated speed by separately exciting the series field winding, is shown in Fig. 2. The resistance of the booster armature circuit is 0.08 Ω and that of the series field is 0.05 Ω.

Determine the potential difference v_L at the load end of the feeder when the rectifier is supplying (a) 20 A, (b) 40 A direct current; and determine the terminal potential difference of the booster generator in each case.

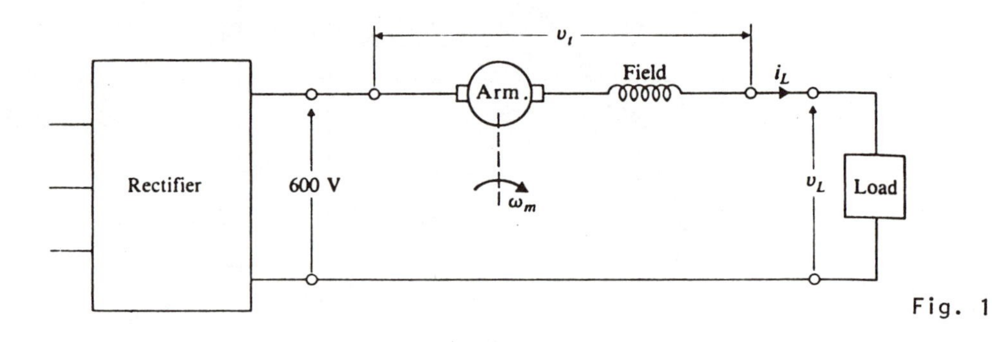

Fig. 1

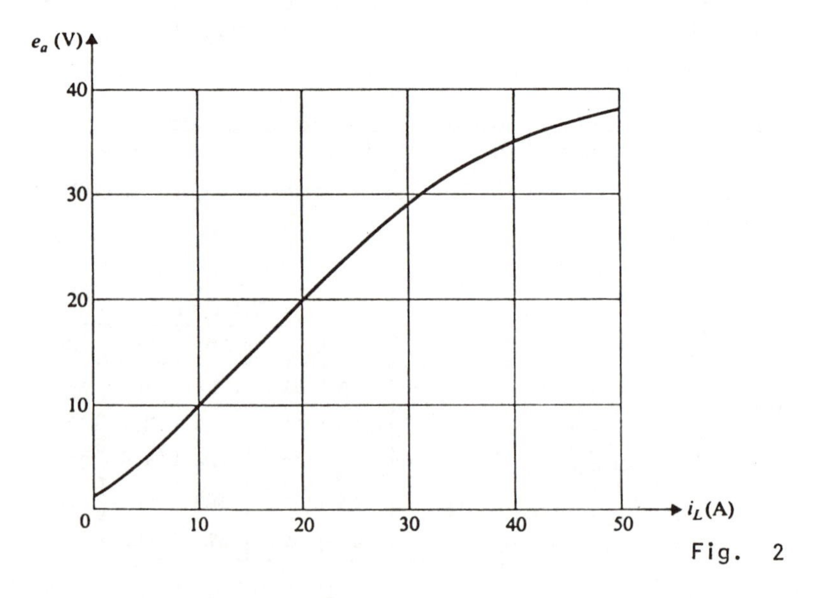

Fig. 2

Solution: The steady-state equivalent circuit of the system is shown in Fig. 3. From that circuit,

$$v_L = v_R + e_a - (2R_F + R_a + R_s)i_L \tag{1}$$

$$v_t = e_a - (R_a + R_s)i_L \tag{2}$$

a) $i_L = 20$ A.

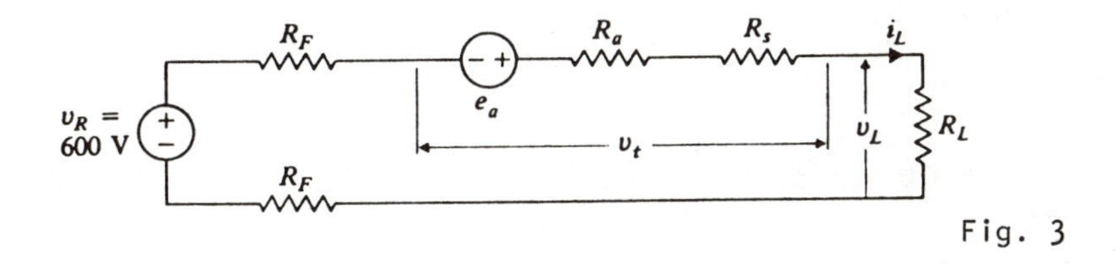

Fig. 3

From Fig. 2, $e_a = 20$ V. Substitution in Eq. (1) gives

$$v_L = 600 + 20 - 20(2 \times 0.41 + 0.08 + 0.05) = 601 \text{ V}$$

Substitution in Eq. (2) gives

$$v_t = 20 - 20(0.08 + 0.05) = 17.4 \quad \text{V}$$

b) $i_L = 40$ A.

From Fig. 2, $e_a = 35$ V.

$$v_L = 600 + 35 - 40(2 \times 0.41 + 0.08 + 0.05) = 597 \text{ V}$$

$$v_t = 35 - 40(0.08 + 0.05) = 29.8 \quad \text{V.}$$

ARMATURE REACTION, DEMAGNETIZING AND CROSS-MAGNETIZING AMPERE-TURNS

• PROBLEM 4-24

An eight-pole generator has a lap winding of 576 active conductors and carries a current of 100 amp. When it is necessary to advance the brushes 15 electrical space degrees, how many demagnetizing and how many cross-magnetizing ampere-turns are present?

Solution: The number of conductors that will be affected will be the conductors in twice the brush shift angle or the conductors in a 30° arc. Since this is a lap winding there are as many parallel paths as poles. There are actually eight paths and, therefore, 240 electrical degrees covered by the demagnetizing conductors.

The whole machine has 1440 electrical degrees; or the part of the armature that is influenced by 240° is in the ratio 240 to 1440. The turns that cause the demagnetizing effect are

$$\frac{2 \times 15° \times 8}{360° \times 4} \times 576 = 96 \text{ conductors}$$

which gives

$$\frac{96}{2} = 48 \text{ turns}$$

and the current per path will be

$$\frac{100}{8} = 12.5 \text{ amp}$$

since the eight-pole lap winding has eight paths in parallel.

The total number of ampere-turns causing a demagnetizing effect are

$$12.5 \times 48 = 600$$

Since the total number of ampere-turns will be

$$\frac{576}{2} \times 12.5 = 3600$$

the cross-magnetizing turns will be equal to 3000.

The same calculation can be made on a basis of ampere-turns per pole, and using space degrees:

$$\frac{15°}{4} \text{ space degrees} = 3.75 \text{ degrees}$$

$$\text{Ampere-turns per pole} = \frac{\alpha}{360°} \times Z \times \frac{I}{p}$$

where α = brush shift space degrees

Z = active conductors on armature

I = output current

p = path in winding

$$NI_D \text{ per pole} = \frac{3.75°}{360°} \times 765 \times \frac{100}{8}$$

$$= 75 \text{ ampere-turns per pole.}$$

Total demagnetizing ampere-turns = 75 × 8 = 600

• PROBLEM 4-25

The armature for an eight-pole generator has a simplex wave winding of 428 inductors and 107 commutator bars. The current rating is 100 amp. Calculate

(a) The cross-magnetizing ampere-turns per pole with the brushes on the geometric neutral.

(b) The demagnetizing ampere-turns per pole if the brushes are shifted forward the width of two commutator bars.

(c) The cross-magnetizing ampere-turns with the brushes shifted as in (b).

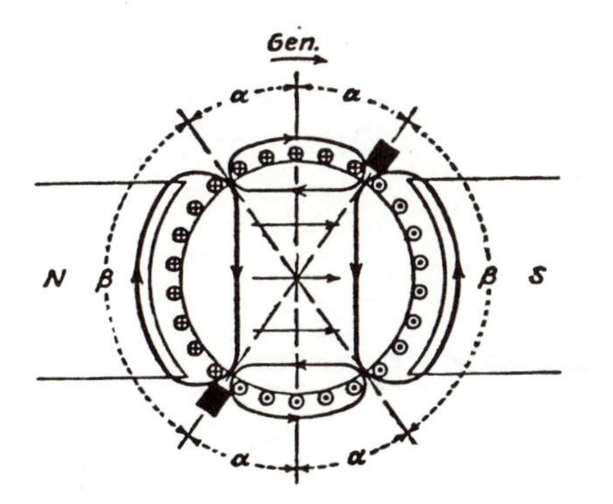

Fig. 1

Armature reaction magnetomotive forces when the brushes are shifted for improvement of commutation in a generator.

Solution: (a) Since, with the brushes on geometric neutral, the entire armature-reaction effect is cross-magnetizing,

$$H_a = \frac{z \times I_a}{2 \times \text{poles} \times \text{paths}} \quad \text{amp-turns per pole}$$

$$= \frac{428 \times 100}{2 \times 8 \times 2} = 1337.5 \text{ NI per pole}$$

(b) It is unnecessary in this case to calculate the value of angle α, since the ratio α/360 will be equal to the ratio of bars shift to total bars; hence,

$$dH_a = \frac{\alpha}{360} \times \frac{zI_a}{\text{Paths}} \quad \text{amp-turns per pole}$$

$$= \frac{2}{107} \times \frac{428 \times 100}{2} = 400 \text{ NI per pole}$$

(c) The cross-magnetizing NI with brushes shifted may be determined by subtracting the demagnetizing NI determined in (b) from the total armature NI determined in (a), as

$$1337.5 - 400 = 937.5 \text{ NI per pole}$$

or this determination may be made through application of the following equation:

$$cH_a = \frac{\beta}{720} \times \frac{zI_a}{\text{Paths}} \text{ amp-turns per pole}$$

In this case, it is necessary to calculate the value of angle β. With eight poles, there will be 16 angles α and 8 angles β.

$$\alpha = \frac{2}{107} \times 360 = 6.73 \text{ deg}$$

therefore

$$\beta = \frac{360 - 16 \times 6.73}{8} = 31.54 \text{ deg}$$

and

$$cH_a = \frac{31.54}{720} \times \frac{428 \times 100}{2} = 937.5 \text{ NI per pole.}$$

LOSSES AND EFFICIENCY

• PROBLEM 4-26

A 250-kw 230-volt compound generator is delivering 800 amp at 230 volts. The shunt-field current is 12 amp. The armature resistance is 0.007 ohm, and the series-field resistance is 0.002 ohm. The stray power at this load is 5,500 watts. The generator is connected long-shunt. Determine generator efficiency at this load.

Solution: Output = 230 × 800 = 184,000 watts.

Shunt-field loss = 230 × 12 = 2,760 watts.

Armature loss = $812^2 \times 0.007$ = 4,615 watts.

Series-field loss = $812^2 \times 0.002$ = 1,319 watts.

Stray power = 5,500 watts.

Stray-load loss, 0.01 × 184,000 = 1,840 watts.

Total loss =16,034 watts.

$$\text{Eff} = \frac{184{,}000}{184{,}000 + 16{,}034} = \frac{184{,}000}{200{,}034} = 0.920, \text{ or } 92.0\%$$

• PROBLEM 4-27

A certain 110-volt shunt generator has an armature and brush resistance of 0.06 ohm at full load of 85 amperes. The resistance of the shunt field is 45 ohms, and the stray power losses are found to be 897 watts. Calculate the full-load efficiency of the generator.

Solution: The total armature current is the sum of the load current plus the field current.

$$\text{The field current is } \frac{110}{45} = 2.44 \text{ amperes}$$

$$\therefore I_a = 85 + 2.44 = 87.44 \text{ amperes}$$

$$I_a^2 R_a = 87.44^2 \times 0.06 = 449 \text{ watts}$$

$$I_f^2 R_f = 2.44^2 \times 45 = 268 \text{ watts}$$

$$\text{Total copper loss} = 717 \text{ watts}$$

$$\text{Stray power loss} = 897 \text{ watts}$$

$$\therefore \text{Total loss} = 1{,}614 \text{ watts}$$

$$\text{Total output} = 85 \times 110 = 9{,}350 \text{ watts}$$

Total output + losses

$$= 9{,}350 + 1{,}614 = 10{,}964 \text{ watts}$$

$$\text{Efficiency} = \frac{\text{output}}{\text{output + losses}} = \frac{9{,}350}{10{,}964} = 85.2 \text{ percent.}$$

• PROBLEM 4-28

A certain 5½-KW 125-volt d-c generator was operated as a shunt motor at no load for determination of core and friction losses. With 135 volts--equal to generated voltage at full load--applied to the armature, and normal speed of 1700 rpm, the current input to the armature was found to be 2.5 amp. Resistance of the armature circuit with full-load current flowing was found to be 0.2 ohm, of the series field 0.025 ohm, and of the shunt-field circuit 100 ohms. Determine the various losses and calculate the full-load efficiency.

Solution: Rotational loss = Armature voltage × Armature current.

135 x 2.5 = 337 watts rotational loss (neglecting the small copper loss involved).

$$\text{Rated load current} = \frac{\text{Power}}{\text{Voltage}}$$

$$= \frac{5500}{125} = 44 \text{ amp}$$

$$\text{Shunt-field current} = \frac{\text{Applied voltage}}{\text{Resistance}}$$

$$= \frac{125}{100} = 1.25 \text{ amp}$$

Armature and shunt-field current (assuming a long-shunt connection) = 44 + 1.25 = 45.25 amp.

$I_a{}^2R_a = \overline{45.25}^2 \times 0.2 \qquad = 410$ watts armature copper loss

$I_a{}^2R_f = \overline{45.25}^2 \times 0.025 \qquad = 51$ watts series-field copper loss

$I_fR_f = 1.25 \times 125 \qquad = 156$ watts shunt-field copper loss

$\therefore$ Total losses = 337 + 410 + 51 + 156 = 954 watts

Output at full load = 5500 watts

Input = output + losses = 5500 + 954 = 6454 watts

Therefore

$$\text{Efficiency} = \frac{\text{output}}{\text{input}} = \frac{5500}{6454} = 0.852 = 85.2\%.$$

• PROBLEM 4-29

A 125-kW, 250-V, 1800-rev/min, cumulative compound d-c generator has the following winding resistances:

$$r_a = 0.025\Omega \qquad r_{se} = 0.010\Omega \qquad r_f = 30\Omega$$

The machine is long-shunt connected. Its stray-power loss at rated voltage and 1800 rev/min is 5000 W. When operated at rated speed, load, and terminal voltage, the shunt-field current is 5A. Find the efficiency and input horsepower requirements under these conditions.

Solution: $\qquad P_{out} = 125{,}000$ W

$$\text{Shunt-field copper loss} = I_f^2 r_f$$

$$= 25 \times 30 = 750 \text{ W.}$$

$$I_a = I_{se} = I_{load} + I_f$$

$$= \frac{125{,}000}{250} + 5 = 505 \text{ A}$$

Series-field copper loss $= I_{se}^2 r_{se}$

$$= 505^2 \times 0.01 = 2550 \text{ W.}$$

$$\text{ACL} = I_a^2 r_a = 505^2 \times 0.025 = 6380 \text{ W.}$$

Brush-drop loss $= 2I_a = 1010$ W.

Stray load loss = 1 percent of 125 kW

= 1250 W.

$P_{rot} \equiv$ stray-power loss = 5000 W.

Total losses = 16,940 W.

$$\text{Efficiency} = \frac{P_{out}}{P_{out} + \text{losses}} = \frac{125{,}000}{141{,}900} = 88.1\%.$$

The drive motor must supply the field rheostat power as well as those losses involved in the efficiency calculation. The total shunt-field-circuit power is 5A × 250 V or 1250 W. Subtracting the field copper loss gives 500 W lost in the rheostat. The input power is, then,

$$P_{in} = P_{out} + \text{efficiency losses} + \text{rheostat losses}$$

$$= 125{,}000 + 16{,}490 + 500 = 141{,}990 \text{ W}$$

or 190.3 hp

For this machine, maximum efficiency would result with an armature current given by the following equation:

Stray-power loss + shunt field copper loss

$$= 5000 + 750 = I_a^2(r_a + r_{se}) = I_a^2 \times 0.035.$$

$$I_a = \sqrt{164{,}000} = 405 \text{ A.}$$

which is about 80 percent of full load.

• PROBLEM 4-30

A 10 kW, 230 V, 1750 rpm shunt generator was run light as a motor to determine its rotational losses at its rated load. The applied voltage across the armature, V_a, computed for the test, was 245 V, and the armature current drawn was 2 A. The field resistance of the generator was 230 ohms and the armature circuit resistance measured 0.2 ohm. Calculate:

(a) The rotational (stray power) losses at full load

(b) The full-load armature circuit loss and the field loss

(c) The generator efficiency at $\frac{1}{4}$, $\frac{1}{2}$, and $\frac{3}{4}$ of the rated load; at the rated load, and at 1¼ times the rated load.

Solution:

(a) Rotational loss $= V_a I_a - I_a^2 R_a$

$$= (245 \times 2) - (2^2 \times 0.2) = 490 - 0.8$$

$$= 489.2 \text{ W}$$

Note that 490 W may be used with negligible error because of negligible electric armature loss.

(b) At the rated load,

$$I_L = \frac{W}{V_t} = \frac{10{,}000 \text{ W}}{230 \text{ V}} = 43.5 \text{ A.}$$

$$I_a = I_f + I_L = \frac{230 \text{ V}}{230 \ \Omega} + 43.5 = 44.5 \text{ A.}$$

The full-load armature loss

$$I_a^2 R_a = (44.5)^2 \times 0.2 = 376 \text{ W.}$$

The field loss

$$V_f I_f = 230 \text{ V} \times 1 \text{ A} = 230 \text{ W.}$$

(c) The efficiency at any load, for a generator is given by the following equation:

$$\eta = \frac{\text{Output at that load}}{\text{Output at the load} + \text{Rotational loss} + \text{Electric loss at that load}}$$

Efficiency at $\frac{1}{4}$ load

$$= \frac{10,000/4}{(10,000/4) + 489.2 + [(376/16) + 230]} \times 100$$

= 77 percent.

Efficiency at $\frac{1}{2}$ load

$$= \frac{10,000/2}{(10,000/2) + 489.2 + [(376/4) + 230]} \times 100$$

= 86.2 percent.

Efficiency at $\frac{3}{4}$ load

$$= \frac{10,000 \times (3/4)}{[10,000(3/4)] + 489.2 + ([376(9/16)] + 230)} \times 100$$

= 89 percent.

Efficiency at full load

$$= \frac{10,000}{10,000 + 489.2 + [376 + 230]} \times 100$$

= 90.1 percent.

Efficiency at $1\frac{1}{4}$ × the rated load (or $\frac{5}{4}$ rated load)

$$= \frac{10,000 \times (5/4)}{[10,000(5/4)] + 489.2 + ([376(25/16)] + 230)} \times 100$$

= 90.6 percent.

• PROBLEM 4-31

A 20-kw 220-volt shunt generator, having an armature resistance of 0.096 ohm, is delivering 85 amp at 222 volts. Its speed is 920 rpm, and the field current is 2.24 amp. Determine: (a) no-load conditions under which generator must be operated (as motor) in order that its no-load stray power may be equal to that which exists under given conditions of operation; (b) value of stray power if armature input at no load is 4.02 amp at 230.4 volts; (c) efficiency of generator under given operating conditions.

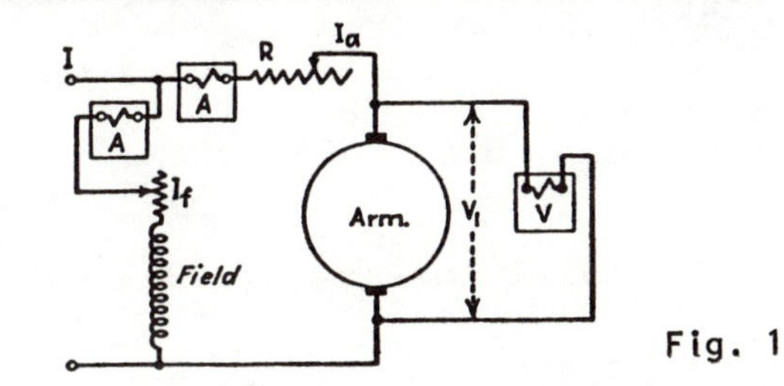

Fig. 1

Connections for stray-power measurement.

Solution: (a) $E = V + (I + I_f) \times R_a$

$$= 222 + (85 + 2.24)0.096 = 230.4 \text{ volts}$$

$$\frac{E}{S} = \frac{230.4}{920} = 0.250.$$

Although the dynamo is a generator, it is operated as a motor to determine its stray power. The connections are shown in Fig. 1. The field rheostat is adjusted until the terminal voltage divided by the speed is equal to 0.250. The armature-resistance drop is negligible at no load. The speed is then adjusted by means of the armature rheostat R until it is 920 rpm. Changing the armature rheostat does not change the flux and hence does not change the ratio E/S.

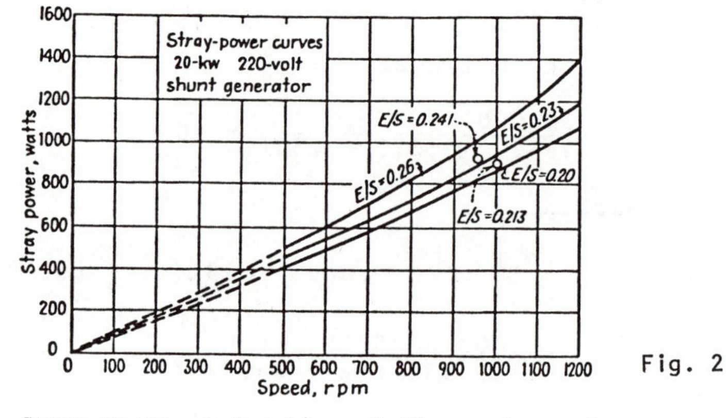

Fig. 2

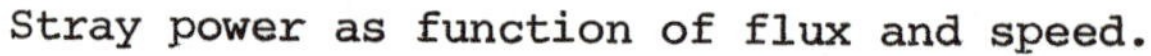
Stray power as function of flux and speed.

(b) Stray Power = $VI_a - I_a^2R_a$

S.P. = $230.4 \times 4.02 - (4.02)^2 \times 0.096 = 926.2$ watts.

(See Fig. 2)

(c) Output = $222 \times 85 = 18{,}870$ watts.

$I_a^2R_a = (85 + 2.24)^2 0.096 = 730.7$ watts.

$VI_f = 222 \times 2.24 = 497.3$ watts.

S.P. $= 926.2$ watts.

Stray-load loss, $0.01 \times 18{,}870 = 188.7$ watts.

Total losses= 2,342.9 watts.

$$\text{Eff } \eta = \frac{18{,}870}{18{,}870 + 2{,}343} = 0.882, \text{ or } 88.2\%.$$

• PROBLEM 4-32

The following data apply to a 100-kW 250-V six-pole 900-rpm compound generator (long-shunt connection):

No-load rotational losses = 3840 W

Armature resistance at 75°C = 0.012Ω

Series-field resistance at 75°C = 0.004Ω

Commutating-pole field resistance at 75°C = 0.004Ω

Shunt-field current = 2.60 A

Assume a stray-load loss equal to 1% of the output and calculate the rated-load efficiency.

Solution: The total resistance of the armature circuit, not including brushes, is the sum of the resistance of the armature, series-field, and commutating-pole field windings:

$$r_a = 0.012 + 0.004 + 0.004 = 0.020\Omega$$

The armature current is the sum of the load current and the shunt-field current, or

$$I_a = \frac{100,000}{250} + 2.6 = 402.6A$$

The losses may be tabulated and totaled as follows:

No-load rotational losses	3840W
Armature-circuit copper losses = $(402.6)^2 \times 0.02$	3240W
Brush-contact loss $(2I_a) = 2 \times 402.6$	805W
Shunt-field circuit copper losses = 250×2.6	650W
Stray-load loss = $0.01 \times 100,000$	1000W
Total losses	9535W

The input at rated load is therefore

100,000 + 9535 = 109,535 W

and the efficiency is given by

$$\text{Efficiency} = 1 - \frac{\text{losses}}{\text{input}}$$

$$= 1 - \frac{9535}{109,535} = 0.915.$$

• PROBLEM 4-33

A 560-kw, n = 375 rpm, 8-pole generator has the following dimensions:

Armature dia.: 39.5 in.
Gross core length: 14.5 in.
Radial vents: 4
Width of a vent: $\frac{3}{8}$ in.
Number of slots: 82
Slot pitch: 1.513 in.
Length of the gap: 0.315 in.
Carter factor: 1.14
Flux density in the gap B_g: 62,500 lines/in.^{2}
Slot width: 0.605 in.
Pole pitch: 15.5 in.
I^2R-loss in the armature: 11,500 watts
Fictitious tooth density at root of the tooth: 141,400 lines/in.^{2}
Pole arc b_p: 9.5 in.
Length of pole: 14.5 in.
Conductors per slot: 6
3 conductors per layer: (u = 3)
Upper conductor: 0.157 × 0.51 in.
Lower conductor: 0.157 × 0.67 in.
Av. height of cond.: 0.59 in.
Av. current density: 2975 amp/in.^{2}
Commutator dia.: 33.5 in.
Commutator pitch: 0.427 in.
Brush width: 1.0 in.
Half mean length turn: 34.4 in.
Net core length: 13 in.

Find (a) pole-face loss (b) copper loss and (c) skin-effect loss.

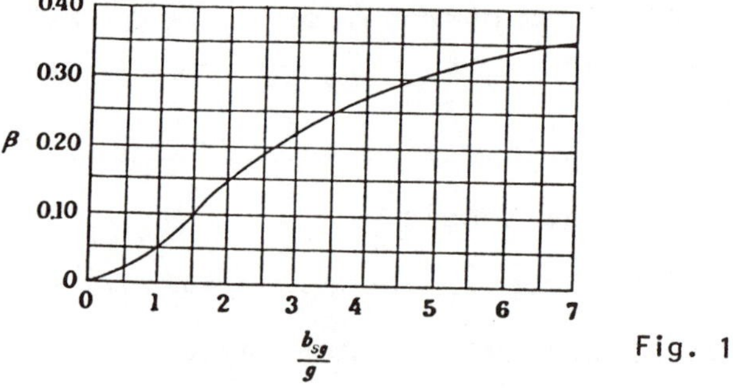

Fig. 1

The quantity β as a function of b_{sg}/g

Solution: (a) Pole-face Loss. $\beta = 0.14$ corresponds to $\frac{b_s}{g} = \frac{0.605}{0.315} = 1.92$ (Fig. 1).

$$B_0 = \beta k_c B_g.$$

$$= 0.14 \times 1.14 \times 62,500 = 10,000 \text{ lines/in.}^2$$

The equation:

$$\ell_s = k_s \left(\frac{Qn}{1000}\right)^{1.5} \left(\frac{B_0 \tau_s}{1000}\right)^2 \text{ watts/in.}^2,$$

yields for 1/16 in. thick steel sheet

$$\ell_s = 6 \times 10^{-4}\left(\frac{82 \times 375}{10{,}000}\right)^{1.5}\left(\frac{10{,}000 \times 1{,}513}{1000}\right)^2$$

$$= 0.745 \text{ watts/in.}^2$$

The total pole-face loss is

$$L_s = p\ell_s b_p \ell_p \text{ watts.}$$

$$= 8 \times 14.5 \times 9.5 \times 0.745 = 820 \text{ watts}$$

(b) Copper Loss Due to the Main Flux. The frequency is

$$f = np/120$$

$$= \frac{8 \times 375}{120} = 25 \text{ c/sec}$$

The equation: $d_M = 16\ \dfrac{hf}{\rho}\left(\dfrac{B'_t}{6450} - 16\right)$ amp/in.^{2}, yields

for $\rho = 0.825$ (at 75° C)

$$d_M = 16\ \frac{0.59 \times 25}{0.825}\left(\frac{141{,}400}{6450} - 16\right) = 1690 \text{ amp/in.}^2$$

λ is equal to 13/34.4 = 0.378. The following equations

$$\varepsilon_M = \left(\frac{d_M}{d_r}\right)^2 \lambda$$

$$L_M = \varepsilon_M I^2 R,$$

yield for this loss

$$\varepsilon_M = \left(\frac{1690}{2975}\right)^2 \times 0.378 = 0.122$$

$$L_M = 0.122 \times 11{,}500 = 1400 \text{ watts}$$

(c) Skin-effect Loss.

$$\xi = 0.316h\sqrt{\frac{b_{cu}}{b_s}\ \frac{f}{\rho}}$$

where h = height of the conductor

b_{cu} = width of all conductors in the slot width

b_s = width of the slot

$f = \dfrac{Pn}{120}$

ρ = resistivity of the conductor material, in micro ohms/in.^{3}

$$\xi = 0.316 \times 0.59 \sqrt{\frac{3 \times 0.157}{0.605} \times \frac{25}{0.825}} = 0.906$$

From the following equations:

$$\sigma = \frac{b_b + (u - 1)\tau_c}{\tau(D_c/D)} \frac{1}{\xi^2}$$

$$F = \frac{0.116}{0.13 + \sigma},$$

$$\sigma = \frac{1 + (3 - 1) \times 0.427}{15.5 \times 33.5/39.5} = 0.172$$

$$F = \frac{0.116}{0.13 + 0.172} = 0.384$$

The equations

$$\varepsilon_{sk} = \frac{4}{3\pi} m^2 \xi^2 \lambda F$$

$$L_{sk} = \varepsilon_{sk} I^2 R,$$

yield for this loss (m = 2)

$$\epsilon_{sk} = \frac{4}{3\pi} \times 4 \times 0.906^2 \times 0.378 \times 0.384 = 0.202$$

and

$$L_{sk} = 0.202 \times 11{,}500 = 2220 \text{ watts}$$

PARALLEL OPERATION OF D. C. GENERATORS

• PROBLEM 4-34

Two compound generators are operated in parallel. If generator A has a series-field resistance of 0.0015 ohm and is adjusted to take 30 percent of the line load while generator B with a series field resistance of 0.001 ohm takes the remainder of the load, determine the current flow in the two series fields and the equalizer when the load is 230 kw at 230 volts.

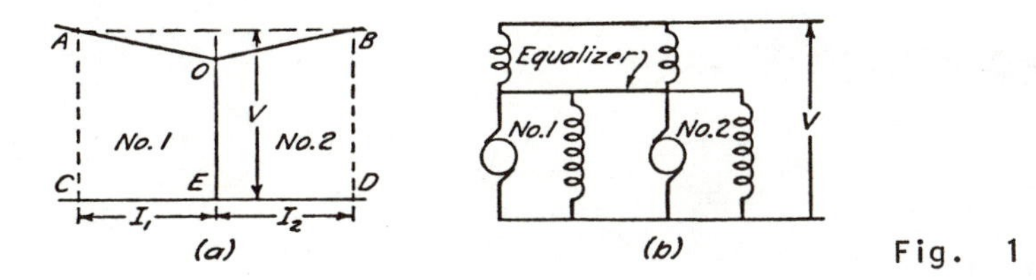

Fig. 1

(a) Characteristic of two over-compound generators in parallel with the same no-load voltage without the equalizer. (b) Circuit diagram of two compound machines showing the equalizer.

Solution: $I = I_A + I_B = \dfrac{230{,}000}{230} = 1000$ amp

Now

$$\frac{I_A}{I_B} = \frac{0.001}{0.0015},$$

$\therefore\ I_A = 400$ amp, series-field current and

$I_B = 600$ amp, series-field current.

$I_A = 0.3 \times 1000 = 300$ amp, machine A armature.

$I_B = 1000 - 300 = 700$ amp, machine B armature.

$\therefore\ I_E = 400 - 300 = 100$ amp in the equalizer after adjustment.

CHAPTER 5

D.C. MOTORS

COUNTER EMF, MOTOR STARTING REQUIREMENTS

• PROBLEM 5-1

A 115-volt shunt motor has an armature whose resistance is 0.22 ohm. Assuming a voltage across the brush contacts of 2 volts, what armature current will flow (a) when the counter emf is 108 volts? (b) if the motor load is increased so that the counter emf drops to 106 volts?

Solution:

$$I_A = \frac{V_A - E_C}{R_A} \text{ amp},$$

where

I_A = armature current

V_A = applied voltage

E_C = counter emf

R_A = armature resistance

(a) $$I_A = \frac{(115 - 2) - 108}{0.22} = 22.7 \text{ amp.}$$

(b) $$I_A = \frac{(115 - 2) - 106}{0.22} = 31.8 \text{ amp.}$$

• PROBLEM 5-2

A six-pole, 30-slot dc commutator generator has a lap-wound armature winding. The armature constant, K_a is 9.55.

The generator is operated with a separately excited field such that the field flux per pole is 0.04 weber/pole. The generator is driven at a speed of 3000 rpm. The no load armature voltage if 418.9 V.

The above generator is now operated as a motor. The flux is 0.04 weber/pole. It is desired to supply a load requiring a torque of 50 n-m at 4000 rpm. Armature circuit resistance is 0.075 ohm. Calculate (a) the back emf, (b) the required armature current and voltage to supply this load.

Solution: (a) The back emf is

$$E = K_a \phi_p V$$

$$= 9.55 \times 0.04 \times 418.9 = 160 \text{ V}$$

(b) The armature current can be found directly from the following equation:

$$T_d = K_a \phi_p I_a \qquad \text{N-m,}$$

which gives

$$I_a = 50/(9.55 \times 0.04) = 130.9 \text{ A}$$

The necessary armature voltage is

$$V = E \pm I_a R_a$$

$$= 160 + 130.9 \times 0.075 = 160 + 9.8 = 169.8 \text{ V}$$

• PROBLEM 5-3

The armature of a 220-volt shunt motor has a resistance of 0.18 ohm. If the armature current is not to exceed 76 amp, calculate: (a) the resistance that must be inserted in series with the armature at the instant of starting; (b) the value to which this resistance can be reduced when the armature accelerates until E_c is 168 volts; (c) the armature current at the instant of starting if no resistance is inserted in the armature circuit. (Assume a 2-volt drop at the brushes.)

Solution:

(a) $I_A = \frac{V_A - E_c}{R_A + R}$ amp.

$$\therefore R = \frac{V_A - E_c}{I_A} - R_A = \frac{(230 - 2) - 0}{76} - 0.18$$

$$= 2.82 \text{ ohms.}$$

(b) $R = \frac{(230 - 2) - 168}{76} - 0.18 = 0.61$ ohm.

(c) $I_A = \frac{(230 - 2)}{0.18} = 1,265$ amp (a very dangerous value).

• PROBLEM 5-4

A 5-hp, 120-volt shunt motor has an armature resistance of 0.10 ohm, and a full-load armature current of 40 amp. Determine the value of the series resistance to add to the armature to limit the initial starting current to 150% of normal.

Solution: $I_a = 40 \times 1.5$ at starting; $V = 120$, the line voltage;

I_a is given by

$$I_a = \frac{V - 2\Delta V}{\Sigma r}$$

Solving for Σr,

$$\Sigma r = \frac{V - 2\Delta V}{I_a}$$

$$= \frac{120 - 2}{60} = 1.966 \text{ ohms.}$$

Therefore, a series resistance of 1.966 - 0.1 = 1.866 would be required.

• PROBLEM 5-5

(i) A 120 V dc shunt motor has an armature resistance of 0.2 ohms and a brush volt drop of 2 V. The rated full-load armature current is 75 A. Calculate the current at the instant of starting, and the percent of full load.

(ii) Calculate the various values (taps) of starting resistance to limit the current in the motor of Part (i) to

(a) 150 percent rated load at the instant of starting.

(b) 150 percent rated load, when the counter emf is 25 percent of the armature voltage, V_a.

(c) 150 percent rated load, when the counter emf is 50 percent of the armature voltage, V_a.

(iii) Find the counter emf at full load, without starting resistance.

Solution:

(i) $I_{st} = \frac{V_a - BD}{R_a} = \frac{120 - 2}{0.2} = 590$ A (counter emf is zero)

Percent full load $= \frac{590 \text{ A}}{75 \text{ A}} \times 100 = 786$ percent

(ii) $I_a = \frac{V_a - (E_c + BD)}{R_a + R_s}$

Solving for R_s,

$$R_s = \frac{V_a - (E_c + BD)}{I_a} - R_a$$

(a) At starting, E_c is zero; $R_s = \frac{V_a - BD}{I_a} - R_a$

$$= \frac{120 - 2}{1.5 \times 75} - 0.2$$

$$= 1.05 - 0.2 = 0.85\ \Omega$$

(b) $R_s = \frac{V_a - (E_c + BD)}{I_a} - R_a = \frac{120 - 30 - 2}{1.5 \times 75} - 0.2 = 0.782 - 0.2$

$= 0.582\ \Omega$

(c) $R_s = \frac{120 - (60 + 2)}{1.5 \times 75} - 0.2 = 0.516 - 0.2 = 0.316\ \Omega$

(iii) $E_c = V_a - (I_a R_a + BD) = 120 - [(75 \times 0.2) + 2]$

$= 103$ V.

• PROBLEM 5-6

Given a 10-hp 230-volt 1000-rpm series motor, having rated-load efficiency of 85.5 percent, armature resistance, including brushes, of 0.28 ohm, and field resistance of 0.15 ohm. (a) Assuming that the field flux varies directly with armature current, what value of resistance should be placed in series with this motor, when starting, in order that the starting current may be limited to a value that will exert a starting torque equal to 1.5 times rated-load torque? Compare the results with those obtained with a constant-flux condition, such as exists in the shunt motor. (b) If the starting resistance of (a) is allowed to remain in the circuit, what will be the resulting speed with rated-load torque being developed? What resistance is required for a speed of 600 rpm?

Solution:

(a) Rated current $= \frac{10 \times 746}{230 \times 0.855} = 38$ amp

$$\frac{T_1}{T_2} = \frac{1}{1.5} = \frac{38^2}{{I_2}^2}$$

from which

$I_2 = 46.5$ amp

$$R = \frac{E}{I} = \frac{230}{46.5} = 4.95 \text{ ohms}$$

Starting resistance $R_s = R - (R_a + R_f)$

$= 4.95 - (0.28 + 0.15) = 4.52$ ohms

If the field flux is assumed to remain constant, at rated-load value, the current required to produce 1.5 times normal torque will be equal to 1.5 times rated-load current or $1.5 \times 38 = 57$ amp. The resistance required then equals $230/57 = 4.03$ ohms.

$4.03 - (0.28 + 0.15) = 3.6$ ohms to be added.

(b) Normal counter emf $= 230 - 38\ (0.28 + 0.15) =$ 213.7 volts. New value of counter emf $= 230 - 38\ (0.28 + 0.15 + 4.52) = 41.9$ volts.

Then

$$\frac{213.7}{41.9} = \frac{1000}{S_2}$$

from which

$$S_2 = \frac{1000 \times 41.9}{213.7} = 196 \text{ rpm.}$$

For a speed of 600 rpm, since the developed torque is assumed to remain the same, the values of I_a and ϕ will be unchanged, and the counter emf will become 600/1000 × 213.7 = 128.2 volts.

$$E = V - I_a(R_a + R_f + R_s)$$

which gives

$$38(0.28 + 0.15 + R_s) = 230 - 128.2 = 101.8 \text{ volts}$$

from which

$$R_s = \frac{101.8 - 16.34}{38} = 2.25 \text{ ohms.}$$

• PROBLEM 5-7

(a) A shunt motor is running at 1200 rpm for a load which requires an armature current of 50 amp from a 230-volt source. At no load the armature current is 5 amp. If the effect of armature reaction has reduced the air-gap flux 2 percent from no load to full load, determine the no-load speed. The armature resistance is 0.15 ohm.

(b) The rated line current of a 230-volt shunt motor is 56 amp. If the shunt-field circuit resistance is 230 ohms and the armature circuit resistance is 0.15 ohm, what would be the line current, assuming that the motor, at standstill, is connected across rated voltage? How much external resistance must be connected in the armature circuit to limit the current at starting to 125 percent full-load armature current?

Solution: (a) Full load:

$$N = \frac{V_T - I_aR_a}{K\phi}$$

$$1200 = \frac{230 - 50(0.15)}{K(0.98\phi)}$$

No load:

$$N = \frac{230 - 5(0.15)}{K\phi}$$

$$\frac{N}{1200} = \frac{\dfrac{230 - 5(0.15)}{K\phi}}{\dfrac{230 - 50(0.15)}{K\phi(0.98)}}$$

$$N = \frac{1200[230 - 5(0.15)]0.98}{230 - 50(0.15)}$$

$$N = \frac{1200(229.25)(0.98)}{222.50} = 1211 \text{ rpm.}$$

(b) $$I_f = \frac{230}{230} = 1 \text{ amp}$$

$$I_a = \frac{230}{0.15} = 1533 \text{ amp}$$

$$I_L = 1534 \text{ amp}$$

Full-load $I_a = 56 - 1 = 55$ amp

125 percent full-load $I_a = 68.75$ amp

Total armature circuit resistance $\frac{230}{68.75} = 3.35$ ohms

$\therefore$ External resistance $= 3.35 - 0.15 = 3.2$ ohms.

• PROBLEM 5-8

The nameplate data for a small dc motor are 125 V, 36 A, 1150 r/min. The armature-circuit resistance R_a is given as 0.370Ω, and the armature power input on no load at rated speed is 325 W. The no-load saturation curve for the motor taken at rated speed is shown in Fig. 1. The motor is to be started by means of the starter of Fig. 2, with the field current i_f set at 2.5 A.

Assuming that the current during starting will be allowed to vary over the range $40 < i_L < 75$ A, determine

a) The total starter resistance required.

b) The speed at which the first reduction in starter resistance takes place.

c) The magnitude of the first reduction in starter resistance.

d) The speed when all starter resistance has been cut out, and $i_L = 36$ A.

e) The no-load speed if the mechanical load is uncoupled.

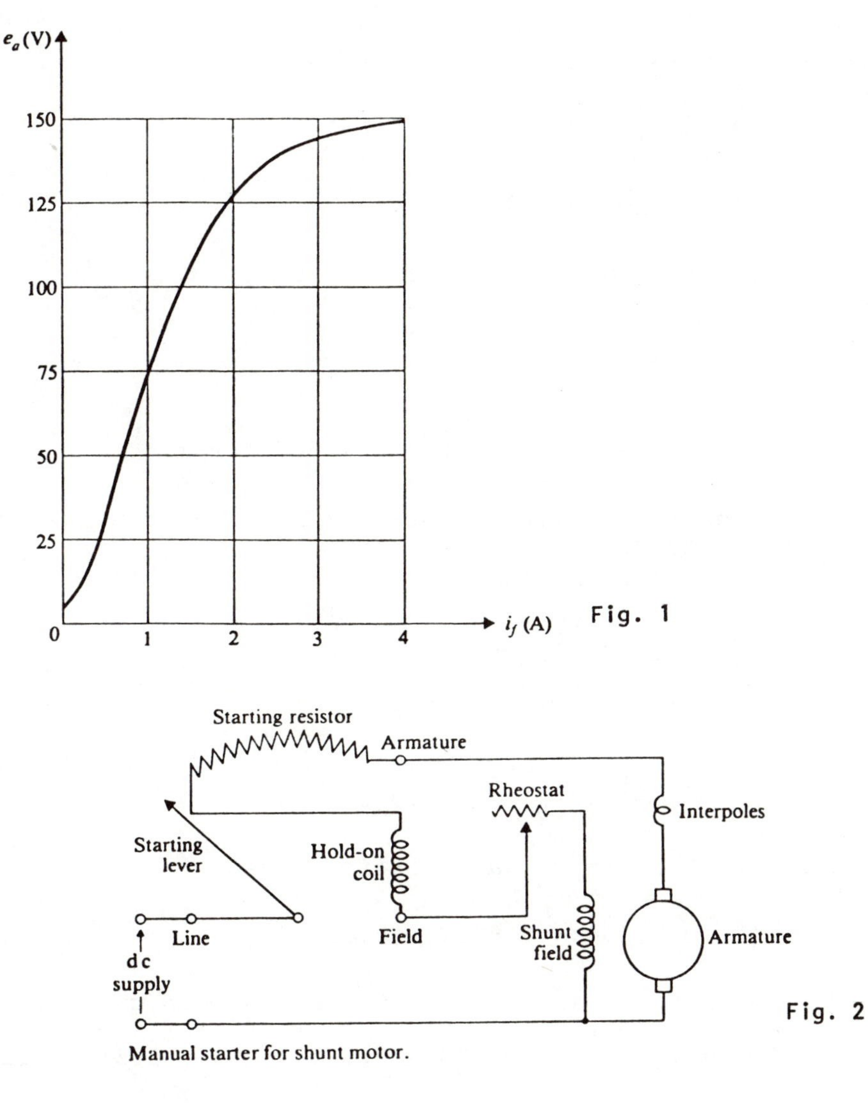

Fig. 1

Fig. 2

Manual starter for shunt motor.

<u>Solution</u>: a) At standstill,

$$i_a = i_L - i_f = 75 - 2.5 = 72.5 \text{ A}$$

$$R_a + R_d = \frac{125}{72.5} = 1.724 \ \Omega$$

$$R_d = 1.724 - 0.370 = 1.354 \ \Omega$$

b) The machine accelerates until

$$i_a = 40 - 2.5 = 37.5 \text{ A}$$

$$e_a = V_t - (R_a + R_d)i_a = 125 - 1.724 \times 37.5 = 60.35 \text{ V}$$

From Fig. 1, for $i_f = 2.5$ A, $e_a = 137$ V. This curve was taken at speed

$$\omega_0 = \frac{2\pi}{60} \times 1150 = 120.4 \text{ rad/s}$$

Thus, at $i_f = 2.5$ A,

$$k\phi = \frac{137}{120.4} = 1.138$$

Hence, for $i_a = 37.5$ A, and $e_a = 60.35$ V,

$$\omega_m = \frac{e_a}{k\phi}$$

$$= \frac{60.35}{1.138} = 53.03 \text{ rad/s} = 506.4 \text{ r/min}$$

c) When the resistance step is removed, the armature current may rise to 72.5A. Thus,

$$R_a + R_{d1} = \frac{V_t - e_a}{i_a}$$

and

$$R_{d1} = \frac{125 - 60.35}{72.5} - 0.370 = 0.5217 \ \Omega$$

The reduction in R_d has therefore been

$$\Delta R_d = 1.354 - 0.5217 = 0.832 \ \Omega$$

d) When all starting resistance has been cut out,

$$e_a = V_t - R_a i_a = 125 - 0.370 \times (36 - 2.5) = 112.6 \text{ V}$$

$$\omega_m = \frac{e_a}{k\phi} = \frac{112.6}{1.138} = 98.95 \text{ rad/s} = 944.9 \text{ r/min}$$

e) On no load, the power input to the armature is (approximately):

$$V_t i_a = 325 \text{ W}$$

$$i_a = \frac{325}{125} = 2.6 \text{ A}$$

$$e_a = V_t - R_a i_a = 125 - 0.370 \times 2.6 = 124.0 \text{ V}$$

$$\text{Speed} = \frac{124.0}{1.138} = 109.0 \text{ rad/s} = 1041 \text{ r/min.}$$

SPEED AND TORQUE CHARACTERISTICS

• PROBLEM 5-9

In a motor the armature resistance is 0.1 ohm. When connnected across 110-volt mains the armature takes 20 amp, and its speed is 1,200 rpm. Determine its speed when the armature takes 50 amp from the same mains, with the field increased 10 percent.

Solution: $S = K \dfrac{V - I_a R_a}{\phi}$

$$\frac{S_2}{S_1} = \frac{K\dfrac{110 - 50 \times 0.1}{\phi_2}}{K\dfrac{110 - 20 \times 0.1}{\phi_1}} = \frac{\dfrac{105}{\phi_2}}{\dfrac{108}{\phi_1}} = \frac{105}{\phi_2} \cdot \frac{\phi_1}{108},$$

$$S_1 = 1{,}200$$

Therefore,

$$S_2 = 1{,}200 \; \frac{105}{108} \cdot \frac{\phi_1}{\phi_2}$$

But

$$\phi_2 = 1.10\phi_1.$$

Therefore,

$$S_2 = 1{,}200 \; \frac{105}{108}\left(\frac{\phi_1}{1.10\phi_1}\right) = 1{,}061 \text{ rpm.}$$

• PROBLEM 5-10

A 25-hp 240-volt series motor takes 93 amp when driving its rated load at 800 rpm. The armature resistance is 0.12 ohm, and the series-field resistance is 0.08 ohm. At what speed will the motor operate if the load is partially removed so that the motor takes 31 amp? Assume that the flux is reduced by 50 percent for a current drop of 66 2/3 percent and that the brush drop is 2 volts at both loads.

Solution:

$$S = \frac{V_a - I_a R_a}{k\phi} \text{ rpm}$$

$$800 = \frac{(240 - 2) - 93(0.12 + 0.08)}{k\phi_{FL}}$$

$$S_x = \frac{(240 - 2) - 31(0.12 + 0.08)}{k(0.5 \times \phi_{FL})}$$

$$\frac{S_x}{800} = \frac{232.8/k(0.5\ \phi_{FL})}{219.4/k\phi_{FL}} = \frac{231.8}{219.4 \times 0.5}$$

Therefore

$$S_x = 800 \times \frac{231.8}{219.4 \times 0.5} = 1,690 \text{ rpm.}$$

• PROBLEM 5-11

A 120 V dc shunt motor having an armature circuit resistance of 0.2 ohm and a field circuit resistance of 60 ohms, draws a line current of 40 A at full load. The brush volt drop is 3 V and rated, full load speed is 1800 rpm. Calculate:

(a) The speed at half load.

(b) The speed at an overload of 125 percent.

Solution: (a) At full load

$$I_a = I_\ell - I_f = 40\text{A} - \frac{120\text{ V}}{60\Omega} = 38\text{ A};$$

$$E_c = V_a - (I_aR_a + BD) = 120 - (38 \times 0.2 + 3)$$

$$= 109.4\text{ V}$$

At the rated speed of 1800 rpm

$$E_c = 109.4\text{ V} \quad \text{and} \quad I_a = 38\text{ A (full load)}$$

Half-load speed

$$I_a = \frac{38\text{ A}}{2} = 19\text{ A};$$

$$E_c = V_a - (I_aR_a + BD) = 120 - (19 \times 0.2 + 3)$$

$$= 113.2\text{ V}$$

Using the ratio method, half-load speed

$$S = S_{orig}\frac{E_{final}}{E_{orig}} = 1800\ \frac{113.2}{109.4} = 1860\text{ rpm.}$$

(b) At 1¼ load

$$I_a = \frac{5}{4}38\text{ A} = 47.5\text{ A};$$

$$E_c = V_a - (I_aR_a + BD) = 120 - (47.5 \times 0.2 + 3)$$

$= 107.5$ V

$$S_{5/4} = 1800 \frac{107.5}{109.4} = 1765 \text{ rpm}$$

• PROBLEM 5-12

(i) A certain 230-volt motor has an armature-circuit resistance of 0.3 ohm and runs at a speed of 1200 rpm, with normal excitation and 50-amp armature current.
(a) If the load is increased until the armature current has risen to 60 amp, what will be the new value of speed?
(b) If the load is reduced until the armature current has dropped to 25 amp, what will the speed become?

(ii) (a) If an attempt is made to start the motor of part (i) without auxiliary resistance, what will be the value of the current at the instant of closing the switch? (b) What series resistance should be inserted in order that the armature current at the moment of starting may be limited to 1.5 times the normal value of 50 amp?

Solution:

(i) Normal $E = V - I_aR_a = 230 - (50 \times 0.3) = 215$ volts

With $I_a = 60$ amp, $I_aR_a = 18$ volts, and $E = 230 - 18$

$= 212$ volts

Since $E = K_1 \phi S$ and ϕ is assumed constant, speed will vary directly,

(a) $\frac{1200}{S_2} = \frac{215}{212}$ from which $S_2 = 1183$ rpm

With $I_a = 25$ amp, $I_aR_a = 7.5$ volts and $E = 230 - 7.5$

$= 222.5$ volts

(b) $\frac{1200}{S_3} = \frac{215}{222.5}$ from which $S_3 = 1242$ rpm

The effects of armature reaction and changes in armature-circuit resistance with current are neglected in this solution.

(ii) (a) $I_a = \frac{V}{R_a} = \frac{230}{0.3} = 766 \frac{2}{3}$ amp

which is 15.3 times normal value.

(b) Permissible starting current (I_a)

$$= 1.5 \times 50 = 75 \text{ amp}$$

Now

$$I_a = \frac{V}{R_a + R_s}$$

$$75 = \frac{230}{0.3 + R_s}$$

from which

$$R_s = 2.77 \text{ ohms.}$$

• PROBLEM 5-13

The armature circuit resistance of a 25 hp, 250 V series motor is 0.1 ohms, the brush volt drop is 3 V and the resistance of the series field is 0.05 ohms. When the series motor takes 85 A, the speed is 600 rpm. Calculate:

(a) The speed when the current is 100 A

(b) The speed when the current is 40 A

Neglect armature reaction and assume that the machine is operating on the linear portion of its saturation curve at all times

(c) Recompute speeds in (a) and (b), using a 0.05 ohm diverter at these speeds.

Solution:

(a) $E_{c2} = V_a - I_a(R_a + R_s) - BD = 250 - 100(0.15) - 3$

$= 232$ V when $I_a = 100$ A

$E_{c1} = 250 - 85(0.15) - 3 = 234.3$ V at a speed of 600 rpm when $I_a = 85$ A

$S = K\frac{E}{\phi}$, assuming ϕ is proportional to I_a (on the linear portion of saturation curve)

$$S_2 = S_1 \frac{E_2}{E_1} \times \frac{\phi_1}{\phi_2} = 600 \frac{232}{234.3} \times \frac{85}{100} = 506 \text{ rpm}$$

(b) $E_{c3} = V_a - I_a(R_a + R_s) - BD = 250 - 40(0.15) - 3$

$= 241$ V at 40A

$$S_3 = S_1\left(\frac{E_{c3}}{E_{c1}}\right) \times \frac{\phi_1}{\phi_3} = 600\left(\frac{241}{234.3}\right) \times \left(\frac{85}{40}\right) = 1260 \text{ rpm}$$

(c) The effect of the diverter is to reduce the series field current (and flux) to half their previous values.

$$E_{c2} = V_a - I_a(R_a + R_{sd}) - BD = 250 - 100(0.125) - 3$$

$$= 234.5 \text{ V at } 100 \text{ A}$$

$$E_{c3} = V_a - I_a(R_a + R_{sd}) - BD = 250 - 40(0.125) - 3$$

$$= 242 \text{ V at } 40 \text{ A}$$

$$S_2 = S_1\left(\frac{E_{c2}}{E_{c1}}\right) \times \frac{\phi_1}{\phi_2} = \frac{234.5}{234.3} \times \frac{85\text{A}}{\left(\frac{100}{2}\right)\text{A}} = 1022 \text{ rpm.}$$

$$S_3 = S_1\left(\frac{E_{c3}}{E_{c1}}\right) \times \frac{\phi_1}{\phi_3} = 600\left(\frac{242}{234.3}\right) \times \frac{85\text{A}}{\left(\frac{40}{2}\right)\text{A}} = 2630 \text{ rpm.}$$

• PROBLEM 5-14

(a) A 10-hp 230-volt shunt-wound motor has rated speed of 1000 rpm and full-load efficiency of 86 percent. Armature-circuit resistance, 0.26 ohm; field-circuit resistance, 225-ohms. If this motor is operating under rated load and the field flux is very quickly reduced to 50 percent of its normal value, what will be the effect upon counter emf, armature current and torque? What effect will this change have upon the operation of the motor, and what will be its speed when stable operating conditions have been regained?

(b) Referring to Part (a) what value of resistance must be inserted in series with the armature of this motor in order that the speed may be reduced to one-half its normal value, torque and excitation remaining at their normal full-load values? What effect will this change have upon the horsepower output and efficiency of the motor?

Solution:

(a) $\text{HP} = \dfrac{2\pi TN}{33{,}000}$

$\therefore$ Normal torque $= \dfrac{10 \times 33{,}000}{2\pi \times 1000} = 52.5$ lb at 1 ft radius or 52.5 lb-ft

Rated load $I_c = \dfrac{\text{HP} \times 746}{V_a \times \eta} - \dfrac{V_a}{R_f}$

$$= \frac{10 \times 746}{230 \times 0.86} - \frac{230}{225} = 36.7 \text{ amp.}$$

From the following equation:

$$E = V - I_c R_a,$$

$$\text{Normal counter emf} = 230 - (36.7 \times 0.26)$$

$$= 220.5 \text{ volts}$$

When flux is reduced, the counter emf is reduced to one-half its former value, or 110.25 volts.

The instantaneous value of $I_a = \frac{230 - 110.25}{0.26} = 460$ amp, which is 12.5 times rated value.

The instantaneous value of torque $= 52.5 \times \frac{50}{100} \times \frac{460}{36.7}$ = 330 lb-ft, or 6.25 times normal value.

If fuses do not blow, the motor speed will rise until equilibrium is resotred. Assuming the torque requirements to be unchanged, the armature current under the new condition of operation will be double its previous value, or 73.4 amp, since $T = K_2\phi\ I_a$, and ϕ has been cut in half. (Note that sudden changes in field current are inadvisable because of the excessive transient armature currents that may result. Slower changes in field current permit speed changes to keep step with excitation, current, and torque changes so that excessive armature currents are avoided.)

The new value of counter emf = 230 - (73.4 × 0.26) = 211 volts

$$\frac{K_1\phi_1S_1}{K_1\phi_2S_2} = \frac{1 \times 1000}{0.5 \times S_2} = \frac{220.5}{211}$$

from which

$$S_2 = 1915 \text{ rpm}$$

(b) From Part (a), rated I_a = 36.7 amp, and normal counter emf = 220.5 volts.

If speed is reduced with no change in excitation, the counter emf will be reduced in direct porportion; therefore

The new counter emf $= 220.5 \times \frac{500}{1000} = 110.25$ volts

Then

$$I_a(R_a + R_s) = 230 - 110.25 = 119.75 \text{ volts}$$

Substituting

$$36.7\ (0.26 + R_s) = 9.5 + 36.7\ R_s = 119.75 \text{ volts}$$

from which

$$R_s = \frac{119.75 - 9.5}{36.7} = 3 \text{ ohms to be added.}$$

Since speed has been cut in half and torque is unchanged, the horsepower output will be one-half its former value, or 5 hp.

Also, since torque and excitation are unchanged, the armature current has the same value as before, making watts input the same, and the efficiency will therefore be one-half its former value, or 43 percent.

• PROBLEM 5-15

A 150-hp, 600-rpm, 250-volt, 485-amp, d-c series motor has the magnetization curve given in Fig. 1. It has an armature resistance including brushes of 0.018 ohm and the series field is 0.002 ohm. One hundred percent quantities for the magnetization curve are; volts = 240, field current = 485, speed = 600 rpm. Effective demagnetizing field amperes at rated current is 30 amp and may be considered linear with armature current for the purposes of this example.

(a) Determine for a 250 line volts and 200 line amperes condition, the speed in revolutions per minute and the developed torque in the motor in foot-pounds and newton-meters.

(b) A resistance, called a diverter resistance, whose ohmic value is 0.002 ohm, is connected across the series field. Determine the quantities as in part (a) for the line voltage and current conditions as in part (a).

Solution: (a) The generated voltage in the armature is

$$E_a = V - I_a R_a$$

$$= 250 - 200(0.018 + 0.002) = 246 \text{ volts}$$

The effective field current is

$$I_f = I_a - \left(\frac{I_a}{I_R}\right) \times I_{edf}$$

$= 200 - (200/485) \times 30 = 187.6$ A

Or field current in percent is $I_f = (187.6/485) \times 100 = 38.7\%$.

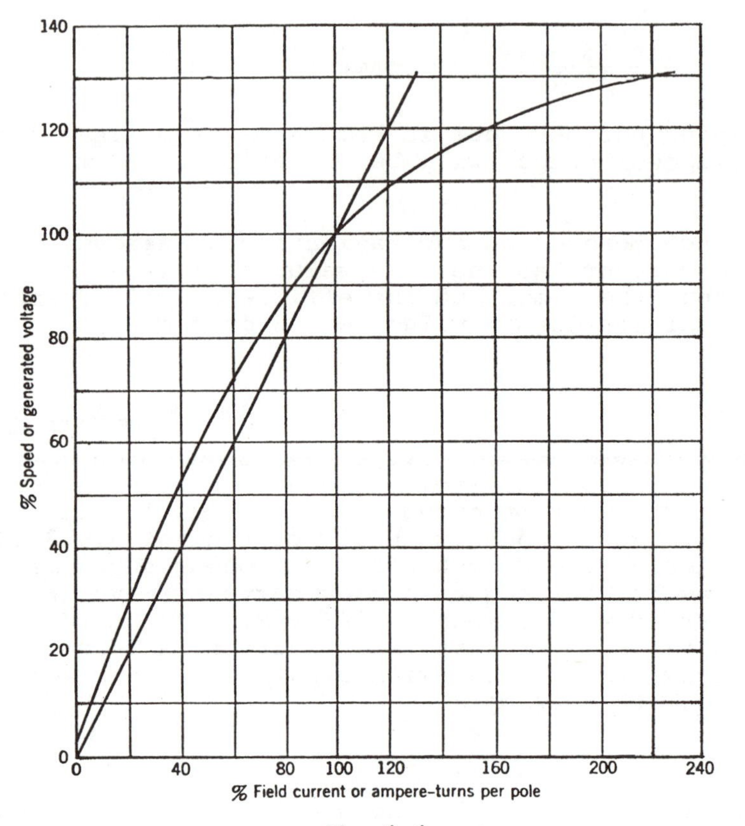

Magnetization curve.

Fig. 1

From the open-circuit characteristic or magnetization curve, the induced voltage at the curve speed, 600 rpm, is read as 51% or $E_a' = 0.51 \times 250 = 127.5$ volts. The 250-volt and 200-amp speed is given by

$$n = 600(246/127.5) = 1160 \text{ rpm}$$

The generator speed is

$$\omega = 1160 \times \pi/60 = 121.3 \text{ rad/sec}$$

The developed torque is

$$T = EI_a/\omega$$

$$= (246 \times 200)/121.3 = 405 \text{ n-m}$$

or

$$T = 405 \times (550/746) = 299 \text{ lb-ft.}$$

(b) The generated voltage in the armature is

$$E_a = 250 - 200\left(0.018 + \frac{0.002 \times 0.002}{0.002 + 0.002}\right) = 246.2 \text{ V}$$

The effective current is

$$I_f = 100 - (200/485) \times 30 = 87.6 \text{ amp}$$

which is $(87.6/485) \times 100 = 18.1\%$.

From the magnetization curve, the generated voltage is 28% or 70 volts at 600 rpm.

$$n = (246.2/70)600 = 2118 \text{ rpm}$$

$$T = (200 \times 246.2 \times 7.04)/2118 = 164 \text{ lb-ft}$$

or

$$T = 164/0.738 = 222 \text{ n-m.}$$

• PROBLEM 5-16

A cumulative compound motor has a varying load upon it which requires a variation in armature current from 50 amp to 100 amp. If the series-field current causes the air-gap flux to change by 3 percent for each 10 amp of armature current, find the ratio of torques developed for the two values of armature current.

Solution: $T = K_T \phi_R I_a$

$$\therefore T_{50} = K_T(115)(50) \text{ and}$$

$$T_{100} = K_T(130)(100)$$

$$\frac{T_{100}}{T_{50}} = \frac{K_T(130)(100)}{K_T(115)(50)} = \frac{260}{115} = \frac{2.26}{1}$$

• PROBLEM 5-17

The magnetization curve of a d-c machine and other data are given in Figure 1. This machine is connected as a shunt motor to a 125-volt d-c line. The field current is adjusted for 0.70 amperes. Find the speed of this machine when it develops a torque of 30 N-m. Assume the windings to be at a temperature of 75° C.

Solution: From the magnetization curve, at 1200 rev/min, and a shunt-field current of 0.70 A,

$$E_g^* = 90.0 \text{ V}$$

and

$$\omega_B = \frac{2\pi N}{60}$$

$$= \frac{1200}{60} \cdot 2\pi = 40\pi \text{ rad/s}$$

Then

$$K_a\phi^* = \frac{E_g^*}{\omega_B} = \frac{90.0}{40\pi} = 0.716.$$

The speed is found from $\omega = E_g/K_a\phi$. The terminal voltage is known, and to find E_g the armature current must be calculated in order to determine the $I_a r_a$ drop. A knowledge of the $K_a\phi$ permits calculation of I_a from the developed torque. The developed torque is given as 30 N-m. Then

$$\tau_d = 30 = K_a\phi^* I_a = 0.716\ I_a$$

$$I_a = \frac{30}{0.716} = 41.9 \text{ A}$$

and

$$E_g = V_T - I_a r_a = 125 - 41.9 \times 0.20$$

$$= 116.6 \text{ V}$$

Finally,

$$\omega = \frac{E_g}{K_a\phi^*} = \frac{116.6}{0.716} = 162.8 \text{ rad/s}$$

or

1555 rev/min.

Two other techniques could have been used in solving this problem:

1. Since the excitation is constant, $K_a\phi$ is constant (assuming no armature reaction). Then

$$\frac{\omega}{\omega_B} = \frac{n}{n_B} = \frac{E_g/K_a\ \phi}{E_g^*/K_a\phi} = \frac{E_g}{E_g^*}$$

E_g and E_g^* are found as before. Then $n = n_B(E_g/E_g^*) = (116.6/90.0) \times 1200 = 1555$ rev/min.

2. The torque could be converted to lb-ft and the English-units equations employed:

$$\tau_d = 30.0 \times 0.738 = 22.1 \text{ lb-ft}$$

$$K_a\phi = \frac{E_g^*}{n_B} = \frac{90.0}{1200} = 0.0750 \text{ V-min/rev}$$

$$\tau_d = 7.04\ K_a\phi I_a = 7.04 \times 0.0750\ I_a = 22.1 \text{ lb-ft}$$

$$I_a = 41.9 \text{ A.}$$

As before

$$E_g = 116.6 \text{ V} = K_a\phi n.$$

$$n = \frac{116.6}{0.0750} = 1555 \text{ rev/min.}$$

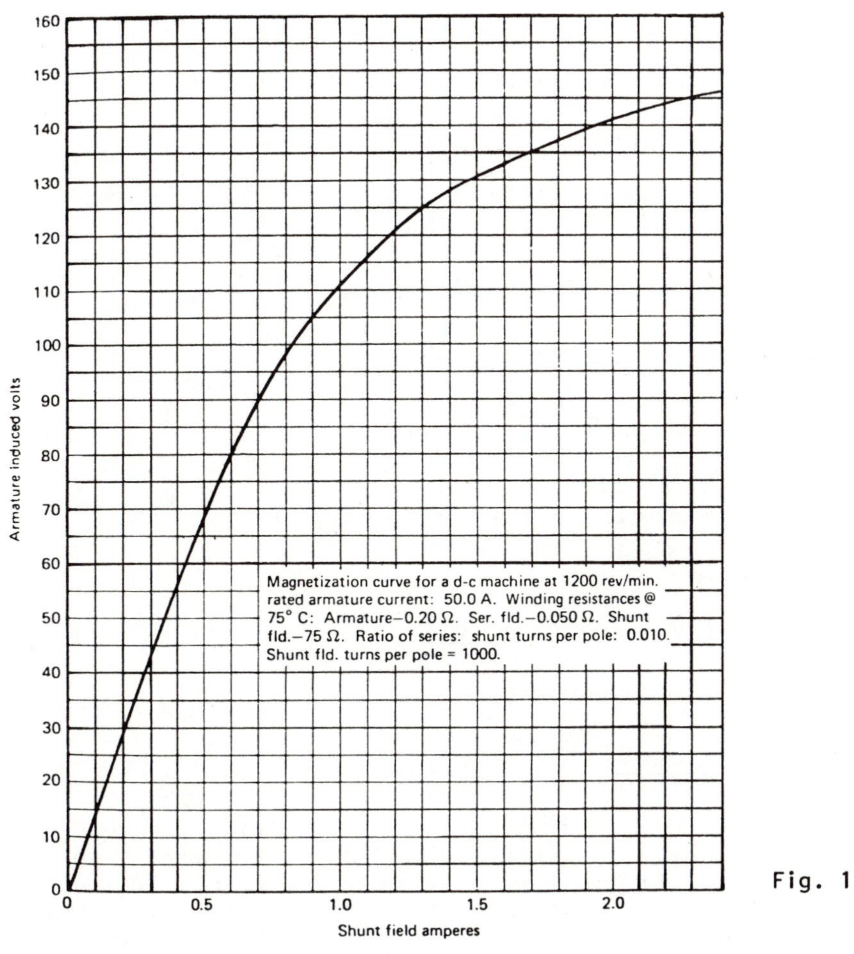

Fig. 1

Magnetization curve for a d-c machine at 1200 rev/min.

• PROBLEM 5-18

(a) A 220-volt shunt motor has an armature resistance of 0.2 ohm. For a given load on the motor, the armature current is 30 amp. What is the immediate effect on the torque developed by the motor if the field flux is reduced by 3 percent?

(b) Consider a 10-hp motor with an armature resistance of 0.5 ohm. This motor was connected directly to a 230-volt supply line.

If the full-load armature current of the motor is 50 amp and it is desired to limit the starting current to 150 percent of this value, find the starting resistance that must be added in series with the armature.

Solution: (a) The torque developed when the armature current is 30 amp is

$$T_1 = K'\phi I_a = K'\phi(30) \quad \text{lb-ft}$$

and the counter emf is

$$E_{g1} = E_t - I_a R_a = 220 - (30 \times 0.2) = 214 \text{ volts}$$

If ϕ is reduced by 3 percent, the value of E_g is also reduced by 3 percent, since $E_g = K\phi n$, and the speed n cannot change instantly. Thus, the new counter emf is

$$E_{g2} = 0.97 \times 214 = 207.58 \text{ volts}$$

The new armature current is

$$I_{a2} = \frac{E_t - E_g}{R_a} = \frac{220 - 207.58}{0.2} = 62.1 \text{ amp}$$

and the new value of torque developed is

$$T_2 = K'(0.97)\phi(62.1)$$

$$= K'\phi 60.24 \quad \text{lb-ft}$$

The torque increase is

$$\frac{T_2}{T_1} = \frac{K'\phi 60.24}{K'\phi 30} = 2.008 \text{ times}$$

Thus, a 3 percent decrease in field flux more than doubles the torque developed by the motor. This increased torque causes the armature speed to increase to a higher value at which the increased counter emf limits the armature current to a value just large enough to carry the load at the higher speed.

(b) Starting resistance (added in series with armature)

$$R_s = \frac{E_t}{I_s} - R_a$$

$$= \frac{230}{40 \times 1.5} - 0.5$$

$$= \frac{230}{60} - 0.5 = 3.33 \text{ ohms.}$$

• PROBLEM 5-19

A 25-h.p., 500-rev/min, d.c. shunt-wound motor operates from a constant supply voltage of 500 V. The full-load armature current is 42 A. The field resistance is 500Ω, the armature resistance is 0.6Ω and the brush drop may be neglected.

Calculate:

(a) the field current required to operate at full-load torque when running at 500 rev/min. What would be the no-load speed with this field current?

(b) Calculate the speed, with this field current, at which the machine must be driven in order to regenerate with full-load armature current.

(c) Calculate the extra field-circuit resistance required to run at 600 rev/min; (i) on no load, (ii) at full-load torque.

(d) Calculate the external armature circuit resistance required to operate at 300 rev/min with full-load torque.

(e) The machine has a cumulative series winding added, its strength at full-load current being equivalent to 0.2 amperes through the shunt turns. By how much will this reduce the full-load speed if the field current is maintained at 0.9 A? What will be the new torque and output horsepower? The additional resistance due to the series turns can be neglected.

Solution: The magnetic characteristic was taken at 400 rev/min.

The problem will be solved in mechanical engineers' units so that $E = k_N N$ and $T = 7.04\ k_N\ I_a$. N is the speed in rev/min and k_N is the generated e.m.f. per rev/min. From the data given below, k_N is calculated immediately and the curve k_N/I_f plotted in Fig. 1.

Field current	0.4	0.6	0.8	1.0	1.2	A
Generated e.m.f.	236	300	356	400	432	V
k_N = e.m.f./400	0.59	0.75	0.89	1.0	1.08	V per rev/min

(a) E at full load = $V - I_aR_a$ = 500 - 42 × 0.6 = 474.8 V

$\therefore$ k_N required = 474.8/500 = 0.9496

From the k_N/I_f curve, this requires 0.9 A and the field circuit resistance must be 500/0.9 = 555 Ω, i.e. an external 55 Ω.

On no load, E = 500 V

$$\therefore \quad N = E/k_N = 500/0.9496 = 527 \text{ rev/min.}$$

(b) When regenerating E = $V + I_aR_a$ = 525.2 V

$$\therefore \quad N = 525.2/0.9496 = 554 \text{ rev/min.}$$

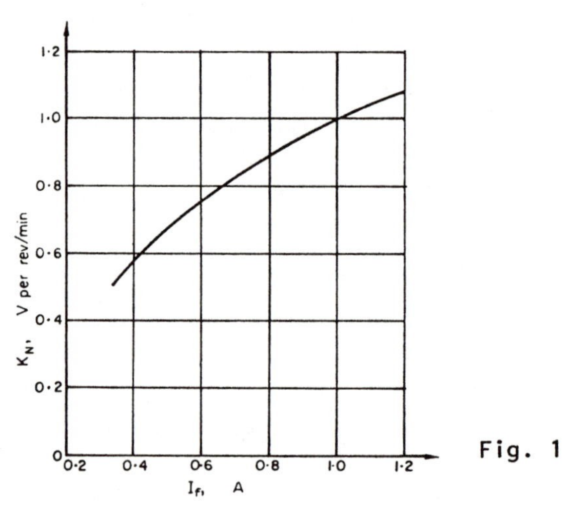

Fig. 1

(c) (i) On no load, E = 500 V $\therefore$ k_N required = 500/600 = 0.833

From k_N/I_f curve this requires 0.7 A as the field current.

The extra field circuit resistance is 500/0.7 - 555 = <u>159 Ω</u>

(ii) Full-load gross torque T = 7.04 × 0.9496 × 42

= 280 lbf-ft

Useful torque at full load, from the rating

$$= 25 \times 746 \times \frac{33{,}000}{2\pi \times 500} = 262.5 \text{ lbf-ft}$$

i.e. there is a mechanical loss torque of 17.5 lbf-ft due to the iron and mechanical losses. Assuming this varies as speed, the total torque developed at 600 rev/min must be 262.5 + 17.5 × 600/500 = 283.5 lbf-ft.

This is only a nominal allowance for the change of loss torque and no great error would follow from assuming it to be constant. Hence, $T = 7.04\ k_N I_a = 283.5$, from which $k_N = 40.2/I_a$.

Further, $E = 500 - 0.6I_a = k_N N = (40.2I_a) \times 600$, from which a quadratic equation in I_a can be obtained; i.e. $I_a^2 - 833\ I_a + 40{,}200 = 0$.

The lower and only practicable value of I_a is found to be 51.5A and this gives $k_N = 40.2/51.5 = 0.78$, and $E = 500 - 0.6 \times 51.5 = 468.8$ V.

I_f from the curve is 0.625 A, and the extra field circuit resistance is $500/0.625 - 555 = 245\,\Omega$.

Note that k_N is very nearly $0.9496 \times 500/600 = 0.79$. This neglects the small change in E due to the additional I_aR_a drop necessary to maintain the torque at full load with reduced flux.

(d) The developed torque must be 262.5 + 17.5 × 300/500 = 273 lbf-ft.

With the flux maintained as is usual for speed reductions, the current becomes 42 × 273/280 = 41 A.

The generated e.m.f. will be $k_N N = 0.9496 \times 300 = 284$ V.

The total resistance drop in the armature circuit must be 216 V.

∴ the external resistance must be $216/41 - 0.6 = 4.66\ \Omega$.

(e) Effective excitation in shunt terms = 0.9 + 0.2 A = 1.1 A

This gives a k_N of 1.04 volts per rev/min

Hence $N = E/k_N = 474.8/1.04 = 455$ rev/min

Developed torque T = 7.04 × 1.04 × 42 = 307 lbf-ft

Useful torque = 307 - 17.5 × 455/500 = 291 lbf-ft

$$\text{Output horsepower} = \frac{2\pi \times 455 \times 291}{33{,}000} = 25.2 \text{ h.p.}$$

• PROBLEM 5-20

A d-c generator and a d-c motor have their armatures directly connected electrically to form a Ward Leonard system. The generator is driven at a constant speed, and the motor field is excited by a constant current. The magnetization curve of the generator is assumed to be linear. The motor is rated at 3 hp and has the following characteristics:

Armature resistance = 0.75 ohm

Inertia of motor plus load = 40 $lb\text{-}ft^2$

Torque per unit armature current = 0.92 lb-ft/amp

Armature inductance is negligible.

The generator is of a size consistent with supplying the 3-hp motor and has the following characteristics:

Armature resistance = 0.25 ohm

Field resistance = 200 ohms

Field inductance = 10 henrys

Generated voltage per unit field current = 1,500 volts/amp

Armature inductance is negligible.

Initially the generator is running but is unexcited, and the motor is excited but stationary. The load torque is negligible.

(a) A constant voltage of 36.6 volts is suddenly impressed on the generator field at t = 0. Find the motor speed $\omega_m(t)$ as a function of time. Approximations are permissible.

(b) Find the time for the speed to reach 98 percent of its final value.

(c) After the motor has reached its final speed, a step of load torque of 8 lb-ft is suddenly applied to the motor at t = 0. Find the speed $\omega_m(t)$.

Solution: With $T_L = 0$, the block diagram of Fig. 1b reduces to Fig. 2a, where $\tau_m = JR_a/K_m^2$. All quantities must be in mks units. From the data,

$$J = \frac{40}{23.7} = 1.69 \text{ kg-m}^2 \qquad R_a = 0.75 + 0.25 = 1.00 \text{ ohm}$$

$$K_m = \frac{0.92}{0.738} = 1.25 \text{ newton-m/amp or volt-sec/radian}$$

$$\tau_m = \frac{(1.69)(1.00)}{(1.25)^2} = 1.08 \text{ sec} \qquad \tau_{fg} = \frac{10}{200} = 0.05 \text{ sec}$$

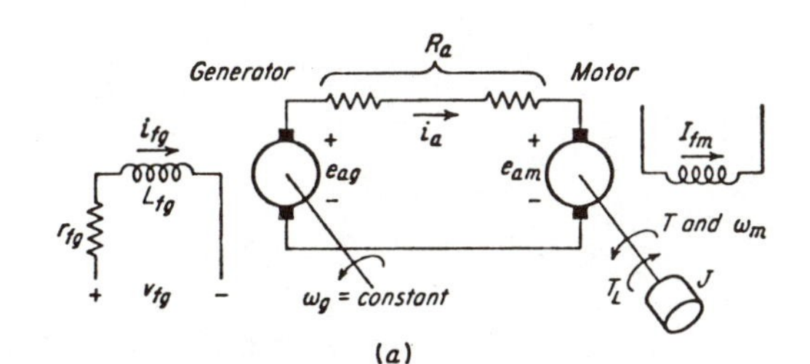

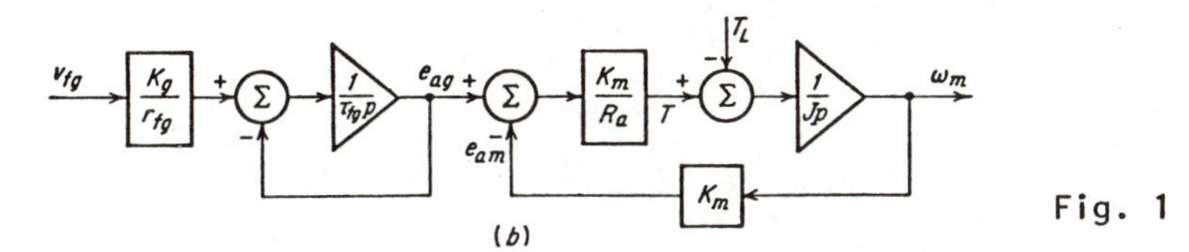

Fig. 1

(a) Circuit diagram and (b) block diagram of an adjustable-armature-voltage, or Ward Leonard, System of speed control.

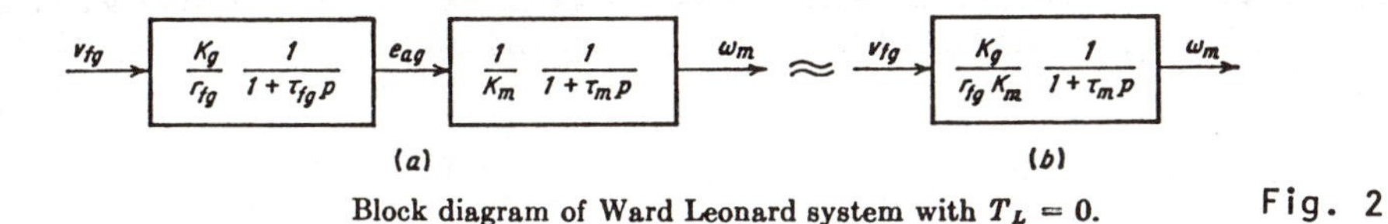

Block diagram of Ward Leonard system with $T_L = 0$. Fig. 2

(a) The block diagram represents two time lags in cascade. But the generator-field time constant τ_{fg} is very much less than the motor-plus-inertia time constant τ_m.

Therefore the generator-field time lag can be neglected with very little error; i.e., the generated emf e_{ag} builds up so rapidly that its effect in the armature circuit is almost that of a step change, and the block diagram reduces to Fig. 2b. For a step input v_{fg} = 36.6 volts, this transfer function represents an exponential build-up to a final value

$$\omega_m(\infty) = \frac{(36.6)(1,500)}{(200)(1.25)} = 219 \text{ rad/sec}$$

or $\omega_m(t) = 219(1 - \varepsilon^{-t/1.08})$

(b) A simple exponential will decay to 0.0183, or approximately 2 percent of its final value in a time equal to four time constants. For the motor to come up to 98 percent of its final speed then requires about (4)(1.08), or 4.3 sec.

(c) The decrease in speed caused by a suddenly applied step of load torque is

$$\Delta\omega_m(\infty) = \frac{T_L R_a}{K_m^2}$$

For $T_L = 8/0.738 = 10.8$ newton-m

$$\Delta\omega_m(\infty) = \frac{(10.8)(1.00)}{(1.25)^2} = 6.9 \text{ rad/sec.}$$

The change is an exponential with time constant τ_m. Therefore the speed is

$$\omega_m(t) = 219 - 6.9(1 - \varepsilon^{-t/1.08})$$

$$= 212 + 7\varepsilon^{-t/1.08} .$$

• PROBLEM 5-21

The following data apply to the position control system of Fig. 1:

Motor armature resistance $r_{am} = 6$ ohms

Generator armature resistance $r_{ag} = 4$ ohms

Motor speed-voltage constant $K_m = 0.9$ volt-sec/rad

Inertia of motor armature alone $J_m = 0.002$ kg-m^2

The motor drives a load of inertia $J_L = 10$ kg-m^2 through ideal gears reducing the speed by the factor k_G.

The system is to be adjusted so that the steady-state errors for a constant-velocity input of 20°/sec or for a steady load torque of 10 newton-m are not to exceed 1°, with all quantities referred to the load shaft.

(a) Find the over-all gain $K_1 K_p$ of the error detector, amplifier, and generator.

(b) Find the gear ratio k_G.

(c) Find the steady-state motor speed, armature voltage, and armature current for a constant-velocity input of 20°/sec with load torque $T_L = 0$.

(d) Find the armature voltage and current at standstill with $T_L = 10$ newton-m.

(e) Check the dynamic performance; i.e., find α, ζ, and ω_n.

(f) For a step input θ_i of 20°, find the initial values of generator emf e_{ag} and armature current i_a.

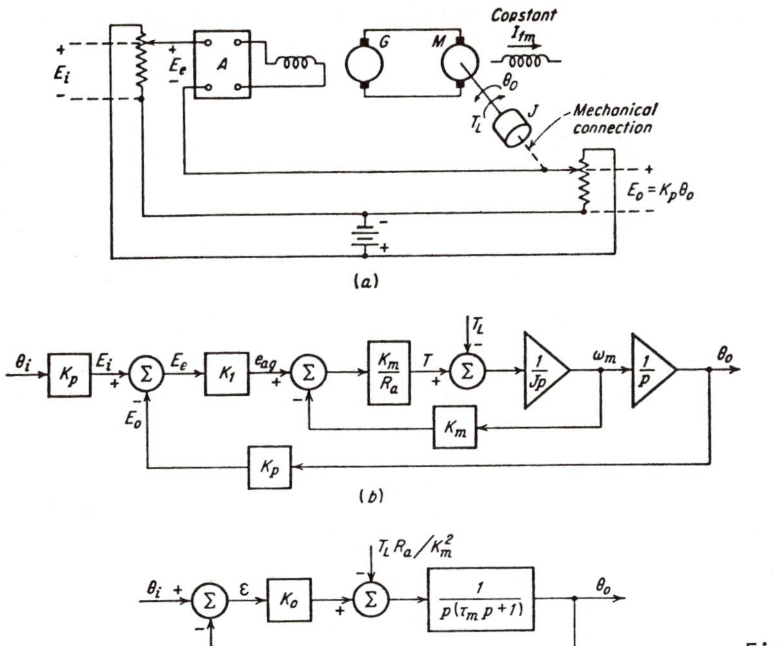

Ward Leonard position-control servo system. (a) Schematic diagram. (b) Block diagram. (c) Simplified block diagram.

Solution: Refer all quantities to the motor shaft; thus

$$\text{Angle at motor} = k_G \text{ (angle at load)}$$

$$\text{Torque at motor} = \frac{\text{torque at load}}{k_G}$$

$$\text{Load inertia referred to motor} = \frac{J_L}{k_G^2}$$

(a) From

$$\varepsilon(\infty) = \frac{\omega_i}{K_0},$$

to meet the velocity-error specification with $T_L = 0$,

$$K_0 = \frac{\omega_i}{\varepsilon(\infty)} = \frac{20k_G}{k_G} = 20$$

$$K_1K_p = K_0K_m = (20)(0.9) = 18 \text{ volts/rad at motor shaft}$$

(b) Referred to the motor shaft,

$$k_G\varepsilon(\infty) = \frac{R_a}{K_m^2K_0}\frac{T_L}{k_G}$$

with error and load torque given at the load shaft. To meet the load-torque specification with $\omega_i = 0$,

$$k_G^2 = \frac{R_a}{K_m^2K_0}\frac{T_L}{\varepsilon(\infty)}$$

with $R_a = r_{ag} + r_{am} = 10$ ohms and $\varepsilon(\infty) = \pi/180$ rad. Thus,

$$k_G^2 = \frac{(10)(10)(180)}{(0.9)^2(20)\pi} = 354 \quad \text{or} \quad k_G = 18.8$$

(c) For $\omega_i = 20°/\text{sec}$ at load, the steady-state motor speed is

$$\omega_m = k_G\omega_i = \frac{(18.8)(20)\pi}{180} = 6.56 \text{ rad/sec} = 63 \text{ rpm}$$

The motor counter emf is

$$e_{am} = K_m\omega_m = 5.9 \text{ volts}$$

With $T_L = 0$ and $i_a = 0$, the generator emf $e_{ag} = 5.9$ volts.

(d) For $T_L = 10$ newton-m at load, the steady-state motor torque is

$$T = \frac{T_L}{k_G} = \frac{10}{18.8} = 0.534 \text{ newton-m}$$

The armature current is

$$i_a = \frac{T}{K_m}\ \frac{0.534}{0.9} = 0.59 \text{ amp}$$

With the motor stationary, $e_{am} = 0$, and

$$e_{ag} = i_aR_a = (0.59)(10) = 5.9 \text{ volts}$$

(e) The motor-plus-load inertia referred to the motor is

$$J = J_m + \frac{J_L}{k_G^2} = 0.002 + \frac{10}{354} = 0.00483 \text{ kg-m}^2$$

The time constant is

$$\tau_m = \frac{JR_a}{K_m^2} = \frac{(0.00483)(10)}{0.81} = 0.06 \text{ sec}$$

$$\alpha = \frac{1}{2\tau_m} = \frac{1}{0.12} = 8.3 \text{ sec}^{-1}$$

$$\omega_n = \sqrt{\frac{K_0}{\tau_m}} = \sqrt{\frac{20}{0.06}} = 18.3 \text{ rad/sec, or } 2.9 \text{ cps}$$

$$\zeta = \frac{\alpha}{\omega_n} = \frac{8.3}{18.3} = 0.45$$

The dynamic performance is satisfactory in so far as damping is concerned.

(f) For a step input θ_i of 20°, or 0.349 rad at the load shaft, the initial value of the generator emf is

$$e_{ag}(0+) = K_1 K_p k_G \varepsilon = (18)(18.8)(0.349) = 118 \text{ volts}$$

With the motor stationary and armature inductances negligible, the initial armature current is

$$i_a(0+) = \frac{e_{ag}(0+)}{R_a} = 11.8 \text{ amp}$$

The initial power generated is $e_{ag} i_a$ = 1,390 watts. The rotating machines must be capable of handling voltages, currents, and power surges of the magnitude just found. The signal-flow analysis alone does not show these requirements directly. Thus there is a lot more to the design of a servo system than merely the proper choice of gains, damping ratios, and so forth. The energy-conversion equipment has to be capable of doing the work required of it by the demands of the signals impressed on it.

• PROBLEM 5-22

The figure shows a portion of the magnetization curve obtained at 1800 rpm on a 25-hp 250-V 84-A shunt motor. The resistance of the shunt-field circuit, including the field rheostat, is 184 Ω and the resistance of the armature circuit, including the commutating or interpole winding and brushes, is 0.082 Ω. The field winding has 3000 turns per pole.

The demagnetizing mmf F_a of armature reaction, at rated armature current, is 0.09 A in terms of the shunt-field current. The no-load losses, which include windage and friction losses as well as core losses, are 1300 W. The stray-load loss is assumed to be 0.01 of the output. The resistance of the field circuit is assumed constant at 184 Ω. Calculate (a) the speed of the motor when it draws 84 A from the line, (b) the electromechanical power, (c) the mechanical power output, (d) the output torque, and (e) the efficiency.

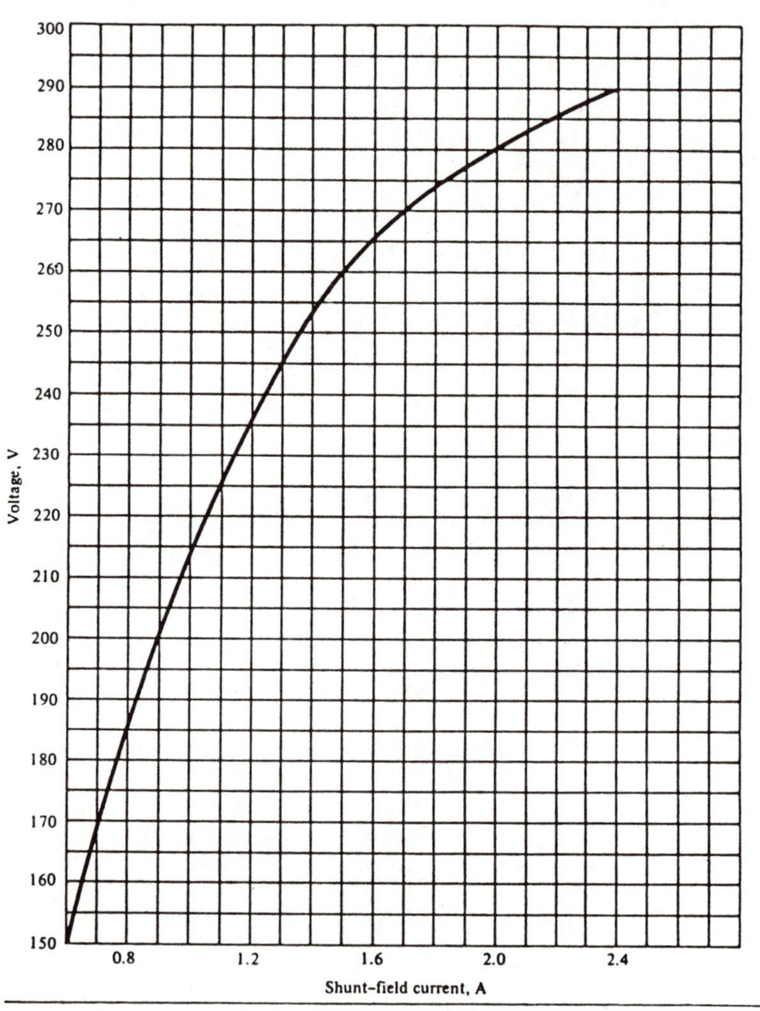

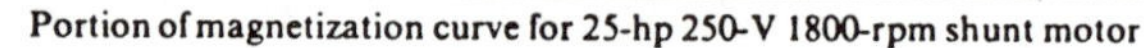

Portion of magnetization curve for 25-hp 250-V 1800-rpm shunt motor

Solution: a. The field current is $I_f = 250 \div 184 = 1.36$ A and the armature current is,

$$I_a = I - I_f = 84 - 1.36 = 82.6 \text{ A}$$

The counter emf is

$$E' = V - r_a I_a = 250 - 0.082 \times 82.6 = 250 - 6.8$$

$$= 243.2 \text{ V}$$

The net field excitation is the field mmf minus the demagnetizing mmf of armature reaction:

$$I_{f(net)} = 1.36 - 0.09 = 1.27 \text{ A.}$$

Since the armature current is near rated value, the value of 0.09 can be used as the demagnetizing mmf with negligible error.

The magnetization curve in the figure shows that a field current of 1.27 A produces a generated voltage of 241.5 V at 1800 rpm.

The motor speed is, therefore,

$$n' = \frac{E'}{E} n = \frac{243.2}{241.5} \times 1800 = 1813 \text{ rpm}$$

b. The electromechanical power is

$$P_{em} = E'I_a = 243.2 \times 82.6 = 20{,}050 \text{ W}$$

c. The mechanical power is the electromechanical power minus the sum of the rotational losses plus the stray-load losses:

$$P_{mech} = P_{em} - (P_{rot} + P_{stray})$$

$$= 20{,}050 - 1300 - 0.01\, P_{mech}$$

$$= 18{,}560 \text{ W} \quad \text{or} \quad \frac{18{,}560}{746} = 24.9 \text{ hp.}$$

d. $\text{Torque} = \dfrac{P_{mech}}{\omega_m} = \dfrac{18{,}560}{2\pi \times 1813/60} = 97.9 \text{ N-m/rad}$

or $97.9 \times 0.738 = 72.2$ lb-ft/rad.

e. Losses

Rotational	=	1,300 W
Stray load = 0.01 × 18,560	=	185 W
$I_a^2 r_a = (82.5)^2 \times 0.082$	=	560 W
$I_f^2 r_f = 250 \times 1.36$	=	340 W
Total losses	=	2,385 W
Output	=	18,560 W
Input		20,945 W

$$\text{Efficiency} = 1 - \frac{\text{losses}}{\text{input}} = 1 - \frac{2385}{20{,}945} = 0.886.$$

It is interesting to note that the full-load speed of the motor in this example is greater than the no-load (1800 rpm) speed due to F_a. This suggests that the motor is somewhat unstable. The motor can be given a drooping speed characteristic by adding a series- or stabilizing-field winding of a few turns.

• PROBLEM 5-23

A 10-HP d-c shunt motor is operated in the three-phase, half-wave thyristor drive shown in Fig. 1. The armature circuit resistance is 0.21 Ω. The armature circuit self-inductance is 0.0105 henry. The moment of inertia is 0.432 kgm^2. The field current is set such that the voltage constant is 1.42 v/(rad/sec) or 1.42 nm/amp. The three-phase source has line-to-neutral voltage of 205 v (rms). v_{an} = 290 sin 377t. The phase sequence is a-b-c. The firing angle is set at 45°. The constant-torque load requires the average electromagnetic torque to be 45 nm. (a) Find the speed. (b) Find the peak-to-peak ripple in the armature current. (c) Investigate the speed dip.

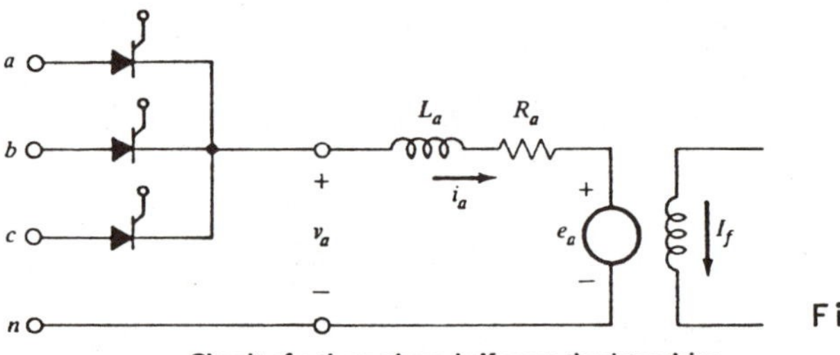

Fig. 1

Circuit of a three-phase, half-wave thyristor drive.

Solution: (a) The average armature terminal voltage is

$$V_a = 0.827\ V_1 \cos\alpha = 0.827(290)\cos 45° = 169.6 \text{ v}$$

The average current is

$$I_a = T/k' = 45/1.42 = 31.7 \text{ amp}$$

The generated voltage is

$$E_a = V_a - R_a I_a = 169.6 - (0.21)(31.7) = 162.9 \text{ v}$$

The speed is

$$\omega_m = E_a/k' = 162.9/1.42 = 114.7 \text{ rad/sec}$$

(b) The armature circuit impedance is

$$\vec{Z}_a = R_a + j\omega L_a = 0.21 + j(377)(0.0105)$$

$$= 0.21 + j3.96 = 3.97 \underline{/87°}\ \Omega$$

The component of the current due to the source is

$$i_1 = (V_1/Z_a) \sin (\omega t - \theta_a)$$

$$= (290/3.97) \sin (377t - 87°)$$

$$= -73 \cos (377t + 3°) \text{ amp}$$

An approximation to the peak-to-peak ripple is given by the following equation:

$$R_{pp} \approx 0.5I_1 \cos(\alpha - \theta_a) = 0.5(73) \cos (45° - 87°)$$

$$= 27.1 \text{ amp}$$

As a matter of interest, the current during a conducting interval is

$$i_a(t) = 73 \sin (377t - 87°) - 775.7 + 862.4e^{-t/0.05}$$

for $75° < \omega t < 195°$.

The minimum current is

$$i_{a\ min} = i_a(t_a) = i_a(t_b) = 13.7 \text{ amp.}$$

The maximum current is

$$i_{a\ max} = i_a(0.00666) = 40.2 \text{ amp.}$$

(c) The speed dip is

$$\Delta\omega_m = k'R_{pp}/J(3\omega) = -1.42(27.1)/(0.432)(1131)$$

$$= -0.079 \text{ rad/sec.}$$

Expressed as a ratio to the speed, the speed dip is

$$\Delta\omega_m/\omega_m = -0.079/114.7 = 0.0007.$$

HORSE POWER

• PROBLEM 5-24

A certain load to be driven at 1750 r/min requires a torque of 60 lb. ft. What horsepower will be required to drive the load?

Solution:

$$\text{Horse power} = \frac{2\pi \times T \times N}{33,000}$$

$$= \frac{2\pi \times 60 \times 1750}{33,000} = 20 \text{ hp.}$$

• PROBLEM 5-25

The field winding of a shunt motor has a resistance of 110 ohms, and the voltage applied to it is 220 volts. What is the amount of power expended in the field excitation?

Solution: The current through the field is

$$I_f = \frac{E_s}{R_f} = \frac{220}{110} = 2 \text{ amperes.}$$

Power expended is

$$E_s I_f = 220 \times 2 = 440 \text{ watts.}$$

The same results will also be obtained by using the equation

$$\frac{E_s^2}{R_f} = P_f = \frac{220^2}{110} = 440 \text{ watts.}$$

• PROBLEM 5-26

Calculate the horsepower output, torque, and efficiency of a shunt motor from the following data:

$I_\ell = 19.8$ amperes

$E_\ell = 230$ volts

Balance reading = 12 lbs. corrected for zero reading

Brake arm = 2 ft.

Speed = 1100 r/min

Solution: The horsepower output of the motor is:

$$\frac{2\pi \times 2 \times 12 \times 1100}{33,000} = 5.03 \text{ hp}$$

Torque = FR = 12 × 2 = 24 lb. ft.

$$\text{Efficiency} = \frac{\text{output}}{\text{input}} = \frac{5.03 \times 746}{230 \times 19.8} = 0.824, \text{ or } 82.4\%$$

• PROBLEM 5-27

(1) The terminal voltage of a motor is 240 volts and the armature current is 60 amp. The armature resistance is 0.08 ohm. The counter emf is 235.2 volts.

What is the power developed by the motor (a) in watts and (b) in horsepower?

(2) The measured speed of a motor is 1,100 rpm. The net force registered on the scale used with a prony brake is 16 lb. If the brake arm measures 28 in., what is the horsepower output of the motor?

Solution:

(1) (a) Developed power = $E_g I_a$ watts

$$= 235.2 \times 60 = 14{,}112 \text{ watts.}$$

(b) $$\text{Horsepower} = \frac{\text{watts}}{746}$$

$$= \frac{14{,}112}{746} = 18.92 \text{ hp.}$$

(2) T = Fr

$$= 16 \times \frac{28}{12} = 37.3 \text{ lb-ft}$$

$$\text{Horsepower output} = \frac{nT}{5{,}252}$$

$$= \frac{1{,}100 \times 37.3}{5{,}252} = 7.82 \text{ hp.}$$

• PROBLEM 5-28

Consider a lap-wound armature 4 feet in diameter, having 12 poles. The winding consists of 240 coils of 4 turns each and the length of the pole face is 10 inches. Sixty percent of the conductors lie under the pole face where the flux density is 60,000 lines per square inch, and 15 percent lie in the pole fringe where the average density is 35,000 lines per square inch. What horsepower is the motor developing if the current flowing into the armature is 480 amperes and the machine is rotating at 200 r.p.m.?

Solution: There are two things to find--the peripheral pull in pounds and the peripheral velocity in feet per minute. The product of these two quantities divided by 33,000 (the number of foot-pounds per minute in 1 horsepower) will give the horsepower of the motor.

As the armature is lap wound and the machine has 12 poles, the winding must have 12 paths. Therefore the current per path = 480 ÷ 12 = 40 amperes, and this is the current in each conductor. The active length of each conductor is 10 inches. The total number of conductors = 240 x 4 x 2 = 1920. Of these, 60 percent (i.e., 1152), lie in a field of 60,000 lines per square inch and 15 percent (i.e., 288) lie in a field of 35,000 lines per square inch.

Therefore, the force in pounds is

$$F = 0.885\ B\ell I \times 10^{-7}$$

$$= 0.885 \times 10^{-7} \times 40[(1152 \times 10 \times 60{,}000) + (288 \times 10 \times 35{,}000)] = 2800;$$

The peripheral speed in feet per minute

$$= 4 \times \pi \times 200 = 2515;$$

The horsepower developed therefore

$$= \frac{2\pi FLN}{33{,}000} = \frac{2515 \times 2800}{33{,}000} = 213.4.$$

• PROBLEM 5-29

A 50-hp, 500-volt shunt motor draws a line current of 4.5 amperes at no load. The shunt field resistance is 250 ohms and the armature resistance, exclusive of brushes, is 0.3 ohm. The brush drop is 2 volts. The full-load line current is 84 amperes. What is the horsepower output and efficiency?

Solution: From the data supplied,

Full-load armature current

$$I_a = I_L - I_f = 84 - \frac{500}{250} = 82 \text{ amperes.}$$

No-load armature current

$$I_a = I_L - I_f = 4.5 - \frac{500}{250} = 2.5 \text{ amperes.}$$

Stray power loss

$$P_{sp} = I_L \times E = 2.5 \times 500 = 1250 \text{ watts.}$$

Brush loss

$$= 2 \times I_a = 2 \times 82 = 164 \text{ watts.}$$

The efficiency of the motor

$$= \frac{500 \times 84 - [(82^2 \times 0.3) + (500 \times 2) + 164 + 1250]}{500 \times 84}$$

$$= \frac{42,000 - 4431}{42,000} = \frac{37,569}{42,000} = 0.8945, \text{ or } 89.45\%$$

Horsepower output

$$= \frac{37,569}{746} = 50.36 \text{ hp.}$$

LOSSES AND EFFICIENCY

• PROBLEM 5-30

A dc motor requires 10 kilowatts to enable it to supply its full capacity of 10 horsepower to its pulley. What is its full-load efficiency?

Solution:

$$\text{Efficiency} = \frac{\text{output}}{\text{input}} = \frac{10 \times 746}{10,000} = 0.746, \text{ or } 74.6\%$$

• PROBLEM 5-31

A 7-hp motor takes 6.3 kilowatts at full load. What is its efficiency?

Solution:

$$\text{Efficiency} = \frac{\text{output}}{\text{input}},$$

$$\text{Output} = 7 \times 746 = 5222 \text{ watts.}$$

$$\text{Input} = 1000 \times 6.3 = 6300 \text{ watts.}$$

$$\text{Efficiency} = \frac{5222}{6300} = 0.829, \text{ or } 83\% \text{ (approx.)}$$

• PROBLEM 5-32

The rating of a certain machine from the name-plate is 110 volts, 38.5 amperes, 5 h.p. Find the input, output and efficiency at full load.

Solution:

The watts input, full load = 110 × 38.5 = 4350 watts

The watts output, full load = 5 × 746 = 3730 watts

Hence the full-load efficiency = 3730 ÷ 4350 = 85.7 percent.

• PROBLEM 5-33

Assume that the dc motor shown in Fig. 1 draws a current of 10 amperes from the line with a supply voltage of 100 volts. If the total mechanical loss (friction, windage, etc.) is 90 watts, calculate the

(a) Copper losses in the field.

(b) Armature current.

(c) Copper losses in the armature.

(d) Total loss.

(e) Motor input.

(f) Motor output.

(g) Efficiency.

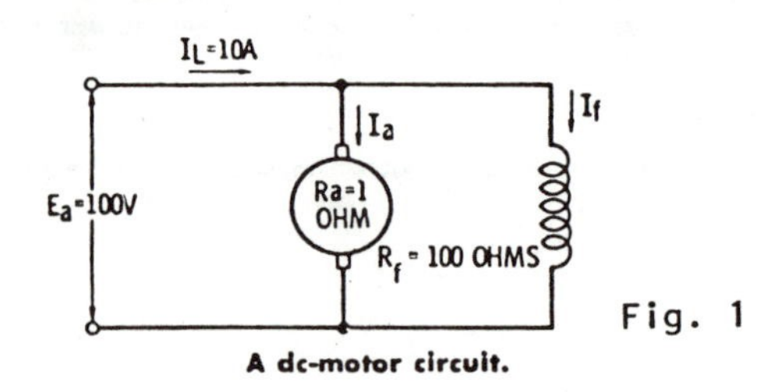

A dc-motor circuit.

Fig. 1

Solution: From the foregoing data,

(a) The copper losses in the field are

$I_f^2 \times R_f = 1^2 \times 100 = 100$ watts.

(b) The armature current is

$10 - 1 = 9$ amperes

(c) Copper losses in the armature are

$I_a^2 \times R_a = 9^2 \times 1 = 81$ watts

(d) The total loss is

$100 + 81 + 90 = 271$ watts

(e) Motor input is

$E_a \times I_L = 100 \times 10 = 1000$ watts

(f) Motor output is

input - losses = 1000 - 271 = 729 watts

(g) The efficiency is

$$\frac{\text{output}}{\text{input}} \times 100 = \frac{729}{1000} \times 100 = 72.9 \text{ percent}$$

• PROBLEM 5-34

A shunt motor with an armature and field resistance of 0.055 and 32 ohms, respectively, is to be tested for its mechanical efficiency by means of a rope brake. When the motor is running at 1400 r/min, the longitudinal pull on the 6-inch-diameter pulley is 57 lbs. Simultaneous readings of the line voltmeter and ammeter are 105 and 35, respectively. Calculate the

(a) Counter emf.

(b) Copper losses.

(c) Efficiency.

Solution: From the foregoing data:

$$I_a = I_L - I_f = 35 - \frac{105}{32} = 31.7 \text{ amperes.}$$

where

I_a = Current of armature

I_L = Current of line

I_f = Current of field

(a) The counter emf is

$$E_{armature} = E_{line} - (I_{arm} \times R_{arm})$$

$$E_a = E_L - (I_aR_a) = 105 - (31.7 \times 0.055)$$

$= 103.26$ volts.

(b) The copper losses are

$$\text{Power Loss for Copper} = (I^2 \times R_f) + (I_a^{\,2} \times R_a)$$

$$P_c = I^2R_f + I_a^{\,2}R_a = (3.3^2 \times 32) + (31.7^2 \times 0.055)$$

$$= 404 \text{ watts.}$$

(c) output

$$\frac{2\pi \times \text{r/min} \times \text{radius of pulley in ft.} \times \text{pull on pulley in lbs.}}{33{,}000}$$

$$= \frac{2\pi \times 1400 \times 3/12 \times 57}{33{,}000} = 3.8 \text{ hp.}$$

$$\text{Input} = \frac{\text{Voltage of line} \times \text{Current of line}}{746 \text{ watts}}$$

$$= \frac{105 \times 35}{746} = 4.93 \text{ hp}$$

$$\text{Efficiency} = \frac{\text{output}}{\text{input}}.$$

$$= \frac{3.8}{4.93} = 0.771, \text{ or } 77\% \text{ (approx.)}.$$

• PROBLEM 5-35

(a) The input current to a 220-volt, long shunt compound motor at no load is 6 amp. The shunt-field circuit resistance is 220 ohms; the armature resistance is 0.10 ohm; and the series-field resistance is 0.08 ohm. What is the stray power loss?

(b) Assume the input rating of the motor in Part (a) to be 56 amp at 220 volts with a no-load speed of 1000 rpm. If the stray power loss varies as the counter-electromotive force and the effects of armature reaction are considered negligible, determine the efficiency at rated input.

Solution:

(a) Motor input $= 6 \times 220 = 1320$ watts.

Shunt-field current is 1 amp. Shunt-field loss $= 1 \times 220 = 220$ watts. Armature current is 5 amp. Armature and series-field loss:

$$I_a^2(R_a + R_s) = (5)^2 \times 0.18 = 25 \times 0.18 = 4.5 \text{ watts}$$

Stray power loss = motors input - (shunt-field loss + armature loss + series - field loss)

$$= 1320 - (220 + 4.5) = 1095.5 \text{ watts.}$$

(b) $$\text{Full-load speed} = \text{no-load speed} \times \frac{E_g \text{ at full load}}{E_g \text{ at no load}}$$

At no load, $E_g = 220 - 5(0.1 + 0.08) = 219.1$ volts.

At full load, $E_g = 220 - 55(0.1 + 0.08) = 210.1$ volts.

$$\therefore \text{Full-load speed} = 1000 \times \frac{210.1}{219.1} = 960 \text{ rpm}$$

$$\text{Stray power loss at full load} = 1095.5 \times \frac{210.1}{219.1} = 1050 \text{ watts}$$

Shunt-field loss = 1 × 220 = 220 watts

Armature loss = $(55)^2 \times 0.1 = 302.5$ watts

Series-field loss = $(55)^2 \times 0.08 = 242$ watts

$$\text{Percentage efficiency} = 100 \times \frac{(\text{input-losses})(\text{watts})}{\text{input (watts)}}$$

$$= 100 \times \left[\frac{(56 \times 220) - (220 + 302.5 + 242 + 1050)}{56 \times 220}\right]$$

$$= 100 \times \frac{12,320 - 1814.5}{12,320}$$

$$= 100 \times \frac{10,505.5}{12,320}$$

Percentage efficiency = 85.3 percent.

• PROBLEM 5-36

The nameplate rating of a shunt motor is 150 hp, 600 volts, 205 amp, 1,700 rpm. The resistance of the shunt field circuit is 240 ohms, and the total armature circuit resistance is 0.15 ohm. The motor has a commutating winding, and armature reaction is neglected.

(a) Calculate the full-load efficiency of the motor and the speed regulation.

(b) Calculate the efficiency and the delivered torque when the motor draws half-load current from the line.

(c) If the motor is to deliver full-load torque at 1,200 rpm, what value of resistance must be added to the armature circuit?

Solution:

(a) At full load

$$\text{Motor input} = \frac{VI}{1{,}000}\ \text{kw}$$

$$= \frac{600 \times 205}{1{,}000} \qquad = 123.0\ \text{kw}$$

$$\text{Motor output} = 150 \times 0.746 \qquad = \underline{111.9\ \text{kw}}$$

$$\text{Motor losses (input - output)} \qquad = 11.1\ \text{kw}$$

$$\text{Efficiency} = \frac{111.9}{123.0} \times 100 \qquad = 91.0\%$$

The losses are

$$\text{Shunt field} = \frac{600 \times \frac{600}{240}}{1{,}000} \qquad = 1.50\ \text{kw}$$

$$\text{Arm. copper} = \frac{I^2R}{1{,}000}\ \text{kw.}$$

$$= \frac{(205 - 2.5)^2 \times 0.15}{1{,}000} = 6.15\ \text{kw}$$

$$\text{Stray power} = \text{remainder} \qquad = \underline{3.45\ \text{kw}}$$

$$\text{Total losses} \qquad = 11.10\ \text{kw}$$

The no-load current to the motor will be that required for the field and the stray-power losses, that is,

$$\text{no-load current} = 2.5 + \frac{3{,}450}{600} = 8.25\ \text{amp}$$

The generated voltages are

$$E_g\ \text{full-load} = 600 - (205 - 2.5) \times 0.15 = 569.6\ \text{volts}$$

$$E_g\ \text{no-load} = 600 - (8.25 - 2.5) \times 0.15 = 599.14\ \text{volts}$$

whence the no-load speed is

$$1{,}700 \times \frac{599.1}{569.6} = 1{,}790\ \text{rpm}$$

and

$$\text{speed regulation} = \frac{1{,}790 - 1{,}700}{1{,}700} \times 100 = 5.3\%$$

The delivered torque at full load is 463 lb-ft, obtained from the horsepower equation. The developed torque is 478 lb-ft and includes the stray-power loss.

(b) At half-load current, the motor draws 102.5 amp, of which 100 amp flow through the armature circuit. The field and stray-power losses are assumed to remain constant, but the new armature copper loss is 1.5 kw. The total losses are 6.45 kw, the motor input is 61.5 kw, the output is 55.05 kw, and the motor efficiency is 89.6%. The generated voltage for the new condition is 585 volts, the speed is 1,746 rpm, and the delivered torque is 222 lb-ft.

(c) With constant flux, full-load torque requires full-load armature current of 202.5 amp. Hence,

$$E_g \text{ at } 1{,}200 \text{ rpm} = \frac{1{,}200}{1{,}700} \times 569.6 = 402 \text{ volts}$$

and the resistance to be added in the armature circuit is

$$\frac{600 - 402}{(205 - 2.5)} - 0.15 = 0.83 \text{ ohm.}$$

• PROBLEM 5-37

Two similar 120-volt 7.5-hp. motors are connected in the manner shown in the figure. The armature resistance of each is 0.12 ohm. The fields are adjusted so that the motor current I_1 is 57 amp, and the generator current I_2 is 45 amp. Under these conditions the power source is supplying a current I of 12 amp at 120 volts. Determine stray power of each machine under these conditions.

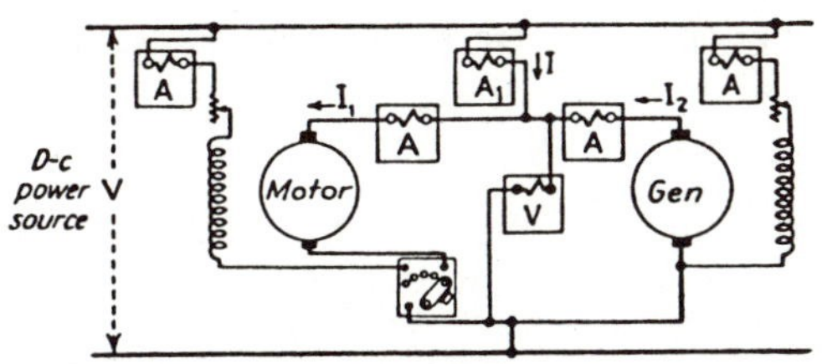

Kapp opposition method for determining losses.

Solution: Power supplied by line,

$$P = 120 \times 12 = 1{,}440 \text{ watts.}$$

$$I_1{}^2R_1 = 57^2 \times 0.12 = 390 \text{ watts.}$$

$$I_2{}^2R_2 = 45^2 \times 0.12 = 243 \text{ watts.}$$

$$\text{Total} = 633 \text{ watts.}$$

Total stray power = 1,440 - 633 = 807 watts.

$$E_1 = 120 - (57 \times 0.12) = 113.2 \text{ volts.}$$

$$E_2 = 120 + (45 \times 0.12) = 125.4 \text{ volts.}$$

The motor stray power

$$P_1 = \frac{113.2}{113.2 + 125.4} 807 = 383 \text{ watts.}$$

The generator stray power

$$P_2 = \frac{125.4}{113.2 + 125.4} 807 = 424 \text{ watts.}$$

• PROBLEM 5-38

(i) A shunt generator when running light as a motor at 1,000 rpm takes 12 amp from 115-volt mains. The field current is 7 amp, and the armature resistance is 0.03 ohm. Determine stray-power loss of the machine at this particular value of flux and speed.

(ii) Assume that the foregoing generator is delivering 100 amp at 110 volts and 1,000 rpm. The field current is 7.0 amp. Determine: (a) stray power under this condition of load; (b) efficiency.

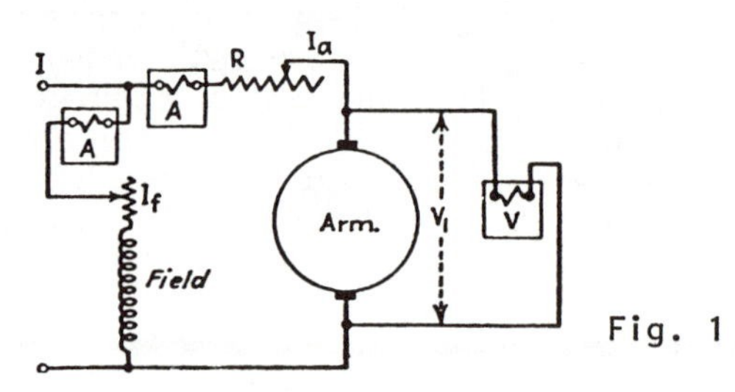

Fig. 1

Connections for stray-power measurement.

Solution:

(i) The armature current $I_a = 12 - 7 = 5$ amp.

The stray power S.P. $= 115 \times 5 - (5)^2\ 0.03$

$= 575 - 0.75 = 574$ watts.

Note that the armature $I_a{}^2R_a$ loss is practically negligible in this case.

It follows from

$$\phi = \frac{1}{K}\left(\frac{E}{S}\right) .$$

where K is a constant

ϕ = flux per pole in maxwells per sq. in.

S = speed.

E = induced emf,

that if E and S under load conditions be duplicated at no load the flux will be the same. Since stray power is a function of flux and speed, it also follows that the stray power under load and no-load conditions will be the same.

(ii) (a) The induced emf

$$E = 110 + (107 \times 0.03) = 113.2 \text{ volts.}$$

$$S = 1{,}000 \text{ rpm.}$$

To make the adjustments of E and S, the generator is run as a motor, connected as in Fig. 1. The rheostat R is first adjusted so that V_1 is equal to 113.2 volts, the small armature voltage drop under these conditions being negligible. The field rheostat is then adjusted to give a speed of 1,000 rpm. The machine is operating at the same value of speed and flux as it did under the given load. Therefore, the stray power is the same in the two cases. The current I_a is 4.8 amp, and V_1 = 113.2 volts. (This neglects the small drop in the armature, 4.8 × 0.03.) The stray power

$$S.P. = 113.2 \times 4.8 - (4.8)^2 0.03 = 542.7 \text{ watts.}$$

(b) For a generator,

$$\eta = \frac{VI}{VI + I_a^2R_a + I_c^2R_c + I_S^2R_S + S.P. + P_{SL}}$$

Output under load = 110 x 100 = 11,000 watts.

$I_a{}^2R_a$ = $(100 + 7)^2 0.03$ = 343.5 watts.

VI_f = 110 × 7 = 770.0 watts.

S.P. = 542.7 watts.

Stray-load loss, P_{SL}

= 0.01 × 11,000 = 110.0 watts.

Total losses = 1,766.2 watts.

$$= \frac{11{,}000}{11{,}000 + 1{,}766} = \frac{11{,}000}{12{,}766} = 0.862, \text{ or } 86.2\%.$$

• **PROBLEM 5-39**

Various losses are shown plotted as curves in Fig. 1. The results are taken from a test of a 5-h.p. shunt motor and are about right for any continuous-current machine of this capacity. The shunt-field loss is 97 watts at all loads; and the stray power is 320 watts at no load and practically the same at full load; the brush-contact resistance loss is 3 watts at no load and 80 watts at full load; the I^2R loss at no-load armature is less than one watt and at full load it is 242 watts. Calculate the efficiency.

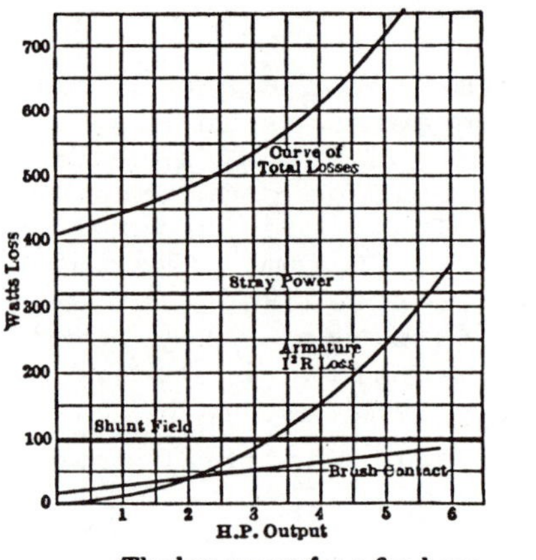

Fig. 1

—The loss curves for a five horse-power continuous-current motor.

Solution:

$$\text{Efficiency} = \frac{\text{Output}}{\text{Output + losses}}$$

At full load the output of the motor is equal to 746 × 5 = 3730 watts.

The input must be

$$3730 + (242 + 97 + 80 + 320) = 4469 \text{ watts.}$$

Hence,

$$\text{The full-load efficiency} = \frac{3730}{4469} = 83.6 \text{ percent.}$$

When the output = 4 h.p.,

$$\text{efficiency} = \frac{\text{output}}{\text{output + losses}} = \frac{2982}{2982 + 627}$$

$$= 82.7 \text{ percent.}$$

When the output = 3 h.p.,

$$\text{efficiency} = \frac{2238}{2238 + 550} = 80.3 \text{ percent.}$$

When the output = 2 h.p.,

$$\text{efficiency} = \frac{1492}{1492 + 495} = 75.1 \text{ percent.}$$

When the output = 1 h.p.,

$$\text{efficiency} = \frac{746}{746 + 455} = 62.1 \text{ percent.}$$

When the output = 0 h.p.,

$$\text{efficiency} = \frac{0}{430} = 0 \text{ percent.}$$

• PROBLEM 5-40

(a) What is the efficiency of the motor whose torque and speed curves are given in Fig. 1, at a speed of 540 r.p.m. and at 230 volts?

(b) Calculate the efficiency of the motor whose tractive effort curve is given in Fig. 2, at a speed of 14 miles per hour.

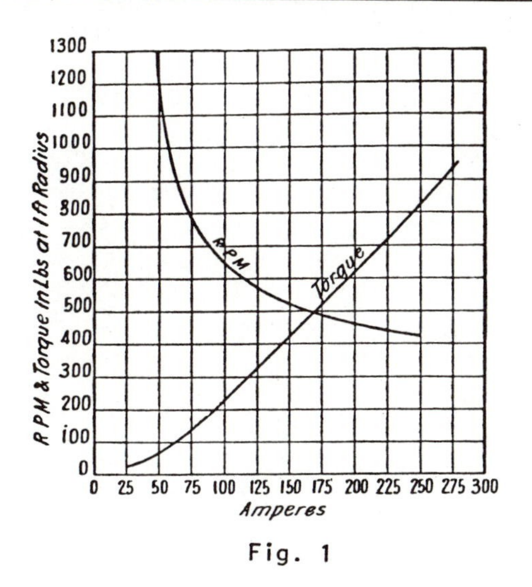

Fig. 1

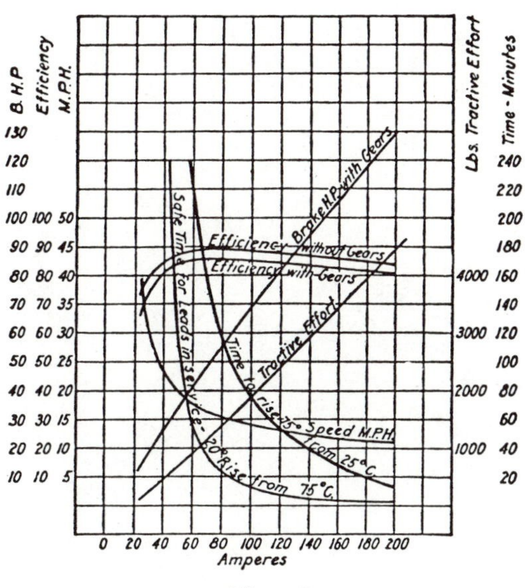

Fig. 2

Solution: (a) At a speed of 540 r.p.m. the current intake is about 140 amperes. The corresponding torque is 360 pound-feet. The efficiency at this speed is, therefore,

$$\eta = \frac{0.142 \quad \eta \; T}{\text{watts input}}$$

$$= \frac{0.142 \times 360 \times 540 \times 100}{140 \times 230}$$

$$= 86 \text{ percent.}$$

(b) At a speed of 14 miles per hour the tractive effort is 1,900 pounds. The current intake at this speed is about 100 amperes.

The output is 1.99 × 1,900 × 14 watts, and the input is 600 × 100 watts. Hence, efficiency is

$$\eta = \frac{1.99 \times \text{tractive effort} \times \text{miles per hour}}{\text{input in watts}}$$

$$\eta = \frac{1.99 \times 1{,}900 \times 14}{600 \times 100} = 88.3 \text{ percent.}$$

which checks very closely with that given in the figure for efficiency without gears.

CHAPTER 6

TRANSFORMERS

PRIMARY AND SECONDARY VOLTAGES AND CURRENTS, POWER FACTOR

• PROBLEM 6-1

The 2,300-volt primary winding of a 60-cycle transformer has 4,800 turns. Calculate: (a) the mutual flux ϕ_m; (b) the number of turns in the 230-volt secondary winding; (c) the maximum flux in the core of a 60-cycle transformer that has 1320 primary turns and 46 secondary turns is 3.76×10^6 maxwells. Also, calculate the primary and secondary induced voltages.

Solution: (a) $E = 4.44 f\, N\, \phi_m \times 10^{-8}$ volts.

$$\therefore \quad \phi_m = \frac{2{,}300 \times 10^8}{4.44 \times 60 \times 4{,}800} = 1.8 \times 10^5 \text{ maxwells.}$$

(b) $E_S = 4.44\, f\, N_S\, \phi_m \times 10^{-8}$ volts.

$$\therefore \quad N_S = \frac{230 \times 10^8}{4.44 \times 60 \times 1.8 \times 10^5} = 480 \text{ turns.}$$

(c) $E_P = 4.44\, f\, N_P\, \phi_m \times 10^{-8}$ volts

$$= 4.44 \times 60 \times 1{,}320 \times 3.76 \times 10^6 \times 10^{-8}$$

$$= 13{,}200 \text{ volts.}$$

$$E_S = 4.44 \; f \; N_S \; \phi_m \times 10^{-8} \text{ volts}$$

$$= 4.44 \times 60 \times 46 \times 3.76 \times 10^6 \times 10^{-8}$$

$$= 460 \text{ volts.}$$

• PROBLEM 6-2

The voltage, v = 100sin377t - 20sin1885t, is applied to a 200-turn transformer winding. Derive the equation for the flux in the core, neglecting leakage flux and winding resistance. Determine the rms values of the voltage and the flux.

Solution: From the following equation:

$$\int e_1 dt = -N_1 \int d\Phi,$$

$$\int e_1 dt = -N_1 \Phi$$

or

$$\Phi = -\frac{1}{N_1}\int e_1 dt$$

$$= \left(\frac{1}{200} - \frac{100}{377}\cos 377t + \frac{20}{1885}\cos 1885t\right)$$

$$= 0.00133\cos 377t - 5.3 \times 10^{-5}\cos 1885t$$

The rms values are

$$\Phi = \sqrt{\frac{(.00133)^2}{2} + \frac{(5.3 \times 10^{-5})^2}{2}} = 0.0094 \text{ Wb}$$

$$V = \sqrt{\frac{100^2}{2} + \frac{20^2}{2}} = 72.1 \text{ V}$$

• PROBLEM 6-3

A certain 10-kva 60-cycle transformer with primary voltage rating 2,400 volts, secondary 240 volts, has a core area of 12.25 sq in., and the length of mean flux path in the core is 23.5 in. The primary is wound with 1,100 turns and the secondary with 110 turns. Calculate the full-load and no-load currents.

Solution: The full-load values of primary and secondary currents are

$$I_1 = \frac{10,000}{2,400} = 4.17 \text{ amp.}$$

$$I_2 = \frac{10,000}{240} = 41.7 \text{ amp.}$$

The maximum flux in the core is

$$\Phi_m = \frac{10^8}{\sqrt{2}} \cdot \frac{E_g}{\pi f n} = \frac{2,400 \times 10^8}{4.44 \times 60 \times 1,100} = 819,000 \text{ maxwells}$$

and so the maximum flux density is

$$B_m = \frac{819,000}{12.25} = 66,700 \text{ maxwells/sq in.}$$

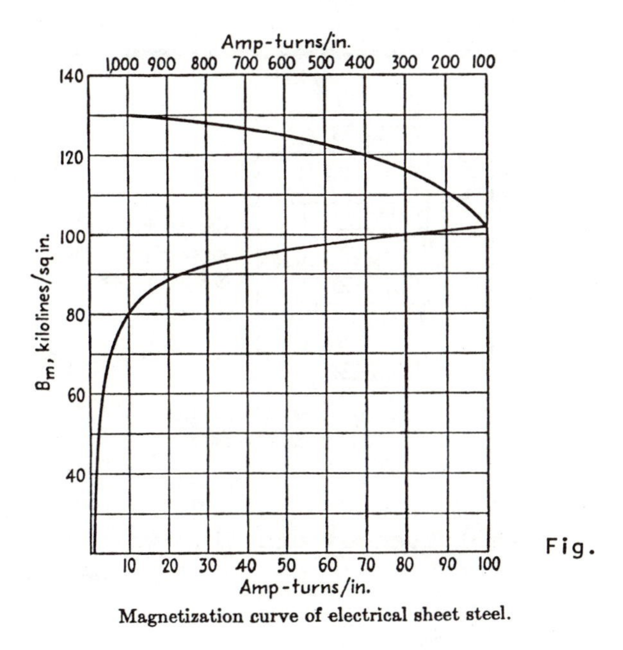

Fig.

Magnetization curve of electrical sheet steel.

Figure 1 shows a magnetization curve for the core material used. Corresponding to this value of B_m the required value of ampere-turns per inch is seen to be 4.8, and so the peak value of ampere-turns required is

$$n_1 I_{0m} = 23.5 \times 4.8 = 113 \text{ amp-turns}$$

If the exciting current is assumed to vary sinusoidally, the effective value of no-load current required is

$$I_0 = \frac{113}{1,100\sqrt{2}} = 0.072 \text{ amp}$$

which is less than 2 per cent of the rated full-load value. It should be mentioned here that the assumption of sinusoidal wave shape for this current may not be correct, but the effective value as computed in this fashion is sufficiently accurate.

• PROBLEM 6-4

Fig. 1 shows the core dimensions and stacking factor of a small experimental transformer. Fig. 3a shows the sinusoidal curve of induced voltage e, replotted from the oscillogram of Fig. 2, and the corresponding sinusoidal flux wave ϕ, lagging the induced voltage by 90 degrees. Calculate the amplitude of the flux wave and the maximum flux density in the iron.

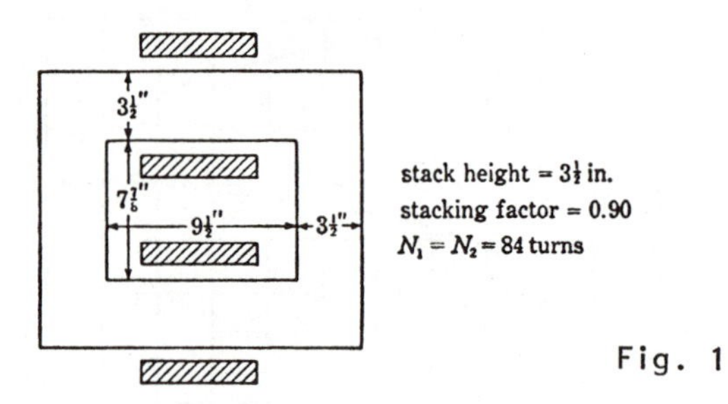

Fig. 1

Core of a small experimental transformer.

Solution: The amplitude of the flux wave is

$$\phi_{max} = \frac{E}{4.44fN}$$

$$= \frac{200}{4.44 \times 60 \times 84}$$

$$= 0.00895 \text{ weber or } 895,000 \text{ maxwells.}$$

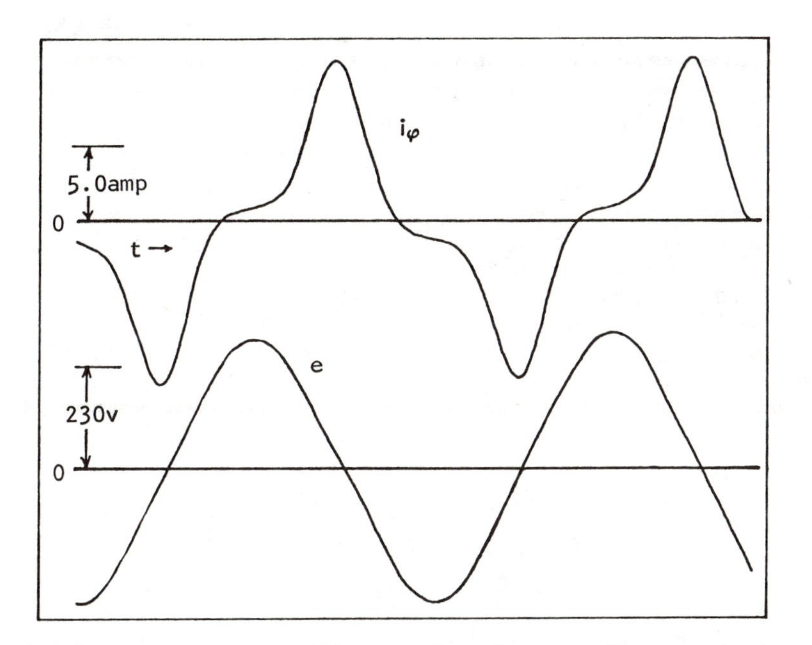

Fig. 2

Exciting-current oscillogram for the transformer of Fig. 1 with rms induced voltage of 200 volts at a frequency of 60 cycles per second.

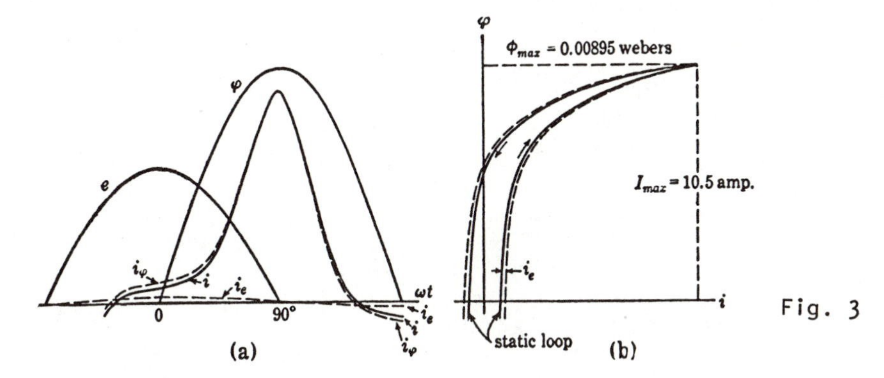

Fig. 3

Graphical construction for determination of the exciting current of the transformer of Fig. 1.

From the core dimensions and the stacking factor of 0.90, the maximum flux density in the iron is

$$B_{max} = \frac{\phi_{max}}{\text{net area}}$$

$$= \frac{895{,}000}{3.50 \times 3.50 \times 0.90} = 81{,}200 \text{ lines/sq in.}$$

$$= \frac{81{,}200}{6.45} = 12{,}600 \text{ gausses, or } 1.26 \text{ webers/sq m.}$$

• PROBLEM 6-5

(i) A transformer with 200 turns on the H winding is to be wound to step the voltage down from 240 to 120 volts. Find the number of turns T_x on the X winding.

(ii) A transformer supplies a load with 30 amp at 240 volts. If the primary voltage is 2,400 volts, find (a) the secondary volt-amperes, (b) the primary volt-amperes, and (c) the primary current.

(iii) What is the rated kilowatt output of a 5-kva 2,400/120-volt transformer at (a) 100 per cent, (b) 80 per cent, and (c) 30 per cent power factor? (d) What is the rated current output?

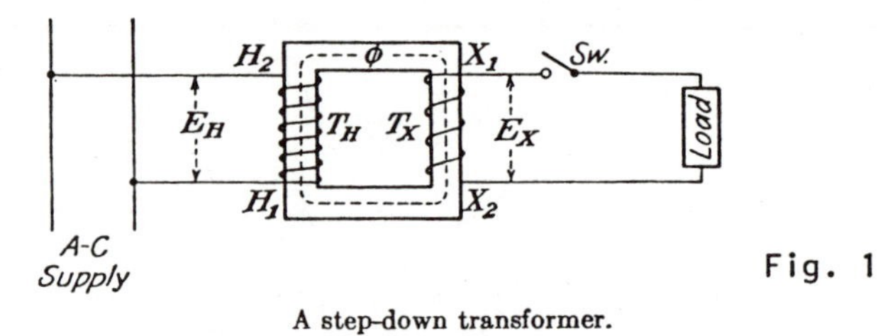

Fig. 1

A step-down transformer.

Solution: (i) From Fig. 1,

$$\frac{E_H}{E_X} = \frac{T_H}{T_X}$$

$$\frac{240}{120} = \frac{200}{T_X}$$

$$\therefore \quad 240T_X = 24,000$$

$$\therefore \quad T_X = 100.$$

(ii) (a) $E_X I_X = 30 \times 240 = 7,200$ va.

(b) $E_H I_H = E_X I_X = 7,200$ va.

(c) $E_H I_H = 7,200$ va

$$2,400 I_H = 7,200 \text{ va}$$

$$\therefore \quad I_H = \frac{7,200}{2,400} = 3 \text{ amp.}$$

(iii) (a) $P = \text{kva} \times \text{power factor} = 5 \times 1.0 = 5$ kw.

(b) $P = 5 \times 0.8 = 4$ kw.

(c) $P = 5 \times 0.3 = 1.5$ kw.

(d) $I = \frac{va}{E} = \frac{5{,}000}{120} = 41.7$ amp.

• PROBLEM 6-6

A 10-kva single-phase transformer designed for 2,000/400 volts has the following constants: $R_1 = 5.5$; $R_2 = 0.2$; $X_1 = 12$; $X_2 = 0.45$. Calculate the approximate value of the secondary terminal voltage at full load, 80 percent power factor (lagging), when the primary supply voltage is 2,000 volts.

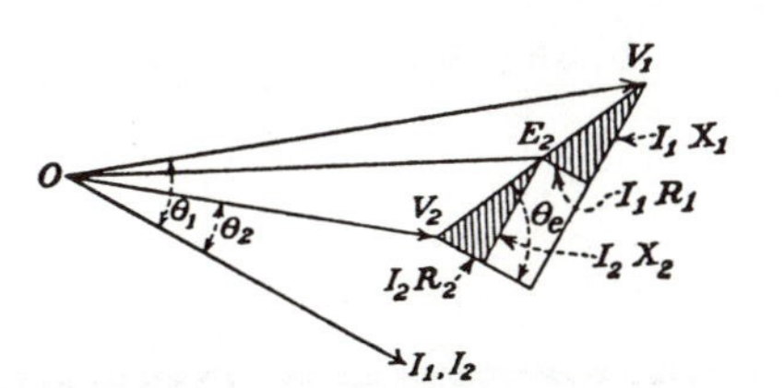

Fig. 1

Simplified phasor diagram of approximate equivalent circuit.

Solution: Referring to Fig. 1, let all quantities be expressed in terms of the primary, so that

$$I_1 = I_2 = \frac{10{,}000}{2{,}000} = 5 \text{ amp.}$$

$$a = \frac{2{,}000}{400} = 5.$$

$$R_1 + R_2 = 5.5 + \left(5^2 \times 0.2\right) = 10.5 \text{ ohms}$$

$$X_1 + X_2 = 12 + \left(5^2 \times 0.45\right) = 23.25 \text{ ohms}$$

Let the current $I_1 = I_2$ in Fig. 1 be taken as the axis of reference, so that in complex notation

$$\vec{I}_1 = \vec{I}_2 = 5 + j0.$$

It follows that

$$\vec{V}_1 = \vec{V}_2 + \vec{I}_2[(R_1 + R_2) + j(X_1 + X_2)].$$

and since by the given conditions

$$\vec{V}_2 = (0.8 + j0.6)V_2,$$

$$\vec{V}_1 = (0.8 + j0.6)V_2 + 5(10.5 + j23.25)$$

or $$\vec{V}_1 = (0.8V_2 + 52.5) + j(0.6V_2 + 116.25)$$

from which,

$$V_1^2 = 2{,}000^2 = (0.8V_2 + 52.5)^2 + (0.6V_2 + 116.25)^2$$

and $V_2 = 1{,}887.25$ volts in terms of the primary and its actual value is

$$V_2 = \frac{1{,}887.25}{5} = 377.45 \text{ volts}$$

• PROBLEM 6-7

A 10-kVA, 2400-240-V, single-phase transformer has the following resistances and leakage reactances. Find the primary voltage required to produce 240 V at the secondary terminals at full load, when the load power factor is

(a) 0.8 power factor lagging
(b) 0.8 power factor, leading.

$r_1 = 3.00\Omega$ $\qquad r_2 = 0.0300\Omega$

$x_1 = 15.00\Omega$ $\qquad x_2 = 0.150\Omega.$

Solution:

$$\vec{Z}_{eq2} = \frac{r_1}{a^2} + r_2 + j\left(\frac{x_1}{a^2} + x_2\right)$$

$$a = \frac{V_{1B}}{V_{2B}} = \frac{2400}{240} = 10$$

$$\vec{Z}_{eq2} = \left(\frac{3.00}{100} + 0.0300\right) + j\left(\frac{15.00}{100} + 0.150\right)$$

$$= 0.0600 + j0.300$$

$$= 0.3059\angle 78.69\ \Omega$$

(a) 0.8 power factor, lagging:

1. $I_{2B} = \frac{S_B}{V_{2B}} = \frac{10{,}000}{240} = 41.7\ A$

With V_2 as a reference, the full-load secondary current is

$$\vec{I}_{2fl} = 41.7\angle -\cos^{-1}0.8$$

$$= 41.7\angle -36.87°\ A$$

2. $\frac{\vec{V}_1}{a} = \vec{V}_2 + \vec{I}_2\vec{Z}_{eq2} = 240\angle 0° + 41.7 \times 0.3059\angle 78.69° + 36.87°$

$$= 240 + j0 + (9.506 + j8.506)$$

$$= 249.506 + j8.506$$

$$= 249.65\angle 1.952°.$$

3. $|\vec{V}_1| = a|\vec{V}_2| = 2496.5\ V.$

(b) 0.8 power factor, leading

1. $\vec{I}_{2fl} = 41.7\angle +36.87°\ A$

2. $\frac{\vec{V}_1}{a} = 240\angle 0° + 41.7 \times 0.3059\angle 78.69° + 36.87°$

$$= 240 + j0 + 12.76\angle 115.56$$

$$= 240 + j0 + (-5.505 + j11.51)$$

$$= 234.50 + j11.51 = 243.78\angle 2.81°.$$

3. $|\vec{V}_1| = a|\vec{V}_2| = 2347.8\ V.$

• PROBLEM 6-8

A certain power transformer is connected between a transmission line and a load. The secondary terminal voltage is 707 sin 377t and the load current is i_2 = 141.4 sin (377t - 30°). The primary winding has 300 turns and a resistance of 2.00Ω. The secondary winding has 30 turns and a resistance of 0.0200Ω. The leakage inductance of the primary is 0.0300H while that of the secondary is 3.00×10^{-4}H. The exciting current of this transformer is 0.707 sin (377t - 80°). Find the turns ratio a, the primary and secondary induced rms voltages E_1 and E_2, and the primary current and terminal voltages. Compare the actual voltage and current ratios with the turns ratio.

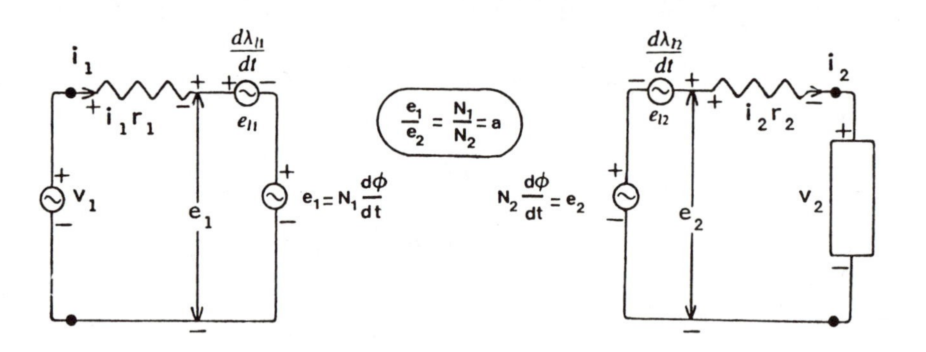

Fig. 1

Primary and secondary transformer voltages.

Solution: The rms phasor notation will be used to avoid involvement in trigonometric identities. Then the secondary voltage and current are:

$$\vec{V}_2 = \frac{707}{\sqrt{2}}\angle 0° = 500\angle 0° = 500 + j0 \text{ V}$$

$$\vec{I}_2 = \frac{141.4}{\sqrt{2}}\angle -30° = 100\angle -30° = 86.6 - j50 \text{ A}$$

The secondary $\vec{I}_2 r_2$ drop is

$$0.0200(86.6 - j50) = 1.732 - j1.00 \text{ V}$$

It has been noted that positive secondary current sets up a negative magnetomotive force. The direction of the secondary leakage flux is in opposition to the core flux. Therefore, the secondary leakage flux linkage is

$$\lambda_{\ell 2} = L_{\ell 2}(-i_2)$$

$$\lambda_{\ell 2} = 3.00 \times 10^{-4} \times (-i_2)$$

$$= -0.0424 \sin (377t - 30°) \text{ Wb turns}$$

The secondary leakage-flux voltage is

$$e_{\ell 2} = \frac{d\lambda_{\ell 2}}{dt} = -0.0424 \times 377 \cos (377t - 30°)$$

$$= -15.98 \cos (377t - 30°)$$

$$= -15.98 \sin (377t - 30° + 90°) \text{ V}$$

Then

$$\vec{E}_{\ell 2} = \frac{-15.98}{\sqrt{2}}\underline{/90° - 30°} = -j11.31\underline{/-30°}$$

$$= -11.31\underline{/+60°} \text{ V}$$

Pause to see what has happened. Note that

$$-\vec{E}_{\ell 2} = 3.00 \times 10^{-4} \times 377 \times jI_2 = j\omega 3.00 \times 10^{-4} I_2$$

where $\omega = 377$ rad/s ($f = ?$) and 3.00×10^{-4} is the secondary leakage inductance, $L_{\ell 2}$. Then $\omega L_{\ell 2} = x_2$ is the secondary leakage reactance. On this basis

$$x_2 = \omega L_{\ell 2} = 377 \times 3.00 \times 10^{-4} = 0.1131\Omega$$

$$-\vec{E}_{\ell 2} = \vec{I}_2 jx_2 = (100\underline{/-30}) \times 0.1131\underline{/90°}$$

$$= 11.31\underline{/+60°} = 5.655 + j9.79 \text{ V}$$

Applying Kirchhoff's Voltage Law to the secondary (Fig. 1),

$$\vec{E}_2 = \vec{V}_2 + \vec{I}_2 r_2 - \vec{E}_{\ell 2}$$

$$\equiv \vec{V}_2 + \vec{I}_2(r_2 + jx_2)$$

$$= 500 + j0 + (1.732 - j1.00) + (5.655 + j9.79)$$

$$= 507.4 + j8.79 = 507.46\angle 0.992° \text{ V}$$

Now

$$a = \frac{N_1}{N_2} = \frac{300}{30} = 10$$

$$\vec{E}_1 = a\vec{E}_2 = 5074.6\angle 0.992°$$

$$= 5074.6 + j87.9 \text{ V}$$

$$\vec{I}_1 = \frac{\vec{I}_2}{a} + \vec{I}_{ex} \ .$$

$$\vec{I}_{ex} = \frac{0.707}{\sqrt{2}}\angle -80° = 0.5\angle -80°$$

$$= 0.0868 - j0.492 \text{ A.}$$

$$\vec{I}_1 = \frac{86.6 - j50}{10} + 0.0868 - j0.492$$

$$= 8.75 - j5.49 = 10.33\angle -32.1° \text{ A.}$$

$$\text{Actual current ratio} = \frac{I_2}{I_1} = \frac{100}{10.33} = 9.68.$$

Compare with $a = 10$.

Now

$$\vec{V}_1 = \vec{E}_1 + \vec{I}_1 r_1 + \vec{E}_{\ell 1}.$$

$$\vec{E}_{\ell 1} = j\omega L_{\ell 1}\vec{I}_1$$

$$= j \times 377 \times 0.300 \times 10.33\angle -32.1°$$

$$= 116.83\angle 90° - 32.1°$$

$$= 62.1 + j99.0 \text{ V.}$$

$$\vec{I}_1 r_1 = (8.75 - j5.49) \times 2.00$$

$$= 17.5 - j11.0 \text{ V.}$$

$$\therefore \quad \vec{V}_1 = (5074.6 + j87.9) + (17.5 - j11.0) + (62.1 + j99.0)$$

$$= 5154.2 + j175.9 = 5157.2 \angle 1.95° \text{ V.}$$

The actual voltage ratio is

$$\frac{V_1}{V_2} = \frac{5157.2}{500} = 10.31.$$

• PROBLEM 6-9

The open-secondary test for a transformer showed an exciting current, I_E, of 5.2 amperes and an input of 185 watts when 110 volts was impressed upon the low-tension side. If the test had been made by applying 2200 volts to the high-tension side of an exciting current of 5.2/20 = 0.26 ampere would have been needed, with the same power unit. Determine the power-factor of the transformer.

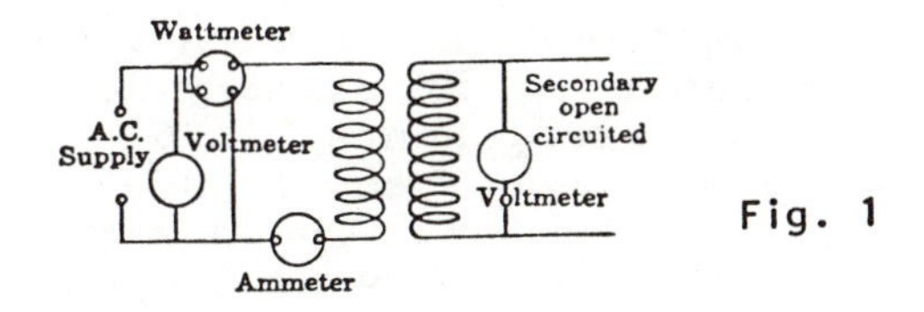

Fig. 1

Proper connections for measuring the core loss of a transformer.

Solution: The components I_M and I_H are determined from the no-load, open-secondary test (Fig. 1).

The component of current supplying the iron losses is then

$$I_H = \frac{P}{V} = 185/2200 = 0.0842 \text{ ampere,}$$

and the magnetizing component of current is

$$I_M = \sqrt{I_E^2 - I_H^2} = \sqrt{0.26^2 - 0.0842^2} = 0.246 \text{ ampere.}$$

The no-load power factor is

$$\cos\phi = \frac{P}{VI} = \frac{185}{2200 \times 0.26} = \frac{185}{572} = 0.323.$$

Full-load secondary current is 25,000/110 = 227.272 amperes, and the load component in the primary is 227.272/20 = 11.3636 amperes. With a secondary power factor of unity, the primary load component is assumed in phase with the voltage, an assumption with negligible error as the IR and IX drops of the transformer are small. Similarly I_H is taken in phase with E, and I_M as 90° behind E_1.

The total primary current, I_1, is then obtained by combining its vertical and horizontal components. Thus

$$I_1 = \sqrt{(I_2' + I_H)^2 + I_M^2}$$

$$= \sqrt{(11.3636 + 0.0842)^2 + (0.246)^2}$$

$$= \sqrt{11.4478^2 + 0.246^2} = 11.4505 \text{ amperes.}$$

The primary power factor is $\cos\phi_1 = \frac{11.4478}{11.4505} = 0.9998.$

To calculate the power factor with a secondary power factor of 0.8 lagging, the primary load component is broken up into its components 11.3636 × 0.8 = 9.0809 amperes in phase with E_1, and 11.3636 × 0.6 = 6.8182 amperes lagging E_1 by 90°. These are combined with the corresponding components of the exciting current.

The primary current is then

$$I_1 = \sqrt{(9.0809 + 0.0842)^2 + (6.8182 + 0.246)^2}$$

$$= \sqrt{9.1651^2 + 7.0642^2} = 11.5716 \text{ amperes,}$$

and the primary power factor is

$$\cos\phi_1 = \frac{9.1651}{11.5716} = 0.7923.$$

The last two calculations were carried out much more

accurately than was necessary with the idea of showing that at full load the primary power factor is practically the same as that of the secondary.

The half-load primary currents, for secondary power factors of unity and 0.8 lagging, respectively, are

$$I_1 = \sqrt{(5.6818 + 0.0842)^2 + (0.246)^2}$$

$$= \sqrt{5.7660^2 + 0.246^2} = 5.7712 \text{ A.}$$

$$I_1 = \sqrt{(5.6818 \times 0.8 + 0.0842)^2 + (5.6818 \times 0.6 + 0.246)^2}$$

$$= \sqrt{4.6296^2 + 3.6551^2} = 5.8985 \text{ A.}$$

The respective power factors are

$$\cos\phi_1 = \frac{5.7660}{5.7712} = 0.9994.$$

$$\cos\phi_1 = \frac{4.6296}{5.8985} = 0.7848.$$

• PROBLEM 6-10

The low-voltage secondary winding of a 100-kva 60 ∿ 12,000 : 2,400-volt distribution transformer is supplying its rated kilovolt-amperes at rated secondary voltage to an inductive load of 0.80 power factor. Determine the primary current, primary voltage, and primary power factor. The resistances and leakage reactances of the windings are

$R_1 = 7.0$ ohms $\qquad R_2 = 0.30$ ohm

$X_{\ell 1} = 19.0$ ohms $\qquad X_{\ell 2} = 0.75$ ohm

The tabulated results of an open-circuit test, taken on the low-voltage side, are

Applied voltage in volts	Exciting current in amperes	Core loss in watts
2,400	1.50	940
2,500	1.67	1,020
2,600	1.87	1,110

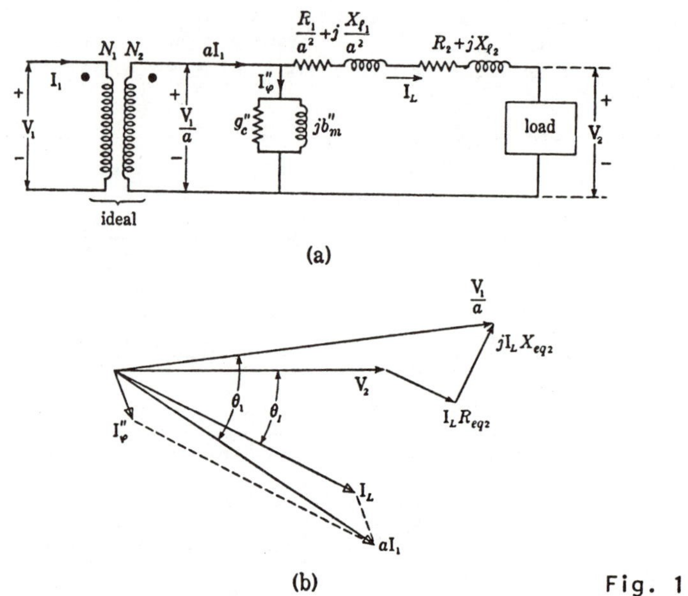

Fig. 1

A simplified equivalent circuit and vector diagram.

Solution: The equivalent resistance and equivalent reactance referred to the secondary are

$$R_{eq2} = \frac{R_1}{a^2} + R_2 = 0.28 + 0.30 = 0.58 \text{ ohms.}$$

$$X_{eq2} = \frac{X_{\ell 1}}{a^2} + X_{\ell 2} = 0.76 + 0.75 = 1.51 \text{ ohms.}$$

The secondary terminal voltage being used as the reference vector, the vector secondary current is,

$$\vec{I}_L = 33.4 - j25.0 \text{ amp.}$$

The primary terminal voltage referred to the secondary is determined by computation of the vector equivalent-resistance and equivalent-reactance voltage drops referred to the secondary and addition of them to the secondary terminal voltage; thus

$$\vec{I}_L R_{eq2} = (33.4 - j25.0)0.58 = 19.4 - j14.5 \text{ v.}$$

$$j\vec{I}_L X_{eq2} = j(33.4 - j25.0)1.51 = 37.8 + j50.5 \text{ v.}$$

$$\vec{V}_2 \qquad\qquad = 2,400.0 + j\,0$$

$$\frac{\vec{V}_1}{a} = \text{vector sum} = 2{,}457.2 + j36.0 \text{ v.}$$

$$= 2{,}458\underline{/0.8°}.$$

The primary terminal voltage is

$$V_1 = 5.00 \times 2{,}458 = 12{,}290 \text{ v.}$$

To calculate the primary current and power factor, the exciting current must be determined. According to the equivalent circuit of Fig. 1a, the exciting current is taken as the no-load current at the primary terminal voltage. Thus the exciting current referred to the secondary is the no-load current at 2,458 volts. By interpolation from the open-circuit data,

$$I''_\phi = 1.60 \text{ amp.}$$

$$P_c = 987 \text{ w.}$$

The no-load power factor is

$$\cos\theta_{n\ell} = \frac{987}{2{,}458 \times 1.60} = 0.251.$$

That is, the exciting current lags the applied voltage by an angle $\theta_{n\ell}$, or 75.5 degrees. Since the angle of $\vec{V}_1$ is +0.8 degree, the vector expression for $\vec{I}''_\phi$ is

$$\vec{I}''_\phi = 1.60\underline{/-74.7°} = 0.42 - j\,1.54 \text{ amp.}$$

$$a\vec{I}'_L = \vec{I}_L = 33.4 - j25.0$$

$$a\vec{I}_1 = \text{sum} = 33.8 - j26.5 \text{ amp.}$$

$$= 43.0\underline{/-38.1°} \text{ amp.}$$

$$\vec{I}_1 = \frac{a\vec{I}_1}{a} = \frac{43.0}{5.0}\underline{/-38.1°}$$

$$= 8.60\underline{/-38.1°} \text{ amp.}$$

The primary current lags the primary terminal voltage by

$$\theta_1 = 38.1 + 0.8 = 38.9°,$$

and therefore the primary power factor is

$$\cos\theta_1 = 0.778, \text{ lagging current.}$$

These results compare very well with the results of the "exact" analysis.

• PROBLEM 6-11

The low-voltage secondary winding of a 100-kva 60 ~ 12,000 : 2,400-volt distribution transformer is supplying its rated kilovolt-amperes at rated secondary voltage to an inductive load of 0.80 power factor. Determine the primary current, primary voltage, and primary power factor using exact vector relations.

The resistances and leakage reactances of the windings are

$$R_1 = 7.0 \text{ ohms} \qquad R_2 = 0.30 \text{ ohm}$$

$$X_{\ell 1} = 19.0 \text{ ohms} \qquad X_{\ell 2} = 0.75 \text{ ohm}$$

The tabulated results of an open-circuit test, taken on the low-voltage side, are

Applied voltage in volts	Exciting current in amperes	Core loss in watts
2,400	1.50	940
2,500	1.67	1,020
2,600	1.87	1,110

Solution: Since the excitation data are given in terms of the low-voltage secondary side it is convenient to refer primary quantities to the secondary.

$$\text{Ratio of transformation } a \equiv \frac{N_1}{N_2} = \frac{12,000}{2,400} = 5.00.$$

The primary constants referred to the secondary are

$$\frac{R_1}{a^2} = \frac{7.0}{25.0} = 0.28 \text{ ohm}$$

$$\frac{X_{\ell 1}}{a^2} = \frac{19.0}{25.0} = 0.76 \text{ ohm.}$$

Note that the primary resistance and leakage reactance referred to the secondary are of the same order of magnitude as the secondary resistance and leakage reactance, as is usually true.

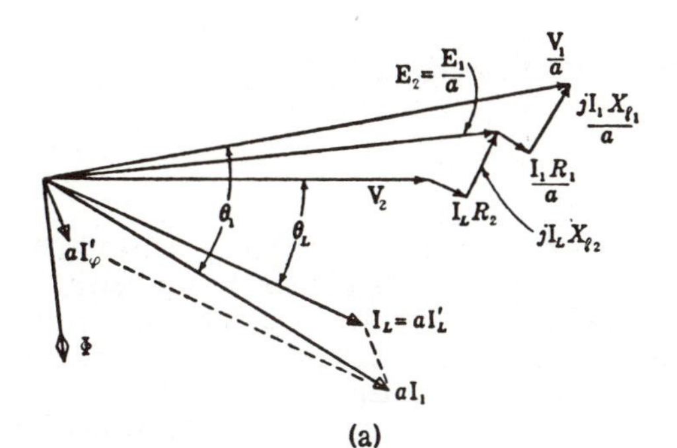

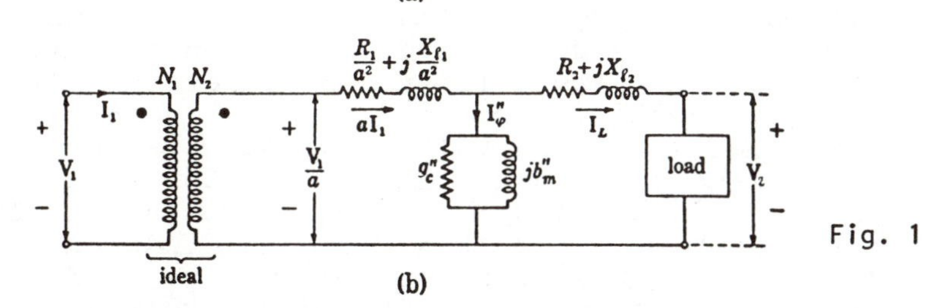

Fig. 1

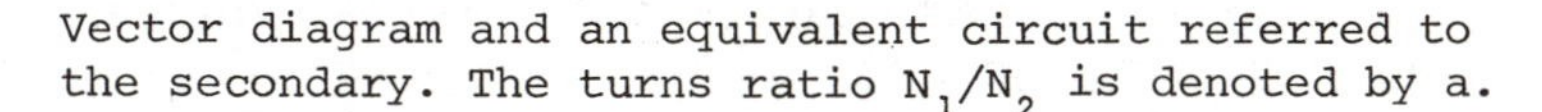

Vector diagram and an equivalent circuit referred to the secondary. The turns ratio N_1/N_2 is denoted by a.

The vector diagram is given in Fig. 1a. At rated output the load current is

$$I_L = \frac{100{,}000}{2{,}400} = 41.7 \text{ amp.}$$

If the secondary terminal voltage is chosen as the reference vector, the vector expression for the load current is

$$\vec{I}_L = 41.7\underline{/\cos^{-1}0.80}\text{, lagging} = 41.7\underline{/-36.9°} \text{ amp.}$$

$$= 41.7\ (0.80 - j0.60) = 33.4 - j25.0 \text{ amp.}$$

The secondary resistance drop is

$$\vec{I}_L R_2 = (33.4 - j25.0)0.30 = 10.0 - j7.5 \text{ v.}$$

The secondary leakage reactance drop is

$$j\vec{I}_L x_{\ell 2} = j(33.4 - j25.0)0.75 = \quad 18.8 + j25.0$$

The secondary terminal voltage is

$$\vec{V}_2 \qquad = 2{,}400.0 + j\ 0$$

$$\vec{E}_2 = \text{vector sum} \qquad = 2{,}428.8 + j17.5 \text{ v.}$$

$$= 2{,}429\underline{/0.4^\circ} \text{ v.}$$

In order to determine the primary current, it is next necessary to determine the exciting current, which then is added to the load component of the primary current. The exciting current referred to the secondary equals the no-load current measured on the low-voltage side for a no-load condition at which the no-load induced voltage equals 2,429 volts--the value of the secondary induced voltage under load. The open-circuit test data show that, to a first approximation, this exciting current is about 1.5 amperes. Since the leakage impedance of the low-voltage winding is about 0.8 ohm, the leakage-impedance voltage drop for this no-load condition is but 1.5 × 0.8 or 1.2 volts, and therefore is negligible. Thus at no load the induced voltage very nearly equals the applied voltage, and therefore the exciting current for an induced voltage of 2,429 volts under load equals the no-load current for very nearly the same applied voltage--say 2,430 volts. By interpolation from the open-circuit data at 2,430 volts

$$I''_\phi = 1.55 \text{ amp.}$$

$$P_c = 963 \text{ w.}$$

The no-load power factor is

$$\frac{963}{2{,}430 \times 1.55} = 0.256.$$

That is, the exciting current lags the induced voltage by $\cos^{-1}0.256$, or 75.2 degrees. Since $\vec{E}_2$ is at an angle of +0.4 degrees, the vector expression for the exciting current is

$$\vec{I}''_\phi = a\vec{I}'_\phi = 1.55\underline{/-74.8^\circ} \quad = \quad 0.41 - j\ 1.50 \text{ amp.}$$

But the load current is

$$\vec{I}_L = a\vec{I}'_L \qquad = \qquad \underline{33.4 - j25.0.}$$

The vector sum is

$$a\vec{I}_1 \qquad = \qquad 33.8 - j26.5 \text{ amp.}$$

$$= \qquad 43.0\underline{/-38.1°} \text{ amp.}$$

The primary terminal voltage now can be determined by adding the primary resistance and leakage-reactance voltage drops to the induced voltage. Referred to the secondary, these voltage drops are

$$a\vec{I}_1\frac{R_1}{a^2} = (33.8 - j26.5)0.28 = \quad 9.5 - j\ 7.4 \text{ v.}$$

$$ja\vec{I}_1\frac{X_{\ell 1}}{a^2} = j(33.8 - j26.5)0.76 = 20.2 + j25.7$$

and

$$\frac{\vec{E}_1}{a} = \vec{E}_2 \qquad = \underline{2{,}428.8 + j17.5}$$

$$\frac{\vec{V}_1}{a} = \text{vector sum} \qquad = 2{,}458.5 + j35.8 \text{ v.}$$

$$= 2{,}459\underline{/0.8°} \text{ v.}$$

Hence the results are

$$I_1 = \frac{aI_1}{a} = \frac{43.0}{5.00} = 8.60 \text{ amp.}$$

$$V_1 = a\ \frac{V_1}{a} = 5.00 \times 2{,}459 = 12{,}290 \text{ v.}$$

The primary current lags the primary terminal voltage by

$$\theta_1 = 38.1 + 0.8 = 38.9°.$$

Hence the primary power factor is

$$\cos\theta_1 = 0.778, \text{ lagging current.}$$

Two transformers are V-connected in accordance with Fig. 1. They supply a balanced three-phase load of 100 kva at a line voltage of 220 volts and a lagging power factor of 0.8. The phase sequence is a-b-c. (This information is pertinent because the transformer circuit, in contrast to the load, is unbalanced.) Find all secondary voltage and current phasors.

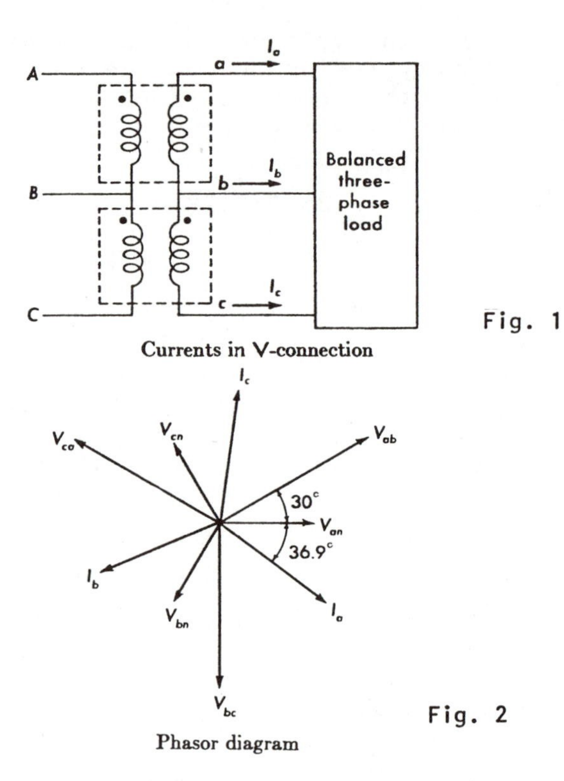

Currents in V-connection

Fig. 1

Phasor diagram

Fig. 2

Solution: In most three-phase circuits involving balanced loads, the most convenient choice of the axis of reference is one of the voltages to neutral. (It is always permissible to assume that the load is wye-connected; if it is actually delta-connected, the delta can be replaced by an equivalent wye.) Then

$$\vec{V}_{an} = V\underline{/0°} = 127\underline{/0°}\ , \quad \vec{V}_{bn} = V\underline{/\pm 120°} = 127\underline{/-120°},$$

$$\vec{V}_{cn} = V\underline{/\pm 120°} = 127\underline{/120°}$$

The line voltages are

$$\vec{V}_{ab} = \sqrt{3}\, V\underline{/\pm 30°} = 220\underline{/30°}\ ,$$

$$\vec{V}_{bc} = \sqrt{3}\, V\underline{/\pm 90°} = 220\underline{/-90°}\ ,$$

$$\vec{V}_{ca} = \sqrt{3}\, V\underline{/\pm 150°} = 220\underline{/150°}.$$

The line current magnitude is obtained from the following equation:

$$|\vec{I}_{L2}| = \frac{1{,}000P_a}{\sqrt{3}|\vec{V}_{L2}|}$$

$|\vec{I}_{L2}| = \dfrac{100{,}000}{\sqrt{3} \times 220} = 264$ amp. Since the power factor of a balanced load is the cosine of the angle between line current and voltage to neutral, the line current phasors are:

$$\vec{I}_a = I\underline{/-\theta} = 264\underline{/-36.9°}\ ,\quad \vec{I}_b = I\underline{/-\theta \pm 120°} = 264\underline{/-156.9°},$$

$$\vec{I}_c = I\underline{/-\theta \pm 120°} = 264\underline{/83.1°}.$$

All these phasors are shown in Fig. 2.

• PROBLEM 6-13

A 2400-volt, 100-kva transformer has an equivalent impedance of 0.708 + j0.92 ohms. Determine the base values for the per-unit system, the per-unit equivalent impedance, and the equivalent IZ voltage at one-half rated current.

Solution: The base values are:

$$E_{base} = 2400 \text{ volts}$$

$$I_{base} = \frac{VA}{E_{base}} = \frac{100{,}000}{2400} = 41.7 \text{ amperes.}$$

$$Z_{base} = \frac{E_{base}}{I_{base}} = \frac{2400}{41.7} = 57.55 \text{ ohms.}$$

The per-unit equivalent impedance:

$$\vec{Z}_{\text{per unit}} = \frac{\vec{Z}}{Z_{\text{base}}} = \frac{0.708 + j0.92}{57.55}$$

$$= 0.0123 + j0.016 \text{ per unit.}$$

At one-half rated current:

$$\vec{I}\vec{Z}_{\text{per unit}} = 0.5(0.0123 + j0.016)$$

$$= 0.00615 + j0.008$$

$$= 0.0101 \text{ per unit.}$$

Therefore, the IZ voltage is 1.01 percent of the rated voltage, or

$$0.0101 \times 2400 = 24.24 \text{ volts.}$$

• PROBLEM 6-14

A 100-kva, 12,000:2,400-volt, 60 ∿ distribution transformer is connected to a single-phase bus of voltage V_B (nominally 12,000 volts) through a high-voltage line of impedance 15.1 + j64.0 ohms, and supplies power to an inductive load through a low-voltage line of impedance 1.27 + j1.76 ohms, as shown in Fig. 1.

Determine the bus voltage V_B necessary to maintain 2,300 volts at the load when the load takes the rated current of the transformer at 0.8 power factor.

The data obtained from a short-circuit test in which the measurements were made on the low-voltage side of the transformer are:

$$V = 123 \text{ volts}$$

$$I = 41.7 \text{ amperes}$$

$$P = 1.05 \text{ kw.}$$

Solution: This problem is conveniently solved by use of per unit quantities. Accordingly, the first step is to express all circuit constants in per unit on the rating of the transformer as a base. If subscripts H and X indicate the high- and low-voltage circuits,

$$\text{Base } V_H = 12{,}000 \text{ v.} \qquad \text{Base } V_X = 2{,}400 \text{ v.}$$

$$\text{Base } I_H = \frac{100{,}000}{12{,}000} \qquad \text{Base } I_X = \frac{100{,}000}{2{,}400}$$

$$= 8.34 \text{ amp.} \qquad = 41.7 \text{ amp.}$$

$$\text{Base } Z_H = \frac{12{,}000}{8.34} \qquad \text{Base } Z_X = \frac{2{,}400}{41.7}$$

$$= 1{,}440 \text{ ohms.} \qquad = 57.5 \text{ ohms.}$$

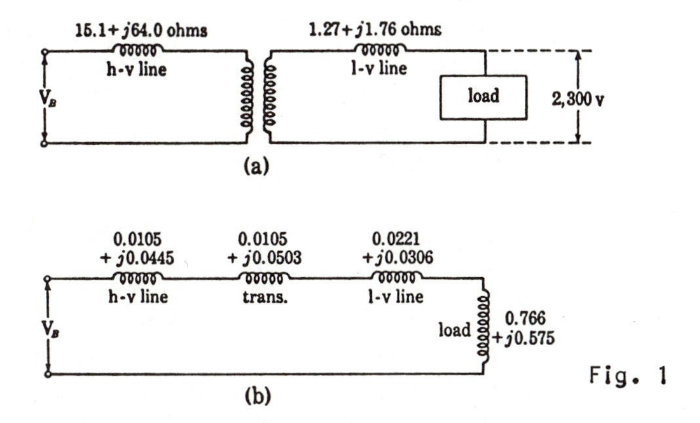

Fig. 1

(a) Circuit diagram. (b) Equivalent circuit on a per unit basis.

From the data and the values of the base impedances, the per unit value of the high-voltage line impedance is

$$\frac{15.1 + j64.0}{1{,}440} = 0.0105 + j0.0445 \text{ per unit impedance,}$$

and of the low-voltage line impedance is

$$\frac{1.27 + j1.76}{57.5} = 0.0221 + j0.0306 \text{ per unit impedance.}$$

Converting the short-circuit data of the transformer to per unit:

$$V = \frac{123}{2{,}400} = 0.0513 \text{ per unit voltage}$$

$$I = 1.00 \text{ per unit current}$$

$$P = \frac{1.05}{100} = 0.0105 \text{ per unit power.}$$

Since the voltage is the impedance drop at normal current, its per unit value equals the per unit equivalent impedance, and also, since the power is the copper loss at normal current, its per unit value equals the per unit equivalent resistance. Hence the per unit equivalent reactance of the transformer is

$$\sqrt{0.0513^2 - 0.0105^2} = 0.0503 \text{ per unit impedance.}$$

The per unit equivalent impedance of the transformer is

$$\vec{Z}_{eq} = 0.0105 + j0.0503 \text{ per unit impedance.}$$

The per unit load voltage is

$$\frac{2,300}{2,400} = 0.958.$$

Since the load current is the rated value, the load is equivalent to an impedance whose per unit magnitude equals the per unit load voltage. Since the power factor is 0.800 inductive, the per unit vector load impedance is

$$0.958(0.800 + j0.600) = 0.766 + j0.575 \text{ per unit impedance.}$$

Since the high-voltage line impedance is only about 0.05 per unit, and a reasonable value of the exciting current of the transformer (not given) is about 0.05 per unit, the magnitude of the impedance drop in the high-voltage line due to the exciting current is about 0.05 × 0.05 or 0.0025 per unit voltage. Therefore negligible error results if the exciting current is neglected and the transformer is represented by its equivalent impedance in series with the feeders, as in Fig. 1b.

The impedance of the whole circuit viewed from the high-voltage bus is the sum of the four components, and is

$$0.809 + j0.700 = 1.07\underline{/\theta} \text{ per unit impedance.}$$

Since the current is the normal value, this per unit impedance also equals the per unit sending-end bus voltage. Hence the bus voltage V_B is

$$V_B = 1.07 \text{ per unit voltage}$$

$$= 1.07 \times 12{,}000 \text{ or } 12.850 \text{ v.}$$

• PROBLEM 6-15

A 50-kva 2,400:240-volt 60-cps distribution transformer has a leakage impedance of 0.72 + j0.92 ohm in the high-voltage winding and 0.0070 + j0.0090 ohm in the low-voltage winding. At rated voltage and frequency the admittance Y_ϕ of the shunt branch accounting for the exciting current is $(0.324 - j2.24) \times 10^{-2}$ mho when viewed from the low-voltage side.

This transformer is used to step down the voltage at the load end of a feeder whose impedance is 0.30 + j1.60 ohms. The voltage V_s at the sending end of the feeder is 2,400 volts.

Find the voltage at the secondary terminals of the transformer when the load connected to its secondary draws rated current from the transformer and the power factor of the load is 0.80 lagging. Neglect the voltage drops in the transformer and feeder caused by the exciting current.

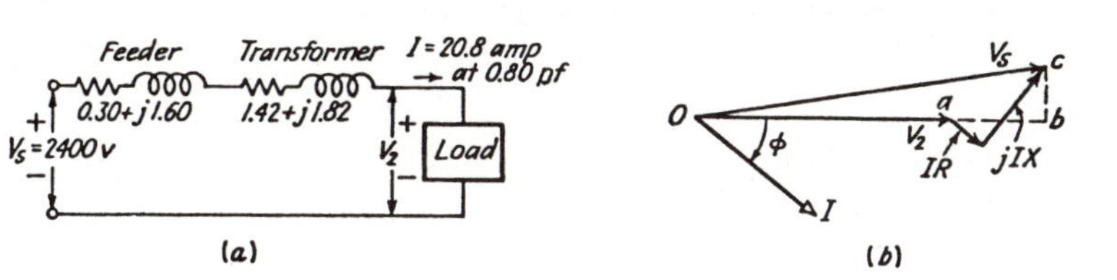

Equivalent circuit and phasor diagram

Fig. 1

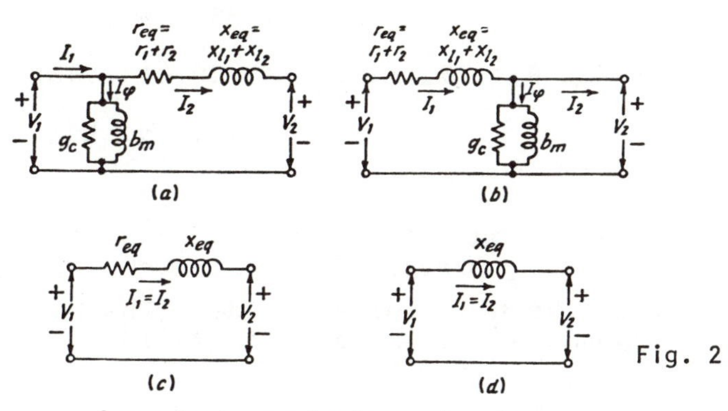

Fig. 2

Aproximate equivalent circuits.

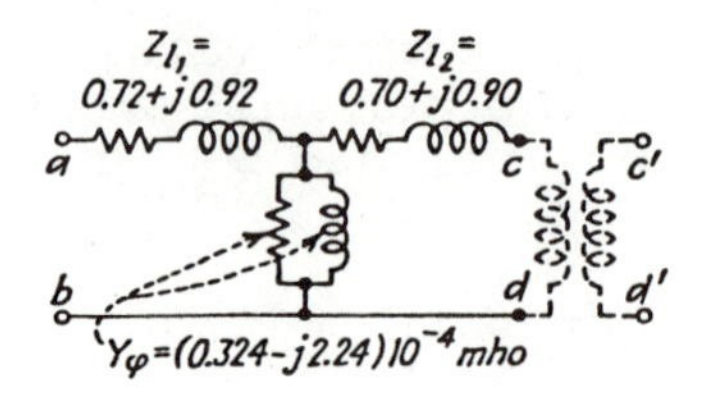

Fig. 3

Equvalent circuit.

Solution: The circuit with all quantities referred to the high-voltage (primary) side of the transformer is shown in Fig. 1a, wherein the transformer is represented by its equivalent impedance, as in Fig. 2. From Fig. 3, the value of the equivalent impedance is $\vec{Z}_{eq}$ = 1.42 + j1.82 ohms, and the combined impedance of the feeder and transformer in series is $\vec{Z}$ = 1.72 + j3.42 ohms. From the transformer rating, the load current referred to the high-voltage side is I = 50,000/2,400 = 20.8 amp.

The phasor diagram referred to the high-voltage side is shown in Fig. 1, from which

$$Ob = \sqrt{V_s^2 - (bc)^2}$$

and

$$V_2 = Ob - \dot{a}b$$

Note that

$$bc = IX \cos \phi - IR \sin \phi$$

$$ab = IR \cos \phi + IX \sin \phi$$

where R and X are the combined resistance and reactance, respectively. Thus

$$bc = (20.8)(3.42)(0.80) - (20.8)(1.72)(0.60)$$

$$= 35.5 \text{ volts}$$

$$ab = (20.8)(1.72)(0.80) + (20.8)(3.42)(0.60)$$

$$= 71.4 \text{ volts}$$

Substitution of numerical values shows that Ob very nearly equals V_s, or 2,400 volts. Then V_2 = 2,329 volts referred to the high-voltage side. The actual voltage at the secondary terminals is 2,329/10, or

$$V_2 = 233 \text{ volts.}$$

• **PROBLEM 6-16**

A 2,000-kw 750-volt d-c six-phase synchronous converter is fed by a bank of transformers connected in delta on the primary side and diametrical on the secondary side. The three-phase primary line voltage is 13,200 volts. Assume that the machine operates at full load at an efficiency of 96 per cent and a power factor of 0.95, and calculate: (a) the direct-current output; (b) the six-phase voltage between adjacent slip rings; (c) the voltage of each transformer secondary; (d) the ratio of transformation of each transformer; (e) the alternating current delivered by each of the transformer secondaries; (f) the current in each of the transformer primary coils; (g) the current in each of the line wires on the primary side.

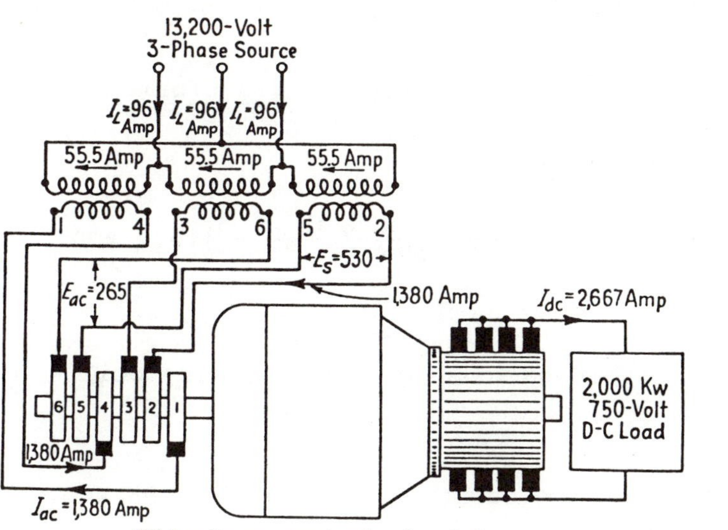

Wiring diagram representing the solution.

Fig. 1

Solution:

(a) $$I_{dc} = \frac{2,000,000}{750} = 2,667 \text{ amp}$$

(b) $$E_{ac} = 0.354 \times 750 = 265 \text{ volts} \quad \text{(see Table 1)}$$

(c) Volts per transformer secondary coil = 2 × 265 = 530 volts. (This follows from the geometry of the six-phase hexagon, in which the voltage across a diagonal = 2 × volts per side.)

(d) $$a = \frac{13,200}{530} = 24.9:1$$

(e) $$I_{ac} = \frac{2,667 \times 0.472}{0.96 \times 0.95} = 1,380 \text{ amp.}$$

(The equation in Table 1 is for power factor = 1 and efficiency = 100 percent.)

Table 1 CONVERSION RATIOS IN SYNCHRONOUS CONVERTERS

	Formula*	2-ring 1-phase	3-ring 3-phase	6-ring 6-phase	12-ring 12-phase
E_{ac}	$\frac{E_{dc}}{\sqrt{2}} \times \sin\frac{\pi}{n}$	$0.707E_{dc}$	$0.612E_{dc}$	$0.354E_{dc}$	$0.182E_{dc}$
I_{ac}	$\frac{2.83}{n} \times I_{dc}$	$1.414I_{dc}$	$0.943I_{dc}$	$0.472I_{dc}$	$0.236I_{dc}$

* n = number of rings; efficiency = 100 per cent; power factor = 1.0.

The transformer secondary current can also be calculated as follows:

$$I_{ac} = \frac{\text{power per transformer}}{\text{volts per transformer} \times \text{PF} \times \text{eff}}$$

$$= \frac{(2{,}000{,}000/3)}{530 \times 0.96 \times 0.95} = 1{,}380 \text{ amp}$$

(f) Current in each transformer primary coil $= \frac{1{,}380}{24.9} =$ 55.5 amp

(g) Primary line current $= 55.5 \times \sqrt{3} = 96$ amp.

The primary line current can also be calculated as follows:

$$\text{Primary } I_L = \frac{P}{\sqrt{3} \times E_L \times \text{PF} \times \text{eff}}$$

$$= \frac{2{,}000{,}000}{\sqrt{3} \times 13{,}200 \times 0.96 \times 0.95} = 96 \text{ amp.}$$

A wiring diagram illustrating the solution to this example, with all currents and voltages shown, is given in Fig. 1.

• **PROBLEM 6-17**

Refer to Fig. 1. The magnitude of the three-phase line voltages is 220 volts; that of the two-phase voltages is 120 volts for each phase. Consider the transformers to be ideal.

(a) Find the turns ratios.

(b) Connect a load consuming 12 kw at unity power factor across terminals a and b. Leave terminals c and d open. Find all currents.

(c) Reconnect the same load to the teaser terminals c and d. Leave terminals a and b open. Find all currents.

(d) Each transformer secondary is supplying a unity-power-factor load of 12 kw. (This constitutes a balanced two-phase load.) Find all currents.

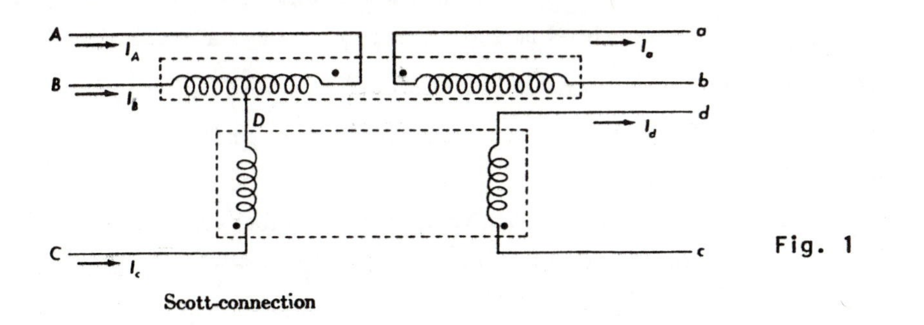

Fig. 1

Solution: (a) For the main transformer

$$a_M = \frac{|\vec{V}_{AB}|}{|\vec{V}_{ab}|} = \frac{220}{120} = 1.83.$$

For the teaser

$$a_T = \frac{|\vec{V}_{CD}|}{|\vec{V}_{cd}|} = \frac{0.866 \times 220}{120} = 1.59.$$

The two values check with the following equation:

$$a_T = 0.8666\ a_M.$$

(b) Using V_{AB} as our axis of reference, $\vec{V}_{AB} = 220\underline{/0°}$, $\vec{V}_{ab} = 120\underline{/0°}$, $\vec{I}_a = \frac{12,000}{120}\underline{/0°} = 100\underline{/0°}$, and $\vec{I}_d = 0$. For the primary currents $\vec{I}_A = \frac{100}{1.83}\underline{/0°} = 54.7\underline{/0°}$,

$\vec{I}_B = 54.7\underline{/180°}$, and $\vec{I}_C = 0$. Actually, the teaser primary carries its exciting current, but this is being neglected as are the other imperfections of the transformers.

(c) This time, $\vec{I}_a = 0$ and $|\vec{I}_d| = \frac{12,000}{120} = 100$ amp. This current is in phase with V_{dc}. Maintaining the phase sequence a-b-c and the choice of V_{AB} as axis of reference, $\vec{V}_{dc} = 120\underline{/-90°}$, thus $\vec{I}_d = 100\underline{/-90°}$ and $\vec{I}_C = -\frac{1}{a_T}\vec{I}_d = -\frac{100}{1.59}\underline{/90°} = 63\underline{/90°}$. (Watch the dots for proper polarities.)

At this point an interesting question arises: This current I_C must find its way back through the main primary, but how can the main primary carry a current if its secondary is open-circuited? The answer is that any amount of current can flow in that primary as long as the total mmf is zero, and this is what happens (since the junction D is at the midpoint of the winding) if the current divides into halves. Thus $\vec{I}_A = -\frac{1}{2}\vec{I}_C = 31.5\underline{/-90°}$ and $\vec{I}_B = -\frac{1}{2}\vec{I}_C = 31.5\underline{/-90°}$.

(d) The secondary currents are $\vec{I}_a = 100\underline{/0°}$ A as in part (b), and $\vec{I}_d = 100\underline{/-90°}$ A as in part (c). The current in the teaser primary is the same as in part (c), thus $\vec{I}_C = 63\underline{/90°}$ A. To find the current in the main primary, we must observe that the total mmf in the main transformer must still be zero (as in every ideal transformer), and that Kirchhoff's current law for the junction D must be satisfied. Thus we must add the solutions of parts (b) and (c), resulting in $\vec{I}_A = 54.7\underline{/0°} + 31.5\underline{/-90°} = 63\underline{/-30°}$ A and $\vec{I}_B = 54.7\underline{/180°} + 31.5\underline{/-90°} = 63\underline{/-150°}$ A. It is worth noting that the three primary line currents form a symmetrical three-phase system, and that their magnitudes and phase angles are the same as if they were flowing to supply a balanced three-phase load whose power and power factor are the same as those of the actual two-phase load.

TRANSFORMER CONSTANTS AND THE EQUIVALENT CIRCUIT

• PROBLEM 6-18

A short-circuit test performed on a 10-kva 2,400:240-volt transformer gave the following readings:

E_1 = 76.8 volts $\quad$ I_2 = 41.7 amp

I_1 = 4.17 amp $\quad$ W = 181 watts

Determine the transformer constants.

Solution: From the given data, the full-load copper loss is 181 watts. There is also a small amount of iron loss in the transformer, but the flux density is so low that this amounts to only a small fraction of a watt and is negligible.

$$\therefore \quad R = \frac{W}{(I_1)^2} = \frac{181}{(4.17)^2} = 10.4 \text{ ohms}$$

$$Z = \frac{E_1}{I_1} = \frac{76.8}{4.17} = 18.4 \text{ ohms}$$

$$X = \sqrt{Z^2 - R^2} = \sqrt{(18.4)^2 - (10.4)^2} = 15.2 \text{ ohms}$$

To get the primary and secondary values separately, the primary and equivalent secondary values are assumed to be equal; thus

$$R_1 = R_2' = \frac{10.4}{2} = 5.2 \text{ ohms}$$

$$\therefore \quad X_1 = X_2' = \frac{15.2}{2} = 7.6 \text{ ohms}$$

$$Z_1 = Z_2' = \frac{18.4}{2} = 9.2 \text{ ohms}$$

$$\therefore \quad R_2 = \frac{5.2}{100} = 0.052 \text{ ohm}$$

$$X_2 = \frac{7.6}{100} = 0.076 \text{ ohm}$$

$$\therefore \quad Z_2 = \frac{9.2}{100} = 0.092 \text{ ohm}$$

A 200-kva, 13,200/2,200-volt, 60-cps, single-phase transformer has the following test data:

	Volts	Amp	Watts	Frequency	Volts
Open circuit	2,200	3.1	1550	60 cps	12,800
Short circuit	210	90.9	2500	60 cps	

Determine the parameters of the T-equivalent circuit when referred to the low voltage winding.

Solution: Turns ratio = 12,800/2,200 = 5.82

The high and low voltage rated currents are respectively

$$I_H = 200 \times 10^3/13,200 = 15.16 \text{ amp.}$$

$$I_L = 200 \times 10^3/2200 = 90.9 \text{ amp.}$$

From this data, it is concluded that both the open-circuit and the short-circuit tests were performed with the instruments on the low voltage side.

$$\therefore \quad R = V_1{}^2/P_{oc} = (2200)^2/1550 = 3120.0 \text{ ohms}$$

$$Y_{oc} = I_\phi/V_1 = 3.1/2200 = 1.41 \times 10^{-3} \text{ mho.}$$

$$|b_{mag}| = \sqrt{Y_{oc}^2 - g_c^2}$$

$$= \sqrt{(1.41 \times 10^{-3})^2 - (0.321 \times 10^{-3})^2}$$

$$= 1.374 \times 10^{-3} \text{ mho.}$$

$$|X_{mag}| = \frac{1}{|b_{mag}|}$$

or $\quad |X_{mag}| \approx 10^3/1.374 = 727.0$ ohms

From the short-circuit test, the leakage impedance components are determined.

$$r_1' + r_2 = R_{sc} = \frac{P_{sc}}{I_{sc}^2}$$

or $r_1' + r_2 = R_{sc} = 2500/(90.9)^2 = 0.303$ ohm.

$$Z_{sc} = \frac{V_{sc}}{I_{sc}}$$

$$|X_{sc}| = \sqrt{Z_{sc}^2 - R_{sc}^2} = \sqrt{(2.31)^2 - (0.303)^2} = 2.28 \text{ ohm.}$$

• PROBLEM 6-20

(1) An open-circuit test on the 240-V winding of the transformer in the figure yielded the following data, corrected for instrument losses:

Volts	Amperes	Watts
240	16.75	580

Calculate the exciting admittance, conductance, and susceptance.

(2) The following data were obtained in a short-circuit test of the transformer in the figure, with its low-voltage winding short-circuited.

Volts	Amperes	Watts
63.0	62.5	1660

Calculate (a) the equivalent primary impedance, (b) the equivalent primary reactance, and (c) the equivalent primary resistance.

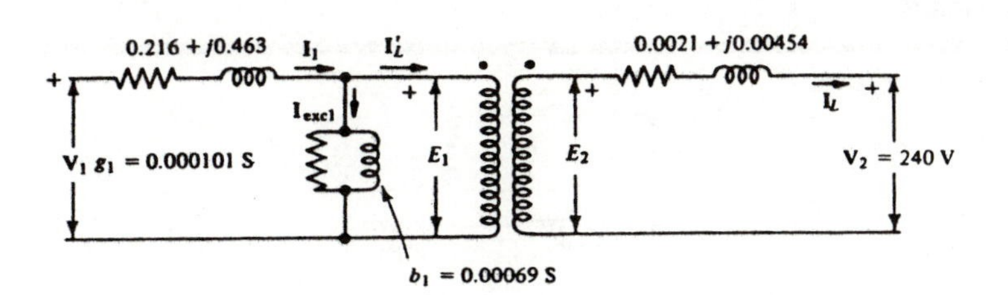

Equivalent circuit.

Solution: (1) The exciting admittance is

$$y \approx \frac{I_{exc}}{V} = \frac{16.75}{240} = 0.0698 \text{ S}$$

$$g \approx \frac{P_{oc}}{V^2} = \frac{580}{(240)^2} = 0.0101 \text{ S}$$

from which the exciting susceptance is found to be

$$b = \sqrt{y^2 - g^2}$$

$$= \sqrt{(0.0698)^2 - (0.0101)^2} = 0.069 \text{ S.}$$

(2)

a) $$Z_{eql} \approx \frac{V_{sc}}{I_{sc}} = \frac{63.0}{62.5} = 1.008\Omega$$

b) $$X_{eql} = \sqrt{Z_{eql}^2 - R_{eql}^2} \approx \sqrt{(1.008)^2 - (0.425)^2} = 0.915\Omega$$

c) $$R_{eql} \approx \frac{P_{sc}}{I_{sc}^2} = \frac{1660}{(62.5)^2} = 0.425\Omega$$

• PROBLEM 6-21

The short-circuit and open-circuit tests on a power transformer rated 60 cycles, 12,000 kva, 132/22 kv are as follows:

	Short-Circuit	Open-Circuit
Voltage, primary	10,900	132,000
Voltage, secondary	0	22,020
Current, primary	90.9	0.95
Current, secondary	545.5	0
Kw, input	71.4	54.5

The primary and secondary d-c resistances at 75°C are: $r_1 = 4.11$, $r_2 = 0.0975$ ohms. Find the transformer constants.

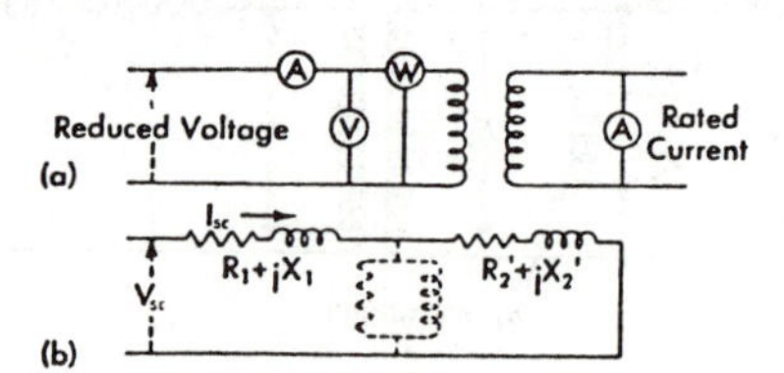

Fig. 1

Short-circuit test.

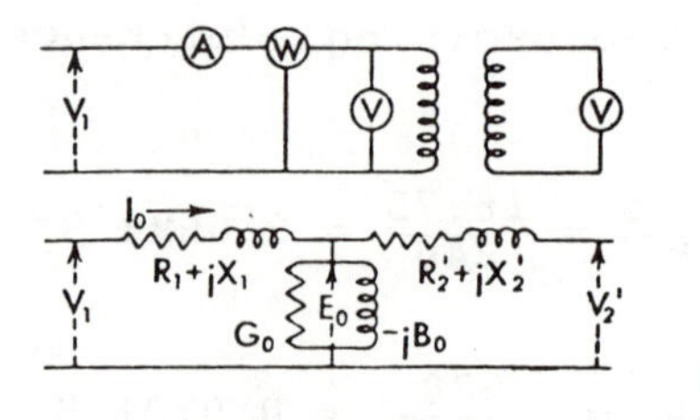

Fig. 2

Open-circuit test.

<u>Solution</u>: (a)

$$\frac{N_1}{N_2} = \left(\frac{I_2}{I_1}\right)_{sc} = \left(\frac{V_1}{V_2}\right)_{oc} = \left(\frac{545.5}{90.9}\right)_{sc} = 6.00,$$

i.e. a = 6.

Also $$\frac{N_1}{N_2} = \left(\frac{132,000}{22,020}\right)_{oc} = 5.99$$

(b) $$\cos\theta_{sc} = \frac{W_{sc}}{V_{sc}I_{sc}} = \frac{71,400}{10,900 \times 90.9} = 0.0721.$$

∴ $$\sin\theta_{sc} = 0.9974.$$

$$\theta_{sc} = -85.85°.$$

(c) $$Z_{sc} = \frac{V_{sc}}{I_{sc}} = \frac{10,900}{90.9} = 120 \text{ ohms}.$$

(d) $$R_{sc} = \frac{W_{sc}}{I^2_{sc}} = \frac{71,400}{90.9^2} = 8.65 \text{ ohms}.$$

(e) $$X_{sc} = \sqrt{Z^2_{sc} - R^2_{sc}} = \sqrt{120^2 - 8.65^2} = 119.6 \text{ ohms}.$$

(f) $$R_1 = \left(\frac{r_1}{r_1 + a^2r_2}\right)R_{sc} = \frac{4.11 \times 8.65}{4.11 + 0.0975 \times 36}$$

$$= 4.67 \text{ ohms}$$

∴ $$R_2' = 8.65 - 4.67 = 3.98 \text{ ohms}$$

$$R_2 = 3.98/36 = 0.1106 \text{ ohms}.$$

$$X_1 = \frac{4.11 \times 119.6}{7.62} = 64.5 \text{ ohms}$$

∴ $$X_2' = 119.6 - 64.5 = 55.1 \text{ ohms}.$$

$$X_2 = 55.1/36 = 1.53 \text{ ohms}.$$

$$\cos\theta_{oc} = \frac{54,500}{132,000 \times 0.95} = 0.435.$$

∴ $$\sin\theta_{oc} = 0.9002 \qquad \theta_{oc} = -64.2°.$$

$$E_0 = 132,000 - (64.7\underline{/85.85°})(0.95\underline{/-64.2°})$$

$$= 131,943 - j23 \text{ V}.$$

(g) $$Y_0 = \frac{T_0}{E_0} = \frac{0.95}{131,944} = 7.20 \times 10^{-6} \text{ mho.}$$

(h) $$G_0 = \frac{W_{oc} - R_1 I_0^2}{E_0^2} = \frac{54,500 - 4.67 \times 0.95^2}{131,944}$$

$$= 3.13 \times 10^{-6} \text{ mho.}$$

(i) $$B_0 = \sqrt{Y_0^2 - G_0^2} = \sqrt{7.20^2 - 3.13^2}$$

$$\times 10^{-6} = 6.49 \times 10^{-6} \text{ mho.}$$

Note how insignificant is the effect of the impedances in determining Y_0 and G_0.

• PROBLEM 6-22

The following measurements were obtained from tests carried out on a 10-kVA, 2300:230-V, 60-Hz distribution transformer:

Open-circuit test, with the low-potential winding excited:

Applied potential difference, V_{oc} = 230 V

Current, I_{oc} = 0.45 A

Input power, P_{oc} = 70 W

Short-circuit test, with the high-potential winding excited:

Applied potential difference, V_{sc} = 120 V

Current, I_{sc} = 4.5 A

Input power, P_{sc} = 240 W

Winding resistances, measured by dc bridge:

$R_{hp} = 5.80\ \Omega$, $R_{lp} = 0.0605\ \Omega$

a) Determine the equivalent circuit of the transformer referred to the low-potential side.

b) Express the exciting current of the transformer as a percentage of the rated full-load current.

Solution: a) For the open-circuit test, the copper loss in the low-potential winding is

$$P_w = R_{1p} I_{oc}^2 = 0.0605 \times 0.45^2 = 0.0163 \quad W$$

This is a very small proportion of the input power on open circuit and may be neglected. The power factor on open circuit is

$$\cos\theta = \frac{P_{oc}}{V_{oc} I_{oc}} = \frac{70}{230 \times 0.45} = 0.676.$$

$$\therefore \quad \theta = 47.4°$$

Fig. 1

Equivalent circuit for open-circuit test.

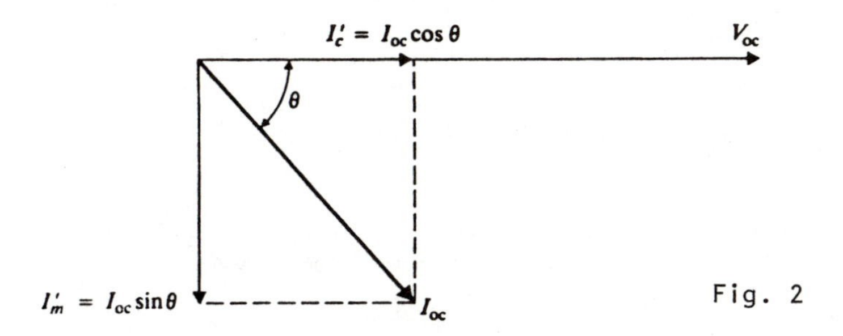

Fig. 2

Phasor diagram for the circuit of Fig. 1.

The phasor diagram for this test, corresponding to the circuit diagram of Fig. 1, is shown in Fig. 2. From this diagram,

$$R'_c = \frac{V_{oc}}{I_{oc}\cos\theta} = \frac{230}{0.45 \times 0.676} = 756 \ \Omega.$$

$$X'_m = \frac{V_{oc}}{I_{oc}\sin\theta} = \frac{230}{0.45 \times 0.737} = 694 \ \Omega.$$

and these parameters are measured on, and therefore referred to, the low-potential side of the transformer.

The circuit diagram for the short-circuit test is that of Fig. 1. From that diagram,

$$R'_{eq} = \frac{P_{sc}}{I_{sc}^2} = \frac{240}{4.5^2} = 11.85 \ \Omega.$$

The transformer impedance on short circuit is

$$Z_{sc} = \frac{V_{sc}}{I_{sc}} = \frac{120}{4.5} = 26.7 \ \Omega.$$

so that

$$X'_{eq} = \left[Z_{sc}^2 - (R'_{eq})^2\right]^{1/2} = [26.7^2 - 11.8^2]^{1/2}$$

$$= 24.0 \ \Omega.$$

and these parameters are measured on, and therefore referred to, the high-potential side of the transformer. Referred to the low-potential side,

$$X''_{eq} = \left(\frac{230}{2300}\right)^2 X'_{eq} = 0.240 \ \Omega.$$

so that

$$X_{\ell 2} = X''_{\ell 1} = \frac{0.240}{2} = 0.120 \ \Omega.$$

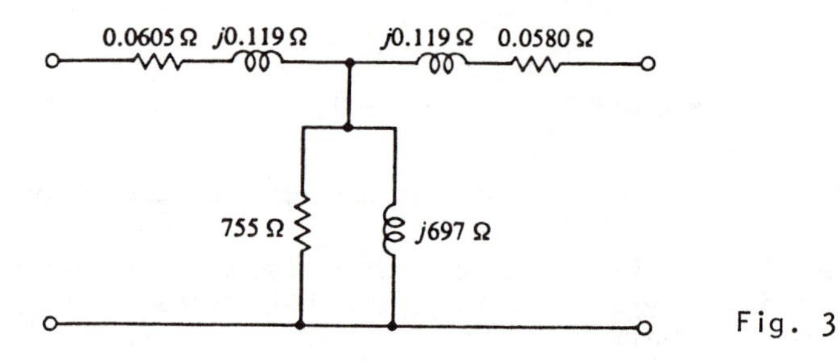

Fig. 3

Equivalent circuit of transformer.

The resistance of the high-potential winding referred to the low-potential side is

$$R''_{hp} = \left(\frac{230}{2300}\right)^2 R_{hp} = 0.058 \ \Omega.$$

The equivalent circuit of the transformer referred to the low-potential side is shown in Fig. 3.

b) The rated output S of the transformer is 10 kVA, thus the rated full-load current in the low-potential winding is

$$I_{1p} = \frac{S}{V_{1p}} = \frac{10 \times 10^3}{230} = 43.5 \text{ A}.$$

The exciting current is I_{oc}; thus, as a percentage of full-load current, the exciting current is

$$\frac{I_{oc}}{I_{1p}} \times 100\% = \frac{0.45}{43.5} \times 100\% = 1.04\%.$$

• PROBLEM 6-23

A 200 kVA single-phase transformer with a voltage ratio 6350/660 V has the following winding resistances and reactances:

$$R_1 = 1.56\Omega \quad R_2 = 0.016\Omega \quad X_1 = 4.67\Omega \quad X_2 = 0.048\Omega.$$

On no load the transformer takes a current of 0.96 A at a power factor of 0.263 lagging. (a) Calculate the equivalent circuit parameters referred to the high-voltage winding. (b) Calculate the voltage regulation at unity power factor (u.p.f.), 0.8 lagging p.f., and 0.8 leading p.f.

Solution:

(a) $$R_{el} = R_1 + R_2(N_1/N_2)^2 = 1.56 + 0.016(635/66)^2 = 3.04\Omega$$

$$X_{el} = X_1 + X_2(N_1/N_2)^2 = 4.67 + 0.048(635/66)^2 = 9.12\Omega$$

For the magnetising branch, the no-load current is split in two components $I_p = I_0\cos\theta$ and $I_m = I_0\sin\theta$.

$$\therefore \quad r_m = \frac{V}{I_0\cos\theta} = \frac{6350}{0.96 \times 0.263} = 25.2 \text{ k}\Omega$$

$$x_m = \frac{V}{I_0\sin\theta} = \frac{6350}{0.96 \times \sqrt{(1 - 0.263^2)}} = 6.85 \text{ k}\Omega$$

(b) Full-load current (f.l.) $= 200 \times 10^3/6350 = 31.5$ A

$I_{f.l.}R = 31.5 \times 3.04 = 96$ V

$I_{f.l.}X = 31.5 \times 9.12 = 288$ V.

$$E - V = IR\cos\theta + IX\sin\theta + (IX\cos\theta - IR\sin\theta)^2/2E \quad (1)$$

u.p.f: $E - V = 96 \times 1 + 288^2/(2 \times 6350) = 96 + 6.4$

$= \underline{102.4 \text{ V}}$

0.8 lag: $E - V = 96 \times 0.8 + 288 \times 0.6$

$+ \dfrac{(288 \times 0.8 - 96 \times 0.6)^2}{2 \times 6350}$

$= 249.7 + 2.3 = 252$ V

0.8 lead: $E - V = 96 \times 0.8 + 288 \times (-0.6)$

$+ \dfrac{(288 \times 0.8 - 96 \times (-0.6))^2}{2 \times 6350.}$

$= -96.3 + 6.5 = -89.8$ V

The corresponding terminal voltages E - (E - V) are:

in primary terms 6248, 6908, 6440 V respectively
in secondary terms 648, 633, 669 V respectively

Note that the last term in Eq.(1) is very small and only affects the answer appreciably as the power factor becomes leading. It will be neglected subsequently.

$\text{Regulation}_{p.u.} = R_{p.u.}\cos\theta + X_{p.u.}\sin\theta$

$R_{p.u.} = 96/6350 = 0.0151$

$X_{p.u.} = 288/6,350 = 0.0453\Omega$

u.p.f: $E - V = 0.0151 \times 1 = 0.0511$ per unit

$= 96$ V

0.8 lag: $E - V = 0.0151 \times 0.8 + 0.0453 \times 0.6$

$= 0.0392 \qquad = 249$ V

$$0.8 \text{ lead: } E - V = 0.0151 \times 0.8 + 0.0453$$

$$\times (-0.6) = -0.0151 \qquad = 96 \text{ V}$$

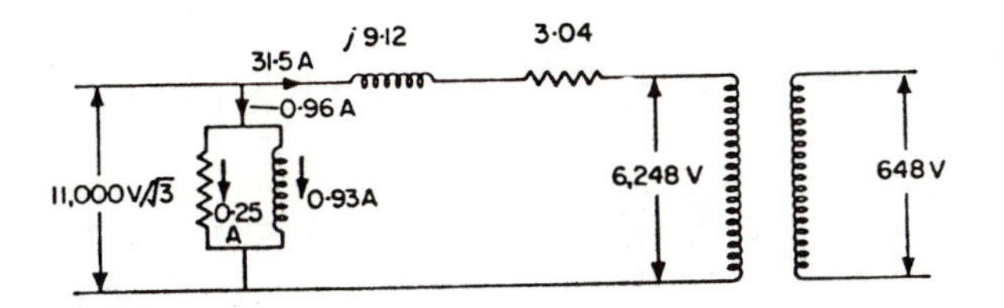

The figure shows the equivalent circuit, currents and voltages at full load, unity power factor.

The figure shows the equivalent circuit, currents and voltages at full load, unity power factor.

• PROBLEM 6-24

> From short-circuit and open-circuit tests on a 60-cycle, 2200/110-volt, 110-kva transformer it is determined that $r_1 = 0.22$ ohm, $x_1 = 2.00$ ohms, $r_2 = 0.0005$ ohm, $x_2 = 0.005$ ohm, $g_{01} = 0.000182$ mho, and $b_{01} = 0.00091$ mho.
>
> Compute: (a) Per-unit values of resistance, reactance, conductance, and susceptance, (b) per-unit values of equivalent resistance and reactance referred to the primary and to the secondary, (c) the general-circuit constants, and (d) the voltage regulation at 0.8 power factor, lagging current, and the primary current when supplying a 0.8 power factor (lagging) load at rated kva.

Solution:

(a) Base Primary Volts = 2200.

Base Volt-amperes = 110,000.

Base Primary Amperes = 110,000/2200 = 50.

Base Secondary Volts = 110.

Base Secondary Amperes = 110,000/110 = 1000.

Base Ohms (Primary) = 2200/50 = 44.

Base Ohms (Secondary) = 110/1000 = 0.11.

Base Mhos = 50/2200 = 1/44

Whence

$$r_1 = \frac{0.22}{44} = 0.005 \text{ per-unit}$$

$$x_1 = \frac{2.00}{44} = 0.0454 \text{ per unit}$$

$$r_2 = \frac{0.0005}{0.11} = 0.00454 \text{ per unit}$$

$$x_2 = \frac{0.005}{0.11} = 0.0454 \text{ per unit}$$

$$g_{01} = \frac{0.000182}{\frac{1}{44}} = 0.00801 \text{ per unit}$$

$$b_{01} = \frac{0.00091}{\frac{1}{44}} = 0.040 \text{ per unit}$$

(b) Referred to Primary

$$R_0 = r_1 + \left(\frac{2200}{110}\right)^2 r_2 = 0.22 + 0.20 = 0.42 \text{ ohm}$$

$$\equiv \frac{0.42}{44} = 0.00954 \text{ per unit}$$

$$X_0 = x_1 + \left(\frac{2200}{110}\right)^2 x_2 = 2.0 + 2.0 = 4.0 \text{ ohm}$$

$$\equiv \frac{4}{44} = 0.0908 \text{ per unit}$$

Referred to Secondary

$$R_0 = r_2 + \left(\frac{110}{2200}\right)^2 r_1 = 0.0005 + 0.00055$$

$$= 0.00105 \text{ ohm} \equiv \frac{0.00105}{0.11} = 0.00954 \text{ per unit}$$

$$X_0 = x_2 + \left(\frac{110}{2200}\right)^2 x_1 = 0.005 + 0.005 = 0.01 \text{ ohm}$$

$$\equiv \frac{0.01}{0.11} = 0.0908 \text{ per unit}$$

Note that in the per-unit system the values of equivalent resistance and reactance are identical, respectively, whether the reference be the primary or the secondary. Also, each equivalent value is the arithmetic sum of the per-unit values of primary and secondary components, i.e., $R_0 = r_1 + r_2$ and $X_0 = x_1 + x_2$ in per-unit values.

(c) Using the simplified values,

$$\vec{A}_0 = a = \frac{\text{Base Primary Volts}}{\text{Base Secondary Volts}} = \frac{1}{1} = 1 \text{ per unit}$$

$$\vec{B}_0 = \frac{Z_0}{a} = Z_0 = R_0 + jX_0 = 0.00954 + j0.0908 \text{ per unit}$$

$$\vec{C}_0 = aY_0 = Y_0 = 0.00801 - j0.040 \text{ per unit}$$

$$\vec{D}_0 = \frac{1}{a} = 1 \text{ per unit}$$

(d)

$$\vec{V}_1 = \vec{A}_0\vec{V}_2 + \vec{B}_0\vec{I}_2\angle\theta_2 \quad (V_2 \text{ is reference})$$

$$= (1\times1) + (0.00954 + j0.0908) \times 1$$

$$\times (\cos\theta_2 - j\sin\theta_2)$$

$$= (1 + 0.007632 + 0.05448) + j(0.07264 - 0.005724)$$

$$= 1.062 + j0.067$$

$$V_1 = 1.0641 \text{ per unit}$$

Whence $\quad$ Percentage of Regulation $= 100 \times \frac{1.0641 - 1}{1}$

$= 6.41\%$

Also, percentage of regulation is given by

$$\text{Percentage of Regulation} = p + \frac{q^2}{200}$$

where $\quad p = R_0\cos\theta + X_0\sin\theta$

$q = X_0\cos\theta - R_0\sin\theta$

(R_0 and X_0 are in per-unit values).

Thus,

$$p = [(0.00954 \times 0.8) + (0.0908 \times 0.6)] \times 100$$

$$= [0.007632 + 0.05448] \times 100$$

$$= 0.0621 \times 100 = 6.21$$

$$q = [(0.0908 \times 0.8) - (0.00954 \times 0.6)] \times 100$$
$$= [0.07264 - 0.005724] \times 100$$
$$= 0.0669 \times 100$$
$$= 6.69$$

And $\text{Percentage of Regulation} = 6.21 + \frac{(6.69)^2}{200}$

$$= 6.43\% \text{ (nearly)}$$

(e) $\vec{I}_1 = \vec{C}_0\vec{V}_2 + \vec{D}_0\vec{I}_2 \angle\theta_2$

$$= (0.00801 - j0.040) + 1 \times (0.8 - j0.6)$$
$$= 0.80801 - j0.64$$

$$I_1 = 1.0307 \text{ per unit}$$
$$= 51.54 \text{ amperes}$$

• PROBLEM 6-25

A four-winding transformer has circuits rated as follows:

	Kva	Volts	Amperes
Coil 1	10,000	132,000	75.75
Coil 2	10,000	66,000	151.50
Coil 3	5,000	13,200	378.75
Coil 4	5,000	6,600	757.50

The short-circuit data from successive short-circuit tests are:

Unit (short-circuit)	Applied Voltage	Current (amperes)	Power Loss (kilowatts)
Winding 2	$V_1 = 13{,}217$	$I_1 = 75.75$	50.3
Winding 3	$V_2 = 2{,}660$	$I_2 = 75.75$	25.2
Winding 4	$V_3 = 525$	$I_3 = 378.8$	30.0
Winding 4	$V_1 = 10{,}700$	$I_1 = 37.88$	19.8
Winding 4	$V_2 = 4{,}640$	$I_2 = 75.75$	24.7
Winding 3	$V_1 = 7{,}925$	$I_1 = 37.88$	20.5

Establish the equivalent circuit for this transformer. Using the equivalent circuit determine the voltage on winding 4 when windings 3 and 4 are open-circuited, winding 2 is short-circuited and a potential of 13,217 volts is impressed on winding 1.

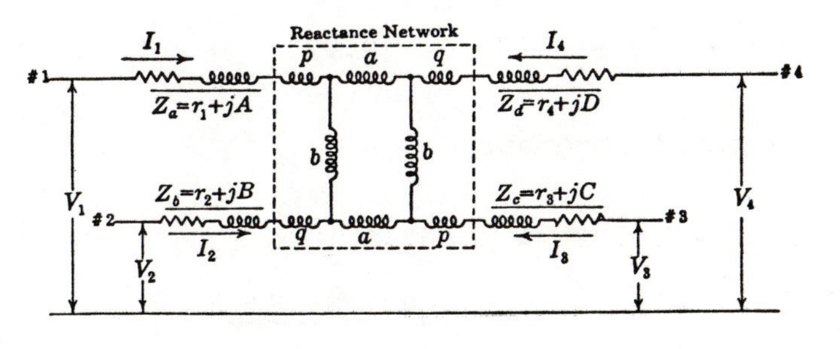

Fig. 1

Solution: Referring to Fig. 1,

$$p = M + \sqrt{MN} = -0.0404 + \sqrt{0.0404 \times 0.0093}$$

$$= -0.021 \text{ per unit}$$

$$q = N + \sqrt{MN} = -0.0093 + \sqrt{0.0404 \times 0.0093}$$

$$= 0.010 \text{ per unit}$$

$$a = -2p = 0.042 \text{ per unit ohms}$$

$$b = -2q = -0.02 \text{ per unit ohms.}$$

The terminal branch impedances are:

$$\vec{Z}_1 = Z_a + jp = 0.00264 + j0.0802\Omega.$$

$$\vec{Z}_2 = Z_b + jq = 0.00481 + j0.0493\Omega.$$

$$\vec{Z}_3 = Z_c + jp = 0.00467 + j0.0287\Omega.$$

$$\vec{Z}_4 = Z_d + jq = 0.00340 + j0.0794\Omega.$$

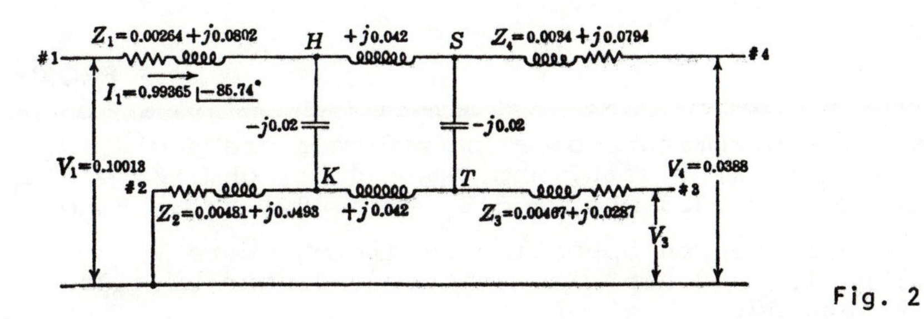

Fig. 2

The equivalent circuit is shown in Fig. 2, in which the equivalent reactance of the parallel network between H and K is

$$j\,\frac{-0.02 \times 0.064}{0.044} = -j0.029.$$

whence the total impedance is

$$Z_1 + Z_2 - j0.029 = 0.00745 + j0.1005$$
$$= 0.10077\angle 85.74°\Omega.$$

$$\vec{I}_1 = \frac{0.10013}{0.10077\angle 85.74°} = 0.99365\angle -85.74°$$
$$= 0.07393 - j0.99087 \text{ A.}$$

$$\vec{V}_{HK} = 0.10013 - I_1(Z_1 + Z_2)$$
$$= 0.10013 - 0.99365\angle -85.74° \times 0.12894\angle 86.70°$$
$$= 0.10013 - 0.1281\angle 0.96°$$
$$= -0.027957 - j0.02152 \text{ V.}$$

$$\vec{V}_{HS} = \frac{0.042}{0.064} \times V_{HK} = -0.01835 - j0.01410 \text{ V.}$$

$$\vec{I}_1\vec{Z}_1 = (0.07393 - j0.99087)(0.00264 + j0.0802)$$
$$= 0.079665 + j0.00329 \text{ V.}$$

$$\vec{V}_4 = 0.10013 - (V_{HS} + I_1Z_1)$$
$$= (0.10013 - 0.06131) - j0.0008$$
$$= 0.03882 - j0.0008$$

$$V_4 = 0.0388 \text{ per unit}$$
$$= 256 \text{ volts.}$$

• PROBLEM 6-26

(a) The parameters of a transformer are $a = 2$, $R_1 = 0.052$ ohm, $X_1 = 0.107$ ohm, $R_2 = 0.011$ ohm, $X_2 = 0.024$ ohm, $G_c = 0.0023$ mho, and $B_m = -0.0041$ mho. Find $\vec{V}_1$ and $\vec{I}_1$ for an operating condition where $V_2 = 115$ volts, $I_2 = 80$ amp, and the power factor on side 2 is 0.9 lagging.

(b) Solve part (a), using the approximate equivalent circuit of Fig. 3.

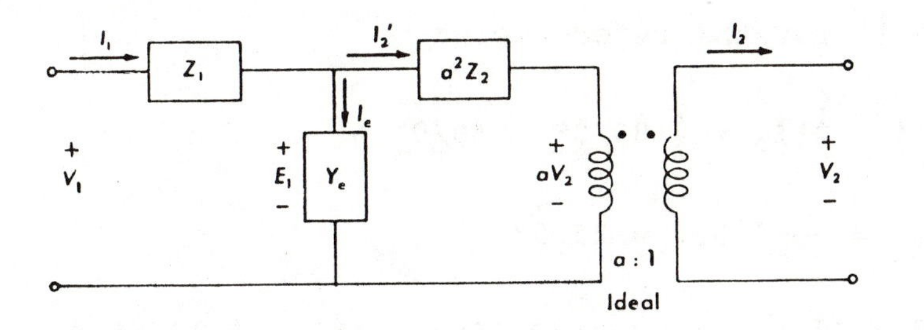

Fig. 1

Equivalent circuit.

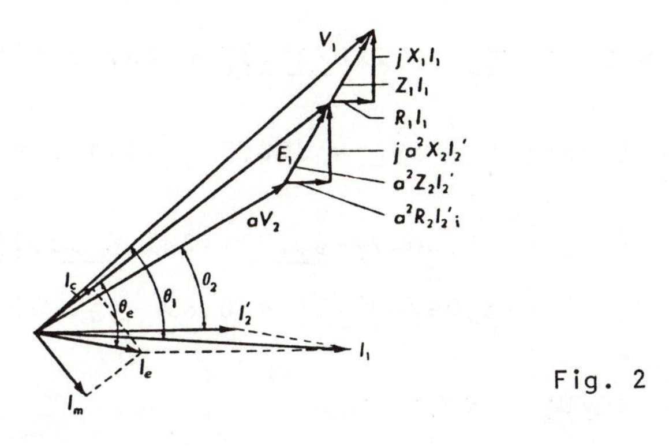

Fig. 2

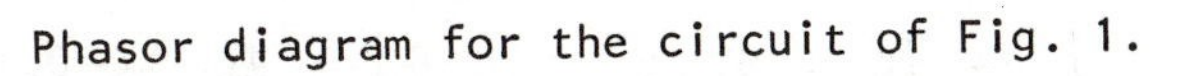
Phasor diagram for the circuit of Fig. 1.

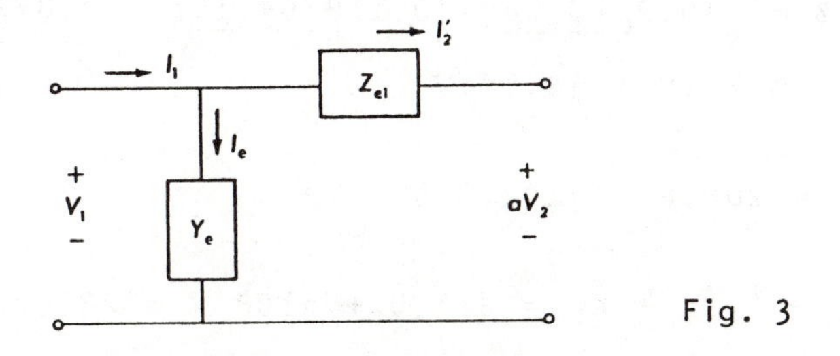

Fig. 3

An approximate equivalent circuit.

Solution:

(a) $\vec{Z}_1 = R_1 + jX_1 = 0.052 + j0.107 = 0.119\underline{/64.1°}\ \Omega$

$\vec{Z}_2 = R_2 + jX_2 = 0.011 + j0.024 = 0.0264\underline{/65.4°}\ \Omega$

$a^2\vec{Z}_2 = 0.106\underline{/65.4°}$

$\vec{Y}_e = G_c + jB_m = 0.0023 - j0.0041$

$= 0.0047\underline{/-60.8°}$ mho.

Choose $\vec{I}_2'$ for the reference.

$$\vec{I}_2' = \frac{1}{a}\,\vec{I}_2 = \frac{1}{2}\,80\underline{/0^\circ} = 40\underline{/0^\circ}\ \text{A}$$

$$\theta_2 = \cos^{-1}0.9 = 25.9^\circ$$

$$\vec{I}_2'(a^2\vec{Z}_2) = (40\underline{/0^\circ})(0.106\underline{/65.4^\circ}) = 4.24\underline{/65.4^\circ}$$
$$= 1.77 + j3.86$$

$$a\vec{V}_2 = 2(115\underline{/\theta_2}) = 230\underline{/25.9^\circ} = 207.0 + j100.4$$

$$\vec{E}_1 = \vec{I}_2'(a^2\vec{Z}_2) + a\vec{V}_2 = 208.8 + j104.3 = 231.6\underline{/25.7^\circ}\ \text{V}$$

$$\vec{I}_e = \vec{Y}_e\vec{E}_1 = (0.0047\underline{/-60.8^\circ})(231.6\underline{/25.7^\circ})$$
$$= 1.09\underline{/-35.1^\circ} = 0.89 - j0.63\ \text{A}$$

$$\vec{I}_2' = 40\underline{/0^\circ} = 40 + j0\ \text{A}$$

$$\vec{I}_1 = \vec{I}_e + \vec{I}_2' = 40.9 - j0.63 = 40.9\underline{/-0.88^\circ}\ \text{A}$$

$$\vec{I}_1\vec{Z}_1 = (40.9\underline{/-0.88^\circ})(0.119\underline{/64.1^\circ}) = 4.87\underline{/63.2^\circ}$$
$$= 2.19 + j4.35\ \text{V}$$

$$\vec{E}_1 = 208.8 + j104.3\ \text{V}$$

$$\vec{V}_1 = \vec{I}_1\vec{Z}_1 + \vec{E}_1 = 211.0 + j108.7 = 237.3\underline{/27.2^\circ}\ \text{V}$$

Figure 2 is an illustrative phasor diagram for this example. The relative magnitudes have been distorted in the figure to show details clearly.

$$\vec{Z}_{el} = \vec{Z}_1 + a^2\vec{Z}_2 = (0.052 + 0.044) + j(0.107 + 0.096)$$
$$= 0.096 + j0.203 = 0.224\underline{/64.7^\circ}\ \Omega$$

$$\vec{I}_2' = 40\underline{/0^\circ}\ \text{A}$$

$$\vec{I}_2'\vec{Z}_{el} = (40\underline{/0^\circ})(0.224\underline{/64.7^\circ}) = 8.96\underline{/64.7^\circ}$$
$$= 3.83 + j8.1\ \text{V}$$

$$a\vec{V}_2 = 230\underline{/25.9°} = 207.0 + j100.4$$

$$\vec{V}_1 = \vec{I}_2'\vec{Z}_{el} + a\vec{V}_2 = 210.8 + j108.5 = 237.1\underline{/27.2°}\ \text{V}$$

$$\vec{I}_e = \vec{Y}_e\vec{V}_1 = (0.0047\underline{/-60.8°})(237.1\underline{/27.2°})$$

$$= 1.11\underline{/-33.6°} = 0.93 - j0.62\ \text{A}$$

$$\vec{I}_2' = 40\underline{/0°} = 40 + j0\ \text{A}$$

$$\vec{I}_1 = \vec{I}_e + \vec{I}_2' = 40.9 - j0.62 = 40.9\underline{/-0.87°}\ \text{A}$$

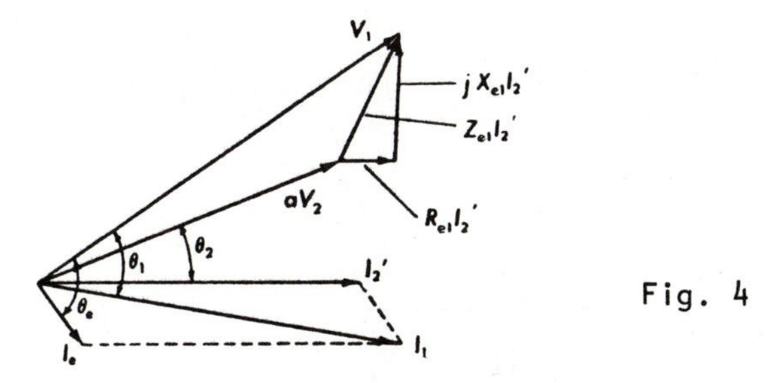

Phasor diagram for the circuit of Fig. 3

Figure 4 can serve to illustrate this solution. The reader is cautioned that the phasor diagram is not drawn to scale. $\vec{I}_2'\vec{Z}_{el}$ and $\vec{I}_e$ are shown much larger than actual to make the details clear.

• PROBLEM 6-27

The parameters of the equivalent circuit of a 100-kva, 2200/220-volt, 60-cycle transformer are given at 75°C as follows:

$r_1 = 0.286$ ohm $\qquad r_2' = 0.319$ ohm

$x_1 = 0.73$ ohm $\qquad x_2' = 0.73$ ohm

$r_m = 302$ ohms

$x_m = 1222$ ohms

If the load impedance on the low-voltage side of the transformer is $Z_L = 0.387 + j0.290$ ohm, solve the equivalent circuit when the voltage V_1 is 2300 volts.

Solution:

$$a = \frac{E_2}{E_1} = \frac{2200}{220} = 10$$

$$\vec{Z}_L' = R_L' + jX_L' = a^2(R_L + jX_L) = 38.7 + j29.0$$
$$= 48.4\underline{/36.8}\ \Omega$$

$$\vec{Z}_2' = (r_2' + R_L') + j(x_2' + X_L')$$
$$= (0.319 + 38.7) + j(0.73 + 29.0)$$
$$= 39.02 + j29.73 = 49.0\underline{/37.3}\ \Omega$$

$$\vec{Z}_m = 302 + j1222 = 1258\underline{/76.1}\ \Omega$$

$$\vec{Y}_m = 1/\vec{Z}_m = 0.795 \times 10^{-3}\underline{/-76.1}\ \text{mho.}$$

Now

$$\vec{I}_1 = \frac{\vec{V}_1}{\dfrac{1}{\vec{Z}_1 + (1/\vec{Z}_2') + \vec{Y}_m}}$$

$$= \frac{2300\underline{/0}}{\left[0.286+j0.73+\dfrac{1}{[1/(49.0\underline{/37.3})]+0.795\times10^{-3}\underline{/-76.1}}\right]}$$

$$= \frac{2300\underline{/0}}{48.2\underline{/39.2}} = 47.7\underline{/-39.2}\ \text{A}$$

$$\vec{I}_2' = \vec{I}_1 \times \frac{\vec{Z}_m}{\vec{Z}_2' + \vec{Z}_m}$$

$$= 47.7\underline{/-39.2} \times \frac{1258\underline{/76.1}}{1258\underline{/76.1} + 49.0\underline{/37.3}}$$

$$= 46.2\underline{/-40.5}\ \text{A}$$

$$\vec{I}_m = \vec{I}_1 \times \frac{\vec{Z}_2'}{\vec{Z}_m + \vec{Z}_2'} = 1.80\underline{/-76.7}\ \text{A}$$

Input power factor = cos 39.2° = 0.775 lagging

Power input = 2300 × 47.7 × 0.775 = 85.0 kw

Power output $= (46.2)^2 \times 38.7 = 82.7$ kw

Primary copper losses $= I_1^2 r_1 = (47.7)^2 \times 0.286$
$= 650$ watts

Secondary copper losses $= (\dot{I}_2')^2 r_2' = (46.2)^2 \times 0.319$
$= 680$ watts

Core losses $= I_m^2 r_m = (1.80)^2 \times 302 = 980$ watts

$$\text{Efficiency } \eta = \frac{\text{output}}{\text{input}} \times 100 = \frac{82.7}{85.0} \times 100 = 97.3\%$$

Voltage across load $= 46.2 \times 48.4 = 2240$ volts.

$$\text{Voltage Regulation} = \frac{V_2\text{no-load} - V_2\text{rated}}{V_2\text{rated}} \times 100$$

$$= \frac{2300 - 2240}{2240} \times 100 = 2.68\%$$

$$= 0.0268 \text{ pu.}$$

• PROBLEM 6-28

A 20-kVA, 2200:220-V, 60-Hz, single-phase transformer has the following equivalent-circuit parameters referred to the high-potential side of the transformer.

$R_1 = 2.51\ \Omega$ $\qquad$ $R_2' = 3.11\ \Omega$

$X_{\ell 1} = 10.9\ \Omega$ $\qquad$ $X_{\ell 2}' = 10.9\ \Omega$

$X_m' = 25,100\ \Omega$

The transformer is supplying 15 kVA at 220 volts and a lagging power factor of 0.85.

Determine the required potential difference at the high-potential terminals of the transformer.

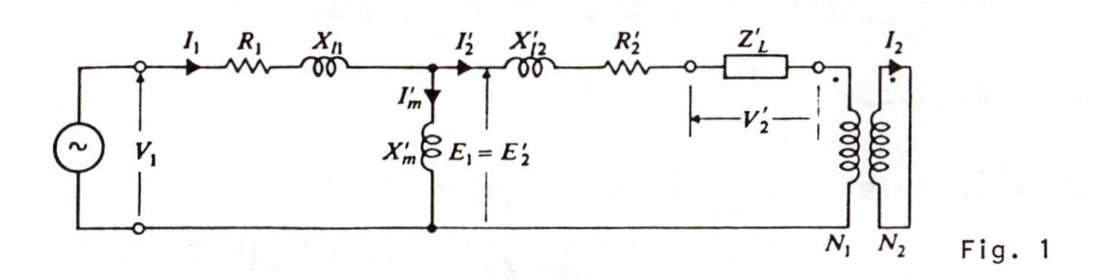

Fig. 1

Equivalent Circuit for Transformers with Sinusoidal Excitation.

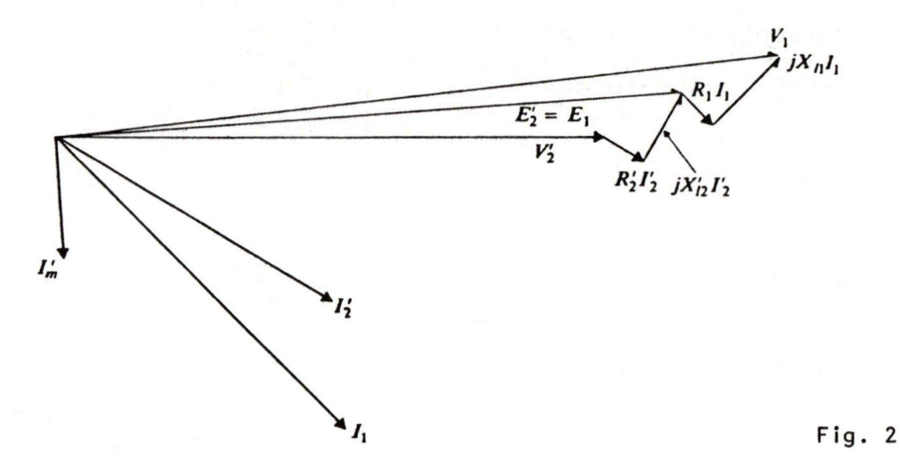

Fig. 2

Phasor diagram for system of Fig. 1.

Solution: The equivalent circuit of Fig. 1 and phasor diagram of Fig. 2 are applicable.

Note that the transformer is not supplying its rated output power. Also note that the rating gives the nominal ratio of terminal potential differences--that is, it gives the turns ratio of the ideal transformer. To a close approximation, this is also the ratio of potential differences on no load--that is, with the low-potential terminals open circuited.

$$\text{Let } \vec{V}_2 = 220\underline{/0} \text{ V.}$$

$$\vec{V}_2' = \frac{2200}{220}\vec{V}_2 = 2200\underline{/0} \text{ V.}$$

$$\therefore \quad I_2 = \frac{15 \times 10^3}{220} = 68.2 \text{ A.}$$

$$\cos^{-1} 0.85 = 31.7°.$$

$$\vec{I}_2 = 68.2\underline{/-31.7°} \text{ A.}$$

$$\vec{I}_2' = \frac{220}{2200} \times 68.2\underline{/-31.7°} = 6.82\underline{/-31.7°} \text{ A.}$$

From Fig. 1,

$$\vec{E}_1 = \vec{V}_2' + (R_2' + jX_{\ell 2}')\vec{I}_2' = 2200\underline{/0} + (3.11 + j10.9)6.82\underline{/-31.7°} = 2260\underline{/1.3°} \text{ V.}$$

$$\vec{I}_m' = \frac{\vec{E}_1}{jX_m'} = \frac{2260\underline{/1.3°}}{25{,}100\underline{/90°}} = 0.090\underline{/-88.7°} \text{ A.}$$

Note that I'_m is very small compared with I'_2.

$$\vec{I}_1 = \vec{I}'_2 + \vec{I}'_m = 6.82\angle{-31.7°} + 0.090\angle{-88.7°}$$
$$= 6.87\angle{-32.3°} \text{ A.}$$

$$\vec{V}_1 = \vec{E}_1 + (R_1 + jX_{\ell 1})\vec{I}_1$$
$$= 2260\angle{1.3°} + (2.51 + j10.9)6.87\angle{-32.3°}$$
$$= 2311\angle{2.6°} \text{ V.}$$

Thus V_1 = 2311 V, as compared with the rated or name-plate value of 2200 V. The additional potential of 111 V is needed to overcome the impedance of the transformer.

REGULATION

• PROBLEM 6-29

(a) A 2300-volt/230-volt, 10-kva transformer has the following constants: 1 percent resistance; 10 percent reactance; 5 percent exciting current. Find the equivalent resistance, reactance, and exciting current referred to both the low voltage and high voltage sides.

(b) Determine the voltage regulation of the transformer given in part (a) for load-power factors of unity, 80 per cent lag, and 80 per cent lead. Draw a vector diagram for these three cases.

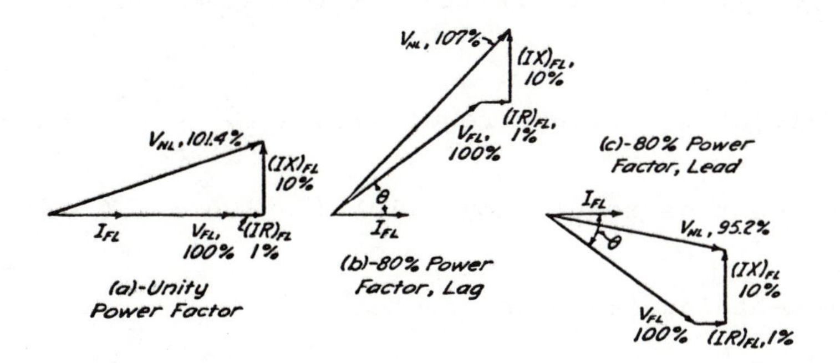

Vector diagrams at full-load and different power factors for the simplified transformer circuit.

Solution: (a) Low voltage side:

$$\text{Full-load current} = \frac{10,000}{230} = 43.5 \text{ amp.}$$

$$IR_{OL} = 230 \times 0.01 = 2.3 \text{ volts,}$$

$$R_{OL} = \frac{2.3}{43.5} = 0.0528 \text{ ohm,}$$

$$IX_{OL} = 230 \times 0.1 = 23 \text{ volts,}$$

$$X_{OL} = \frac{23}{43.5} = 0.528 \text{ ohm,}$$

$$I_{OL} = 43.5 \times 0.05 = 2.175 \text{ amp.}$$

High voltage side:

$$\text{Full-load current} = \frac{10,000}{2300} = 4.35 \text{ amp}$$

$$IR_{OH} = 2300 \times 0.01 = 23 \text{ volts,}$$

$$R_{OH} = \frac{23}{4.35} = 5.28 \text{ ohms,}$$

$$IX_{OH} = 2300 \times 0.1 = 230 \text{ volts,}$$

$$X_{OH} = \frac{230}{4.35} = 52.8 \text{ ohms,}$$

$$I_{OH} = 4.35 \times 0.05 = 0.2175 \text{ amp.}$$

(b) Unity power factor:

$$V_{NL} = \sqrt{(100 + 1)^2 + (10)^2} = 101.4 \text{ per cent}$$

Now

$$\text{Percentage voltage regulation} = \frac{V_{NL} - V_{FL}}{V_{FL}} \times 100$$

$\therefore$ Percentage regulation = 101.4 - 100 = 1.4 per cent

80 per cent lagging power factor:

$$V_{NL} = \sqrt{(100 \times 0.8 + 1)^2 + (100 \times 0.6 + 10)^2}$$

$$= \sqrt{(81)^2 + (70)^2} = 107 \text{ per cent}$$

$\therefore$ Percentage regulation = 107 - 100 = 7 per cent

80 per cent leading power factor:

$$V_{NL} = \sqrt{(100 \times 0.8 + 1)^2 + (100 \times 0.6 - 10)^2}$$
$$= \sqrt{(81)^2 + (50)^2} = 95.2 \text{ per}$$

$\therefore$ Percentage regulation = 95.2 - 100 = -4.8 per cent

• PROBLEM 6-30

A 2300/230 V, 20 kVA step-down transformer is connected as shown in the figure, with the low voltage side short-circuited. Short-circuit data obtained for the high voltage side are:

wattmeter reading = 250 W

voltmeter reading = 50 V

ammeter reading = 8.7 A

Calculate:
(a) equivalent impedance, reactance and resistance referred to the high side voltage
(b) equivalent impedance, reactance and resistance referred to the low side voltage
(c) voltage regulation at unity power factor
(d) voltage regulation at 0.7 PF lagging.

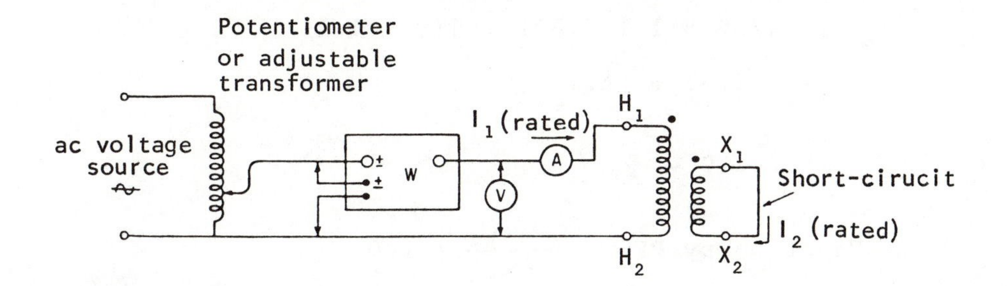

Solution:

(a) $$Z_{el} = \frac{V_{sc}}{I_{sc}} = \frac{50 \text{ V}}{8.7 \text{ A}} = 5.75 \ \Omega$$

$$R_{el} = \frac{P_{sc}}{(I_{sc})^2} = \frac{250}{(8.7)^2} = 3.3 \ \Omega$$

For X_{el}, $\theta = \text{arc cos } \frac{R_{el}}{Z_{el}} = \text{arc cos } \frac{3.3}{5.75} = 55°$

$$X_{e1} = Z_{e1} \sin\theta = 5.75 \sin 55° = 4.71\ \Omega$$

(b) $$Z_{e2} = \frac{Z_{e1}}{\alpha^2} = \frac{5.75\ \Omega}{10^2} = 0.0575\ \Omega,$$

where α is the input to output voltage ratio.

$$R_{e2} = \frac{R_{e1}}{\alpha^2} = \frac{3.3\ \Omega}{10^2} = 0.033\ \Omega$$

$$X_{e2} = \frac{X_{e1}}{\alpha^2} = \frac{4.71\ \Omega}{10^2} = 0.0471\ \Omega$$

(c) Rated secondary load current,

$$I_2 = \frac{\text{kVA} \times 1000}{V_2} = \frac{20 \times 10^3}{230\ \text{V}} = 87\ \text{A}$$

$$I_2 R_{e2} = 87\ \text{A} \times 0.033\ \Omega = 2.87\ \text{V}$$

$$I_2 X_{e2} = 87\ \text{A} \times 0.0471\ \Omega = 4.1\ \text{V}$$

Full load secondary induced emf at unity PF is

$$\vec{E}_2 = (V_2\cos\theta_2 + I_2 R_{e2}) + j(V_2 \sin\theta_2 + I_2 X_{e2})$$

$$= (230 \times 1 + 2.87) + j(0 + 4.1)$$

$$= 232.87 + j4.1$$

$$E_2 = 232.9\ \text{V}$$

$$\text{VR at unity PF} = \frac{E_2 - V_2}{V_2} \times 100$$

$$= \frac{232.9 - 230}{230} \times 100 = \frac{2.9}{230} \times 100$$

$$= 1.26$$

(d) Full load secondary induced emf at 0.7 PF lagging

$$\vec{E}_2 = (V_2\cos\theta_2 + I_2 R_{e2}) + j(V_2 \sin\theta_2 + I_2 X_{e2})$$

$$= (230 \times 0.7 + 2.87) + j(230 \times 0.713 + 4.1)$$

$$= (161 + 2.87) + j(164 + 4.1) = 163.9 + j168.1$$

$E_2 = 235$

$$\text{VR at 0.7 PF lagging} = \frac{E_2 - V_2}{V_2} = \frac{235 - 230}{230}$$

$$= 2.175\%.$$

• PROBLEM 6-31

A short-circuit test on a 15-kva, 2400/240-volt, 60-cycle transformer yields the following result: 6.25 amperes, 131 volts, and 214 watts. Determine the regulation of the transformer for a load of 0.8 lagging power factor by means of the approximate equivalent circuit.

Solution: (a) Using actual values of voltage, current, and impedance:

$$\text{Power factor for short-circuit conditions} = \frac{214}{131 \times 6.25}$$

$$= 0.261 \text{ lagging}$$

Therefore, the phase angle for short-circuit conditions

$$= 74°52' \text{ lagging}$$

$$\vec{Z}_{eq} = \frac{\vec{E}_1}{\vec{I}_1} = \frac{131\angle 0°}{6.25\angle -74°52'} = 20.96\angle 74°52' \text{ ohms}$$

$$R_{eq} = 20.96 \times \cos 74°52' = 5.49 \text{ ohms}$$

$$X_{eq} = 20.96 \times \sin 74°52' = 19.97 \text{ ohms}$$

Therefore,

$$\vec{E}_1 = 2400(0.8 + j0.6) + 6.25(5.49 + j19.97)$$

$$= 1920 \quad + j1440$$

$$+ 34.3 + j\ 124.8$$

$$= 1954.3 + j1564.8$$

$$= 2502.2 \text{ volts}$$

$$\text{Percent regulation} = \frac{E_1\text{no load} - E_1\text{full load}}{E_1\text{full load}} \times 100$$

$$= \frac{2502.2 - 2400}{2400} \times 100$$

$$= 4.26 \text{ per cent.}$$

(b) Using per-unit values:

Base value of primary voltage = 2400 volts.

Base value of primary current $= \frac{15{,}000}{2400} = 6.25$ amperes.

Base value of equivalent impedance referred to primary side $= \frac{2400}{6.25} = 384$ ohms.

$$\text{Power factor for short-circuit conditions} = \frac{214}{131 \times 6.25}$$

$$= 0.261 \text{ lagging}$$

Therefore, the phase angle for short-circuit conditions = 74°52' lagging

$$131 \text{ volts} \equiv \frac{131}{2400} = 0.0546 \text{ per unit volts}$$

$$6.25 \text{ amperes} \equiv \frac{6.25}{6.25} = 1.0 \text{ per-unit amperes}$$

$$\vec{Z}_{eq} = \frac{0.0546\underline{/0°}}{1.0\underline{/-74°52'}} = 0.0546\underline{/74°52'} \text{ per-unit ohm}$$

$$R_{eq} = 0.0546 \times \cos 74°52' = 0.01425 \text{ per-unit ohm}$$

$$X_{eq} = 0.0546 \times \sin 74°52' = 0.0506 \text{ per-unit ohm}$$

Therefore, $\vec{E}_1 = 1.0(0.8 + j0.6) + 1.0(0.01425 + j0.0506)$

$$= 0.8 \quad + j0.6$$

$$+ \; 0.01425 + j0.0506$$

$$= 0.81425 + j0.6506$$

$$= 1.0426 \text{ per-unit volts}$$

$$\therefore \text{Per cent regulation} = \frac{1.0426 - 1.0}{1.0} \times 100$$

$$= 4.26 \text{ per cent}$$

• PROBLEM 6-32

A 5-kva transformer has a nominal voltage rating of 1,100/110 volts. With the low-voltage winding short-circuited, it is found by experiment that 33 volts is required to circulate rated full-load current, and the corresponding power input is 85 watts. Find the percent regulation when the load takes rated current at a power factor of 80 percent, lagging.

Solution: (a) V_2 can be determined from the following equation:

$$V_1^2 = [V_2 + V_e \cos(\theta_e - \theta_2)]^2 + [V_e \sin(\theta_e - \theta_2)]^2 \quad (1)$$

From the given data,

$$I_1 = \frac{5,000}{1,100} = 4.55 \text{ amp}$$

$$\cos\theta_e = \frac{85}{33 \times 4.55} = 0.567$$

$$\therefore \quad \theta_e = 55°29'$$

$$\cos\theta_2 = 0.8$$

$$\therefore \quad \theta_2 = 36°52'$$

and $V_e \cos(\theta_e - \theta_2) = 33 \cos 18°37' = 31.27.$

$V_e \sin(\theta_e - \theta_2) = 33 \sin 18°37' = 10.53.$

Substituting in Eq. (1),

$$1,100^2 = (V_2 + 31.27)^2 + 10.53^2$$

$$\therefore \quad V_2 = 1,068.7$$

$$\text{Regulation} = \frac{V_1 - V_2}{V_1} \times 100\%$$

$$\text{Regulation} = \frac{1,100 - 1,068.7}{1,100} \times 100 = 2.85\%$$

(b) Another method: Alternatively, V_2 can be found from the following equation:

$$V_1^2 = (V_2 + I_1 R_e \cos\theta_2 + I_1 X_e \sin\theta_2)^2 + (I_1 X_e \cos\theta_2 - I_1 R_e \sin\theta_2)^2 \qquad (2)$$

$$I_1 = \frac{5,000}{1,100} = 4.55 \text{ amp}$$

$$Z_e = \frac{V_e}{I_1} = \frac{33}{4.55} = 7.25 \text{ ohms}$$

$$R_e = \frac{P_e}{I_1^2} = \frac{85}{4.55^2} = 4.125 \text{ ohms.}$$

$$X_e = \sqrt{Z_e^2 - R_e^2} = \sqrt{7.25^2 - 4.125^2} = 5.96 \text{ ohms.}$$

Substituting in Eq.(2),

$$1,100^2 = (V_2 + 4.55 \times 4.125 \times 0.8 + 4.55 \times 5.96 \times 0.6)^2 + (4.55 \times 5.96 \times 0.8 - 4.55 \times 4.125 \times 0.6)^2$$

$$1,100^2 - 10.44^2 = (V_2 + 15.0 + 16.3)^2$$

$$V_2 = 1,068.7$$

$$\therefore \quad \text{Regulation} = \frac{1,100 - 1,068.7}{1,100} \times 100 = 2.85\%$$

• PROBLEM 6-33

The following measurements were obtained from tests carried out on a 10-kVA, 2300:230-V, 60-Hz distribution transformer:

Open-circuit test, with the low-potential winding excited:

Applied potential difference, $V_{oc} = 230$ V

Current, $I_{oc} = 0.45$ A

Input power, $P_{oc} = 70$ W

Short-circuit test, with the high-potential winding excited:

Applied potential difference, $V_{sc} = 120$ V

Current, $I_{sc} = 4.5$ A

Input power, $P_{sc} = 240$ W

Winding resistances, measured by dc bridge:

$$R_{hp} = 5.80\ \Omega, \qquad R_{lp} = 0.0605\ \Omega$$

The transformer is supplying full load at 230 volts and a power factor of 0.8 lagging. Determine

a) The primary potential difference required.
b) The power factor at the primary terminals.
c) The transformer regulation.
d) The approximate change in turns ratio required if the primary potential difference is fixed at 2300 V.

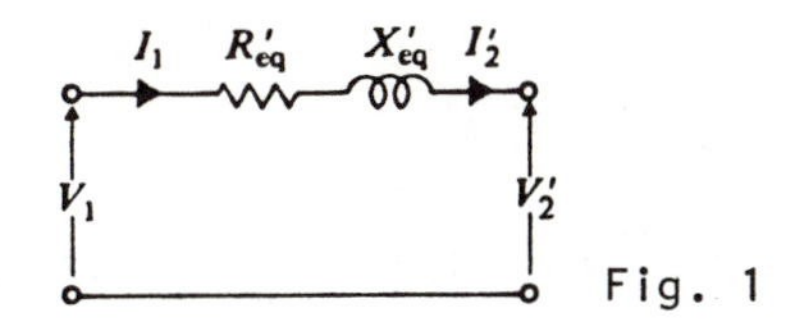

Fig. 1

Approximate transformer equivalent circuits.

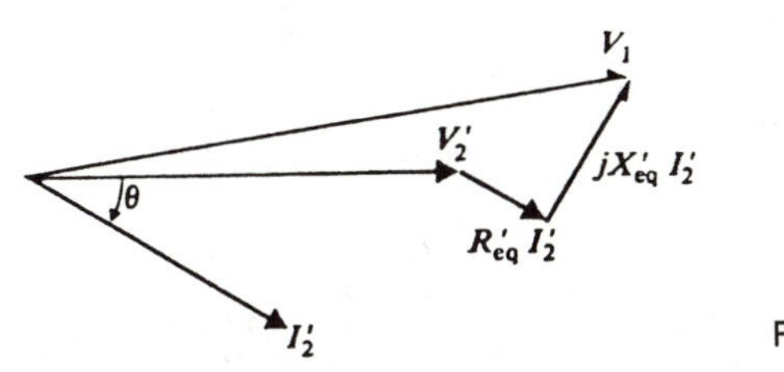

Fig. 2

Phasor diagrams for the approximate equivalent circuit of Fig. 1 for lagging power factor.

Solution: Since $\cos^{-1} 0.8 = 36.8°$, the equivalent circuit and phasor diagram for the specified condition of operation are as shown in Fig. 1 and 2, except the quantities will be referred to the secondary winding.

a) Since

$$\vec{V}_2 = 230\angle 0°, \text{ then } \vec{I}_2 = 43.5\angle -36.8° \text{ V.}$$

$$\vec{V}_1'' = \vec{V}_2 + (R''_{eq} + jX''_{eq})\vec{I}_2$$

$$R'_{eq} = \frac{P_{sc}}{I^2_{sc}} = \frac{240}{(4.5)^2} = 11.85\ \Omega$$

$$R''_{eq} = \left(\frac{230}{2300}\right)^2 R'_{eq} = 0.118\ \Omega$$

$$Z_{sc} = \frac{V_{sc}}{I_{sc}} = \frac{120}{4.5} = 26.7\ \Omega$$

$$X'_{eq} = [Z^2_{sc} - (R'_{eq})^2]^{1/2}$$

$$= [(26.7)^2 - (11.8)^2]^{1/2} = 24.0\ \Omega.$$

$$X''_{eq} = \left(\frac{230}{2300}\right)^2 X'_{eq} = 0.240\ \Omega.$$

$$R''_{eq} + jX''_{eq} = 0.118 + j0.240 = 0.266\underline{/63.5^\circ}\ \Omega$$

Thus

$$\vec{V}''_1 = 230\underline{/0^\circ} + 0.266\underline{/63.5^\circ} \times 43.5\underline{/-36.8^\circ}$$

$$= 240.4\underline{/1.24^\circ}\ \text{V}.$$

In Fig. 2 the angle between $\vec{V}''_1$ and $\vec{V}_2$ is 1.24°.

$$V_1 = \frac{2300}{230} \times V''_1 = 10 \times 240.4 = 2404\ \text{V}.$$

b) The power factor at the primary terminals is $\cos(36.8^\circ + 1.24^\circ) = 0.788$.

c) When load is removed from the transformer, V_2 will rise to the value of V''_1 on full load; that is, V_2 = 240.4 volts. Thus

$$\text{Percentage regulation} = \frac{240.4 - 230}{230} \times 100\% = 4.52\%$$

d) The result in (c) shows that the turns ratio of the transformer must be reduced on load by 4.52% if the secondary potential difference is to remain unaltered. That is,

$$\frac{N_1}{N_2} = \frac{100 - 4.52}{100} \times 10 = 9.548.$$

• PROBLEM 6-34

The following are the constants of a 300 kva 11,000: 2,300-volt 60-cycle power transformer.

$r_1 = 1.28$ ohms $\qquad r_2 = 0.0467$ ohm

$x_1 = 4.24$ ohms $\qquad x_2 = 0.162$ ohm

The core loss and the no-load current when 2,300 volts at 60 cycles are impressed on the low-voltage winding are

$P_n = 2,140$ watts $\qquad I_n = 3.57$ amp

Assume that an inductive load of 300 kva at 0.8 power factor is connected to the low-voltage terminals. Calculate the voltage which it is necessary to impress on the high-voltage terminals in order to maintain 2,300 volts across the load. The vector diagram of the transformer is shown in the figure.

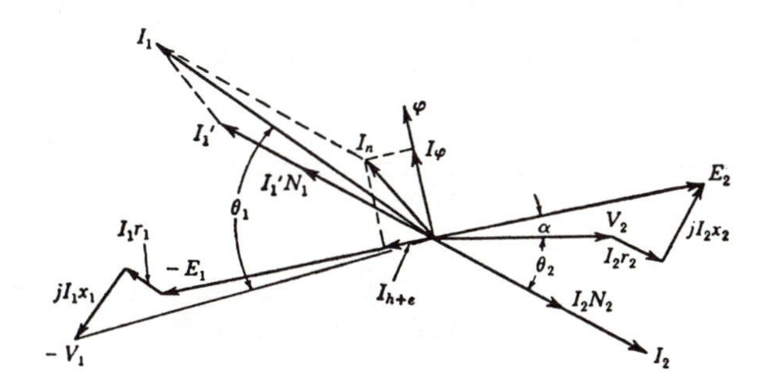

Solution:

$$I_2 = \frac{300,000}{2,300} = 130.4 \text{ amp.}$$

$$\vec{Z}_2 = r_2 + jx_2 = 0.0467 + j0.162\ \Omega.$$

Take the secondary voltage as reference axis.

$$\vec{V}_2 = 2,300(1 + j0) \text{ volts.}$$

$$\vec{I}_2 = 130.4(0.8 - j0.6) = 104.3 - j78.2 \text{ amp}$$

$$\vec{E}_2 = \vec{V}_2 + \vec{I}_2\vec{Z}_2 = 2,300(1 + j0) + (104.3 - j78.2) \times (0.0467 + j0.162)$$

$$= 2,317.5 + j13.45 \text{ volts.}$$

$$a = \frac{11,000}{2,300} = 4.783.$$

$$\vec{I}_1' = -\frac{1}{a}\vec{I}_2 = -\left(\frac{104.3 - j78.2}{4.783}\right) = -21.81 + j16.35 \text{ amp.}$$

As an approximation assume that I_n is fixed by the terminal voltage instead of by E_1. Referred to the high-voltage side of the transformer the exciting current is

$$I_n = \frac{3.57}{4.783} = 0.746 \text{ amp.}$$

$$I_{h+e} = \frac{2,140}{2,300} \times \frac{1}{a} = \frac{2,140}{11,000} = 0.195 \text{ amp.}$$

As an approximation assume that I_ϕ is sinusoidal.

$$I_\phi = \sqrt{(0.746)^2 - (0.195)^2} = 0.720 \text{ amp}$$

$$\vec{I}_1 = \vec{I}_1' + \vec{I}_n = \vec{I}_1' + I_{h+e}(-\cos\alpha - j\sin\alpha) + jI_\phi(\cos\alpha + j\sin\alpha) \text{ A.}$$

$$\cos\alpha = \frac{2,318}{\sqrt{(2,318)^2 + (13.4)^2}} = 1.000$$

$$\sin\alpha = \frac{13.4}{\sqrt{(2,318)^2 + (13.4)^2}} = 0.0058$$

$$\therefore \quad \vec{I}_1 = (-21.81 + j16.35) + 0.195(-1.000 - j0.0058) + j0.720(1.000 + j0.0058)$$

$$= (-21.81 + j16.35) + (-0.195 - j0.001) + (j0.720 - 0.004)$$

$$= -22.01 + j17.07 \text{ amp}$$

$$\vec{E}_1 = a\vec{E}_2 = 4.783(2317.5 + j13.45) \text{ volts}$$

$$-\vec{V}_1 = -\vec{E}_1 + \vec{I}_1(r_1 + jx_1)$$

$$= -4.783(2317.5 + j13.45) + (-22.01 + j17.07)(1.28 + j4.24)$$

$$= (-11,087 - j64.3) + (-100.6 - j71.5)$$

$$= -11,188 - j135.8 \text{ volts}$$

$$V_1 = \sqrt{(11,188)^2 + (135.8)^2} = 11,189 \text{ volts}$$

The open-circuit secondary voltage corresponding to a given primary impressed voltage is given by

$$V_2' = \frac{V_1 - I_n(r_1\cos\theta_n + x_1\sin\theta_n)}{a}$$

The no-load primary copper loss is negligible and is omitted when finding the cosine and sine of the no-load power-factor angle.

$$\cos\theta_n = \frac{2,140}{2,300 \times 3.57} = 0.261$$

$$\therefore \quad \sin\theta_n = 0.966$$

$$I_n(\text{on the high-voltage side}) = \frac{3.57}{4.783} = 0.746 \text{ amp}$$

$$V_2' = \frac{11,189 - 0.746(1.28 \times 0.261 + 4.24 \times 0.966)}{4.783}$$

$$= \frac{11,186}{4.783} = 2,338.7 \text{ volts}$$

$$\therefore \quad \text{Regulation} = \frac{V_2' - V_2}{V_2} \times 100 = 1.68 \text{ per}$$

• PROBLEM 6-35

A 10-kva 2,400:240-volt 60-cycle transformer has the following data:

$R_1 = 5.2$ ohms $\qquad$ $R_2 = 0.052$ ohm

$X_1 = 7.6$ ohms $\qquad$ $X_2 = 0.076$ ohm

Assume it is operating at rated full load, 0.8 power factor lagging, with rated output voltage, and with input voltage adjusted to give rated output voltage under this condition of load. Determine (a) the input voltage E_1, (b) the generated voltage E_{2g}, (c) the regulation (i) at 0.8 power factor lagging, (ii) at unity power factor and (iii) at 0.8 power factor leading.

Solution:

$$I_2 = \frac{10,000}{240} = 41.7 \text{ amp.}$$

$$I_2' = \frac{41.7}{10} = 4.17 \text{ amp} = I_1.$$

$$R_2' = 100 \times 0.052 = 5.2 \text{ ohms}.$$

Note that the equivalent resistance of the secondary is usually about equal to that of the primary, which means that the secondary copper loss is about equal to that in the primary. This represents a properly balanced design since the volume and weight of copper in the primary and secondary coils are approximately equal. So the loss per pound or per cubic inch is the same in both windings. Similarly the equivalent secondary leakage reactance is about equal to the primary value:

$$R = R_1 + R_2' = 10.4 \text{ ohms}.$$

$$X_2' = 100 \times 0.076 = 7.6 \text{ ohms}.$$

$$X = X_1 + X_2' = 15.2 \text{ ohms}.$$

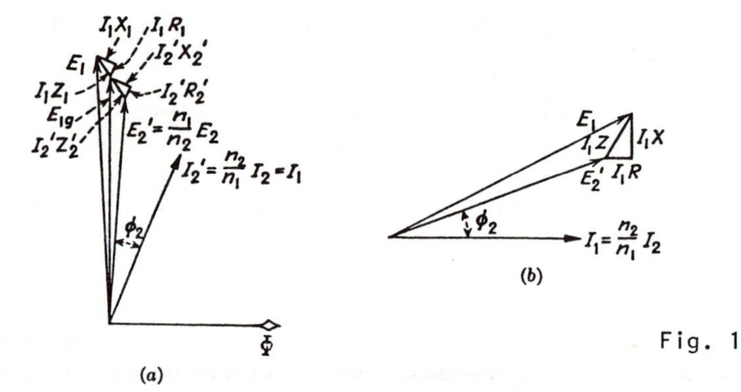

Fig. 1

Simplified vector diagrams of transformer with lagging power-factor load.

(a) The problem of finding E_1 will be solved by using components in phase and in quadrature with the current; thus

$$E_2' = 10 \times 240 = 2{,}400 \text{ volts}$$

$$E_2' \cos \phi_2 = 2{,}400 \times 0.8 = 1{,}920 \text{ volts}$$

$$E_2' \sin \phi_2 = 2{,}400 \times 0.6 = 1{,}440 \text{ volts}$$

$$IR = 4.17 \times 10.4 = 43.4 \text{ volts}$$

$$IX = 4.17 \times 15.2 = 63.4 \text{ volts}$$

$$E_1 = \sqrt{(E_2' \cos \phi_2 + IR)^2 + (E_2' \sin \phi_2 + IX)^2}$$

$$= \sqrt{(1,920 + 43.4)^2 + (1,440 + 63.4)^2} = 2,473 \text{ volts}$$

which is about 3 percent higher than the value required at no load to give rated secondary voltage. Conversely, if the primary voltage is held at its rated value of 2,400 volts, the secondary voltage will fall under load to a value about 3 percent less than 240, or about 233 volts.

The value of the primary power factor may be computed as

$$\cos \phi_1 = \frac{1,963.4}{2,473} = 0.794$$

which is very nearly equal to the load power factor.

(b) In order to maintain E_2 at its rated value, the flux and generated voltage must be increased slightly. E_{2g} is found from the following equation:

$$E_{2g} = \sqrt{(E_2 \cos \phi_2 + I_2R_2)^2 + (E_2 \sin \phi_2 + I_2X_2)^2}$$

and $E_2 \cos \phi_2 = 192$ volts.

$E_2 \sin \phi_2 = 144$ volts.

$I_2R_2 = 41.7 \times 0.052 = 2.17$ volts

$I_2X_2 = 41.7 \times 0.076 = 3.17$ volts

$$\therefore \quad E_{2g} = \sqrt{(192 + 2.17)^2 + (144 + 3.17)^2} = 243.7 \text{ volts}$$

which indicates that ϕ must be increased about 1.5 percent above the value required at no load.

(c) (i) The regulation for 0.8 power factor lagging is

$$\text{Reg} = \frac{2,473 - 2,400}{2,400} = \frac{73}{2,400} = 3.04 \%$$

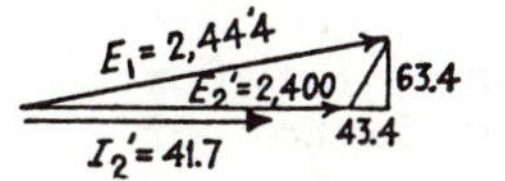

(a) Unity power-factor load

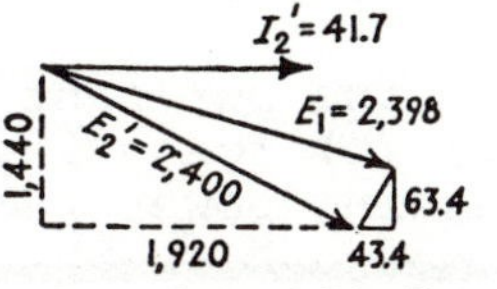

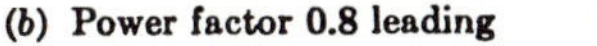

(b) Power factor 0.8 leading

Fig. 2

Transformer-regulation diagrams.

(ii) For the same transformer with unity power-factor load (see Fig. 2a)

$$E_1 = \sqrt{(2,400 + 43.4)^2 + (63.4)^2} = 2,444 \text{ volts.}$$

$$\text{Reg} = \frac{2,444 - 2,400}{2,400} = \frac{44}{2,400} = 1.83\%$$

(iii) For a 0.8 leading power-factor load (see Fig. 2b)

$$E_1 = \sqrt{(1,920 + 43.4)^2 + (1,440 - 63.4)^2}$$

$$= 2,398 \text{ volts.}$$

$$\text{Reg} = \frac{2,398 - 2,400}{2,400} = \frac{-2}{2,400} = -0.01\%$$

For this value of power factor the regulation is practically zero. For a lower value of leading power factor, the regulation is definitely negative, which means that the terminal voltage will rise as the transformer load is increased.

• PROBLEM 6-36

A 25-kva 2,300/230-volt distribution transformer has the following resistance and leakage-reactance values (ohms): $R_p = 0.8$; $X_p = 3.2$; $R_s = 0.009$; $X_s = 0.03$.
(1) Calculate the equivalent values of resistance, reactance, and impedance: (a) in secondary terms; (b) in primary terms.

(2) Calculate the equivalent resistance and reactance voltage drops for a secondary load current of 109 amp; (a) in secondary terms; (b) in primary terms.

(3) Calculate the per cent regulation: (a) for unity power factor; (b) for a lagging power factor of 0.8; (c) for a leading power factor of 0.866.

Solution:

(a) Ratio of transformation $a = \frac{2,300}{230} = 10.$

$$\left.\begin{aligned} R_e &= R_S + \frac{R_P}{a^2} \\ &= 0.009 + \frac{0.8}{100} = 0.017\Omega \\ X_e &= X_S + \frac{X_P}{a^2} \\ &= 0.03 + \frac{3.2}{100} = 0.062\Omega \\ \therefore \quad Z_e &= \sqrt{(0.017)^2 + (0.062)^2} = 0.0642\Omega \end{aligned}\right\} \text{in secondary terms}$$

(b)

$$\left.\begin{aligned} R_e &= a^2 R_S + R_P \\ &= (100 \times 0.009) + 0.8 = 1.7\Omega \\ X_e &= a^2 X_S + X_P \\ &= (100 \times 0.03) + 3.2 = 6.2\Omega \\ \therefore \quad Z_e &= \sqrt{(1.7)^2 + (6.2)^2} = 6.42\Omega \end{aligned}\right\} \text{in primary terms}$$

(2)

(a) $$\left.\begin{aligned} I_S R_e &= 109 \times 0.017 = 1.85 \text{ volts} \\ I_S X_e &= 109 \times 0.062 = 6.75 \text{ volts} \end{aligned}\right\} \text{in secondary terms}$$

(b) $$\left.\begin{aligned} I_P R_e &= \frac{109}{10} \times 1.7 = 18.5 \text{ volts} \\ I_P X_e &= \frac{109}{10} \times 6.2 = 67.5 \text{ volts} \end{aligned}\right\} \text{in primary terms}$$

(3)

(a) $V_P = \sqrt{(2,300 + 18.5)^2 + (67.5)^2} = 2,320\text{V}$

Per cent regulation (power factor = 1) =

$$\frac{2,320 - 2,300}{2,300} \times 100 = 0.87.$$

(b) $V_P = \Big\{[(2,300 \times 0.8) + 18.5]^2 +$

$[(2,300 \times 0.6) + 67.5]^2\Big\}^{1/2}$

$= 2,360V$

Per cent regulation (power factor = 0.8 lag) =

$$\frac{2,360 - 2,300}{2,300} \times 100 = 2.61$$

(c) $V_P =$

$$\sqrt{[(2,300 \times 0.866) + 18.5]^2 + [(2,300 \times 0.5) - 67.5]^2}$$

$= 2,280V$

Per cent regulation (power factor = 0.866 lead) =

$$\frac{2,280 - 2300}{2,300} \times 100 = - 0.87.$$

• PROBLEM 6-37

A 10-kva 2,400:240-volt 60-cycle transformer has the following data:

$R_1 = 5.2$ ohms $\qquad R_2 = 0.052$ ohm

$X_1 = 7.6$ ohms $\qquad X_2 = 0.076$ ohm

The iron loss was found to be 61 watts.

Calculate (a) the per-unit values of various impedances, (b) regulation using per-unit system, (c) efficiency using per-unit system.

Solution:

$E_1 = 2{,}400$ volts $= 1.0$ per unit V

$E_2 = 240$ volts $= 1.0$ per unit V

$I_1 = 4.17$ amp $= 1.0$ per unit A

$I_2 = 41.7$ amp $= 1.0$ per unit A

P_1 or $P_2 = 10{,}000$ watts $= 1.0$ per unit W

Also

10,000 va = 1.0 per unit.

(a) To get the per-unit values of the various impedances,

$$R_1 = 5.2 \text{ ohms} = \frac{5.2 \times 4.17}{2{,}400} = 0.00905 \text{ per unit.}$$

$$R_2 = 0.052 \text{ ohm} = \frac{0.052 \times 41.7}{240} = 0.00905 \text{ per unit}$$

$$R = 10.4 \text{ ohms} = 0.0181 \text{ per unit}$$

$$X_1 = 7.6 \text{ ohms} = 0.0132 \text{ per unit} = X_2$$

$$X = 0.0264 \text{ per unit}$$

$$Z = \sqrt{(0.0181)^2 + (0.0264)^2} = 0.032 \text{ per unit}$$

(b) To find regulation in the per-unit system, E_1 is obtained by vector addition

$$E_1 = E_2 + I_1 Z$$

If this is computed for rated secondary voltage and current, i.e.,

$$E_2 = 1.0\text{V} \quad \text{and} \quad I_2 = I_1 = 1.0\text{A}$$

then the regulation is

$$\text{Reg} = E_1 - 1.0$$

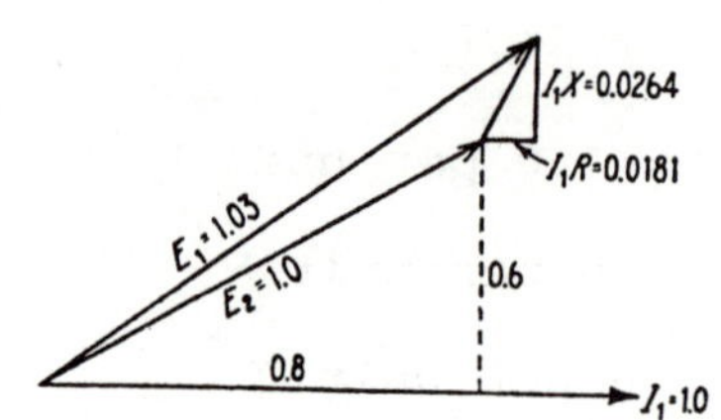

Transformer-regulation diagram using per-unit quantities.

From the figure, using per-unit quantities to compute regulation for 0.8 power factor lagging,

$$I_1R = R = 0.0181\Omega.$$

$$I_1X = X = 0.0264\Omega.$$

$$E_1 = \sqrt{(0.8 + 0.0181)^2 + (0.6 + 0.0264)^2} = 1\ 0304\text{V}.$$

Hence, Reg. $= 1.0304 - 1.0 = 0.0304.$

(c) To compute efficiency using the per-unit system, the computation for full load, 0.8 power factor, will be repeated.

$$\text{Iron loss} = \frac{61}{10{,}000} = 0.0061.$$

$$\text{Copper loss} = R = 0.0181.$$

$$\text{Output} = 0.8.$$

$$\text{Eff} = \frac{0.8}{0.8 + 0.0061 + 0.0181} = \frac{0.8}{0.8242} = 0.971$$

Note that in the per-unit system, the resistance, the IR drop, and the I^2R loss are all equal at full load. It will be found that per-unit notation has other advantages than the simplification of computations. The effects of various design factors upon performance are almost immediately apparent when expressed in per unit.

• PROBLEM 6-38

The following are the constants of a 300 kva 11,000:2,300-volt 60-cycle power transformer.

$r_1 = 1.28$ ohms $\qquad r_2 = 0.0467$ ohm

$x_1 = 4.24$ ohms $\qquad x_2 = 0.162$ ohm

Calculate the regulation using the per-unit method.

Solution: The primary voltage is given by

$$V_1 = V_2(1 + j0) + I_2(\cos\theta_2 - j\sin\theta_2) \times (r_e - jx_e) \qquad (1)$$

Using rated values as the base values for voltage and kilovolt-amperes, the base value of V_2 is 2,300 and the per unit value is 2,300/2,300 = 1. The base value of V_1 is 11,000. The base value of current is computed from the base kilovolt-amperes and for the secondary is

$$\frac{(300 \times 1{,}000)}{2{,}300} = 130.4.$$

The base amperes for the primary is similarly

$$\frac{(300 \times 1{,}000)}{11{,}000} = 27.3.$$

I_2 has a per-unit value of 130.4/130.4 = 1. The base ohms for the primary is 11,000/27.3 = 402 and for the secondary is 2,300/130.4 = 17.64. The transformer resistances and reactances have the following per-unit values:

$$r_1 = \frac{1.28}{402} = 0.00318 \qquad x_1 = \frac{4.24}{402} = 0.01054$$

$$r_2 = \frac{0.0467}{17.64} = 0.00265 \qquad x_2 = \frac{0.162}{17.64} = 0.00920$$

$$r_e = r_1 + r_2 = 0.00583 \qquad x_e = x_1 + x_2 = 0.01974.$$

Therefore, from Eq. (1),

$$V_1 = 1.0(1 + j0) + 1.0(0.8 - j0.6)(0.00583 + j0.01974)$$
$$= 1.0 + j0 + 0.01652 + j0.01228$$
$$= 1.01652 + j0.01228$$

The magnitude of V_1 is 1.0166. The primary voltage required is therefore

$$1.0166 \times 11{,}000 = 11{,}183 \text{ volts.}$$

and the regulation is

$$\frac{V_1 - V_2}{V_2} = \frac{1.0166 - 1.0}{1.0} = 0.0166 \text{ or } 1.66 \text{ per cent.}$$

LOSSES AND EFFICIENCY

• PROBLEM 6-39

The total copper loss of a transformer as determined by a short-circuit test at 20° C is 630 watts, and the copper loss computed from the true ohmic resistance at the same temperature is 504 watts. What is the load loss at the working temperature of 75° C?

Solution:

Eddy-current loss at 20° C = 630 - 504 = 126 watts.

True copper loss at 75° C can be found using the following formula:

True copper loss at temp. t_2 =

$$\text{True copper loss at temp. } t_1 \cdot \left[\frac{t_2 + 234.5}{t_1 + 234.5}\right]$$

∴ True copper loss at 75° C =

$$504 \times \frac{75 + 234.5}{20 + 234.5} = 613 \text{ watts.}$$

Eddy-current loss at 75° C can be found using the following formula:

Eddy-current loss at temp. t_2 =

$$\text{Eddy-current loss at temp. } t_1 \times \left[\frac{t_1 + 234.5}{t_2 + 234.5}\right]$$

$\therefore$ Eddy-current loss at 75° C =

$$126 \times \frac{20 + 234.5}{75 + 234.5} = 104 \text{ watts.}$$

Load loss at 75° C = 613 + 104 = 717 watts.

• PROBLEM 6-40

A 4,400-volt 60-cycle transformer has core loss of 840 watts, of which one-third is eddy-current loss. Determine the core loss when the transformer is connected (a) to a 4,600-volt 60-cycle source, (b) to a 4,400-volt 50-cycle source, and (c) to a 4,600-volt 50-cycle source.

Solution:

$$P_e = \frac{1}{3} \times 840 = 280 \text{ watts.} \quad P_h = 840 - 280 = 560 \text{ watts.}$$

$$P_c = \text{hysteresis loss} + \text{eddy-current loss} = P_h + P_e$$

where

$$P_h = k_1\left(\frac{E^{1.6}}{f^{0.6}}\right);$$

$$P_e = k_2 E^2,$$

(a) $$P_c = 560 \times \left[\frac{\left(\frac{4600}{4400}\right)^{1.6}}{\left(\frac{60}{60}\right)^{0.6}}\right] + 280 \left(\frac{4600}{4400}\right)^2$$

$$= 560 \times \left(\frac{4,600}{4,400}\right)^{1.6} + 280 \times \left(\frac{4,600}{4,400}\right)^2$$

$$= (560 \times 1.073) + (280 \times 1.093) =$$

$$601 + 306 = 907 \text{ watts.}$$

(b) $P_c = 560 \times \left(\frac{60}{50}\right)^{0.6} + 280 = (560 \times 1.121) + 280$

$= 624 + 280 = 904$ watts

(c) $P_c = 560 \times \left[\left(\frac{4,600}{4,400}\right)^{1.6} \left(\frac{60}{50}\right)^2\right] + \left[280 \times \left(\frac{4,600}{4,400}\right)^2\right]$

$= (560 \times 1.073 \times 1.121) + (280 \times 1.093)$

$= 674 + 306 = 980$ watts.

• PROBLEM 6-41

A 25-kva, 2200- to 110-volt, 60-cycle distribution transformer is tested for losses and ratio. With connections as in Fig. 1, the high-tension winding being open, the wattmeter read 185 watts, the ammeter 5.2 amperes, and primary voltmeter 110 volts.

With connections as in Fig. 2, the low-tension winding being short-circuited, the primary voltage was adjusted until rated secondary current of 25,000/110 = 227.2 amperes was flowing. The primary wattmeter then read 500 watts, and the primary voltmeter 190 volts. Find the losses and equivalent resistances of the transformer.

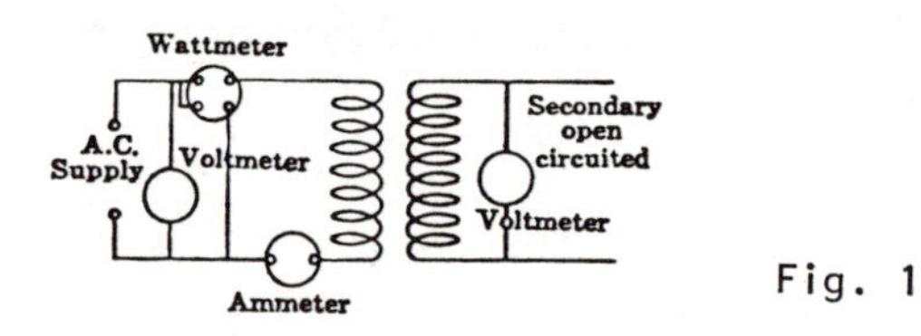

Fig. 1

Proper connections for measuring the core loss of a transformer.

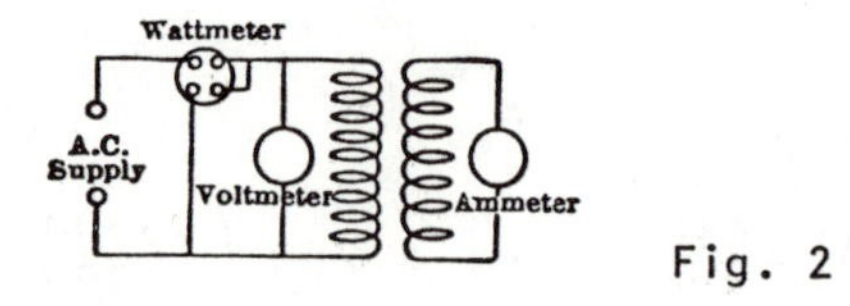

Fig. 2

Proper connections for measuring the copper loss of a transformer.

Solution: The core or iron loss of the transformer is evidently 185 watts, and the no-load primary power factor is equal to

$$\frac{P}{VI} = 185/(110 \times 5.2) = 0.323.$$

The ratio of the transformer, u, is 2200/110 = 20.

With 227.2 amperes rated current, flowing in the short-circuited secondary, the rated primary current is

$$I_1 = I_2/u = 227.2/20 = 11.36 = 25{,}000 \text{ watts}/2200 \text{ volts}.$$

The primary equivalent resistance is

$$R_1' = \frac{P_{sc}}{I_1^2} = \frac{500}{11.36^2} = 3.874 \text{ ohms}.$$

The secondary equivalent resestance is

$$R_2' = \frac{P_{sc}}{I_{sc}^2} = \frac{500}{227.2^2} = 0.00968 \text{ ohm},$$

$$= R_1'/u^2 = 3.874/400 = 0.00968 \text{ ohm}.$$

If the primary and secondary copper losses of the transformer were equal and each therefore equal to 250 watts, the actual primary resistance would be $250/11.36^2 = 1.937$ ohms and that of the secondary $250/227.2^2 = 0.00484$ ohm.

The primary equivalent resistance would then be

$$R'_1 = R_1 + u^2R_2 = 1.937 + (400 \times 0.00484) =$$

$$3.873 \text{ ohms},$$

and the secondary equivalent resistance would be

$$R'_2 = R_1/u^2 + R = (1.937/400) + 0.00484 =$$

$$0.00968 \text{ ohm}.$$

Number 29 gage sheet steel designated as U.S.S. Transformer 52 yields the following data:

Frequency	Flux density, kilogauss	Core loss, watts per lb
30	8	0.135
60	12	0.75
30	12	0.31

Calculate the hysteresis and eddy-current losses for each of the frequencies shown in the table.

Solution: The losses are determined using the following equations:

$$\text{Hysteresis loss} \quad P_h = k_h f B_m^x \tag{1}$$

$$\text{Eddy-current loss} \quad P_e = k_e f^2 B_m^2 \tag{2}$$

From the given data,

$$x = \frac{\log \left[\dfrac{B_2{}^2 (P_2 - a^2 P_3)}{(P_2 - aP_3)\ B_1{}^2 - a(a-1) P_1 B_2{}^2} \right]}{\log (B_2/B_1} = 2.06$$

where

$$a = f_2/f_1$$

$$k_h = \frac{P_2 - a^2 P_3}{f_2 (1-a) B_2^x} = 489 \times 10^{-7}$$

$$k_e = \frac{P_2 - a\ P_3}{f_2{}^2 B_2{}^2} \left(\frac{a}{a-1}\right) = 501 \times 10^{-9}$$

P_h and P_e are computed by substituting these values in Eqs. (1) and (2), and are shown in the following tabulation:

f	B, kilogauss	P_h, watts per lb	P_e, watts per lb
30	8	0.106	0.029
60	12	0.490	0.260
30	12	0.245	0.065

The results of these calculations show that the eddy-current loss is the smaller of the two parts of the core loss; it is usually between 20 and 50 per cent of the total core loss.

• PROBLEM 6-43

A transformer has the following losses corresponding to three conditions:

B_1 = 25,800 lines per sq in., $f_1 = 30$, $P_1 = 270$

B_2 = 64,500 lines per sq in., $f_2 = 60$, $P_2 = 2920$

B_2 = 64,500 lines per sq in., $f_1 = 30$, $P_3 = 1250$

Find the total core loss

Solution:

$$a = \frac{\log\left[\dfrac{1 - \beta^2 P_3/P_2}{\beta(1-\beta)(P_1/P_2) + (B_1/B_2)^2(1-\beta P_3/P_2)}\right]}{\log(B_2/B_1)}$$

Then, since $\beta = 2$,

$$a = \frac{\log\left[\dfrac{1 - 4 \times 0.428}{2(1-2)0.0925 + 0.4^2(1 - 2 \times 0.428)}\right]}{\log 2.50} = 1.615$$

$$k' = \frac{f_1^2 P_2 - f_2^2 P_3}{f_1 f_2 (f_1 - f_2) B_2^a}$$

$$= \frac{30 \times 30 \times 2920 - 60 \times 60 \times 1250}{30 \times 60(30 - 60)64{,}500^{1.615}} = 0.60 \quad 10^{-6}.$$

$$k'' = \frac{f_1P_2 - f_2P_3}{f_1f_2(f_2 - f_1)B_2^2}$$

$$= \frac{30 \times 2920 - 60 \times 1250}{30 \times 60(60 - 30)64{,}500^2} = 0.56 \times 10^{-10}$$

Hence the core loss for this transformer is

$$P = k'f\,B_{max}^{a} + k''f^2B_{max}^2$$

$$= \frac{0.60}{10^6}\,f\,B_{max}^{1.615} + \frac{0.56}{10^{10}}\,f^2B_{max}^2 \text{ W.}$$

• PROBLEM 6-44

The following data were obtained when a short-circuit test was performed upon a 100-kva 2,400/240-volt distribution transformer: E_{SC} = 72 volts; I_{SC} = 41.6 amp; P_{SC} = 1,180 watts. All instruments
high side, and the low side was short-circuited.

(1) Calculate: (a) the equivalent resistance, impedance, and reactance; (b) the per cent regulation at a power factor of 0.75 lagging.

(2) Calculate the copper losses when the load is (a) 125 kva, (b) 75 kva, (c) 85 kw at a power factor of 0.772.

Solution:

(a) $$R_e = \frac{P_{SC}}{I_{SC}^2}$$

$$= \frac{1{,}180}{(41.6)^2} = 0.682 \text{ ohm.}$$

$$Z_e = \frac{E_{SC}}{I_{SC}}$$

$$= \frac{72}{41.6} = 1.73 \text{ ohms.}$$

$$X_e = \sqrt{Z_e^2 - R_e^2}$$

$$= \sqrt{(1.73)^2 - (0.682)^2} = 1.59 \text{ ohms.}$$

(b)

$$\text{Rated current} = \frac{100,000}{2,400} = 41.6 \text{ amp.}$$

$$\therefore \qquad IR_e \text{ drop} = 41.6 \times 0.682 = 28.4 \text{ volts}$$

$$IX_e \text{ drop} = 41.6 \times 1.59 = 66.2 \text{ volts.}$$

Now,

$$V_P = \sqrt{[V_{in} \cos\phi + IR_e]^2 + [V_{in} \sin\phi + IX_e]^2}$$

$$= \left\{ [2,400 \times 0.75) + 28.4]^2 + [(2,400 \times 0.66) + 66.2]^2 \right\}^{1/2}$$

$$= 2,460 \text{ volts.}$$

$$\text{Per cent regulation} = \frac{V_{NL} - V_{FL}}{V_{FL}} \times 100$$

$$= \frac{2,460 - 2,400}{2,400} \times 100 = 2.5.$$

(2)

(a) $\text{Copper loss} = \left(\frac{125}{100}\right)^2 \times 1, \quad = 1,845 \text{ watts.}$

(b) $\text{Copper loss} = \left(\frac{75}{100}\right)^2 \times 1,180 = 663 \text{ watts.}$

(c) $\text{Copper loss} = \left(\frac{85/0.772}{100}\right)^2 \times 1,180 = 1,430 \text{ watts.}$

Test data for a 10-kva, 2,300/230-volt, 60-cycle-per-second distribution transformer are as follows: open-circuit test, with input to low side, 230 volts, 0.62 amp, 69 watts; short-circuit test, with input to high side, 127 volts, 4.9 amp, 263 watts; ohmic resistance test with input to low side 1.89 volts (d-c), 41.0 amp (d-c), and with input to high side, 23.4 volts (d-c), 4.60 amp (d-c).

aV_2 = 2,300 volts, $I'_{2\ rated}$ = 4.35 amp,

R_{el} = 10.9 ohms, and P_c = 69 watts.

For the above transformer, find the efficiency when it is operated at rated frequency and rated voltage for: (a) rated kva at pf = 1; (b) one half of rated kva at pf = 0.6 lead.

Solution:

(a) Rated kva, pf = 1.

$$P_{out} = aV_2I'_2 \quad \cos\theta_2 = 2{,}300(4.35)(1) = 10{,}000 \text{ watts}$$

$$P_{cu} = I'^2_2 R_{el} = (4.35)^2 10.9 = 206 \text{ watts}$$

$$P_c = 69 \text{ watts}$$

$$\eta = 1 - \frac{P_c + P_{cu}}{P_{in}}$$

$$= 1 - \frac{P_c + I'^2_2 \ R_{el}}{aV_2I'_2 \cos\theta_2 + P_c + I'^2_2 \ R_{el}}$$

$$= 1 - \frac{69 + 206}{10{,}000 + 69 + 206} = 1 - 0.0268 =$$

$$0.973 \equiv 97.3\%$$

(b) One half rated kva, pf = 0.6 lead.

$$I'_2 = \frac{1}{2} I'_{2\ rated} = \frac{1}{2}(4.35) = 2.17 \text{ amp}$$

$$P_{out} = aV_2I'_2 \cos\theta_2 = 2{,}300(2.17)(0.6) = 3{,}000 \text{ watts}$$

$$P_{cu} = (2.17)^2 10.9 = 51.5 \text{ watts}$$

$$P_c = 69 \text{ watts}$$

$$\eta = 1 - \frac{69 + 51.5}{3{,}000 + 69 + 51.5} = 1 - 0.0385 = 0.961$$

$$\equiv 96.1\%$$

• PROBLEM 6-46

A 50-kva 2,200/220 volt 60-cycle transformer has a core loss, determined by the open-circuit test, of 350 watts and a copper loss, at rated current, of 630 watts, determined by the short-circuit test. Find the efficiency (a) at full load, unity power factor; (b) at three-fourths load, unity power factor; (c) at full kva rating; 80 per cent power factor; (d) at three-fourths of rated kva, 80 per cent power factor.

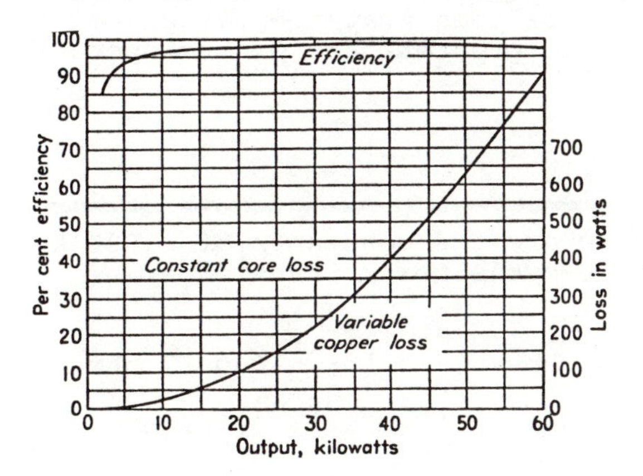

Efficiency curves, 50-kva transformer.

Solution:

(a) Efficiency η is given by

$$\eta = \frac{\text{output}}{\text{output + losses}} \times 100$$

$$\therefore \qquad \eta = \frac{50{,}000 \times 100}{50{,}000 + 350 + 630} = 98.1\%$$

(b)

$$\eta = \frac{3/4\ (50{,}000) \times 100}{3/4\ (50{,}000) + 350 + (3/4)^2 630} = 98.2\%$$

(c)

$$\eta = \frac{50{,}000 \times 0.80 \times 100}{50{,}000 \times 0.80 + 350 + 630} = 97.6\%$$

(d)

$$\eta = \frac{3/4\ (50{,}000 \times 0.80) \times 100}{3/4\ (50{,}000 \times 0.80) + (3/4)^2 630} = 97.7\%$$

The variation of efficiency with load is illustrated in the figure, which likewise shows the constant and variable components of the total loss. The abscissas in this figure represent output in kilowatts, and since kilowatt output is proportional to the current, the curve showing the variable copper loss (I^2R) is a parabola. The intersection of this curve with the constant-loss line occurs at a point where the efficiency is a maximum. It will be seen that the efficiency changes very little over the greater part of the operating range.

• PROBLEM 6-47

The following are the constants of a 300 kva 11,000: 2,300-volt 60-cycle power transformer.

$r_1 = 1.28$ ohms $\qquad r_2 = 0.0467$ ohm

$x_1 = 4.24$ ohms $\qquad x_2 = 0.162$ ohm

When this transformer is operated at no load, with 2,300 volts at 60 cycles impressed on its low-voltage winding, the current and power are $I_n = 3.57$ amp and $P_n = 2{,}140$ watts. When the high-voltage winding is short-circuited and 48.4 volts at 60 cycles are impressed on the low-voltage winding, the input current and power are $I_{SC} = 132.6$ amp and $P_{SC} = 1{,}934$ watts. The d-c resistances r_1 and r_2 were measured at a temperature of 25°C and the temperature during the short-circuit test was also 25°C. Calculate the efficiency for a full kva, 0.8 power-factor load.

Solution: Full load secondary current is

$I_2 = 300{,}000/2{,}300 = 130.5$ amp.

$$r_e = \text{equivalent resistance} = \frac{1{,}934}{(132.6)^2} = 0.110 \text{ ohm}$$

$$Z_e = \text{equivalent impedance} = \frac{48.4}{132.6} = 0.365 \text{ ohm}$$

$$X_e = \text{equivalent reactance} = \sqrt{(0.365)^2 - (0.110)^2} = 0.347 \text{ ohm}$$

All three of the above constants are referred to the low-voltage side since the voltage was impressed on the low-voltage side when making the short-circuit test.

$$r_{e(dc)} \text{ (referred to low-voltage side)} = \frac{r_1}{a^2} + r_2$$

where

$$a = \text{ratio of transformation} = \frac{11{,}000}{2{,}300} = 4.783$$

Thus,

$$r_{e(dc)} \text{ at } 25°C = \frac{1.28}{(4.783)^2} + 0.0467$$

$$= 0.103 \text{ ohm.}$$

Now,

$$r_e \text{ at } 75°C = 0.103 \frac{234.5 + 75}{234.5 + 25} + (0.110 - 0.103) \times \frac{234.5 + 25}{234.5 + 75}$$

$$= 0.122 + 0.006 = 0.128 \text{ ohm.}$$

The no-load copper loss is negligible. Therefore the power input when the secondary is open is the core loss. The core loss at 60 cycles, P_{h+e}, is therefore 2,140 watts. The efficiency for a full kva load at 0.8 power factor is given by

$$\text{Efficiency} = \frac{\text{output}}{\text{input}} = \frac{V_2 I_2 \cos\theta_2}{V_2 I_2 \cos\theta_2 + P_{h+e} + I_2{}^2 r_e}$$

$$= \frac{300{,}000 \times 0.8}{300{,}000 \times 0.8 + 2{,}140 + (130.5)^2 \times 0.128}$$

$$= \frac{240{,}000}{240{,}000 + 4{,}320} = 1 - \frac{4{,}320}{240{,}000 + 4{,}320}$$

= 0.982 or 98.2 per cent.

• PROBLEM 6-48

An open-circuit test on a 150-kVa 2400/240-V 60-Hz transformer yielded the following data, corrected for instrument losses:

Volts	Amperes	Watts
240	16.75	580

The following data were obtained in a short-circuit test of the transformer, with its low-voltage winding short-circuited.

Volts	Amperes	Watts
63.0	62.5	1660

Calculate the efficiency of the transformer (a) at rated load 0.80 power factor and (b) at one-half rated load 0.60 power factor.

Solution: (a) The real power output is

$$\text{output} = 150{,}000 \times 0.80 = 120{,}000 \text{ W}.$$

The load loss at rated load equals the real power measured in the short-circuit test at rated current:

$$\text{load loss} = P_{SC} = 1660 \text{ W.}$$

The core loss is taken as the no-load loss (i.e., the real power measured in the open-circuit test):

$$\text{core loss} = P_{OC} = 580 \text{ W}$$

$$\text{losses} = 1660 + 580 = 2240 \text{ W}$$

Real power input = output + losses

$$= 120{,}000 + 2240 = 122{,}240 \text{ W.}$$

The rated-load efficiency at 0.80 power factor is therefore

$$\text{Efficiency} = 1 - \frac{\text{losses}}{\text{input}} = 1 - \frac{2240}{122{,}240}$$

$$= 1 - 0.0183 = 0.9817.$$

(b) The real power output at one-half rated load and 0.60 power factor is:

$$\text{output} = \frac{1}{2} \times 150{,}000 \times 0.60 = 45{,}000 \text{ W.}$$

The load loss, being I^2R_{eq}, varies as the current squared, and at one-half rated load the rated current is one-half rated value. The load loss is therefore $\left(\frac{1}{2}\right)^2 = \frac{1}{4}$ of that at rated current value; hence,

$$\text{load losses} = \frac{1}{4} P_{SC} = \frac{1}{4} \times 1660 = 415 \text{ W.}$$

The core losses are considered unaffected by the load, as long as the secondary terminal voltage is at its rated value. Therefore,

$$\text{core losses} = P_{OC} = 580 \text{ W.}$$

$$\text{total losses} = 415 + 580 = 995 \text{ W.}$$

and

$$\text{input} = 45{,}000 + 995 = 45{,}995 \text{ W.}$$

The efficiency at one-half rated load and 0.60 power factor is

$$\text{Efficiency} = 1 - \frac{995}{45{,}995} = 0.9784.$$

• PROBLEM 6-49

Compute the regulation and efficiency at full load, 0.80 power factor, lagging current, of the 15-kva, 2,400: 240-volt, 60 ~ distribution transformer to which the following data apply. (Subscript H means high-voltage, subscript X means low-voltage winding.)

Short-circuit test	Open-circuit test
$V_H = 74.5$ v	$V_X = 240$ v
$I_H = 6.25$ amp	$I_X = 1.70$ amp
$P_H = 237$ watts	$P_X = 84$ watts
Frequency = 60 ~ $\theta = 25$ C	Frequency = 60 ~

Direct-current resistances measured at 25° C

$R_{dcH} = 2.80$ ohms $\qquad R_{dcX} = 0.0276$ ohm

The data given above have been corrected for instrument losses where this correction was necessary.

Solution: The rated current of the high-voltage winding is

$$I_H = \frac{15{,}000}{2{,}400} = 6.25 \text{ amp.}$$

The short-circuit test was taken at rated current, as is usually done.

From the short-circuit data,

$$R_{eq\,H} = \frac{P_H}{I_H^2} = \frac{237}{(6.25)^2} = 6.07 \text{ ohms at } 25° \text{ C.}$$

For computation of the equivalent resistance at 75° C, the direct-current component of the equivalent resistance must be determined. The ratio of transformation is 10, and from the direct-current resistances

$$R_{dcH} + a^2 R_{dcX} = 2.80 + 2.76 = 5.56 \text{ ohms at } 25° \text{ C.}$$

The equivalent resistance at 75° C is given by

$$R_{eqH(75)} = R_{dcH(\theta)} \cdot \frac{309.5}{234.5 + \theta}$$

$$+ [R_{eqH(\theta)} - R_{dcH(\theta)}] \frac{234.5 + \theta}{309.5}$$

$$= 5.56 \times \frac{309.5}{234.5 + 25}$$

$$+ [6.07 - 5.56] \quad \times \frac{234.5 + 25}{309.5}$$

$$= 7.05 \text{ ohms.}$$

The conventional efficiency now can be determined as follows:

$I^2 R_{eqH(75)} = (6.25)^2 \times 7.05 =$		276 watts
Core loss	=	84 watts
Total loss = sum	=	360 watts
Output = 0.80 × 15,000	=	12,000 watts
Input = sum	=	12,360 watts

$$\frac{\text{Losses}}{\text{Input}} = \frac{360}{12,360} = 0.0291.$$

$$\text{Efficiency} = 1 - \frac{\text{Losses}}{\text{Input}} = 1 - 0.0291$$

$$= 0.9709.$$

Computing the voltage regulation requires that the equivalent reactance be found. From the short-circuit data,

$$Z_{eqH} = \frac{V_H}{I_H} = \frac{74.5}{6.25} = 11.92 \text{ ohms}$$

$$X_{eqH} = \sqrt{Z^2_{eqH} - R^2_{eqH}}$$

Note that the value of R_{eqH} at the temperature of the test should be used in determining X_{eqH}. Hence

$$X_{eqH} = \sqrt{11.92^2 - 6.07^2} = 10.27 \text{ ohms.}$$

Since the constants are referred to the high-voltage side, this winding may be considered the secondary when the regulation is computed. The regulation is then given by

$$\text{Regulation} = \left(\frac{I_L R_{eq2} \cos\theta_L + I_L X_{eq2} \sin\theta_L}{V_2}\right)$$

$$+ \frac{1}{2}\left(\frac{I_L X_{eq2} \cos\theta_L - I_L R_{eq2} \sin\theta_L}{V_2}\right)^2 \quad (1)$$

$\cos\theta_L = 0.800$ $\sin\theta_L = +0.600$

Simplifying the notation, in Eq. 1,

$IR \cos\theta = 35.3$	$IX \cos\theta = 51.4$
$IX \sin\theta = 38.5$	$IR \sin\theta = 26.4$
Sum = 73.8	Difference = 25.0

$$\frac{73.8}{2{,}400} = 0.0308 \qquad \left(\frac{1}{2}\,\frac{25.0}{2{,}400}\right)^2 = \text{negligible.}$$

Hence,

Regulation = 0.0308, or 3.08%.

• PROBLEM 6-50

A 100-kva, 2200/220-volt, 60-cycle transformer is tested on open-circuit at no-load and short-circuit at 25°C and the following readings are recorded:

Open circuit (H.V. winding open)	Short circuit (L.V. winding short-circuited)
$V_1 = 220$ volts	$V_{SC} = 70$ volts
$I_0 = 18$ amp	$I_{SC} = 45.5$ amp $= I_1$ rated
$P_0 = 980$ watts	$P_{SC} = 1050$ watts

The resistance of the high-voltage winding after the short-circuit test was 0.24 ohm, and the temperature was 25°C. a = 2200/220 = 10.

Find the parameters of the equivalent circuit, the regulation at full-load unity p.f. and 0.80 p.f. lagging, and the efficiency at unity p.f. and 0.80 p.f. lagging.

Solution: $I_1 \text{(rated)} = \frac{100{,}000}{2200} = 45.5 \text{ amp.}$

$$Z_e(25°C) = \frac{V_{SC}}{I_{SC}} = \frac{70}{45.5} = 1.54 \text{ ohms}$$

$$R_e(25°C) = \frac{P_{SC}}{I_{SC}^2} = \frac{1050}{(45.5)^2} = 0.507 \text{ ohm.}$$

$$X_e = \sqrt{Z_e^2 - R_e^2}$$

$$= \sqrt{(1.54)^2 - (0.507)^2} = 1.46 \text{ ohms.}$$

$$R_e(75°C) = 0.507 \times \frac{234.5 + 75}{234.5 + 25} = 0.605 \text{ ohm.}$$

$$r_1(75°C) = 0.24 \times \frac{234.5 + 75}{234.5 + 25} = 0.286 \text{ ohm.}$$

$$r_2' = R_e - r_1 = 0.605 - 0.286 = 0.319 \text{ ohm}$$

$$x_1 = x_2' = 0.73 \text{ ohm}$$

$$P_{h+e} = 980 - (1.80)^2 \times 0.24 = 979 \text{ watts}$$

$$E_1 = 2200 - 1.80 \times 0.73 = 2200 \text{ volts}$$

$$x_m = \frac{2200}{1.80} = 1222 \text{ ohms}$$

$$g_m = \frac{P_{h+e}}{E_2^2} = \frac{979}{(2200)^2} = 0.000202 \text{ mho}$$

$$r_m = g_m x_m^2 = 0.000202 \times (1222)^2 = 302 \text{ ohms}$$

Unit voltage = 220 × 10 = 2200 volts (V_2')

Unit current = 45.5 amp

Unit impedance = 48.4 ohms

$$e_r = \frac{0.605}{48.4} = 0.0125 \text{ V}$$

$$e_x = \frac{1.46}{48.4} = 0.0302 \text{ V}$$

Regulation at unity p.f.

$$\text{Regulation} = e_r \cos \phi_2 \pm e_x \sin \phi_2 + \frac{(e_x \cos \phi_2 \pm e_r \sin \phi_2)^2}{2}$$

$$= 0.0125 + \frac{(0.0302)^2}{2}$$

$$= 0.0125 + 0.000455 = 0.0130$$

Regulation at 0.80 p.f. lagging

$$\text{Reg.} = 0.80 \times 0.0125 + 0.60 \times 0.0302 + \frac{(0.80 \times 0.0302 - 0.60 \times 0.0125)^2}{2}$$

$$= 0.0100 + 0.0181 + 0.000139$$

$$= 0.0282.$$

Efficiency:

$$\text{Load losses } (75°\text{C}) = 1050 \times \frac{234.5 + 75}{234.5 + 25} = 1252 \text{ watts.}$$

Efficiency at rated load unity p.f.:

$$\eta = \frac{\text{output}}{\text{output + losses}}$$

$$= \frac{100}{100 + 0.980 + 1.25} = 0.977.$$

Efficiency at rated load 0.80 p.f. lagging:

$$\eta = \frac{100 \times 0.80}{100 \times 0.80 + 0.980 + 1.25} = 0.973.$$

Efficiency at $\frac{3}{4}$ load 0.80 p.f. lagging:

$$\eta = \frac{100 \times \frac{3}{4} \times 0.80}{100 \times \frac{3}{4} \times 0.80 + 0.980 \times (\frac{3}{4})^2 + 1.25} = 0.972.$$

• PROBLEM 6-51

A 10-kva 2,400:240-volt 60-cycle transformer has the following data:

$R_1 = 5.2$ ohms $\qquad R_2 = 0.052$ ohm

$X_1 = 7.6$ ohms $\qquad X_2 = 0.076$ ohm

The iron loss was found to be 61 watts, by measuring the input with no load at rated voltage, and this value is assumed to remain constant at all values of load. Determine the efficiency at (a) full load, unity power factor, (b) half load, unity power factor, (c) full load, 80 per cent power factor.

Solution:

$$\text{Primary current } I_1 = \frac{10,000}{2,400} = 4.17 \text{ amp.}$$

$$\text{Secondary current } I_2 = 10,000/240 = 41.7 \text{ amp.}$$

$$\text{Primary copper loss} = I_1^2R_1 = (4.17)^2 \times 5.2 = 90.5 \text{ watts}$$

$$\text{Secondary copper loss} = I_2^2R_2 = (41.7)^2 \times 0.052 = 90.5 \text{ watts}$$

$$\text{Total copper loss} = 181 \text{ watts.}$$

(a) The output at full load, unity power factor, is 10,000 watts, and so the efficiency at this load is

$$\eta = \frac{10,000}{10,000 + 61 + 181} = \frac{10,000}{10,242} = 97.6\%$$

(b) At half load, unity power factor, the output is 5,000 watts, and the copper loss is 45.3 watts; hence

$$\eta = \frac{5,000}{5,000 + 61 + 45.3} = \frac{5,000}{5,106.3} = 97.9\%$$

(c) Full load at 80 percent power factor means full-load current or kva, and so the output is only 8,000 watts. The iron and copper losses are 61 watts and 181 watts, respectively, the same as for full-load unity power factor; therefore

$$\eta = \frac{8,000}{8,000 + 242} = \frac{8,000}{8,242} = 97.1\%$$

Curves of efficiency plotted against output are shown in the figure for 100 per cent and 80 per cent power factor load.

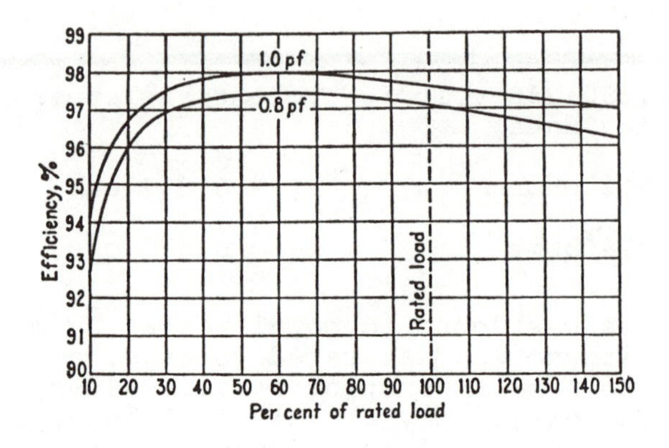

Efficiency curves of 10-kva distribution transformer.

The efficiency is the same for 80 percent power factor leading and 80 percent power factor lagging. Best efficiency is obtained at 100 percent power factor but not at 100 percent of rated load. For this transformer, the best efficiency occurs at about 58 per cent load, where the copper loss is equal to the iron loss. By designing the transformer in this way, it is made to have a very flat efficiency curve over the whole working range, and the high efficiency is maintained down to very light load.

• PROBLEM 6-52

A 10kVA distribution transformer has a 240V secondary winding. The equivalent internal resistance of this transformer, referred to that winding, is 0.048Ω. The core loss of this transformer is 75W. At what kVA load will this transformer operate at maximum efficiency?

Solution:

$$P_k = I_2^2 R \ ,$$

(I_2 is the current in the secondary winding)

$$75\text{W} = I_2^2 \times 0.048$$

$$I_2 = \sqrt{75/0.048} = 39.5 \text{ A}$$

Then at this current, the output volt-amperes of the transformer would be

$|S| = V_2 I_2 = 240 \times 39.5 = 9480$ VA or 95% of rated load.

• PROBLEM 6-53

A 5-kva transformer is supplying a lighting load; it would, in a normal day's run, operate at full load $1\frac{1}{2}$ hours and perhaps at half load $1\frac{1}{2}$ hours; during the rest of the day there would be no load on the transformer. The iron loss is 200 watts and the full-load copper loss (I^2R) is 200 watts. Find the all-day efficiency of the transformer.

Solution: The copper loss at half load would be equal to $(\frac{1}{2})^2 \times 200$ watts = 50 watts.

As the transformer is connected to its supply line all day, the core loss for all day = 24 x 200 = 4800 watthours.

The copper loss for the $1\frac{1}{2}$ hours full load = $1\frac{1}{2} \times 200$ = 300 watthours and during the $1\frac{1}{2}$ hours half load = $1\frac{1}{2} \times 50$ = 75 watthours, so that the total copper loss for the day's run is 375 watthours. The total loss in the transformer for 24 hours = 4800 + 375 = 5175 watthours.

The energy output in one day = $1\frac{1}{2} \times 5000 + 1\frac{1}{2} \times 2500$ = 11,250 watthours.

The energy input in one day = 11,250 + 5175 = 16,425 watthours.

The all-day efficiency = $\frac{\text{output}}{\text{input}} = \frac{11,250}{16,425}$ = 68.5 percent.

The efficiency of this transformer in the ordinary sense $= \frac{5000}{5400}$ = 92.7 percent.

• PROBLEM 6-54

A 200 kVA single-phase transformer with a voltage ratio 6350/660 V has the following winding resistances and reactances:

$$R_1 = 1.56\Omega \quad R_2 = 0.016\Omega \quad X_1 = 4.67\Omega \quad X_2 = 0.048\Omega .$$

On no load the transformer takes a current of 0.96 A at a power factor of 0.263 lagging.

Using $R_{eql} = 3.04\Omega$, calculate the efficiency of the transformer at full load and half load when the power factor is unity and also when the power factor is 0.8. Also calculate the maximum efficiency.

The transformer is in circuit continuously. For a total of 8 hours, it delivers a load of 160 kW at 0.8 p.f.

For a total of 6 hours it delivers a load of 80 kW at u.p.f.

For the remainder of the 24-hour cycle it is on no load.

What is the all-day efficiency?

Solution: In the following, to avoid complications, which serve no purpose at this stage, the effects of regulation will be neglected. For example, at 0.8 lagging and 0.8 leading power-factor, the terminal voltages and hence the outputs are different at full-load current. The efficiency will be based on an output equal to the power factor, lagging or leading, multiplied by the rated kVA or a fraction of it as required. Further, the full-load copper loss will be taken as $I^2_{f.l.} R_e$ though, in fact, the current in the primary resistance should include the small no-load current. Actually, R_e is determined in the first place from the measured copper loss on short circuit with full-load current in the windings, so the error which is very small, is in the value of R_e only.

$$\text{Iron loss} = VI \times \text{p.f.}$$

$$= 6350 \times 0.96 \times 0.263 = 1.6 \text{ kW.}$$

$$\text{The copper loss at full load} = I^2_{f.l.} R_{eql}$$

$$= (31.5)^2 \times 3.04$$

$$= 3.02 \text{ kW.}$$

The copper loss at half full-load

$= (0.5)^2 \; I^2_{f.l.} \; R_{eql}$

$= (0.5)^2 \times 3.02 = 0.775$ kW.

$\therefore$ Total loss at full load = 1.6 + 3.02 = 4.62 kW

Total loss at half load = 1.6 + 0.755 = 2.36 kW

Unity Power-factor

F.l. output = 200 kW; Input = 204.62 kW;

$$\eta = 1 - \frac{\text{losses}}{(\text{output + losses})} = 1 - \frac{\text{losses}}{\text{input}}$$

$$= 1 - \frac{4.62}{204.62} = 97.74\%$$

Half f.l. output = 100 kW; Input = 102.36 kW;

$$\eta = 1 - \frac{\text{losses}}{(\text{output + losses})} = 1 - \frac{\text{losses}}{\text{input}}$$

$$= 1 - \frac{2.36}{102.36} = 97.7\%$$

0.8 p.f.

F.l. output = 200 x 0.8 = 160 kW; Input = 164.62 kW;

$$\eta = 1 - \frac{4.62}{164.62} = 97.19\%$$

Half f.l. output = 100 x 0.8 = 80 kw; Input = 82.36

$$\eta = 1 - \frac{2.36}{82.36} = 97.14\%$$

The maximum efficiency occurs when per unit load current,

$$I_{pu} = \sqrt{\frac{\text{iron loss}}{\text{full load copper loss}}}$$

$$= \sqrt{\left(\frac{1.6}{3.02}\right)} = 0.73 \text{ A.}$$

At unity power-factor, the absolute maximum efficiency occurs when the output is 0.73 x 200 = 146 kW

and $\eta_{max} = 1 - \frac{1.6 + 1.6}{146 + 3.2} = 97.86\%$

Note, in working out the efficiencies in per unit, the losses would first be expressed as a fraction of the rated kVA.

All-day Efficiency.

At 160 kW, 0.8 p.f. Copper loss = 3.02 kW,

Total loss = 4.62 kW.

At 80 kW, u.p.f., copper loss = $3.02(80/200)^2 = 0.48$.

Total loss = 2.08 kW.

On no load Total loss = 1.6 kW.

For 8 hours, output = 160 x 8 = 1,280 kWh;

loss = 4.62 x 8 = 37 kWh

For 6 hours, output = 80 x 6 = 480 kWh;

loss = 2.08 x 6 = 12.5 kWh

For 10 hours, output = 0

loss = 1.6 x 10 = 16 kWh

In 24 hours, total output = 1760 kWh;

total loss = 65.5 kWh

All-day efficiency

$$= 1 - \frac{(\text{losses in kWh})}{(\text{output in kWh + losses in kWh})}$$

$$= 1 - \frac{65.5}{1760 + 65.5} = 96.41\%$$

• PROBLEM 6-55

Consider the transformer for which the approximate equivalent circuit (referred to the low-voltage side) is shown in the figure. The secondary load is 12,000 kva at 0.80 pf lag, and 22 kv.

(a) What is the regulation?

(b) Suppose this transformer has a daily load cycle in which it carries 20 percent overload for 2 hours, full load for 6 hours, 75 percent load for 8 hours, and 50 percent load for 8 hours. What is its all-day efficiency?

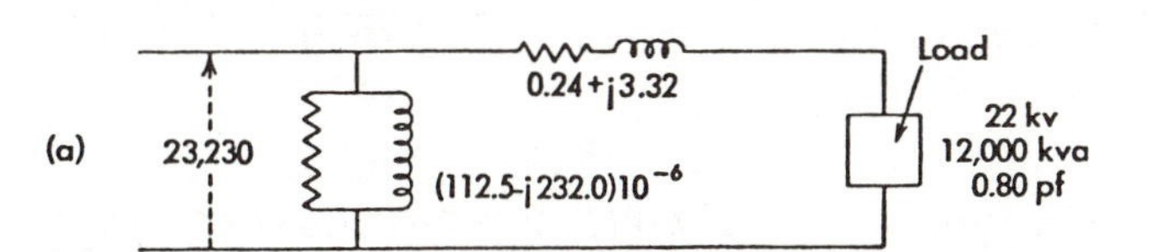

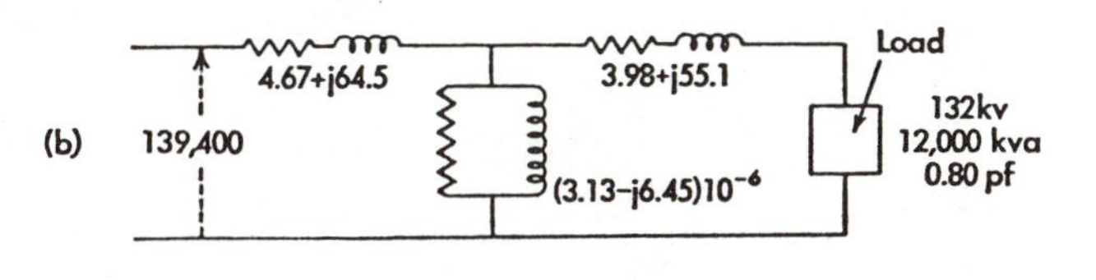

Solution: (a) Full load current is

$$I_2 = \frac{12{,}000}{22} = 545.5 \text{ amp}$$

The regulation is

$$\text{Reg} = \frac{RI_2 \cos\theta_2 + XI_2 \sin\theta_2}{V_2}$$

$$+ \frac{1}{2}\left(\frac{XI_2 \cos\theta_2 - RI_2 \sin\theta_2}{V_2}\right)^2$$

$$= \frac{0.24 \times 545.5 \times 0.80 + 3.32 \times 545.5 \times 0.60}{22{,}000}$$

$$+ \frac{1}{2}\left(\frac{3.32 \times 545.5 \times 0.80 - 0.24 \times 545.5 \times 0.60}{22{,}000}\right)^2$$

$$= 0.0561$$

The losses are

$$\frac{0.24 \times 545.5^2}{1000} + \frac{112.5}{10^6} \times \frac{22{,}000^2}{1000} = 71.4 + 54.4$$

$$= 125.8 \text{ kw}$$

The output is

$$= 12{,}000 \times 0.80 = 9600 \text{ kw}$$

The efficiency then is

$$\eta = 1 - \frac{125.8}{9600 + 125.8} = 0.987$$

(b) $$\eta = \frac{V_2 I \cos\theta_2 (\alpha_1 t_1 + \alpha_2 t_2 + \alpha_3 t_3 + \ldots)}{V_2 I \cos\theta_2 (\alpha_1 t_1 + \alpha_2 t_2 + \ldots) + RI^2(\alpha_1^2 t_1 + \alpha_2^2 t_2 + \ldots) + 24W}$$

$$= \frac{22{,}000 \times 545.5 \times 0.80 (1.20 \times 2 + 1.00 \times 6 + 0.75 \times 8 + 0.50 \times 8)}{\left\{\begin{matrix} 22{,}000 \times 545.5 \times 0.80 (1.20 \times 2 + 1.00 \times 6 + 0.75 \times 8 + 0.50 \times 8) \\ + 24 \times 54{,}500 \\ + 0.24 \times 545.5^2 (1.20^2 \times 2 + 1.00^2 \times 6 + 0.75^2 \times 8 + 0.50^2 \times 8) \end{matrix}\right\}}$$

$$= 0.9865$$

The primary current is

$$\vec{I}_1 = \frac{1}{6}\left[545.5(0.8 - j0.6) + \left(\frac{112.5 - j232.0}{10^6}\right) 23{,}230\right]$$

$$= 73.0 - j55.3 \text{ A.}$$

The primary voltage is

$$\vec{V}_1 = 6\,[22{,}000 + (0.24 + j3.32)\,545.5(0.8 - j0.6)]$$

$$= 139{,}150 + j8220 = 139{,}400 \;\underline{/3.4^\circ} \text{ V.}$$

By way of comparison, it is instructive to carry out detailed calculations for the exact equivalent circuit, Fig. b.

$$\vec{I}_2' = \frac{12,000}{132}(0.8 - j0.6) = 72.7 - j54.5$$

$$= 90.8 \underline{/-36.9°} \text{ A.}$$

$$\vec{E}_0 = 132,000 + (72.7 - j54.5)(3.98 + j55.1)$$

$$= 135,290 + j3793 = 135,300 \underline{/1.6°} \text{ V.}$$

$$\vec{I}_0 = \left(\frac{3.13 - j6.45}{10^6}\right)(135,290 + j3793) = 0.45 - j0.86$$

$$= 0.97 \underline{/-62.4°} \text{ A.}$$

$$\vec{I}_1 = (72.7 - j54.5) + (0.45 - j0.86) = 73.2 - j55.4$$

$$= 91.9 \underline{/-37.1°} \text{ A.}$$

$$\vec{V}_1 = (135,290 + j3793) + (4.67 + j64.5)(73.2 - j55.4)$$

$$= 139,207 + j8254 = 139,400 \underline{/3.4°} \text{ V.}$$

$$\text{Reg} = \frac{139,400 - 132,000}{132,000} = 0.0561.$$

$$\text{Losses} = \frac{4.67 \times 91.9^2}{1000} + \frac{3.98 \times 90.8^2}{1000} + \frac{3.13}{10^6} \times \frac{135,300^2}{1000}$$

$$= 129.7 \text{ kw.}$$

$$\text{Efficiency} = 1 - \frac{129.7}{12,000 + 129.7} = 0.989.$$

Determine the energy efficiency of a 15-kva, 2400:240 volt, 60 cycles/sec. transformer operating on the load cycle shown in the figure.

The core loss of the transformer is 84 watts. The hot and cold values of the equivalent resistance are $R_{eqH(75)} = 7.05$ ohms and $R_{eqH(25)} = 6.07$ ohms.

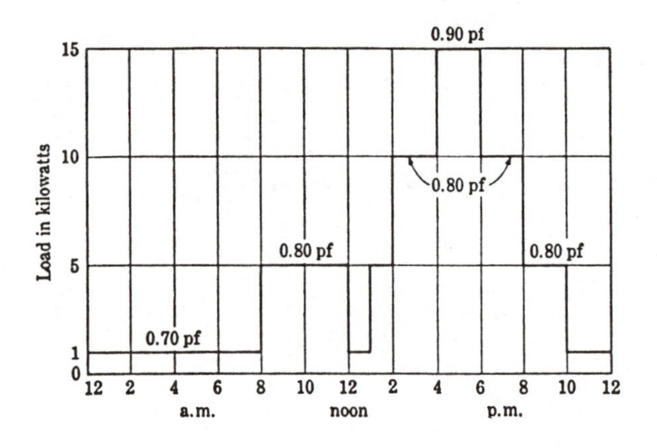

Simplified daily load curve.

Solution: Examination of the figure shows that the daily load cycle can be summarized as follows:

Hours	*Kilowatts output*	*Power factor*
2	15	0.90
4	10	0.80
7	5	0.80
11	1	0.70

The losses are the constant core loss, which must be supplied continuously, and the load loss, which varies with the square of the current. Thus the energy dissipated in core loss is

$$\frac{24 \times 84}{1,000} = 2.02 \text{ kwhr daily.}$$

The daily output and load loss can be computed and, for convenience, arranged as in the table below. For computing the load loss. the equivalent resistance is taken as the average of its hot and cold values; that is,

$$R_{eqH} = \frac{6.07 + 7.05}{2} = 6.56 \text{ ohms.}$$

The current in the high-voltage winding is given by

$$I_H = \frac{kva}{2,400}\ A$$

Total output = 116 kwhr   Total load loss = 1.676 kwhr

Hours	Output				I_H	$I_H^2 R_{eqH}$ Kw	Load loss Kwhr
	Kw	Kwhr	P.f.	Kva			
2	15	30	0.90	16.67	6.94	0.316	0.632
4	10	40	0.80	12.5	5.20	0.177	0.709
7	5	35	0.80	6.25	2.60	0.0443	0.310
11	1	11	0.70	1.43	0.595	0.0023	0.025

The energy efficiency now can be computed from the energy output and losses, as follows:

Total load loss = 1.68 kwhr daily

Total core loss = 2.02

Total loss = 3.70

Output = 116.00

Input = 119.70

Hence

$$\text{Energy efficiency} = 1 - \frac{\text{losses}}{\text{input}} = 1 - \frac{3.70}{119.7} = 0.9691$$

$$= 96.91\%$$

AUTOTRANSFORMERS AND MULTICIRCUIT TRANSFORMERS

• PROBLEM 6-57

A 100-kva, 2300- to -230- volt, 60-cycle, 2-winding transformer is used as an autotransformer having a single winding in order to step up the voltage of a 2300-volt line by 10%. If the transformer has 2% losses, a 2.2% regulation, and a 3.3% impedance (Z_e) as a 2-winding transformer, find its characteristics as a 2300/2530-volt autotransformer.

Solution:

Primary voltage = 2300 volts.

Load voltage = 2530 volts.

Ratio of transformation = $\frac{10}{11}$

Now

I_1 (as 2-winding transformer) $= \frac{100,000}{2300} = 43.5$ amp.

I_2 (as 2-winding transformer) $= \frac{100,000}{230} = 435$ amp.

I_2 (as autotransformer) = 435 amp.

I (as autotransformer) = 43.5 amp.

I_1 (as autotransformer) = 478.5 amp.

Output (as autotransformer) $= 2530 \times 435$

$= 1100$ kva.

Losses $= \frac{1}{11} \times 0.02$

$= 0.00182$

Regulation $= \frac{1}{11} \times 0.022$

$= 0.002.$

Impedance $= \frac{1}{11} \times 0.033$

$= 0.003.$

Assume 230-volt winding to be insulated for 2300 volts.

• PROBLEM 6-58

A 10-kva 2,300/230 volt two-winding transformer is connected as an autotransformer with the L.T. winding additively in series with the H.T. winding as in the figure. A potential difference of 2,530 volts is impressed upon terminals ac, and a load is connected to terminals ab so that the current in winding bc is equal to the rated current of the L.T. winding. Compare the volt-ampere rating of the auto-transformer with that of the same transformer when connected as a simple two-winding transformer.

Solution: The rated current in the 230-volt winding is

$$I_1 = \frac{10,000}{230} = 43.48 \text{ amp}$$

and since there are ten times as many turns in ab as in bc , the balance of ampere-turns (ignoring the exciting current) requires that the current in the 2,300-volt winding shall be

$$I_3 = I_2 - I_1 = \frac{10,000}{2,300} = 4.35 \text{ amp}$$

which is the rated current of that winding.

The current supplied to the load is

$$I_2 = I_1 + I_3 = \frac{10,000}{230} + \frac{10,000}{2,300} = \frac{10,000}{230} \times \frac{11}{10}$$

$$= 47.83 \text{ amp}$$

The ratio of transformation of the auto-transformer is

$$a = \frac{V_1}{V_2} = \frac{2,300 + 230}{2,300} = 1.1$$

it follows that

$$I_2 = aI_1 = 1.1 \times \frac{10,000}{230} = 47.83 \text{ amp.}$$

The total volt-amperes supplied from the source is

$$P = V_1I_1 = 2,530 \times \frac{10,000}{230} = 11 \times 10,000 \text{ VA}$$

while the volt-amperes supplied inductively is

$$P_i = V_2(I_2 - I_1) = P\frac{a - 1}{a} = P \times \frac{1}{11} = 10,000 \text{ VA}$$

which is the volt-ampere rating of the original transformer.

The power supplied conductively by way of the L.T. winding bc is

$$P_c = \frac{P}{a} = 100,000 \text{ VA}$$

The conclusion may then be drawn that a two-winding transformer connected as a step-down autotransformer will have a volt-ampere rating $a/(a - 1)$ times its rating as a simple transformer. This increase becomes larger as the ratio of transformation a approaches unity; but in the limiting case when $a = 1$, the power transferred conductively is equal to the entire input, and the power transferred inductively, p_i, becomes zero, which means that there is no need for the autotransformer.

Otherwise expressed, if an autotransformer is to be designed to have a rating P, it will consist of a two-winding transformer having a rating $P_i = \frac{a - 1}{a} P = \frac{V_1 - V_2}{V_1} P$, which becomes smaller as V_2 approaches V_1. When designed in this way, the material in the autotransformer will be utilized to the maximum extent.

• PROBLEM 6-59

A 20-kva load is to be supplied at 500 v. An ideal step-up autotransformer is used to connect this load to a 400-v source. Find (a) the voltage and current of the series winding, (b) the voltage and current of the common winding, (c) the kva rating of this transformer if it were used as a two-winding transformer.

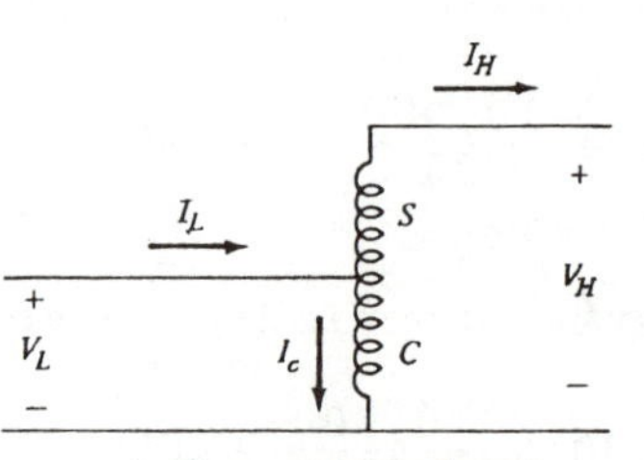

Step-up autotransformer.

Solution: (a) The circuit shown in the figure is used. The autotransformer must have high side voltage, $V_H = 500$ V, and low side voltage, $V_L = 400$ volts. The series winding voltage is

$$V_S = V_H - V_L = 500 - 400 = 100 \text{ v}$$

The series winding current is the load current

$$I_S = I_H = I_{Load} = 20,000/500 = 40 \text{ amp}$$

(b) The voltage of the common winding is the low side voltage

$$V_C = V_L = 400 \text{ v}$$

The current in the common winding is

$$I_C = (N_S/N_C) I_S = (V_S/V_C) I_S = (100/400)(40) = 10 \text{ amp}$$

An alternative method to find I_C is to find I_L first:

$$I_L = 20,000/400 = 50 \text{ amp}$$

Then

$$I_C = I_L - I_H = 50 - 40 = 10 \text{ amp}$$

(c) The apparent power associated with the two-winding transformer is the product of the voltage and current of one winding

$$P_{aT} = V_C I_C = (400 \text{ v})(10 \text{ amp}) = 4000 \text{ volt amp} = 4 \text{ kva.}$$

If it were used as a two-winding transformer, its rating would be 4 kva, 400/100 volts.

• PROBLEM 6-60

A 10-kva, 440/110-volt, 60-cps, single-phase transformer has an efficiency of 97.5% at a rated load of unity power factor. The full-load copper loss is 150 watts, and the core loss is 100 watts. Determine (a) the kilovolt-ampere rating as a 550/440-volt autotransformer, and (b) the efficiency when supplying full load at 80% lagging power factor.

Solution: (a) The 110-volt current rating is 10,000/110 = 90.8. This current is the high-voltage, full-load current, thus the kilovolt-ampere rating is given by 90.8 x 550/1000 = 50 kva.

(b) The load power at a power factor of 0.8 lag is 50 x 0.8 = 40 kw. The power in is the power out plus losses, therefore,

$$P_{in} = 40 + 0.150 + 0.100 = 40.25 \text{ kw.}$$

$$\therefore \text{ Efficiency } \eta = 1 - \frac{\text{losses}}{P_{in}} = 1 - \frac{0.25}{40.25} = 0.9938$$

$$= 99.38\%.$$

The efficiency at full load is 99.38%. Of course, the reason that the efficiency is so high is that only 10 kva is being transformed as it would in the two-winding transformer.

• PROBLEM 6-61

The following measurements were obtained from tests carried out on a 10-kVA, 2300:230-V, 60-Hz distribution transformer:

Open-circuit test, with the low-potential winding excited:

Applied potential difference, $V_{OC} = 230$ V

Current, $I_{OC} = 0.45$ A

Input power, $P_{OC} = 70$ W

Short-circuit test, with the high-potential winding excited:

Applied potential difference, $V_{SC} = 120$ V

Current, $I_{SC} = 4.5$ A

Input power, $P_{SC} = 240$ W

Winding resistances, measured by dc bridge:

$R_{hp} = 5.80\ \Omega, \quad R_{lp} = 0.0605\ \Omega$

The transformer has its two windings connected in series to form an autotransformer giving a small reduction to potential difference from a 2300-volt line. Determine

a) The transformation ratio on open circuit.

b) The permissible output of the transformer if the winding currents are not to exceed those for full-load operation as a two-winding transformer.

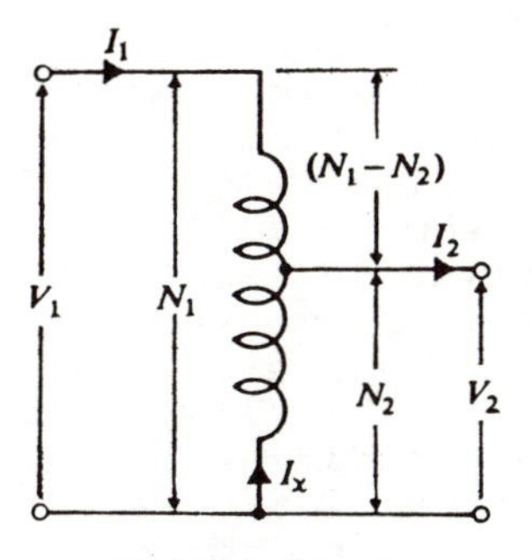

Autotransformer.

Solution: a) The connection for operation as an autotransformer is as shown in the figure, where $V_1 = 2300$ volts. The original primary winding is now N_2, and the original secondary winding is $(N_1 - N_2)$. The turns ratio of the transformer is thus

$$\frac{V_1}{V_2} = \frac{2300 + 230}{2300} = \frac{2530}{2300} = 1.1$$

b) On open circuit,

$$V_2 = \frac{2300}{2530} \times 2300 = 2090 \text{ V}$$

Winding $(N_1 - N_2)$ is rated at 43.5 A. Winding N_2 is rated at 4.35 A. Neglecting the exciting current,

$$I_x = I_2 - I_1$$

That is,

$$4.35 = I_2 - 43.5$$

and

$$I_2 = 47.8 \text{ A}$$

Rating of the autotransformer is thus

$$V_2 I_2 = 2090 \times 47.8 = 99.9 \text{ kVA.}$$

Compare this with its 10 kVA rating as a two-winding transformer.

(i) A 60-cycle-per-second, two-winding transformer is rated 3 kva, 220/110 volts. This transformer is reconnected as a step-up autotransformer to deliver 330 volts to a resistive load when the input is from a 220-volt source. Assume that the transformer is ideal. Find (a) the value of the load resistance for which rated current will flow in each winding, (b) the load power for the condition of (a), (c) the power delivered by transformer action and the power delivered by conduction, (d) the input impedance looking into the low side.

(ii) A 60-cycle-per-second, two-winding transformer is rated 3 kva, 220/110 volts. This transformer is connected as a step-down autotransformer to deliver 110 volts to a load impedance of (3 + j2) ohms when the input is from a 330-volt source. Assume that the transformer is ideal. Find (a) the load current, the input current, and the current in each winding, (b) the load power, (c) the power delivered by transformer action and the power delivered by conduction, (d) the input impedance looking into the high side.

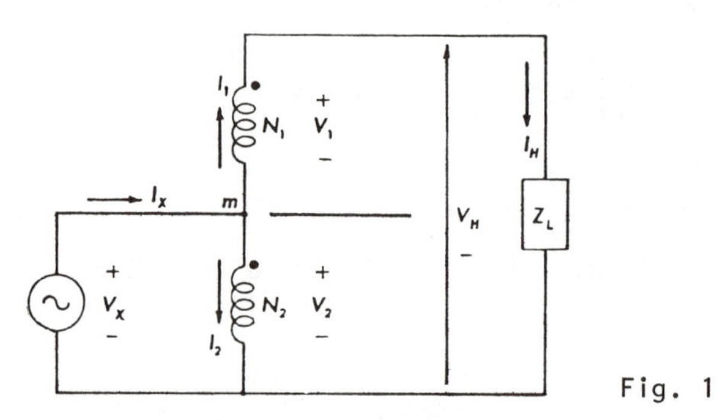

Fig. 1

Ideal step-up autotransformer.

Solution: (i) Refer to Fig. 1. The 110-volt winding is winding 1. Thus

$$a = 110 \text{ volts}/220 \text{ volts} = 1/2$$

$$I_H = I_1 = \frac{3{,}000 \text{ volt-amperes (va)}}{110 \text{ volts}} = 27.3 \text{ amp}$$

(a) $$Z_L = R = \frac{330 \text{ volts}}{27.3 \text{ amp}} = 12.1 \text{ ohms}$$

$$P_{aH} = V_H I_H = 330 \text{ volts} \times 27.3 \text{ amp} = 9{,}000 \text{ va}$$

Since the load power factor is unity,

(b) $$P_H = 9{,}000 \text{ watts}$$

(c) $P_{aT} = V_1 I_1 = 110 \text{ volts} \times 27.3 \text{ amp} = 3{,}000 \text{ va}$

$P_T = 3{,}000 \text{ watts}$

$P_{aC} = V_X I_H = 220 \text{ volts} \times 27.3 \text{ amp} = 6{,}000 \text{ va}$

$P_C = 6{,}000 \text{ watts}$

(d) $I_2 = aI_1 = \frac{1}{2} \times 27.3 \text{ amp} = 13.6 \text{ amp}$

$I_X = I_1 + I_2 = 27.3 \text{ amp} + 13.6 \text{ amp} = 40.9 \text{ amp}$

$$Z_{in} = \frac{V_X}{I_X} = \frac{220 \text{ volts}}{40.9 \text{ amp}} = 5.37 \text{ ohms}$$

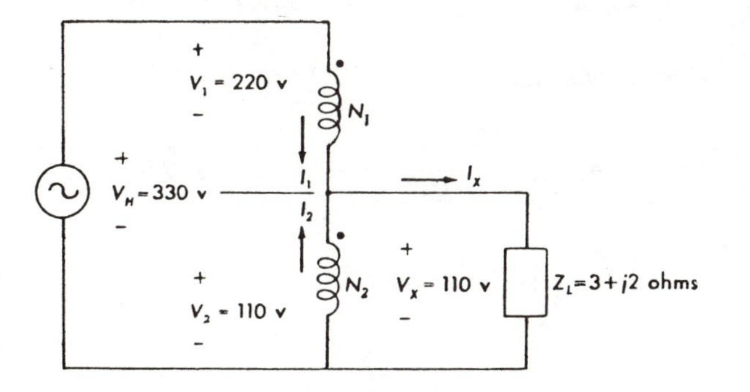

Fig. 2

Ideal step-up autotransformer.

(ii) The circuit is shown in Fig. 2. Choose the arrow directions for the currents opposite to part (i). This choice is more satisfying to one's intuition in regard to the direction of energy flow. The 220-volt winding is winding 1. Thus

$$a = 220 \text{ volts}/110 \text{ volts} = 2.$$

Choose $\vec{V}_H$ as the reference.

$$\vec{V}_H = 330 \ \underline{/0^\circ}$$

$$\vec{V}_X = 110 \ \underline{/0^\circ}$$

$$\vec{Z}_L = 3 + j2 = 3.61 \ \underline{/33.7^\circ} \text{ ohms.}$$

(a) $$\vec{I}_X = \frac{\vec{V}_X}{\vec{Z}_L} = \frac{110 \ \underline{/0^\circ}}{3.61 \ \underline{/33.7^\circ}} = 30.5 \ \underline{/-33.7^\circ} \text{ amp.}$$

$$\vec{I}_H = \frac{1}{a+1}\vec{I}_X = \frac{1}{2+1} 30.5 \underline{/-33.7°} = 10.17 \underline{/-33.7°} \text{ amp}$$

$$= 10.17 \underline{/-33.7°} \text{ amp.}$$

$$\vec{I}_1 = \vec{I}_H = 10.17 \underline{/-33.7°} \text{ amp.}$$

$$\vec{I}_2 = a\vec{I}_1 = 20.33 \underline{/-33.7°} \text{ amp.}$$

(b) $P_L = V_X I_X \cos\theta = 110 \times 30.5 \times \cos 33.7°$

$= 2{,}800$ watts.

(c) $P_T = V_X I_2 \cos\theta = 110 \times 20.33 \times \cos 33.7°$

$= 1{,}866$ watts.

$P_C = V_X I_H \cos\theta = 110 \times 10.17 \times \cos 33.7°$

$= 933$ watts.

(d) $$\vec{Z}_{in} = \frac{\vec{V}_H}{\vec{I}_H} = \frac{330 \underline{/0°}}{10.17 \underline{/-33.7°}} = 32.5 \underline{/33.7°}$$

$= (27 + j18)$ ohms.

• PROBLEM 6-63

A two-winding transformer is rated 2.2 kva, 220/110 volts, 60 cycles per second. Winding A is the 220-volt winding and has an impedance of 0.24 + j0.40 ohms. Winding B is the 110-volt winding and has an impedance of 0.05 + j0.09 ohms. The core loss is 28 watts for rated voltage and frequency.

The transformer is connected as an autotransformer to supply a load at 220 volts from a source of 330 volts. For a load current of 30 amp at 0.9 pf lagging, find: (a) the voltage regulation; (b) the efficiency.

Solution: Winding B will be the series winding (winding 1), and winding A will be the common winding (winding 2).

$$a = 110 \text{ volts}/220 \text{ volts} = 1/2$$

$$\vec{Z}_1 = \vec{Z}_B = 0.05 + j0.09 \text{ ohms}$$

$$a^2\vec{Z}_2 = a^2\vec{Z}_A = (½)^2(0.24 + j0.40) = 0.06 + j0.10 \text{ ohms}$$

$$\vec{Z}_{eH} = \vec{Z}_1 + a^2\vec{Z}_2 = 0.11 + j0.19 \text{ ohms}$$

$$I'_X = \frac{1}{a+1} I_X = \frac{1}{\frac{1}{2}+1}(30) = \frac{2}{3}(30) = 20 \text{ amp.}$$

$$Pf = 0.9 \text{ lag}, \quad \cos\theta = 0.9, \quad \sin\theta = 0.437.$$

(a) $$\varepsilon \approx \varepsilon_1 = \frac{I_{1\text{ rated}}}{V_{L\text{ rated}}}(R_{el}\cos\theta_2 + X_{el}\sin\theta_2).$$

$$= \frac{20 \text{ amp}}{330 \text{ volts}}(0.11 \times 0.9 + 0.19 \times 0.437)$$

$$= 0.011$$

$$P_C = 28 \text{ watts}$$

$$P_{cu} = I'^2_X R_{eH} = (20)^2 0.11 = 44 \text{ watts}$$

$$P_{out} = V_X I_X \cos\theta_X = 220 \times 30 \times 0.9 = 5{,}940 \text{ watts}$$

(b) $$\eta = 1 - \frac{P_c + P_{cu}}{P_{out} + P_c + P_{cu}}$$

$$= 1 - \frac{28 + 44}{5{,}940 + 28 + 44} = 1 - 0.012 = 0.988$$

• PROBLEM 6-64

The results of three short-circuit tests on a 7,960:2,400:600-volt 60-cps single-phase transformer are as follows:

Test	Winding excited	Winding short-circuited	Applied voltage, volts	Current in excited winding, amp
1	1	2	252	62.7
2	1	3	770	62.7
3	2	3	217	208

Resistances may be neglected. The rating of the 7,960-volt primary winding is 1,000 kva, of the 2,400-volt secondary is 500 kva, and of the 600-volt tertiary is 500 kva.

(a) Compute the per-unit values of the equivalent-circuit impedances of this transformer on a 1,000-kva rated-voltage base.

(b) Three of these transformers are used in a 3,000-kva Y-Δ-Δ 3-phase bank to supply 2,400-volt and 600-volt auxiliary power circuits in a generating station. The Y-connected primaries are connected to the 13,800-volt main bus. Compute the per-unit values of the steady-state short-circuit currents and of the voltage at the terminals of the secondary windings if a 3-phase short circuit occurs at the terminals of the tertiary windings with 13,800 volts maintained at the primary line terminals. Use a 3,000-kva 3-phase rated-voltage base.

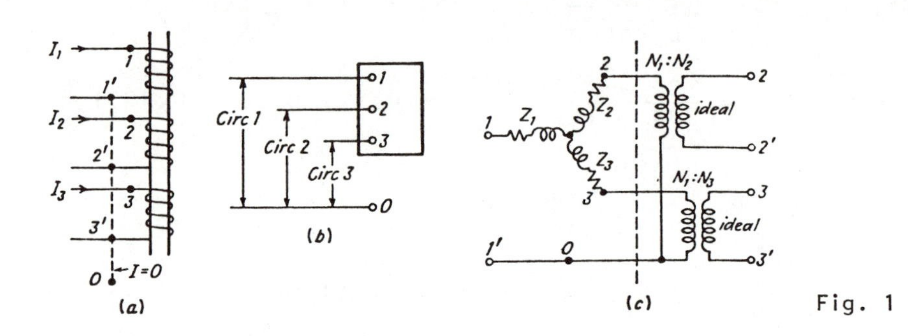

Fig. 1

(a) Elementary 3-winding transformer. (b and c) Steps in the development of its equivalent circuit.

Solution: (a) First convert the short circuit data to per unit on 1,000 kva per phase.

For primary: $V_{base} = 7,960$ volts

$$I_{base} = \frac{1,000}{7.96} = 125.4 \text{ amp}$$

For secondary: $V_{base} = 2,400$ volts.

$$I_{base} = \frac{1,000}{2.4} = 416 \text{ amp.}$$

Conversion of the test data to per unit then gives

Test	Windings	V	I
1	1 and 2	0.0316	0.500
2	1 and 3	0.0967	0.500
3	2 and 3	0.0905	0.500

From test 1, the short-circuit impedance Z_{12} is

$$Z_{12} = \frac{0.0316}{0.500} = 0.0632 \text{ per unit}$$

Similarly, from tests 2 and 3,

$$Z_{13} = \frac{0.0967}{0.500} = 0.1934 \text{ per unit.}$$

$$Z_{23} = \frac{0.0905}{0.500} = 0.1910 \text{ per unit.}$$

The equivalent-circuit constants are given by:

$$Z_1 = \tfrac{1}{2}(Z_{12} + Z_{13} - Z_{23})$$

$$Z_2 = \tfrac{1}{2}(Z_{23} + Z_{12} - Z_{13})$$

$$Z_3 = \tfrac{1}{2}(Z_{13} + Z_{23} - Z_{12}).$$

Thus,

$$\vec{Z}_1 = jX_1 = j0.0378 \text{ per unit.}$$

$$\vec{Z}_2 = jX_2 = j0.0254 \text{ per unit.}$$

$$\vec{Z}_3 = jX_3 = j0.1556 \text{ per unit.}$$

(b) Base line-to-line voltage for the Y-connected primaries is $\sqrt{3}$ (7,960) = 13,800 volts, or the bus voltage is 1.00 per unit. From the equivalent circuit with a short circuit on the tertiaries,

$$I_{sc} = \frac{V_1}{Z_1 + Z_3} = \frac{V_1}{Z_{13}} = \frac{1.00}{0.1934} = 5.18 \text{ per unit.}$$

(Note, however, that this current is 10.36 per unit on the rating of the tertiaries.) If the voltage drops caused by the secondary load current are neglected in comparison with those due to the short-circuit current, the secondary terminal voltage equals the voltage at the junction of the three impedances Z_1, Z_2, and Z_3 in Fig. 1, whence

$$V_2 = I_{sc}Z_3 = (5.18)(0.1556) = 0.805 \text{ per unit.}$$

• PROBLEM 6-65

Given a single-phase three-winding transformer with the following characteristics:

Ratings:

Coil 1, 1000 kva at 6600 volts (151.5 amperes at full load)

Coil 2, 1000 kva at 1100 volts (909 amperes at full load)

Coil 3, 500 kva at 220 volts (2273 amperes at full load)

Combined open- and short-circuit test:

Coil 3 open, reactance from coil 1 to 2 = 0.11 per unit (based on 1000 kva)

Coil 2 open, reactance from coil 1 to 3 = 0.10 per unit (based on 500 kva)

Coil 1 open, reactance from coil 2 to 3 = 0.07 per unit (based on 500 kva)

Copper loss with coil 3 open = 8 kw

Copper loss with coil 2 open = 5.5 kw

Copper loss with coil 1 open = 6.6 kw

(a) When coil 1 is used as primary, if the load on coil 2 is 500 amperes at 0.8 power factor (lagging current) and the load on coil 3 is 600 amperes at 0.9 power factor (leading current), using primary voltage as reference, what is the kva load and power factor of coil 1?

(b) If rated voltage is applied to coil 1, what are the voltages of coils 2 and 3 when the two secondary coils are loaded in this manner?

(c) With 6600 volts applied to coil 1, what is the open-circuit voltage of coil 3 when 2 is short-circuited?

(d) With 6600 volts applied to coil 1, what current will flow in coil 1 when coils 2 and 3 are both short-circuited? What is the value of the current in coil 2 and in coil 3 under these conditions?

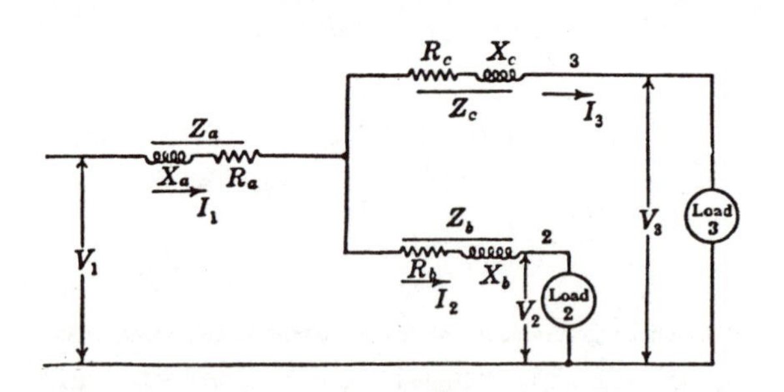

Equivalent Circuit of Three-winding Transformer.

Solution: First, it is necessary to determine the per-unit impedance values between the various windings with the third winding open-circuited. To do this, the per-unit resistances must be found, and, finally, the per-unit impedances must all be referred to the same base. The per-unit resistances referred to a given base are equal to the copper losses in kilowatts at the base kva, divided by the base kva. Hence, the per-unit impedances are,

Coil 3 open, $\vec{Z}_{ab} = \frac{8}{1000} + j0.11 = 0.008 + j0.11$
(based on 1000 kva)

Coil 2 open, $\vec{Z}_{ca} = \frac{5.5}{500} + j0.10 = 0.011 + j0.10$
(based on 500 kva)

Coil 1 open, $\vec{Z}_{bc} = \frac{6.6}{500} + j0.07 = 0.0132 + j0.07$
(based on 500 kva)

Since the primary coil is chosen as reference, all values should be based on 1000 kva. Hence $\vec{Z}_{ab} = 0.008 + j0.11$, $\vec{Z}_{bc} = 0.0264 + j0.14$, $\vec{Z}_{ca} = 0.022 + j0.20$.

$\vec{Z}_a$, $\vec{Z}_b$ and $\vec{Z}_c$ are given by

$$\vec{Z}_a = \tfrac{1}{2}\,[\vec{Z}_{ab} + \vec{Z}_{ca} - \vec{Z}_{bc}]$$

$$\vec{Z}_b = \tfrac{1}{2}\,[\vec{Z}_{bc} + \vec{Z}_{ab} - \vec{Z}_{ca}]$$

$$\vec{Z}_c = \tfrac{1}{2}\,[\vec{Z}_{ca} + \vec{Z}_{bc} - \vec{Z}_{ab}]$$

Substituting numerical values,

$$\vec{Z}_a = \tfrac{1}{2}[(0.008 + 0.022 - 0.0264) + j(0.11 + 0.20 - 0.14)] = 0.0018 + j0.085$$

$$\vec{Z}_b = \tfrac{1}{2}[(0.0264 + 0.008 - 0.022) + j(0.14 + 0.11 - 0.20)] = 0.0062 + j0.025$$

$$\vec{Z}_c = \tfrac{1}{2}[(0.022 + 0.0264 - 0.008) + j(0.20 + 0.14 - 0.11)] = 0.0202 + j0.115$$

(a) Actual Kva on Coil 2 $= 500 \times (0.8 - j0.6) \times 1.1$

$= (440 - j330) = 550\ \underline{/-36.9^\circ}$

Actual Kva on Coil 3 $= 600 \times (0.9 + j0.436) \times 0.22$

$= (118.8 + j57.55) = 132\underline{/25.9^\circ}$

Per-unit $\vec{I}_2 = \frac{440 - j330}{1000} = (0.44 - j0.33)$

$= 0.55\ \underline{/-36.9^\circ}$

Per-unit $\vec{I}_3 = \frac{118.8 + j57.55}{1000} = (0.1188 + j0.0576)$

$= 0.132\ \underline{/25.9^\circ}$

Referring to the figure, it is seen that $\vec{I}_1 = \vec{I}_2 + \vec{I}_3$

Whence $\quad$ Per-unit $\vec{I}_1 = 0.5588 - j0.2726$

$$= 0.62 \angle{-26°}$$

and $\quad$ Actual Kva on Coil 1 $= 1000 \times 0.62 = 620 \angle{-26°}$

Power Factor of Load on Coil 1 $= \dfrac{0.5588}{0.62} = 0.9$, lagging

(b) From the equivalent circuit of the figure

$$\vec{V}_2 = \vec{V}_1 - \vec{I}_1\vec{Z}_a - \vec{I}_2\vec{Z}_b$$

Therefore, using per-unit values,

$$\vec{V}_2 = 1 - (0.5588 - j0.2726)(0.0018 + j0.085)$$
$$- (0.44 - j0.33)(0.0062 + j0.025)$$
$$= 1 - (0.0242 + j0.0469) - (0.011 + j0.009)$$
$$= (1 - 0.0352) - j0.0559$$
$$= 0.9648 - j0.0559$$
$$= 0.9655 \angle{-3.32°} \text{ per-unit volts}$$

$$\vec{V}_3 = 1 - (0.5588 - j0.2726)(0.0018 + j0.085) - (0.1188$$
$$+ j0.0576)(0.0202 + j0.115)$$
$$= (1 - 0.02) - j0.0617$$
$$= 0.9800 - j0.0617$$
$$= 0.9819 \angle{-3.6°} \text{ per-unit volts}$$

In volts,

$$\vec{V}_2 = 1100 \times 0.9655 \angle{-3.32°} = 1062 \angle{-3.32°}$$
$$\vec{V}_3 = 220 \times 0.9819 \angle{-3.6°} = 216.0 \angle{-3.6°}$$

(c) When coil 2 is short-circuited, $\vec{V}_2 = 0$ and the impedance to short-circuit current is

$$\vec{Z}_a + \vec{Z}_b = 0.008 + j0.11$$

Therefore, the per-unit value of the short-circuit current is

$$\frac{\vec{V}}{\vec{Z}_a + \vec{Z}_b} = \frac{1}{0.008 + j0.11} = 0.657 - j9.043$$

The potential drop in coil 1 due to the short-circuit current is

$$\vec{I}_s\vec{Z}_a = (0.657 - j9.043)(0.0018 + j0.085)$$

$$= 0.7687 + j0.0397$$

Since coil 3 is on open circuit, there is no potential drop in it and, consequently, the per-unit voltage of coil 3 is

$$\vec{V}_3 = 1 - 0.7687 - j0.0397$$

$$= 0.2313 - j0.0397$$

$$= 0.234 \angle -9.76°$$

In volts,

$$\vec{V}_3 = 220 \times 0.234 \angle -9.76° = 51.48 \angle -9.76°$$

(d) When the secondary coils are both short-circuited, impedances $\vec{Z}_b$ and $\vec{Z}_c$ are in parallel, and the total impedance to short-circuit current is this parallel combination plus $\vec{Z}_a$. The per-unit value of short-circuit current is equal to the reciprocal of the total per-unit impedance. Hence,

$$\text{Per-unit } \vec{I}_s = \frac{1}{\vec{Z}_a + \dfrac{1}{\dfrac{1}{\vec{Z}_b} + \dfrac{1}{\vec{Z}_c}}} = \frac{\vec{Z}_b + \vec{Z}_c}{\vec{Z}_a\vec{Z}_b + \vec{Z}_b\vec{Z}_c + \vec{Z}_c\vec{Z}_a}$$

$$\vec{Z}_a = 0.0018 + j0.085 = 0.0851 \angle 87.3°$$

$$\vec{Z}_b = 0.0062 + j0.025 = 0.0258 \angle 76.1°$$

$$\vec{Z}_c = 0.0202 + j0.115 = 0.1165 \angle 80.0°$$

$$\vec{Z}_b + \vec{Z}_c = 0.0264 + j0.140 = 0.1424 \angle 79.3°$$

$$\vec{Z}_a\vec{Z}_b = 0.00219 \angle 163.4° = -0.0021 + j0.000625$$

$$\vec{Z}_b\vec{Z}_c = 0.0030 \angle 156.1° = -0.002745 + j0.001215$$

$$\vec{Z}_c\vec{Z}_a = 0.00991 \angle 167.3° = -0.00965 + j0.00218$$

$$\vec{Z}_a\vec{Z}_b + \vec{Z}_b\vec{Z}_c + \vec{Z}_c\vec{Z}_a = -0.0145 + j0.0040 = 0.015 \angle 164.0°$$

Whence

$$\vec{I}_s = \frac{0.1424 \angle 79.3°}{0.015 \angle 164.0°} = 9.42 \angle -84.7° \text{ per unit}$$

Therefore, in amperes, $I_s = 151.5 \times 9.42 = 1426$ amperes

Referring to the figure, it is seen that with the loads short-circuited, the primary current will divide inversely as the impedances $\vec{Z}_b$ and $\vec{Z}_c$. Hence,

$$\text{Per-unit Amperes in Coil 2} = \frac{0.1165}{0.1427} \times 9.42 = 7.70.$$

$$\text{Per-unit Amperes in Coil 3} = \frac{0.0258}{0.1427} \times 9.42 = 1.70.$$

Therefore, in amperes,

Current in Coil 2 = 7.70 x 909 x 2 = 14,000

Current in Coil 3 = 1.7 x 2273 x 2 = 7720

The factor, 2, is essential because the actual base is 500 kva.

PARALLEL AND POLYPHASE OPERATION OF TRANSFORMERS

• PROBLEM 6-66

A 500 kVA 1-ph transformer with 0.010 p.u. resistance and 0.05 p.u. leakage reactance is to share a load of 750 kVA at p.f. 0.80 lagging with a 250 kVA transformer with per-unit resistance and reactance of 0.015Ω and 0.04Ω. Find the load on each transformer (a) when both secondary voltages are 400 V, and (b) when the open-circuit secondary voltages are respectively 405 V and 415 V.

Solution: (a) The per-unit impedances expressed on a common base of 500 kVA are

$$\vec{Z}_1 = 0.010 + j0.05 = 0.051 \angle 79°$$

$$\vec{Z}_2 = 2(0.015 + j0.04) = 0.085 \angle 69°$$

$$\vec{Z}_1 + \vec{Z}_2 = 0.04 + j0.13 = 0.136 \angle 73°$$

The load is $\vec{S} = 750(0.8 - j0.6) = 750 \angle -37°$ KVA.

Applying the following formulae:

$$\vec{S}_1 = \vec{S} \frac{\vec{Z}_2}{\vec{Z}_1 + \vec{Z}_2}$$

$$\vec{S}_2 = \vec{S} \frac{\vec{Z}_1}{\vec{Z}_1 + \vec{Z}_2},$$

$$\vec{S}_1 = 750 \angle -37^\circ \frac{0.085 \angle 69^\circ}{0.136 \angle 73^\circ} = 471 \angle -40^\circ$$

$$= 359 - j305 \text{ kVA}$$

$$\vec{S}_2 = 281 \angle -31^\circ = 241 - j305 \text{ kVA}$$

The total active power is 359 + 241 = 600 kW (= 750 × 0.8), and the total reactive power is 450 kvar (=750 × 0.6). The 250 kVA transformer operates with a 12½% overload because of its smaller per-unit leakage impedance.

(b) With the Millman theorem it is necessary to work in ohmic values of impedance

$$\vec{Z}_1 = 0.0032 + j0.0160 = 0.0163 \angle 79^\circ \Omega$$

$$\vec{Z}_2 = 0.0096 + j0.0256 = 0.0275 \angle 69^\circ \Omega$$

It is also necessary to estimate the load impedance $\vec{Z}$: assuming an output voltage on load of 395 V, then $\vec{Z} = 0.208 \angle 37^\circ \Omega$

$$I = \frac{\vec{E}_1}{\vec{Z}_1} + \frac{\vec{E}_2}{\vec{Z}_2} = \frac{405 \angle 0^\circ}{0.0163 \angle 79^\circ} + \frac{415 \angle 0^\circ}{0.0275 \angle 69^\circ}$$

$$= 39{,}700 \angle -75.2^\circ \text{ A}$$

$$\frac{1}{\vec{Z}_0} = \frac{1}{0.208 \angle 37^\circ} + \frac{1}{0.0163 \angle 79^\circ} + \frac{1}{0.0275 \angle 69^\circ}$$

$$= \frac{1}{0.0099 \angle 73.5^\circ} \text{ mho}$$

The secondary terminal voltage is

$$\vec{V} = \vec{I} \vec{Z}_0$$

$$= (39{,}700 \angle -75.2^\circ)(0.0099 \angle 73.5^\circ) = 393 \angle -1.7^\circ$$

$$= 393 - j12 \text{ V}$$

The internal volt drop in the first transformer is $\vec{E}_1 - \vec{V} = 405 - (393 - j12) = 17 \angle 45^\circ$ V, and in the second is $22 - j12 = 25 \angle 29^\circ$ V, whence

$$\vec{I}_1 = \frac{17 \angle 45^\circ}{0.0163 \angle 79^\circ} = 1{,}040 \angle -34^\circ \text{ A}$$

and $\vec{I}_2 = 910 \angle -40^\circ$ A

The loads are $\vec{S}_1 = \vec{V}\vec{I}_1$ and $\vec{S}_2 = \vec{V}\vec{I}_2$, giving

$$\vec{S}_1 = 340 - j220 = 410 \text{ kVA and } \vec{S}_2 = 270 - j220$$

$$= 350 \text{ kVA}$$

The combined load is $\vec{S} = 610 - j445$ kVA ($\simeq 600 - j450$) and the 250 kVA transformer is overloaded by 40%. The secondary circulating current on no load is

$$(\vec{E}_1 - \vec{E}_2)/(\vec{Z}_1 + \vec{Z}_2) = 230 \text{ A}$$

corresponding to about 95 kVA and a considerable waste in I^2R loss.

• PROBLEM 6-67

Two 50-kva, single-phase transformers are connected in parallel on both the high- and low-tension sides. Their constants are given in the following tabulation:

Unit	Open-circuit Voltages		Resistance		Reactance	
	High-tension	Low-tension	High-tension	Low-tension	High-tension	Low-tension
1	22,500	2,310	61.6	0.661	110	1.16
2	22,400	2,320	61.6	0.661	110	1.16

These transformers supply a combined load of 93 kw at a power factor of 0.89 on the low-tension side with lagging current at a terminal voltage of 2300. What is the applied primary voltage under load and at no load? What is the load current of each transformer? What is the value of the circulating current at no load?

Solution:

For Transformer 1

$$a_1 = \frac{22,500}{2310} = 9.74$$

$$R_{01} = 0.661 + \frac{61.6}{(9.74)^2}$$

$$= 0.661 + 0.651$$

$$= 1.312 \text{ ohms}$$

$$X_{01} = 1.16 + \frac{110}{94.7}$$

$$= 1.16 + 1.16$$

$$= 2.32 \text{ ohms}$$

$$\vec{Z}_{01} = 1.312 + j2.32$$

$$= 2.6552 \angle 60.5° \text{ ohms}$$

$$\frac{1}{\vec{Z}_{01}} = 0.3766 \angle -60.5° \text{ mho}$$

$$\frac{1}{a_1\vec{Z}_{01}} = 0.03866 \angle -60.5°$$

For Transformer 2

$$a_2 = \frac{22,400}{2320} = 9.65$$

$$R_{02} = 0.661 + \frac{61.6}{(9.65)^2}$$

$$= 0.661 + 0.661$$

$$= 1.322 \text{ ohms}$$

$$X_{02} = 1.16 + \frac{110}{93.1}$$

$$= 1.16 + 1.18$$

$$= 2.34 \text{ ohms}$$

$$\vec{Z}_{02} = 1.322 + j2.34$$

$$= 2.6876 \angle 60.5° \text{ ohm.}$$

$$\frac{1}{\vec{Z}_{02}} = 0.3720 \angle -60.5° \text{ mho.}$$

$$\frac{1}{a_2\vec{Z}_{02}} = 0.03856\angle{-60.5°}$$

$$\frac{1}{\vec{Z}_t} = 0.3766\angle{-60.5°} + 0.3720\angle{-60.5°} \ldots\ldots = 0.7486\angle{-60.5°}$$

$$\sum \frac{1}{a_n\vec{Z}_{0n}} = 0.03866\angle{-60.5°} + 0.3856\angle{-60.5°}$$

$$\ldots = 0.07722\angle{-60.5°}$$

$$\vec{I}_2 = \frac{93{,}000}{2300 \times 0.89} = 45.45\angle{-27.1°}\text{ A.}$$

The primary voltage $\vec{V}_1$ is given by

$$\vec{V}_1 = \frac{\vec{I}_2 + \frac{\vec{V}_2}{\vec{Z}_t}}{\sum \frac{1}{a_n\vec{Z}_{0n}}}$$

$$\vec{V}_1 = \frac{45.45\angle{-27.1°} + (2300 \times 0.7486)\angle{-60.5°}}{0.07722\angle{-60.5°}}$$

$$= \frac{45.45(0.89 - j0.4555) + 1721.78(0.4925 - j0.87)}{0.07722\angle{-60.5°}}$$

$$= \frac{888.88 - j1518.55}{0.07722 -\angle{60.5°}}$$

$$= \frac{1758.7\angle{-59.7°}}{0.07722\angle{-60.5°}} = 22{,}782\angle{0.8°}\text{ volts.}$$

At no-load, $\vec{I}_2 = 0$, hence,

$$\vec{V}_1 = \frac{\vec{V}_2}{\vec{Z}_t \sum \frac{1}{a_n\vec{Z}_{0n}}}$$

$$\vec{V}_1 = \frac{2300 \times 0.7486\angle{-60.5°}}{0.07722\angle{-60.5°}} = \frac{1721.78}{0.07722}\angle{0°} = 22{,}303\angle{0°}\text{ volts}$$

$\vec{I}_{21}$ and $\vec{I}_{22}$ are given by

$$\vec{I}_{21} = \frac{1}{\vec{Z}_{01}} \left[\frac{\vec{V}_1}{a_1} - \vec{V}_2\right]$$

$$\vec{I}_{22} = \frac{1}{\vec{Z}_{02}} \left[\frac{\vec{V}_1}{a_2} - \vec{V}_2\right]$$

Thus, under load,

$$\vec{I}_{21} = 0.3766\angle{-60.5°}\left[\frac{22,782\angle{0.8}}{9.74} - 2300\right]$$

$$= 0.3766\angle{-60.5°}[39.01 + j32.75]$$

$$= 0.3766\angle{-60.5°} \times 50.9\angle{40°}$$

$$= 19.22\angle{-20.5°} \text{ amperes}$$

$$\vec{I}_{22} = 0.3720\angle{-60.5°}\left[\frac{22,782\angle{0.8}}{9.65} - 2300\right]$$

$$= 0.3720\angle{-60.5°}[60.83 + j33.05]$$

$$= 0.3720\angle{-60.5°} \times 68.55\angle{28.6°}$$

$$= 25.55\angle{-31.9°} \text{ amperes}$$

At no load,

$$\vec{I}_{21} = 0.3766\angle{-60.5°}\left[\frac{22,303\angle{0°}}{9.74} - 2300\right]$$

$$= 0.3766\angle{-60.5°} \times (-10)$$

$$= 3.766\angle{119.5°} \text{ amperes}$$

$$\vec{I}_{22} = 0.372\angle{-60.5°}\left[\frac{22,303\angle{0°}}{9.65} - 2300\right]$$

$$= 0.372\angle{-60.5°} \times 10.15$$

$$= 3.766\angle{-60.5°}$$

Since there are but two transformers in parallel, it is clear that the same circulating current must flow in each transformer at no load and that the currents are 180 degrees out of phase. The results obtained bear out this relation. This example emphasizes the undesirability of operating transformers in parallel when there is an appreciable difference in the ratios of transformation.

Summarizing, it can be said in regard to parallel operation, that for satisfactory results,

(a) The transformers should have equal voltage ratings.

(b) The ratios of transformation should be the same.

(c) The ratio of equivalent resistance to equivalent reactance should be the same for each transformer.

(d) The equivalent impedances of the transformers should be inversely proportional to the current ratings.

• PROBLEM 6-68

Two 10-kva transformers A and B are operated in parallel to supply a 20-kva load. The per-unit values of internal impedance (on a 10-kva base) are

$$R_A = 0.0181$$

$$X_A = 0.0264$$

$$\vec{Z}_A = 0.0181 + j0.0264$$

$$= 0.032\underline{/55.6}$$

$$R_B = 0.015$$

$$X_B = 0.036$$

$$\vec{Z}_B = 0.015 + j0.036$$

$$= 0.039\underline{/67.4}$$

The total current is

$$I = 2.0$$

(a) Determine the current in each transformer.

(b) Determine the maximum kva of the load that can be supplied by these transformers without overloading either of the transformers.

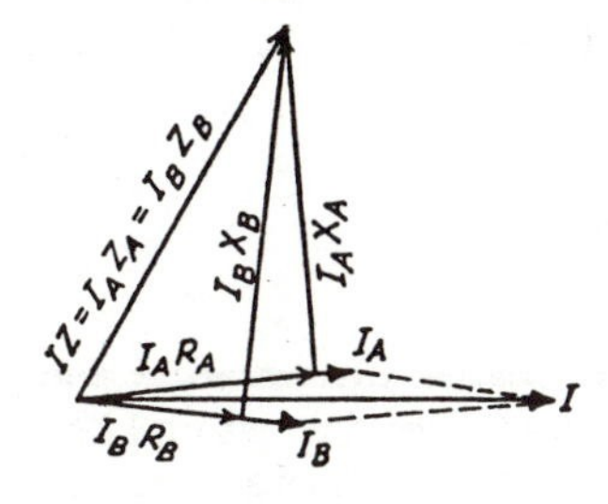

Vector diagram of load division with transformers in parallel.

Solution: (a) The total parallel impedance is

$$Z = \frac{\vec{Z}_A \times \vec{Z}_B}{\vec{Z}_A + \vec{Z}_B} = \frac{0.001247\underline{/123}}{0.0331 + j0.0624} = \frac{0.001247\underline{/123}}{0.0706\underline{/62.1}}$$

$$= 0.01766\underline{/60.9}$$

and so the IZ drop for each transformer is

$$\vec{I}\vec{Z} = 0.03532\underline{/60.9}$$

Therefore the currents of transformers A and B are, respectively,

$$\vec{I}_A = \frac{0.03532\underline{/60.9}}{0.032\underline{/55.6}} = 1.103\underline{/5.3}$$

$$\vec{I}_B = \frac{0.03532\underline{/60.9}}{0.039\underline{/67.4}} = 0.906\underline{/-6.5}$$

The vector relationships are shown in the figure.

(b) The load on the transformers is not split equally, and transformer A is overloaded more than 10 per cent.

To avoid overloading transformer A, the total load must be limited to 18.1 kva, or about 90 per cent of the total rated kva. Also, the two transformers operate at different power factors. If the load power factor in the problem is unity, the phase angle for transformer A is 5.3 deg leading, and that for B is 6.5 deg lagging. If the total load power factor were 0.8 lagging, the load on transformer A would be 9.4 kw at 0.852 power factor, while that on B would be only 6.6 kw at 0.727 power factor.

• PROBLEM 6-69

Two transformers, connected in parallel on both the high- and low-voltage sides, are characterized by the following data, where the impedances and resistances are given in terms of the low-voltage sides:

Unit	Kva rating	Voltage	Impedance, ohms	Resistance, ohms
A	100	4,600/230	0.027	0.008
B	200	4,610/225	0.013	0.003

The load, connected to the low-voltage side, takes 150 kw at a lagging power factor of 0.85, and the terminal voltage is 235 volts. Find the primary voltage and the current supplied by each transformer.

Solution:

Transformer A

$a_1 = 4{,}600/230 = 20$

$\vec{Z}' = 0.008 + j0.027\ \Omega.$

$\vec{Y}' = 10.08 - j34.05$ mho.

$\frac{\vec{Y}'}{a_1} = 0.504 - j1.702$

Transformer B

$a_2 = 4{,}610/225 = 20.49$

$\vec{Z}'' = 0.003 + j0.013\ \Omega.$

$\vec{Y}'' = 16.85 - j73.03$ mho.

$$\frac{\vec{Y}''}{a_2} = 0.822 - j3.563$$

$$\sum \vec{Y}_k = 26.93 - j107.08 \text{ mho}$$

$$\sum \frac{\vec{Y}_k}{a_k} = 1.326 - j5.265 \ .$$

From the given data,

$$I = \frac{150,000}{225 \times 0.85} = 784 \text{ amp.}$$

Assume that V_2 is taken as the axis of reference so that

$$\vec{V}_2 = V_2 + j0 = 225 \text{ V.}$$

and since I lags behind V_2 by the angle $\cos^{-1} 0.85$ or $\sin^{-1} 0.527$,

$$\vec{I} = 784(0.85 - j0.527) = 666.4 - j413.17 \text{ A.}$$

The primary voltage $\vec{V}_1$ is given by

$$\vec{V}_1 = \frac{\vec{V}_2 \sum \vec{Y}_k + \vec{I}}{\sum(\vec{Y}_k/a_k)}$$

$$\vec{V}_1 = \frac{225(26.93 - j107.08) + 784(0.85 - j0.527)}{1.326 - j5.265}$$

$$= \frac{6,725.6 - j24,506}{1.326 - j5.265} = \frac{25,411\angle -74°39.16'}{5.428\angle -75°51.85'}$$

$$= 4,681\angle 1°12.7' \text{ V.}$$

The primary current supplied by transformer A is given by

$$\vec{I}' = \frac{\vec{I} + \vec{V}_1 [(\sum \vec{Y}_k/a_1) - \Sigma(\vec{Y}_k/a_k)]}{\vec{Z}' \sum \vec{Y}_k}$$

$$= \left\{784(0.85 - j0.527) + 4{,}681\underline{/1^\circ 12.7'} \left[\frac{26.93 - j107.08}{20} - (1.326 - j5.265)\right]\right\} \div \left\{(0.008 + j0.027)(26.93 - j107.08)\right\}$$

$$= 363.8\underline{/-44^\circ 38'} = 258.9 - j255.7 \text{ A.}$$

and therefore

$$\vec{I}'' = \vec{I} - \vec{I}' = 407.5 - j157.47 = 436.9\underline{/-21^\circ 7.7'} \text{ A.}$$

These results show that transformer A is loaded to 83.5 per cent of its current rating, while transformer B is loaded only to 49.3 per cent of its rated current.

• PROBLEM 6-70

Two 10-kva 2,200/220-volt transformers each has an impedance

$$\vec{Z} = R + jX = 8 + j12 \text{ ohms.}$$

They are connected in open Δ to a noninductive load as in Fig. 1, where

$$\vec{Z}_\ell = R_\ell + j0 = 500 \text{ ohms,}$$

all quantities being expressed in terms of the primary.

Find the current in each branch of the circuit.

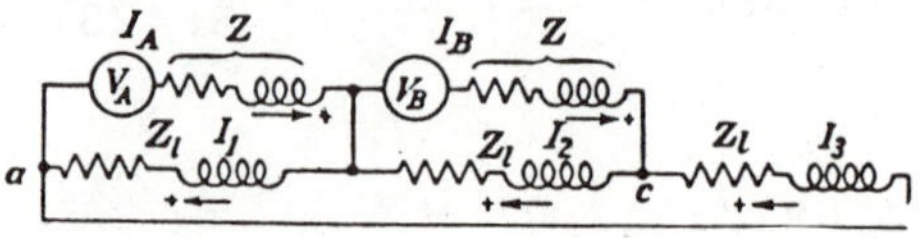

Fig. 1

Equivalent circuit, transformers in open Δ, loads in Δ.

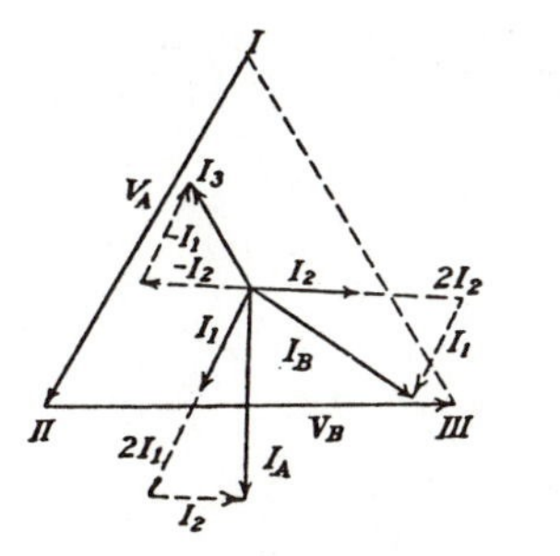

Fig. 2

Phase relations of voltages and currents, open-Δ connection.

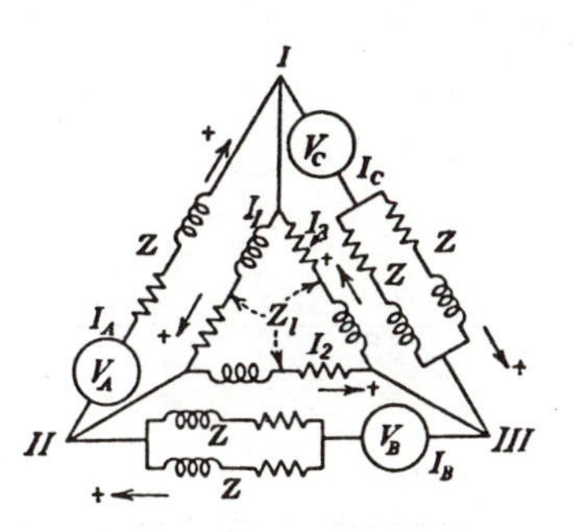

Fig. 3

Equivalent circuit, closed- Δ and open-Δ transformers in parallel.

Solution:

$$\vec{V}_B = \vec{V}_A \underline{/120°} = \vec{V}_A(-0.5 + j0.866) \text{ V.}$$

$$\vec{I}_1 = \frac{\vec{V}_A(\vec{Z}_\ell + 2\vec{Z}) - \vec{V}_B\vec{Z}}{(\vec{Z}_\ell + 3\vec{Z})(\vec{Z}_\ell + \vec{Z})}$$

$$= V_A \frac{(500 + 16 + j24) - (-0.5 + j0.866)(8 + j12)}{(500 + 24 + j36)(500 + 8 + j12)}$$

$$= V_A \frac{530.39 + j23.07}{265{,}760 + j24{,}576} = \frac{V_A}{502.72} \underline{/-2°47.6'} \text{ A.}$$

$$\vec{I}_2 = \frac{\vec{V}_B(\vec{Z}_\ell + 2\vec{Z}) - \vec{V}_A\vec{Z}}{(\vec{Z}_\ell + 3\vec{Z})(\vec{Z}_\ell + \vec{Z})}$$

$$= V_A \frac{(-0.5 + j0.866)(500 + 16 + j24) - (8 + j12)}{265{,}760 + j24{,}576}$$

$$= \frac{\vec{V}_A}{522.36} \underline{/118°51.7'} \text{ A.}$$

$$\vec{I}_3 = -\vec{I}_1 - \vec{I}_2 = \frac{\vec{V}_A}{525.2} \underline{/236°4.2'} \text{ A.}$$

$$\vec{I}_A = \frac{\vec{V}_A(2\vec{Z}_\ell + 3\vec{Z}) + \vec{V}_B\vec{Z}_\ell}{(\vec{Z}_\ell + 3\vec{Z})(\vec{Z}_\ell + \vec{Z})}$$

$$= \vec{V}_A \frac{(1{,}000 + 24 + j36) + (-0.5 + j0.866)500}{265{,}760 + j24{,}576}$$

$$= \frac{\vec{V}_A}{294.9} \underline{/25°55.8'} \text{ A.}$$

$$\vec{I}_B = \frac{\vec{V}_A\vec{Z}_\ell + \vec{V}_B(2\vec{Z}_\ell + 3\vec{Z})}{(\vec{Z}_\ell + 3\vec{Z})(\vec{Z}_\ell + \vec{Z})}$$

$$= \vec{V}_A \frac{500 + (-0.5 + j0.866)(1{,}024 + j36)}{265{,}760 + j24{,}576}$$

$$= \frac{\vec{V}_A}{306.8} \underline{/87°33.7'} \text{ A.}$$

These results are plotted in Fig. 2, which shows clearly the severe unbalancing of the currents, as to phase relations, in the supply lines.

• PROBLEM 6-71

What should be the ratings and turns ratio of a three-phase transformer to transform 10,000 KVA from 230 KV to 4160 V' if the transformer is to be connected (a) Y-Δ, (b) Δ-Y, (c) Δ-Δ?

Solution:

Rated primary line current $\equiv I_{L1B} = \dfrac{S_B}{\sqrt{3}V_{L1B}}$

$$= \frac{10,000,000}{\sqrt{3} \times 230,000} = 25.1 \text{ A}$$

Rated secondary line current $\equiv$

$$I_{L2B} = \frac{10,000,000}{\sqrt{3} \cdot 4160} = 1388 \text{ A}$$

(a) Y-Δ:

Rated kVA = $S_B/1000$ = 10,000 kVA.

Rated $I_1 \equiv I_{1B} = I_{L1B} = 25.1$ A.

Rated $I_2 \equiv I_{2B} = I_{L2B}/\sqrt{3} = 801$ A.

Rated V_{L1} = 230 kV; Rated V_{L2} = 4160 V.

Rated $V_1 \equiv V_{1B} = 230/\sqrt{3} = 132.8$ kV.

Rated $V_2 \equiv V_{2B} = 4160$ V.

Turns ratio = $V_{1B}/V_{2B} = 132.8 \times 10^3/4160 = 31.9$.

kVA per phase = 10,000/3 = 3333 kVA

(b) Δ-Y:

Rated kVA = 10,000

kVA per phase = 3333.

$V_{1B} = V_{L1B} = 230$ kV.

$V_{2B} = V_{L2B}/\sqrt{3} = 4160/\sqrt{3} = 2400$ V.

$I_{1B} = I_{L1B}/\sqrt{3} = 14.5$ A.

$I_{2B} = I_{L2B} = 1338$ A.

$\therefore \quad a = V_{1B}/V_{2B} = 95.8.$

(c) Δ-Δ:

Rated kVA = 10,000.

kVA/phase = 3333.

$V_{1B} = V_{L1B} = 230$ kV.

$V_{2B} = V_{L2B} = 4160$ V.

$I_{1B} = I_{L1B}/\sqrt{3} = 14.5$ A.

$I_{2B} = I_{L2B}/\sqrt{3} = 801$ A.

$$\therefore \quad a = \frac{V_{1B}}{V_{2B}} = 55.3.$$

• PROBLEM 6-72

A 5,000-kva 3-phase 60 ~ 3-winding 14,400:2,400:575-volt transformer is to be used as a station-service transformer to supply turbine and boiler auxiliaries in a generating station. The 14,400-volt winding is rated 5,000 kva; the 2,400- and 575-volt windings are each rated 2,500 kva. All three windings are Δ-connected.

To avoid excessive interrupting duty on the 575-volt switchgear, the transformer is designed so that a solid, symmetrical, three-phase short circuit directly at the terminals of the 575-volt winding, with rated voltage sustained on the 14,400-volt winding, will cause steady-state currents limited to 25,000 amperes in the lines emanating from the terminals of the 575-volt winding. When the transformer is designed to meet these requirements, the short-circuit reactance of the 14,400:2,400-volt windings is found to be 6.0 per cent on a 5,000-kva base, and the short-circuit reactance of the 2,400:575-volt windings is 10.0 per cent on a 2,500-kva base. Winding resistances are small enough to be neglected.

Under these special conditions, determine the full-load voltages of the 575- and 2,400-volt buses. These voltages are to be computed for 2,500-kva loads at 0.85 power factor, lagging, on each bus and with rated voltage impressed on the 14,400-volt winding.

Solution: The constants of the equivalent circuit will first be found. Let the 14,400-, 2,400-, and 575-v windings be numbered 1, 2, and 3, respectively. Then, in per unit on a 2,500-kva base,

$$X_{(12)} = 0.06 \times \frac{2,500}{5,000} = 0.03$$

$$X_{(23)} = 0.10.$$

On a 2,500-kva base, unit current in the 575-v lines is

$$\frac{2,500 \times 1,000}{\sqrt{3} \times 575}, \text{ or } 2,510 \text{ amp.}$$

Hence the 25,000-amp short-circuit current is

$$\frac{25,000}{2,510}, \text{ or } 9.96 \text{ per unit,}$$

and

$$X_{(31)} = \frac{1}{9.96} = 0.1003.$$

The branch impedances of the Y-equivalent circuit are given by

$$Z_1 = \frac{1}{2}\,[Z_{(12)} + Z_{(13)} - Z_{(23)}] \tag{1}$$

$$Z_2 = \frac{1}{2}\,[Z_{(21)} + Z_{(23)} - Z_{(13)}] \tag{2}$$

$$Z_3 = \frac{1}{2}\,[Z_{(31)} + Z_{(32)} - Z_{(12)}] \tag{3}$$

Since winding resistances are negligible, Eqs. (1), (2) and (3) yield

$$X_1 = \frac{1}{2}\,(0.03 + 0.1003 - 0.10) = 0.0151$$

$$X_2 = \frac{1}{2}\,(0.03 + 0.10 - 0.1003) = 0.0149$$

$$X_3 = \frac{1}{2}\,(0.1003 + 0.10 - 0.03) = 0.0851.$$

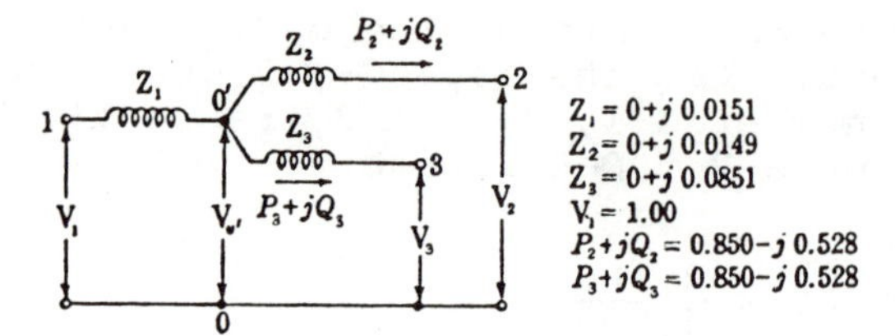

Equivalent circuit for the three-winding transformer, showing conditions of the problem. All quantities are expressed in per unit on a 2,500-kva base.

The equivalent circuit of the transformer is given in the figure, where the conditions of the problem are also presented. Inspection of the figure shows that in a straightforward solution of the problem the transformer currents would have to be taken as unknowns and an unwieldy set of simultaneous equations (some quadratics) would have to be established. To avoid the excessive labor and possible inaccuracy of such a solution, a method of successive approximations will be adopted.

Recognizing that all windings carry somewhere near their rated currents, one can roughly estimate a voltage drop of 1.5% in Z_2, and 8.5% in Z_3. The current in Z_1 is approximately twice the unit current (on a 2,500-kva base), and therefore the voltage drop in Z_1 is approximately 3%. The voltage drops subtract vectorially from the rated voltage applied to winding 1, and consequently the terminal voltages of windings 2 and 3 are somewhat greater than the arithmetic differences between the applied voltage V_1 and the voltage drops. Approximate values of the terminal voltages V_2 and V_3 can be obtained, however, if the vector relations are ignored; thus

$$V_2 \simeq 1.00 - 0.03 - 0.01 = 0.96 \text{ V}$$

$$V_3 \simeq 1.00 - 0.03 - 0.08 = 0.89 \text{ V}$$

On this basis, the winding currents are

$$I_2 \simeq \frac{1.00}{0.96} = 1.04 \text{ A}$$

$$I_3 \simeq \frac{1.00}{0.89} = 1.12 \text{ A}$$

and

$$I_1 \simeq 1.04 + 1.12 = 2.16 \text{ A.}$$

The vector power input to winding 1 can now be estimated if the inductive reactive power (I^2X) absorbed by the transformer leakage reactances is added to the vector power outputs of windings 2 and 3. The reactive power absorbed by the leakage reactances is

$$\Sigma I^2X = I_1^2X_1 + I_2^2X_2 + I_3^2X_3$$

$$\simeq (2.16)^2(0.0151) + (1.04)^2(0.0149) + (1.12)^2(0.0851)$$

$$\simeq 0.19 \text{ VAR.}$$

By convention, the algebraic sign associated with inductive reactive power is negative, and therefore, the vector power input $P_1 + jQ_1$ to winding 1 is

$$P_1 + jQ_1 = P_2 + jQ_2 + P_3 + jQ_3 - j\Sigma I^2X$$

$$\simeq 2(0.850 - j0.528) - j0.19$$

$$\simeq 1.70 - j1.25.$$

A solution for V_2 and V_3 can be formulated from this input, giving values which, compared with those estimated above, will serve as a guide for a second approximation if one is necessary. Since V_1 equals 1.00 per unit, the per unit vector current I_1 in winding 1 equals the per unit vector power $P_1 + jQ_1$ when $\vec{V}_1$ is the reference vector. Thus

$$\vec{I}_1 = 1.70 - j1.25 \text{ A}$$

The vector voltage $V_{0'}$ of the common point 0' in the figure is

$$\vec{V}_{0'} = \vec{V}_1 - j\vec{I}_1\vec{X}_1$$

$$= 1.00 + j0 - j0.0151(1.70 - j1.25)$$

$$= 0.981 - j0.0256 \text{ V.}$$

The vector terminal voltages of windings 2 and 3 now can be determined.

$$\vec{V}_2 = \vec{V}_{0'} - j\vec{I}_2\vec{X}_2$$

$$= 0.981 - j0.0256 - j0.0149\left(\frac{1.70 - j1.25}{2}\right)$$

$$= 0.972 - j0.0383 \text{ V}$$

The magnitude of $\vec{V}_2$ is

$$V_2 = 0.972 \text{ per unit, or } 2{,}330 \text{ v.}$$

The vector voltage $\vec{V}_3$ is

$$\vec{V}_3 = \vec{V}_{0'} - j\vec{I}_3\vec{X}_3$$

$$= 0.981 - j0.0256 - j0.0851\left(\frac{1.70 - j1.25}{2}\right)$$

$$= 0.928 - j0.0979 \text{ V.}$$

The magnitude of $\vec{V}_3$ is

$$V_3 = 0.930 \text{ per unit, or } 535 \text{ V.}$$

Carrying through this process again with these voltages as second approximations yields practically identical results. The full-load voltages of the transformer are, then, 2,330 and 535 v.

• PROBLEM 6-73

Consider the circuit shown in the schematic diagram of Fig. 1, comprising a Y-Δ bank of three 1,000-kva 63,500: 33,000-volt single-phase transformers connected on their primary sides to a balanced three-phase source (the 110,000-volt bus in a substation). The secondaries of this bank supply power to a Δ-Δ bank of three 1,000-kva 33,000:13,200-volt transformers through a three-phase transmission line.

Determine what voltage is required at the substation bus in order to maintain the rated line-to-line voltage of 13,200 volts at the secondary terminals of the Δ-Δ bank when this bank supplies a balanced three-phase load of 3,000 kva at unity power factor.

Data: The impedance of the 33,000-volt transmission line is

$$\vec{Z}_{line} = 7.3 + j18.2 \text{ ohms per phase.}$$

The equivalent impedance $\vec{Z}_R$ of each of the Δ-Δ transformers at the receiving end of the line is

$$\vec{Z}_R = 1.71 + j9.33 \text{ ohms}$$

referred to the low-voltage side. Their core loss is

$$P_c = 5.6 \text{ kw, each transformer,}$$

and their magnetizing reactive kva is

$$(VI)_{mag} = 51 \text{ kvar, each transformer.}$$

The average results of single-phase open-circuit and short-circuit tests taken on the sending-end transformers are:

Open-circuit test	Short-circuit test
V = 33000 v	V = 2,640 v
I = 1.24 amp	I = 30.3 amp
P = 5.30 kw	P = 9.81 kw

In both these tests, the measurements were made on the low-voltage (33,000-volt) sides at rated frequency (60 cycles per second).

Solution: The first step is to reduce the circuit to a single phase of a Y-connected equivalent, as shown in Fig. 1b, in which the load and source are each considered as one phase of a Y-connected circuit, and each transformer bank is represented by one phase of a Y-Y equivalent comprising an ideal transformer in combination with a series impedance and a shunt admittance.

The neutral points n of all the Y's can be considered connected, as shown by the broken line in Fig. 1b.

Although the exciting currents of the transformers ordinarily would be neglected in the solution of a problem of this kind, they will be included in the following solution, wherever they have any effect, for the sake of completeness.

The exciting currents may be accounted for by shunt admittances connected on either side of the transformers. It is convenient to connect the exciting admittances of the load-end transformers across the secondary terminals, since the excitation, given as

$$P_c - j(VI)_{mag} = 5.6 - j51 \text{ vector kva per phase,}$$

then can be combined directly in parallel with the load, given as 1,000 + j0 vector kva per phase. Thus the load and the excitation of the load-end transformers absorb a combined vector power of

1,006 - j51 vector kva per phase.

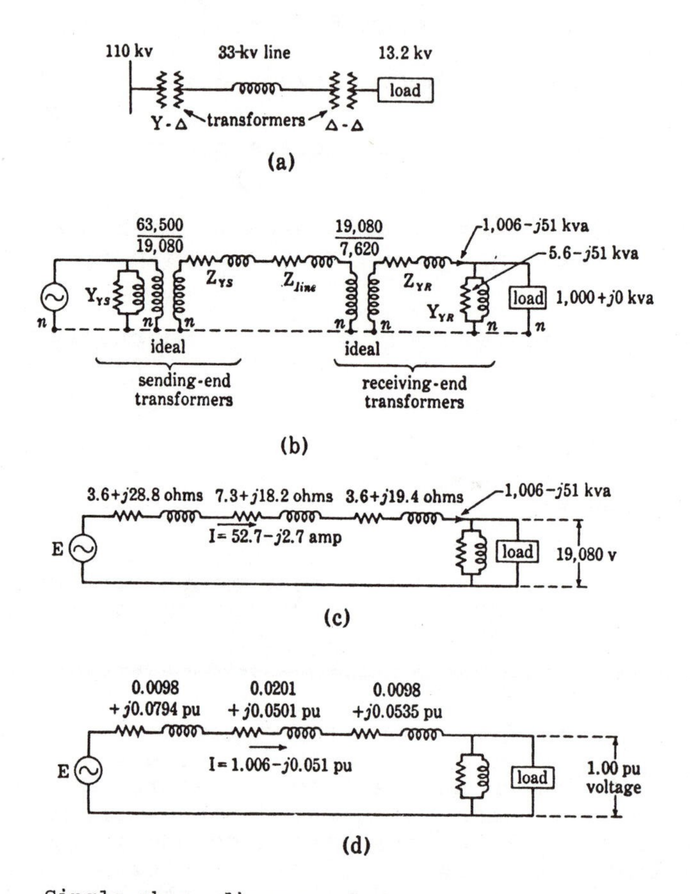

Fig. 1

Single-phase diagram of the circuit, and steps in the simplification of the circuit.

However, if the exciting admittances of the sending-end transformers are connected across their primary terminals, as in Fig. 1b, their exciting currents have no effect on the source voltage required to maintain rated voltage at the load, although they do affect the current taken from the source. When treated in this manner, the exciting admittances of the sending-end transformers thus do not enter into the solution of this problem, since the source current is not required.

The ratios of transformation of the Y-Y connected ideal transformers in Fig. 1b are such that they produce the same open-circuit voltages as the bank they represent. The equivalent Y voltages to neutral for the 33,000: 13,200-v, Δ-Δ bank at the load end are

$$\frac{33{,}000}{\sqrt{3}} = 19{,}080 \text{ v}$$

$$\frac{13{,}200}{\sqrt{3}} = 7{,}620 \text{ v}$$

and the equivalent Y voltages to neutral for the Y-Δ bank at the sending end are

$$\frac{110{,}000}{\sqrt{3}} = 63{,}500 \text{ v}$$

and

$$19{,}080 \text{ v.}$$

Thus the ratios of transformation of the ideal transformers are as marked in Fig. 1b.

The equivalent impedance of each of the Δ-Δ transformers at the receiving end is given as

$$\vec{Z}_{\Delta R} = 1.71 + j9.33 \text{ ohms}$$

referred to the low-voltage side. The equivalent Y-connected impedance is

$$\vec{Z}_{YR} = \frac{1}{3}\,(1.71 + j9.33) = 0.57 + j3.11 \text{ ohms}$$

Referred to the primary side of the receiving-end ideal transformer, this equivalent impedance is

$$a^2\vec{Z}_{YR} = \left(\frac{19{,}080}{7{,}620}\right)^2 (0.57 + j3.11)$$

$$= 3.6 + j19.4 \text{ ohms.}$$

From the short-circuit data, the short-circuit impedance of each sending-end transformer referred to the Δ-connected secondary side is

$$Z_{\Delta S} = \frac{V}{I} = \frac{2{,}640}{30.3} = 87.1 \text{ ohms}$$

$$R_{\Delta S} = \frac{P}{I^2} = \frac{9{,}810}{(30.3)^2} = 10.7 \text{ ohms}$$

$$X_{\Delta S} = \sqrt{Z_{\Delta S}^2 - R_{\Delta S}^2} = 86.4 \text{ ohms.}$$

The equivalent Y-connected impedance is

$$\vec{Z}_{YS} = \frac{1}{3}\,\vec{Z}_{\Delta S}$$

$$= \frac{1}{3}\,(10.7 + j86.4) = 3.6 + j28.8 \text{ ohms}$$

If the load and the equivalent impedance and exciting current of the receiving-end transformer are all referred to the primary side, and the source voltage is referred to the secondary side of the sending-end transformer, the ideal transformers may be omitted, and the circuit of Fig. 1b reduces to that shown in Fig. 1c, in which a load of 1,006 - j51 vector kva per phase is supplied at a voltage of 19,080 v per phase through a series impedance whose value equals the sum of its three components, or

$$\vec{Z} = \vec{Z}_{YS} + \vec{Z}_{line} + a^2\vec{Z}_{YR} = 14.5 + j66.4 \text{ ohms}$$

If the load voltage is taken as the reference vector, the vector current in the transmission line is

$$I = \frac{\text{kva per phase}}{\text{kv per phase Y}} = \frac{1{,}006 - j51}{19.08}$$

$$= 52.7 - j2.7 \text{ amp per phase Y.}$$

The required sending-end voltage now can be determined by adding the impedance drop in the system to the load voltage; thus

$$\vec{I}\vec{Z} = (52.7 - j2.7)(14.5 + j66.4)$$

$$= 940 + j3{,}460 \text{ v Y}$$

$$\vec{V} = 19{,}080 + j\quad 0$$

$$\vec{E} = \text{sum} = 20{,}020 + j3{,}460$$

$$E = 20{,}300 \text{ vY}$$

The voltage E is the substation voltage to neutral, referred to the low-voltage secondary sides of the sending-end transformers. The actual substation voltage is

$$\frac{63,500}{19,080} \times 20,300 = 67,500 \text{ v to neutral}$$

or

$$\sqrt{3} \times 67,500 = 117,000 \text{ v line-to-line,}$$

which is the sending-end voltage required to maintain rated voltage at the load end.

• PROBLEM 6-74

Three 10-kVA, 1330:230-V, 60-Hz transformers are connected wye-delta to supply at 230 volts line-to-line a heating load of 2 kW per phase and a three-phase induction-motor load of 21 kVA. The power factor of the induction-motor load is 0.8.

In this system the loads are connected to the transformers by means of a common three-phase feeder whose impedance is 0.003 + j0.010 Ω per phase. The transformers themselves are supplied from a constant-potential source by means of a three-phase feeder whose impedance is 0.75 + j5.0 Ω per phase. The equivalent impedance of one transformer referred to the low-potential side is 0.118 + j0.238 Ω. Determine the required source potential difference if that at the load is to be 230 V.

Solution: Since the secondaries of the transformer bank are connected in delta, it is convenient to refer their equivalent impedances to the primary wye-connected side before determining the equivalent wye-wye bank of ideal transformers. Thus,

$$R'_{eq} + jX'_{eq} = \left[\frac{1330}{230}\right]^2 (0.118 + j0.238) = 3.94 + j7.96 \ \Omega$$

The required turns ratio of the ideal transformers in the equivalent wye-wye bank is given by the ratio of the line-to-line potential differences of the ideal wye-delta bank:

$$n' = \frac{1330\sqrt{3}}{230} = 10$$

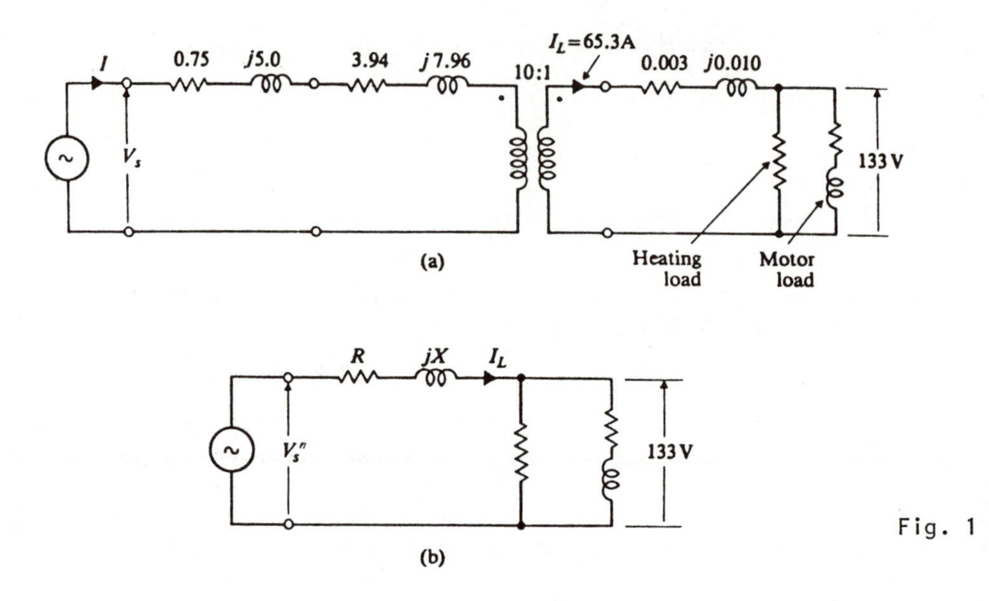

Fig. 1

Per-phase equivalent circuit.

The per-phase equivalent circuit of the system is therefore that shown in Fig. 1(a). The quantities on the primary side of the ideal transformer may now be referred to the secondary and the impedances of the transformer and both feeders combined to give the circuit shown in Fig. 1(b), where

$$R = 0.003 + \frac{1}{10^2}(0.75 + 3.94) = 0.050\ \Omega$$

$$X = 0.010 + \frac{1}{10^2}(5.0 + 7.96) = 0.140\ \Omega$$

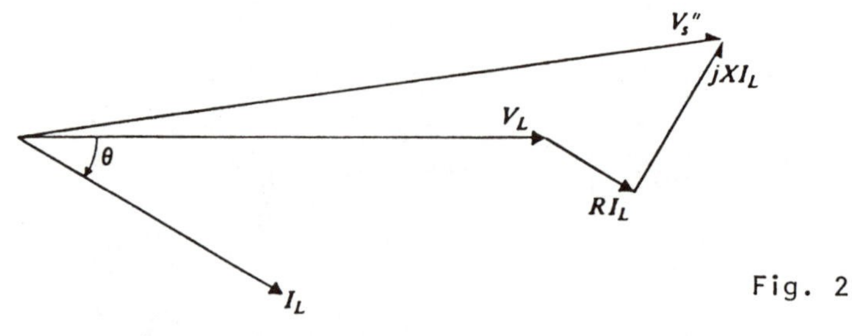

Fig. 2

Phasor diagram for circuit of Fig. 1.

Figure 2 shows a phasor diagram for the circuit of Fig. 1(b), in which the reference phasor is

$$\vec{V}_L = \left(230/\sqrt{3}\right)\underline{/0} = 130\underline{/0}\ \text{V}$$

The apparent power per phase delivered to the load is

26/3 = 8.67 kVA, made up of 7.6 kW of active power and

4.2 kVAR of reactive power. The load current is therefore

$$I_L = \frac{8.67 \times 10^3}{133} = 65.3 \text{ A}$$

at angle

$$\theta = -\cos^{-1} \frac{7.6}{8.67} = -28.7°.$$

Now

$$\vec{V}''_S = \vec{V}_L + (R + jX)\, \vec{I}_L$$

$$= 133\angle 0 + (0.050 + j0.140) \times 65.3\angle -28.7°$$

$$= 140\angle 2.64° \text{ V}$$

so that the line-to-neutral potential difference at the source must be

$$10 \times 140 = 1400 \text{ V}$$

and the line-to-line potential difference is

$$1400\sqrt{3} = 1730 \text{ V}$$

Alternative Method of Solution: In the foregoing solution, the equivalent impedances of the transformers were referred to the wye-connected side of the bank so that the equivalent impedance of one transformer might be shown in Fig. 1(a). If the problem dealt with a delta-delta bank, it would not be possible to do this, and an alternative method of showing the transformer equivalent impedance in the per-phase equivalent circuit would have to be employed. This method involves the wye-delta transformation of Fig. 3.

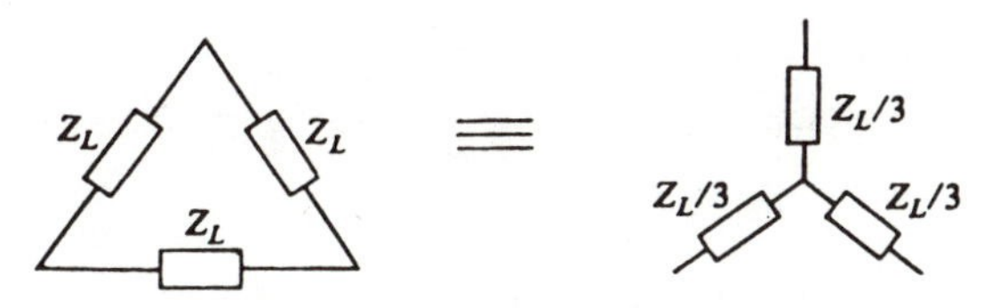

Fig. 3

Wye-delta transformation.

The equivalent impedances of the transformers that are connected in delta, may be transformed to their equivalent wye values. Thus

$$\vec{Z}''_{eq}\ (Y) = \frac{1}{3}\,\vec{Z}''_{eq}(\Delta) = \frac{1}{3}\ (0.118 + j0.238)$$

$$= 0.0393 + j0.0793\ \Omega$$

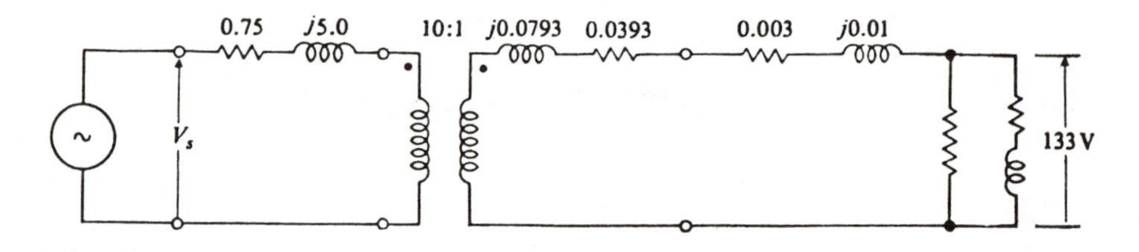

Alternative equivalent circuit to that of Fig. 1 (a). Fig. 4

The per-phase equivalent circuit of the system is therefore that shown in Fig. 4, which is equivalent to Fig. 1(a).

• PROBLEM 6-75

Three identical transformers are connected in Y on both H.T. and L.T. sides. Each transformer is rated at 100 kva, 11,500/230 volts, 60 cycles. On open-circuit test, each unit consumes 560 watts at a power factor of 0.155. On short-circuit test, 217.5 volts impressed upon the H.T. winding circulates 8.7 amp, and the power consumed is 1,135 watts.

A balanced three-phase voltage of 15,000 volts, line to line, is impressed upon the H.T. side of the Y-connected group. The L.T. terminals are connected to a star-connected set of three reactors, each having an impedance of 0.6 ohm, but one of them is noninductive, another has a lagging power factor of 0.866, and the third a leading power factor of 0.500.

Find the current and voltage in each branch of the load.

Solution: From the short-circuit data,

$$\text{Power factor} = \cos\theta_e = \frac{1{,}135}{217.5 \times 8.7} = 0.6$$

$$\therefore \quad \sin\theta_e = 0.8$$

$$\text{Ratio of transformation} = a = \frac{11,500}{230} = 50.$$

The equivalent impedance, resistance, and reactance of each transformer are then

in terms of H.T. side	In terms of L.T. side
$Z = \frac{217.5}{8.7} = 25\ \Omega$	$Z = \frac{25}{a^2} = 0.010\ \Omega$
$R = Z \cos \theta_e = 15\ \Omega$	$R = \frac{15}{a^2} = 0.006\ \Omega$
$X = Z \sin \theta_e = 20\ \Omega$	$X = \frac{20}{a^2} = 0.008\ \Omega$

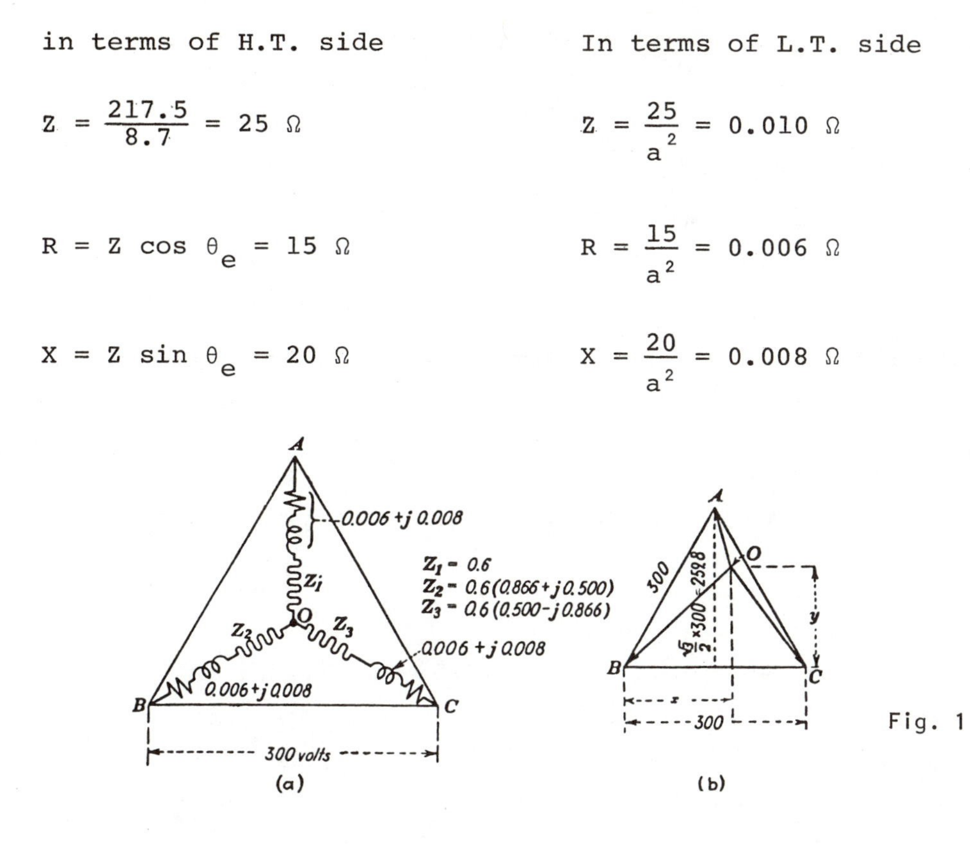

Equivalent circuit, unbalanced load.

The primary (H.T.) impressed voltage of 15,000 volts, line to line, is equivalent to a L.T. line voltage of 300 volts between lines. Consequently, the equivalent circuit, neglecting the exciting current, is of the type shown in Fig. 1a, all quantities being expressed in terms of the L.T. side.

In Fig. 1b, the phase voltages OA, OB, OC depend upon the potential of the common junction point O, which is defined, relative to point B, by the unknown coordinates x and y. Thus,

$$\vec{E}_A = -(x - 150) + j(259.8 - y)\ \text{V}.$$

$$\vec{E}_B = -x - jy\ \text{V}.$$

$$\vec{E}_C = (300 - x) - jy \text{ V.}$$

The phase currents are therefore

$$\vec{I}_A = \frac{-(x - 150) + j(259.8 - y)}{0.606 + j0.008} \text{ A}$$

$$\vec{I}_B = \frac{-x - jy}{0.526 + j0.308} \text{ A.}$$

$$\vec{I}_C = \frac{(300 - x) - jy}{0.306 - j0.512} \text{ A.}$$

On rationalizing these expressions for the three currents, the results are

$$\vec{I}_A = -1.649x - 0.0218y + 253.01 + j(425.14 - 1.649y + 0.0218x) \text{ A.}$$

$$\vec{I}_B = -1.415x - 0.829y + j(0.829x - 1.415y) \text{ A.}$$

$$\vec{I}_C = -0.860x + 1.439y + 258 + j(-1.439x - 0.860y + 431.7) \text{ A.}$$

Since $\vec{I}_A + \vec{I}_B + \vec{I}_C = 0$, the real and the imaginary terms of the summation must each be identically zero; hence,

$$-3.924x + 0.588y + 511.01 = 0$$

$$-0.588x - 3.924y + 856.84 = 0$$

from which

$$x = 159.37$$

$$y = 114.36$$

$\vec{I}_A = -12.28 + j240.03 = 240.34\underline{/92°56'}$ A.

$\vec{I}_B = -320.31 - j29.70 = 321.7\underline{/185°18'}$ A.

$\vec{I}_C = 285.51 + j104 = 303.86\underline{/20°1'}$ A.

$\vec{E}_A = -9.37 + j115.44 = 115.82\underline{/94°38'}$ V.

$\vec{E}_B = -159.37 - j114.36 = 196.16\underline{/215°40'}$ V.

$\vec{E}_C = 140.63 - j114.36 = 181.26\underline{/320°53'}$ V.

TRANSFORMER DESIGN

• PROBLEM 6-76

Determine the core dimensions for a core-type 2-kVA, 220:55-V, 60-Hz transformer of the configuration shown in the figure. Assume the following parameters:

Permissible peak core flux density, B = 1T

Stacking factor, $k_i = 0.9$

Winding space factor, $k_w = 0.25$

Permissible rms current density $J = 2 \times 10^6 A/m^2$.
Volts/turn, $V_t = 1$ V.

The core should be square in cross section. The window should be approximately twice as high as it is wide. The transformer is to be air-cooled by convection.

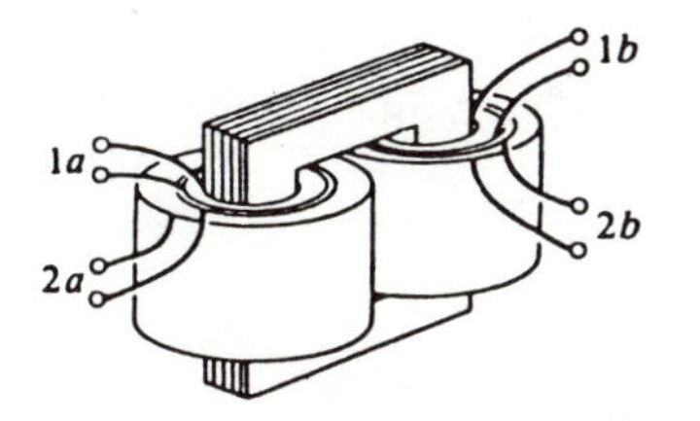

Winding and core arrangement. Core-type with core of laminated sheets.

Solution: The rms flux density $B = 1/\sqrt{2} = 0.707$ T. The low value of winding space factor is to allow ample space for cooling air to circulate round the winding. The rms current density is made low to reduce copper loss in view of the cooling method.

Assume that this is a step-down transformer, so that $V_1 = 220$ V. Then since $V_t = 1$ V,

$$N_1 = 220, \qquad N_2 = 55$$

The gross core cross-section area is

$$\frac{A_i}{k_i} = \frac{V_1}{2\pi f N_1 \, B k_i} = \frac{220}{2\pi \times 60 \times 220 \times 0.707 \times 0.9}$$

$$= 0.00417 \text{ m} .$$

Length of core side $= (A_i/k_i)^{\frac{1}{2}} = 64.6$ mm. The window area is

$$\frac{A_w}{k_w} = \frac{S}{\pi f B k_i A_i k_w J}$$

$$= \frac{2 \times 10^3}{\pi \times 60 \times 0.707 \times 0.9 \times 0.00417 \times 0.25 \times 2 \times 10^6}$$

$$= 0.00798 \text{ m}^2$$

Let d = window width; then

$$\frac{A_w}{k_w} = 2d^2 = 0.00798$$

from which

$$d = 63.2 \text{ mm}$$

$\therefore$ Window height $= 2d = 126.4$ mm.

CHAPTER 7

A.C. MACHINES

A. C. GENERATOR - WAVE FORM AND FREQUENCY OF GENERATED EMF

• **PROBLEM 7-1**

What is the equation of a 25-cycle-current sine wave, having an rms value of 30 amp, and what is the value of the current when the time is 0.005 sec? What fraction of a cycle has the wave gone through in 0.005 sec ? Assume that the wave crosses the time axis in a positive direction when the time is equal to zero.

Solution:

$$I_m = 30\sqrt{2} = 42.4 \text{ amp.}$$

$$\omega = 2\pi 25 = 157$$

The equation of the current sine wave is then

$$i = 42.4 \sin 157t.$$

The value of the current when t = 0.005 sec is

$$\begin{aligned} i &= 42.4 \sin 157 \times 0.005 \\ &= 42.4 \sin 0.785 \text{ radian} \\ &= 30 \text{ amp.} \end{aligned}$$

As the wave completes 360° in 1/25, or 0.04 sec, in 0.005 sec, it will have completed 0.005/0.040 = 1/8 cycle.

• PROBLEM 7-2

(a) A 60-cycle current 9 sin ωt is added to a 60-cycle current

$$8 \cos \omega t, \text{ where } \omega = 2\pi 60.$$

Determine the resultant current i_3.

(b) Two 25-cycle emfs differing in phase by 60° are given by

$$e_1 = 120 \sin (\omega t - 30°)$$

and

$$e_2 = 100 \sin (\omega t - 90°),$$

where $\omega = 2\pi 25$, (see figure). Determine their sum e_3.

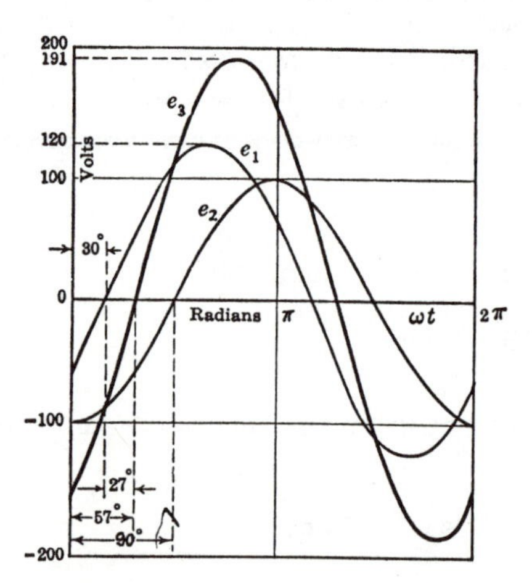

Addition of sine waves differing in phase by 60°.

Solution: (a) Using the trigonometric relation

$$A \sin x + B \cos x = \sqrt{A^2 + B^2} \sin (x + \tan^{-1} B/A),$$

$$i_3 = \sqrt{9^2 + 8^2} \sin (\omega t + \theta),$$

$$\tan \theta = 8/9 = 0.888 \qquad \theta = 46.1°.$$

$$i_3 = 12.05 \sin (\omega t + 46.1°).$$

(b)

$$e_3 = 120 \sin(\omega t - 30°) + 100 \sin(\omega t - 90°) \quad (1)$$

Expanding the terms on the right hand side of Eqn. (1),

$$e_3 = 120(\sin \omega t \cos 30° - \cos \omega t \sin 30°) +$$
$$100(\sin \omega t \cos 90° - \cos \omega t \sin 90°),$$

$\cos 30° = 0.866$; $\sin 30° = 0.5$; $\cos 90° = 0$; $\sin 90° = 1$

$$e_3 = 104 \sin \omega t - 60 \cos \omega t + 0 - 100 \cos \omega t$$

$$= 104 \sin \omega t - 160 \cos \omega t$$

Using the trigonometric relation

$$A \sin x + B \cos x = \sqrt{A^2 + B^2} \sin(x + \tan^{-1} B/A),$$

$$e_3 = \sqrt{104^2 + 160^2} \sin\left(\omega t + \tan^{-1} \frac{-160}{104}\right)$$

$$= 191 \sin(\omega t - 57.0°),$$

$-57.0° = \tan^{-1}(-160/104) = \tan^{-1}(-1.538)$. Since the numerator is negative and the denominator positive, θ must be negative and in the fourth quadrant. The angle between the 120-volt wave e_1 and the resultant wave e_2 is $57.0° - 30.0° = 27.0°$.

• PROBLEM 7-3

(a) A 60-cycle alternator has 2 poles. What is the speed of the alternator?

(b) A 60-cycle alternator has a speed of 120 rpm. How many poles has it?

Solution: (a) $f = \frac{PS}{120}$ cycles/sec (1)

where f = frequency, P = number of poles, S = rpm

$$\therefore \quad S = \frac{120f}{P}$$

$$= \frac{120 \times 60}{2}$$

$$= 3600 \text{ rpm}$$

(b) This may be solved without using Eqn. (1) directly. The 2-pole 60-cycle alternator rotates at 3,600 rpm. Therefore the 60-cycle, 120 rpm alternator must have

$$\frac{3,600}{120} 2 = 60 \text{ poles}$$

• PROBLEM 7-4

An a-c generator has six poles and operates at 1,200 rpm.

(a) What frequency does it generate? (b) At what speed must the generator operate to develop 25 cycles? 50 cycles? (c) How many poles are there in a generator that operates at a speed of 240 rpm and develops a frequency of 60 cycles?

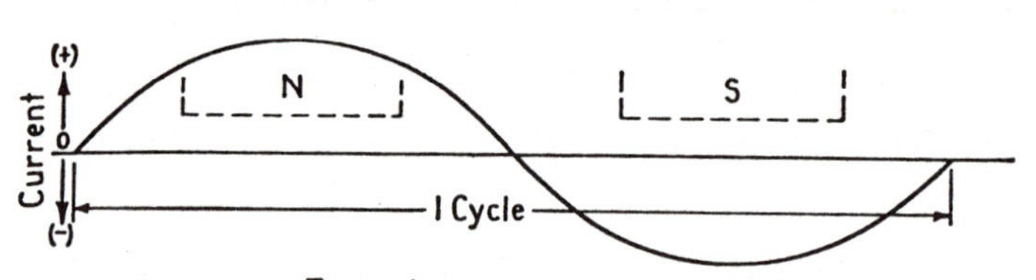

(a) Two poles - one cycle per revolution.

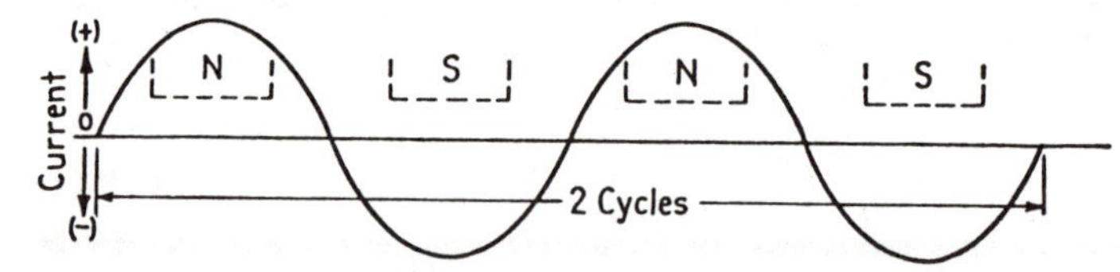

(b) Four poles - two cycles per revolution.

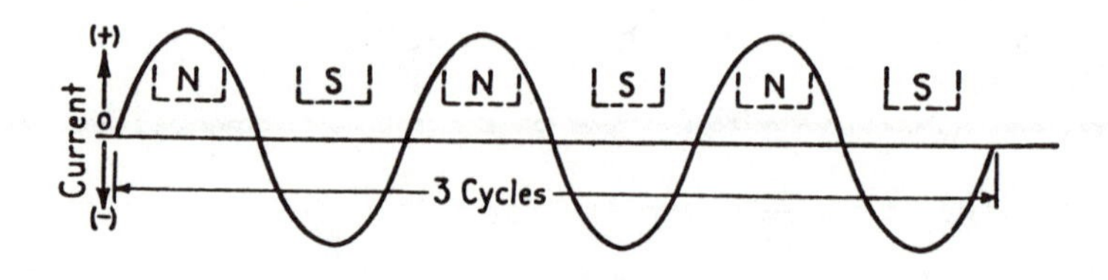

(c) Six poles - three cycles per revolution.

Sketch illustrating the relation between the number of poles and the generated a-c frequency in cycles per revolution.

Solution: (a)

$$f = \frac{P \times rpm}{120}$$

$$f = \frac{6 \times 1{,}200}{120} = 60 \text{ cycles/sec.}$$

(b)

$$rpm_{25} = \frac{120 \times 25}{6} = 500.$$

$$rpm_{50} = \frac{120 \times 50}{6} = 1{,}000.$$

(c)

$$P = \frac{120 \times f}{rpm} = \frac{120 \times 60}{240} = 30 \text{ poles}$$

The diagrams in the figure represent graphically the manner in which the number of cycles per revolution increases with an increase in the number of poles.

GENERATED VOLTAGE AND TERMINAL VOLTAGE

• PROBLEM 7-5

The effective voltage for a 5-turn coil on a 6-pole generator is 10.6 volts. The flux per pole is 0.00795 weber. Calculate the frequency and the speed of the machine, and write the equation for the voltage per coil assuming it to be sinusoidal.

Solution: The average voltage per conductor is

$$E_{avg} = \frac{E_{eff}}{\text{number of turns} \times 2 \times 1.11};$$

also

$$E_{avg} = \frac{\phi}{t} \text{ volts per conductor}$$

Thus,

$$E_{avg} = \frac{10.6}{5 \times 2} \times \frac{1}{1.11} = \frac{0.00795}{t} \text{ volts per conductor}$$

$$t = 0.00833 \text{ second per half cycle}$$

$$f = \frac{1}{2t} = 60 \text{ cycles per second}$$

Now,

$$f = \frac{P}{2} \times \frac{rpm}{60} \text{ cycles per second.}$$

$$= \frac{6 \times rpm}{2 \times 60} \text{ cycles per second}$$

Therefore,

$$\text{speed} = 1{,}200 \text{ rpm}$$

and

$$e = 10.6\sqrt{2} \sin 2\pi 60t$$

$$= 15 \sin 377t \text{ volts}$$

If the average voltage per turn is used, the time taken to change the flux through the loop from maximum to zero is 0.00417 sec, corresponding to 1/4 cycle. Whereas one conductor moves one pole pitch of 180 electrical degrees in cutting the flux of one pole, the coil (of two conductors) need move only 90 electrical degrees to cut the same flux.

An alternate method of solution is to use the expression for the maximum voltage generated. The maximum voltage generated in a coil of N turns is

$$E_{max} = N_{\omega}\Phi$$

$$\sqrt{2}(10.6) = 5 \times 2\pi f(0.00795)$$

$$f = \frac{15}{0.25} = 60 \text{ cycles per second}$$

• PROBLEM 7-6

The machine in Fig. 1 has a cylindrical rotor with radius of 0.11 m and length (into the page) of 0.26 m. The distance across the air gap is 0.003 m. The reluctance of the steel parts may be neglected. The stator coil, a - a', has 80 turns. The rotor coil, b - b', has 200 turns. Assume the rotor mmf produces a sinusoidally distributed flux density in the air gap. The rotor turns clockwise at a speed of 3600 rpm (i.e., $\alpha = 377t$). Let the rotor current, i_b, be 15 amp (d-c) with the polarity shown. Find the voltage function in the open-circuited stator coil. Find the rms value of the voltage.

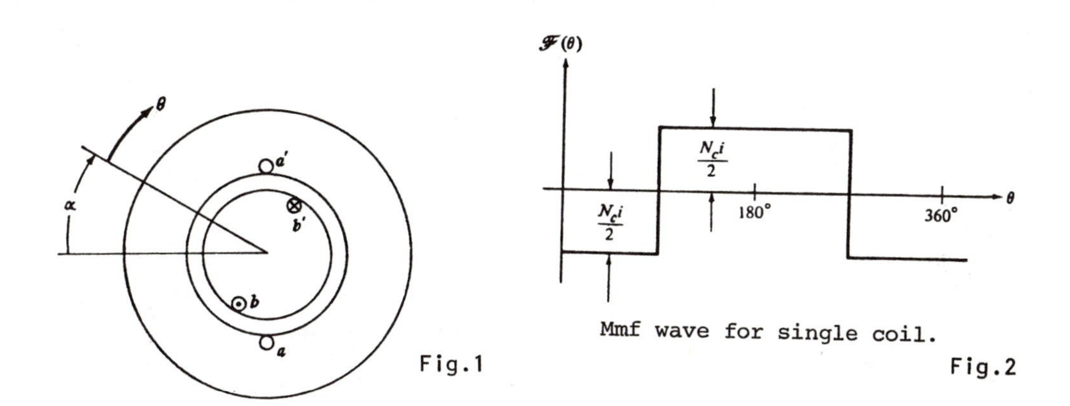

Fig.1

Mmf wave for single coil.

Fig.2

Solution: The mmf in the air gap has the rectangular wave shape of Fig. 2. The peak mmf of the sinusoid for the fundamental component is

$$F_{peak} = (4/\pi)(N_c i/2)$$

The corresponding peak value of the flux density is found by using the magnetic circuit properties of the air gap

$$B_{peak} = \left(\frac{\mu_0}{g}\right) F_{peak} = \left(\frac{4\pi \times 10^{-7}}{0.003}\right)\left(\frac{4}{\pi}\right)\left(\frac{200 \times 15}{2}\right) =$$

$$0.8 \text{ weber/m}^2$$

$$\phi = \frac{4\ell r}{p} B_{peak} = \frac{4(0.26)(0.11)}{2} (0.8) =$$

$$0.0458 \text{ weber}$$

The flux linkages of the stator coil are given by

$$\lambda = N_a \phi \cos \alpha = 80 \times 0.0458 \cos 377t = 3.66 \cos 377t$$

The instantaneous value of voltage in winding a is

$$e_{a'a} = d\lambda/dt = -1380 \sin 377t \text{ v}$$

$$e_{aa'} = 1380 \sin 377t \text{ v}$$

The rms value of this voltage is

$$E_{rms} = E_{max}/\sqrt{2} = 976 \text{ v}$$

We could also find the rms value of voltage by using the following equation:

$$E = 4.44\ N_c \phi f = 4.44 \times 80 \times 0.0458 \times 60 = 976 \text{ v.}$$

• PROBLEM 7-7

(a) Assume the cross-sectional area of the poles in the figure to be 800 square centimeters, and that the flux density is 5,000 lines of induction per square centimeter. What is the pressure between the brushes when the single loop armature makes 3,600 rotations per minute?

(b) How many inductors will have to be connected in series if the electromotive force in part (a) is to be 48 volts?

(c) A six-pole generator operating at a speed of 900 revolutions per minute has 960 inductors on the armature. What will be the electromotive force developed if a two-circuit winding is used and if the flux emanating from one pole is 1,200,000 lines?

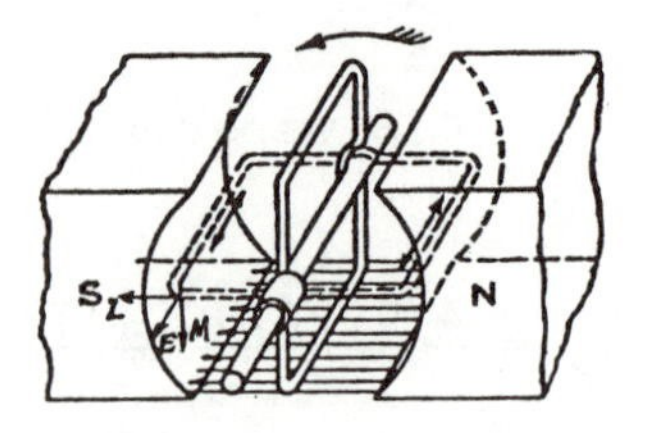

Solution:

(a) $$E = \frac{2nZ\ \Phi}{60 \times 10^8} \text{ volts}$$

$$n = 3{,}600 \text{ rpm}$$

$$Z = 2$$

$$\Phi = 5{,}000 \times 800 = 4{,}000{,}000 = 4 \times 10^6 \text{ maxwells.}$$

Then

$$E = \frac{4 \times 10^6 \times 3{,}600 \times 2 \times 2}{60 \times 10^8}$$

$$= 9.6 \text{ volts.}$$

(b) Two inductors in series generate 9.6 volts. Hence one inductor generates 4.8 volts. To generate 48 volts will require $48 \div 4.8 = 10$ inductors, or a loop of 5 turns.

(c)

$$E = \frac{p\Phi Zn}{q60 \times 10^8}$$

$$p = 6$$

$$q = 2$$

$$\Phi = 1{,}200{,}000 = 1.2 \times 10^6$$

$$n = 900 \text{ rpm}$$

$$Z = 960$$

Then

$$E = \frac{6 \times 1.2 \times 10^6 \times 960 \times 900}{2 \times 60 \times 10^8}$$

$$= 518 \text{ volts.}$$

• **PROBLEM 7-8**

A 60-kva 220-volt 60-cycle alternator has an effective armature resistance of 0.016 ohm and an armature leakage reactance of 0.070 ohm. Determine induced emf when the machine is delivering rated current at a load power factor of unity.

Solution: The current

$$I = \frac{60,000}{220} = 273 \text{ amp,}$$

Armature resistance drop is

$$IR = 273 \times 0.016 = 4.37 \text{ volts,}$$

Armature leakage reactance drop is

$$IX = 273 \times 0.070 = 19.1 \text{ volts,}$$

The induced emf is then given by

$$E_a = \sqrt{(V + IR)^2 + (IX)^2}$$

$$= \sqrt{(220 + 4.4)^2 + (19.1)^2} = 225 \text{ volts.}$$

• PROBLEM 7-9

(i) A 1000 kVA, 4600 V, three-phase, wye-connected alternator has an armature resistance of 2 ohms per phase and a synchronous armature reactance, X_s, of 20 ohms per phase. Find the full-load generated voltage per phase at:

(a) Unity power factor

(b) A power factor of 0.75 lagging

(ii) Repeat part (i) to determine the generated voltage per phase at full load with:

(a) A leading load of 0.75 PF

(b) A leading load of 0.4 PF.

Solution: (i)

$$V_p = \frac{V_L}{\sqrt{3}} = \frac{4600 \text{ V}}{1.73} = 2660 \text{ V;}$$

$$I_p = \frac{\text{kVA} \times 1000}{3V_p} = \frac{1000 \times 1000}{3 \times 2600} = 125 \text{ A.}$$

$$I_aR_a \text{ drop/phase} = 125 \text{ A} \times 2\Omega = 250 \text{ V.}$$

I_aX_s drop/phase = 125 A × 20Ω = 2500 V.

(a) At unity power factor,

$$\vec{E}_g = (V_p + I_aR_a) + jI_aX_s = (2660 + 250) + j2500$$

$$= 2910 + j2500$$

$$E_g = 3845 \text{ V/phase}$$

(b) At 0.75 PF lagging,

$$\vec{E}_g = (V_p \cos\theta + I_aR_a) + j(V_p \sin\theta + I_aX_s)$$

$$= (2660 \times 0.75 + 250) + j(2660 \times 0.676 + 2500)$$

$$= 2250 + j4270$$

$$E_g = 4820 \text{ V/phase}$$

(ii)

From part (i),

I_aR_a/phase = 250 V.

I_aX_s/phase = 2500 V.

(a) At 0.75 PF leading

$$\vec{E}_g = (V_p \cos\theta + I_aR_a) + j(V_p \sin\theta - I_sX_a)$$

$$= [(2660 \times 0.75) + 250] + j[(2660 \times 0.676) - 2500]$$

$$= 2250 - j730$$

$$E_g = 2360 \text{ V/phase}$$

(b) At 0.40 PF leading

$$\vec{E}_g = [(2660 \times 0.4) + 250] + j[(2660 \times 0.916) - 2500]$$

$$= 1314 - j40$$

$$E_g = 1315 \text{ V/phase}$$

Note that the generated voltage is less than the terminal voltage at both power factors, and decreases as the power factor becomes more leading.

• PROBLEM 7-10

Each of two alternator coils Oa and Ob, Fig. (a), is generating an emf of 160 volts. These voltages differ in phase by 90°. Determine the voltage across their open ends if they are connected together at O as shown.

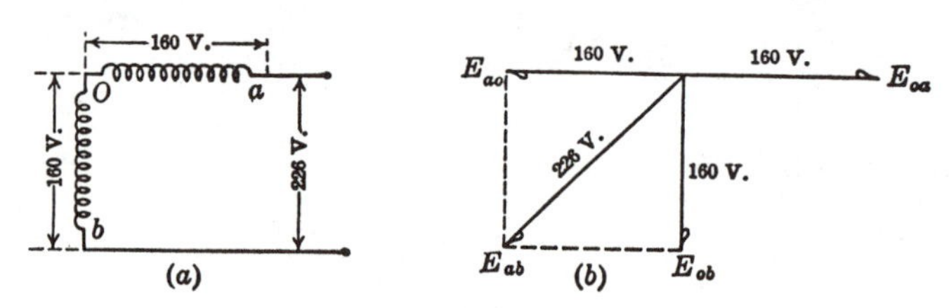

Vector addition of two equal voltages having a 90° phase difference.

Solution: Let E_{oa} and E_{ob}, Fig. (b), represent the voltages across coils Oa and Ob. Let the voltage across the open ends a and b be denoted by E_{ab}. To obtain the voltage $\vec{E}_{ab}$, it is necessary to use $\vec{E}_{ao}$, displaced 180° from $\vec{E}_{oa}$. Then, vectorially, $\vec{E}_{ab} = \vec{E}_{ao} + \vec{E}_{ob}$. Combining $\vec{E}_{ao}$ and $\vec{E}_{ob}$ vectorially, the voltage $\vec{E}_{ab}$ is obtained. As E_{ao} and E_{ob} are at right angles, their resultant, which is the hypotenuse of a right triangle, is

$$E_{ab} = \sqrt{E_{ao}^2 + E_{ob}^2} = \sqrt{160^2 + 160^2} = 226 \text{ volts.}$$

• PROBLEM 7-11

(a) Two single-phase alternators are in parallel and operating in synchronism. One machine has a sine wave effective voltage of 100 volts, the second has a square flat-top wave form of 100 volts effective. What is the maximum instantaneous circulating current between the two machines, if the synchronous impedance of each machine is 3.68 ohms?

(b) Two single-phase alternators, each having a synchronous reactance of 3.678 ohms and an effective resistance of 0.0554 ohm, are connected in parallel when the induced voltages are 6650 and 6550 volts, respectively, and in exact synchronism. Determine the circulating current and its power factor with respect to the terminal voltage.

Solution: (a)

Maximum voltage of machine 1 = 100 × 1.414

= 141.4 volts

Maximum voltage of machine 2 = 100 volts

Resultant maximum instantaneous voltage = 100 volts

$$\text{Maximum instantaneous current} = \frac{V}{2Z_s}$$

where Z_s = synchronous impedance of each machine

$$\therefore \text{ Maximum instantaneous current} = \frac{100}{2 \times 3.68} = 13.6 \text{ amp.}$$

(b) Resultant voltage = 6650 − 6550 = 100 volts

$$\text{Impedance of the system} = \sqrt{x^2 + r^2}$$

$$= 2\sqrt{3.678^2 + 0.0554^2} = 7.3568 \text{ ohms}$$

$$\therefore \quad \text{Circulating current} = \frac{V}{Z} = \frac{100}{7.3568} = 13.59 \text{ amp}$$

$$\text{Power factor} = \frac{0.1108}{7.3568} = 0.015 \text{ lagging.}$$

• PROBLEM 7-12

A six-pole generator has a lap winding. If there are 300 inductors on the surface of the armature, and if the flux is 900,000 lines per pole, what will be the voltage between the brushes at a speed of 1,500 revolutions per minute?

(b) What voltage will the generator in part (a) develop if the winding be wave connected, other conditions remaining as before?

Solution:

(a)

$$E = \frac{pnZ\Phi}{q \times 60 \times 10^8}.$$

$p = 6.$

$n = 1,500$ rpm.

$\Phi = 900,000 = 9 \times 10^5.$

$q = 6.$

$Z = 300.$

Then

$$E = \frac{6 \times 1,500 \times 9 \times 10^5 \times 300}{6 \times 60 \times 10^8}$$

$$= 67.5 \text{ volts.}$$

(b) All the conditions are the same as before, except q which is 2 instead of 6, hence

$$E = \frac{6 \times 1,500 \times 9 \times 10^5 \times 300}{2 \times 60 \times 10^8}$$

$$= 202.5 \text{ volts.}$$

• PROBLEM 7-13

(i) A four-pole generator has a total of 500 inductors on its armature and is designed to have 2×10^6 lines of magnetic flux per pole crossing its air gap with normal excitation. What voltage will be generated at a speed of 1800 rpm (a) if the armature is simplex wave wound, (b) if the armature is simplex lap wound? (c) If the allowable current is 5 amp per path, what will be the kilowatts generated by the machine in each case?

(ii) A six-pole generator is driven at a speed of 1200 rpm. The flux per pole is 5×10^6 maxwells. Calculate the inductors necessary to generate 250 volts when the armature is (a) simplex lap wound and (b) duplex wave wound. (c) If the normal generation capacity is 100 kw, what must be the current capacity per path for each type of winding?

Solution: (i)

$$E = \frac{Z}{\text{Paths}} \times \phi \times \text{poles} \times \frac{\text{rpm}}{60} \times 10^{-8} \text{ volts.}$$

where Z = Number of inductors

(a) $$E = \frac{500}{2} \times 2 \times 10^6 \times 4 \times \frac{1800}{60} \times 10^{-8}$$

= 600 volts, for wave-wound machine

(b) $$E = \frac{500}{4} \times 2 \times 10^6 \times 4 \times \frac{1800}{60} \times 10^{-8}$$

= 300 volts for lap-wound machine

(c)

Total current = 5 × 2 = 10 amp for wave winding

$$\text{Kilowatts} = \frac{10 \times 600}{1000} = 6$$

Total current = 5 × 4 = 20 amp for lap winding

$$\text{Kilowatts} = \frac{20 \times 300}{1000} = 6$$

(ii)

(a) $$250 = \frac{Z}{6} \times 5 \times 10^6 \times 6 \times \frac{1200}{60} \times 10^{-8} \text{ volts}$$

from which

$$Z = 250 \text{ inductors}$$

(b) For the duplex wave winding, there being four paths, only 167 inductors will be required for the same voltage per path.

(c) Line current = 100,000 ÷ 250 = 400 amp

For lap winding

$$\text{Current capacity per path} = \frac{400}{6} = 66\tfrac{2}{3} \text{ amp}$$

For wave winding

$$\text{Current capacity per path} = \frac{400}{4} = 100 \text{ amp}$$

Note that the kilowatt values here given would not represent the power that the generator could supply to an external load, since an appreciable part of the power generated would be absorbed in overcoming internal losses in the machine itself.

• PROBLEM 7-14

(i) (a) A triplex lap-wound armature is used in a 14-pole machine with fourteen brush sets, each spanning three commutator bars. Calculate the number of paths in the armature.

(b) Repeat (a) for a triplex wave-wound armature having two such brush sets and 14 poles.

(ii) Calculate the generated emf in each of the above problems if the flux per pole is 4.2×10^6 lines, the generator speed is 60 rpm, and there are 420 coils on the armature, each coil having 20 turns.

Solution: (i) For a lap winding, $a = mP$

where a is the number of parallel paths in the armature

m is the multiplicity of the armature

P is the number of poles.

Thus

(a) $a = 3 \times 14 = 42$ paths

(b) For a wave winding,

$$a = 2m = 2 \times 3 = 6 \text{ paths}$$

(ii) (a) $Z = 420$ coils $\times$ 20 turns/coil $\times$ 2 conductors/turn $= 16{,}800$ conductors.

$$E_g = \frac{\phi ZSP}{60a} \times 10^{-8} \text{ V}$$

$$= \frac{4.2 \times 10^6 \times 16{,}800 \times 60 \times 14}{60 \times 42} \times 10^{-8}$$

$$= 235.2 \text{ V}$$

(b) $$E_g = \frac{4.2 \times 10^6 \times 16{,}800 \times 60 \times 14}{60 \times 6 \times 10} = 1646.4 \text{ V}$$

• PROBLEM 7-15

A 72-slot three-phase stator armature is wound for six poles, using double-layer lap coils having 20 turns per coil with a 5/6 pitch. The flux per pole is 4.8×10^6 lines, and the rotor speed is 1200 rpm. Calculate:

(a) The generated effective voltage per coil of a full-pitch coil.

(b) The total number of turns per phase.

(c) The distribution factor.

(d) The pitch factor.

(e) The total generated voltage per phase from (a), (c), and (d) above, and by the following equation:

$$E_{gp} = 4.44\ \phi\ N_p f K_p k_d \times 10^{-8} \text{ V} \qquad (1)$$

Solution: (a) $E_{g/coil} = 4.44 \Phi N_c f \times 10^{-8}$ V

$$= 4.44(4.8 \times 10^6)(20)\left(\frac{6 \times 1200}{120}\right) \times 10^{-8}$$

$$= 256 \text{ V/coil}$$

(b)

$$N_p = \frac{CN_c}{p} = \frac{\text{total armature coils} \times \text{turns/coil}}{\text{number of phases}}$$

$$= 72 \text{ coils/3 phase} \times 20 \text{ turns/coil}$$

$$= 480 \text{ turns/phase}$$

(c)

$$k_d = \frac{\sin(n\alpha/2)}{n \sin(\alpha/2)},$$

where

$$n = 72 \text{ slots/(3 phase} \times 6 \text{ poles)}$$

$$= 4 \text{ slots/pole-phase}$$

and

$$\alpha = (6 \text{ poles} \times 180° \text{/pole})/72 \text{ slots} = 15° \text{/slot}$$

$$k_d = \frac{\sin[(4 \times 15)/2]}{4 \sin\left(\frac{15}{2}\right)} = \frac{\sin 30°}{4 \sin 7.5°} = 0.958$$

(d)

$$k_p = \sin \frac{p°}{2} = \sin\left(\frac{5}{6} \times \frac{180}{2}\right) = \sin 75° = 0.966$$

(e)

$$E_{gp} = 4.44 \times 480 \text{ turns/phase} \times 4.8 \times 10^6 \times 60 \times 0.966 \times 0.958 \times 10^{-8}$$

$$= 5680 \text{ V [from Eq. (1)]}$$

$$E_{gp} = 256 \text{ V/coil} \times 24 \text{ coils/phase} \times 0.966 \times 0.958$$

$$= 5680 \text{ V/phase [from (a), (c), and (d)].}$$

• **PROBLEM 7-16**

A 6-pole 3-phase 60-cycle alternator has 12 slots per pole and four conductors per slot. The winding is five-sixths pitch. There are 2,500,000 maxwells (= 0.025 weber) entering the armature from each north pole, and this flux is sinusoidally distributed along the air gap. The armature coils are all connected in series. The winding is Y-connected. Determine the open-circuit emf of the alternator.

Solution: The total number of slots is 72.

The series conductors per phase, therefore, are

$$Z = \frac{4 \times 72}{3} = 96$$

Slots per pole per phase = 72/(6 × 3) = 4.

$$k_b = \frac{\sin(n\alpha/2)}{n \sin(\alpha/2)}$$

where n = number of slots per pole per phase,

α = Electrical angle between slots.

$$n = 4, \ \alpha = \frac{180°}{12} = 15°$$

$$k_b = \frac{\sin \frac{4 \times 15}{2}}{4 \sin (15°/2)} = \frac{\sin 30°}{4 \sin 7.5°} = \frac{0.5}{0.522}$$

$$= 0.958.$$

$$k_p = \cos \frac{180° (1 - p)}{2}$$

where p is the pitch, expressed as a fraction.

With five-sixths pitch,

$$k_p = \cos \frac{180° (1 - 5/6)}{2}$$

$$= \cos 15° = 0.966.$$

Alternatively, k_b and k_p can be obtained from the following tables:

Values of Breadth Factor k_b

Slots per hole per phase	Single-phase	2-phase	3-phase
1	1.000	1.000	1.000
2	0.707	0.924	0.966
3	0.667	0.910	0.960
4	0.653	0.907	0.958

Values of Pitch Factor k_p

Pitch	9/10	6/7	5/6	4/5	3/4	2/3
k_p	0.988	0.974	0.966	0.951	0.924	0.866

The total induced emf per phase is

$$E = 2.22\ k_b k_p\ Z\phi f\ 10^{-8}\ \text{volts}$$

$$= 2.22 \times 0.958 \times 0.966 \times 96 \times 2{,}500{,}000 \times 60 \times 10^{-8}$$

$$= 296\ \text{volts.}$$

As the winding is Y-connected, the terminal voltage is

$$296\sqrt{3} = 513\ \text{volts.}$$

• PROBLEM 7-17

(1) Calculate the coil pitches (coil spans) and indicate the slots into which the first coils should be placed for the following armature windings: (a) 28 slots, four poles; (b) 39 slots, four poles· (c) 78 slots, six poles; (d) 121 slots, eight poles; (e) 258 slots, 14 poles.

(2) Calculate the commutator pitches Y_c for the following pole and commutator segment combinations: (a) six poles, 34 segments; (b) eight poles, 63 segments; (c) 10 poles, 326 segments. In each case, trace the winding around the commutator once; start at segment 1 and show that after one trip around the commutator, a segment is reached that is one behind or one ahead of the starting segment.

Solution:

(1) $$Y_s = \frac{S}{P} - K$$

(a) $Y_s = \frac{28}{4} - 0 = 7$ slots 1 and 8

(b) $Y_s = \frac{39}{4} - \frac{3}{4} = 9$ slots 1 and 10

(c) $Y_s = \frac{78}{6} - 0 = 16$ slots 1 and 17

(d) $Y_s = \frac{121}{8} - \frac{1}{8} = 15$ slots 1 and 16

(e) $Y_s = \frac{258}{14} - \frac{6}{14} = 18$ slots 1 and 19

(2) $$Y_c = \frac{C \pm 1}{P/2}$$

(a) $$Y_c = \frac{34 - 1}{3} = \frac{33}{3} = 11$$

Tracing 1-12-23-34 (one behind segment 1).

(b) $$Y_c = \frac{63 + 1}{4} = \frac{64}{4} = 16$$

Tracing 1-17-33-49-2 (one ahead of segment 1).

(c) $$Y_c = \frac{326 - 1}{5} = \frac{325}{5} = 65$$

Tracing, 1-66-131-196-261-326 (one behind segment 1).

• PROBLEM 7-18

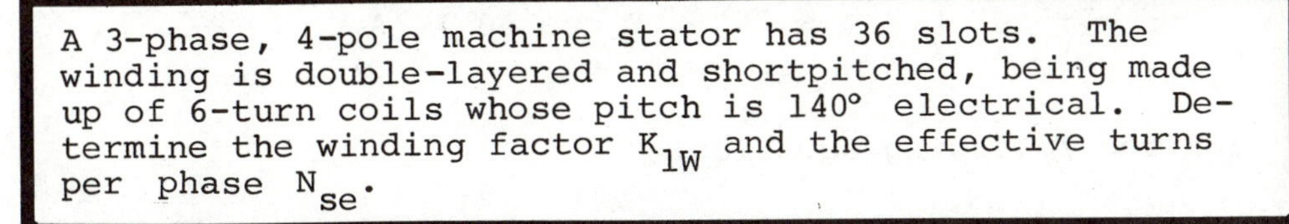

A 3-phase, 4-pole machine stator has 36 slots. The winding is double-layered and shortpitched, being made up of 6-turn coils whose pitch is 140° electrical. Determine the winding factor K_{1W} and the effective turns per phase N_{se}.

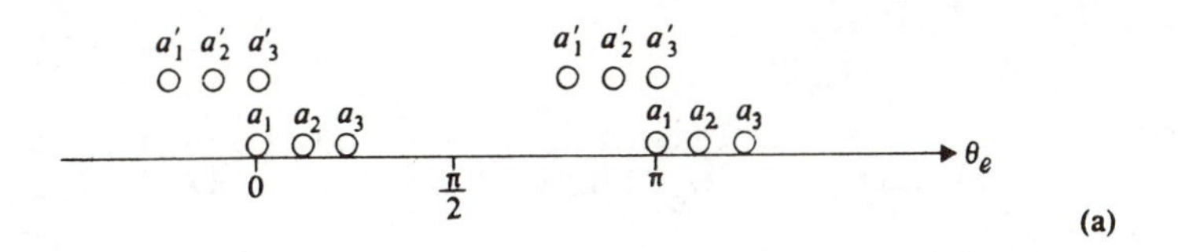

(a)

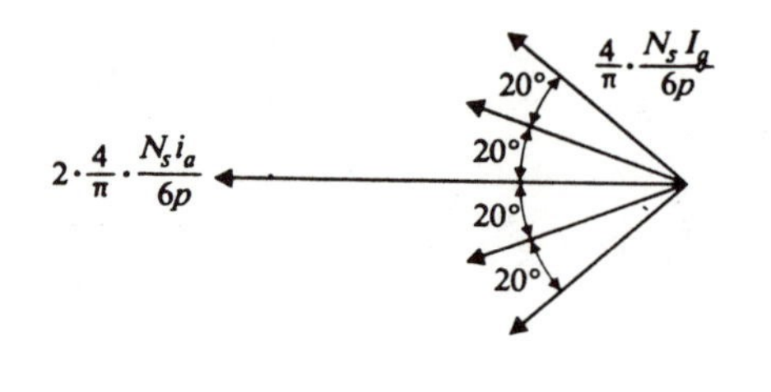

(b)

Solution: Since there are 36 slots with two coil-sides per slot, there will be 36 coils in the winding, giving 12 coils per phase. Thus

$$N_s = 6 \times 12 = 72 \text{ turns}$$

The winding is 4-pole; there will, therefore, be three coils per pole per phase and three slots in a phase belt. The angle subtended by one slot pitch at the rotor axis is

$$\frac{2\pi}{36} \text{ radians} = \frac{4\pi}{36} \text{ electrical radians} = 20° \text{ electrical}$$

The coil pitch is 140° electrical, so that each coil spans seven slot pitches. The arrangement of the coil sides of phase "a" in the slots will therefore be as shown in Fig. (a).

The peak amplitude of the stepped wave produced by this winding $N_s i_a/p$. Since 6 coils produce this amplitude, the amplitude of the rectangular wave produced by one coil is $N_s i_a/6p$. The amplitude of the fundamental component produced by one coil is thus $(4/\pi)(N_s i_a/6p)$ A. The vector diagram from which the resultant fundamental component may be obtained is shown in Fig. (b), where it is seen that the fundamental component of the stepped wave produced by the whole phase "a" winding is

$$\frac{4}{\pi}\,\frac{N_s i_a}{6p}\,(2 + 2\cos 20° + 2\cos 40°) \quad A$$

From the following equation:

$$F_{sg1} = \frac{4}{\pi}\,\frac{N_s}{p}\,i_s \quad A$$

it is seen that the fundamental mmf amplitude for a concentrated winding of N_s turns would be $(4/\pi)(N_s i_a/p)$. Thus

$$K_{1W} = \frac{\text{Fundamental mmf amplitude for distributed winding of } N_s \text{ turns}}{\text{Fundamental mmf amplitude for concentrated winding of } N_s \text{ turns}} .$$

$$= \frac{1}{6} (2 + 2 \cos 20° + 2 \cos 40°) = 0.902$$

and

$$N_{se} = K_{1W} \frac{4}{\pi} N_s = 0.902 \times \frac{4}{\pi} \times 72 = 83.5 \text{ turns.}$$

This paradoxical result, namely that $N_{se} > N_s$, is due to the fact that N_{se} is calculated from the amplitude of the fundamental component.

• PROBLEM 7-19

Calculate the effect of armature reaction in a machine that is designed as follows:

Series conductors per phase	= 240
Slot width, a_3	= 0.504 in.
Slot depth, b_c	= 2.5 in.
Pole pitch, τ	= 6.024 in.
Axial length of stacking, L	= 4.5 in.
Number of stator slots, S_1	= 144
Phases, m	= 3
Poles, P	= 24
Pitch, p	= $\frac{5}{6}$ or 0.833
Slot constant, K_s	= $b_c/3a_3$ = 1.65
Frequency	= 60 cycles

The armature winding is made up of 10 conductors, each of 4 parallel straps per pole, connected for 2 parallel paths.

<u>Solution</u>: For full pitch,

$$X_{slot} = 2\pi fC^2 m \cdot 10^{-8} \left(\frac{4.2\ L\ k_s}{S_1}\right)$$

$$= 2\pi \times 60 \times 240^2 \times 3 \times 10^{-8} \times$$

$$\left(\frac{4.2 \times 4.5 \times 1.65}{144}\right) = 0.140.$$

$$k_s = (0.625p + 0.375) = 0.895.$$

$$X_{slot} = 0.140 \times 0.895 = 0.1255 \text{ ohm for } \frac{5}{6} \text{ pitch}$$

For the end-connection leakage,

$$X_{end} = 2\pi fC^2 \times \frac{1}{P}\left(6 + \frac{\tau - 2}{2.4}\right) K_e \times 10^{-8}$$

$$= 2\pi fC^2 \times \frac{1}{24}\left(6 + \frac{6.024 - 2}{2.4}\right) \times 0.79 \times 10^{-8}$$

$$= 0.545 \text{ ohm.}$$

The value 0.79 is the correction for pitch:

$$K_e = (1.25 \times \frac{5}{6} - 0.25) = 0.79\ .$$

The total leakage reactance then is 0.18 ohm. Potier reactance is usually larger than leakage reactance in salient-pole machines although the results are doubtless influenced by neglecting tooth-tip reactance in these calculations.

• PROBLEM 7-20

If the windings of a 2500-kva, 3-phase, 60-cycle, 32-pole, 2300-volt alternator making 225 rpm are connected to a 3-phase supply, rated current will flow when 85 volts are applied and the power supplied will be 31 kw. Determine the effective resistance and armature leakage reactance for the machine if wye-connected; if delta-connected. Ohmic resistance between terminals is 0.03149 ohm. What is the ratio of effective to ohmic resistance? What is the per cent R_a, X_a, and Z_a?

Solution:

$$I = \frac{2500 \times 1000}{\sqrt{3} \times 2300} = 627.57 \text{ line current in amps.}$$

$$I_p = 627.57/\sqrt{3} = 362.32 \text{ (delta-connected) amps.}$$

$$V = 2300 \text{ line voltage} = V_\phi \text{(delta-connected) volts.}$$

$$V_p = 2300/\sqrt{3} = 1327.94 \text{ (wye-connected) volts.}$$

$$\text{Power per phase} = \frac{31,000}{3} = 10,333 \text{ watts per phase}$$

$$R_a = \frac{W}{I_a^2} = \frac{10,333}{(627.57)^2} = 0.02624 \text{ ohm for wye}$$

$$R_a = \frac{W}{I_a^2} = \frac{10,333}{\left(\frac{627.57}{\sqrt{3}}^2\right)} = 0.07872 \text{ ohm for delta}$$

$$R_0 = \frac{R_t}{2} = \frac{0.03149}{2} = 0.01574 \text{ ohm for wye}$$

$$R_0 = \frac{3}{2} R_t = \frac{3}{2} \times 0.03149 = 0.04721 \text{ ohm for delta}$$

$$\%R_a = \frac{0.02624 \times 627.57 \times 100}{1327.94} = 1.24\%.$$

$$\text{Ratio } \frac{R_a}{R_0} = \frac{0.02624}{0.01574} = \frac{0.07872}{0.04721} = 1.67.$$

$$Z_a = \frac{\frac{85}{\sqrt{3}}}{627.57} = 0.0780 \text{ ohm for wye}$$

$$Z_a = \frac{85}{362.32} = 0.23460 \text{ ohm for delta}$$

$$\%Z_a = \frac{0.07820 \times 627.57}{1327.94} \times 100 = 3.70\%.$$

$$X_a = \sqrt{Z_a^2 - R_a^2}$$

$$X_a = \sqrt{(0.07820)^2 - (0.02624)^2} = 0.07367 \text{ ohm for wye}$$

$$X_a = \sqrt{(0.2346)^2 - (0.07872)^2} = 0.22100 \text{ ohm for delta}$$

$$\%X_a = \frac{0.07367 \times 627.57}{1327.94} \times 100 = 3.48.$$

SUMMARY

Per Phase

	V	I	R_0	R_a	$\%R_a$	X_a	$\%X_a$	Z_a	$\%Z_a$
Wye	1328	628	0.0157	0.0262	1.24	0.0737	3.48	0.0782	3.7
Delta	2300	362	0.0472	0.0787	1.24	0.2210	3.48	0.2346	3.7

REGULATION AND EFFICIENCY

• PROBLEM 7-21

A 50-kva 550-volt single-phase alternator has an open-circuit emf of 300 volts when the field current is 14 amp. When the alternator is short-circuited through an ammeter, the armature current is 160 amp, the field current still being 14 amp. The ohmic resistance of the armature between terminals is 0.16 ohm. The ratio of effective to ohmic resistance may be taken as 1.2. Determine (a) synchronous impedance; (b) synchronous reactance; (c) regulation at 0.8 power factor, current lagging.

Solution: The rated current I = 50,000/550 = 91 amp.

(a) The synchronous impedance Z_s is given by

$$Z_s = \frac{E_1}{I_1'}$$

where E_1 is the open circuit emf and I_1' is the short-circuit current.

$$\therefore \qquad Z_s = 300/160 = 1.87 \text{ ohms.}$$

The effective resistance $R = 1.2 \times 0.16 = 0.192$ ohm.

(b) $X_s = \sqrt{Z_s^2 - R^2} = \sqrt{(1.87)^2 - (0.192)^2} = 1.86$ ohms.

(c) $\cos\theta = 0.8$, $\sin\theta = 0.6$.

The armature induced emf,

$$E_a = \sqrt{(V\cos\theta + IR)^2 + (V\sin\theta + IX)^2}$$

$$= \left\{[(550 \times 0.8) + (91 \times 0.192)]^2 + [(550 \times 0.6) + (91 \times 1.86)]^2\right\}^{\frac{1}{2}}$$

$$= \sqrt{209{,}000 + 249{,}000} = 677 \text{ volts.}$$

As the synchronous reactance was used in computing E, the armature reaction was taken into consideration, so that the no-load voltage of the alternator is presumably 677 volts. The regulation, therefore, is given by

$$\text{Regulation} = \frac{E_a - V}{V} \times 100$$

$$= \frac{677 - 550}{550}\, 100 = \frac{127}{550}\, 100 = 23.1 \text{ per cent.}$$

• PROBLEM 7-22

The figure shows the open- and short-circuit characteristics of a 1,500-kva 2,300-volt 60-cycle alternator. Terminal volts and line current are plotted as ordinates with values of field current as abscissas. Assume that the machine is Y-connected. The resistance between each pair of terminals as measured with direct current is 0.12 ohm. Assume that the effective resistance is 1.5 times the ohmic resistance. Determine the synchronous reactance of the alternator and its regulation at 0.85 power factor, current lagging.

Solution: From the figure, the maximum value of the short-circuit current is 1,400 amp, which is equal to the coil current, since the Y-connection is assumed. This corresponds to 240 amp in the field, and at 240 amp field current the open-circuit terminal emf is 2,180 volts. The corresponding coil emf is

$$\frac{2,180}{\sqrt{3}} = 1,260 \text{ volts,}$$

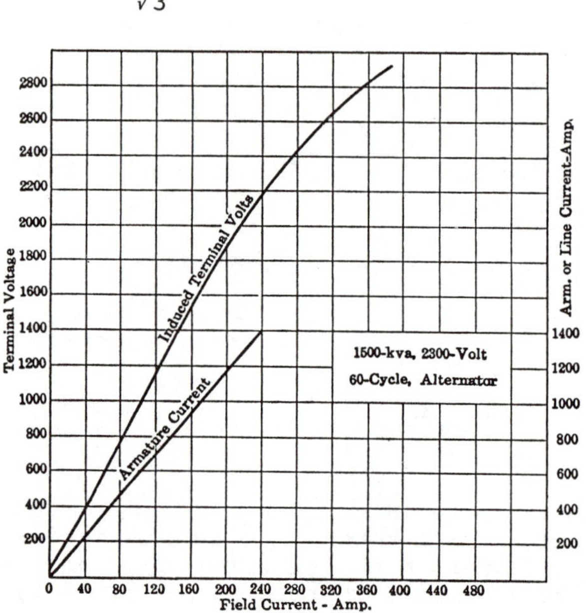

Open-and short- circuit characteristics of 1,500-kva alternator.

The synchronous impedance per coil is then

$$Z_s \text{(per coil)} = \frac{1,260}{1,400} = 0.90 \text{ ohm} = X_s, \text{ nearly.}$$

If the resistance between terminals is 0.12 ohm, it includes two coils in series, as the Y-connection is assumed, so that the ohmic resistance per coil is

$$\frac{0.12}{2} = 0.06 \text{ ohm.}$$

The effective resistance per coil is equal to

$$1.5 \times 0.06 = 0.09 \text{ ohm.}$$

$$\text{Rated current } I = \frac{1,500,000}{2,300\sqrt{3}} = 376 \text{ amp per terminal.}$$

Rated emf per coil, $V = \frac{2,300}{\sqrt{3}} = 1,330$ volts.

$\cos\theta = 0.850, \qquad \theta = 31.8°, \qquad \sin\theta = 0.527.$

No-load emf per coil is given by:

$$E = \sqrt{(V\cos\theta + IR)^2 + (V\sin\theta + IX_s)^2}$$

$$= \left\{[(1,330 \times 0.850) + (376 \times 0.09)]^2 + [(1,330 \times 0.527) + (376 \times 0.90)]^2\right\}^{\frac{1}{2}}$$

$$= 1,560 \text{ volts.}$$

$$\text{Percentage regulation per coil} = \frac{1,560 - 1,330}{1,330}\,100$$

$$= 17.4 \text{ per cent.}$$

Open-circuit terminal emf $= 1,560\sqrt{3} = 2,700$ volts.

$$\text{Percentage regulation using this value} = \frac{2,700 - 2,300}{2,300}\,100$$

$$= 17.4 \text{ per cent.}$$

• PROBLEM 7-23

For the data given on Fig. 1 calculate the regulation of the alternator by the A.S.A. method for an 80 per cent power-factor lagging load.

Solution: From the data the machine is delta-connected.

$$R_a = 0.07872 \text{ ohm per phase}$$

$$I_a = \frac{627.57}{\sqrt{3}} \text{ amp.}$$

$$I_aR_a = \frac{627.57}{\sqrt{3}} \times 0.07872 = 28.48 \text{ volts}$$

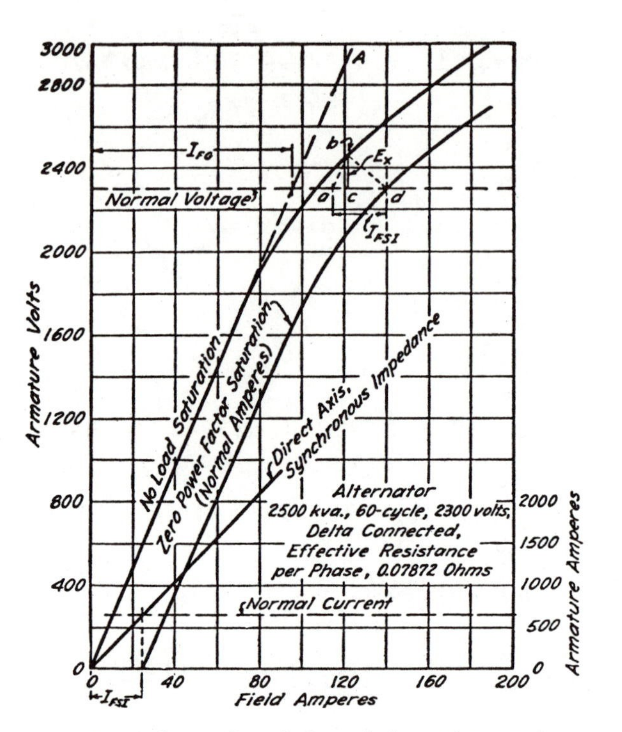

Test data for determining the Potier reactance for an alternator.

Fig.1

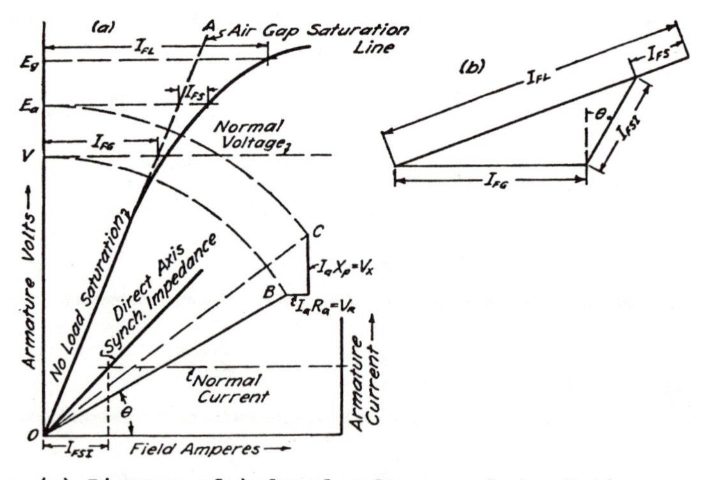

(a) Diagram of induced voltage and regulation.
(b) Determination of load field current.

Fig.2

E_x = 160 volts by construction. By construction shown graphically in Fig. 2 a,

$$I_{FG} = 95.5 \text{ amp}$$

$$I_{FS} = 18.0 \text{ amp}$$

$$I_{FSI} = 25 \text{ amp}$$

To determine the value of I_{FL} it is necessary to add the vector quantities as shown in Fig. 2 b.

$$\vec{I}_{FL} = \vec{I}_{FG} + \vec{I}_{FSI} + \vec{I}_{FS}$$

$$\theta = \cos^{-1} 0.8$$

From Fig. 2 b,

$$I_{FL} = \sqrt{(95.5 + 25 \times 0.6)^2 + (25 \times 0.8)^2} + 18$$

$$= 130.3 \text{ amp.}$$

From Fig. 1,

for $I_{FL} = 130.3$, $E_g = 2540$ volts

$$\text{Percentage regulation} = \frac{\text{no load voltage-rated voltage}}{\text{rated voltage}} \times 100$$

$$= \frac{2540 - 2300}{2300} \times 100 = 10.4 \text{ per cent.}$$

• PROBLEM 7-24

Fig. 1 shows the no-load and low-power-factor characteristics for a 10,000-kva 6,900-volt 514-rpm Y-connected 0.8-power-factor 60-cycle water-wheel alternator. The curves give voltages to neutral. The voltages DB and C'C are the terminal voltages to neutral, 3,980 volts.

The effective resistance of the armature is 0.06 ohm per phase, and the field voltage is 240 volts. Determine by means of the Potier diagram (a) armature leakage reactance; (b) armature reaction in terms of field current; (c) induced emf E_a at 0.8 power factor, lagging current; (d) regulation at 0.8 power factor, lagging current.

Solution:

$$\text{Rated current } I = \frac{10,000}{6,900\sqrt{3}} = 837 \text{ amp}$$

(a) Rated terminal voltage to neutral

$$V = \frac{6,900}{\sqrt{3}} = 3,980 \text{ volts;}$$

Distance AC = 500 volts; therefore, the IX-voltage drop = 500 volts, and

$$X = \frac{IX}{I} = \frac{500}{837} = 0.597 \text{ ohm.}$$

(b) Distance BC = Demagnetizing armature mmf,

$$A = 107 \text{ amp.}$$

The armature induced emf,

$$\vec{E}_a = V + I(\cos\theta - j\sin\theta)(R + jX)$$

$$= 3,980 + 837(0.8 - j0.6)(0.06 + j0.597)$$

$$= 3,980 + 40.2 - j30.1 + j399.5 + 299.5$$

$$= 4,320 + j369.4 \text{ volts,}$$

$$E_a = \sqrt{(4,320)^2 + (369.4)^2} = 4,330 \text{ volts, or } 4.330 \text{ kv}$$

(shown at A_2).

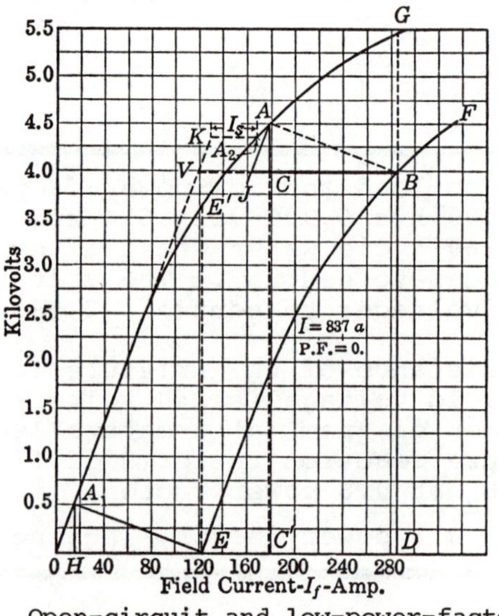

Open-circuit and low-power-factor characteristics of alternator.

Fig.1

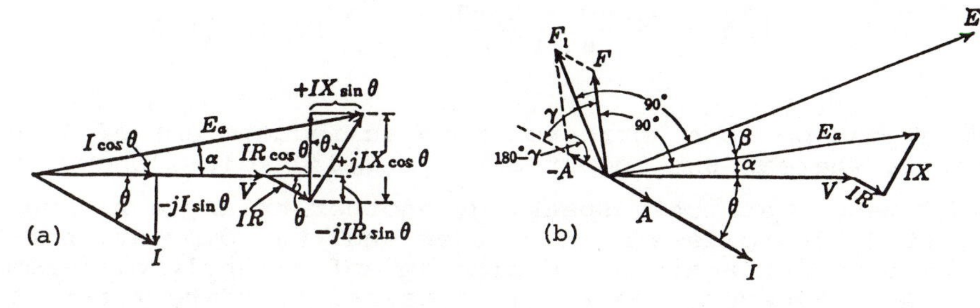

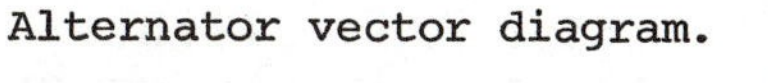
Alternator vector diagram.

Fig.2

From Fig. 1, the field current corresponding to 4,330 volts = 167 amp = F_1

From Fig. 2(b),

$$\tan \alpha = \frac{IX \cos\theta - IR \sin\theta}{V + IR \cos\theta + IX \sin\theta}$$

$$= \frac{369.4}{4{,}320} = 0.0855;$$

$$\alpha = 4.9°;$$

$\cos \theta = 0.80;$ $\theta = 36.9°;$ $\theta + \alpha = 41.8°.$

Referring to Fig. 2(a),

$$\gamma + 90° + \alpha + \theta = 180°;$$

$$\gamma = 90° - (\alpha + \theta) = 90° - 41.8° = 48.2°.$$

Applying the law of cosines to the mmf diagram, Fig. 2, $-A$ being considered a positive magnitude,

$$F_1^2 = F^2 + A^2 - 2FA \cos (180° - \gamma)$$

$$= 167^2 + 107^2 - 2 \times 167 \times 107 \cos 131.8°$$

$$= 27{,}890 + 11{,}450 + 35{,}740 \sin 41.8° = 66{,}000.$$

$$F_1 = 251 \text{ amp}.$$

From the no-load characteristic, Fig. 1, for I_f = 251 amp, E = 5,230 volts, or 9,060 terminal volts.

$$\text{Regulation} = \frac{9,060 - 6,900}{6,900} = 0.313 \text{ or } 31.3\%$$

Note that point A corresponds to an induced emf of 4,500 volts, whereas the computed E_a is 4,330 volts, shown at A_2. Hence, strictly speaking, another triangle having A_2 at 4,330 volts should be used and the computation repeated, which again would give an emf slightly different for A_2. However, the result obtained from the first recomputation will differ only slightly from that obtained originally, and usually the general precision of the method does not warrant recomputation.

• PROBLEM 7-25

A 2,000-kva 2,300-volt three-phase alternator operates at rated kilovolt-amperes at a power factor of 0.85. The d-c armature-winding resistance at 75° C between terminals is 0.08 ohm. The field takes 72 amp at 125 volts from exciter equipment. Friction and windage loss is 18.8 kw, iron losses are 37.6 kw, and stray-load losses are 2.2 kw. Calculate the efficiency of the alternator. (Assume that the effective armature-winding resistance is 1.3 times the d-c value.)

Solution:

$$\text{Output} = 2,000 \times 0.85 = 1,700 \text{ kw.}$$

$$I_L = \frac{2,000,000}{\sqrt{3} \times 2,300} = 503 \text{ amp.}$$

$$R_A \text{ (per phase)} = \frac{0.08}{2} \times 1.3 = 0.052 \text{ ohm.}$$

Losses	Kilowatts
Friction and windage	18.8
Iron	37.6
Field winding = (125 × 72) ÷ 1,000	9.0
Armature winding = 3 × $(503)^2$ × 0.052	39.4
Stray load	2.2
Total	107.0

Per cent efficiency

$$= \left[1 - \frac{\text{kw losses}}{(\text{kva output} \times \text{PF}) + (\text{kw losses})}\right] \times 100$$

$$= \left(1 - \frac{107}{1,700 + 107}\right) \times 100 = 94.1.$$

A. C. MOTOR CHARACTERISTICS

• PROBLEM 7-26

When the field rheostat is cut out, a 230-volt shunt motor generates a counter emf of 220 volts at no load. The resistance of the armature is 2.3 ohms and that of the field is 115 ohms. Calculate the

(a) Current through the armature when the field rheostat is cut out.

(b) Current through the armature when sufficient external resistance has been inserted in the field circuit to make the field current one-half as great.

Solution: (a) The armature current when the field rheostat is cut out is

$$I_a = \frac{E_s - E_a}{R_a} = \frac{230 - 220}{2.3} = 4.35 \text{ amperes.}$$

(b) The current through the field without external resistance is

$$\frac{230}{115} = 2 \text{ amperes.}$$

When the field current has been made half as great by inserting external resistance, the field flux and therefore the counter emf will become half as great, or 110 volts. The armature current in this particular case is therefore

$$I_a = \frac{230 - 110}{2.3} = 52.2 \text{ amperes.}$$

• **PROBLEM 7-27**

A 15-hp, 220-volt, 1800-r/min shunt motor has an efficiency of 87 per cent at full load. The resistance of the field is 440 ohms. Calculate the

(a) Full-load armature current.

(b) Torque of the machine.

Solution: (a) The armature current is

$$I_a = I_\ell - I_f = \frac{15 \times 746}{0.87 \times 220} - \frac{220}{440}$$

$$= 58.46 - 0.5 = 57.96 \text{ amps.}$$

(b) The torque is

$$T = \frac{\text{hp} \times 5252}{N}$$

$$= \frac{15 \times 5252}{1800} = 43.77 \text{ lb. ft.}$$

• **PROBLEM 7-28**

A single-phase motor is taking 20 amperes from a 400-volt, 50-hertz supply, the power factor being 80% lagging. What value of capacitor connected across the circuit will be necessary to raise the power factor to unity?

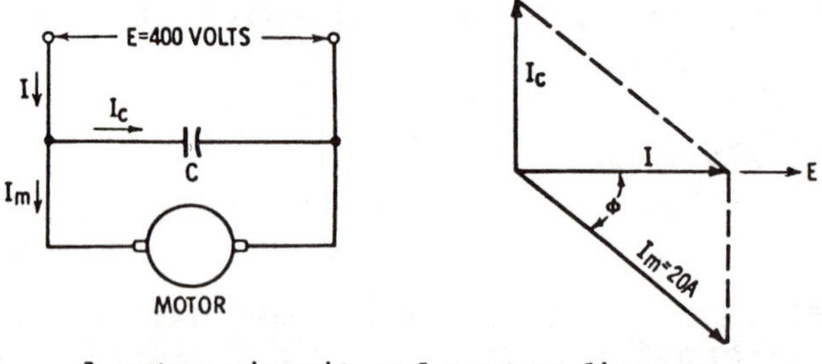

A motor circuit and vector diagram.

Solution: With reference to the vector diagram in the figure, it is evident that

$$I_c = I_m \sin\phi$$

Since

$$I_m = 20, \text{ and } \sin\phi = \sqrt{1 - 0.8^2} = 0.6$$

then

$$I_C = 20 \times 0.6 = 12 \text{ amperes.}$$

and

$$X_C = \frac{E}{I_C} = \frac{400}{12} = 33.3 \text{ ohms.}$$

The equation for the capacitive reactance is

$$X_c = \frac{10^6}{2\pi fC} \text{ ohms.}$$

or

$$C = \frac{10^6}{2\pi f X_C}$$

Inserting values and simplifying,

$$C = \frac{300}{\pi} = 95.5 \text{ microfarads.}$$

NO-LOAD AND LOCKED-ROTOR TESTS

• PROBLEM 7-29

Determine the parameters of a 110 volts, 1/6-HP, 6-pole, 60-cycle split-phase motor from the no-load and locked-rotor test data given below.

(a) No-load test (starting winding open) shows

$V_1 = 110$ volts, $P_0 = 63.0$ watts, $I_0 = 2.70$ amp

r_1 after the test = 2.65 ohms; $P_{F+W} = 3.0$ watts (ball-bearing motor)

(b) Locked-rotor test (starting winding open) shows

$V_L = 110$ volts; $P_L = 851$ watts; $I_L = 11.65$ amp

r_1 after the test = 2.54 ohms.

(The difference in the magnitude of r_1 is due to the fact that the tests were made at different times and at different ambient temperatures.)

Solution: Assume $k_2 = 1.05$; then

$$x_m \simeq \frac{V_1}{I_0} \frac{k_2}{2k_2^2 - 1} \quad ;$$

$$= \frac{110}{2.70} \times \frac{1.05}{2 \times (1.05)^2 - 1} = 35.5 \text{ ohms}$$

r_m can be determined when the other parameters are known.

From the following equations:

$$Z_L = \frac{V_L}{I_L} \; ; \qquad R_L = \frac{P_L}{I_L^2} \quad ;$$

$$X_L = \sqrt{Z_L^2 - R_L^2} \quad ,$$

$$Z_L = \frac{110}{11.65} = 9.44 \text{ ohms;} \qquad R_L = \frac{851}{(11.65)^2} = 6.27 \text{ ohms;}$$

$$X_L = \sqrt{(9.44)^2 - (6.27)^2} = 7.03 \text{ ohms}$$

$$x_1 \simeq 2x_2' = \frac{X_L}{2} = \frac{7.03}{2} = 3.51 \text{ ohms}$$

$$x_2' = 1.75 \text{ ohms}$$

Checking

$$k_2 = 1 + \frac{x_2'}{x_m}$$

$$= 1 + \frac{1.75}{35.5} = 1.0493$$

which is very close to the assumed value of $k_2 = 1.05$.

$$r_2' = \frac{R_L - r_1}{2} k_2^2$$

$$= \frac{6.27 - 2.54}{2} \times (1.05)^2 = 2.06$$

The five parameters as determined from the no-load and locked-rotor tests are (in ohms)

$$r_1 = 2.54 \qquad r_2' = 2.06$$

$$x_m = 35.5$$

$$x_1 = 3.51 \qquad x_2' = 1.75$$

For the calculation of the performance of the motor the hot resistances of the windings at full-load should be used; if full-load data are not available the resistances at 75° C must be used. For the motor considered, r_1 at full-load was 2.85 ohms, while in the locked-rotor test the resistance was 2.54 ohms. It can be assumed that the resistance of the rotor is increased by the load in the same ratio as the stator resistance. Thus at full-load $r_2' = 2.06(2.85/2.54) = 2.31$, and the motor parameters for performance calculations are (in ohms)

$$r_1 = 2.85 \qquad r_2' = 2.31$$

$$x_m = 35.5$$

$$x_1 = 3.51 \qquad x_2' = 1.75$$

r_m can now be determined. The no-load losses of the motor are 63 watts. They consist of copper losses in both windings, iron losses (due to the main flux + rotational losses), and friction and windage losses.

The latter losses are 3.0 watts. The resistance of the stator winding after the no-load test was 2.65 ohms, and thus the loss in the stator copper is

$$P_{co,1} = 2.65 \times (2.7)^2 = 19.3 \text{ watts}$$

From the locked-rotor test, $r_1 = 2.54$ and $r_2' = 2.06$. Since at no-load $r_1 = 2.65$, it can be assumed that at no-load

$$r_2' = 2.06 \, \frac{2.65}{2.54} = 2.15 \text{ ohms}$$

The rotor current at no-load is approximately equal to I_0 ($I'_{2b} \simeq I_0$). Therefore the losses in the rotor winding at no-load are

$$P_{co,2} = 2.15 \times (2.7)^2 = 15.7 \text{ watts}$$

The total iron losses are then

$$P_{ir} = 63.0 - (3.0 + 19.3 + 15.7) = 25.0 \text{ watts}$$

The rotational iron losses at no-load are relatively larger in the single-phase motor than in the polyphase motor. As in the latter motor, the slot openings cause tooth-surface and tooth-pulsation losses in the single-phase motor. However, contrary to the polyphase motor, the single-phase motor has a large no-load current which does not differ very much from the full-load current. Consequently, the harmonics produce additional tooth-surface and tooth-pulsation losses at no-load. All of these additional iron losses are caused by the rotation of the rotor and must be supplied by the rotor. It can be assumed that the losses due to the main flux are approximately half of the total iron losses, i.e.,

$$P_{h+e} \simeq 13 \text{ watts.}$$

This loss determines the magnitude of r_m.

From the following equations:

$$E_1 \simeq V_1 - I_0 x_1$$

$$C = \frac{x_m}{\sqrt{(r'_2/2)^2 + x'^2_2}}$$

$$E'_{2f} \simeq E_1 \frac{C}{1 + C},$$

$$E_1 = 110 - 2.70 \times 3.51 = 100.6$$

$$C = \frac{35.5}{\sqrt{(2.15/2)^2 + (1.75)^2}} = 17.3$$

$$E'_{2f} = 100.6 \frac{17.3}{18.3} = 95.2$$

$$g_m = \frac{P_{h+e}}{E'^2_{2f}} = \frac{13}{(95.2)^2} = 0.0014 \text{ mho.}$$

$$r_m = g_m x_m^2 = 0.0014 \times (35.5)^2 = 1.77 \text{ ohms}$$

The six parameters for performance calculations are (in ohms)

$r_1 = 2.85$ $\quad r_m = 1.77$ $\quad r'_2 = 2.31$

$x_1 = 3.51$ $\quad x_m = 35.5$ $\quad x'_2 = 1.75$

Note.--In general the saturation of the leakage paths should be taken into account when evaluating the locked-rotor test. The saturation of the leakage paths has been neglected here.

POWER, TORQUE AND EFFICIENCY

• PROBLEM 7-30

Calculate the full-load torque, in pound-feet and ounce-inches, of a 1/50-hp 1,500-rpm shaded-pole motor.

(b) A 1/20-hp 1,550-rpm shaded-pole motor has a maximum torque of 38 oz-in. and a starting torque of 12 oz-in. What percentages are these torques with respect to the full-load torque?

Solution:

(a) $$T = \frac{33{,}000}{2\pi} \times \frac{hp}{rpm} = \frac{5{,}250 \times hp}{rpm} \text{ lb-ft}$$

$$= \frac{5{,}250 \times 0.02}{1{,}500} = 0.07 \text{ lb-ft}$$

$$T = \frac{5,250 \times 12 \times 16 \times hp}{rpm} = \frac{hp}{rpm} \times 10^6 \quad \text{oz-in.}$$

$$= \frac{0.02}{1,500} \times 10^6 = 13.4 \text{ oz-in.}$$

(b)

$$T = \frac{0.05}{1,550} \times 10^6 = 32.3 \text{ oz-in.} \qquad \text{(full-load torque)}$$

$$\text{Per cent maximum torque} = \frac{38}{32.3} \times 100 = 118\%$$

$$\text{Per cent starting torque} = \frac{12}{32.3} \times 100 = 37\%$$

• PROBLEM 7-31

A machine with two coils has inductances as follows: (on rotor) $L_1 = 0.1$ henry, (on stator) $L_2 = 0.5$ henry, $M = 0.2 \cos \theta$ henry, where θ is the angle of the rotor coil axis displaced counterclockwise with respect to the stator coil axis. Coil 1 (on rotor) is short circuited. Coil 2 (on stator) is energized from a 60-Hz sinusoidal voltage source of 110 v. Resistances of the coils may be neglected. Assume the circuit operates in sinusoidal steady state. θ is set at 30°. (a) Find an expression for the instantaneous torque on the rotor. (b) Find the value of the average torque on the rotor. (c) Determine the direction of this torque.

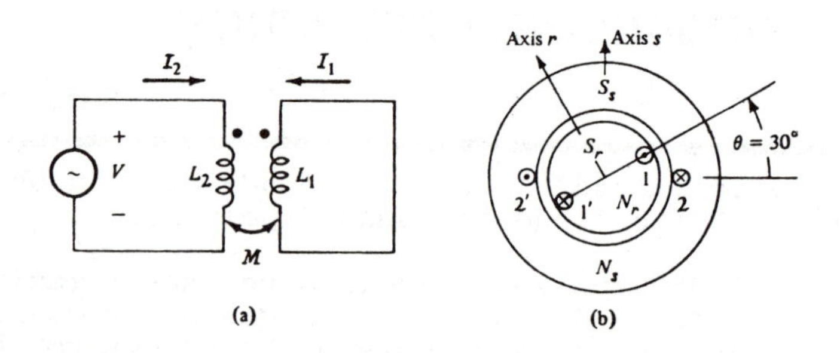

Solution: (a) In order to find the developed torque, one must first determine the values of the currents in the coils. The equivalent circuit is shown in Fig. a. Write Kirchhoff's voltage equations

$$\vec{V} = +j\omega L_2 \vec{I}_2 + j\omega M \vec{I}_1$$

$$0 = +j\omega M \vec{I}_2 + j\omega L_1 \vec{I}_1$$

These yield $\vec{I}_1 = -(M/L_1)\vec{I}_2$, $\vec{I}_2 = \vec{V}/j\omega(L_2 - M^2/L_1)$ A

From the given data, $M = 0.2 \cos 30° = 0.1732$ henry

Choose V for the reference. $\vec{V} = 110\underline{/0°}$ V

Then the currents are $\vec{I}_2 = -j1.46$ A and $\vec{I}_1 = +j2.53$ A

The current functions are

$$i_2(t) = 1.46\sqrt{2} \sin 377\ t,$$

$$i_1(t) = -2.53\sqrt{2} \sin 377\ t$$

Torque is given by

$$T = \frac{1}{2} i_1^2 dL_1/d\theta + \frac{1}{2} i_2^2 dL_2/d\theta + i_1 i_2 dM/d\theta$$

$$dL_1/d\theta = 0, \quad dL_2/d\theta = 0, \quad dM/d\theta = -0.2 \sin \theta$$

$$T = i_1 i_2 dM/d\theta =$$

$$(2.06 \sin 377t)(-3.58 \sin 377t)(-0.2 \sin 30°)$$

$$= 0.74 \sin^2 377t = 0.74\left[\frac{1}{2} - \frac{1}{2}\cos 2(377)t\right]$$

$$= 0.37 - 0.37 \cos 754t \text{ nm.}$$

(b) Since the average value of the cosine function is zero, we can conclude

$$T_{ave} = 0.37 \text{ nm.}$$

(c) Figure b illustrates the conditions of this problem. At an instant of time when the stator current produces a magnetic field polarity, N_s, S_s, as shown, the rotor current has the direction shown. The rotor field results in a repelling action that makes the torque on the rotor to be counterclockwise.

• PROBLEM 7-32

The following constants are for a $\frac{1}{4}$ - hp 60-Hz 115-V four-pole capacitor-start motor:

$r_1 = 2.15\Omega$ $\qquad r_2 = 4.45\Omega$

$x_1 = 3.01\Omega$ $\qquad x_2 = 2.35\Omega$

$x_M = 70.5\Omega$

Core loss = 26.0 W, windage and friction loss = 14.0 W. Calculate for a slip of 0.05 the (a) current, (b) power factor, (c) output, (d) torque, and (e) efficiency.

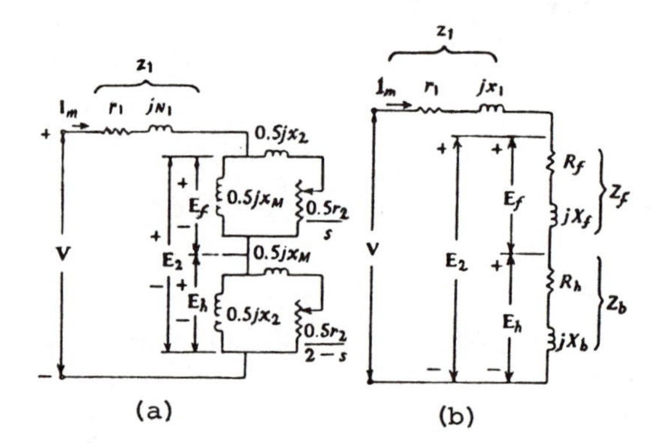

Equivalent circuit of single-phase induction (main winding only). (a) Running, detailed circuit. (b) Simplified circuit.

Solution: (a) The current based on the equivalent circuit in Fig. (b) is

$$\vec{I}_m = \frac{\vec{V}}{\vec{Z}_1 + \vec{Z}_f + \vec{Z}_b}$$

where $\vec{Z}_1 = r_1 + jx_1 = 2.15 + j3.01$,

and where $\vec{Z}_f$ and $\vec{Z}_b$ are found by making use of Fig. (a) as follows:

$$\vec{Z}_f = R_f + jX_f = \frac{[(0.5r_2/s) + j0.5x_2)]j0.5x_M}{(0.5r_2/s) + j(0.5x_2 + 0.5x_M)}$$

$$= \frac{(44.5 + j1.175)j35.25}{44.5 + j36.43} = 27.2\underline{/\ 52.1°}$$

$$= 16.72 + j21.45\Omega$$

$$\vec{Z}_b = R_b + jX_b = \frac{\{[0.5r_2/(2 - s)] + j0.5x_2\}j0.5x_M}{[0.5r_2/(2 - s)] + j(0.5x_2 + 0.5x_M)}$$

$$= \frac{(1.14 + j1.175)j35.25}{1.14 + j36.43} = 1.58\underline{/\ 47.6°}$$

$$= 1.06 + j1.17\Omega$$

and

$$\vec{Z}_1 + \vec{Z}_f + \vec{Z}_b = 2.15 + j3.01 + 16.72 + j21.45 + 1.06 + j1.17$$

$$= 19.93 + j25.63 = 32.5\underline{/\ 52.1°}$$

$$\vec{I}_M = \frac{115}{32.5\underline{/\ 52.1°}} = 3.54\underline{/\ -52.1°}\ \text{A}$$

(b) The power factor is cos 52.1° = 0.615.

(c) The developed power is

$$P_{em} = (1 - S)I_m^2(R_f - R_b)$$

$$= (0.95)(3.54)^2(16.72 - 1.06) = 186\ \text{W}$$

The mechanical power output is

$$P_{out} = P_{em} - P_{rot}$$

and the rotational losses are

$$P_{rot} = 26 + 14 = 40\ \text{W}$$

Hence,

$$P_{out} = 186 - 40 = 146\ \text{W or } 0.196\ \text{hp}$$

(d)

$$\text{Torque} = \frac{P_{out}}{(1 - s)\omega_{syn}} = \frac{146}{0.95[(2\pi \times 1800)/60]}$$

$$= 0.816 \text{ N-m/rad}$$

(e)

$$\text{Efficiency} = \frac{\text{output}}{\text{input}}$$

$$\text{Input} = VI_m \cos\theta = 115 \times 3.54 \times 0.615 = 250 \text{ W}$$

$$\text{Efficiency} = \frac{146}{250} = 0.584.$$

• PROBLEM 7-33

The locked readings of a certain 1/2-hp 220-volt 2-phase 60-cycle 1,800-rpm motor are E = 220 volts, I = 12.1 amp, W = 3.2 kw, T = 5 lb-ft. If the motor is now connected as shown in the figure with an external resistance R in phase B of 20 ohms, calculate the currents I_A and I_b in phases A and B and also the new locked torque.

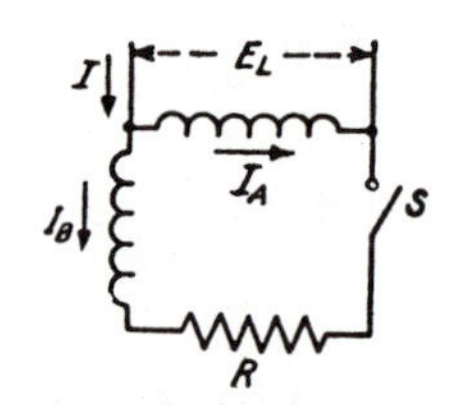

Connections of 2-phase motor for split-phase starting.

Solution:

$$I_A = 12.1 \text{ amp.}$$

$$\cos\phi_A = \frac{1,600}{220 \times 12.1} = 0.6.$$

and so

$$\phi_A = 53.1°.$$

To find I_B, the total equivalent locked resistance and

reactance per phase are computed.

$$Z_A = Z_B = \frac{220}{12.1} = 18.2 \text{ ohms.}$$

$$R_A = \frac{1{,}600}{(12.1)^2} = 10.9 \text{ ohms}$$

$$X_A = X_B = \sqrt{Z_A{}^2 - R_A{}^2} = \sqrt{331 - 119} = \sqrt{212} = 14.55 \text{ ohms.}$$

The value of the external resistance in phase B is 20 ohms, the total resistance of this phase is then

$$R_B = 30.9 \text{ ohms.}$$

$$Z_B = \sqrt{R_B{}^2 + X_B{}^2} = \sqrt{955 + 212} = 34.2 \text{ ohms.}$$

$$I_B = \frac{220}{34.2} = 6.44 \text{ amp.}$$

$$\cos \phi_B = \frac{30.9}{34.2} = 0.905,$$

$$\phi_B = 25.2°$$

and

$$\alpha = 53.1 - 25.2 = 27.9°$$

To find the locked torque,

$$I_B \sin \alpha = 6.44 \times 0.468 = 3.02.$$

The torque is given by

$$T = K\, n_A I_A n_B I_B \sin \alpha \qquad (1)$$

where

n_A = Effective series turns of phase A

n_B = Effective series turns of phase B

α = Phase difference between I_A and I_B

Without the external resistance R, the torque was 5 lb-ft, I_A and I_B were each 12.1 amps and angle α was 90°. Substituting these values in Eq. (1),

$$T_1 = 5 = K\, n_A n_B (12.1)(12.1)(1.0) \tag{2}$$

With the external resistance R, I_A remains same,

$$I_B \sin \alpha = 3.02.$$

Substituting these values in Eq. (1),

$$T_2 = K\, n_A n_B (12.1)(3.02) \tag{3}$$

Eq. (3) divided by Eq. (2) gives

$$\frac{T_2}{5} = \frac{3.02}{12.1}$$

$$T_2 = \frac{3.02}{12.1} \times 5 = 1.25 \text{ lb-ft}$$

It is seen from the above that a 2-phase motor operated on a single-phase line in this fashion would give very poor starting performance.

CHAPTER 8

SYNCHRONOUS GENERATORS

SYNCHRONOUS REACTANCE AND SYNCHRONOUS IMPEDANCE

• **PROBLEM** 8-1

The following data are taken from the open-circuit and short-circuit characteristics of a 45-kva 3-phase Y-connected 220-volt (line to line) 6-pole 60-cps synchronous machine:

From open-circuit characteristic:

Line-to-line voltage = 220 volts

Field current = 2.84 amp

From short-circuit characteristic:

Armature current, amp	118	152
Field current, amp.	2.20	2.84

From air-gap line:

Field current = 2.30 amp

Line-to-line voltage = 202 volts

Compute the unsaturated value of the synchronous reactance, and the short-circuit ratio. Express the synchronous reactance in ohms per phase and also in per unit on the machine rating as a base.

Solution: At a field current of 2.20 amp, the voltage to neutral on the air-gap line is

$$E_{f(ag)} = \frac{202}{\sqrt{3}} = 116.7 \text{ volts}$$

and, for the same field current, the armature current on short circuit is

$$I_{a(sc)} = 118 \text{ amp}$$

$$x_{s(ag)} = \frac{E_{f(ag)}}{I_{a(sc)}} = \frac{116.7}{118} = 0.987 \text{ ohm per phase}$$

Note that rated armature current is $45{,}000/\sqrt{3}(220) = 118$ amp. Therefore $I_{a(sc)} = 1.00$ per unit. The corresponding air-gap-line voltage is

$$E_{f(ag)} = 202/220 = 0.92 \text{ per unit}$$

In per unit,

$$x_{s(ag)} = \frac{0.92}{1.00} = 0.92 \text{ per unit}$$

From the open-circuit and short-circuit characteristics and the equation,

$$x_s = \frac{V_t}{I'_{a(sc)}},$$

$$x_s = \frac{220}{\sqrt{3}\ (152)} = 0.836 \text{ ohm per phase}$$

In per unit, $I'_{a(sc)} = 152/118 = 1.29$, and from the equation above,

$$x_s = \frac{1.00}{1.29} = 0.775 \text{ per unit}$$

From the open-circuit and short-circuit characteristics,

$$\text{SCR} = \frac{2.84}{2.20} = 1.29.$$

• PROBLEM 8-2

The following data are for a three-phase 13,800-V wye-connected 60,000-kVA 60-Hz synchronous generator:

P = 2; stator slots = 36, stator coils = 36, turns in each stator coil = 2

Stator coil pitch = $\frac{2}{3}$, rotor slots = 28

Spacing between rotor slots = $\frac{1}{37}$ of circumference

Rotor coils = 14

Turns in each rotor coil = 15

I.D. of stator iron = 0.948 m

O.D. of rotor iron = 0.865 m

Net axial length of stator iron = 3.365 m, stator coil connection is two-circuit (a = 2) series rotor-coil connection.

Assume that g_e = 1.08 g and calculate the unsaturated self-inductance of the field winding based on the fundamental component of the air-gap flux.

Solution:

$$L_{ff} \cong \frac{3.2 D_g L}{g_e}\left(\frac{k_{wf} N_f}{P}\right)^2 \times 10^{-6}$$

where

$$D_g = \frac{0.948 + 0.865}{2} = 0.9065 \text{ m}$$

$$L = 3.365 \text{ m}$$

$$g_e = \frac{1.08\ (0.948 - 0.865)}{2} = 0.0449 \text{ m}$$

$$N_f = 14 \times 15 = 210 \text{ turns}$$

$$P = 2;$$

The field winding is equivalent to a full-pitch distributed winding and the winding factor k_{wf} is

$$k_{wf} = \frac{\sin\ (\beta_f/2)}{n \sin\ (\gamma_f/2)}$$

where

$\beta_f = \frac{28\pi}{37} = 2.38$ rad or 136.3°, the angle occupied by the field winding under one pole.

$\gamma_f = \frac{2\pi}{37} =$ 0.0541 rad or 9.73°, the angle between adjacent slots.

n = 14 slots per pole.

$$k_{wf} = \frac{\sin\ (136.3/2)}{14 \sin\ (9.73/2)} = \frac{0.928}{14 \times 0.085} = 0.78.$$

Hence,

$$L_{ff} = \frac{3.2 \times 0.9065 \times 3.365}{0.0449}\left(\frac{0.78 \times 210}{2}\right)^2 \times 10^{-6}$$

$$= 1.46 \text{ H}.$$

• PROBLEM 8-3

Determine the synchronous reactances in both axes for the following generator:

Output: 875 kva, $\cos\phi = 0.80$, 3 phases, 60 cycles, 24 poles, 2300 volts, $I_a = 220$ amp.

Stator winding: 180 slots, 6 conductors per slot, coil pitch 6 slot pitches, pole pitch $\tau = 9.52$ in., width of the pole arc $b_p = 6.0$ in., no-load mmf at 2300 volts $M_{f0} = 3610$ AT per pole and $M_{f0g} = 2930$ AT per pole for the gap only. The leakage reactance is 0.826 ohm per phase. For the ratio $b_p/\tau = 6.0/9.52 = 0.63$, $C_d = 0.86$, $C_q = 0.41$.

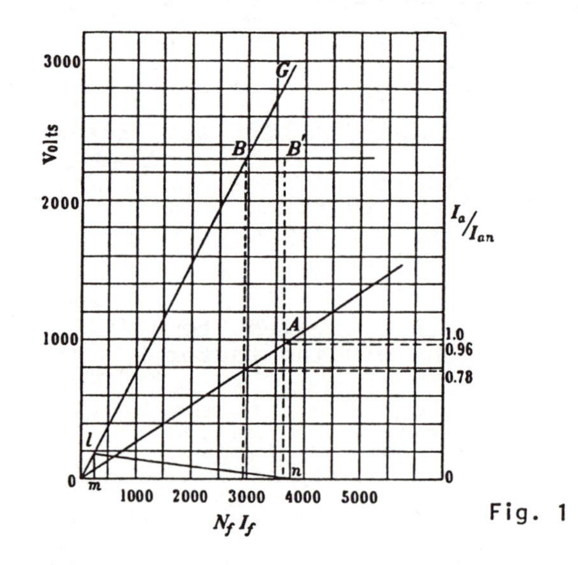

Fig. 1

Solution: The number of slots per pole per phase $q = 180/(3 \times 24) = 2.5$, $k_d = 0.955$, $k_p = \sin[(6/7.5) \times \pi/2] = 0.951$, $k_{dp} = 0.908$. the number of turns per phase

$$N_a = \frac{6 \times 180}{2 \times 3} = 180$$

The total armature mmf in the direct axis at $\psi = 90°$ (short circuit) is

$$M_{ad} = \frac{0.9\ m\ N_a I_a k_{dp}}{p}$$

$$= 0.9 \times 3 \times \frac{180}{24} \times 0.908 \times 220 \times 0.86$$

$$= 3480 \text{ AT per pole}$$

From the data given, the air gap line OG in Fig. 1 may be drawn. Construct the Potier triangle with $\ell m = I_a x_\ell = 220 \times 0.826 = 182$ volts and $mn = 3480$ AT. To point n (3730 AT) corresponds unit armature current (nA) on the short-circuit characteristic--draw this latter characteristic as OA. Corresponding in $M_{f0g} = 2930$ and $M_{f0} = 3610$ (points B and B' respectively) the unsaturated SCR may be determined as 0.78. Hence (in p.u.):

$$x_d = \frac{1}{0.78} = 1.28$$

$$x_\ell = 0.137$$

$$x_{ad} = x_d - x_\ell = 1.28 - 0.137 = 1.14$$

$$x_{aq} = x_{ad} \frac{C_q}{C_d} = 1.14 \times \frac{0.41}{0.86} = 0.543$$

$$x_q = x_\ell + x_{aq} = 0.137 + 0.543 = 0.680$$

In this example the value of q is 2.5. Such a winding is classified as a fractional slot winding.

$$k_d = \frac{E_r}{qE_c}$$

for the distribution factor does not apply to this kind of winding.

• PROBLEM 8-4

The curves of Fig. 1 are for a 500-kva 2,300-volt three-phase alternator. If the average d-c resistance of the three armature-winding phases between pairs of terminals R_t is 0.8 ohm, (a) calculate the values of Z_S and X_S. (Assume that the effective a-c resistance is 1.5 times the d-c resistance.)

(b) A 500-kva 2,300-volt three-phase star-connected alternator has a full-load armature-resistance drop per phase of 50 volts and a combined armature-reactance (E_X) plus armature-reaction (E_{AR}) drop of 500 volts per phase.

For the above data, using the values of R_A and X_S from (a), calculate the percent regulation for a full-load power factor of 0.8 lagging.

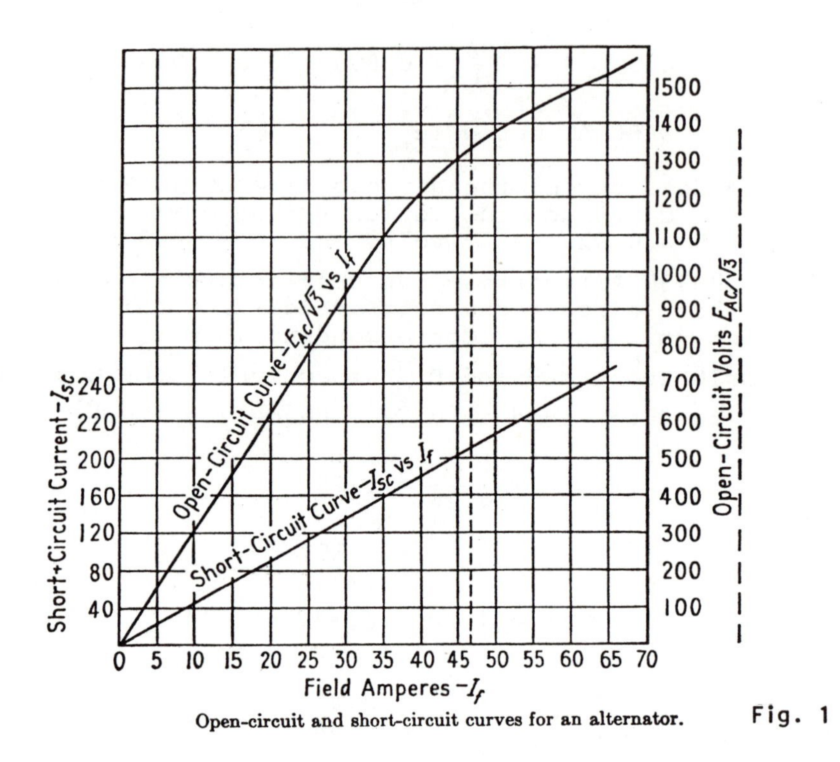

Open-circuit and short-circuit curves for an alternator. Fig. 1

<u>Solution</u>: (a) From the curves of Fig. 1, at $E_{ac}/\sqrt{3} = 2{,}300/\sqrt{3} = 1{,}330$ volts,

$$I_{SC} = 204 \text{ amp}$$

$$Z_S = \frac{1{,}330}{204} = 6.52 \text{ ohms}$$

$$R_A = 1.5 \times \frac{0.8}{2} = 0.6 \text{ ohm}$$

$$Z_S = \sqrt{R_A^2 + X_S^2}\ ,$$

$$\therefore\ X_S = \sqrt{(6.52)^2 - (0.6)^2} = 6.5 \text{ ohms}$$

(b) $$I_L = \frac{500{,}000}{\sqrt{3} \times 2300} = 126$$

$$E_R = 126 \times 0.6 = 75 \text{ volts}$$

$$I_L X_S = 126 \times 6.5 = 8.20$$

$$E_G = \sqrt{\left[(1{,}330 \times 0.8) + 75\right]^2 + \left[(1{,}330 \times 0.6) + 820\right]^2}$$

$$= \sqrt{(1,139)^2 + (1,618)^2} = 1,980 \text{ volts}$$

$$\text{Percent regulation} \doteq \frac{1,980 - 1,330}{1,330} \times 100 = 48.8$$

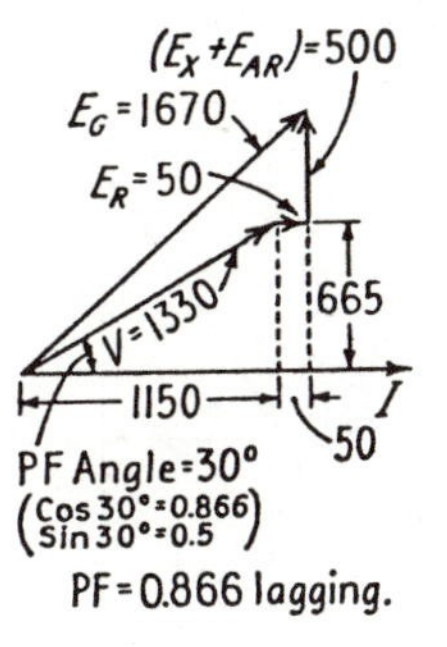

Phasor diagram illustrating part (b).

• PROBLEM 8-5

A 60-cps, 2300-volt (line to line), Y-connected, 1000-kva, 1800-rpm, three-phase synchronous generator has open-circuit, short circuit and zero-lagging power-factor as shown in Fig. 1.

Determine:

(a) Unsaturated value of synchronous reactance

(b) Saturated value of synchronous reactance at the rated terminal voltage condition.

(c) The voltage regulation for a 0.8 lagging power factor load using the value of synchronous reactance determined in part b.

Solution: The rated quantities are determined as

$$\text{Phase voltage} = \frac{2300}{\sqrt{3}} = 1327 \text{ volts}$$

$$\text{Phase current} = \frac{1000(10^3)}{\sqrt{3}\ (2300)} = 251 \text{ amperes}$$

(a) The unsaturated value of synchronous reactance is determined by the air-gap line and the short circuit characteristic. Referring to Fig. 1, the point E on the air-gap line is arbitrarily chosen. The percent voltage is 112% or 1486 volts. The percent field current is 100. The value of short circuit current corresponding to 100% field current is 100% or 251 amperes. The unsaturated value of synchronous reactance is

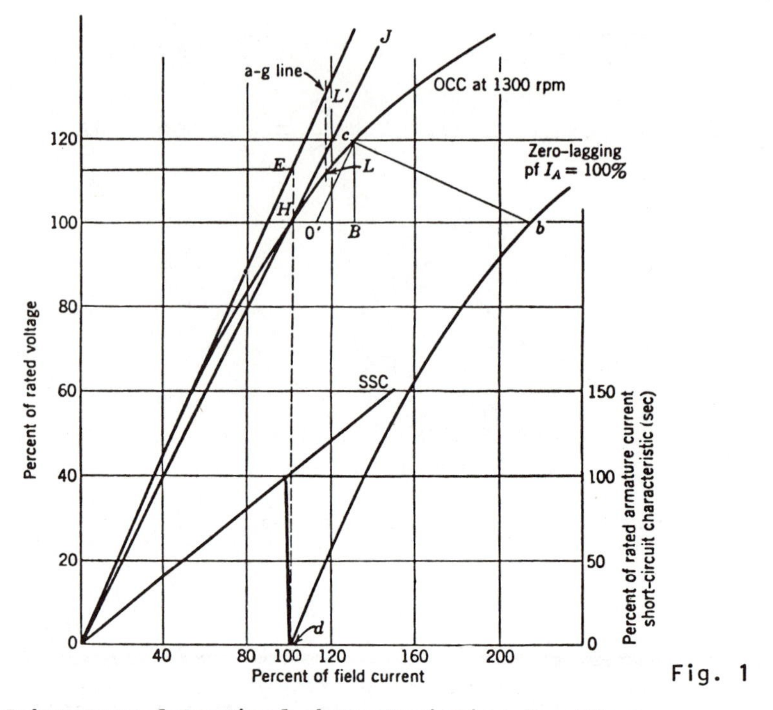

Fig. 1

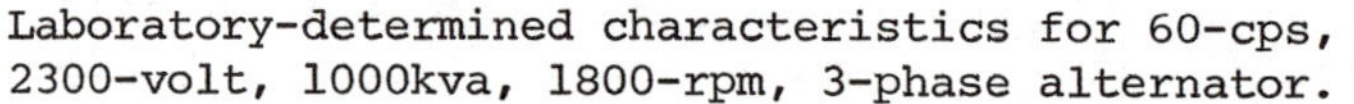
Laboratory-determined characteristics for 60-cps, 2300-volt, 1000kva, 1800-rpm, 3-phase alternator.

$$x_{s(ag)} = \frac{E_{f(ag)}}{I_{a(sc)}}$$

$$= \frac{1486}{251} = 5.92 \text{ ohms per phase}$$

(b) The saturated value of synchronous reactance determined by rated voltage is

$$x_s = \frac{V_A}{I_{a(sc)}}$$

$$= \frac{1327}{251} = 5.28 \text{ ohms per phase}$$

(c) To determine the voltage regulation, the linearized open circuit characteristic, which is the straight line passing through the origin and point H, is used to determine the required field current at the rated load conditions. The no-load terminal voltage is then determined from the actual open circuit characteristic for the same value of the field current. Thus,

$$\vec{E}_f = \vec{V}_A + j\,\vec{I}_a\,X_s$$

$$= 1327 + j0 + 251(0.8 - j0.6)(j5.28)$$

$$= 1327 + j0 + 792 + j1060 = 2370\ \underline{/26.6°}\ \text{V.}$$

In percent of rated voltage,

$$E_f = \frac{2370}{1327}(100) = 179\%$$

From the curve OHJ, the percent field current is determined at 179%. The voltage from the open circuit characteristic is 139.5% or 1850 volts. The voltage regulation is

$$\text{Reg. }(\%) = \frac{1850 - 1327}{1327}(100) = 39.4\%$$

ZERO-POWER-FACTOR TEST AND POTIER'S TRIANGLE

• PROBLEM 8-6

The open-circuit characteristic of a 13,529-kVA 13.8-kV 60-Hz three-phase two-pole turbine generator is shown in the figure.

The zero-power-factor test data follow:

Line to Line (V)	Field Current (A)
0	188
13,800	368

(a) Draw the Potier triangle and determine the Potier reactance x_p and the component of field current A to overcome the mmf of armature reaction.

(b) Calculate the saturation factor k and the regulation of the synchronous generator, when delivering rated load at 0.85 power factor current lagging.

Solution: (a) Rated current $I = \dfrac{13{,}529}{\sqrt{3} \times 13.8} = 566$ A

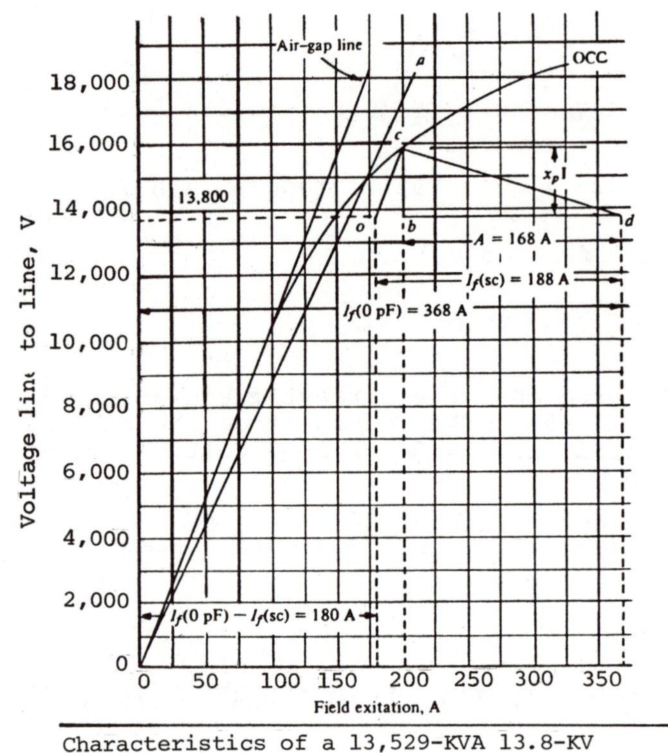

Characteristics of a 13,529-KVA 13.8-KV 60-Hz three-phase two-pole turbine generator.

Distance bd in Fig. 1 is 168 amp = A, the component of field current to overcome the mmf of armature reaction.

$$\text{Potier reactance } x_p = \frac{2000}{\sqrt{3} \times 566} = 2.04\ \Omega,$$

$$x_p \text{ in per unit} = \frac{2000}{13,800} = 0.145$$

(b) Terminal volts per phase, $V = 13,800 \div \sqrt{3} = 7960$

$$\text{Rated current amp per phase, } I = \frac{13,529}{\sqrt{3} \times 13.8} = 566$$

Potier reactance ohms per phase = 2.04 from part (a)

$$\vec{E}_p \cong \vec{V} + jx_p \vec{I}$$

$$\vec{E}_p = 7960 + j2.04 \times 566(0.85 - j0.527)$$

$$= 8570 + j981 = 8630 \underline{/6.65°} \text{ V/phase}$$

The line-to-line voltage is

$$\sqrt{3}\, E_p = 14,950 \text{ V}$$

which requires a field current of R = 173 A as determined on the OCC characteristic in the figure. The same value of voltage requires an mmf of R_{ag} = 142 A on the air-gap line in the figure. Hence,

$$k = \frac{R}{R_{ag}} = \frac{173}{142} = 1.22$$

The unsaturated synchronous reactance is found from the voltage on the air-gap line produced by a field current of 188 A, which is required for the rated short-circuit armature current of 566 A. This voltage is found to be 19,800 V line to line when the air-gap line is extended. So

$$x_{du} = \frac{19,800}{\sqrt{3} \times 566} = 20.2\ \Omega$$

and since x_p = 2.04 Ω, the saturated synchronous reactance is

$$x_d = x_p + \frac{x_{du} - x_p}{k} = 2.04 + \frac{20.2 - 2.04}{1.22} = 16.9\ \Omega$$

and the induced voltage E_k, when r_a is neglected, is

$$\vec{E}_k \cong \vec{V} + jx_d\,\vec{I} = 7960 + j16.9 \times 566(0.85 - j0.527)$$

$$= 13,000 + j8130 = 15,320\ \underline{/32.0°}\ V$$

The line-to-line magnitude of this voltage is

$$\sqrt{3}\,E_k = 26,600\ V$$

The field current obtained for this value on the line oa in the figure is

$$F = 315\ A,$$

which produces a line-to-line voltage on the open-circuit characteristic of 18,300 V. The regulation is therefore,

$$\text{regulation} = \frac{18.3 - 13.8}{13.8} = 0.326.$$

The test results on a 3-phase, 10-MVA, 11-kV, star-connected synchronous generator are as follows:

Open-circuit test.

Line voltage	4	6	8	10	11	12	13	14 kV
Field Current	60	90	126	171	203	252	324	426 A

Short-circuit test.

Armature current 566 A, field current 200 A.

Zero power-factor test.

Armature current 396 A, field current 480 A, line voltage 11 kV.

Calculate the field current:

(a) when supplying 6000 kVA at zero leading power-factor to an unloaded transmission line (line charging).

(b) when generating at full load 0.9 p.f. lagging

(c) when motoring at the same current and power factor as (b)

(d) when running as an unloaded motor taking 6000 kVA at zero p.f. leading.

Neglect armature resistance.

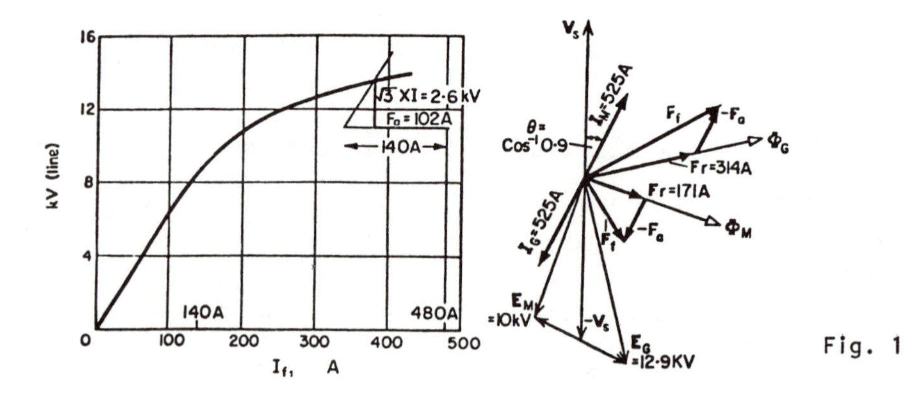

Fig. 1

Solution: The Potier construction is carried out on the accompanying figure after correcting the short circuit information to the same armature current as the zero p.f. point, i.e. at 396 A on s.c. the field current will be (396/566) x 200 = 140 A.

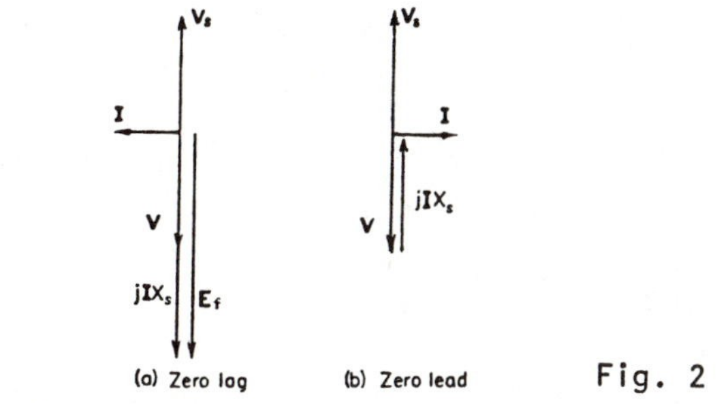

Fig. 2

Effect of power factor on excitation requirements.

At 396 A and 11 kV the field current is 480 A for zero p.f. lagging. This simple correction to the s.c. information is possible because the relationship between field and armature s.c. current is linear. Note that the o.c. curve is plotted in terms of line voltage for convenience so that the Potier reactance drop per phase at 396 A is scaled up by a factor of $\sqrt{3}$.

From Fig. 1, $\sqrt{3}$ XI = 2.6 kV at 396 armature amperes.

$$F_a = 102 \text{ A at } 396 \text{ armature amperes.}$$

(a) At 6000 kVA, $I = \dfrac{6000}{\sqrt{3} \times 11} = 315$ A

$$F_a = (315/396) \times 102 = 81 \text{ A.}$$

$$\sqrt{3}XI = (315/396) \times 2.6$$

$$= 2.06 \text{ kV}$$

At zero p.f. leading (see Fig. 2(b)), E = V - XI = 11- 2.06

= 8.94 kV

From the o.c. curve, this requires a net excitation current

$$F_r = 145 \text{ A}$$

The armature reaction ampere-turns are completely magnetising, Fig. 4c, so:

$$F_f = F_r - F_a = 145 - 81 = 64 \text{ A}$$

(b) Full-load current at 10 MVA = (10/6) x 315 = 525 A

F_a = (525/396) x 102 = 135 A. $\sqrt{3}$XI = (525/396) x 2.6

$$= 3.45 \text{ kV}$$

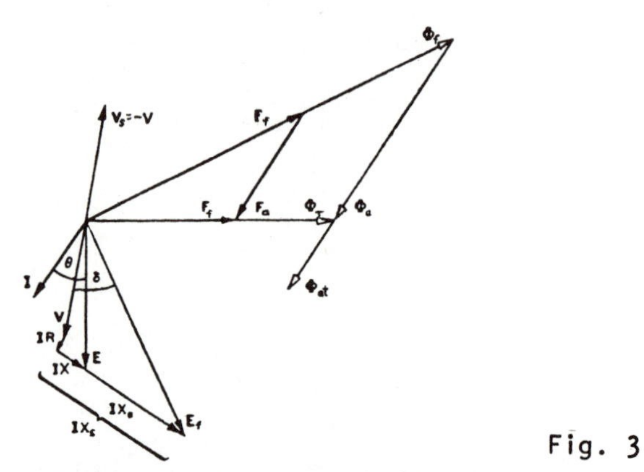

Fig. 3

Generator vector diagram, unsaturated machine.

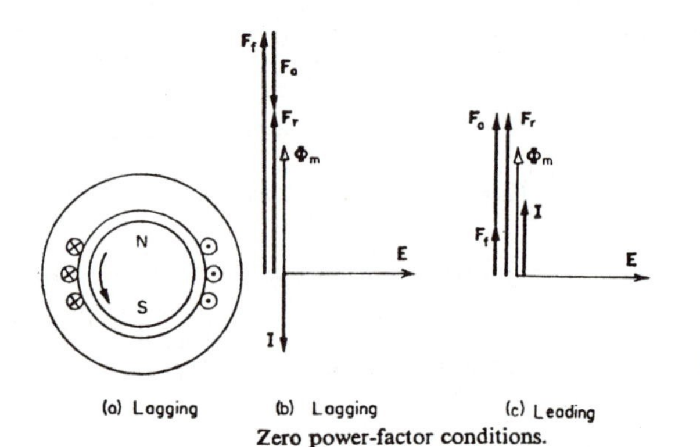

Fig. 4

Zero power-factor conditions.

From the generator diagram constructed in accordance with Fig. 3, E_g = 12.9 kV requiring a net F_r of 314 A and by construction, F_f = 415 A.

With the load removed, the terminal voltage on o.c. would rise to 13.9 kV with this field current maintained.

(c) The motor vector diagram uses the same values of F_a and IX as in (b) but with a relative reversal because the current is opposed to the e.m.f. E_m is 10 kV, F_r is 171 A and F_f = 202 A.

(d) In this case, the armature m.m.f. is completely de-magnetising (see Fig. 4b). Further, E = V + XI, see Fig. 2(a). See also Fig. 5.

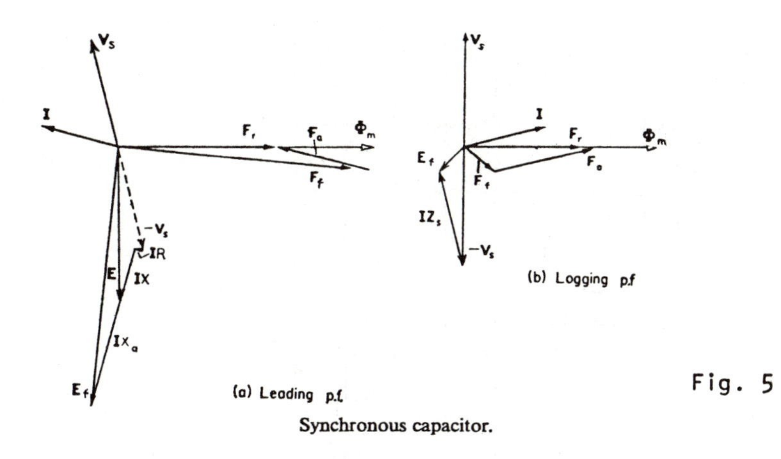

Synchronous capacitor.

Substituting the same values of F_a and $\sqrt{3}$ XI as in (a)

$$E = 11 + 2.06 = 13.06 \text{ kV}$$

From the o.c. curve this requires F_r = 328 A

$$\text{so } F_f = F_r + F_a = 328 + 81 = 409 \text{ A}$$

SYNCHRONOUS GENERATOR CHARACTERISTICS

• PROBLEM 8-8

A 2-pole generator running on no load and excited to normal busbar voltage is dead-short-circuited on all three phases at the instant that the axes of the field and phase A are in coincidence. Find the current in phase A and in the field winding. Per-unit reactances: x_{ad} = 1.50; x_{aq} = 0.60, x = 0.10, x_f = 0.13 p.u. Resistances: all negligible.

Solution: Here v_q = 1.0 p.u. and θ_0 = 0. The d-axis transient reactance is

$$x_d' = \frac{x_{ad}x_f}{x_{ad} + x_f} + x = \frac{(1.50)(0.13)}{1.50 + 0.13} + 0.10 = 0.22 \text{ p.u.}$$

and $x_{sq} = x_{aq} + x$ = 0.70 p.u. Using the solutions above,

$$i_a = 4.5 \cos \omega_1 t - 3.0 - 1.5 \cos 2\omega_1 t$$

$$i_f = 4.2(1 - \cos \omega_1 t) + 1.0 = 5.2 - 4.2 \cos \omega_1 t$$

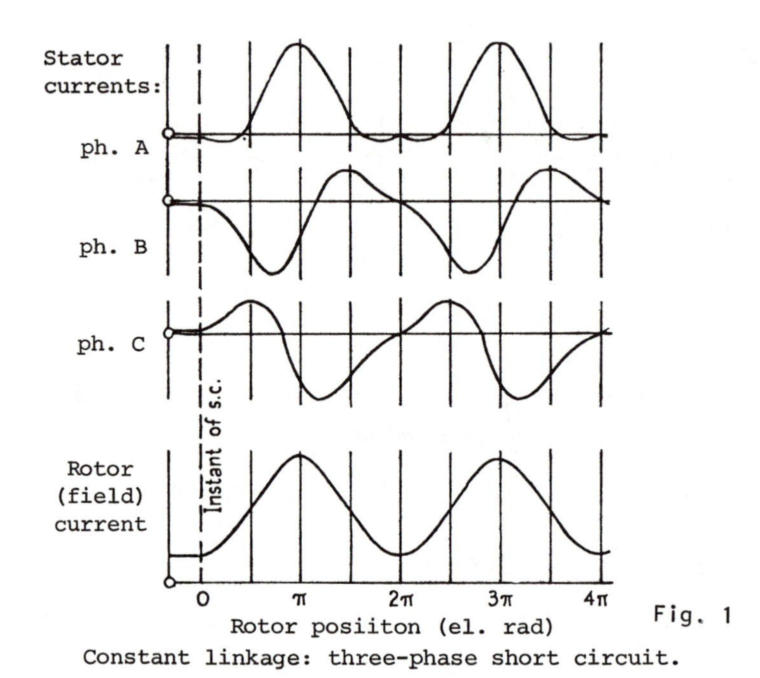

Fig. 1

Constant linkage: three-phase short circuit.

the unity being added to i_f for the initial value i_{f0} = 1.0 p.u. The values correspond to those plotted in Fig. 1.

• PROBLEM 8-9

A three-phase synchronous generator has the following per-unit constants:

Synchronous reactance $x_s = 0.4$

Effective armature resistance $r_e = 0.02$

Neglecting the effect of saturation on the synchronous reactance, calculate the per-unit and percent open-circuit phase voltage which corresponds to the field excitation that produces rated terminal voltage when the generator operates with a full kva inductive load of 0.8 power factor.

Solution: The per-unit open-circuit phase voltage is

$$\vec{E}'_a = 1.00(0.8 + j0.6) + 1.00(1 + j0)(0.02 + j0.4)$$

$$= 0.8 + j0.6 + 0.02 + j0.4$$

$$= 0.82 + j1.00 \text{ V.}$$

$$E'_a = \sqrt{(0.82)^2 + (1.00)^2} = 1.29 \text{ per unit}$$

The percent values are merely the per-unit values multiplied by 100. In the above example the no-load voltage is 129 percent.

• PROBLEM 8-10

The flux-distribution curve of a synchronous machine is represented at no-load by

$$B_x = 100 \sin \frac{x}{\tau}\pi - 14 \sin 3 \frac{x}{\tau}\pi - 20 \sin 5 \frac{x}{\tau}\pi + 1 \sin 7 \frac{x}{\tau}\pi.$$

Determine the emfs induced at no-load by the harmonics (as fractions of the fundamental).

Solution: Let $\Phi_1 = 100$. Then, since $\frac{B_3}{B_1} = 0.14$ and $\frac{\tau_3}{\tau_1} = \frac{1}{3}$, $\Phi_3 = \frac{1}{3}$, $\Phi_0 = \frac{1}{3} \times 14 = 4.67$.

Accordingly

$$\Phi_5 = \frac{1}{5} \times 20 = 4, \; \Phi_7 = \frac{1}{7} \times 1 = 0.143.$$

Further, $f_1 = 60$, $f_3 = 3 \times 60 = 180$, $f_5 = 5 \times 60 = 300$, $f_7 = 7 \times 60 = 420$. The distribution factors are given by the following equation:

$$k_{dv} = \frac{\sin [vq(\alpha_s/2)]}{q \sin [v(\alpha_s/2)]}, \; v = 1, 3, 5, 7$$

$$k_{d1} = +0.966, \; k_{d3} = +0.707, \; k_{d5} = +0.259,$$

$$k_{d7} = -0.259,$$

and the pitch factors are given by the following equation:

$$k_{pv} = \sin [v \frac{w}{\tau} \frac{\pi}{2}], \; v = 1, 3, 5, 7$$

$$k_{p1} = +0.966, \; k_{p3} = \sin\left(3 \times \frac{5}{6} \times 90\right) = -0.707,$$

$$k_{p5} = \sin\left(5 \times \frac{5}{6} \times 90\right) = +0.259,$$

$$k_{p7} = \sin\left(7 \times \frac{5}{6} \times 90\right) = +0.259.$$

The ratios of the emf's are

$$\frac{E_3}{E_1} = \frac{\Phi_3}{\Phi_1} \cdot \frac{f_3}{f_1} \cdot \frac{k_{dp_3}}{k_{dp_1}} = \frac{4.67}{100} \times 3 \times \frac{0.707 \times 0.707}{0.934}$$

$$= 0.075,$$

$$\frac{E_5}{E_1} = \frac{\Phi_5}{\Phi_1} \cdot \frac{f_5}{f_1} \cdot \frac{k_{dp_5}}{k_{dp_1}} = \frac{4}{100} \times 5 \times \frac{0.259 \times 0.259}{0.934}$$

$$= 0.01435,$$

$$\frac{E_7}{E_1} = \frac{0.143}{100} \times 7 \times \frac{0.259 \times 0.259}{0.934} = 0.00072.$$

• PROBLEM 8-11

The following data are for a three-phase 13,800-V wye-connected 60,000-kVA 60-Hz synchronous generator:

P = 2: stator slots = 36, stator coils = 36, turns in each stator coil = 2

Stator coil pitch = $\frac{2}{3}$, rotor slots = 28

Spacing between rotor slots = $\frac{1}{37}$ of circumference

Rotor coils = 14

Turns in each rotor coil = 15

I.D. of stator iron = 0.948 m,

O.D. of rotor iron = 0.865 m

The magnetising reactance x_{ad} = 3.03Ω/phase, k_w = 0.826

Net axial length of stator iron = 3.365 m, stator coil connection is two-circuit (a = 2) series rotor-coil connection.

The generator is delivering rated load at 0.80 power factor, current lagging. The leakage reactance is 0.12 times the magnetizing reactance, and the armature resistance is negligible for this problem.

Neglecting saturation calculate (a) the synchronous reactance in ohms per phase and in per unit; (b) E_{ag}, the voltage behind the leakage impedance; (c) E_{af}, the voltage due to the field current; (d) the resultant flux linking the armature; (e) the flux that links the armature due to the field current; and (f) the flux produced by the armature current.

Show a phasor diagram of the current, voltage, and flux phasors.

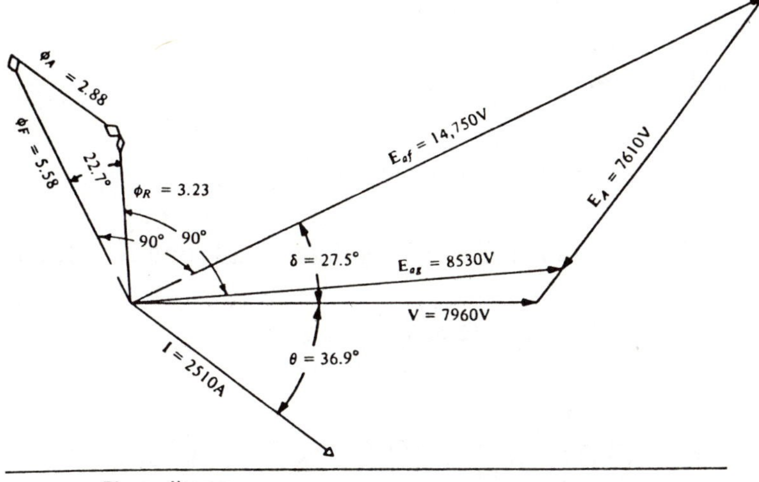

Phasor diagram

Solution:

a. $x_\ell = 0.12x_{ad} = 0.12 \times 3.03 = 0.36\Omega$ /phase

$$x_d = x_{ad} + x_\ell$$

$$= 3.03 + 0.36 = 3.39\ \Omega/\text{phase}$$

$$x_{d(\text{per unit})} = x_{d(\text{ohms})} \frac{VA_{base}}{(\text{volts}_{base})^2}$$

$$= \frac{3.39 \times 60 \times 10^6}{(13.8 \times 10^3)^2} = 1.07$$

b. $\vec{E}_{ag} = \vec{V} + (r_a + jx_\ell)\vec{I}$

For convenience let $\vec{V}$ lie on the axis of reals:

$$\vec{V} = \frac{13{,}800}{\sqrt{3}}(1 + j0) = 7960 + j0\ \text{V/phase}$$

The rated armature current is

$$I = \frac{VA}{\sqrt{3}\ V_{L-L}} = \frac{60 \times 10^6}{\sqrt{3} \times 13.8 \times 10^3} = 2510\ \text{A/phase}$$

and when expressed as a phasor,

$$\vec{I} = 2510(0.80 - j0.60) = 2010 - j1510$$

Since $r_a \cong 0$,

$$\vec{E}_{ag} = 7960 + j0.36(2010 - j1510)$$

$$= 7960 + 544 + j724 = 8504 + j724$$

$$= 8530 \underline{/4.86°}$$

c. $$\vec{E}_{af} = \vec{V} + (r_a + jx_d)\vec{I}$$

$$= 7960 + j3.39(2010 - j1510)$$

$$= 7960 + 5120 + j6820 = 13{,}080 + j6820$$

$$= 14{,}750 \underline{/27.5°}$$

d. $$\phi_R = \frac{aE_{ag}}{4.44fk_wN_{ph}} \quad \text{(where } k_w = k_pk_b\text{)}$$

$$= \frac{2 \times 8530}{4.44 \times 60 \times 0.826 \times 24} = 3.23 \text{ Wb/pole}$$

ϕ_R leads $\vec{E}_{ag}$ by 90°; hence,

$$\phi_R = 3.23 \underline{/90° + 4.8°} = 3.23\underline{/94.8°}$$

e. $$\phi_F = \frac{E_{af}}{E_{ag}} \phi_R = \frac{14{,}750}{8{,}530} \times 3.23 = 5.58 \text{ Wb/pole}$$

$$\vec{\phi}_F = 5.58 \underline{/90° + 27.5°} = 5.58 \underline{/117.5°}$$

f. $$E_A = X_{ad}I = 3.03 \times 2510 = 7610 \text{ V/phase}$$

$$\phi_A = \frac{E_A}{E_{ag}}\phi_R = \frac{7610}{8530} \times 3.23 = 2.88 \text{ Wb/pole}$$

Also

$$\vec{I} = 2510(0.80 - j0.60) = 2510 \underline{/-36.9°}$$

and since the armature flux is in phase with armature current,

$$\phi_A = 2.89 \underline{/-36.9°}$$

The phasor diagram is shown in the figure.

• PROBLEM 8-12

A 3-phase, 2300-V, 100-kVA, 60-Hz, 6-pole synchronous generator has a stator leakage reactance $X_{\ell s}$ of 7.9 Ω and a magnetizing reactance X_{ms} of 56.5 Ω. The winding resistances are negligible. Immediately after the machine was synchronized onto the bus, the field current was 23 A and the power input at the prime-mover coupling was 3.75 kW.

a) Determine the field current required when the generator is supplying rated output at 0.9 lagging power factor (that is, the load on the generator is inductive).

b) Determine the internal torque developed by the machine when it is delivering an output current of 15 A, and the field current is 20 A.

c) Determine the power input at the shaft required to drive the machine out of synchronism at the 20-A field current.

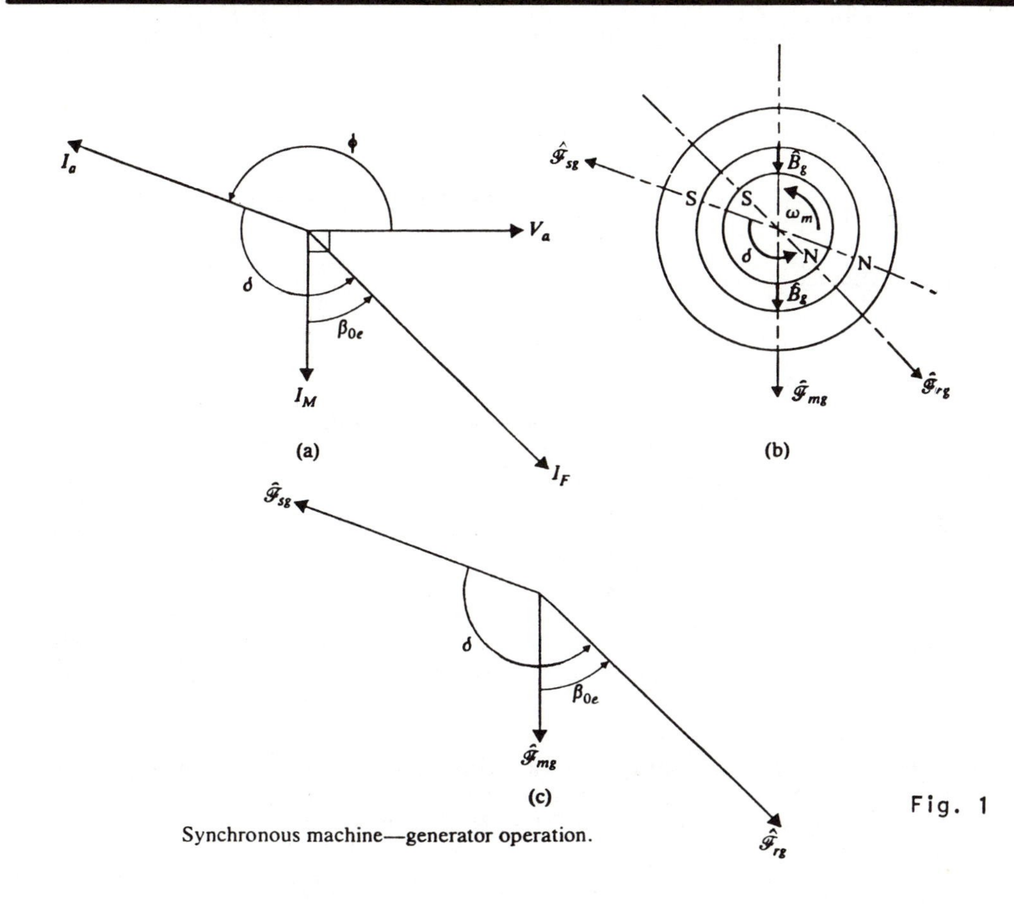

Fig. 1

Synchronous machine—generator operation.

Solution: The synchronous reactance of the machine is

$$X_s = 7.9 + 56.5 = 64.4\ \Omega.$$

On on load,

$$V_a = \frac{2300}{\sqrt{3}} = 1328 \text{ V.}$$

and

$$I_F = I_M = \frac{V_a}{X_s} = \frac{1328}{64.4} = 20.62 \text{ A.}$$

Thus

$$n'' = \frac{I_F}{i_f} = \frac{20.62}{23} = 0.897$$

a) For the condition of operation described, the phasor diagram will have the form shown in Fig. 1, where

$$\vec{V}_a = 1328\underline{|0°} \text{ V}$$

$$\phi = 180° - \cos^{-1}0.9 = 154.2°$$

$$I_a = \frac{100 \times 10^3}{3 \times 1328} = 25.10 \text{ A}$$

$$\vec{I}_M = \frac{\vec{V}_a}{jX_s} = \frac{1328\underline{|0°}}{j64.4} = 20.62\underline{|-90°} \text{ A.}$$

$$\vec{I}_F = \vec{I}_M - \vec{I}_a = 20.62\underline{|-90°} - 25.10\underline{|154.2°} = 41.91\underline{|-49.8°} \text{ A}$$

$$i_f = \frac{41.91}{0.897} = 46.7 \text{ A.}$$

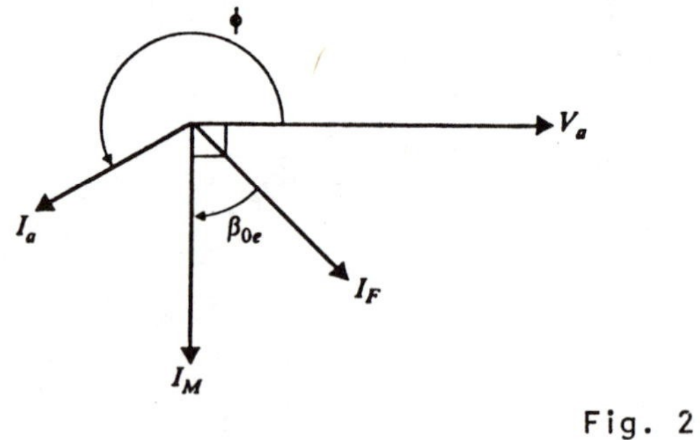

Fig. 2

Diagram for the problem.

b) Since the field current is less than that on synchronization, the phasor diagram corresponding to this operating condition has the form shown in Fig. 2. As in (a),

$$\vec{V}_a = 1328\underline{|0°} \text{ V}$$

$$\vec{I}_M = 20.62\underline{|-90°} \text{ A}$$

$$I_F = 0.897 \times 20 = 17.94 \text{ A}$$

$$I_a = 15 \text{ A}$$

From Fig. 2

$$\cos \beta_{0e} = \frac{I_F^2 + I_M^2 - I_a^2}{2 I_F I_M} = \frac{17.94^2 + 20.62^2 - 15^2}{2 \times 17.94 \times 20.62} = 0.706$$

$$\beta_{0e} = 45.12°$$

$$T = -\frac{p}{2} \times \frac{3}{\omega_s} \times I_F V_a \sin \beta_{0e}$$

$$= -\frac{6}{2} \times \frac{3}{120\pi} \times 17.94 \times 1328 \sin 45.12° = 403 \text{ N.m}$$

c) $$T_{max} = -\frac{6}{2} \times \frac{3}{120\pi} \times 17.94 \times 1328 = 569 \text{ N.m}$$

Air-gap power is

$$P = \frac{2}{p} \omega_s T = \frac{2}{6} \times 120\pi \times 569 = 71.50 \text{ kW.}$$

Rotational loss

$$P_{rot} = 3.75 \text{ kW}$$

The required input at the shaft is therefore

$$P_{shaft} = 71.50 + 3.75 = 75.25 \text{ kW.}$$

• PROBLEM 8-13

A 100-MVA, 16-kV, three-phase, 60-Hz synchronous alternator is delivering rated voltage, open circuited, when a balanced, three-phase short occurs at its terminals. The machine has the following constants, the reactances expressed in per unit on a 100-MVA base:

$$x_d = 1.0 \qquad x'_d = 0.3 \qquad x''_d = 0.2$$

$$T''_d = 0.03 \text{ s} \qquad T'_d = 1.00 \text{ s}$$

Neglecting d-c and double-frequency components of the current,

(a) Find the initial current.

(b) Find the current at the end of two cycles, and at the end of 10 s.

Solution: (a) Initial current = I" = $\frac{E_{f0}}{x_d''}$

In per unit, since E_{f0} = rated value,

$$I''_{pu} = \frac{1}{0.2} = 5 \text{ per unit}$$

$$I_B = \frac{100 \times 10^6}{\sqrt{3} \times 16{,}000} = 3610 \text{ A}$$

Then I" = 18,050 A.

(b) $$I' = \frac{E_{f0}}{x_d'} = \frac{1}{0.3} = 3.33 \text{ per unit}$$

$$I_{ss} = \frac{E_{f0}}{x_d} = 1.0 \text{ per unit}$$

$$I(t) = (I'' - I')e^{-t/T_d''} + (I' - I_{ss})e^{-t/T_d'} + I_{ss}$$

$$= 1.67e^{-t/0.03} + 2.33e^{-t} + 1 \text{ per unit}$$

At $t = \frac{2}{60}$ s:

$$I(0.0333) = 167e^{-1.11} + 2.33e^{-0.0333} + 1$$

$$= 0.55 + 2.25 + 1 = 3.80 \text{ per unit}$$

or 13,720 A

At t = 10 seconds

$$I(t) = 1.67e^{-333} + 2.33e^{-10} + 1$$

$$= 1.0001 \text{ per unit}$$

$$= 3610 \text{ A}$$

• PROBLEM 8-14

Consider a two-pole three-phase Y-connected turbo-generator rated at 2,500 kva, 6,600 volts between terminals, at 3,000 rpm. The frequency is f = pn/120 = 50 cycles. The armature winding is distributed in 60 slots, with four conductors per slot, and its effective resistance and reactance per phase are 0.073 and 0.87 ohm, respectively. The open-circuit characteristic, shown in Fig. 1 is plotted in terms of ampere-turns per pole, with the ordinates representing volts per phase, i.e., from line to neutral. The rotor, illustrated in Fig. 2, has a winding distributed in 10 slots per pole.

Compute the field excitation necessary to develop rated

terminal voltage and rated current when the power factor of the load is (a) 80 per cent, lagging; (b) 100 per cent; (c) zero; and in each case determine the voltage regulation.

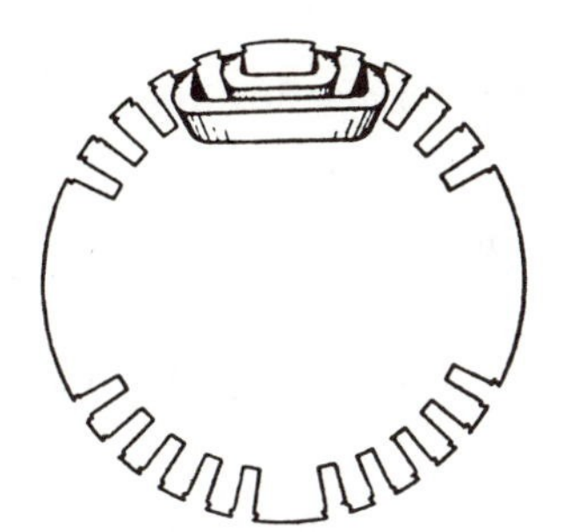

Fig. 2

Bipolar radial-slot rotor.

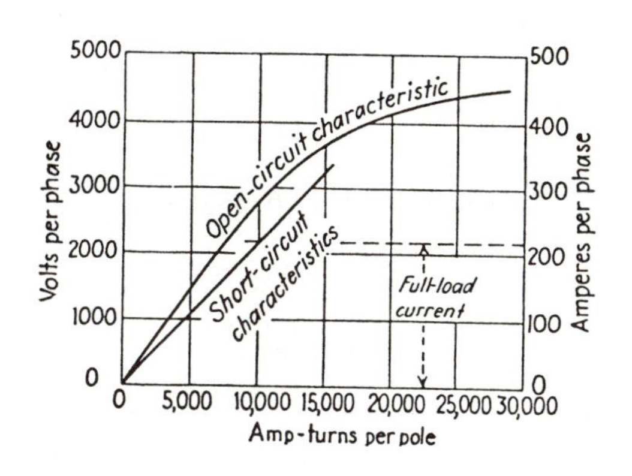

Open and short-circuit characteristics.

Fig. 1

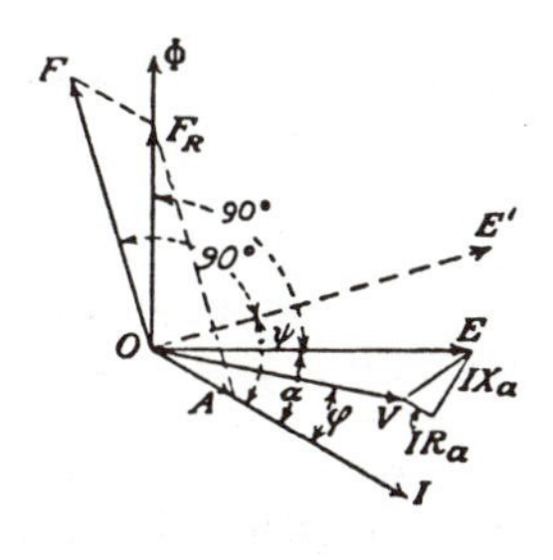

Fig. 3

General (Potier) phasor diagram of alternator with nonsalient poles.

Solution: (a) Load Power Factor = $\cos \phi = 0.8$. From the given data, the rated terminal voltage per phase is

$$V = \frac{6,600}{\sqrt{3}} = 3,810 \text{ V.}$$

and the rated current per phase is

$$I = \frac{2,500,000}{3 \times 3,810} = 219 \text{ amp.}$$

The ohmic and reactive drops per phase are

$$IR_a = 219 \times 0.073 = 16 \text{ volts.}$$

$$IX_a = 219 \times 0.87 = 190 \text{ volts.}$$

For a power factor of 80 per cent ($\cos \phi = 0.8$, $\sin \phi$

= 0.6), the complex expression for V, referred to the current I as axis of reference, is

$$\vec{V} = 3{,}810\ (\cos\phi + j\sin\phi) = 3{,}048 + j2{,}286$$

and the impedance drop is

$$\vec{I}(R_a + jX_a) = (I + j0)(R_a + jX_a) = 16 + j190$$

The induced emf is therefore

$$\vec{E} = \vec{V} + \vec{I}(R_a + jX_a) = (3{,}048 + 16) + j(2{,}286 + 190)$$

$$= 3{,}064 + j2{,}476$$

and

$$E = \sqrt{3{,}064^2 + 2{,}476^2} = 3{,}940 \text{ volts}$$

Referring to the open-circuit characteristic, Fig. 1, it is seen that this voltage corresponds to an excitation of 17,500 amp-turns per pole, which when added geometrically to -A gives the field excitation F. To compute A, it is to be noted, that the entire winding comprises 60 × 4 = 240 conductors, or 120 turns, so that N_p = turns per pole per phase = 120/(3 × 2) = 20. Since q = number of slots per pole per phase = 60/(2 × 3) = 10 and γ = angle between slots = 360/60 = 6°,

$$k_{b_1} = \frac{\sin(q\gamma/2)}{q\sin(\gamma/2)} = \frac{\sin 30°}{10\sin 3°} = 0.955$$

and therefore,

$$A = \frac{2\sqrt{2}}{\pi} k_{b1} k_{p1}\, m\, N_p\, I$$

$$= \frac{2\sqrt{2}}{\pi} \times 3 \times 0.955 \times 20 \times 219 = 11{,}300 \text{ amp-turns per pole}$$

From Fig. 3, $\vec{A}$ is in phase with $\vec{I}$, so that

$$\vec{A} = 11{,}300 + j0$$

The resultant excitation $\vec{F}_R$ is 90° ahead of $\vec{E}$, and $\vec{E}$ is ahead of $\vec{I}$ by an angle α such that

$$\cos\alpha = \frac{3{,}064}{3{,}940} = 0.778$$

$$\sin \alpha = \frac{2,476}{3,940} = 0.628$$

Therefore

$$\vec{F}_R = F_R e^{j(\alpha+90)} = F_R(-\sin\alpha + j\cos\alpha)$$

$$= 17,500\ (-0.628 + j0.778)$$

$$= -10,990 + j13,615$$

Hence,

$$\vec{F} = \vec{F}_R - \vec{A} = -22,290 + j13,615$$

$$F = \sqrt{22,290^2 + 13,615^2} = 26,120$$

The open-circuit voltage corresponding to this excitation, determined from Fig. 1, is 4,450 volts; hence, the regulation is

$$\frac{4,450 - 3,810}{3,810} \times 100 = 16.8\%$$

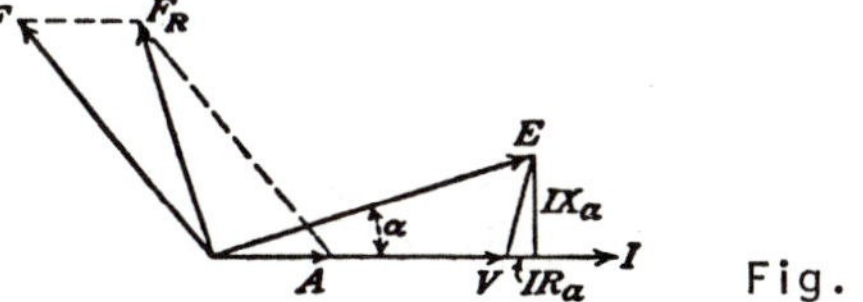

Fig. 4

General phasor diagram, unity power factor.

b. Load Power Factor = cos ϕ = 1.0. The general phasor diagram, Fig. 3, takes the form of Fig. 4 when $\phi = 0$. Taking the current $\vec{I}$ as the axis of reference,

$$\vec{I} = 219 + j0\ \text{A.}$$

$$\vec{V} = 3,810 + j0\ \text{V}$$

and $\vec{E} = \vec{V} + \vec{I}(R_a + jX_a) = 3,810 + 16 + j190$

$$= 3,826 + j190$$

whence $E = \sqrt{3,826^2 + 190^2} = 3,830$

This emf leads the current by an angle α such that

$$\cos \alpha = \frac{3,826}{3,830} = 0.9989$$

$$\sin \alpha = \frac{190}{3,830} = 0.0496$$

and corresponding to E is the resultant excitation F_R = 16,500 amp-turns per pole, from Fig. 1. Since $\vec{F}_R$ leads $\vec{E}$ by 90°, it leads the current by 90 + α; hence,

$$\vec{F}_R = F_R[\cos(90 + \alpha) + j \sin(90 + \alpha)]$$

$$= F_R(-\sin \alpha + j \cos \alpha)$$

$$= 16{,}500 \ (-0.0496 + j0.9989)$$

$$= -818 + j16{,}482$$

The field excitation is therefore

$$\vec{F} = \vec{F}_R - \vec{A} = -12{,}118 + j16{,}482$$

$$F = \sqrt{12{,}118^2 + 16{,}482^2} = 20{,}457$$

and corresponding open-circuit voltage is 4,150. The regulation is therefore

$$\frac{4{,}150 - 3{,}810}{3{,}810} \times 100 = 8.9\%$$

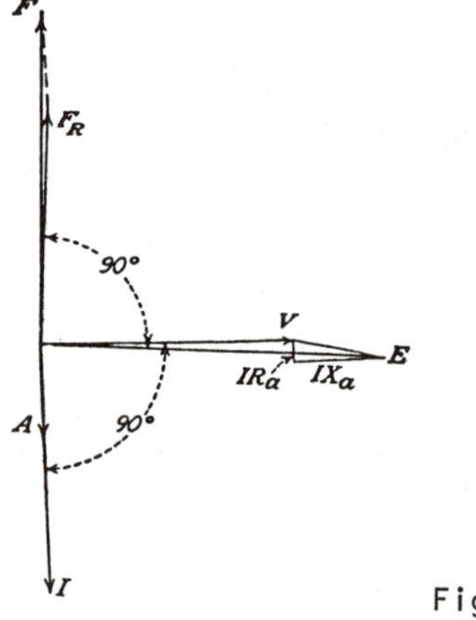

Fig. 5

General phasor diagram, zero power factor.

c. Load Power Factor = cos ϕ = 0. The phasor diagram for this case, Fig. 5 shows that it is sufficiently accurate to write

$$E = V + IX_a = 3{,}810 + 190 = 4{,}000 \text{ V.}$$

whence the value of F_R from Fig. 1 is 18,000 amp-turns per pole, from which

$$F = 18{,}000 + 11{,}300 = 29{,}300 \text{ amp-turns per pole}$$

Corresponding open-circuit voltage is 4,500 volts. The regulation is then

$$\frac{4,500 - 3,810}{3,810} \times 100 = 18.1\%$$

REGULATION AND FIELD CURRENT

• **PROBLEM** 8-15

Calculate the voltage regulation by the unsaturated synchronous-impedance method for operation of the machine of Fig. 1 as a generator with a 0.80 lagging power-factor load.

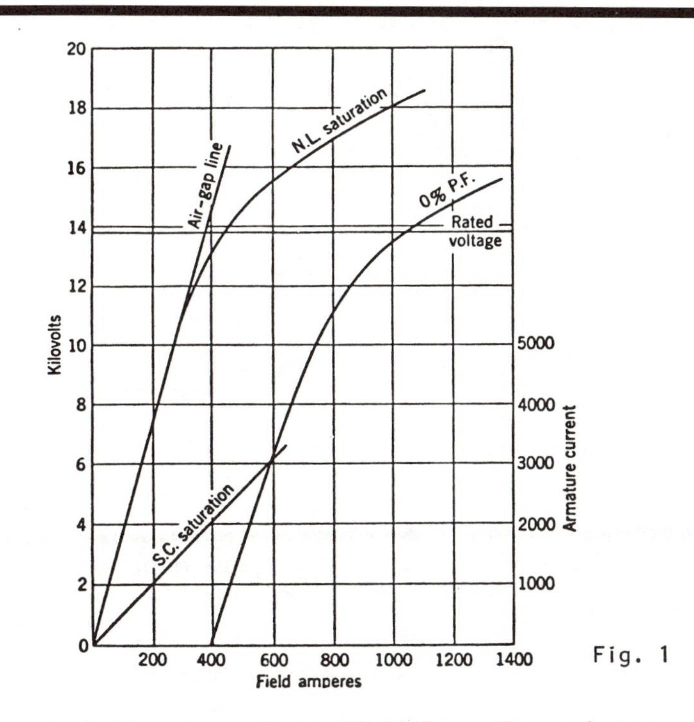

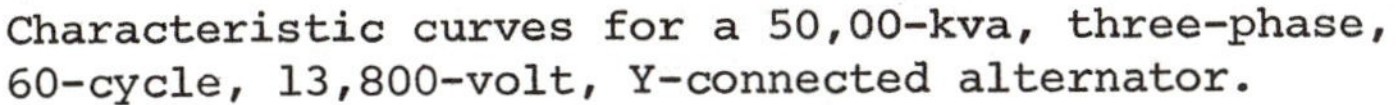
Characteristic curves for a 50,00-kva, three-phase, 60-cycle, 13,800-volt, Y-connected alternator.

Solution:

$$I_{a\ rated} = \frac{50,000}{\sqrt{3} \times 13,800} = 2090 \text{ amperes}$$

From the short-circuit curve of Fig. 1, a field current of 405 amperes is required to produce rated armature current of 2090 amperes.

From the no-load saturation curve, a field current of 405 amperes will produce a no-load terminal voltage of

13,100 volts.

Therefore,

$$Z_{\text{syn unsaturated}} = \frac{13,100}{\sqrt{3} \times 2090} = 3.62 \text{ ohms}$$

Neglect armature resistance since it is very small compared to the synchronous reactance.

Then,

$$X_{\text{syn unsaturated}} = 3.62 \text{ ohms}$$

and $I_a Z_{syn}$ = 7570 volts and is leading the armature current by 90 degrees.

$$\vec{E}_{NL} = \frac{13,800}{\sqrt{3}}(0.80 + j0.6) + j7570$$

$$= 6382 + j4786 + j7570$$

$$= 6382 + j12,356$$

$$E_{NL} = 13,900 \text{ volts}$$

$$\text{Percent regulation} = \frac{E_{NL} - E_{FL}}{E_{FL}} \times 100$$

$$= \frac{13,900 - 7977}{7977} \times 100 = 74\%.$$

• PROBLEM 8-16

A three-phase alternator with the open-circuit and short-circuit characteristics, shown in Fig. 1, has the following rating :

164 kva	Y-connected armature
3600 rpm	2200 volts
2 poles	60 cycles

(a) Calculate synchronous reactance by the Synchronous Impedance method. (b) Calculate the regulation at unity pf and at 0.80 pf.

Solution (a) The full-load current per terminal

$$I_a = \frac{164,000}{\sqrt{3} \times 2200} = 43 \text{ amperes}$$

Rated volts to neutral:

$$V_n = \frac{2200}{\sqrt{3}} = 1270 \text{ volts}$$

The maximum voltage on the open-circuit curve is 3200. At this voltage the field current is 31 amperes, and this same excitation produces a short-circuit current of 70 amperes. These values will be used to calculate X_s.

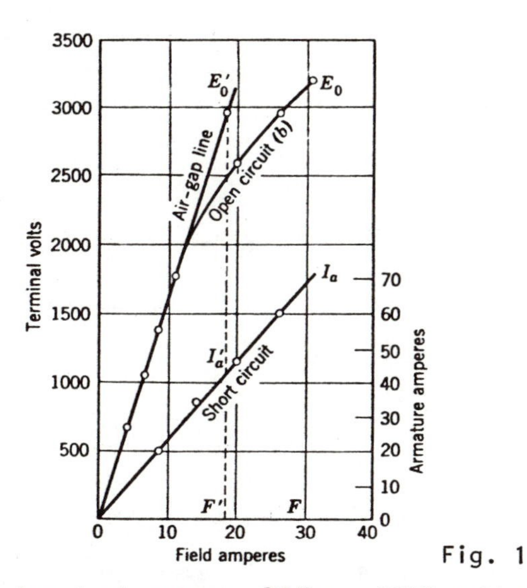

Fig. 1

Alternator test curves. 164kva, 2200 volts, 3600 rpm, 60 cycles, three-phase, Y connected.

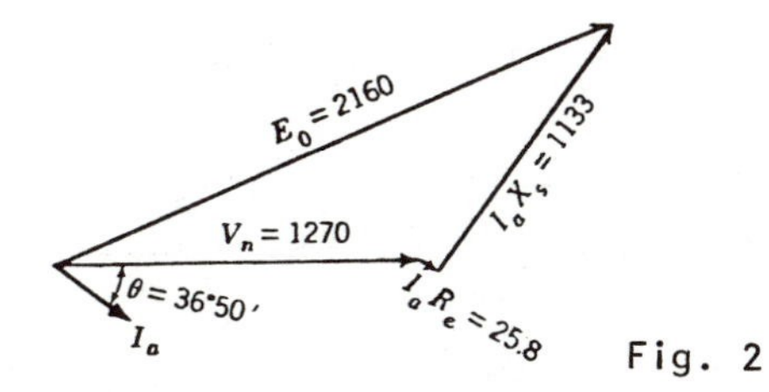

Fig. 2

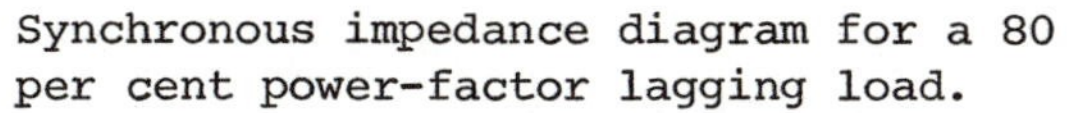
Synchronous impedance diagram for a 80 per cent power-factor lagging load.

$$3200/\sqrt{3} = 1846 \text{ volts per phase}$$

$$Z_s = 1846/70 = 26.37 \text{ ohms per phase}$$

The effective resistance per phase is 0.60 ohm at 75 C. This will be neglected in determining X_s.

Hence,

$$X_s = \sqrt{Z_s^2 - R_e^2}$$

$$X_s \approx 26.37 \text{ ohms per phase.}$$

(b) To calculate the regulation at unity pf, lay off as reference vector the phase value of the terminal voltage V_n. Solve for E_0.

$$\vec{E}_0 = V_n + I_a R_e + jI_a X_s \quad \text{(See Fig. 2)}$$

$$= 1270 + 43 \times 0.60 + j43 \times 26.37$$

$$E_0 = \sqrt{1295.8^2 + 1133^2} = 1721 \text{ volts per phase}$$

$$\text{Regulation} = \frac{E_0 - V_n}{V_n} 100$$

$$= \frac{1721 - 1270}{1270} 100$$

$$= 35.5 \text{ percent at unity pf}$$

The regulation at 0.80 pf, lagging, is calculated below. Refer to Fig. 2 for the vector diagram. The $I_a R_e$ and $I_a X_s$ drops are rotated with the current. The no-load voltage then becomes

$$\vec{E}_0 = V_n + I_a(R_e + jX_s)(\cos\theta - j\sin\theta) \text{ volts per phase}$$

The operator $(\cos\theta - j\sin\theta)$ rotates the current vector clockwise through θ degrees. Of course the voltage-drop triangle rotates similarly. Then,

$$\vec{E}_0 = 1270 + 43(0.60 + j26.37)(0.8 - j0.6)$$

$$= 1270 + 699 + j892.5$$

$$E_0 = 2160 \text{ volts per phase}$$

$$\text{Regulation} = \frac{2160 - 1270}{1270} \times 100$$

$$= 70.1 \text{ percent of 0.80 pf, lagging.}$$

Complex quantities can be avoided in the solution of these problems by the use of trigonometry or accurate diagrams from which the values can be scaled.

• PROBLEM 8-17

The no-load saturation curve for a 10-kva, 1,200 rpm, 3-phase, 230 volts, 25 amperes, 60 cycles synchronous generator is shown in Fig. 1. A short-circuit test with full-load current performed on the generator gave the value of the field current I_f, as 1.05 amp. The measured resistance per phase was 0.20 ohm. Calculate the regulation, using the adjusted synchronous impedance method, at full-load, 0.8 power-factor lagging.

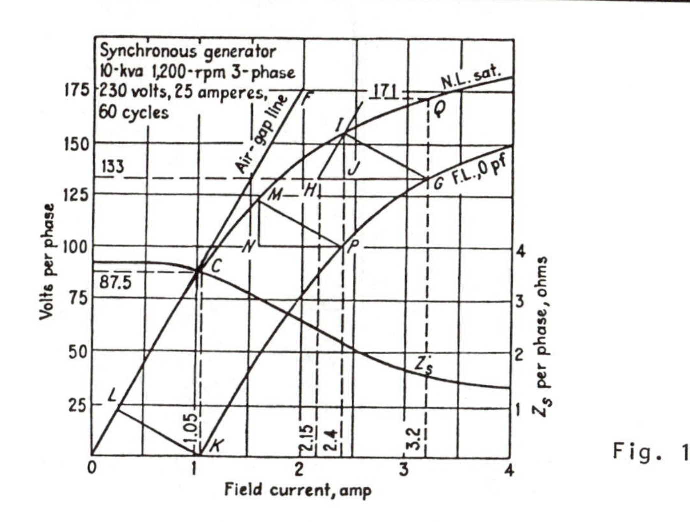

Fig. 1

Full-load zero-power-factor saturation and adjusted synchronous impedance.

Solution: In addition to the no-load saturation curve, a full-load zero-power-factor saturation curve is needed. The point K on the full-load zero-power-factor saturation curve is obtained using the value of Z_f found from the short-circuit test. The point G on curve KPG was determined experimentally by connecting the generator to a 3-phase line with wattmeters and ammeters installed so that field can be adjusted to give full-load armature current with zero wattmeter reading, giving the following values:

E_T = 133 volts (230 between terminals)

I = 25 amp W = 0 I_f = 3.2 amp

Other points on the curve are computed using points K and G.

Curve KPG, as shown in Fig. 1, is substantially the same shape and parallel to the no-load saturation curve OMQ. The first part of the no-load curve is practically a straight line. If this part is extended to form line OF, it is called the air-gap line, since it indicates the field required for the gap only and is not affected by saturation. Point H is determined by making HG = OK. Through H, line HI is drawn parallel to OF, intersecting curve OMQ at point I. Now if OL is made equal to HI, triangles KOL and GHI will be equal. Line IG shows the distance and direction between corresponding points on the two curves. Line IJ is drawn perpendicular to HG; then starting with any point such as M on the full-load curve, measure down MN = IJ and to the right NP = JG to obtain the corresponding point P on the full-load curve.

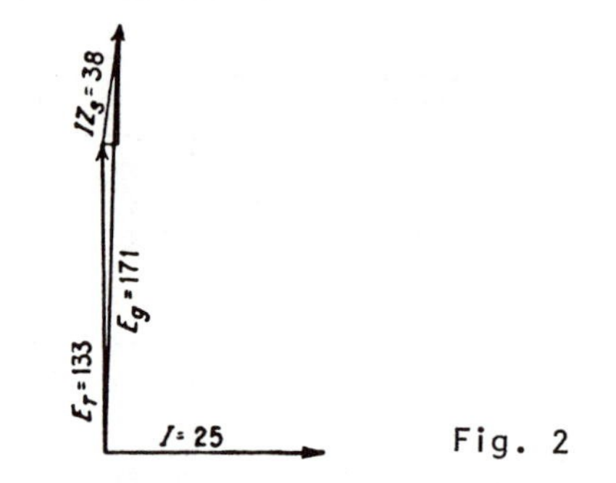

Vector diagram for full-load zero power factor.

At full load, zero power factor lagging, as shown in Fig. 2, E_g and E_T are practically in phase, and the arithmetic difference is IZ_s. For $I_f = 3.2$ the values are $E_g = 171$ at point Q and $E_T = 133$ at point G; therefore

$$IZ_s = 171 - 133 = 38 \text{ volts}$$

and

$$Z_s = 38/25 = 1.52 \text{ ohms}$$

Similarly for other values of field current, the values of Z_s are computed, and this curve is also plotted in Fig. 1. These values of Z_s adjusted for excitation give fairly good results in computations.

The value of excitation at rated voltage for a particular power-factor load is not known, and so it is necessary to pick a point which seems about right. To explain the method, the no-load and full-load curves of Fig. 1 have been redrawn in Fig. 3. For a power factor of 0.8 lagging, assume a full-load value of excitation I_f = 2.5 amp, for which $E_g = 158$ at point U. At point X on the full-load curve, the reading is 106.5 volts; hence

$IZ_s = 158 - 106.5 = 51.5$ volts

The value of IR is

$IR = 25 \times 0.2 = 5$ volts

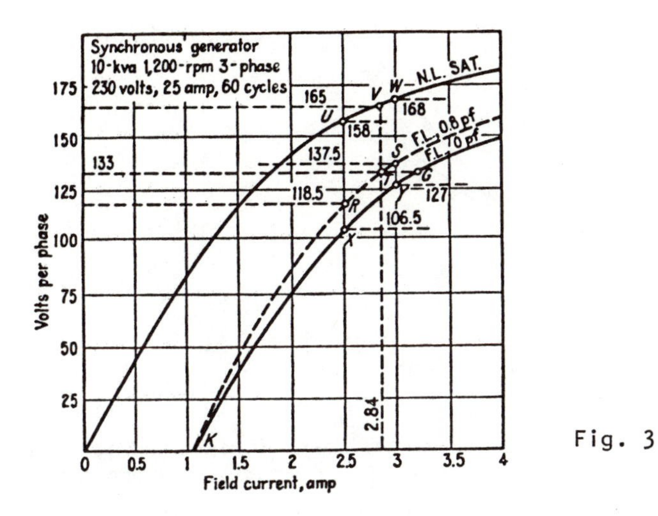

Fig. 3

Construction of full-load 0.8-power-factor saturation curve.

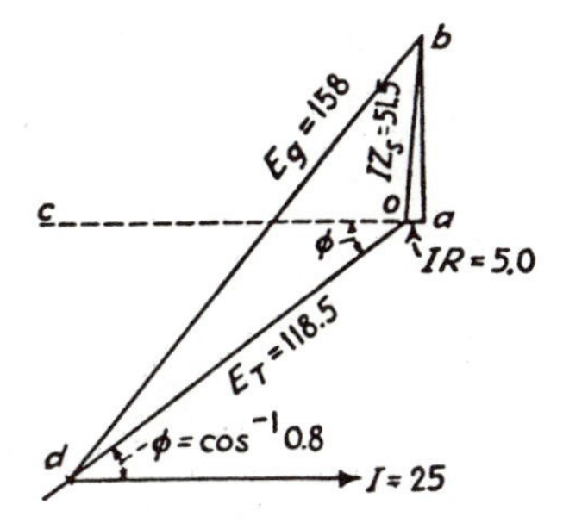

Fig. 4

Computation of terminal voltage for $I_f = 2.5$ amp.

These values are plotted in Fig. 4. From point O the IR is drawn horizontal (Oa = 5). Line ab is drawn vertical, and the IZ_s is laid off, making Ob = 51.5. From O line Od is drawn at an angle of $\phi = \cos^{-1} 0.8$ with the horizontal. From b a radius bd is laid off with length $E_g = 158$ volts, cutting line Od at point d. The terminal voltage then scales 118.5 volts, which is too low, and so a higher value of field current must be used. However, this point is first spotted on Fig. 3 at R.

For the next trial, take $I_f = 3.0$, which gives E_g =

168 at point W and

$$IX_s = 168 - 127 = 41$$

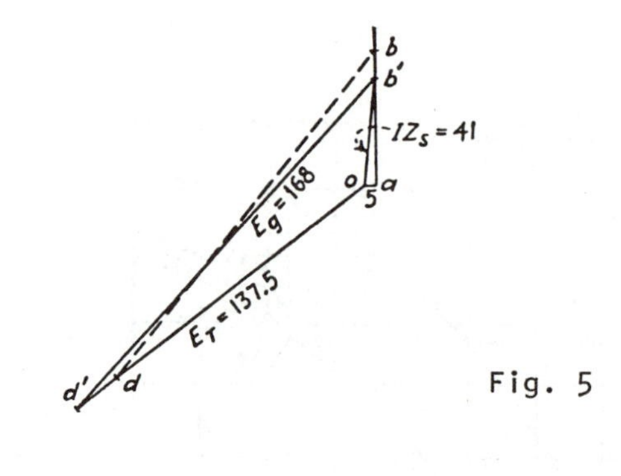

Fig. 5

Computation of terminal voltage for I_f = 3.0 amp.

These values are plotted in Fig. 5 using the same procedure, and E_T comes out 137.5 volts. This value is too high, but it is spotted on Fig. 3 at S. To save time in this graphical computation, the two computations are combined in one diagram. The dotted lines of Fig. 5 show the computation of Fig. 4. Now three points R, S, and K are known on the full-load 0.8-power factor curve. This curve can be roughed in, and point T at rated voltage is obtained, giving I_f = 2.84 amp and E_g = 165 volts. The regulation then is

$$\text{Reg} = \frac{165 - 133}{133} = \frac{32}{133} = 24\%.$$

• PROBLEM 8-18

Consider a 93,750-kva three-phase synchronous generator, whose constants and characteristic curves are given on Fig. 1. An inductive, 93,750-kva load at 0.8 power factor at rated frequency and voltage is used. Calculate the field current and regulation using saturated synchronous reactance.

Solution: Referring to Fig. 1, the distances EG and GF on the figure correspond to 1,750 volts and 426 field amperes. Since the generator is Y-connected and the zero-power-factor curve is for an armature current of 4,100 amp. per terminal, the Potier reactance of this generator is

$$x_a = \frac{1750}{\sqrt{3} \times 4,100} = 0.248 \text{ ohm per phase.}$$

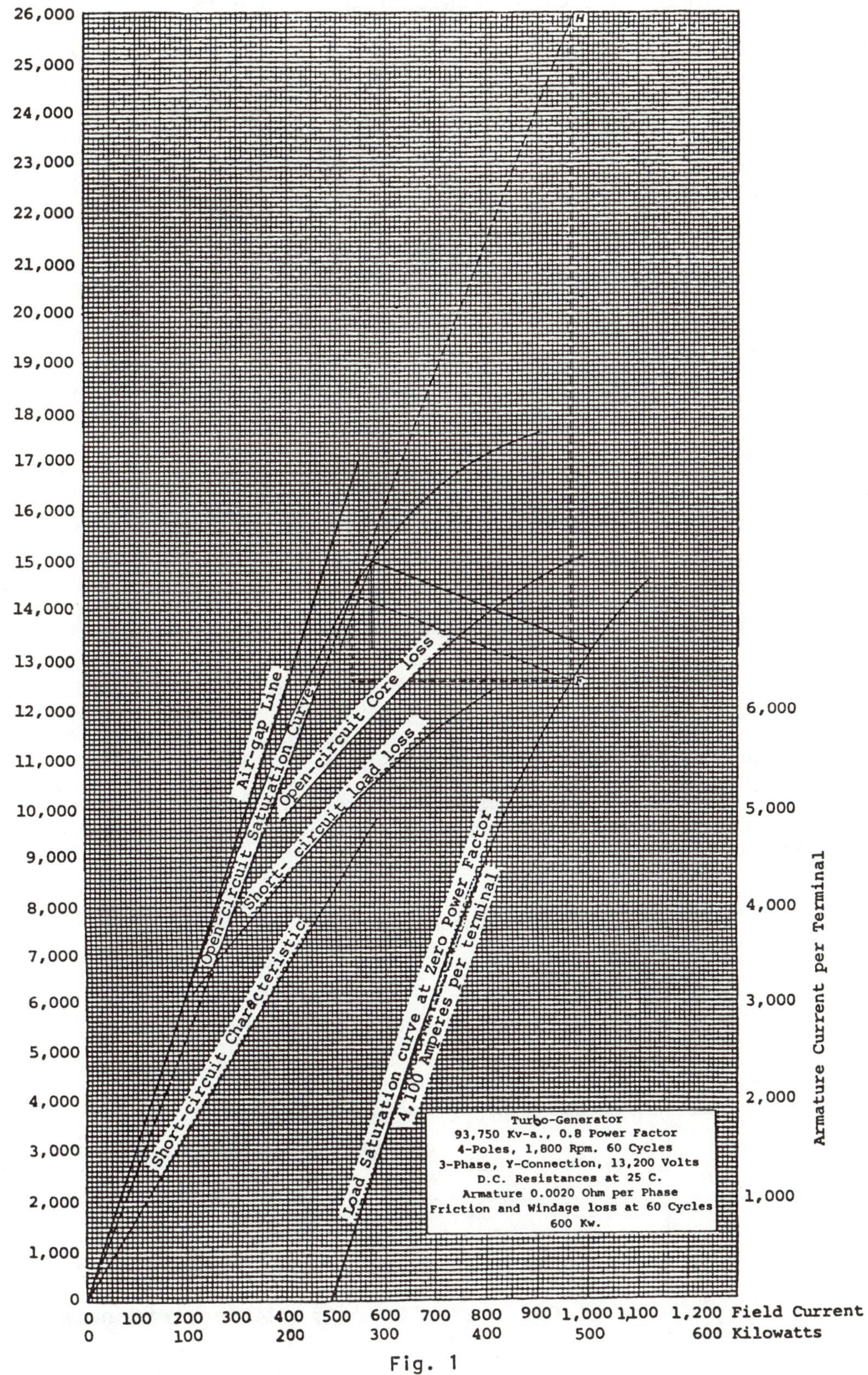

26,000
25,000
24,000
23,000
22,000
21,000
20,000
19,000
18,000
17,000
16,000
15,000
14,000
13,000
12,000
11,000
10,000
9,000
8,000
7,000
6,000
5,000
4,000
3,000
2,000
1,000
0
Volts between Terminals
6,000
5,000
4,000
3,000
2,000
1,000
Armature Current per Terminal
H
Air-gap Line
Open-circuit Saturation Curve
Open-circuit Core loss
Short-circuit load loss
Short-circuit Characteristic
Load Saturation curve at Zero Power Factor
4,100 Amperes per terminal
Turbo-Generator
93,750 Kv-a., 0.8 Power Factor
4-Poles, 1,800 Rpm. 60 Cycles
3-Phase, Y-Connection, 13,200 Volts
D.C. Resistances at 25 C.
Armature 0.0020 Ohm per Phase
Friction and Windage loss at 60 Cycles
600 Kw.
0 100 200 300 400 500 600 700 800 900 1,000 1,100 1,200 Field Current
0 100 200 300 400 500 600 Kilowatts

Fig. 1

The armature effective resistance at 75°C is

$$r_{oh}(\text{at } 25°\text{ C})\ \frac{1 + 0.0042 \times 75}{1 + 0.0042 \times 25}$$

$$+ r_{oh}\ (\text{at } 25°\text{ C})\ (1.82 - 1)$$

$$= 0.0020 \times 1.19 + 0.0020 \times 0.82$$

$$= 0.00402 \text{ ohm per phase}$$

The air-gap voltage for the given load is

$$\vec{E}_a = V + I(r_e + jx_a)$$

$$= \frac{13,200}{\sqrt{3}}\ (1 + j0) + 4,100(0.8 - j0.6)(0.00402 + j0.248)$$

$$= 7,620\ (1 + j0) + (623 + j804)$$

$$= 8,243 + j804 \text{ V.}$$

$$E_a = \sqrt{(8,243)^2 + (804)^2} = 8,280 \text{ volts per phase}$$

$$\sqrt{3}E_a = \sqrt{3} \times 8,280 = 14,340 \text{ volts between terminals}$$

In Fig. 1 draw a straight line through O and the point corresponding to 14,340 volts on the open-circuit saturation curve. This line represents the saturation at which the generator operates for the given load. Construct a new Potier triangle, shown dotted on Fig. 1, with its upper angle at the voltage 14,340 volts. The saturated synchronous reactance at which the generator operates is

$$x_s(\text{saturated}) = \frac{13,270}{\sqrt{3} \times 4,000} = 1.867 \text{ ohms per phase}$$

The excitation voltage is

$$\vec{E}'_a = 7,620\ (1 + j0) + 4,100(0.8 - j0.6)(0.00402 + j1.867)$$

$$= 7,620\ (1 + j0) + (4,605 + j6,110)$$

$$= 12,225 + j6,110$$

$$E'_a = \sqrt{(12,225)^2 + (6,110)^2}$$

$$= 13,650 \text{ volts per phase}$$

$$\sqrt{3}E'_a = \sqrt{3} \times 13,650 = 23,620 \text{ volts between terminals}$$

The load field current is the field current corresponding to 23,620 volts on the saturation line OH, Fig. 1. This current is 890 amp.

The open-circuit voltage for this field current, from the open-circuit saturation curve, is 17,500 volts between terminals. The regulation is

$$\frac{17{,}500 - 13{,}200}{13{,}200} \times 100 = 32.6 \text{ percent}$$

• PROBLEM 8-19

Consider a three-phase cylindrical-rotor Y-connected 60-cycle 13,200-volt four-pole 1,800-rpm synchronous generator, rated at 93,750 kva at 0.8 power factor.

The armature has a two-circuit winding with 72 slots and 2 inductors per slot. The winding pitch is 2/3.

The field has a spiral winding with two circuits in parallel. There are 10 slots per pole and 25 conductors per slot. The distribution of the field coils is given below.

Coil 1 (inner coil) has a spread of 60 electrical degrees

Coil 2 has a spread of 86 2/3 electrical degrees

Coil 3 has a spread of 113 1/3 electrical degrees

Coil 4 has a spread of 140 electrical degrees

Coil 5 (outer coil) has a spread of 166 2/3 electrical degrees

The developed field winding is shown in Fig. 1.

The armature resistance at 25° C is 0.00200 ohm and the armature leakage reactance is 0.248 ohm, both per phase. The ratio of the a-c to the d-c armature resistance at 25° C and 60 cycles is 1.82.

The resistance of the field winding between terminals is 0.156 at 25° C.

The complete characteristic curves of the generator at rated frequency are given in Fig. 2.

Calculate the field current and regulation when the generator carries a full kva inductive load at 0.8 power-factor.

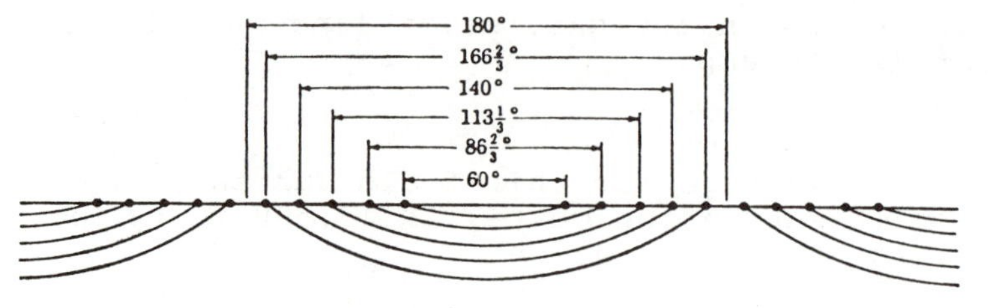

Fig. 1

Solution: The full-load phase current is $\frac{93,750 \times 1,000}{\sqrt{3} \times 13,200}$ = 4,100 amp and the phase voltage is $13,200/\sqrt{3}$ = 7,620 volts. The armature reaction for a balanced load at rated current is found from the equation

$$A = 0.90\ NI\ k_b\ k_p$$

The current in this equation is the current per turn which in this case is one-half the phase current, since the armature has a two-circuit winding.

$$N = \frac{72 \times 2}{2 \times 4} = 18 \text{ turns per pole (in all phases)}$$

$$k_p = \sin\frac{\rho}{2} = \sin\frac{120°}{2} = 0.866$$

$$k_b = \frac{\sin\ (n\alpha/2)}{n\ \sin\ (\alpha/2)}$$

$$n = \frac{72}{4 \times 3} = 6 \text{ slots per phase per pole}$$

$$\alpha = \frac{360° \times 2}{72} = 10 \text{ deg between slots}$$

$$k_b = \frac{\sin\frac{6 \times 10°}{2}}{6\ \sin\frac{10°}{2}} = 0.956$$

$$A = 0.90 \times 18 \times 0.956 \times 0.866 \times \frac{4,100}{2}$$

$$= 27,500 \text{ amp-turns per pole}$$

A is the magnitude of the vector which represents the fundamental of the armature reaction. It is shown on the vector diagram given in Fig. 3. All harmonics are neglected.

The amplitude of the vector which represents the fundamental of the field ampere-turns per ampere of field current is given by $4/\pi\ N_f\ I_f\ k_b\ k_p$.

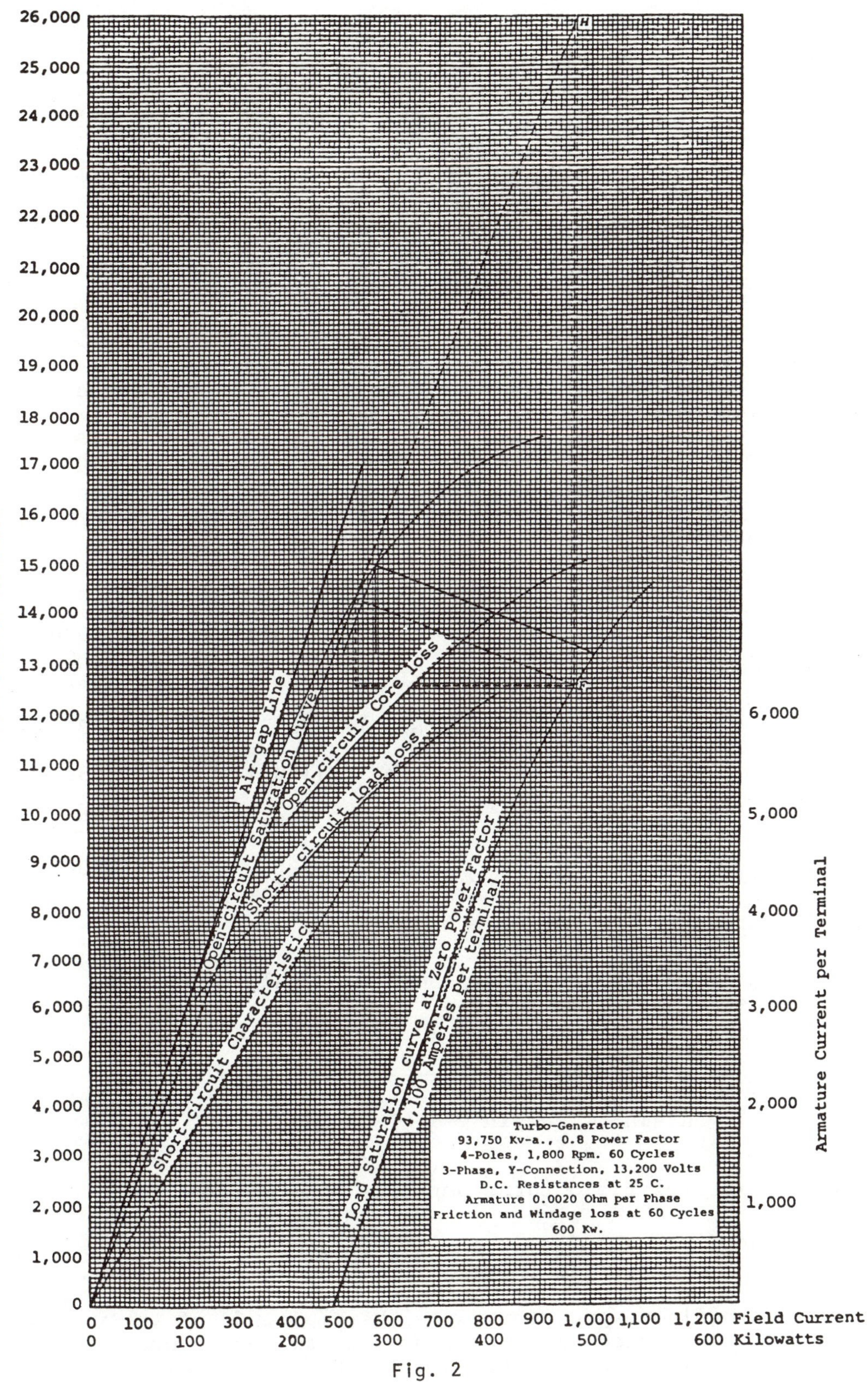
Volts between Terminals
26,000
25,000
24,000
23,000
22,000
21,000
20,000
19,000
18,000
17,000
16,000
15,000
14,000
13,000
12,000
11,000
10,000
9,000
8,000
7,000
6,000
5,000
4,000
3,000
2,000
1,000
0
H
Air-gap Line
Open-circuit Saturation Curve
Open-circuit Core loss
Short-circuit load loss
Short-circuit Characteristic
Load Saturation curve at Zero Power Factor
4,100 Amperes per terminal
Armature Current per Terminal
6,000
5,000
4,000
3,000
2,000
1,000
Turbo-Generator
93,750 Kv-a., 0.8 Power Factor
4-Poles, 1,800 Rpm. 60 Cycles
3-Phase, Y-Connection, 13,200 Volts
D.C. Resistances at 25 C.
Armature 0.0020 Ohm per Phase
Friction and Windage loss at 60 Cycles
600 Kw.
0 100 200 300 400 500 600 700 800 900 1,000 1,100 1,200 Field Current
0 100 200 300 400 500 600 Kilowatts

Fig. 2

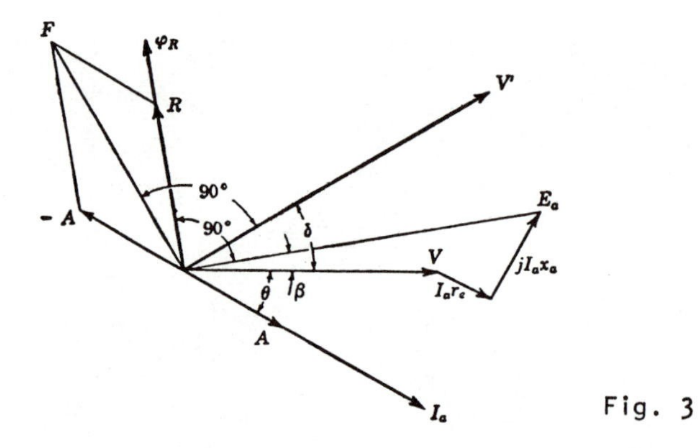

Fig. 3

Since the field has a spiral winding, the axes of all field coils for any given pole coincide. In this case the breadth factors are unity. Since all coils carry equal currents, the average pitch factor for the coils may be used for k_p. The same result is obtained if the actual field winding is assumed to be replaced by a full-pitch distributed lap winding with 10 slots in a belt of conductors.

Coil spreads	Pitch factors	
60°	sin 1/2 (60°)	= 0.5000
86 2/3°	sin 1/2 (86 2/3°)	= 0.6862
113 1/3°	sin 1/2 (113 1/3°)	= 0.8354
140°	sin 1/2(140°)	= 0.9397
166 2/3°	sin 1/2 (166 2/3°)	= 0.9933
		3.9546

$$\text{Average } k_p = \frac{3.9546}{5} = 0.791$$

If the field winding is replaced by a full-pitch distributed winding, $k_p = 1$ and

$$k_b = \frac{\sin \frac{n\alpha}{2}}{n \sin\frac{\alpha}{2}} = \frac{\sin 10 \frac{13\ 1/3°}{2}}{10 \sin \frac{13\ 1/3°}{2}} = \frac{0.9182}{10 \times 0.1161} = 0.791$$

$$\text{The number of field turns per pole} = \frac{10 \times 25}{2} = 125$$

$$\frac{4}{\pi} Nk_pI_f = \frac{4}{\pi} Nk_bI_f = \frac{4}{\pi} 125 \times 0.791\ I_f = 125.8I_f \qquad (1)$$

The foregoing expression gives the amplitude of the vector which can be used on the vector diagram to represent the fundamental of the impressed field ampere-turns. I_f in the equation is the current per turn and for the generator considered is one-half the field current per terminal, since the field winding has two circuits in parallel.

When the effective resistance of the armature of a generator is given at one temperature and is used at another temperature, it is usually assumed that the part of the effective resistance which is due to local losses, skin effect, etc., is not changed by a change in temperature. This is not strictly true but is nearly enough true so far as the resistance drop in the armature is concerned, since the percent resistance drop in the armature of a large generator is small. Assume an operating temperature of 75°C. The armature effective resistance at 75° C is

$$r_{oh}(\text{at } 25^\circ \text{ C}) \frac{1 + 0.0042 \times 75}{1 + 0.0042 \times 25}$$

$$+ r_{oh}(\text{at } 25\%\text{C})(1.82 - 1)$$

$$= 0.0020 \times 1.19 + 0.0020 \times 0.82$$

$$= 0.00402 \text{ ohm per phase}$$

Take V on Fig. 3 as the reference axis.

$$\vec{E}_a = V(1 + j0) + I(\cos\theta - j\sin\theta)(r_e + jx_a)$$

$$= 7{,}620\ (1 + j0) + 4{,}100(0.8 - j0.6)(0.00402 + j0.248)$$

$$= (7{,}620 + j0) + (623 + j804)$$

$$= 8{,}243 + j804 \text{ V}$$

$$E_a = \sqrt{(8{,}243)^2 + (804)^2} = 8{,}280 \text{ volts per phase}$$

$$\sqrt{3}E_a = \sqrt{3} \times 8{,}280 = 14{,}340 \text{ volts between terminals}$$

R is the air-gap field expressed in terms of the equivalent field current and is the current corresponding to the voltage $\sqrt{3}\ E_a = 14{,}340$ on the open-circuit saturation curve (see Fig. 2).

$$R \text{ for } 14{,}340 \text{ volts} = 535 \text{ amp}$$

$$\vec{F} = A\ (-\cos\theta + j\sin\theta) + R(-\sin\beta + j\cos\beta)$$

$$\sin\beta = \frac{840}{8{,}280} = 0.0972 \qquad \cos\beta = \frac{8{,}243}{8{,}280} = 0.996$$

Since R is expressed in terms of the field current, A must be in terms of the equivalent field current. This equivalent current is twice the equivalent current found by dividing A in ampere-turns per pole by the field ampere-turns per pole per ampere of field current.

These field ampere-turns are 125.8 as seen from Eq. (1).

$$A = 2 \frac{27,500}{125.8} = 437 \text{ equivalent field amperes}$$

$$\vec{F} = 437(-0.8 + j0.6) + 535(-0.0972 + j0.996)$$

$$= (-349.6 + j262.2) + (-52.0 + j532)$$

$$= -402 + j794$$

$$F = \sqrt{(402)^2 + (794)^2} = 890 \text{ amp}$$

A field current of 890 amp is required to maintain rated voltage at rated frequency when the generator carries an inductive full kva load at 0.8 power factor.

The voltage found from the open-circuit saturation curve (Fig. 2) corresponding to a field current of 890 amp

is 17500 volts and is the terminal voltage the generator would have at no load if the field current and frequency were kept constant at their full-load values when the load is removed. This voltage is the voltage V' on the vector diagram given in Fig. 3.

The regulation of the generator for the given load is therefore

$$\text{Regulation} = \frac{17,500 - 13,200}{13,200} \times 100 = 32.6 \text{ percent.}$$

• PROBLEM 8-20

The full-load 0.8 power-factor saturation curve for a 10-kva, 1,200 rpm, 3-phase, 230 volts, 25 amperes, 60 cycles synchronous generator, is shown in Fig. 1.

The synchronous impedance curve is shown in Fig. 2. The measured resistance per phase is 0.20 ohm.

(a) Calculate the per-unit values of the resistance per phase R and the synchronous impedance Z_s.

(b) Calculate the regulation using per unit values.

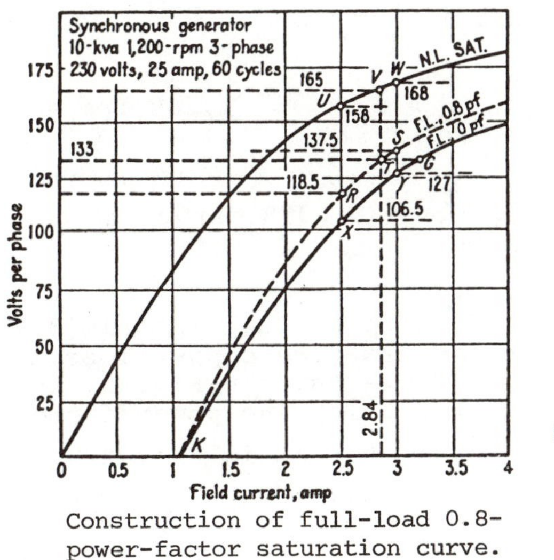

Fig. 1

Construction of full-load 0.8-power-factor saturation curve.

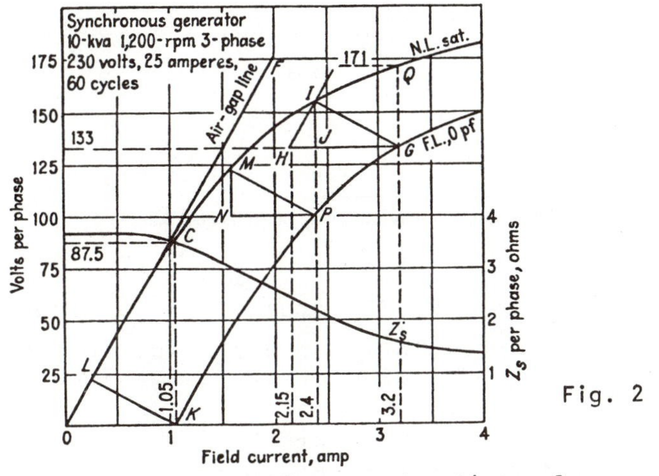

Fig. 2

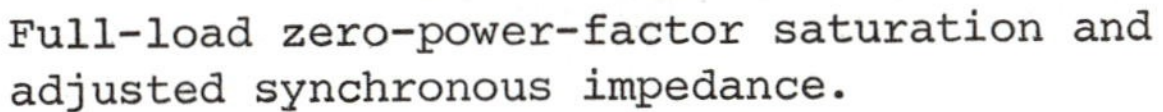
Full-load zero-power-factor saturation and adjusted synchronous impedance.

Solution: The per-unit values on this machine are as follows.

Voltage (per phase)	133 volts	= 1 per unit
Current	25 amp	= 1 per unit
Impedance	133/25	= 5.32 ohms
		= 1 per unit
Kw or kva (per phase) . . .	3.33	= 1 per unit

Kw or kva (total) . . . . . 10 = 1 per unit

Speed . . . . 1,200 rpm = 1 per unit

(a) $R = 0.2 \text{ ohm} = \frac{0.2}{5.32} = 0.0375$ per unit

This shows that the IR drop at full load is 3.75 percent, also that the full-load armature copper loss is 3.75 percent of 10 kw, or 375 watts. In Fig. 1, the value of I_f for full load at 0.8 lagging power factor was 2.84 amp. From the curve for Z_s in Fig. 2,

$$Z_s = 1.68 \text{ ohms} = \frac{1.68}{5.32} = 0.316 \text{ per unit}$$

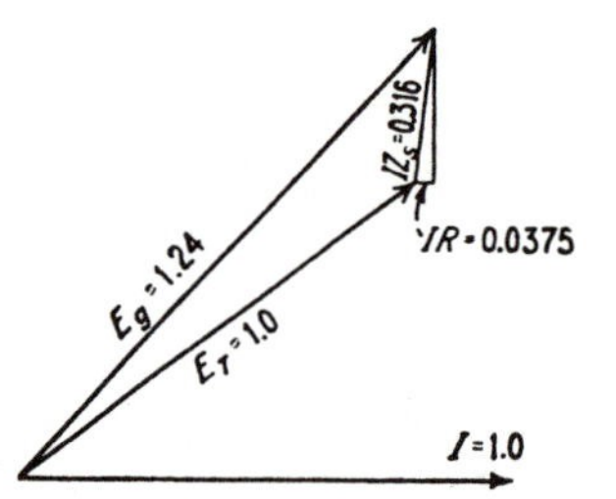

Generator-regulation diagram using per-unit notation.

Fig. 3

(b) From part (a), the IZ_s drop is 31.6 percent of rated voltage. A regulation diagram for this case is shown in Fig. 3. The value of E_g is obtained as

$$E_g = 1.24 \text{ per unit} = 165 \text{ volts.}$$

and the regulation is

$$\text{Reg} = \frac{1.24 - 1.0}{1.0} = 0.24.$$

• PROBLEM 8-21

The no-load saturation curve and the full-load zero-power-factor saturation curve for a 10-kva, 1,200 rpm, 3-phase, 230 volts, 25 amperes, 60 cycles synchronous generator are shown in Fig. 1. The measured resistance per phase is 0.20 ohm. Calculate the regulation, using the AIEE method, at full-load, 0.8 power-factor lagging.

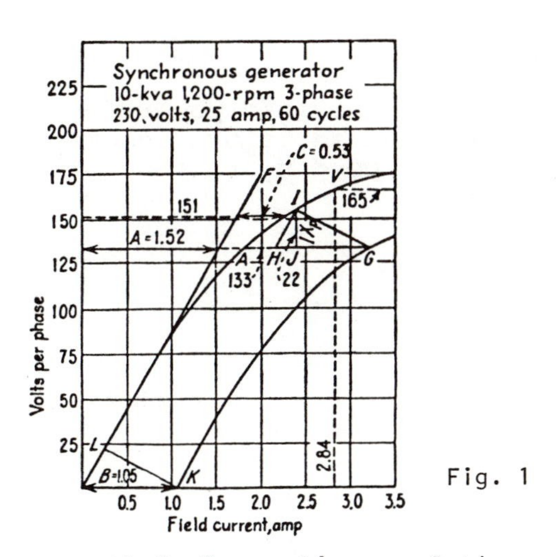

Fig. 1

AIEE method of computing regulation.

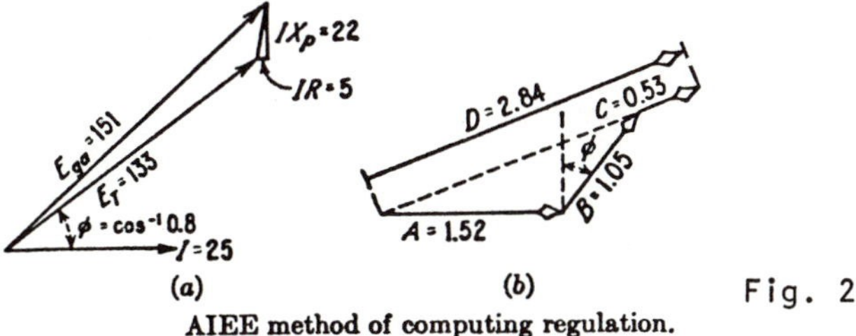

Fig. 2

AIEE method of computing regulation.

Solution: From Fig. 1,

$$IX_p = \text{distance IJ} = 22 \text{ volts}$$

The terminal voltage E_T = 133 volts, leads the current I = 25 amperes, by an angle $\phi = \cos^{-1} 0.8$.

The resistance drop in phase with I is

$$IR = 5 \text{ volts.}$$

These voltages are drawn in Fig. 2(a), giving

$$E_{ga} = 151 \text{ volts}$$

at full load, 80 per cent lagging power factor. The method of finding the field current necessary to obtain this internal voltage under load is shown in Fig. 2(b). The air-gap mmf corresponding to rated voltage

$$A = 1.52 \text{ amp}$$

is added vectorially to the value of field required for full-load current on short circuit,

$$B = 1.05 \text{ amp.}$$

The angle ϕ shown is the load power-factor angle. The additional field required to overcome saturation in the core with flux corresponding to the actual generated voltage E_{ga} is

$$C = 0.53 \text{ amp}$$

which is added arithmetically to the vector sum of A and B, as shown, to give the total field current required,

$$D = 2.84 \text{ amp.}$$

Entering this value of I_f on the no-load magnetization curve of Fig. 1 at point V,

$$E_g = 165 \text{ volts}$$

and so the regulation is

$$\text{Reg} = \frac{165 - 133}{133} = \frac{32}{133} = 24\%$$

• PROBLEM 8-22

Consider a 93,750-kva synchronous generator whose constants and characteristic curves are given in Fig. 1. Calculate the field current and regulation by the A.S.A. method for an inductive 0.8 power-factor load with rated armature current and rated terminal voltage.

Solution: The field currents for rated voltage (13,200 volts) and rated short-circuit armature current (4,100 amp) from the air-gap line and the short-circuit characteristic are 427 and 490 amp. Refer to Fig. 2.

$$\vec{I}'_f = I_y(1 + j0) + I_x(\sin\theta + j\cos\theta)$$

$$= 427(1 + j0) + 490(0.6 + j0.8)$$

$$= 427 + 294 + j392 = 721 + j392$$

$$I'_f = \sqrt{(721)^2 + (392)^2} = 821 \text{ amp}$$

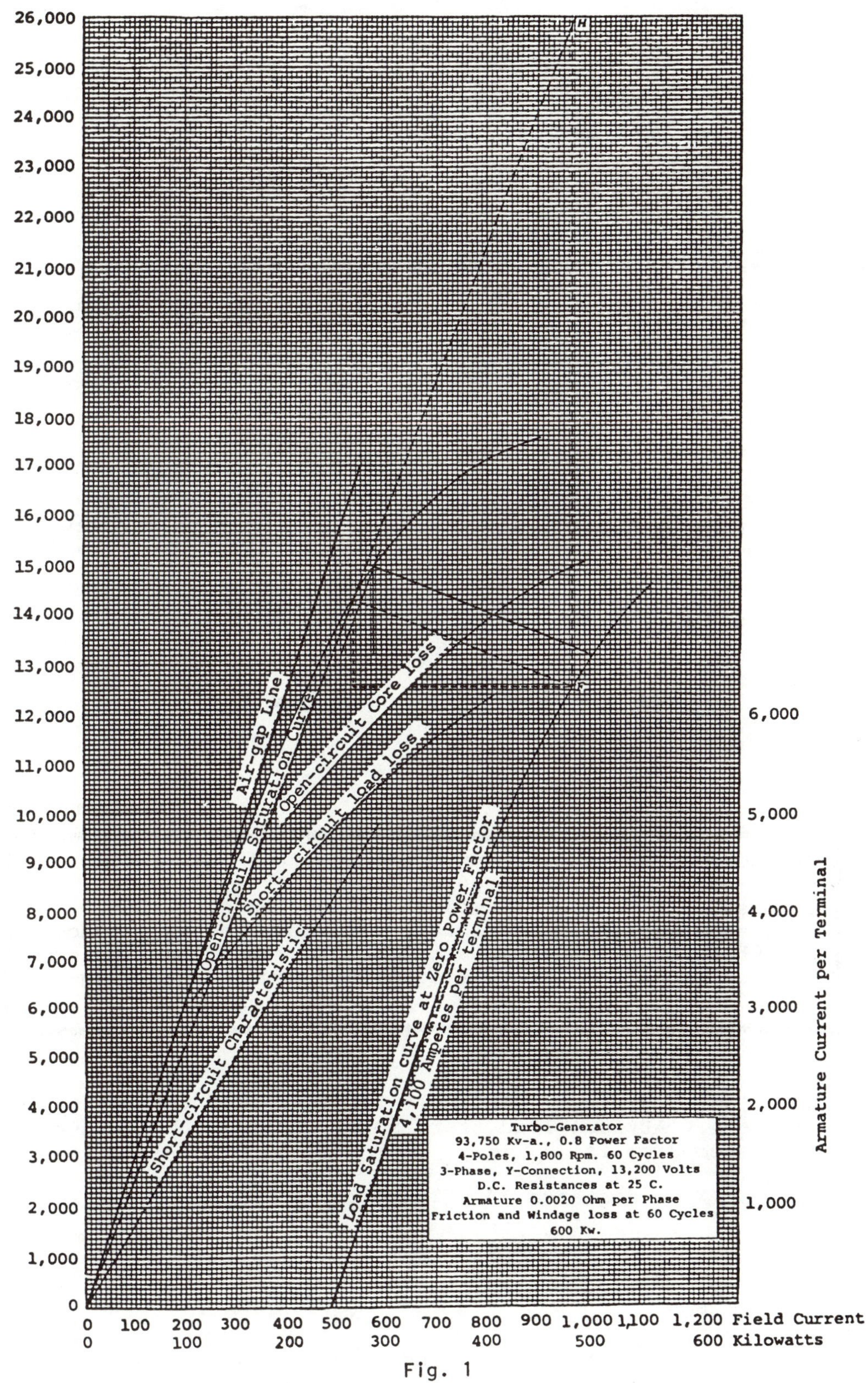

Volts between Terminals
26,000
25,000
24,000
23,000
22,000
21,000
20,000
19,000
18,000
17,000
16,000
15,000
14,000
13,000
12,000
11,000
10,000
9,000
8,000
7,000
6,000
5,000
4,000
3,000
2,000
1,000
0
Armature Current per Terminal
6,000
5,000
4,000
3,000
2,000
1,000
Air-gap Line
Open-circuit Saturation Curve
Open-circuit Core loss
Short-circuit load loss
Short-circuit Characteristic
Load Saturation curve at Zero Power Factor
4,100 Amperes per terminal
H
Turbo-Generator
93,750 Kv-a., 0.8 Power Factor
4-Poles, 1,800 Rpm. 60 Cycles
3-Phase, Y-Connection, 13,200 Volts
D.C. Resistances at 25 C.
Armature 0.0020 Ohm per Phase
Friction and Windage loss at 60 Cycles
600 Kw.
0 100 200 300 400 500 600 700 800 900 1,000 1,100 1,200 Field Current
0 100 200 300 400 500 600 Kilowatts

Fig. 1

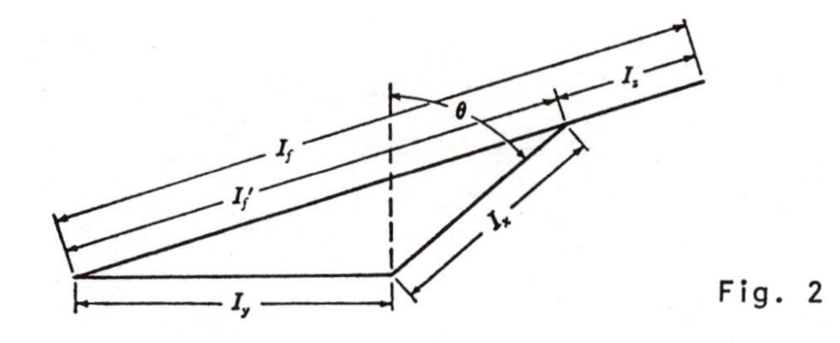

Fig. 2

I_s from Fig. 1 for an air-gap voltage E_a = 14,340 volts between terminals, which is the air-gap voltage found for the given load is

$$I_s = 535 - 465 = 70 \text{ amp}$$

$$I_f = I_f' + I_s = 821 + 70 = 891 \text{ amp}$$

The open-circuit voltage corresponding to this current on the open-circuit saturation curve, Fig. 1, is 17,530 volts between terminals.

The regulation is

$$\frac{17{,}530 - 13{,}200}{13{,}200} \times 100 = 32.6 \text{ per cent}$$

• PROBLEM 8-23

The no-load saturation curve and the full-load zero-power-factor saturation curve for a 10-kva, 1200 rpm, 3-phase, 230 volts, 25 amperes, 60 cycles synchronous generator are shown in Fig. 1. The measured resistance per phase is 0.20 ohm. Calculate the regulation, using the Potier method, at full load, 0.8 power-factor lagging.

Solution: In Fig. 1, triangle IJG is called the Potier triangle. At full load, zero power factor, the leakage reactance drop is

$$IX_p = \text{distance IJ} = 22 \text{ volts}$$

and so

$$X_p = 22/25 = 0.88 \text{ ohm}$$

The leakage reactance X_p, also called Potier reactance,

is caused by the leakage flux. Since the path of this flux is largely in air, X_p is almost independent of excitation or saturation. The armature reaction effect, evaluated in equivalent-field amperes, is

$$JG = 0.8 \text{ amp}$$

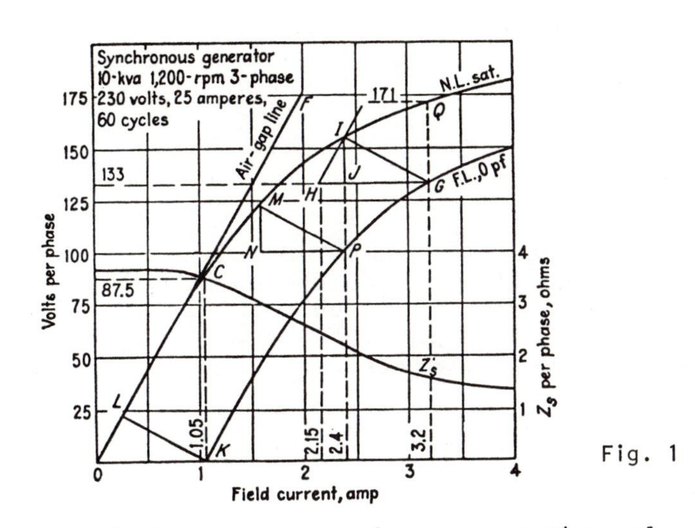

Fig. 1

Full-load zero-power-factor saturation and no-load saturation curves.

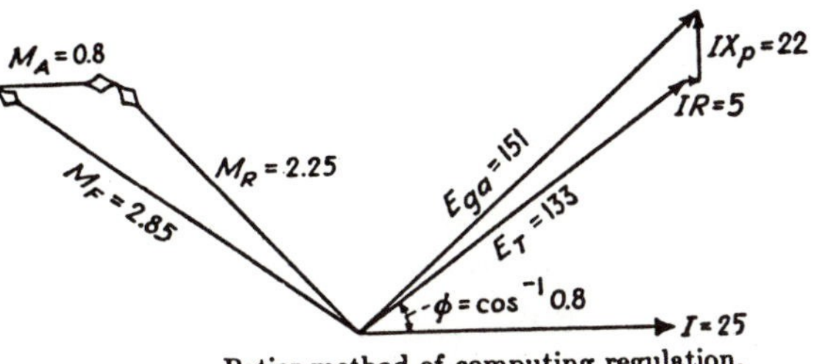

Fig. 2

Potier method of computing regulation.

The method of using these quantities to compute regulation is shown in Fig. 2, for an 80 per cent lagging power-factor load. The terminal voltage per phase

$$E_T = 133 \text{ volts}$$

is drawn at an angle of

$$\phi = \cos^{-1} 0.8$$

ahead of the current

$$I = 25 \text{ amp}$$

To this is added the resistance drop in phase with I,

$$IR = 5 \text{ volts}$$

and the leakage reactance drop in quadrature,

$$IX_p = 22 \text{ volts}$$

to give the internal voltage

$$E_{ga} = 151 \text{ volts}$$

which is the actual generated voltage caused by the total air-gap flux due to the combined field and armature mmfs. The value of this resultant mmf, in equivalent-field amperes, is found from the no-load saturation curve of Fig. 1 as

$$M_R = 2.25 \text{ amp}$$

and this is drawn 90 deg ahead of E_{ga}. The armature mmf, in the same units,

$$M_A = 0.8 \text{ amp}$$

is in phase with the current I. The field excitation required is, therefore,

$$M_F = 2.85 \text{ amp}$$

At no load, with this value of excitation, the generated voltage would be

$$E_g = 165 \text{ volts}$$

and so the regulation, computed by this method, is

$$\text{Reg} = \frac{165 - 133}{133} = 24\%$$

LOSSES AND EFFICIENCY

• **PROBLEM 8-24**

Consider a synchronous machine with negligible armature resistance and leakage reactance and negligible losses to be connected to an infinite bus (i.e., to a system so large that its voltage and frequency remain constant regardless of the power delivered or absorbed). The field current is kept constant at the value which causes the armature current to be zero at no load. Data are given in Fig. 1 with respect to the losses of the 45 kva synchronous machine. Compute its efficiency when running as a synchronous motor at a terminal voltage of 230 volts and with a power input to its armature of 45 kw at 0.80 power factor, leading current. The field current measured in a load test taken under these conditions is I_f(test) = 5.50 amp.

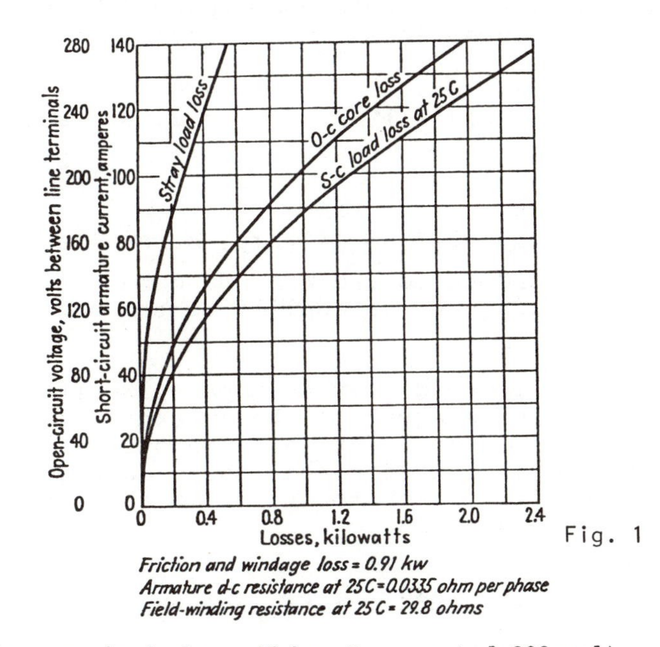

Losses in 3-phase 45-kva Y-connected 220-volt 60-cps 6-pole synchronous machine.

Solution: For the specified operating conditions, the armature current is

$$I_a = \frac{45,000}{\sqrt{3}(230)(0.80)} = 141 \text{ amp}$$

The copper losses are to be computed on the basis of the d-c resistances of the windings at 75°C. Correcting the winding resistances by means of the equation,

$$\frac{r_T}{r_t} = \frac{234.5 + T}{234.5 + t},$$

gives

Field-winding resistance r_f at 75°C = 35.5 ohms

Armature d-c resistance r_a at 75°C = 0.0399 ohm per phase

The field copper loss is

$$I^2_f r_f = (5.50)^2 (35.5) = 1,070 \text{ watts, or } 1.07 \text{ kw}$$

According to the ASA Standards, field-rheostat and exciter losses are not charged against the machine. The armature copper loss is

$$3I_a^2 r_a = (3)(141)^2 (0.0399) = 2,380 \text{ watts, or } 2.38 \text{ kw}$$

and from Fig. 1, at I_a = 141 amp, stray load loss = 0.56 kw. According to the ASA Standards, no temperature correction is to be applied to the stray load loss.

The core loss is read from the open-circuit core-loss curve at a voltage equal to the internal voltage behind the resistance of the machine. The stray load loss is considered to account for the losses caused by the armature leakage flux. For motor action this internal voltage is, as a phasor,

$$\vec{V}_t - \vec{I}_a r_a = \frac{230}{\sqrt{3}} - 141(0.80 + j0.60)(0.0399)$$

$$= 128.4 - j3.4$$

The magnitude is 128.4 volts per phase, or 222 volts between line terminals. From Fig. 1, open-circuit core loss = 1.20 kw. Also, friction and windage loss = 0.91 kw. All losses have now been found.

Total losses =

$$1.07 + 2.38 + 0.56 + 1.20 + 0.91 = 6.12 \text{ kw}$$

The power input is the sum of the a-c input to the

armature and the d-c input to the field, or

$$\text{Input} = 46.07 \text{ kw}$$

Therefore,

$$\text{efficiency} = 1 - \frac{\text{losses}}{\text{input}} = 1 - \frac{6.12}{46.1}$$

$$= 0.867.$$

• **PROBLEM 8-25**

Consider a 93,750-kva 13,200-volt 60-cycle synchronous generator whose characteristic curves and other data are given on Fig. 1. The ventilation loss is 350 kw at rated load. The friction and windage loss is 600 kw. Calculate the efficiency when the generator carries a full-kva inductive load at 0.8 power-factor. The field-current for this load is given as 890 amp.

Solution: The armature effective resistance at 75°C is

$$r_{oh}(\text{at } 25°\text{C}) \frac{1 + 0.0042 \times 75}{1 + 0.0042 \times 25} +$$

$$r_{oh}(\text{at } 25°\text{C})(1.82 - 1)$$

$$= 0.0020 \times 1.19 + 0.0020 \times 0.82$$

$$= 0.00402 \text{ ohm per phase}$$

The armature d-c resistance at 75°C is

$$r_{dc}(\text{at } 75°\text{C}) = 0.0020 \times \frac{1 + 0.0042 \times 75}{1 + 0.0042 \times 25}$$

$$= 0.00238 \text{ ohm per phase}$$

The rated full-load armature current is

$$I = \frac{93,750 \times 1,000}{\sqrt{3} \times 13,200} = 4,100 \text{ amp per phase}$$

From Fig. 1, the load loss at rated frequency at full-load current is 183 kw.

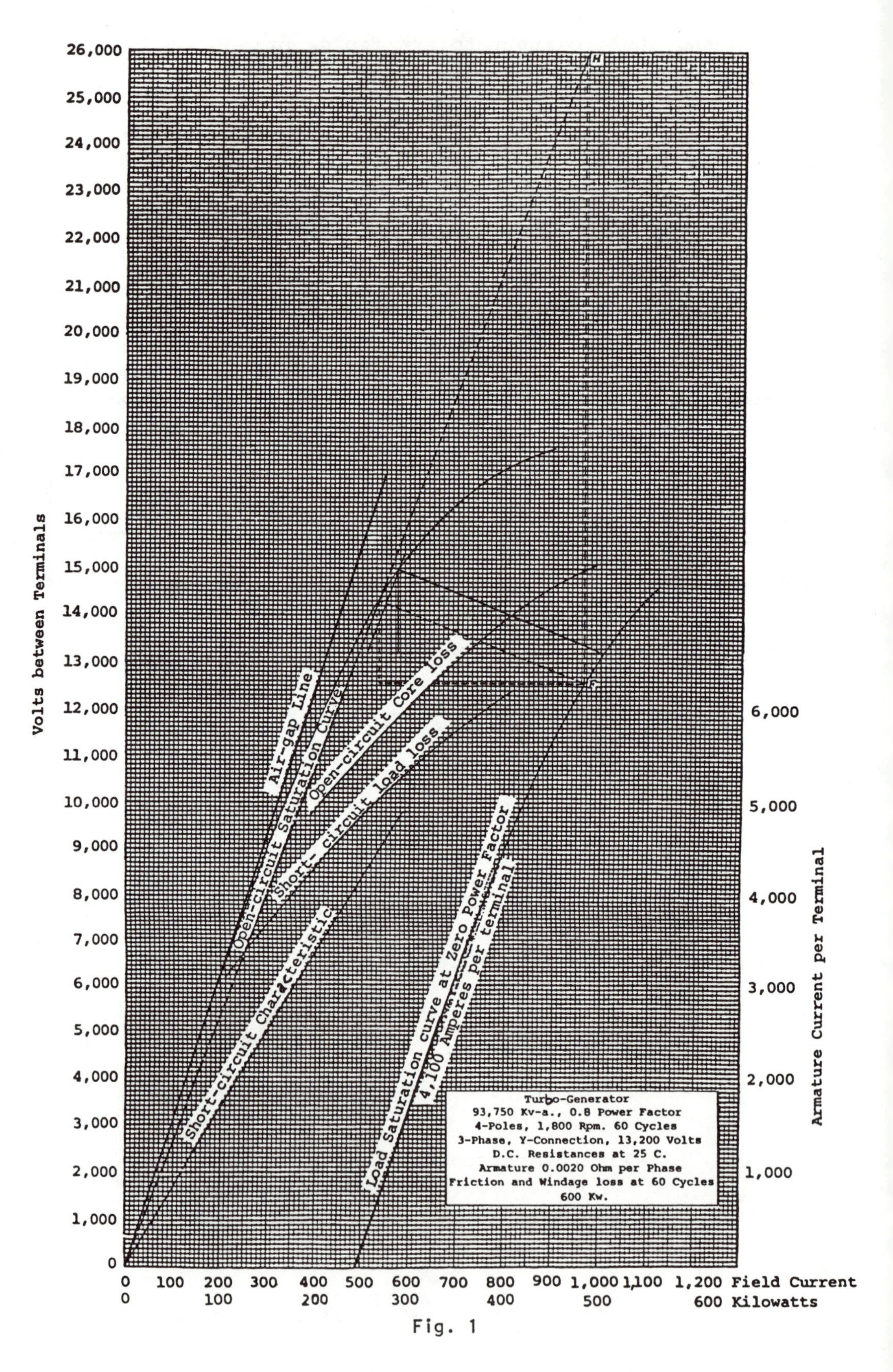
Volts between Terminals
26,000
25,000
24,000
23,000
22,000
21,000
20,000
19,000
18,000
17,000
16,000
15,000
14,000
13,000
12,000
11,000
10,000
9,000
8,000
7,000
6,000
5,000
4,000
3,000
2,000
1,000
0
H
Air-gap Line
Open-circuit Saturation Curve
Open-circuit Core loss
Short-circuit load loss
Open-circuit Saturation Curve
Short-circuit Characteristic
Load Saturation curve at Zero Power Factor
4,100 Amperes per terminal
Turbo-Generator
93,750 Kv-a., 0.8 Power Factor
4-Poles, 1,800 Rpm. 60 Cycles
3-Phase, Y-Connection, 13,200 Volts
D.C. Resistances at 25 C.
Armature 0.0020 Ohm per Phase
Friction and Windage loss at 60 Cycles
600 Kw.
Armature Current per Terminal
6,000
5,000
4,000
3,000
2,000
1,000
0 100 200 300 400 500 600 700 800 900 1,000 1,100 1,200 Field Current
0 100 200 300 400 500 600 Kilowatts

Fig. 1

The stray-load loss at rated armature current is

$$P_s = 183 - 3 \times (4{,}100)^2 \times 0.0020 \times \frac{1}{1{,}000}$$

$$= 82 \text{ kw}$$

In the foregoing calculation of the stray-load loss, the armature resistance at 25°C is used instead of the resistance at 75°C as it is probable that the temperature at which the curves on Fig. 1 were obtained was more nearly 25° than 75°.

$$\vec{E} = \vec{V} + \vec{I}r_e = \frac{13{,}200}{\sqrt{3}}(1 + j0)$$

$$+ 4{,}100(0.8 - j0.6)(0.00402)$$

$$E = 7{,}633 \text{ volts per phase}$$

$$\sqrt{3}\,E = 13{,}220 \text{ volts between terminals}$$

From Fig. 1 the core loss corresponding to the above voltage is 357 kw.

$$\text{Output} = \sqrt{3} \times 13{,}200 \times 4{,}100 \times 0.8$$

$$= 75{,}000{,}000 \text{ watts}$$

Losses (watts)

Armature copper loss = $3(4{,}100)^2 \times 0.00238$ = 120,000

Core loss.............................. = 357,000

Stray-load loss........................ = 82,000

Field loss = $(890)^2 \times 0.186$............... = 147,300

Friction and windage loss.............. = 600,000

Ventilation loss....................... = 350,000

1,656,300 watts

$$\text{Efficiency} = \frac{\text{output}}{\text{output + losses}}$$

$$= \frac{75{,}000}{75{,}000 + 1{,}656} =$$

0.978 or 97.8 per cent

PARALLEL OPERATION OF SYNCHRONOUS GENERATORS

• PROBLEM 8-26

Two parallel-running synchronous generators have e.m.f.s of 1,000 V/ph, and their phase impedances are

$$\vec{z}_1 = (0.1 + j2.0)\Omega$$

and

$$\vec{z}_2 = (0.2 + j3.2)\Omega.$$

They supply a common load of impedance (2.0 + j1.0)Ω/ph. Find their terminal voltage, load currents, active power outputs and no-load circulating current for a phase divergence of 10 deg(e.). The governor characteristics are identical.

Solution: Taking $\vec{E}_1 = (1,000 + j0)V$, then

$$\vec{E}_2 = 1,000(\cos 10° - j\sin 10°)$$

$$= (985 - j174)V.$$

Applying the Millman (parallel-generator) theorem with a common load impedance $\vec{Z}$,

$$\vec{V}\left[\frac{1}{\vec{Z}} + \frac{1}{\vec{z}_1} + \frac{1}{\vec{z}_2}\right] = \frac{\vec{V}}{\vec{Z}_0} = \frac{\vec{E}_1}{\vec{z}_1} + \frac{\vec{E}_2}{\vec{z}_2}$$

gives the common terminal voltage $\vec{V}$. The admittance summation yields $\vec{Z}_0 = 0.905 \angle 66.3°\ \Omega$, and the right-hand side gives the total short-circuit current $\vec{I}_{SC} = 808 \angle -90.7° A$. Hence

$$\vec{V} = \vec{I}_{SC}\vec{Z}_0 = 730 \angle -24.4° = (667 - j303)V$$

is the common terminal voltage. The individual load currents are

$$\vec{I}_1 = (\vec{E}_1 - \vec{V})/\vec{z}_1 = 224 \underline{/-44.9^\circ} = (160 - j159)A$$

$$\vec{I}_2 = (\vec{E}_2 - \vec{V})/\vec{z}_2 = 106 \underline{/-64.3^\circ} = (46 - j96)A$$

The corresponding active powers per phase are

$$P_1 = 155 \text{ kW} \quad \text{and} \quad P_2 = 60\text{kW}.$$

so that machine 1 is retarded and machine 2 advanced. With the machines operating in parallel on zero load,

the circulating current is

$$\vec{I}_S = (\vec{E}_1 - \vec{E}_2)/(\vec{z}_1 + \vec{z}_2) = (34 - j1)A$$

which is substantially an active-power current, corresponding to a generated power of 34 kW/ph in machine 1.

Unless the machines have an exceptionally large inertia, and respond only slowly to synchronizing power, the impedances z_1 and z_2 are not the synchronous impedances, but have lower (dynamic) values.

• PROBLEM 8-27

Two 3-phase, 6.6-kV, star-connected alternators supply a load of 3000 kW at 0.8 p.f. lagging. The synchronous impedance per phase of machine A is 0.5 + j10Ω and of machine B is 0.4 + j12Ω. The excitation of machine A is adjusted so that it delivers 150 A at a lagging power-factor, and the governors are so set that the load is shared equally between the machines. Determine the current, power factor, induced e.m.f. and load angle of each machine.

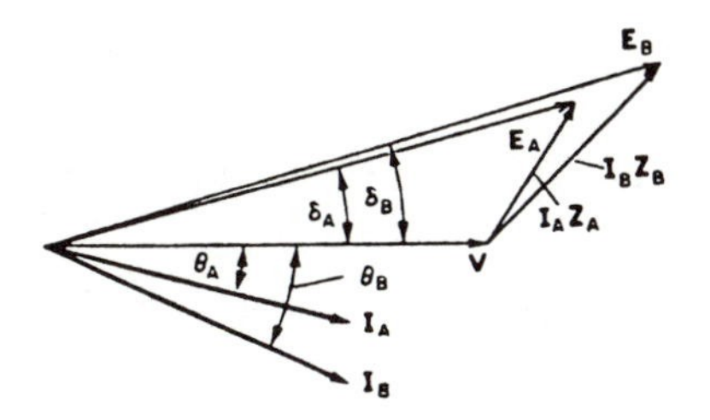

Solution: For machine A,

$$\cos\theta_A = \frac{1500}{\sqrt{3} \times 6.6 \times 150} = 0.874,$$

$$\therefore \quad \theta_A = 29°$$

$$\sin\theta_A = 0.485.$$

$$\text{Total current} = \frac{3000}{\sqrt{3} \times 6.6 \times 0.8} = 328A = 328(0.8 - j0.6)$$

$$= 262 - j195A$$

$$\vec{I}_A = 150(0.874 - j0.485) \qquad = 131 - j72.6 \text{ A}$$

$$\vec{I}_B = 131 - j124.4 \text{ A}$$

$$I_B = 181 \text{ A}$$

$$\therefore \quad \cos\theta_B = \frac{131}{181} = 0.723 \text{ lagging}$$

Taking $\vec{V}$ as reference vector and working in phase values:

$$\vec{E}_A = \vec{V} + \vec{I}_A\vec{Z}_A = (6.6/\sqrt{3}) + (131 - j72.6)(0.5 + j10) \times 10^{-3}$$

$$= 4.6 + j1.27 \text{ kV}$$

$$\text{Load angle } \delta_A = \tan^{-1} 1.27/4.6 = 15.4°$$

$$\text{Line value of e.m.f.} = \sqrt{3}\sqrt{[(4.6)^2 + (1.27)^2]} = 8.26 \text{ kV}$$

$$\vec{E}_B = \vec{V} + \vec{I}_B\vec{Z}_B = (6.6/\sqrt{3}) + (131 - j124.4)(0.4 + j12) \times 10^{-3}$$

$$= 5.35 + j1.52 \text{ kV}$$

Load angle $\delta_B = \tan^{-1} 1.52/5.35 = 15.9°$

Line value of e.m.f. = 9.6 kV

• PROBLEM 8-28

Consider two equal non-salient-pole three-phase 60-cycle Y-connected 6,600-volt 10,000-kva synchronous generators, each with an effective resistance of 0.063 ohm per phase and a synchronous reactance of 1.52 ohms per phase. Assume that the resistances and reactances of the generators are constant. These generators in parallel carry an inductive load of 18,000 kva at 0.9 power factor.

(a) Find the excitation voltages of the generators and their power angles when they share the kw load equally and operate at the same power factor.

(b) Assume that the power load on generator A is increased by adjusting the governors on the prime movers, keeping the frequency the same, until generator A carries two-thirds of the load on the system. The excitation of generator A is kept constant, but the excitation of generator B is adjusted to keep the terminal voltage of the system 6,600 volts. Under these new conditions find the excitation voltage of generator B and the new power angles.

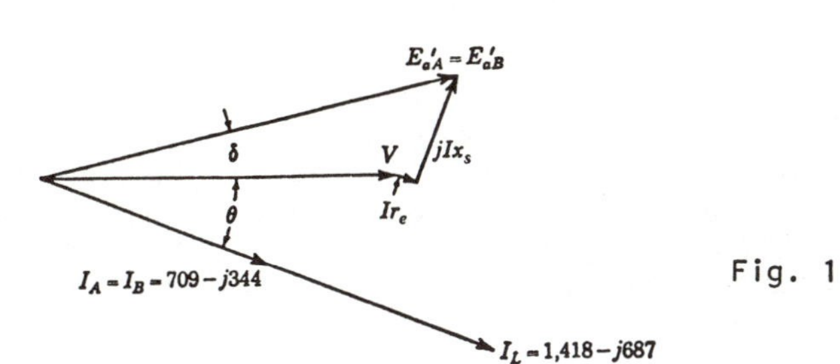

Fig. 1

Solution: (a) The vector diagram for the first condition is shown in Fig. 1. This diagram is not drawn to scale but illustrates the conditions under which the generators are operating. Since the generators are carrying equal loads and operate at the same power factor, they must deliver equal currents.

$$\therefore \quad V = \frac{6,600}{\sqrt{3}} = 3,810 \text{ volts per phase}$$

$$\text{Load current} = I_L = \frac{18,000,000}{\sqrt{3} \times 6,600} = 1,576 \text{ amp per phase}$$

$$\cos\theta = 0.9$$

$$\therefore \quad \sin\theta = 0.4359$$

$$\vec{I}_A = \vec{I}_B = \frac{I_L}{2} = \frac{1,576}{2}(0.9 - j0.4359) = 709 - j344$$

amp per phase

$$\vec{E}'_a \text{ (for each generator)} = 3,810(1 + j0) + (709 - j344)(0.063 + j1.52)$$

$$= (3,810 + j0) + (568 + j1,056)$$

$$= 4,378 + j1,056 \text{ volts per phase}$$

$$E'_a = \sqrt{(4,378)^2 + (1,056)^2} = 4,504 \text{ volts per phase}$$

$$\sqrt{3}\,E'_a = \sqrt{3} \times 4,504 = 7,800 \text{ volts between terminals}$$

$$\tan\delta = \frac{1,056}{4,378} = 0.2413$$

$$\therefore \quad \delta = 13.57 \text{ deg}$$

(b) Neglecting the effect of armature resistance, the maximum power occurs at a power angle of 90 deg. If the terminal voltage and frequency of one of the generators could be maintained constant, the power angle could increase (90 - 13.57) deg before the generator would develop its maximum power. This maximum power is

$$3 \times \frac{E'_a V}{x_s}\sin\delta = 3 \times \frac{4,504 \times 3,810}{1.52} \times 1.00$$

$$= 33,870 \text{ kw.}$$

and is greater than three times the rating of the generator. This neglects the change in synchronous reactance with the change in load and also neglects the effect of the armature resistance. The generator can swing beyond a power angle of 90 deg and still be stable, provided that at this greater angle the power output of the generator plus losses is larger than the power input from its prime mover, since under this condition the generator would swing back to its initial phase position.

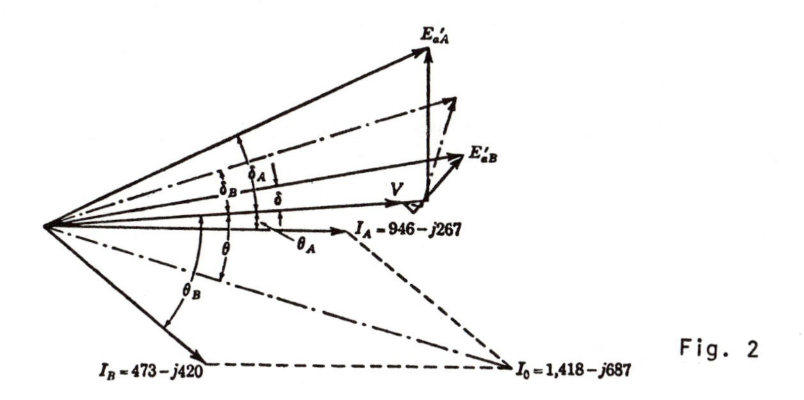

Fig. 2

The conditions when the load on generator A has been increased to two-thirds of the entire load on the system are shown in Fig. 2, which is not drawn to scale. The dot-and-dash lines on this figure show the initial conditions.

$I_A(\text{active}) = 2/3 \times 1.576 \times 0.9 = 946$ amp per phase

$I_B(\text{active}) = 1/3 \times 1{,}576 \times 0.9 = 473$ amp per phase

Generator A

$$4{,}504\angle\delta_A = 3{,}810(1 + j0) + (946 + jI_r)(0.063 + j1.52)$$

where I_r is the reactive component of the armature current.

$$4{,}504\angle\delta_A = 3{,}810 + (59.60 - 1.52I_r) + j(1{,}438 + 0.063I_r)$$

$$(4{,}504)^2 = (3{,}870 - 1.52I_r)^2 + (1{,}438 + 0.063I_r)^2$$

$$I_r = +5{,}273 \text{ or } -267 \text{ amp}$$

Use the smaller value. Then

$$\vec{I}_A = 946 - j267 \text{ amp}$$

$$\vec{E}'_a = 3{,}810(1 + j0) + (946 - j267)(0.063 + j1.52)$$

$$= 4{,}275 + j1.421$$

$$E'_a = \sqrt{(4{,}275)^2 + (1{,}421)^2} = 4{,}503 \text{ volts per phase}$$

which checks with the value 4,504 volts previously found for generator A.

$$\tan \delta_A = \frac{1{,}421}{4{,}275} = 0.3325$$

$$\therefore \qquad \delta_A = 18.39 \text{ deg}$$

Generator A has swung ahead in phase from its initial phase position by (18.39 - 13.57) = 4.82 deg

Generator B

$$\vec{I}_B = \vec{I}_L - \vec{I}_A$$

$$= 1{,}576(0.900 - j0.4359) - (946 - j267)$$

$$= 472 - j420 \text{ amp per phase}$$

$$\vec{E}'_a = 3{,}810(1 + j0) + (472 - j420)(0.063 + j1.52)$$

$$= 4{,}478 + j692 \text{ volts per phase}$$

$$E'_a = \sqrt{(4{,}478)^2 + (692)^2} = 4{,}531 \text{ volts per phase}$$

$$\tan \delta_B = \frac{692}{4{,}478} = 0.1545$$

$$\therefore \ \delta_B = 7.64 \text{ deg}$$

Generator B has swung behind its initial phase position by

$$(7.64 - 13.57) = -5.93$$

electrical degrees. Generator A has swung ahead of its initial phase position. The currents delivered by the two generators can be brought into phase without changing the terminal voltage of the system by increasing the excitation of generator A and decreasing the excitation of generator B.

CHAPTER 9

SYNCHRONOUS MOTORS

SYNCHRONOUS MOTOR CHARACTERISTICS

• PROBLEM 9-1

(i) The inside diameter of the stator of a small, three-phase, six-pole synchronous machine is 0.300 m. and the length of the stator stack is 0.250 m. If the air-gap flux density is sinusoidally distributed at the inside surface of the stator, and has a peak value of 0.96 T. find the flux per pole.

(ii) The machine of Part-(i) has 36 slots in its stator; that is, there are 6 slots per pole. Since each pole corresponds to 180 electrical degrees, there are 180°/6 or 30 electrical degrees per slot. (a) If each coil has a span of 5 slots, what are the coil pitch, p, and the pitch factor, k_p?

(b) What is the maximum flux linkage with each coil, if the coils have two turns?

(c) If the machine is running at 1000 rev/min, what rms voltage is induced in each coil?

Solution:

(i)

$$\phi = \frac{4LrB_{max}}{p} = \frac{4 \times 0.250 \times 0.150 \times 0.96}{6} = 0.024 \text{ Wb}$$

(ii)

(a) $p = 5 \times 30 = 150°$ $\quad k_p = \sin\frac{p}{2} = \sin 75° = 0.9659$

(b)

$$\lambda_{c\,max} = N_c k_p \phi \; .$$

From Part (i)

$$\phi = 0.024 \text{ Wb/pole.}$$

Then

$$\lambda_{c\ max} = 2 \times (\sin 75°) \times 0.024 = 0.046 \text{ Wb turns.}$$

(c)

$$f = \frac{pn_s}{120} = \frac{6 \times 1000}{120} = \frac{1000}{20} = 50 \text{ Hz}$$

Then

$$E_c = \sqrt{2}\pi f N_c k_p \phi = \sqrt{2}\pi \times 50 \times 0.046 = 10.3 \text{ V.}$$

• PROBLEM 9-2

(i) A 20 pole, 40 hp, 660 V, 60 Hz, three-phase, wye-connected, synchronous motor is operating at no-load with its generated voltage per phase exactly equal to the phase voltage applied to its armature. At no-load, the rotor is retarded 0.5 mechanical degree from its synchronous position. The synchronous reactance is 10 ohms, and the effective armature resistance is 1 ohm per phase. Calculate:

(a) The rotor shift from the synchronous position, in electrical degrees

(b) The resultant emf across the armature, per phase

(c) The armature current, per phase

(d) The power per phase, and the total power drawn by the motor from the bus

(e) The armature power loss, and the developed horsepower.

(ii) Repeat part-(i) with a mechanical displacement of 5° between rotor and synchronous position.

Solution:

(i)

(a) $$\alpha = P\left(\frac{\beta}{2}\right) = 20\left(\frac{0.5}{2}\right) = 5°$$

(b) $V_p = \frac{V_L}{\sqrt{3}} = \frac{660}{1.73} = 381$ V;

$E_{gp} = 381$ V also, as given

$\vec{E}_r = (V_p - E_{gp} \cos \alpha) + j(E_{gp} \sin \alpha)$

$= (381 - 381 \cos 5°) + j(381 \sin 5°)$

$= 1.54 + j33.2 = 33.2 \angle 87.3°$ V/phase.

(c) $\vec{Z}_s = R_a + jX_s = 1.0 + j10 = 10\angle 84.3°$ Ω/phase

$\vec{I}_a = \frac{\vec{E}_r}{\vec{Z}_p} = \frac{33.2\angle 87.3°}{10\angle 84.3°} = 3.32\angle 3.0°$ A/phase

(d) $P_p = V_p I_a \cos \theta = 381 \times 3.32 \cos 3° =$

$381 \times 3.32 \times 0.999$

$= 1265$ W/phase

$P_t = 3P_p = 3 \times 1265$ W $= 3795$ W

(e) $3 \times I_a^2 R_a = 3 \times (3.32)^2 \times 1.0 = 33$ W.

$\text{Horsepower} = \frac{3795 - 33 \text{ W}}{746 \text{ W/hp}} = 5.3$ hp.

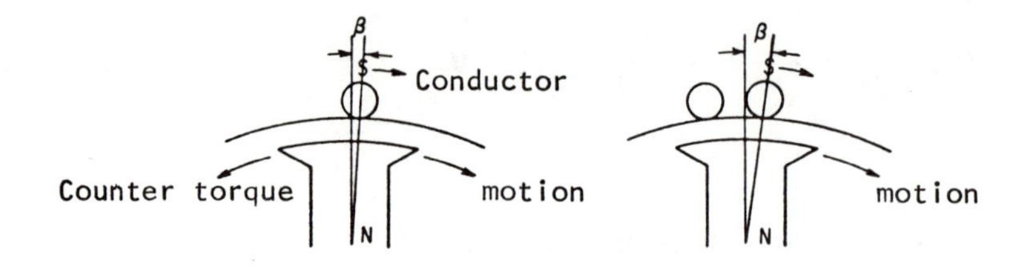

(a) No load. (b) Increased load.

Effect of load on rotor position.

(ii)

(a) $$\alpha = \frac{P\beta}{2} = \frac{20 \times 5}{2} = 50° \text{ (electrical degrees)}.$$

(b) $$\vec{E}_r = (V_p - E_{gp} \cos \alpha) + j(E_{gp} \sin \alpha)$$

$$= 381 - 381 \cos 50° + j381 \sin 50°$$

$$= 141 + j292 = 334\angle 64.2° \text{ V/phase}.$$

(c) $$\vec{I}_a = \frac{\vec{E}_r}{\vec{Z}_p} = \frac{324\angle 64.2° \text{ V}}{10\angle 84.3° \ \Omega} = 32.4 \angle -20.1° \text{ A}.$$

(d) $$P_p = V_p I_a \cos \theta = 381 \times 32.4 \cos 20.1° = 11{,}600 \text{ W}.$$

$$P_t = 3P_p = 3 \times 11{,}600 = 34{,}800 \text{ W}.$$

(e) $$3I_a^2 R_a = 3 \times (32.4)^2 \ 1.0 = 3150 \text{ W}$$

$$\text{Horsepower} = \frac{34{,}800 - 3150 \text{ W}}{746 \text{ W/hp}} = 42.5 \text{ hp}.$$

• **PROBLEM 9-3**

The following test results were obtained from a 3-phase, 60-Hz, 6-pole, 15-kVA, 220-V synchronous motor:

Open-Circuit Test
Field current = 6.7 A
Terminal potential difference = 220 V

Short-Circuit Test
Field current = 6.7 A
Average line current = 57 A

The motor is designed to operate at a full-load leading power factor of 0.8. Determine

(a) The required full-load field current.

(b) The reactive power absorbed by the motor on full load.

<u>Solution</u>:

(a) Per-phase synchronous reactance is

$$X_s = \frac{220}{\sqrt{3} \times 57} = 2.23 \ \Omega .$$

$$n'' = \frac{I_F}{i_f} = \frac{57}{6.7} = 8.51 .$$

Full-load current is

$$I_a = \frac{15 \times 10^3}{220\sqrt{3}} = 39.4 \text{ A.}$$

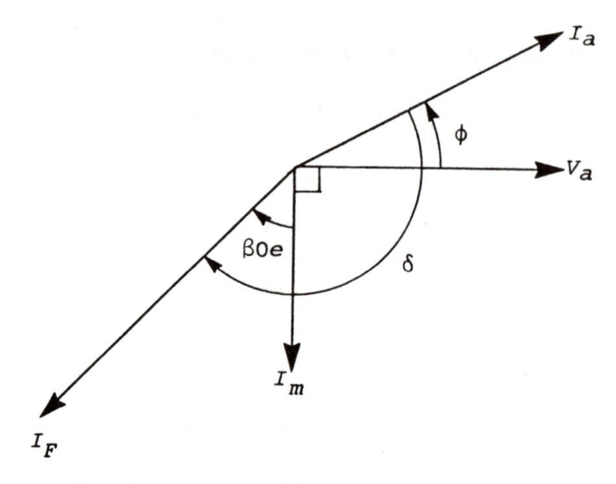

Synchronous machine- Increase effect of varying rotor excitation.

Fig.1

On full load, the phasor diagram for the motor will have the form illustrated in Fig. 1, where

$$\phi = \cos^{-1} 0.8 = 36.9° .$$

If

$$V_a = \frac{220}{\sqrt{3}}\angle 0° \text{V}.$$

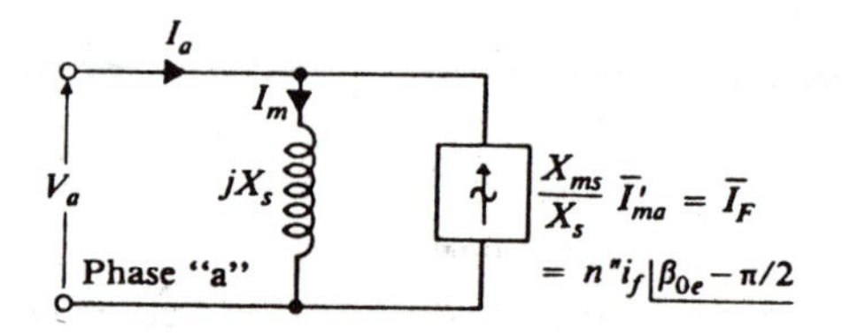

Simplified steady-state equivalent circuit of a cylindrical-rotor synchronous machine.

Fig.2

then from Fig. 2

$$\vec{I}_M = \frac{220\angle 0°}{\sqrt{3} \times 2.23\angle 90°} = 57.0\angle -90° \text{ A}.$$

In the phasor diagram,

$$\vec{I}_f = \vec{I}_M - \vec{I}_a = 57.0\angle -90° - 39.4\angle 36.9° = 86.6\angle -111° \text{ A}.$$

Thus

$$i_f = \frac{86.6}{8.51} = 10.0 \text{ A}$$

(b) Reactive power absorbed by the motor is

$$Q = 3 \times \frac{220}{\sqrt{3}} \times 39.4 \sin 36.9° = 9.00 \text{ kVAR.}$$

• PROBLEM 9-4

The characteristic curves of a 45-kva six-pole three-phase Y-connected 220-volt 60-cycle salient-pole synchronous motor are given in Fig. 1. The zero-power-factor curve in the figure is for 118 amp per terminal. Other data for the motor are:

Field resistance = 30 ohms at 75°C

Friction and windage loss = 810 watts at 60 cycles

Stray-load loss = 540 watts at 60 cycles and rated armature current

Direct-current armature resistance = 0.050 ohm per phase at 75°C

Unsaturated quadrature synchronous reactance = 0.53 ohm per phase

Assume that the motor is operating at rated kva load, with rated voltage and frequency impressed on its terminals, with leading current and power factor at 0.8.

Calculate the field current and efficiency of the motor using the double-reactance method.

Solution:

$$\text{Rated current} = I_a = \frac{45{,}000}{\sqrt{3} \times 220} = 118 \text{ amp per phase.}$$

$$r_e = \text{effective resistance} = 0.050 + \frac{540}{3(118)^2} = 0.063$$

ohm per phase at 75°C and rated current

As an approximation, use the Potier reactance in place of the leakage reactance for finding the air-gap voltage.

Potier reactance = x_p, assumed equal to x_a,

$$= \frac{271 - 220}{\sqrt{3} \times 118} = 0.250 \text{ ohm per phase}$$

$$\vec{E}_a = \text{air-gap voltage} = V + I_a(r_e + jx_a)$$

$$= \frac{220}{\sqrt{3}}(1 + j0) + 118(-0.8 - j0.6)(0.063 + j0.250)$$

$$= (127.1 + j0) + (11.8 - j28.1)$$

$$= 138.9 - j28.1 \text{ V.}$$

$$|\vec{E}_a| = \sqrt{(138.9)^2 + (28.1)^2} = 141.7 \text{ volts per phase}$$

$$\sqrt{3}\,|\vec{E}_a| = \sqrt{3} \times 141.7 = 245.5 \text{ volts between terminals}$$

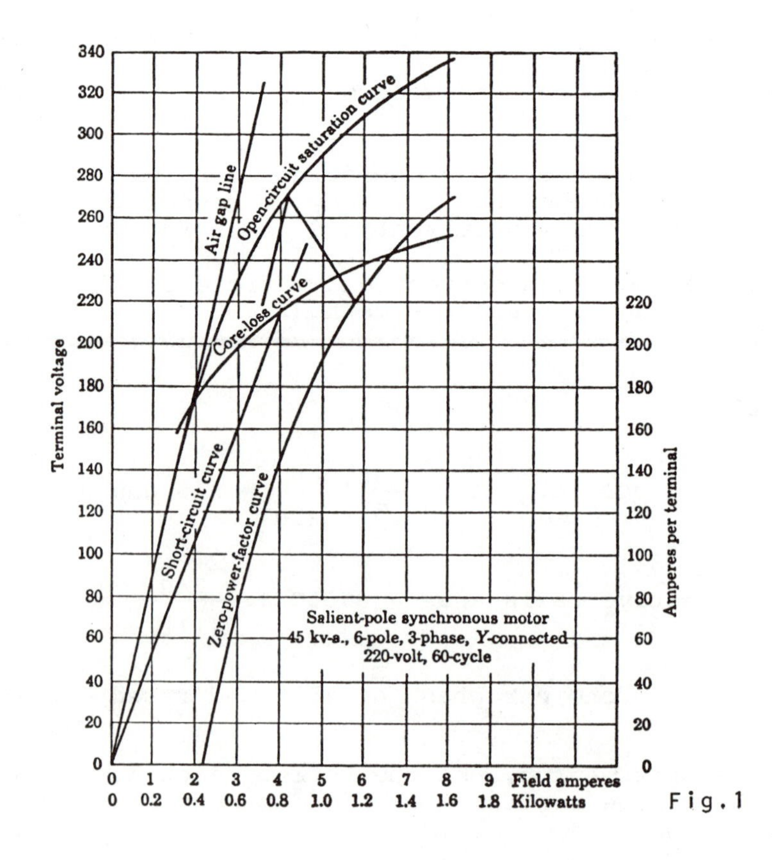

Fig.1

From the open-circuit saturation curve on Fig. 1, the field current corresponding to a voltage of 245.5 volts between terminals is $I_f = 3.40$ amp. From the air-gap line, the voltage corresponding to $I_f = 3.40$ amp is 305 volts between terminals.

$$\text{Saturation factor, } k = \frac{305}{246} = 1.24$$

The unsaturated direct-axis synchronous reactance from the short-circuit curve and the air-gap line is

$$x_d'(\text{unsaturated}) = \frac{198.0}{\sqrt{3} \times 118.0} = 0.969 \text{ ohm per phase}$$

$$x_d(\text{saturated}) = (x_d' - x_a)\frac{1}{k} + x_a$$

$$= (0.969 - 0.25)\frac{1}{1.24} + 0.25$$

$$= 0.83 \text{ ohm per phase}$$

Assume as an approximation that the saturation factor for the quadrature-axis synchronous reactance is the same as the saturation factor for the direct-axis synchronous reactance.

$$x_q(\text{saturated}) = (0.53 - 0.25)\frac{1}{1.24} + 0.25$$

$$= 0.476 \text{ ohm per phase}$$

The power angle δ is given by

$$\tan\delta = \frac{-I_a(r_e \sin\theta' + x_q \cos\theta')}{V + I_a(x_q \sin\theta' - r_e \cos\theta')}$$

$$= \frac{-118(0.063 \times 0.6 + 0.476 \times 0.8)}{127.1 + 118(0.476 \times 0.6 - 0.063 \times 0.8)}$$

$$= \frac{-118 \times 0.4186}{127.1 + 118 \times 0.2360} = -0.3191$$

$$\delta = -17.7 \text{ deg.}$$

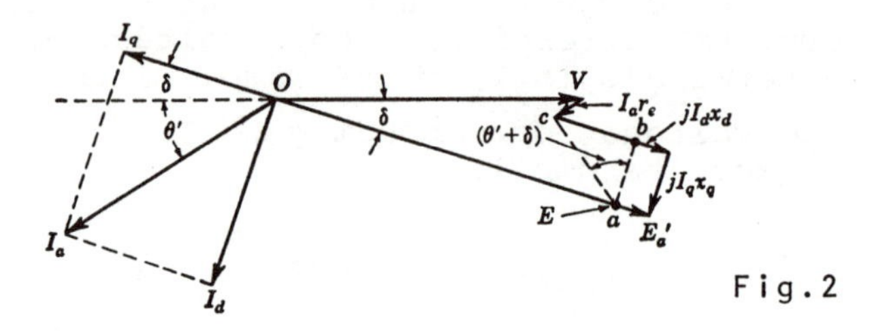

Fig.2

Refer to Fig. 2. The angle between I_a and I_q is $(\theta' + \delta)$.

$$\theta' = \tan^{-1} 0.8 = 36.87 \text{ deg}$$

$$(\theta' + \delta) = 36.87 + 17.7 = 54.57 \text{ deg}$$

$$\sin \delta = 0.3040 \qquad \cos \delta = 0.9527$$

$$\sin (\theta' + \delta) = 0.8148 \qquad \cos (\theta' + \delta) = 0.5797$$

$\vec{E}_a'$ is given by

$$\vec{E}_a'(1 + j0) = V(\cos \delta + j \sin \delta) + I_a[-\cos (\theta' + \delta) - j \sin (\theta' + \delta)]r_e - I_a \cos (\theta' + \delta)jx_q - jI_a \sin (\theta' + \delta)jx_d$$

$$= 127.1(0.9527 + j0.3040) + 118.0(-0.5797 - j0.8148)0.063 - 118.0(0.5797)(j0.476) - j118.0(0.8148)(j0.830)$$

$$= (121.0 + j38.6) + (-4.31 - j6.06) - (j32.5) + (79.8)$$

$$= 196.5 + j0$$

$$|\vec{E}_a'| = 196.5 \text{ volts per phase}$$

$$\sqrt{3}\, E_a' = \sqrt{3} \times 196.5 = 340.3 \text{ volts between terminals}$$

The voltage 340.3 is the excitation voltage at load saturation. The required field current for the assumed load is the field current found from the saturation line corresponding to 340.3 volts. By extrapolation this field current is

$$I_f = \frac{340.3}{245.5} \times 3.40 = 4.71 \text{ amp}$$

According to the Standardization Rules of the AIEE and also the ASA Standards for Rotating Electrical Machinery, the proper core loss to use when finding the efficiency is that corresponding to a voltage equal to the terminal voltage corrected for the armature resistance drop. This voltage is

$$\vec{V} + \vec{I}_a r_e = 127.1 + 118.0(-0.8 - j0.6)(0.063)$$

$$= 121.1 - j4.46 \text{ volts per phase}$$

$$|\vec{V} + \vec{I}_a r_e| = \sqrt{(121.1)^2 + (4.46)^2} = 121.2 \text{ volts per phase}$$

$$\sqrt{3}\ |\vec{V} + \vec{I}_a r_e| = \sqrt{3} \times 121.2 = 210.0 \text{ volts between terminals}$$

From the core-loss curve on Fig. 1, the core loss corresponding to 210.0 volts is 740 watts.

Losses	Watts
Armature copper $3(118)^2 0.050$............	2,089
Stray-load............................	540
Core..................................	740
Field copper $(4.71)^2 30$..................	666
Friction and windage....................	810
Total............................	4,845

$$\text{Efficiency} = \frac{\text{Input} - \text{losses}}{\text{Input}}$$

$$= \frac{45{,}000 \times 0.8 - 4{,}845}{45{,}000 \times 0.8}$$

$$= 1 - \frac{4{,}845}{36{,}000} = 0.866 \text{ or } 86.6 \text{ per cent}$$

TORQUE, TORQUE ANGLE, HUNTING OSCILLATION

• PROBLEM 9-5

A 2,000-hp 1.0-power-factor 3-phase Y-connected, 2,300-volt 30-pole 60-cps synchronous motor has a synchronous reactance of 1.95 ohms per phase. For the purposes of this problem all losses may be neglected.

(a) Compute the maximum torque in pound-feet which this motor can deliver if it is supplied with power from a constant-voltage constant-frequency source, commonly called an infinite bus, and if its field excitation is constant at the value which would result in 1.00 power factor at rated load.

(b) Instead of the infinite bus of part a, suppose that the motor were supplied with power from a 3-phase Y-connected 2,300-volt 1,750-kva 2-pole 3,600-rpm turbine-generator whose synchronous reactance is 2.65 ohms per phase. The generator is driven at rated speed, and the field excitations of generator and motor are adjusted so that the motor runs at 1.00 power factor and rated terminal voltage at full load. The field excitations of both machines are then held constant, and the mechanical load on the synchronous motor is gradually increased. Compute the maximum motor torque under these conditions. Also compute the terminal voltage when the motor is delivering its maximum torque.

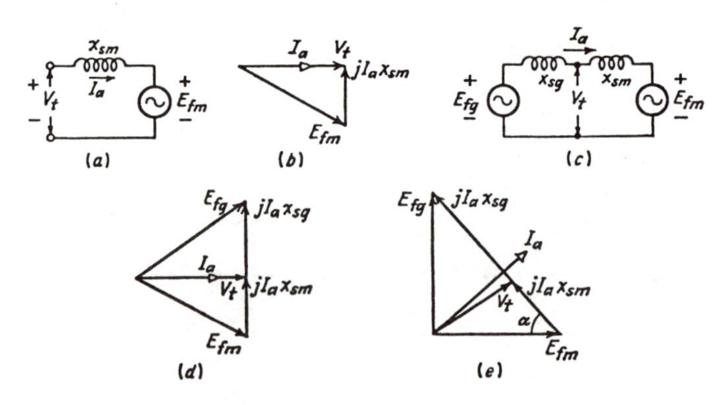

Equivalent circuits and phasor diagrams.

Solution: Although this machine is undoubtedly of the salient-pole type, we shall solve the problem by simple cylindrical-rotor theory. The solution accordingly neglects reluctance torque. The machine actually would develop a maximum torque somewhat greater than the computed value.

(a) The equivalent circuit is shown in Fig. 1a and the

phasor diagram at full load in Fig. 1b, wherein E_{fm} is the excitation voltage of the motor and x_{sm} is its synchronous reactance. From the motor rating with losses neglected,

Rated kva = 2,000 × 0.746 = 1,492 kva, 3-phase

= 497 kva per phase.

$$\text{Rated voltage} = \frac{2{,}300}{\sqrt{3}} = 1{,}330 \text{ volts to neutral.}$$

$$\text{Rated current} = \frac{497{,}000}{1{,}330} = 374 \text{ amp per phase Y}$$

$$I_a x_{sm} = 374 \times 1.95 = 730 \text{ volts per phase}$$

From the phasor diagram at full load

$$E_{fm} = \sqrt{V_t^2 + (I_a x_{sm})^2} = 1{,}515 \text{ volts}$$

When the power source is an infinite bus and the field excitation is constant, V_t and E_{fm} are constant. Substitution of V_t for E_1, E_{fm} for E_2, and x_{sm} for X in the equation,

$$P_{max} = E_1 E_2 / X$$

gives

$$P_{max} = \frac{V_t E_{fm}}{x_{sm}}$$

$$= \frac{1{,}330 \times 1{,}515}{1.95} = 1{,}030 \times 10^3 \text{ watts per phase}$$

$$= 3{,}090 \text{ kw for 3 phases}$$

(In per unit, P_{max} = 3,090/1,492 = 2.07.) With 30 poles at 60 cps, synchronous speed = 4 rev/sec.

$$T_{max} = \frac{P_{max}}{\omega_s} = \frac{3{,}090 \times 10^3}{2\pi \times 4} = 123 \times 10^3 \text{ newton-m}$$

$$= 0.738(123 \times 10^3) = 90{,}600 \text{ lb-ft.}$$

(b) When the power source is the turbine-generator, the equivalent circuit becomes that shown in Fig. 1c, wherein E_{fg} is the excitation voltage of the generator and x_{sg} is its synchronous reactance. The phaser diagram at full motor load, 1.00 power factor, is shown in Fig. 1d. As before,

$$V_t = 1{,}330 \text{ volts at full load}$$

$$E_{fm} = 1{,}515 \text{ volts.}$$

The synchronous-reactance drop in the generator is

$$I_a x_{sg} = 374 \times 2.65 = 991 \text{ volts}$$

and from the phasor diagram

$$E_{fg} = \sqrt{V_t^2 + (I_a x_{sg})^2} = 1{,}655 \text{ volts.}$$

Since the field excitations and speeds of both machines are constant, E_{fg} and E_{fm} are constant. Substitution of E_{fg} for E_1, E_{fm} for E_2, and $x_{sg} + x_{sm}$ for X in the equation,

$$P_{max} = E_1 E_2 / X_1$$

gives

$$P_{max} = \frac{E_{fg} E_{fm}}{x_{sg} + x_{sm}}$$

$$= \frac{1{,}655 \times 1{,}515}{4.60} = 545 \times 10^3 \text{ watts per phase}$$

$$= 1{,}635 \text{ kw for 3 phases}$$

(In per unit, $P_{max} = 1{,}635/1{,}492 = 1.095$.)

$$T_{max} = \frac{P_{max}}{\omega_s} = \frac{1{,}635 \times 10}{2\pi \times 4} = 65 \times 10^3 \text{ newton-m}$$

$$= 48{,}000 \text{ lb-ft}$$

Synchronism would be lost if a load torque greater than this value were applied to the motor shaft. The motor would stall, the generator would tend to overspeed, and the circuit would be opened by circuit-breaker action.

With fixed excitations, maximum power occurs when E_{fg} leads E_{fm} by 90°, as shown in Fig. 1e. From this phasor diagram

$$I_a(x_{sg} + x_{sm}) = \sqrt{E_{fg}^2 + E_{fm}^2} = 2,240 \text{ volts}$$

$$I_a = \frac{2,240}{4.60} = 488 \text{ amp}$$

$$I_a x_{sm} = 488 \times 1.95 = 951 \text{ volts}$$

$$\cos \alpha = \frac{E_{fm}}{I_a(x_{sg} + x_{sm})} = \frac{1,515}{2,240} = 0.676$$

$$\sin \alpha = \frac{E_{fg}}{I_a(x_{sg} + x_{sm})} = \frac{1,655}{2,240} = 0.739$$

The phasor equation for the terminal voltage is

$$V_t = E_{fm} + jI_a x_{sm} = E_{fm} - I_a x_{sm} \cos \alpha + jI_a x_{sm} \sin \alpha$$

$$= 1,515 - 643 + j703 = 872 + j703$$

The magnitude of V_t is

$$V_t = 1,120 \text{ volts to neutral}$$

$$= 1,940 \text{ volts, line to line}$$

When the source is the turbine-generator, as in part b, the effect of its impedance causes the terminal voltage to decrease with increasing load, thereby reducing the maximum power from 3,090 kw in part a to 1,635 kw in part b.

• PROBLEM 9-6

A 75-HP, three-phase, six-pole, 60-Hz, Y-connected, cylindrical-rotor synchronous motor has synchronous reactance of 9.6 Ω per phase. Its rated terminal voltage is 500 v per phase. (a) Find the value of excitation voltage that makes maximum torque to be 120 percent of rated torque. (b) The machine is operated with the excitation voltage set as in part (a). For rated load torque, find the armature current, the power factor, and the torque angle.

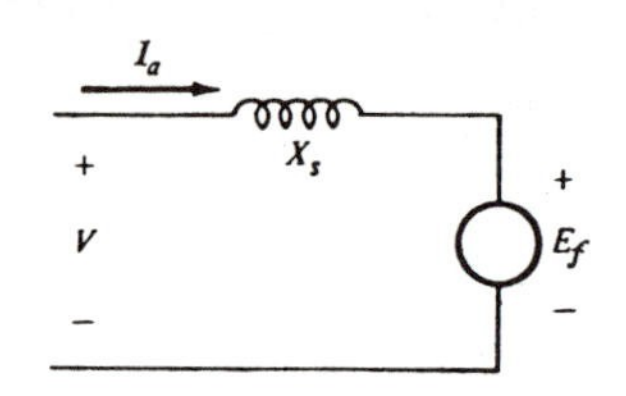

Current direction choice for motor operation.

Solution: (a) Synchronous speed is

$$\omega_s = 4\pi f/p = 4\pi(60)/6 = 40\pi \text{ rad/sec}$$

$$P_{rated} = 75 \times 746 = 56{,}000 \text{ w}$$

$$T_{rated} = 56{,}000/40\pi = 446 \text{ nm}$$

The excitation voltage can be found from the following equation:

$$T_{max} = \frac{1}{\omega_{x_m}} \frac{3VE_f}{X_s}$$

$$\therefore \quad T_{max} = 3(500)(E_f)/40\pi(9.6) = 1.2(446)$$

$$\therefore \quad E_f = 430 \text{ v}$$

(b) The torque angle can be found from the following equation:

$$P = \frac{3VE_f}{X_s} \sin\delta$$

$$P = \frac{3(500)(430)}{9.6} \sin\delta = 56{,}000$$

$\delta = 56.5°$

For the equivalent circuit shown in the figure, Kirchhoff's voltage equation is given by the following equation:

$$\vec{E}_f = \vec{V} - jX_s\vec{I}_a \qquad (1)$$

Use the terminal voltage as the reference

$$\vec{V} = 500\angle 0° \text{ v.}$$

For motor operation, the excitation voltage lags behind the terminal voltage by angle δ

$$\vec{E}_f = 430\angle -56.5° \text{ v.}$$

Solve Eq. 1 for the armature current

$$\vec{I}_a = \frac{\vec{V} - \vec{E}_f}{jX_s} = \frac{500\angle 0° - 430\angle -56.5°}{j9.6}$$

$$= 53.3\angle -45.3° \text{ amp.}$$

Thus, I_a = 53.3 amp. The power factor is P.F. = cos 45.3° = 0.7 lagging.

• PROBLEM 9-7

For the motor described below, (a) determine the power input and power developed, (b) determine the maximum torque angle (pull-out point) and the stability factor.

200 horsepower

Unity pf

8 poles 60 cycles 440 volts three phase

R_e = 0.0285 ohm per phase

X_d = 1.27 ohms per phase

X_q = 0.774 ohm per phase

Full-load amperes = 210 amperes per terminal

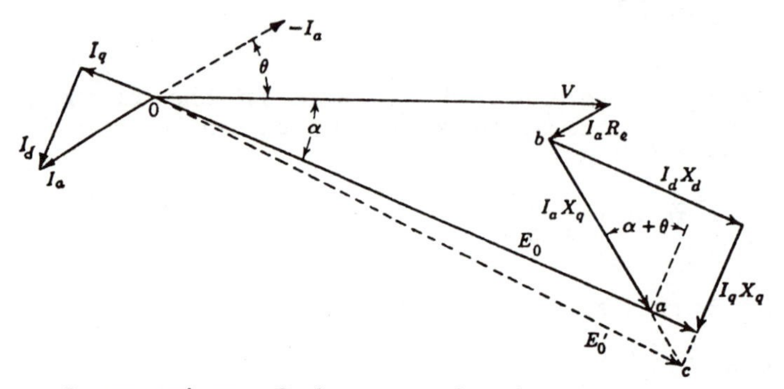

A comparison of the open-circuit voltage as obtained by the two-reaction method versus the synchronous-impedance method.

Solution: (a) From the figure,

$$\tan\alpha = \frac{-I_aX_q}{V - I_aR_e} = \frac{-210 \times 0.774}{254 - 210 \times 0.0285}$$

Then

$$\alpha = -33°13' \text{ as the full-load torque angle}$$

$$\therefore \quad \sin\alpha = -0.548$$

$$\cos\alpha = 0.837$$

The excitation voltage necessary to produce unity pf is, using E_0 as reference vector ($\sin\alpha$ is positive),

$$E_0 = (V - I_aR_e)\cos\alpha + I_aX_d \sin\alpha$$

$$= (254 - 210 \times 0.0285)0.837 + 210 \times 1.27 \times 0.548$$

$$= 354 \text{ volts}$$

$$P_{input} = \frac{V^2\sin\alpha}{X_q}\cos\alpha + \frac{VE_0 - V^2\cos\alpha}{X_d}\sin\alpha$$

$$= \frac{254 \times 354}{1.27}(-0.548) + \frac{254^2(1.27 - 0.774)}{2 \times 1.27 \times 0.774}(-0.9166)$$

$$= -38{,}800 - 14{,}800 = -53{,}300 \text{ watts per phase}$$

The input is 159.9 kilowatts for three phases.

The power developed is input minus armature copper losses, or

$$159{,}900 - 3 \times 210^2 \times 0.0285 = 155{,}420 \text{ watts}$$

The current components:

$$I_q = I_a \cos \alpha = 210 \times 0.837 = 175.5$$

$$I_d = I_a \sin \alpha = 210 \times 0.548 = 115.0$$

Check power input by

$$P_{input} = E_0 I_q - I_q I_d \ (X_d - X_q)$$

$$P_{input} = 354 \times 175.5 - 175.5 \times 115(1.27 - 0.774)$$

$$= 53{,}500 \text{ watts per phase}$$

(b)

$$B = \frac{V(X_d - X_q)}{X_q} \equiv \frac{254(1.27 - 0.774)}{0.774} = 162$$

$$\cos \alpha \approx -\frac{354}{4 \times 162} \pm \sqrt{\frac{1}{2} + \left(\frac{354}{4 \times 162}\right)^2}$$

$$= -0.544 \pm 0.891 = 0.347.$$

The positive sign gives useful values. Hence the torque angle at pull out is 69°40'. Note how widely this differs from the cylindrical rotor assumption of approximately 90%.

The stability factor:

$$P_r = \frac{dP}{d\alpha} \approx \frac{254 \times 354}{1.27} \times 0.837$$

$$+ \frac{254^2(1.27 - 0.774)}{1.27 \times 0.774} \times 0.3998$$

$$= 72{,}500 \text{ watts per phase per electrical radian}$$

$$= 217.5 \text{ kilowatts for 3 phases}$$

In the synchronous motor, synchronous watts and watt output are represented by the same figure.

$$\text{Torque in pound-feet} = \frac{7.04 \times \text{synchronous watts}}{\text{Synchronous rpm}}$$

$$= \frac{7.04 \times 217{,}500}{900} = 1720$$

Hence, in terms of pound-feet of torque per electrical radian, the stability factor is 1720; or, per mechanical radian

$$1720 \times P/2 = 6880$$

This is, of course, approximate, being based on watt input rather than output.

The maximum input, at the point of pull out is

$$P_{max} = \frac{254 \times 354}{1.27} \sin(-69°40') + \frac{254^2(1.27 - 0.774)}{2 \times 1.27 \times 0.774} \times \sin(-139°20')$$

$$= -77{,}400 \text{ watts per phase or } 232.2 \text{ kilowatts for 3 phases}$$

It will be recalled that these values are based on input, not output. Neglecting this error, and, since the values obtained are also proportional to torque,

$$\frac{\text{Maximum torque}}{\text{Full-load torque}} \approx \frac{232.2 \text{ kilowatts}}{162.75 \text{ kilowatts}} = 1.43$$

(This motor does not meet A.S.A. or N.E.M.A. Standards, which require that a 200-horsepower motor of this speed have a maximum torque of at least 1.75 times the full-load value.)

• PROBLEM 9-8

(a) Determine the resultant current when a three-phase, 2500-kva, 2300-volt synchronous motor with R_a = 1.24 per cent and X_s = 24.9 per cent is operating idle at 20 per cent overexcitation. The no-load losses are to be neglected. Assume wye connection and E_c out of phase with V by 180°.

(b) The machine in Part-(a) has normal excitation, and the shaft load causes the induced electromotive force to fall back 8°. Determine the angles α, Δ, γ, and θ, and also the power delivered (including the friction and windage), the power required, and the copper losses, not neglecting the no-load losses.

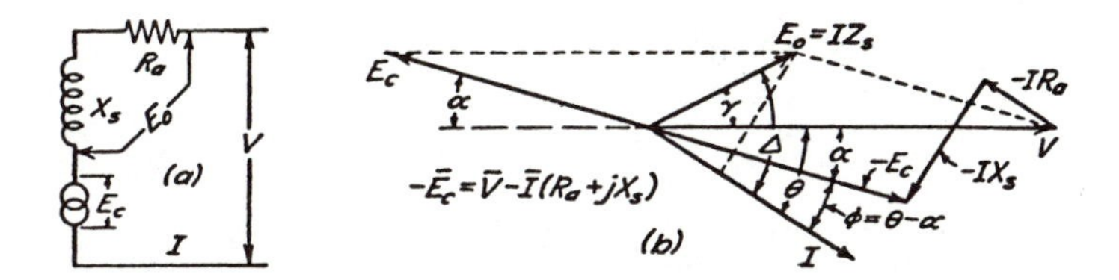

The equivalent circuit and vector diagram for a synchronous motor; line to neutral for a polyphase motor.

Solution: (a)

$$I_L = \frac{2500 \times 1000}{\sqrt{3} \times 2300} = 627.57 \text{ amp}$$

$$V_p = 2300 \div \sqrt{3} = 1327.94 \text{ volts}$$

$$X_s = \frac{V_p}{I_L}$$

$$= \frac{0.249 \times 1327.94}{627.57} = 0.5269 \text{ ohm}$$

$$\vec{E}_0 = (\vec{V} + \vec{E}_c) = (1327.94 + j0) + (-1327.94 + j0) \times 1.2$$

$$= -265.58 + j0 \text{ V}$$

$$\vec{I} = \frac{\vec{E}_0}{X_s} = \frac{265.58 + j0}{0 + j0.5269} \text{ (R neglected)}$$

$$\vec{I} = 0 + j504$$

$$I = 504 \text{ amp.}$$

(b) $\cos 8° = 0.9903$ $\qquad$ $\sin 8° = 0.1392$

$$\vec{E}_c = -(0.9903 \times 1327.94) + j(0.1392 \times 1327.94)$$
$$= -1315.06 + j184.85 \text{ V}$$

$$\vec{V} = 1327.94 + j0 \text{ V.}$$

$$\vec{E}_0 = \vec{E}_c + \vec{V} = 12.88 + j184.85 \text{ V.}$$

$$\vec{I} = \frac{12.88 + j184.85}{0.02624 + j0.5269} = 351.18 - j6.96 \text{ amp.}$$

Angle (Fig. 1b)		Sin	Cos	
α	8°	0.1392	0.9903	$V = 1327.94$ volts
γ	86.1°	0.9976	0.0695	$E_0 = 185.30$ volts
Δ	87.2°	0.9988	0.0488	$E_c = 1327.94$ volts
θ	1.1°	0.0198	0.9998	$I = 351.25$ amp

$$P_{cu} = \frac{3 \times 185.3 \times 351.25 \times 0.0488}{1000} = 9.5 \text{ kw (copper losses).}$$

$$P = \frac{\sqrt{3} \times 2300 \times 351.25 \times 0.9998}{1000} = 1399 \text{ kw (input).}$$

$$P_{del} = \frac{3V_pI \cos(\theta \pm \alpha)}{1000} \text{ kw}$$

$$= \frac{3 \times 1327.94 \times 351.25 \times \cos(\theta \pm \alpha)}{1000} = 1389 \text{ kw}$$

(delivered).

• **PROBLEM 9-9**

An ideal 1.5 MW, 50 Hz, 3-ph, 8-pole synchronous machine with a reaction reactance of 1.25 p.u. runs on 6.6 kV infinite busbars. The moment of inertia of the rotor and its mechanical attachments is 3100 kg-m^2. The machine operates

(a) on no load with 1.0 p.u. excitation;

(b) as a compensator with 2.0 p.u. excitation;

(c) as a motor on full load (1.0 p.u.) at power factor 0.87 leading;

(d) as a generator on 0.75 p.u. active power at an output power factor of 0.78 lagging.

For each case estimate the load angle, synchronizing power and torque, and the natural frequency of hunting oscillation; for case (d) find the voltage regulation.

The electrical load diagram in Fig. 2 is drawn for x_a = 1.25 p.u.; the power/load-angle curves in Fig. 1 and the V-curves in Fig. 3 also apply here. Using per-unit values based on the full rating (1.5 MW) enables calculations to be made for the whole machine direct instead of per phase.

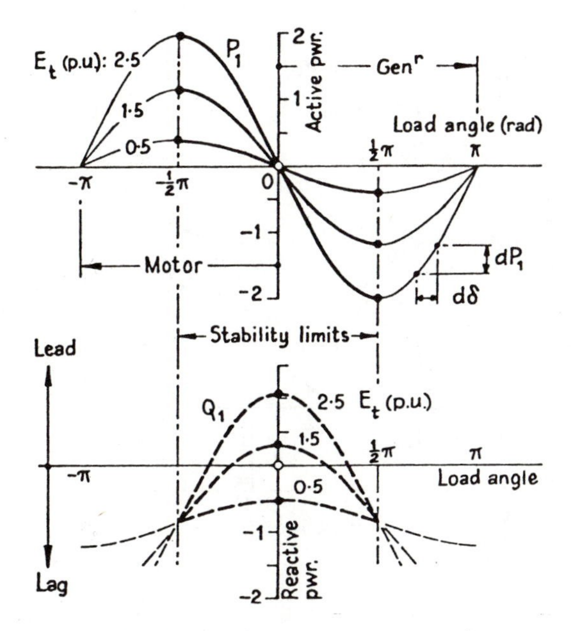

Active and reactive power/
load-angle relations.

Fig.1

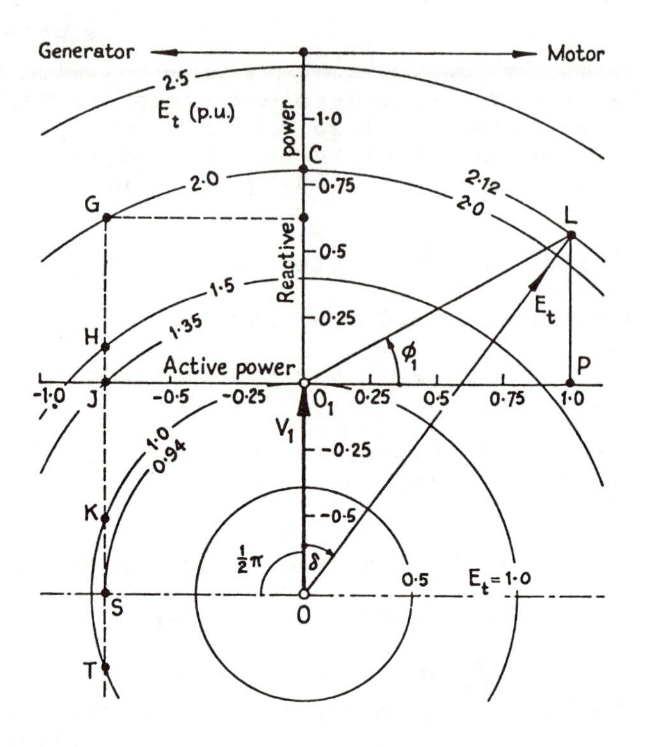

Ideal synchronous machine: electrical load diagram. The scales are in per-unit values for a reaction reactance of 1.25 p.u.

Fig.2

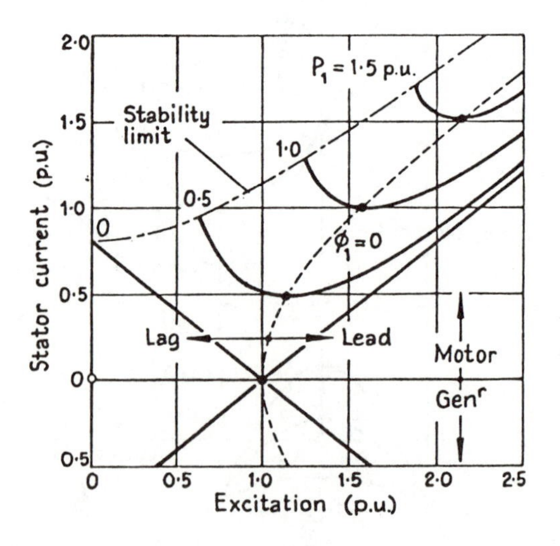

V-curves. The scales are in per-unit values for a reaction reactance of 1.25 p.u.

Fig.3

Solution: (a) No Load: The no-load operating point with $E_t = 1.0$ p.u. is O_1 in Fig. 2. The stator current and the load angle are both zero. The synchronizing power is

$$p_s = (V_1/x_a)E_t \cos \delta$$

$$= (1.0/1.25)1.0 = 0.80 \text{ p.u./rad(e.)} =$$

$$1.20 \text{ MW/rad(e.)}.$$

As 1 rad(m.) = 4 rad(e.), then also

$$p_s = 0.80 \times 4 = 3.2 \text{ p.u./rad(m.)}$$

$$= 0.056 \text{ p.u./deg (m.)}$$

The synchronous speed is $\omega_s = 2\pi 50/4 = 78.6$ rad(m)/s so that the corresponding synchronizing torque is

$$M_s = p_s/\omega_s$$

$$= 3.2 \times 1.5 \times 10^3/78.5 = 61 \text{ kN-m/rad(m.)}$$

The natural angular frequency of hunting oscillation is

$$\omega_0 = \sqrt{(M_s/J)}$$

$$= \sqrt{(61{,}000/3{,}100)} = 4.44 \text{ rad(m.)/s}$$

that is, an oscillation frequency $f_0 = 0.71$ Hz, or one complete swing in 1.4 s.

(b) Compensator: Raising E_t to 2.0 p.u. shifts the operating point to C. The load angle is still zero (there being no active power concerned) but the machine takes a reactive input of 0.80 p.u. = 1.2 Mvar leading as a synchronous capacitor. The values required are

$$p_s = (1.0/1.25)2.0 = 1.60 \text{ p.u./rad(e.)}$$

$$= 6.4 \text{ p.u./rad(m.)}$$

$$M_s = 6.4 \times 1.5 \times 10^3/78.5 = 122 \text{ kN-m/rad(m.)}$$

$$\omega_0 = \sqrt{(122{,}000/3{,}100)} = 6.28 \text{ rad/s}, f_0 = 1.0 \text{ Hz}.$$

(c) Motor: The operating point is L, with 1.0 p.u. active power (1.5 MW), 0.57 p.u. leading reactive power (0.85 Mvar) and 1.15 p.u. apparent power (1.72 MVA). The necessary excitation is readily calculated, or measured from the load diagram: it is $E_t = 2.12$ p.u. The load angle is $-36° = -0.63$ rad(e.). The synchronizing power and torque are

$$p_s = (1.0/1.25) \times 2.12 \times \cos(-36°) \times 4$$

$$= 5.49 \text{ p.u./rad(m.)}$$

$$M_s = 5.49 \times 1.5 \times 10^3/78.5 = 105 \text{ kN-m/rad(m.)}$$

and the natural frequency is $\omega_0 = 5.8$ rad(m.)/s, $f_0 = 0.92$ Hz.

(d) Generator: The conditions are given by G, with OG $= E_t = 2.0$ p.u., and inputs of -0.75 p.u. active and $+ 0.60$ p.u. reactive. The output of the machine as a generator is therefore $+ 0.75$ p.u. = 1.12MW and -0.60 = 0.9 Mvar lagging. By calculation or measurement, $\delta = +28°$ $= +0.49$ rad(e.). Then

$$p_s = 5.64 \text{ p.u./rad(m.)}, \quad M_s = 108 \text{ kN-m/rad(m.)},$$

$$\omega_0 = 5.9 \text{ rad(m.)/s}, \quad f_0 = 0.94 \text{ Hz}$$

The regulation is $\varepsilon = (E_t - V_1)/V_1 = 1.0$ p.u., because obviously the open-circuit terminal voltage is double the busbar voltage.

CIRCLE CHART AND V-CURVES

• PROBLEM 9-10

The no-load saturation curve and the curve of Z_s values as a function of field current, for a 10-kva, 1,200 rpm, 3-phase, 230 volts, 60 cycles synchronous motor, are shown in Fig. 1. The measured mechanical and iron losses are 560 watts. The measured resistance per phase is 0.20 ohms.

(a) Draw the circle chart for this motor.

(b) Compute the V-curve data for the motor using the circle chart.

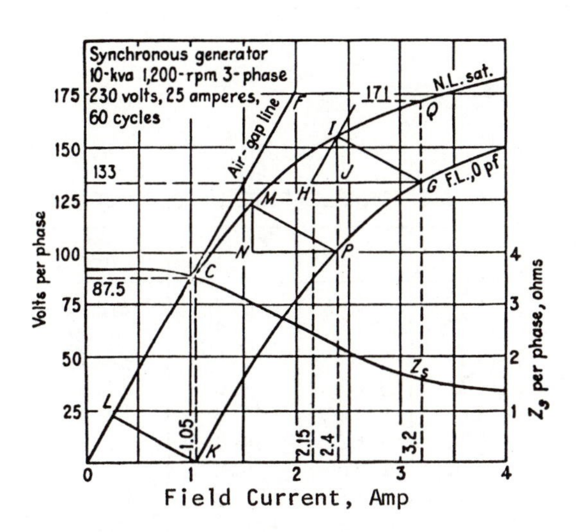

Full-load zero-power-factor saturation and adjusted synchronous impedance.

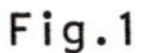
Fig.1

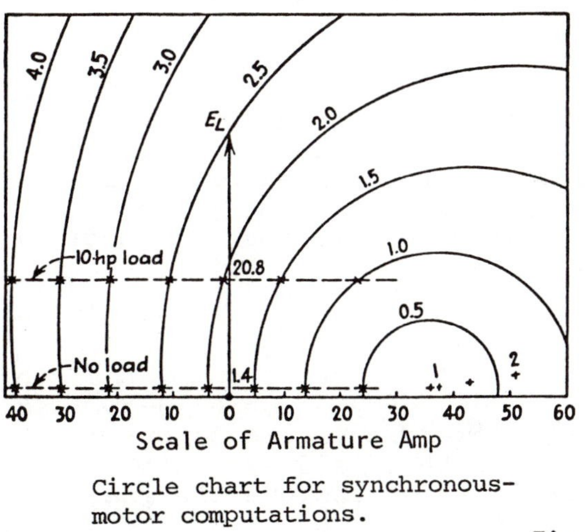

Circle chart for synchronous-motor computations.

Fig.2

Solution: (a) The values of E_g and Z_s for various values of field current I_f are obtained from Fig. 1. The data for drawing the circle chart are computed using the following equations:

Radius of the circle = E_g/Z_s

Distance to the center = E_L/Z_s

$$= \frac{E_g/Z_s \times 100}{\text{Per cent excitation}}$$

Slope of the line on which the center is located is

$$\theta = \cos^{-1} \frac{R}{Z_s}$$

VALUES FOR PLOTTING CIRCLES OF FIG.2

(1)	(2)	(3)	(4)	(5)	(6)
I_f	E_g	Z_s	E_g/Z_s	E_L/Z_s	$\cos\theta$
0.5	44	3.68	12.0	36.1	0.054
1.0	84	3.55	23.6	37.5	0.056
1.5	117	3.10	37.8	42.9	0.065
2.0	142	2.60	54.7	51.2	0.077
2.5	158	2.10	75.2	63.3	0.095
3.0	168	1.65	101.9	80.6	0.121
3.5	176	1.45	121.3	91.8	0.138
4.0	182	1.32	138.0	100.9	0.152

Table 1

The computed data are shown in Table 1. The circle chart is drawn using these data and is shown in Fig. 2.

The data for columns (2) and (3) were obtained from Fig. 1. Column (4) gives the radius of the circle, and (5) gives the distance to the center. Column (6) gives the slope of the line on which the center is located. The figures on the circles are the values of field current, shown in column (1).

(b) After the circle chart is completed, the computation of V curves is very easy. For a 10-hp load the a-c watts input at unity power factor is computed by adding losses to output. Since the field excitation is supplied separately, this does not include the field loss. The armature copper loss is figured by a cut-and-try computation, using the value of 0.2 ohm for R.

	Watts
Output..................................	7,460
Mechanical and iron loss..............	560
Armature copper = $3 \times (20.8)^2 \times 0.2$.....	260
	8,280

And finally, the current is computed to check the assumed value:

$$I = \frac{8,280}{\sqrt{3} \times 230} = 20.8 \text{ amp}$$

A horizontal line is drawn 20.8 amp above the base line, as shown in Fig. 2. The points where this line intersects the circles can then be measured to find the armature currents corresponding to the various values of field current. A locus line for no load is also plotted with vertical component of current,

$$I = \frac{560}{\sqrt{3} \times 230} = 1.4 \text{ amp}$$

Values of field current and line current for both computations are tabulated in Table 2.

COMPUTED V-CURVE DATA

I_f	Armature current at	
	No load	Full load
0	36.1	
0.5	24.0	
1.0	13.7	30.8
1.5	4.8	21.7
2.0	3.8	20.8
2.5	12.0	23.2
3.0	21.2	29.6
3.5	30.0	36.6
4.0	38.0	43.5

Table 2

POWER, POWER FACTOR AND EFFICIENCY

• PROBLEM 9-11

A 2300-V, three-phase synchronous motor driving a pump is provided with a line ammeter and a field rheostat. When the rheostat is adjusted so that the a-c line current is a minimum, the ammeter reads 8.8A. Approximately what horsepower is being delivered to the pump? How should the rheostat be adjusted so that the motor is operating at 0.8 power-factor leading? How many kVARs is the motor supplying to the system at 0.8 power factor leading?

Solution: At minimum line current, the power factor is unity. The power drawn from the line is thus

$$P = \sqrt{3}\, V_L I_L = \sqrt{3} \times 2300 \times 8.8 = 35 \text{ kW}$$

Neglecting losses, HP $\cong$ 35,000/746 = 47 hp. The a-c power is practically independent of field current. Then at 0.8 power factor,

$$|S| = \frac{P}{\text{p.f.}} = \frac{35{,}000}{0.8} = \frac{(\sqrt{3} \times 2300) \times 8.8}{0.8}$$

$$= (\sqrt{3} \times 2300) \times 11.0 \quad \text{VA}$$

Thus the line current should be 11.0A. To make sure the power factor is leading, the d-c field current should be increased until the a-c ammeter reads 11.0 amperes. This is accomplished by decreasing the field rheostat resistance.

The kVARs supplied by the motor are given by

$$Q = |S| \sin \theta_{\phi m} = |S| \sin \cos^{-1} (\text{p.f.})$$

$$= \frac{35 \text{ kW}}{0.8} \sin \cos^{-1} 0.8$$

$$= \frac{35}{0.8} \times 0.6 = 26.25 \text{ kVAR}$$

• PROBLEM 9-12

A three-phase, Y-connected load takes 50 A current at 0.707 lagging power factor at 220 V between the lines. A three-phase Y-connected round-rotor synchronous motor, having a synchronous reactance of 1.27 ohm per phase, is connected in parallel with the load. The power developed by the motor is 33 kW at a power angle of 30°. Neglecting the armature resistance, calculate (a) the reactive kVA of the motor, and (b) the overall power factor of the motor and the load.

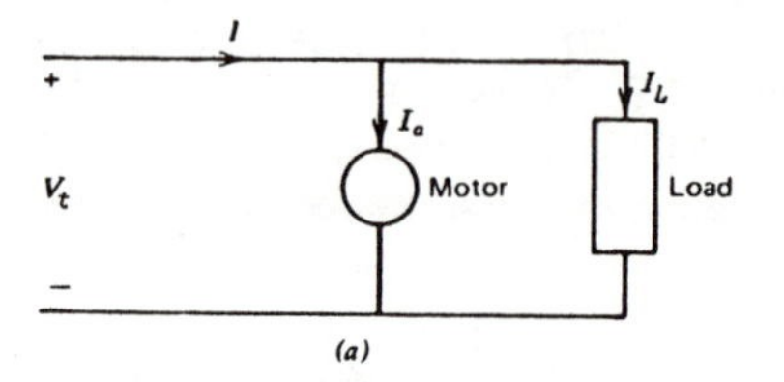

(a)

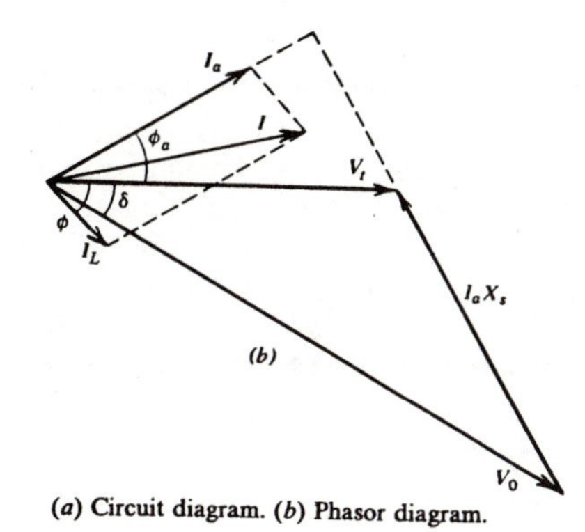

(a) Circuit diagram. (b) Phasor diagram.

Solution: The circuit and the phasor diagram, on a per-phase basis, are shown in the figure.

$$P_d = \frac{V_0 V_t}{X_s} \sin \delta$$

$$= \frac{1}{3} \times 33{,}000 = \frac{220}{\sqrt{3}} \frac{V_0}{1.278} \sin 30°$$

which yields V_0 = 220 volt. From the phasor diagram,

$$I_a X_s = 127$$

or

Ia = 127/1.27 = 100 A

and

$$\phi_a = 30°.$$

The reactive kVA of the motor =

$$\sqrt{3} \times V_0 I_a \sin \phi_a = \sqrt{3} \times \frac{220}{1000} \times 100 \times \sin 30$$

$$= 19 \text{ kvar}$$

The overall power-factor angle ϕ is given by

$$\tan \phi = \frac{I_a \sin \phi_a - I_L \sin \phi_L}{I_a \cos \phi_a + I_L \cos \phi_L} = 0.122 ,$$

or $\phi = 7°$ and $\cos \phi = 0.992$ leading.

• PROBLEM 9-13

A 3-phase, 100-h.p., 440-V, star-connected synchronous motor has a synchronous impedance per phase of 0.1 + j1Ω. The excitation and torque losses are 4 kW and may be assumed constant. Calculate the line current, power factor and efficiency when operating at full load with an excitation equivalent to 400 line volts.

Solution:

$$\alpha = \sin^{-1} R/Z_s = \sin^{-1} 0.1/\sqrt{[(0.1)^2 + 1^2]}$$

$$= \sin^{-1} 0.1/1.005 = 5.7°$$

Gross output = 100 × 746 + 4,000 = 78,600 watts.

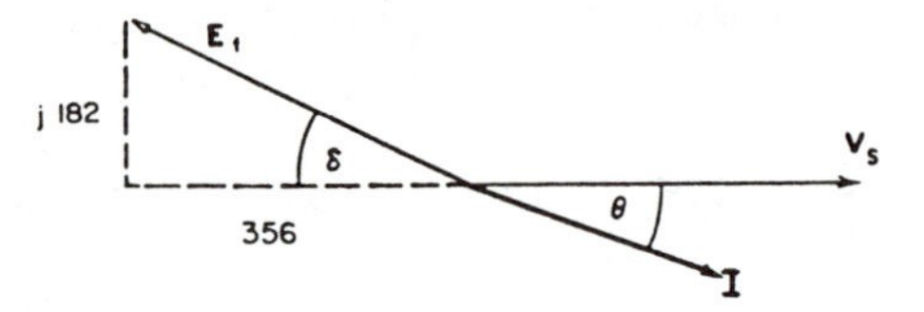

δ can be found by using the figure and the following equation:

$$E_f I \cos(\delta + \theta) = \frac{V_s E_f}{Z_s} \sin(\delta + \alpha) - \frac{E_f^2 R}{Z_s^2}$$

$$78{,}600 = 3\left[\frac{(440/\sqrt{3})(400/\sqrt{3})}{1.005} \sin(\delta + 5.7)° - \frac{(400/\sqrt{3})^2 \times 0.1}{(1.005)^2}\right]$$

$$= \frac{400 \times 440}{1.005}\left[\sin(\delta + 5.7)° - \frac{400 \times 0.1}{440 \times 1.005}\right]$$

$$0.45 = \sin(\delta + 5.7)° - 0.0905$$

from which $\delta = (\sin^{-1} 0.5405) - 5.7° = 27.1°$

$$\vec{E}_f = 400(-\cos\delta + j\sin\delta) = 400(-0.89 + j0.455)$$

$$= -356 + j182 \text{ V}$$

$$\vec{I} = \frac{(\vec{V}_s + \vec{E}_f)}{\vec{Z}_s} = \frac{440 - 356 + j182}{0.1 + j1} = 109 - j38.1 =$$

$$115.5\text{A}$$

$$\cos\theta = 109/115.5 = 0.945 \text{ lagging}$$

The answer can be checked by calculating $\sqrt{3}VI \cos\theta$ - copper losses and comparing with the gross mechanical output

i.e. $\sqrt{3} \times 440 \times 109 - 3 \times 115.5^2 \times 0.1 = 83{,}000 - 4000$

$$= 79{,}000 \text{ watts.}$$

The error is only about 0.5% and is within slide rule accuracy.

$$\text{Efficiency} = \frac{746 \times 100}{\sqrt{3} \times 440 \times 109} = 89.6\% .$$

• PROBLEM 9-14

(i) A 100-hp 600-volt 1,200-rpm 3-phase Y-connected synchronous motor has an armature resistance of 0.052 ohm per phase and a leakage reactance of 0.42 ohm per phase. At rated load and 0.8 power factor, leading current, determine (a) induced armature emf per phase E_a at rated load; (b) angle α between current and E_a; (c) mechanical power developed within armature at rated load. Under the foregoing conditions, the motor has a rated-load efficiency, excluding field loss, of 0.92.

(ii) Repeat the example, with lagging current.

Solution: (i)

(a)

$$\text{Motor input} = \frac{100 \times 746}{0.92} = 81{,}100 \text{ watts.}$$

$$\text{Current } I = \frac{81{,}100}{\sqrt{3} \times 600 \times 0.80} = 97.6 \text{ amp.}$$

$$\text{Voltage per phase} = \frac{600}{\sqrt{3}} = 346 \text{ volts.}$$

$$E_a = \sqrt{(V\cos\theta - IR)^2 + (V\sin\theta + IX)^2}.$$

$$E_a = \left\{\left[(346 \times 0.80)-(97.6 \times 0.052)\right]^2 + \left[(346 \times 0.60)+(97.6 \times 0.42)\right]^2\right\}^{\frac{1}{2}} = 368.3 \text{ volts}$$

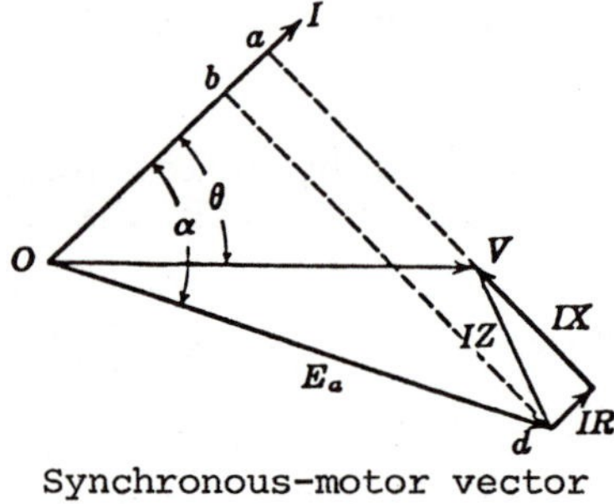

Synchronous-motor vector diagram, leading current.
Fig.1

(b) A study of Fig. 1 shows that

$$\tan\alpha = \frac{bd}{Ob} = \frac{V\sin\theta + IX}{V\cos\theta - IR} = \frac{248.6}{271.7} = 0.915.$$

$$\alpha = 42.5°.$$

$$E_a = V - IZ$$

$$= V - I(\cos\theta + j\sin\theta)(R + jX)$$

$$E_a = 346 - 97.6(0.80 + j0.60)(0.052 + j0.42)$$

$$= 346 - (-20.5 + j35.9) = 366.5 - j35.9.$$

$$|E_a| = \sqrt{(366.5)^2 + (35.9)^2} = 368.3 \text{ volts}$$

(c) The mechanical power developed is equal to the product of the induced emf, the current, and the cosine of the angle between them.

$$P_m = 3 \times 368.3 \times 97.6 \times \cos 42.5° = 79{,}600 \text{ watts.}$$

This power P_m is also equal to the power input minus the armature resistance loss.

$$P_m = 81{,}100 - 3 \times (97.6)^2 \times 0.052 = 79{,}600 \text{ watts.}$$

The power developed at the pulley is less than P_m by the rotational losses, namely, friction, windage, and rotational core losses.

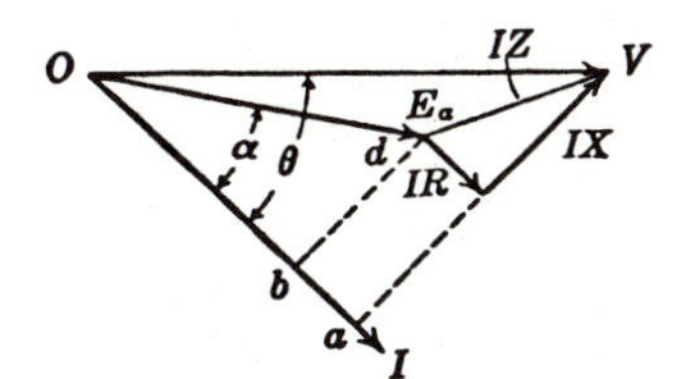

Synchronous-motor vector diagram, lagging current.

Fig.2

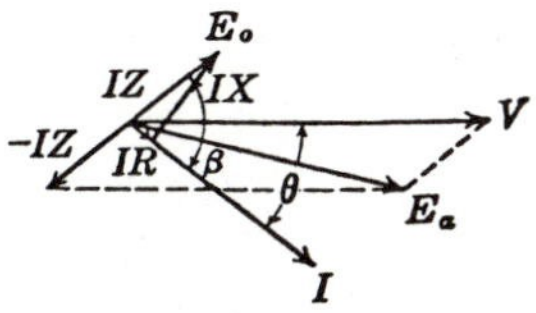

Vector diagram for counter emf less than terminal voltage.

Fig.3

Lagging Current.--In Fig. 2, which corresponds to Fig. 3, E_a and IR are projected on the current vector. Then

$$E_a = \sqrt{(V \cos \theta - IR)^2 + (V \sin \theta - IX)^2}. \tag{1}$$

The solution by complex notation is

$$\begin{aligned} \vec{E}_a &= \vec{V} - \vec{I}\vec{Z} \\ &= V - I(\cos \theta - j \sin \theta)(R + jX) \end{aligned} \tag{2}$$

(ii) (a) Using Eq. (1),

$$E_a = \Big\{[(346 \times 0.80) - (97.6 \times 0.052)]^2 + [(346 \times 0.60) - (97.6 \times 0.42)]^2\Big\}^{\frac{1}{2}} = 319 \text{ volts.}$$

Using Eq. (2),

$$\vec{E}_a = 346 - 97.6(0.80 - j0.60)(0.052 + j0.42)$$
$$= 346 - (28.7 + j29.8) = 317.3 - j29.8.$$

$$|\vec{E}_a| = \sqrt{(317.3)^2 + (29.8)^2} = 319 \text{ volts (check)}.$$

From Fig. 2,

(b) $$\tan \alpha = \frac{bd}{Ob} = \frac{V \sin \theta - IX}{V \cos \theta - IR} = \frac{166.6}{271.7} = 0.613.$$

$$\alpha = 31.5°.$$

(c) $$P_m = 3 \times 319 \times 97.6 \times \cos 31.5° = 79{,}600 \text{ watts}.$$

(This is the same value as before, which it should be.)

These values of E_a are the induced emfs in the armature. Since X is the armature leakage reactance, the values of E_a are the actual emfs induced in the armature under the given conditions of load. If the motor were disconnected from the line but driven mechanically, with the frequency and excitation unchanged, the emf at the terminals would be the no-load, or excitation, emf E. The excitation emf may be computed by using in the foregoing equations the synchronous reactance X_s, rather than the leakage reactance X.

• PROBLEM 9-15

A manufacturing plant takes 200 kw, at 0.6 power factor, from a 600-volt 60-cycle 3-phase system. It is desired to raise the power factor of the entire system to 0.9 by means of a synchronous motor, which at the same time is to drive a direct-current shunt generator, requiring that the synchronous motor take 80 kw from the line. What should be the rating of the synchronous motor in volts and amperes?

Solution: The vector diagram is shown in the figure. Assume that the system is Y-connected. The problem will be worked for one phase only.

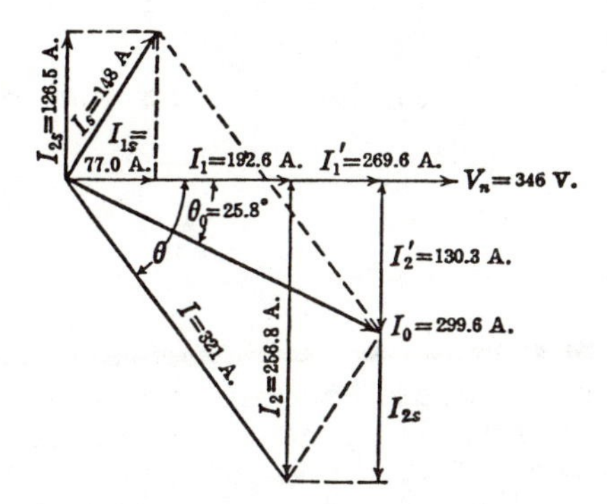

Vector diagram for synchronous motor which raises system power factor to cos θ_0.

Voltage to neutral $V_n = \dfrac{600}{\sqrt{3}} = 360$ volts.

Current per phase $I = \dfrac{200{,}000}{\sqrt{3} \times 600 \times 0.60} = 321$ amp.

Energy current of load $I_1 = I \cos\theta = 321 \times 0.6$

$= 192.6$ amp.

Quadrature current of load $I_2 = I \sin\theta = 321 \times 0.8$

$= 256.8$ amp.

At 0.9 power factor, the resultant power-factor angle $\theta_0 = 25.8°$.

Energy current of the synchronous motor

$$I_{1_s} = \frac{80{,}000}{\sqrt{3} \times 600} = 77.0 \text{ amp.}$$

Total energy current $I_1' = I_1 + I_{1_s} = 192.6 + 77.0 = 269.6$ amp.
Quadrature current of the system

$$I_2' = 269.6 \tan 25.8\% = 269.6 \quad 0.4834 = 130.3 \text{ amp.}$$

Quadrature current of the synchronous motor

$$I_{2_s} = I_2 - I_2' = 256.8 - 130.3 = 126.5 \text{ amp.}$$

Total synchronous-motor current

$$I_{2_s} = \sqrt{(I_{1_s})^2 + (I_{2_s})^2} = \sqrt{(77.0)^2 + (126.5)^2}$$

$$= \sqrt{21{,}930} = 148 \text{ amp.}$$

The synchronous motor will then be rated at 600 volts, 148 amp, or will have a rating of 154 kva.

The resultant current I_0, the vector sum of I and I_s, is shown in the figure.

• PROBLEM 9-16

The load of an industrial concern is 400 kva at a power factor of 75 per cent, lagging. An additional motor load of 100 kw is needed. Find the newkilovolt-ampere load and the power factor of the load, if the motor to be added is (a) an induction motor with a power factor of 90 per cent, lagging, and (b) an 80 per cent power factor (leading) synchronous motor.

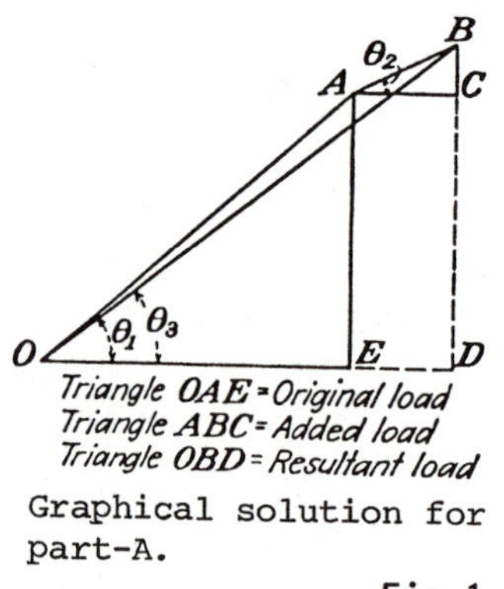

Graphical solution for part-A.

Fig.1

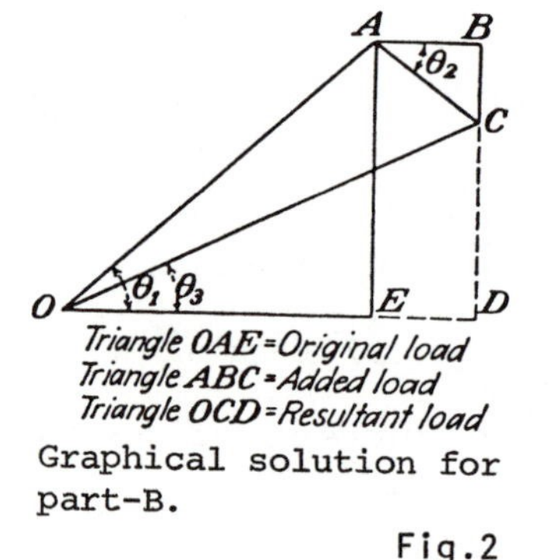

Graphical solution for part-B.

Fig.2

Solution: Original load, 400 kva at 75 per cent power factor.

$$\text{kw} = 400 \times 0.75 = 300 \text{ kw}$$

$$(\text{kva})^2 = (\text{kw})^2 + (\text{kvar})^2$$

$$\text{kvar} = \sqrt{(\text{kva})^2 - (\text{kw})^2}$$

$$= \sqrt{400^2 - 300^2} = 264.6 \text{ kvar, lagging}$$

(a) Induction motor, 100 kw at 90 per cent lagging power factor

$$\text{kva} = \frac{\text{kw}}{\text{power factor}} = \frac{100}{0.9} = 111.1 \text{ kva}$$

$$\text{kvar} = \sqrt{(111.1)^2 - (100)^2} = 48.4 \text{ kvar, lagging}$$

Resultant kilowatt load = 300 + 100 = 400 kw

Resultant kilovar load = 264.6 + 48.4 = 313 kvar

Resultant kilovolt-ampere load $= \sqrt{(\text{kw})^2 + (\text{kvar})^2}$

$= \sqrt{(400)^2 + (313)^2}$

= 507.8 kva

Resultant power factor $= \frac{\text{kw}}{\text{kva}} = \frac{400}{507.8}$

= 0.787 = 78.7 per cent, lagging

Since the relation between kilowatts, kilovars, and kilovolt-amperes of any load may be represented by a right triangle, the solution of the above problem may be made graphically as shown in Fig. 1. The triangle OAE represents the original load; OE = 300 kw, EA = 264.6 kvar, and OA = 400 kva. The angle θ_1 is the power-factor angle of the original load. Triangle ABC represents the added induction-motor load, sides AC, CB, and AB representing 100 kw, 48.4 kvar, and 111.1 kva, respectively, with the phase angle represented as θ_2. Triangle OBD represents the resultant load with sides OD, DB, and OB representing 400 kw, 313 kvar, and 507.8 kva, respectively. The resultant power-factor angle is θ_3.

(b) Synchronous motor, 100 kw at 80 per cent leading power factor.

$$\text{kva} = \frac{\text{kw}}{\text{power factor}} = \frac{100}{0.8} = 125 \text{ kva}$$

$$\text{kvar} = \sqrt{(125)^2 - (100)^2} = 75 \text{ kvar leading}$$

Resultant kilowatt load = 300 + 100 = 400 kw

Resultant kilovar load = 264.6 (lagging) - 75 (leading)

= 189.6 kvar (lagging)

Resultant kilovolt-ampere load $= \sqrt{(\text{kw})^2 + (\text{kvar})^2}$

$= \sqrt{(400)^2 + (189.6)^2} = 442.5$ kva

Resultant power factor =

$$\frac{\text{kw}}{\text{kva}} = \frac{400}{442.5} = 0.903 = 90.3 \text{ per cent, lagging}$$

The graphical solution is shown in Fig. 2, with triangles OEA, ABC, and OCD representing the original load, the added load, and the resultant load, respectively.

SYNCHRONOUS MOTOR DYNAMICS

• PROBLEM 9-17

A 200-hp 2,300-volt 3-phase 60-cps 28-pole 257-rpm synchronous motor is directly connected to a large power system. The motor has the following characteristics:

Wk^2 = 10,500 lb-ft^2 (motor plus load)

Synchronizing power P_s = 11.0 kw/elec deg

Damping torque = 1,770 lb-ft/mech rad/sec

(a) Investigate the mode of electrodynamic oscillation of the machine.

(b) Rated mechanical load is suddenly thrown on the motor shaft at a time when it is operating in the steady state, but unloaded. Study the electrodynamic transient which will ensue.

Solution: (a) Throughout this solution, the angle δ will be measured in electrical degrees rather than radians. This fact must be recognized in obtaining P_j, P_d, and P_s from the given data.

The inertia is given as Wk^2 (weight times square of radius of gyration) in English units, a common practice for large machines. In mks units,

$$J = \frac{Wk^2}{23.7} = \frac{10,500}{23.7} = 444 \text{ kg-m}^2$$

From the following equation:

$$P_j = J \frac{2}{\text{Poles}} \frac{2\pi n}{60} \quad ,$$

with the factor $\pi/180$ inserted to convert angular measurement to degrees,

$$P_j = 444 \times \frac{2}{28} \times \frac{2\pi(257)}{60} \times \frac{\pi}{180}$$

$$= 14.9 \text{ watts/elec deg/sec}^2$$

The remaining motor constants in the appropriate units are

$$P_d = 2\pi \times 257 \times 1{,}770 \times \frac{746}{33{,}000} \times \frac{\pi}{180} \times \frac{2}{28}$$

$$= 80.6 \text{ watts/elec deg/sec}$$

$$P_s = 11.0 \times 1{,}000 = 11{,}000 \text{ watts/elec deg}$$

The force-free equation which determines the mode of oscillation is then

$$a\frac{d^2x}{dt^2} + b\frac{dx}{dt} + cx = d$$

or

$$14.9\frac{d^2\delta}{dt^2} + 80.6\frac{d\delta}{dt} + 11{,}000\delta = 0$$

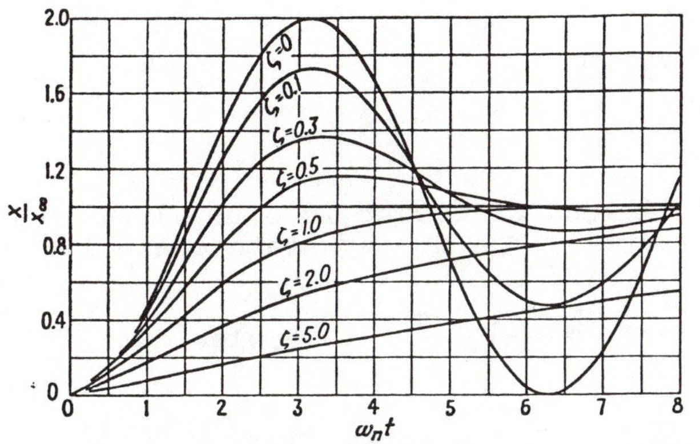

Normalized solutions of second-order linear differential equation for initial-rest conditions.

From the following equations:

$$\frac{c}{a} = \omega_n^{\,2}$$

$$\frac{b}{2\sqrt{ac}} = \rho\,,$$

the undamped angular frequency and damping ratio are, respectively,

$$\omega_n = \sqrt{\frac{11{,}000}{14.9}} = 27 \text{ rad/sec}$$

$$\zeta = \frac{80.6}{2\sqrt{14.9 \times 11{,}000}} = 0.10$$

The magnitude of ζ places the transient response decidedly in the oscillatory region as it does for all synchronous machines. Any operating disturbance will be followed by a relatively slowly damped oscillation, or swing, of the rotor before steady operation at synchronous speed is resumed. A large disturbance may, of course, be followed by complete loss of synchronism. The damped angular velocity of the motor is

$$\omega_d = \omega_n\sqrt{1-\rho^2}$$

$$= 27\sqrt{1 - (0.10)^2} = 26.9 \text{ rad/sec}$$

corresponding to a damped oscillation frequency of

$$f_d = \frac{26.9}{2\pi} = 4.3 \text{ cps}$$

(b) The full load of 200 hp is equivalent to 200 × 746 = 149,200 watts. The steady-state operating angle is

$$\delta_\infty' = \frac{149{,}200}{11{,}000} = 13.6 \text{ elec deg.}$$

In accordance with the following equation:

$$\frac{x}{x_\infty} = 1 - \frac{1}{\sqrt{1 - \rho^2}} e^{-\rho\omega_n t} \sin(\sqrt{1 - \rho^2}\, \omega_n t + \phi),$$

the angular excursions are characterized by the equation

$$\delta = 13.6°[1 - 1.004e^{-2.7t} \sin(26.9t + 84.3°)]$$

The oscillation is depicted graphically by the curve for $\zeta = 0.1$ in the figure, where one unit on the vertical scale is taken as 13.6°.

CHAPTER 10

INDUCTION MACHINES

STARTING AN INDUCTION MOTOR

• PROBLEM 10-1

Compensators are supplied with several secondary voltage taps to permit the user to select the turns ratio (Fig. 1). A 100-hp, 1750-rev/min, 2300-V, three-phase motor draws 150 A and produces a starting torque of 120 percent full-load value, when started at 2300 V. If a compensator is to be used, and the 80 percent voltage tap is selected, find the starting line current and torque.

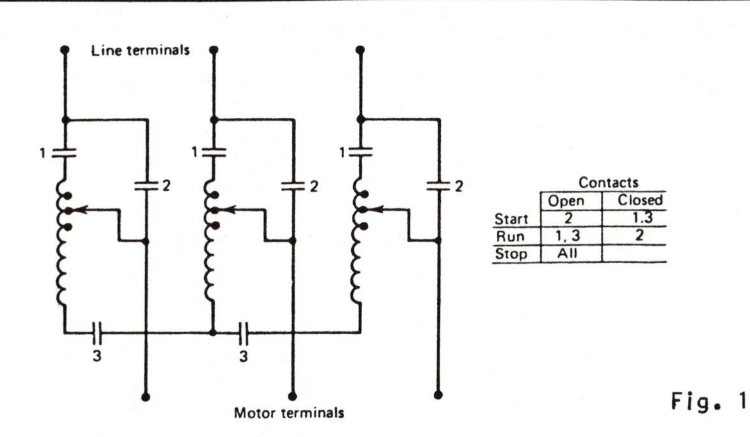

Fig. 1

Starting compensator for induction motors or synchronous motors with damper windings.

Solution: The compensator turns ratio is

$$a = \frac{V_L}{0.8V_L} = \frac{1}{0.8} = 1.25$$

The voltage applied to the motor at starting is

$$\frac{2300}{a} = 0.8 \times 2300 = 1840 \text{ V}$$

and since the motor phase impedance is practically independent of voltage at s = 1,

$$I_{1\ start} = \frac{1840}{\sqrt{3}|Z_{in}|} = \frac{1840}{2300} \times 150 \text{ A} = \frac{150}{a}$$

$$= 0.8 \times 150 = 120 \text{ A.}$$

Then, by the transformer action, the line current is given by

$$I_L = \frac{I_{1\ start}}{a} = \frac{120}{a} = \frac{150}{a^2} = (0.8)^2 \times 150 = 96 \text{ A}$$

In general,

$$I_L = \frac{\text{Starting current at full voltage}}{a^2}$$

The starting voltage applied to the motor is V_L/a, or 1840 V, but the torque is proportional to the voltage squared. Then

Starting torque with compensator

$$= (1/a^2) \times (\text{Starting torque at line voltage})$$

Full-load torque of the motor is given by

$$\tau_{fl} = \frac{5252\text{hp}}{\text{rev/min}} = \frac{525{,}200}{1750} = 300 \text{ lb ft.}$$

Starting torque at rated voltage is then 360 lb ft. Then

$$\tau_{st} = \frac{360}{a^2} = (0.8)^2 \times 360 = 230 \text{ lb ft.}$$

• PROBLEM 10-2

A 50-hp, 2200-V, 13.5-A, 1160r/min., 60-Hz, wound-rotor induction motor has the following equivalent-circuit parameters:

$$R_S = 2.22\ \Omega \qquad X_L = 14.2\ \Omega$$

$$R'_R = 2.97\ \Omega \qquad X_M = 324.0\ \Omega$$

The effective turns ratio is $N_{se}/N_{re} = 1.22$ and $k = 0.979$

a) Determine the external rotor-circuit resistance per phase required to hold down the starting current to three times the rated value.
b) Determine the speed at which the motor would develop an internal torque equal to 125% of the rated torque with this starting resistance still in circuit.
c) Determine the line current that would result if the starting resistance were shorted out of circuit at the speed determined in (b).

Solution: a) Permissible starting current is

$$I_a = 3 \times 13.5 = 40.5\ \text{A}.$$

The corresponding value if I'_A at standstill may be obtained from a circle diagram. From the equivalent circuit of Fig. 2,

$$I_M = \frac{2200}{324.0\sqrt{3}} = 3.920\ \text{A}.$$

The radius of the circle diagram is:

$$\frac{V_a}{2X_L} = \frac{2200}{2 \times 14.2\sqrt{3}} = 44.72\ \text{A}.$$

The circle diagram for starting conditions is shown in Fig. 1, from which

$$\cos\alpha = \frac{40.5^2 + 48.6^2 - 44.7^2}{2 \times 40.5 \times 48.6}$$

$$= \frac{40.5^2 + 3.92^2 - (I'_A)^2}{2 \times 40.5 \times 3.92}$$

From which

$$I'_A = 38.7 \text{ A}$$

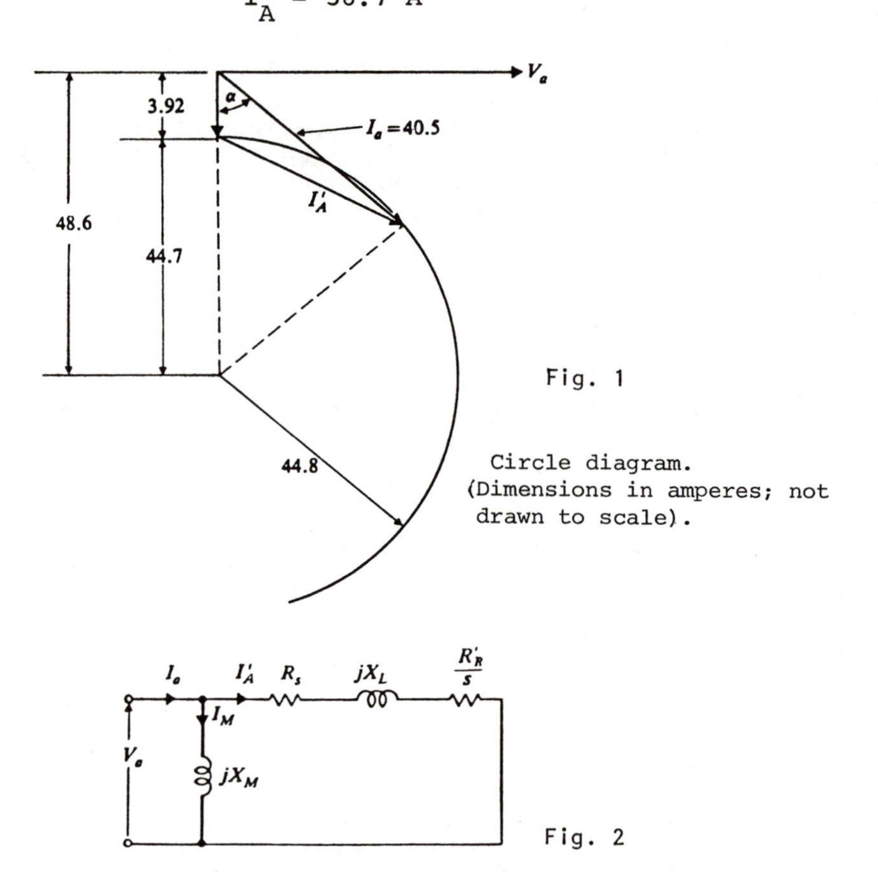

Fig. 1

Circle diagram. (Dimensions in amperes; not drawn to scale).

Fig. 2

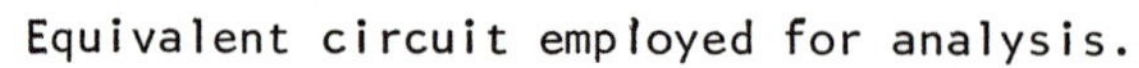

Equivalent circuit employed for analysis.

The impedance at standstill of the right-hand branch of the circuit of Fig. 2 is

$$\vec{Z} = 2.22 + j\,14.2 + 2.97 + R'_{EX}$$

$$= (5.19 + R'_{EX}) + j\,14.2$$

Thus

$$Z^2 = (5.19 + R'_{EX})^2 + 14.2^2 = \left(\frac{2200}{38.7\sqrt{3}}\right)^2$$

From which

$$R'_{EX} = 24.4\ \Omega.$$

Thus

$$R_{ex} = k^2\left(\frac{N_{re}}{N_{se}}\right)^2 R'_{EX} = \frac{0.979^2}{1.22^2} \times 24.5 = 15.8\ \Omega.$$

b) Rated speed $= \frac{2\pi}{60} \times 1160 = 121.5$ rad/s

Rated torque $= \frac{50 \times 746}{121.5} = 307.1$ N•m.

$$T = \frac{3}{\omega_s} \times \frac{p}{2} \times \frac{R'_R}{s} \frac{V_a^2}{[R_s + (R'_R/s)]^2 + X_L{}^2} \quad \text{N.m}$$

$R'_R + R'_{EX} = 27.4$ must be substituted for R'_R; thus

$$1.25 \times 307.1 = \frac{3}{120\pi} \times \frac{6}{2} \times \frac{27.4}{s} \times \frac{(2200/\sqrt{3})^2}{(2.22 + 27.4/s)^2 + 14.2^2}$$

From which s = 0.293, and

$$\omega_m = (1 - s)\ \frac{2}{p}\ \omega_s = (1 - 0.293)\frac{2}{6} \times 120\pi = 88.8 \text{ rad/s}$$
$$= 848 \text{ r/min.}$$

c) Impedance of the right-hand branch of the equivalent circuit at s = 0.293, and with R_{ex} shorted out is

$$\vec{Z} = 2.22 + \frac{2.97}{0.293} + j\ 14.2 = 12.36 + j\ 14.2$$
$$= 18.82\angle 48.96°\ \Omega.$$

$$\vec{I}'_A = \frac{2200\angle 0}{\sqrt{3} \times 18.82\angle 48.96°} = 67.49\angle -48.96°$$

$$= 44.3 - j50.9 \text{ A.}$$

$$\vec{I}_a = \vec{I}'_A + \vec{I}_M = 44.3 - j50.9 - j3.92 = 44.2 - j54.8 \text{ A.}$$

from which

$$I_a = 70.4 \text{ A}$$

Since this is more than three times the rated current, the whole external resistance may not be shorted out. It must be reduced in steps.

• PROBLEM 10-3

A 3-phase, Y-connected, 60-cycle, 220-volt, 5-hp, 1800-rpm, wound-rotor induction motor has the following constants: $R_1 = 0.545$, $R_2' = 0.541$, $X_1 + X_2' = 1.68$. The slip at full load is $s = 0.05$. The line current during the starting cycle is not to exceed 3 times the normal current. The load torque equation is $T(s) = 10 + 6.1(1-s)^2$ lb-ft. Investigate the different starting methods, neglecting the exciting current.

Solution: 1. Line start

$$\text{Normal Current } I = \frac{127}{\sqrt{(0.545 + 0.541/0.05)^2 + 1.68^2}}$$

$$= 11.07 \text{ amps}$$

$$\text{Normal torque } T = \frac{33000 \times 3}{2\pi 1800 \times 746} \frac{0.541}{0.05} 11.07^2 = 15.6 \text{ lb-ft}$$

$$\text{Starting current } I_{(0)} = \frac{127}{\sqrt{(0.545 + 0.541)^2 + 1.68^2}}$$

$$= 63.5 \text{ amps}$$

$$\text{Starting torque } T_{(0)} = 0.01175 \times 0.541 \times 63.5^2$$

$$= 25.6 \text{ lb-ft}$$

The starting current is nearly twice the allowable line current, which precludes this method of starting.

2. Compensator starting

The maximum permissible line starting current is $I'_{max} = 3 \times 11.07 = 33.2$ amperes. The initial transformer tap is

$$a_1 = \sqrt[4]{\frac{1}{(R_1 + R_2')^2 + (X_1 + X_2')^2}} \sqrt{\frac{V_1}{I'_{max}}}$$

$$= \left(\frac{127}{33.2}\right)^{1/2} \frac{1}{[(0.545 + 0.541)^2 + 1.68^2]^{1/4}}$$

$$= 1.38$$

giving an initial motor current of $1.38 \times 33.2 = 45.8$ amperes and a starting torque of

$$T_{(0)} = 0.01175 \times 0.541 \times 45.8^2 = 13.4 \text{ lb-ft}$$

From the following equations:

$$T(s_k) = \frac{33000 m_1}{2\pi n_1 746} \frac{R_2'}{s_k} \frac{(V_1/a_k)^2}{(R_1 + R_2'/s_k)^2 + (X_1 + X_2')^2}$$

$$T(s) = 10 + 6.1(1 - s^2) \text{ lb-ft},$$

the slip at which the tap is changed satisfies the equation,

$$10 + 6.1(1 - s_1)^2 = 0.01175 \frac{0.541}{s_1}$$

$$\times \frac{(127/1.38)^2}{(0.545 + 0.541/s_1)^2 + 1.68^2}$$

from which it is found that $s_1 = 0.108$, and the minimum current is

$$I_{min(k)} = \frac{V_1}{a_k\sqrt{(R_1 + R_2'/s_k)^2 + (X_1 + X_2')^2}}$$

$$= \frac{127/1.38}{\sqrt{(0.545 + 0.541/0.108)^2 + 1.68^2}}$$

$$= 15.80 \text{ amps.}$$

Hence the new transformer ratio is

$$a_{k+1} = \sqrt{a_k \frac{I_{min(k)}}{I'_{max}}}$$

$$a_2 = \sqrt{1.38 \frac{15.80}{33.20}} = 0.81 ,$$

which, being less than unity, means that the motor would be thrown directly on the line at this instant. The line current at the instant after change-over is

$$I = 1.38 \times I_{min} = 21.80 \text{ amps}$$

and the torque jumps to

$$T_{(1)}' = 0.01175 \times \frac{0.541}{0.108} 21.80^2 = 28 \text{ lb-ft}$$

3. Wye-delta starting

Suppose that a motor drawing the same line current and with the same torque characteristics as specified in the example can be designed for a delta running connection, but may be started wye-connected. The starting current per phase in delta with 220 volts across the winding is 63.5/1.73 = 36.6 amperes. If connected in wye, the voltage across each phase is reduced by

$\sqrt{3}$, and the new phase and line current is 36.6/1.73 = 21.2 amps. This is below the permissible maximum line current. The torque, however, being proportional to the square of the current is one-third of the delta starting torque, or 25.6/3 = 8.5 lb-ft. This is below that required by the load, and therefore wye-delta starting is not suitable for this case.

4. Wound rotor (current limitations)

Maximum starting current is 33.2 amperes. This will require a total rotor resistance (referred to the stator) such that

$$33.2 = \frac{127}{\sqrt{(0.545 + r_1)^2 + 1.68^2}} \text{ or } r_1 = 2.88 \text{ ohms}$$

At this current and resistance the starting torque is

$$T_{(0)} = 0.01175 \times 2.88 \times 33.2^2 = 37.3 \text{ lb-ft.}$$

which is several times that required by the load.

If it is desired to have the fewest possible steps in the external resistance, then minimum current may be determined by finding the slip at which the torque with r_1 = 2.88 ohms in the rotor is just equal to the load torque. However, by taking a higher value for the minimum current, more steps, with faster acceleration and smoother starting are possible. In the present case, try I_{min} = 20 amperes. Then from the following equation:

$$\frac{r_{k+1}}{r_k} = \frac{\sqrt{(V_1/I_{max})^2 - X^2} - R_1}{\sqrt{(V_1/I_{min})^2 - X^2} - R_1}, \tag{1}$$

$$= \frac{\sqrt{(127/33.2)^2 - 1.68^2} - 0.545}{\sqrt{(127/20)^2 - 1.68^2} - 0.545} = 0.52$$

And from the following equation:

$$\frac{r_{k+1}}{s_k} = \frac{r_k}{s_{k-1}} \quad (2)$$

starting with $s_0 = 1$, the slip at which the current reaches a minimum is

$$s_1 = 0.52 \times 1 = 0.52$$

At this slip the torque is

$$T_{(1)} = 0.01175 \times \frac{2.88}{0.52} \times 20^2 = 26 \text{ lb-ft}$$

Repeated use of (1) and (2) and the torque equation yield in succession

k	0	1	2	3	4	
r_k	2.88	2.88	1.50	0.78	0.541	(rotor resistance)
s_k	1.00	0.52	0.27	0.14	0.05	(full load slip)
T_k		26	26	26	15.6	(full load torque)
T_k'	37.3	37.3	37.3	33.2		(peak torque)

At the third step, when the rotor is short-circuited, the current jumps to

$$I_{(3)} = \frac{127}{\sqrt{(0.545 + 0.541/0.14)^2 + 1.68^2}} = 26.9 \text{ amps}$$

and the torque to

$$T_{(3)} = 0.01175 \frac{0.545}{0.14} 26.9^2 = 33.2$$

5. Wound-rotor (maximum torque at start)

The rotor resistance for maximum torque at start is

$$r_1 = \sqrt{0.545^2 + 1.68^2} = 1.76 \text{ ohms}$$

and starting current is

$$\frac{127}{\sqrt{(0.545 + 1.76)^2 + 1.68^2}} = 44.7 \text{ amperes}$$

This exceeds the allowable starting current of 33.2 amperes. Nevertheless, the calculations will be completed. Maximum torque is

$$T_{max} = \frac{m_1 V_1^{\,2}}{2(R_1 + \sqrt{R_1^{\,2} + X^2})}$$

$$= 0.01175 \frac{127^2}{2(0.545 + \sqrt{0.545^2 + 1.68^2})}$$

$$= 41 \text{ lb-ft.}$$

Let minimum torque be taken as $T_{min} = 30$ lb-ft. Then, from the following equation:

$$\frac{T_{max}}{T_{min}} = \frac{(R_1 + r_1/s_1)^2 + X^2}{2(R_1 + \sqrt{R_1^{\,2} + X^2})(r_1/s_1)},$$

$$\frac{41}{30} = \frac{(0.545 + 1.76/s_1)^2 + 1.68^2}{2(0.545 + \sqrt{0.545^2 + 1.68^2})(1.76/s_1)}$$

which is satisfied by $s_1 = 0.39$. The new resistance for the rotor at this slip is

$$r_2 = r_1 s_1^{\,2}$$

$$= 1.76 \times 0.39^2 = 0.268 \text{ ohm}$$

Since this is less than the rotor resistance of 0.541 ohms, the rotor would be short-circuited at this step, and the torque jumps to

$$T_{(2)}' = 0.01175 \frac{0.541}{0.39} \frac{127^2}{(0.545 + 0.541/0.39)^2 + 1.68^2}$$

$$= 40.2 \text{ lb-ft.}$$

The abrupt jump in torque could have been avoided, according to the following equation,

$$s' = \sqrt{\frac{rR_2'}{R_1{}^2 + X^2}}$$

by short-circuiting the rotor at a slip

$$s' = \sqrt{\frac{1.76 \times 0.541}{0.545^2 + 1.68^2}} = 0.552$$

DETERMINATION OF EQUIVALENT CIRCUIT PARAMETERS

• PROBLEM 10-4

Tests on a 75-hp three-phase 60-Hz 440-V 88-A six-pole 1170-rpm deep-bar induction motor (class B) yielded results as follows:

a. No Load. 440V line to line, 24.0 A line current, 2.56 kW three-phase power.
b. Locked-Rotor Test at 15 Hz. 28.5 V line to line, 90.0 A line current, 2.77 kW three-phase power.
c. Average Value of DC Resistance Between Stator Terminals. 0.0966 Ω terminal to terminal.
d. Locked-Rotor Test at Rated Voltage, 60 Hz. 440 V line to line, 503 A line current, 150.0 kW three-phase power.

Calculate (a) the constants of the equivalent circuit in Fig. 1 for the running range, and (b) the electromagnetic torque on full-voltage starting.

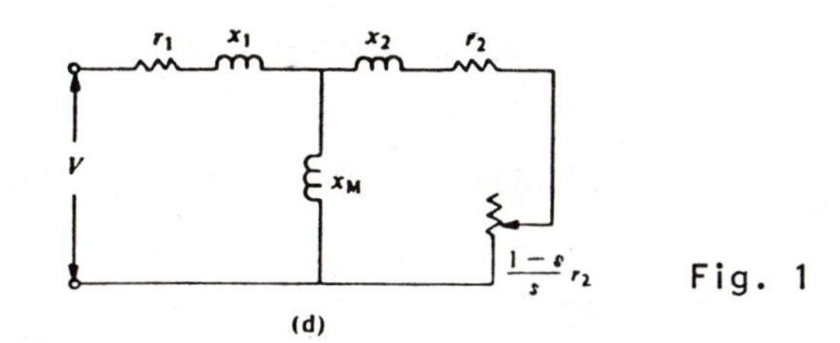

Fig. 1

Simplified equivalent circuit for motor under load.

Solution: a. From the no-load test and the equations,

$$V = \frac{V_0}{\sqrt{3}}, \quad Z_0 = \frac{V}{I_0}, \quad P_0 = 3I_0^2 r_0, \quad x_0 = \sqrt{z_0^2 - r_0^2},$$

one obtains

$$z_0 = \frac{440}{\sqrt{3}(24.0)} = 10.57\ \Omega$$

$$r_0 = \frac{2560}{3(24)^2} = 1.48\ \Omega$$

$$x_0 = \sqrt{(10.57)^2 - (1.48)^2} = 10.45\ \Omega$$

From the dc test,

$$r_1 = 0.0966 \div 2 = 0.0483\ \Omega$$

Table 1
Empirical Proportions of Induction-Motor Leakage Reactances

Type of Motor	Class A	Class B	Class C	Class D	Wound Rotor
x_1	$0.5x_L$	$0.4x_L$	$0.3x_L$	$0.5x_L$	$0.5x_L$
x_2	$0.5x_L$	$0.6x_L$	$0.7x_L$	$0.5x_L$	$0.5x_L$

From the blocked-rotor test at 15 Hz and the equations,

$$z_L = \frac{V_L}{\sqrt{3}I_L}, \quad r_L = r_1 + R_2 = \frac{P_L}{3I_L^2},$$

$$x_L = x_1 + X_2 = \sqrt{z_L^2 - r_L^2},$$

$$z_{L15} = \frac{28.5}{\sqrt{3}(90.0)} = 0.183\Omega \quad r_{L15} = \frac{2770}{3(90.0)^2} = 0.114\Omega$$

$$x_{L15} = \sqrt{(0.183)^2 - (0.114)^2} = 0.1435\Omega$$

The corresponding 60-Hz value is

$$x_L = \frac{60}{15} x_{L15} = \frac{60}{15} \times 0.1435 = 0.574 \ \Omega.$$

The leakage reactances are determined from Table 1 to be

$$x_1 = 0.4x_L = 0.4(0.574) = 0.230 \ \Omega.$$

$$x_2 = 0.6x_L = 0.6(0.574) = 0.344 \ \Omega.$$

The magnetizing reactance is

$$x_M = x_0 - x_1 = 10.45 - 0.23 = 10.22 \ \Omega$$

and the rotor resistance referred to the stator is

$$r_2 = (r_{L15} - r_1)\left(\frac{x_2 + x_M}{x_M}\right)^2 = 0.0657\left(\frac{0.343 + 10.22}{10.22}\right)^2$$

$$= 0.0697 \ \Omega$$

b. Full-voltage starting torque. The equation,

$$T_{em} = \frac{P_{em}}{\omega_m} = \frac{m_1 I_2^2[(1 - s)/s]r_2}{2\pi(1 - s)(n_{syn}/60)} = \frac{m_1 I_2^2 (r_2/s)}{2\pi n_{syn}/60},$$

shows the electromagnetic torque to equal the result of dividing the real-power input to the rotor by the synchronous angular velocity. The power input to the rotor is taken as the difference between the real-power input to the stator and the stator copper loss. At full-voltage starting,

Power input to stator = 150,000 W

$$\text{Stator copper loss} = 3I_1^2 r_1 = 3(503)^2(0.0483)$$

$$= 36{,}700 \text{ W}$$

Power input to rotor = 150,000 - 36,700 = 113,300 W

$$T_{em} = \frac{113,300}{2\pi n_{syn}/60} \quad \text{and} \quad n_{syn} = 1200 \text{ rpm}$$

$$T_{em} = \frac{113,300}{40\pi} = 900 \text{ N-m/rad}$$

The measured torque is lower than the calculated value above because in computing the power input to the rotor winding, the core losses and the stray load losses were not taken into account.

• PROBLEM 10-5

Determine the six parameters of a 3-HP, 440/220-volt, 3-phase, 60-cycle, 4-pole squirrel-cage induction motor from a no-load and a locked-rotor test. The stator resistance at 25°C is 2.26 ohms, and the friction and windage loss is 44 watts. Stray load loss = 48 watts.

No-load test—75°C

$V_1 = 440$ volts

$I_0 = 2.36$ amp

$P_0 = 211$ watts

Locked-rotor test (full voltage)—75°C

$V_L = 440$ volts $= V_1$

$I_L = 29.1$ amp

$P_L = 13.92$ kw

Locked-rotor test (reduced voltage)

$V_L = 76$ volts

$I_L = 4.25$ amp

The skin-effect factor for r_2' is 1.30 and for x_2' is 0.97 (see Fig. 1).

Solution: Saturated reactances

$$Z_L = \frac{440}{\sqrt{3} \times 29.1} = 8.73 \text{ ohms.}$$

$$R_L = \frac{13,920}{3 \times (29.1)^2} = 5.48 \text{ ohms.}$$

$$X_L = \sqrt{Z_L^2 - R_L^2} = 6.80 \text{ ohms.}$$

$$r_1 (\text{at } 75°\text{C}) = 2.26 \times \frac{234.5 + 75}{234.5 + 25} = 2.69 \text{ ohms.}$$

$$\therefore \quad r_2' = 5.48 - 2.69 = 2.79 \text{ ohms}$$

$$x_1 = x_2' = 3.40 \text{ ohms}$$

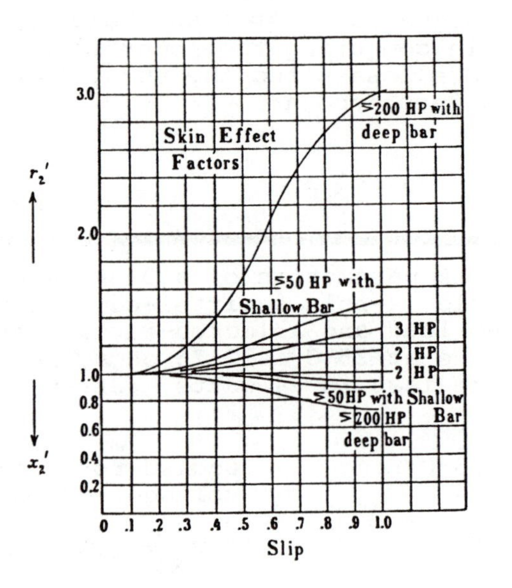

Fig. 1

Skin effect factors for problems.

Unsaturated reactances

$$Z_L = \frac{76}{\sqrt{3} \times 4.25} = 10.3 \text{ ohms}$$

$$X_L = \sqrt{(10.3)^2 - (5.48)^2} = 8.72 \text{ ohms}$$

$$x_1 = \frac{8.72}{2} = 4.36 \text{ ohms}$$

The ratio of unsaturated leakage reactance to saturated leakage reactance (4.36/3.40) is called the saturation factor, and equals 1.28.

In order to determine the running performance from the equivalent circuit it is necessary to correct r_2' and x_2' for skin-effect since the unsaturated parameters above were taken at 60 cycles.

$$\therefore \quad r_2' \text{ (corrected)} = \frac{2.79}{1.30} = 2.14 \text{ ohms}$$

$$x_2' \text{ (corrected)} = \frac{4.36}{0.97} = 4.50 \text{ ohms}$$

$$x_m = \frac{254 - 2.36 \times 4.36}{2.36} = \frac{243.7}{2.36} = 103 \text{ ohms}$$

From the following equations:

$$P_0' = m_1 I_0'^2 r_1 + P_{h+e}$$

and

$$P_{h+e} + P_{ir.rot} = P_0 - 3(I_0)^2 \times r_1$$
$$- \text{(friction and windage loss)},$$
$$= 211 - 3(2.36)^2 \times 2.69 - 44$$
$$= 122 \text{ watts}$$

It will be assumed that one half of the iron losses are due to the main flux so that:

$$P_{h+e} = 61 \text{ watts}$$

$$g_m = \frac{P_{h+e}}{m_1 E_1^2} = \frac{61}{3 \times (243.7)^2} = 3.43 \times 10^{-4} \text{ mho.}$$

$$r_m = g_m x_m^2 = 3.43 \times 10^{-4} \times (103)^2 = 3.66 \text{ ohms.}$$

To check the computed value of r_m, obtain

$$\text{(check)} \quad m_1 I_0^2 r_m = 3 \times (2.36)^2 \times 3.66 = 61 \text{ watts.}$$

$$r_m = \frac{61}{3 \times (2.36)^2} = 3.66 \text{ ohms.}$$

• PROBLEM 10-6

The results of the no-load and blocked-rotor tests on a three phase, Y-connected induction motor are as follows:

No-load test:	line-to-line voltage	=	220 V
	total input power	=	1000 W
	line current	=	20 A
	friction and windage loss	=	400 W
Blocked-rotor test:	line-to-line voltage	=	30 V
	total input power	=	1500 W
	line current	=	50 A

Calculate the parameters of the approximate equivalent circuit shown in Figure 1.

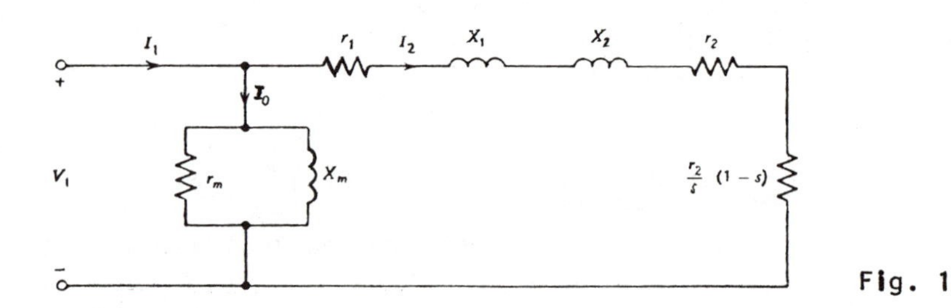

Fig. 1

An approximate equivalent circuit of an induction motor.

Solution:

$$V_0 = \frac{220}{\sqrt{3}} = 127 \text{ V.}$$

$$I_0 = 20 \text{ A.}$$

$$\therefore \quad P_0 = \frac{1}{3}(1000 - 400) = 200 \text{ W.}$$

$$r_m = \frac{V_0^2}{P_0} = \frac{127^2}{200} = 80.5 \ \Omega.$$

$$\phi_0 = \cos^{-1}\frac{P_0}{V_0 I_0} = \cos^{-1}\frac{200}{20 \times 127} = 86°.$$

$$X_m = \frac{V_0}{I_0 \sin\phi_0} = \frac{127}{20 \times 0.99} = 6.4 \ \Omega.$$

Now

$$V_s = \frac{30}{\sqrt{3}} = 17.32 \text{ V.}$$

$$I_s = 50 \text{ A.}$$

$$P_s = \frac{1500}{3} = 500 \text{ W.}$$

Therefore

$$r_e = \frac{P_s}{I_s^2} = \frac{500}{50^2} = 0.2\ \Omega.$$

$$\phi_s = \cos^{-1}\frac{P_s}{V_s I_s} = \cos^{-1}\frac{500}{17.32 \times 50} = 54°.$$

$$X_e = \frac{V_2 \sin\phi}{I_s} = 17.32 \times \frac{0.8}{50} = 0.277\ \Omega.$$

• PROBLEM 10-7

A 3-phase, wye-connected, 60-cycle, 220 volt, 5-hp, 1800 rpm induction motor has the following test data:

Running light		Blocked rotor
220 (127/phase)	voltage	68 (39.3/phase)
4.35	current	20
244 (81.3/phase)	watts	1265 (422/phase)
65 (21.7 phase)	windage and friction	

Stator resistance at 75 C and including skin effect, R_1 = 0.545 ohms per phase. Determine the constants.

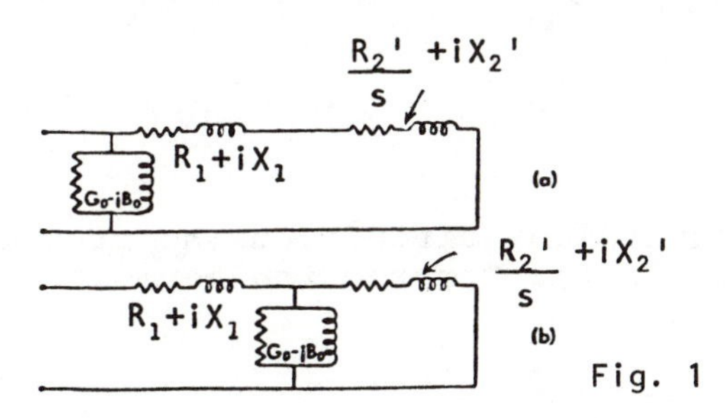

Fig. 1

Solution: For the approximate equivalent circuit,

$$G_0 = \frac{W_0 - F}{V_0^2} = \frac{81.3 - 21.7}{127^2} = 0.0037 \text{ mho,}$$

$$B_0 = \frac{I_0}{V_0}\sqrt{1 - \left(\frac{W_0}{V_0 I_0}\right)^2}$$

$$= \frac{4.35}{127}\sqrt{1 - \left(\frac{81.3}{127 \times 4.35}\right)^2} = 0.0339 \text{ mho}$$

$$\cos\theta_b = \frac{W_b}{V_b I_b} = \frac{422}{39.3 \times 20} = 0.537,$$

$$\sin\theta_b = 0.843.$$

$$R_1 + R_2' = \frac{W_b - G_0 V_b^2}{I_b^2} = \frac{422 - 0.0037 \times 39.3^2}{20^2}$$

$$= 1.040 \text{ ohms},$$

$$\therefore \quad R_2' = 1.040 - 0.545 = 0.495 \text{ ohm}$$

$$X_1 + X_2' = \left\{\frac{V_b^2}{(I_b\cos\theta_b - G_0 V_b)^2 + (I_b\sin\theta_b - B_0 V_b)^2} - (R_1 + R_2')^2\right\}^{1/2}$$

$$= \left\{\frac{39.3^2}{(20\times0.537-0.0037\times39.3)^2 + (20\times0.843-0.0339\times39.3)^2} - 1.040^2\right\}^{1/2}$$

$$= 1.816 \text{ ohm}$$

For the exact equivalent circuit, using previous values of the constants in the "correction terms,"

$$G_0 = \frac{W_0 - F - R_1 I_0^2}{(V_1 - X_1 I_0)^2}$$

$$= \frac{81.3 - 21.7 - 0.545 \times 4.35^2}{(127 - 0.908 \times 4.35)^2} = 0.0033 \text{ mho.}$$

$$Y_0 = \frac{I_0}{V_1 - X_1 I_0} = \frac{4.35}{127 - 0.908 \times 435} = 0.0353 \text{ mho}$$

$$B_0 = \sqrt{Y_0^2 - G_0^2} = \sqrt{0.0353^2 - 0.0033^2} = 0.0352 \text{ mho}$$

$$R_1 + R_2' = \frac{V_b}{I_b}\cos\theta_b + 2R_2'X_2'B_0 + (R_2'^2 - X_2'^2)G_0$$

$$= \frac{39.3}{20} \times 0.537 + 2 \times 0.495 \times 0.908 \times 0.0352$$

$$+ (0.495^2 - 0.908^2)0.0033$$

$$= 1.085 \text{ ohm.}$$

$$X_1 + X_2' = \frac{V_b}{I_b}\sin\theta_b + 2R_2'X_2'G_0 - (R_2'^2 - X_2'^2)B_0$$

$$= \frac{39.3}{20}\, 0.843 + 2 \times 0.495 \times 0.908 \times 0.0033$$

$$- (0.495^2 - 0.908^2)0.0352$$

$$= 1.671 \text{ ohm.}$$

Recapitulating, the results for the different assumptions are tabulated:

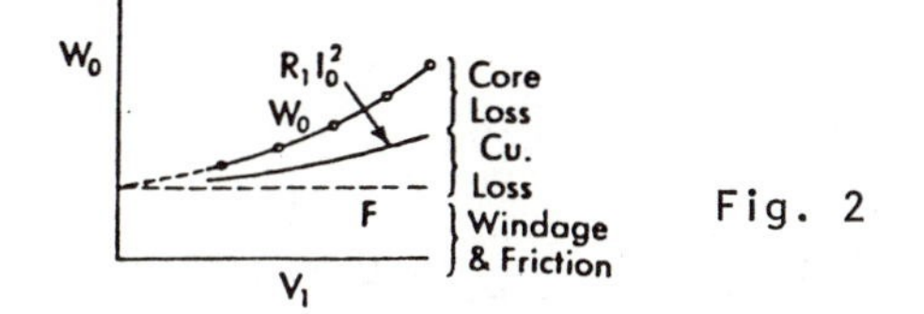

Fig. 2

Constants	Approx. equiv. circuit		Exact equiv. circuit	
	1st approx.	corrected	1st approx.	corrected
G_0	0.0050	0.0037	0.0031	0.0033
B_0	0.0339	0.0339	0.0341	0.0352
$R_1 + R_2'$	1.054	1.040	1.054	1.085
$X_1 + X_2'$	1.656	1.816	1.656	1.671

• **PROBLEM 10-8**

The stator winding of a cage-type induction motor shown in Figure 1 has 24 turns-per-phase. Calculate the factor by which the rotor standstill resistance must be multiplied to refer it to the stator.

Solution: Note from Figure 1 that the stator has three phases, or, $m_1 = 3$. Also, it has four poles, or, $p = 4$; the number of slots per-pole per-phase, $q = 36/4\times3 = 3$, and the slot angle

$$\alpha = \frac{180p}{Q} = \frac{180}{mq} = \frac{180}{3\times3} = 20. \quad \text{Therefore,}$$

$$k_{d1} = \frac{\sin(q\ \alpha/2)}{q \sin \alpha/2}$$

$$= \frac{\sin(3 \times 20/2)}{3\sin(20/2)} = 0.96.$$

Again, from Figure 1, $\tau = 9$ slots and $\beta = 8$ slots. Thus,

$$k_{p1} = \sin \frac{\pi\beta}{2\tau} = \sin \frac{8\pi}{18} = \sin 80° = 0.985.$$

The winding factor for the stator

$$k_{w1} = k_{d1}k_{p1} = 0.945.$$

For the rotor,

$$k_{w2} = 1, \quad N_2 = \frac{p}{4} = 1, \quad \text{and } m_2 = \frac{2Q_2}{p} = 2\times\frac{28}{4} = 14.$$

Substitution of these values in the following equation:

$$r_2' = \frac{m_1}{m_2}\left(\frac{k_{w1}N_1}{k_{w2}N_2}\right)^2 r_2$$

yields

$$r_2' = \frac{3}{14}\left(\frac{0.945 \times 24}{1 \times 1}\right)^2 r_2 = 110r_2$$

or the required factor = 110.

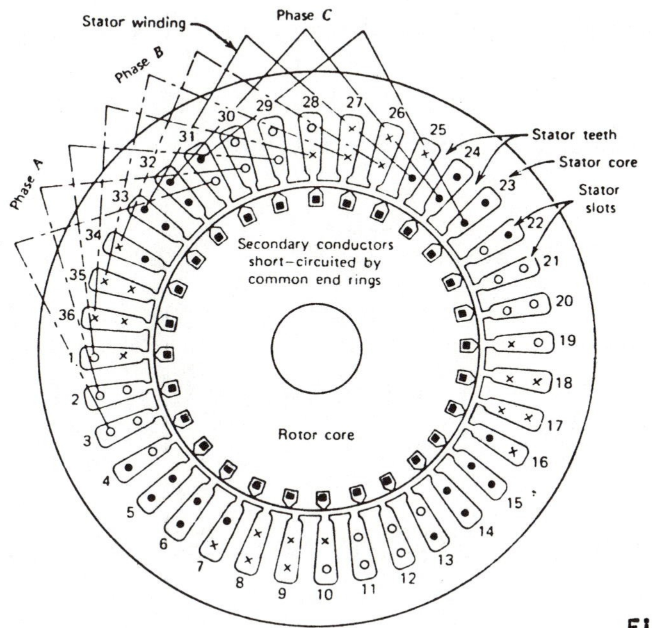

Fig. 1

Stator and rotor windings. Key: O, phase A; X, phase B; and ●, phase C.

SYNCHRONOUS SPEED, SLIP, TORQUE-SLIP CURVES, PULL-OUT TORQUE AND PLUGGING

• PROBLEM 10-9

(a) Calculate the synchronous speed of an eight-pole induction motor when supplied with power from a 60-cycle source.

(b) Calculate the synchronous speed of 60-cycle induction motors having four poles.

(c) The rotor speed of a six-pole 50-cycle induction motor is 960 rpm. Calculate the percent slip.

(d) Calculate the speed of a 60-cycle, 14-pole motor if the slip s is 0.05.

Solution:

(a) $$\text{rpm}_{\text{syn}} = \frac{120 \times f}{p} = \frac{120 \times 60}{8} = 900 \quad \text{(for 60 cycles)}$$

(b) $$rpm_{syn} = \frac{120 \times 60}{4} = 1{,}800 \quad \text{(for four poles)}$$

(c) $$rpm_{syn} = \frac{120 \times 50}{6} = 1{,}000$$

$$\text{Percent slip} = \frac{rpm_{syn} - rpm_{rotor}}{rpm_{syn}} \times 100$$

$$= \frac{1{,}000 - 960}{1{,}000} \times 100 = 4.$$

(d) $$rpm_{rotor} = \frac{120f}{p}(1 - s) = \frac{120 \times 60}{14}(1 - 0.05)$$

$$= 488.$$

• PROBLEM 10-10

Two 60-cycle motors are connected in concatenation to drive a load. If machine 1 has six poles and machine 2 has eight poles, (a) calculate the speed of the combination if the slip is 0.075.

(b) At what other speeds can the load be driven if each motor is operated separately, assuming the same value of slip?

Solution:

(a) $$rpm = \frac{120f_1}{p_1 + p_2}(1 - s_1)$$

$$= \frac{120 \times 60}{6 + 8} \times (1 - 0.075) = 475.$$

(b) $$rpm_{p_1} = \frac{120f_1}{p_1}(1 - s_1)$$

$$rpm_6 = \frac{120 \times 60}{6} \times (1 - 0.075) = 1{,}110.$$

Similarly,

$$rpm_8 = \frac{120 \times 60}{8} \times (1 - 0.075) = 832.$$

• PROBLEM 10-11

For the 15-hp, 4-pole motor described below, find slip at which maximum torque occurs, maximum torque, slip at which maximum power occurs, maximum power, starting torque and starting current.

220 volts	Three-phase	15hp	1725rpm
$R_1 = 0.15$	$R_2 = 0.18$	$X_1 = 0.31$	$_{ss}X_2 = 0.31$

Volts per phase= 127

Friction and windage= 240 watts

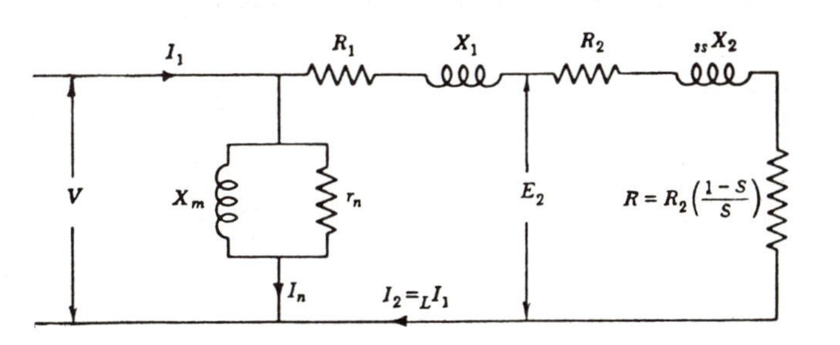

The modified circuit diagram of the induction motor.

Solution: Slip at which maximum torque occurs:

$$S_{mt} = \frac{R_2}{\sqrt{R_1^2 + (X_1 + {}_{ss}X_2)^2}} = \frac{0.18}{\sqrt{0.15^2 + 0.62^2}} = 0.282$$

Maximum torque in synchronous watts

$$T_{max} = \frac{mV^2}{2[R_1 + \sqrt{R_1^2 + (X_1 + {}_{ss}X_2)^2}]}$$

$$= \frac{3 \times 127^2}{2[0.15 + \sqrt{0.15^2 + 0.62^2}]}$$

$$= 30{,}700$$

$$T_{max} = 7.04 \times \frac{30{,}700}{1800} = 120 \text{ lb-ft (118 by test)}$$

Slip at which maximum power occurs:

$$S_{mo} = \frac{R_2}{R_2 + Z_e} = \frac{0.18}{0.18 + 0.703} = 0.204$$

Maximum power in watts:

$$P_m = \frac{mV^2}{2(R_e + Z_e)} = \frac{3 \times 127^2}{2(0.33 + 0.703)} = 23,400$$

$$23,400 \text{ watts} = 31.4 \text{ hp}$$

Starting torque in synchronous watts:

$$\text{Starting torque} = \frac{mV^2 R_2}{Z_e^{\ 2}} = \frac{3 \times 127^2 \times 0.18}{0.703^2} = 17,580$$

$$\text{Starting torque} = 7.04 \times \frac{17,580}{1800} = 68.8 \text{ lb-ft (96 by test)}$$

Starting current:

$$I_2 = \frac{V}{Z_e} = \frac{127}{0.703} = 181 \text{ amperes (182 by test)}$$

Check on starting torque:

$$\text{Starting torque} = mI_2^{\ 2}R_2 = 3 \times 181^2 \times 0.18$$

$$= 17,580 \text{ synchronous watts}$$

• PROBLEM 10-12

A four-pole, three-phase, 220-V, 60-Hz, 10-hp induction motor has the following model impedances:

$r_1 = 0.39\ \Omega \qquad x_1 = 0.35\ \Omega \qquad X_m = 16.0\ \Omega$

$r_2 = 0.14\ \Omega \qquad x_2 = 0.35\ \Omega$

Rotational losses are 350 W.

Find the pull-out torque and the slip at which it occurs.

Solution: "Pull-out" and maximum torque are synonymous. There are at least three procedures available for its calculation: (a) find V_{Th} and Z_1, then apply the formula

$$\tau_{max} = \frac{3}{\omega_s} \frac{V_{Th}^2}{2(R_1 + \sqrt{R_1^2 + (X_1 + x_2)^2})} ;$$

(b) find Z_1, then s_M by

$$s_M = \frac{r_2}{\sqrt{R_1^2 + (X_1 + x_2)^2}}$$

and then find the torque; or (c) find s_M, use the Thevenin-modified model to find I_2, and then $\tau_{max} = (3/\omega_s) I_2^2 (r_2/s_M)$. The first two of these methods will be applied for comparison.

$$\text{(a)} \quad V_{Th} \cong V'_{1m} = V_{1m}\left(\frac{X_m}{x_1 + X_m}\right) = \frac{220}{\sqrt{3}}\left(\frac{16}{16.35}\right)$$

$$= 127.0 \times 0.9736 = 124.3 \text{ V}$$

$$X_1 \cong x_1 = 0.35 \ \Omega \qquad R_1 \cong r_1\left(\frac{Xm}{x_1 + Xm}\right)^2 = 0.39 \times 0.9786^2$$
$$= 0.373\Omega$$

$$\tau_{max} = \frac{3}{188.5}\left(\frac{124.3^2}{2(0.373 + \sqrt{0.373^2 + (0.35 + 0.35)^2}}\right)$$

$$= 105 \text{ N-m}$$

$$\text{(b)} \quad s_M = \frac{r_2}{\sqrt{R_1^2 + (X_1 + x_2)^2}} = \frac{0.14}{\sqrt{0.373^2 + (0.7)^2}} = 0.177$$

$$\frac{r_2}{s_M} = \frac{0.14}{0.177} = 0.791 \ \Omega$$

$$\vec{Z}_f = \frac{jX_m[(r_2/s) + jx_2]}{(r_2/s) + j(x_2 + X_m)} = \frac{j16 \times (0.791 + j0.35)}{0.791 + j16.35}$$

$$= R_f + jX_f = 0.756 + j0.379\ \Omega.$$

$$\vec{Z}_{in} = z_1 + Z_f = 1.146 + j0.729 = 1.358\underline{/32.4°}\ \Omega$$

$$I_1 = \frac{220/\sqrt{3}}{1.358} = 93.5\ A$$

$$P_g = 3I_1^2R_f = 19{,}800\ W.$$

$$\tau_{max} = \frac{P_g}{\omega_s} = \frac{19{,}800}{188.5} = 105\ N\text{-}m.$$

Converting to lb-ft, $\tau_{max} = 105 \times 0.738 = 77.5$ lb-ft.

• PROBLEM 10-13

When operated at rated voltage and frequency with its rotor windings short-circuited, a 500-hp wound-rotor induction motor develops its rated full-load output at a slip of 1.5 per cent. The maximum torque which this motor can develop is 200 per cent of full-load torque. The Q of its Thévenin equivalent circuit is 7.0. For the purposes of this example, rotational and stray load losses may be neglected. Determine:
(a) The rotor I^2R loss at full load, in kilowatts
(b) The slip at maximum torque
(c) The rotor current at maximum torque
(d) The torque at a slip of 20 per cent
(e) The rotor current at a slip of 20 per cent
Express the torque and rotor currents in per unit based on their full-load values.

Solution: (a) Rotor I^2R at Full Load. The power P_{gl} absorbed from the stator divides between mechanical power P and rotor I^2R in the ratio $(1 - s)/s$. Consequently, at full load (neglecting rotational and stray load losses)

$$P_{gl} = \frac{P}{1 - s} = \frac{(500)(0.746)}{0.985} = 379\ kw.$$

$$\text{Rotor } I^2R = sP_{gl} = (0.015)(379) = 5.69\ kw.$$

Parts b to e can readily be solved by means of the normalized curves (Figs. 1 and 2).

(b) Slip at Maximum Torque. From the data, $T_{fl}/T_{max} = 0.50$, where the subscripts fl indicate full load. From Fig. 1 at $Q = 7.0$ and $T/T_{max} = 0.50$,

$$\frac{s}{s_{max\ T}} = \frac{s_{fl}}{s_{max\ T}} = 0.25$$

whence $$s_{max\ T} = \frac{s_{fl}}{0.25} = \frac{0.015}{0.25} = 0.060.$$

(c) Rotor Current at Maximum Torque. From Fig. 2 at $Q = 7.0$ and a slip ratio $s/s_{max\ T} = 0.25$ at full load, the corresponding current ratio is

$$\frac{I_2}{I_{2max\ T}} = \frac{I_{2fl}}{I_{2max\ T}} = 0.355.$$

whence $$I_{2max\ T} = \frac{I_{2fl}}{0.355} = 2.82 I_{2fl}.$$

(d and e) Torque and Rotor Current at $s = 0.20$. The slip ratio is

$$\frac{s}{s_{max\ T}} = \frac{0.20}{0.060} = 3.33.$$

The corresponding torque and current ratios can be read from the curves of Figs. 1 and 2 at $Q = 7.0$ and $s/s_{max\ T} = 3.33$. From Fig. 1

$$\frac{T}{T_{max}} = 0.60 \quad \text{or} \quad T = 0.60 T_{max} = 1.20 T_{fl}$$

From Fig. 2,

$$\frac{I_2}{I_{2\max T}} = 1.40 \quad \text{or} \quad I_2 = 1.40 I_{2\max T}$$

and from (c)

$$I_2 = (1.40)(2.82 I_{2fl}) = 3.95\ I_{2fl}.$$

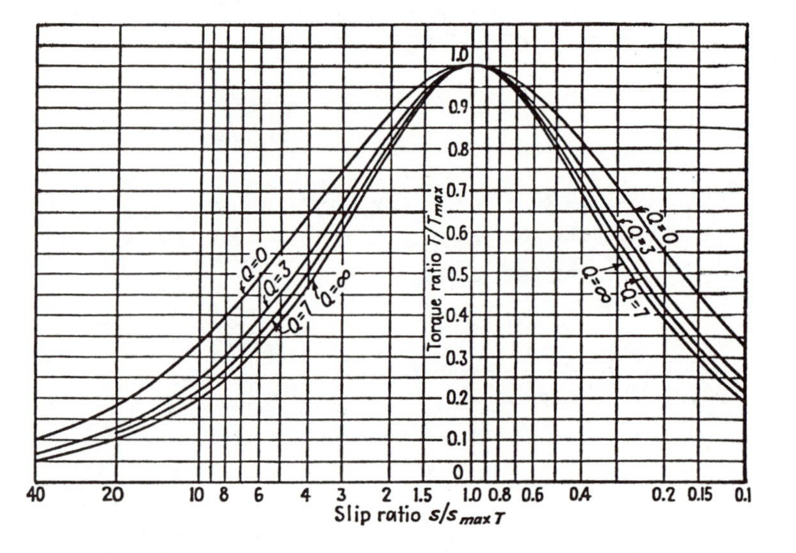

Fig. 1

Normalized torgue-slip curves for polyphase induction motors.

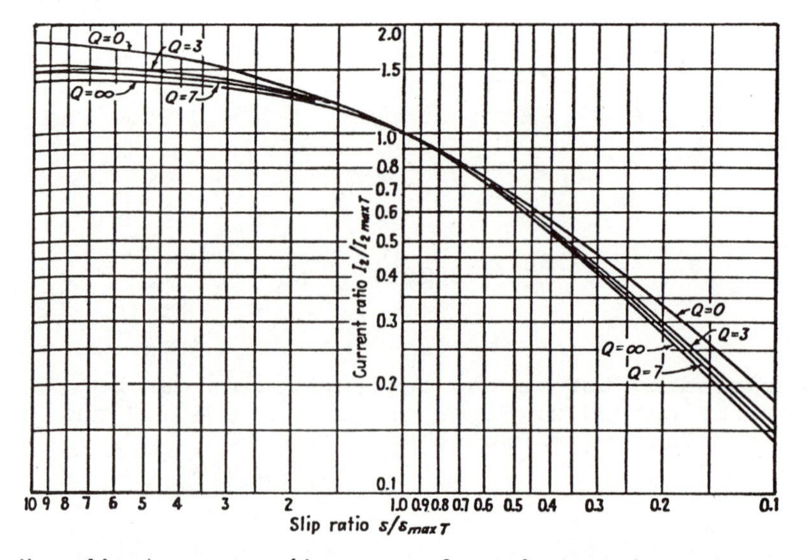

Fig. 2

Normalized current-slip curves for polyphase induction motors.

• **PROBLEM 10-14**

(i) An eight-pole, 60 Hz SCIM is deliberately loaded to a point where pull-out or stalling will occur. The rotor resistance per phase is 0.3 ohm, and the motor stalls at 650 rpm. Calculate:
 (a) The breakdown slip, s_b
 (b) The locked-rotor reactance (the standstill reactance)
 (c) The rotor frequency at the maximum torque point.

(ii) The induced voltage per phase in the rotor of the induction motor is 112 V. Using the data of the part (i) induction motor, determine
 (a) The added rotor resistance
 (b) The rotor power factor which will produce the same starting torque (twice the rated torque) as with the rotor short-circuited
 (c) The starting current.

Solution: (i)

(a) $$S = \frac{120f}{P} = \frac{120 \times 60}{8} = 900 \text{ rpm}$$

$$s_b = \frac{S - S_r}{S} = \frac{900 - 650}{900} = 0.278$$

(b) $$X_{\ell r} = \frac{R_r}{s_b} = \frac{0.3}{0.278} = 1.08\ \Omega$$

(c) $$f_r = sf = 0.278 \times 60 = 16.7 \text{ Hz.}$$

(ii) The new and the original conditions may be summarized in the following table.

(a) $$T_0 = K''_t \left[\frac{R_r}{R_r^2 + X_{\ell r}^2}\right] = K''_t \left[\frac{0.3}{(0.3)^2 + (1.08)^2}\right]$$

$$= K''_t \left(\frac{0.3}{1.25}\right) = K''_t \times 0.24$$

$$T_n = T_0 = K''_t \left[\frac{R_r + R_x}{(R_r + R_x)^2 + X_{\ell r}^2}\right]$$

$$= K''_t \left[\frac{0.3 + R_x}{(0.3 + R_x)^2 + (1.08)^2} \right] = K''_t \times 0.24$$

Simplifying, $0.3 + R_x = 0.24[(0.3 + R_x)^2 + (1.08)^2]$

Expanding and combining the terms yields

$$0.24R_x^2 - 0.856\ R_x = 0.$$

This is a quadratic equation having two roots, which may be factored as $R_x(0.24R_x - 0.856) = 0$, yielding

$$R_x = 0 \quad \text{and} \quad R_x = \frac{0.856}{0.24} = 3.57\ \Omega$$

Note that this solution shows that the original torque is produced with an external resistance of either zero or twelve times the original rotor resistance. Therefore, $R_T = R_r + R_x = 0.3 + 3.57 = 3.87\ \Omega$.

(b) $\vec{Z}_T = R_T + jX_{\ell r} = 3.87 + j1.08 = 4.02\angle 15.6°\ \Omega$

$$\cos\theta = \frac{R_T}{Z_T} = \frac{3.87}{4.02} = \cos 15.6° = 0.963$$

(c) $I_r = \dfrac{E_{\ell r}}{Z_r} = \dfrac{112\ \text{V}}{4.02\ \Omega} = 28\ \text{A}.$

• PROBLEM 10-15

A 420 V, 50 Hz, 3-phase, star-connected, 6-pole induction motor having stator and referred-rotor leakage impedances both given by (1.0 + j2.0) Ω/ph, drives a pure-inertia load at 960 r/min. The total inertia of the drive is $J = 3.0\ \text{kg-m}^2$. Estimate the time to bring the drive to standstill by plugging. Neglect friction, windage and load torques other than inertial, and the effect of transient switching torques.

Solution: Immediately after switching, the speed is given by $\omega_r = \omega_s(1 - s)$ and the angular acceleration by $\alpha = d\omega_r/dt = -\omega_s(ds/dt) = M/J$, where M is the motor torque over the range $s_1 = 1.96$ p.u. initially to $s_2 = 1.0$ at rest. Then $dt = -\omega_s(J/M)ds$. Taking the torque per phase as the approximation in the following equation:

$$M = \frac{1}{\omega_s} \cdot \frac{V_1^2(r_2/s)}{(r_1 + r_2/s)^2 + X_1^2},$$

then

$$dt = -\frac{J\omega_s^2}{3V_1^2}\left[\frac{(r_1 + r_2/s)^2 + X_1^2}{r_2/s}\right] ds$$

Integrating the term in square brackets gives

$$[(r_1^2 + X_1^2)/2r_2](s_2^2 - s_1^2) + 2r_1(s_2 - s_1) + r_2 \ln(s_2/s_1)$$

$$= -24.1 - 1.9 - 0.7 = -26.7$$

with $r_1 = r_2 = 1.0$ and $X_1 = (x_1 + x_2) = 4.0\ \Omega$; then the total time is

$$t = 3.0 \times 105^2 \times 26.7/420^2 = 5.0\text{s}$$

in which $\omega_s = 2\pi \times 1{,}000/60 = 105$ rad/s.

POWER, POWER FACTOR, TORQUE AND EFFICIENCY

• PROBLEM 10-16

A 220-volt, 60-hertz, 10-hp, single-phase induction motor operates at an efficiency of 86% and a power factor of 90%. What capacity should be placed in parallel with the motor so that the feeder supplying the motor will operate at unity power factor?

Solution: The current taken by the motor is

$$hp \times 746 = VI \cos\phi \times \eta$$

$$\therefore \quad I = \frac{hp \times 746}{V \times \cos\phi \times \eta}$$

$$= \frac{10 \times 746}{220 \times 0.9 \times 0.86} = 43.81 \text{ amperes (lagging)}$$

Current taken by the capacitor is

$$I_C = I \times \sin\phi$$

$$= 43.81 \times 0.4352 = 19 \text{ amperes.}$$

The capacitive reactance of the capacitor is

$$X_C = \frac{220}{19} = 11.58 \text{ ohms.}$$

Capacity necessary is

$$C = \frac{10^6}{2\pi \times 60 \times 11.58} = 229 \text{ microfarads.}$$

• PROBLEM 10-17

For a 25-hp, 440-volt, 60-cycle, 1200-rpm, wye-connected induction motor, determine the no-load current and power factor; also, the power factor, current, and speed at 5 per cent slip. The stator resistance per phase is 0.25 ohm and the reactance 0.7 ohm; the rotor resistance and reactance referred to the stator are 0.5 ohm and 0.7 ohm, respectively. The exciting conductance is 0.008 mho and the exciting susceptance is 0.055 mho.

Solution:

$$V_1 = \frac{440}{\sqrt{3}} = 254 \text{ volts}$$

$$\vec{Y}_0 = 0.008 - j0.055 = |0.0555| \text{ mho.}$$

$$\vec{I}_0 = V_1\vec{Y}_0 = (254 + j0)(0.008 - j0.055)$$

$$= 2.03 - j13.97 = |14.12| \text{ amp.}$$

$$\cos\theta_0 = \frac{0.008}{0.0555} = 0.144, \quad \text{power factor} = 14.4 \text{ per cent.}$$

$$\vec{Z}_{01} = r_1 + a^2r_2 + a^2r_2\left[\frac{1-s}{s}\right] + j(X_1 + X_2)$$

$$= 0.25 + 0.5 + 0.5\left(\frac{1 - 0.05}{0.05}\right) + j(0.7 + 0.7)$$

$$= 10.25 + j1.4\ \Omega.$$

$$\vec{Y}_{01} = 0.0958 - j0.0131 \text{ mho.}$$

$$\vec{Y}_1 = \vec{Y}_{01} + \vec{Y}_0 = 0.1038 - j0.0681 = |0.1241| \text{ mho.}$$

$$\cos\theta_1 = \frac{0.1038}{0.1241} = 0.836, \text{ power factor} = 83.6 \text{ per cent.}$$

$$\vec{I}_1 = V_1\vec{Y}_1 = (254 + j0)(0.1038 - j0.0681)$$

$$= 26.37 - j17.30 = |31.5| \text{ amp}$$

Speed = 1200(1 - 0.05) = 1140 rpm.

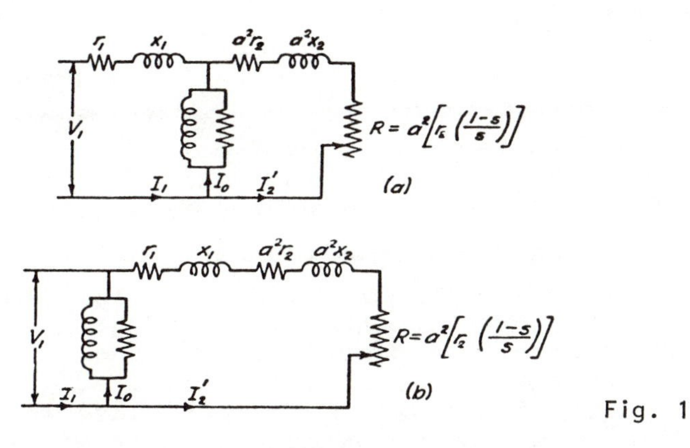

Fig. 1

The equivalent circuit of the polyphase induction motor, showing one line to neutral: (a) accurate equivalent circuit; (b) approximate equivalent circuit.

• **PROBLEM 10-18**

A 3-phase, 15-hp, 440-volt, 17.5-amp, 1,155-rpm squirrel-cage motor drives a pump. Calculate the starting torque of the motor at 65 per cent rated voltage from the following test information.

no load: 462 volts, 9.2 amp, 816 watts, 1,196 rpm

blocked rotor: 223 volts, 74.8 amp, 16.0 kw

resistance between stator terminals = 0.93 ohm.

At 65 per cent rated voltage (286 volts), the stator current is 95.9 amp and the motor input is 26.3 kw. The stator copper loss is 12.84 kw.

Solution: Fixed losses are 698 watts, which modified by the square of the ratio of the starting voltage to the test voltage, are 268 watts. Hence, the rotor input at 65 per cent rated voltage is

26,300 - 12,840 - 268 = 13,192 watts (13.19 synchronous kw, or 17.69 synchronous hp)

$$\text{Starting torque } T_{st} = \frac{7.05}{N} \times (\text{rotor input in watts})\,\text{lb-ft}$$

$$= \frac{13{,}192 \times 7.05}{1{,}200} = 77.4 \text{ lb-ft}$$

Normal torque = 68.3 lb-ft.

Starting torque = 113.5% of rated torque at 65% voltage.

• **PROBLEM 10-19**

A 500-V, 3-phase, 50-cycles/sec, 8-pole, star-connected induction motor has the following equivalent circuit parameters: $R_1 = 0.13\Omega$, $R_2 = 0.32\Omega$, $X_1 = 0.6\Omega$, $X_2 = 1.48\Omega$; magnetizing branch admittance $\vec{Y}_m = 0.004 - j0.05$ mho referred to primary side. The full-load slip is 5%. (a) Determine the full-load gross torque, stator input current and power factor using both approximate and "exact" equivalent circuits. The effective stator/rotor turns ratio per phase is 1/1.57. (b) Using the approximate circuit of (a), calculate the output of the machine when driven at a speed of 780 rev/min.

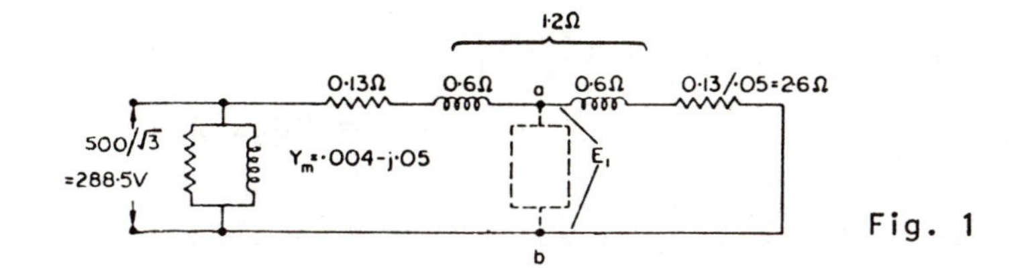

Fig. 1

Solution: (a) $R_2' = 0.32 \times (1/1.57)^2 = 0.13\ \Omega$,

$X_2' = 1.48 \times (1/1.57)^2 = 0.6\ \Omega$

The equivalent circuit for one phase is shown in Fig.-1 for a slip of 0.05. Using the approximate circuit

$$\vec{I}_2' = \frac{\vec{V}_1}{\vec{Z}} = \frac{288.5}{(0.13 + 2.6) + j1.2} = 88.8 - j39 = 97\ \text{A}$$

$$\vec{I}_p = \vec{V}_1\vec{G} = 288.5 \times 0.004 = 1.15 - j0 = 1.15\text{A}$$

$$\vec{I}_m = \vec{V}_1\vec{B} = 288.5 \times -j0.05 = -j14.4 = 14.4\text{A}$$

$$\vec{I}_1 = \vec{I}_2' + \vec{I}_p + \vec{I}_m = 89.95 - j53.4 = 105\ \text{A}$$

Input power factor = 89.95/105 = 0.856

$$\text{Torque} = \frac{3P_2}{\omega_s} = \frac{3I_2'^2R_2'/s}{2\pi f/p} = \frac{3 \times 97^2 \times 2.6}{2\pi \times 50/4} = 935\ \text{N-m.}$$

Using the "exact" circuit, the admittance between a and b is:

$$\vec{Y} = 0.004 - j0.05 + \frac{1}{2.6 + j0.6} = 0.369 - j0.1345\ \text{mho,}$$

the corresponding impedance, $1/\vec{Y} = \vec{Z}_{ab} = 2.4 + j0.872$
$= 2.55\ \Omega$

Adding $R_1 + jX_1$, total impedance $= \vec{Z}' = 2.53 + j1.472$
$= 2.92\ \Omega$.

$$\text{Total input current} = \frac{288.5}{2.53 + j1.472} = 85.5 - j49.6$$

$$= 98.5 \text{ A.}$$

Input power factor = 85.5/98.5 = 0.858.

To find the e.m.f. E_1, since the same current I_1 flows through Z and Z_{ab} in series, the voltage across each impedance is proportional to the impedance modulus, i.e.

$$E_1 = V_1(Z_{ab}/Z') = 288.5(2.55/2.92) = 252 \text{ V},$$

hence

$$I_2'^2 = \frac{252^2}{(2.6)^2 + (0.6)^2} = 8880 \quad \text{and}$$

$$I_2' = \sqrt{8880} = 94.3 \text{ A} \quad \text{and}$$

$$\text{Torque} = 935 \times (94.3/97)^2 = 886 \text{ N-m}$$

$$I_p = E_1 G = 252 \times 0.004 = 1.01 \text{ A.}$$

$$I_m = E_1 B = 252 \times 0.05 = 12.6 \text{ A.}$$

(b) Synchronous speed = 60f/p = 60 × 50/4 = 750 rev/min.

$$\therefore \quad \text{Slip} = (750-780)/780 = -0.04.$$

Apparent rotor circuit resistance = 0.13/-0.04

$$= -3.25 \ \Omega.$$

$$\vec{I}_2' = \frac{288.5}{(0.13 - 3.25) + j1.2} = \frac{288.5}{-3.12 + j1.2}$$

$$= -80.5 - j31 \text{ A.}$$

From the approximate circuit of (a), $\vec{I}_0 = 1.15 - j14.4$,

$$\therefore \quad \vec{I}_1 = \vec{I}_2' + \vec{I}_0 = -79.35 - j45.4 = 91.5 \text{ A}$$

Output kVA = $\sqrt{3} \times 500 \times 91.5 = 79.3$ kVA

Power factor = -79.35/91.5 = 0.865 leading.

Tests on the main winding of a one-hp, 4-pole, 220-V, 60-Hz, single-phase induction motor gave the following results:

$$V_{NL} = 220 \text{ V} \qquad V_{\ell k} = 80 \text{ V}$$

$$I_{NL} = 4.2 \text{ A} \qquad I_{\ell k} = 10.4 \text{ A}$$

$$P_{NL} = 192 \text{ W} \qquad P_{\ell k} = 404 \text{ W}$$

$$R_s = 1.54\ \Omega$$

Determine the line current, power factor, shaft torque, and efficiency of the motor at a speed of 1725 r/min.

Solution: The equivalent circuit of the motor must first be determined.

$$\frac{X_M}{2} \simeq \frac{V_{NL}}{I_{NL}} = \frac{220}{4.2} = 52.4\ \Omega.$$

$$P_{rot} \simeq P_{NL} = 192 \text{ W}.$$

The equivalent circuit and phasor diagram for the locked-rotor test are shown in Fig. 1.

$$\phi_{\ell k} = \cos^{-1} \frac{P_{\ell k}}{V_{\ell k} I_{\ell k}} = \cos^{-1} \frac{404}{80 \times 10.4} = 60.9°$$

Let $\qquad \vec{V}_{\ell k} = 80\angle 0 \text{ V}.$

Then

$$\vec{I}_{\ell k} = 10.4\angle{-60.95°} = 5.05 - j9.09 \text{ A}.$$

$$\vec{I}_M = \frac{\vec{V}_{\ell k}}{jX_M} = \frac{80\angle 0}{j104.8} = -j0.763 \text{ A}.$$

$$\vec{I}_s' = \vec{I}_{\ell k} - \vec{I}_M = 5.05 - j8.33 = 9.74\angle{-58.8°} \text{ A}.$$

$$Z_R = \frac{V_{\ell k}}{I_s'} = \frac{80}{9.74} = 8.21\ \Omega.$$

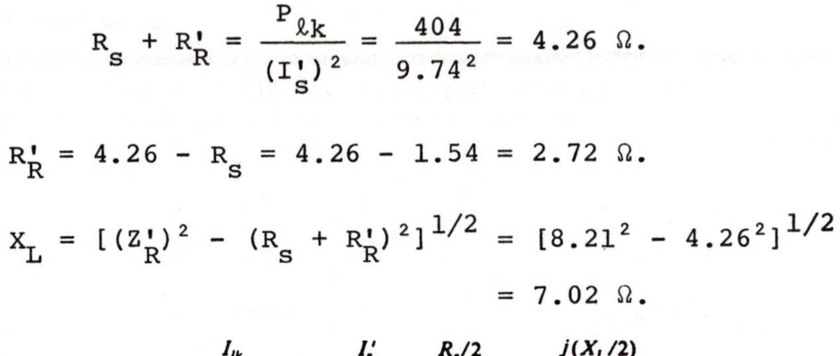

$$R_s + R_R' = \frac{P_{\ell k}}{(I_s')^2} = \frac{404}{9.74^2} = 4.26\ \Omega.$$

$$R_R' = 4.26 - R_s = 4.26 - 1.54 = 2.72\ \Omega.$$

$$X_L = [(Z_R')^2 - (R_s + R_R')^2]^{1/2} = [8.21^2 - 4.26^2]^{1/2}$$

$$= 7.02\ \Omega.$$

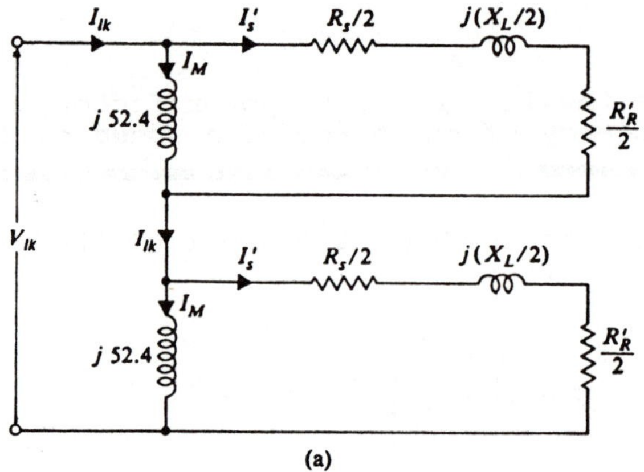

(a)

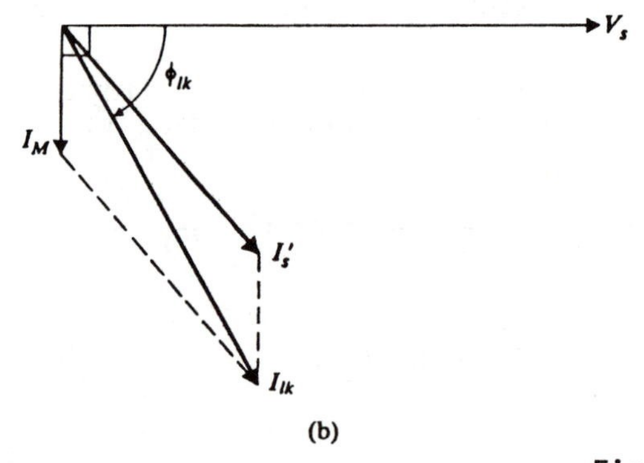

(b)

Fig. 1

The equivalent circuit of the motor is shown in Fig. 2. At 1725 r/min,

$$s_f = \frac{1800 - 1725}{1800} = 0.0417.$$

$$\vec{Z}_{Rf} = 0.770 + \frac{1.36}{0.0417} + j3.51 = 33.4 + j3.51$$

$$= 33.6 \angle 6.0^\circ\ \Omega.$$

$$\vec{Z}_f = \frac{j(X_M/2)\vec{Z}_{Rf}}{j(X_M/2) + \vec{Z}_{Rf}} = \frac{j52.4 \times 33.6\angle 6.0°}{j52.4 + 33.4 + j3.51}$$

$$= \frac{52.4\angle 90° \times 33.6\angle 6.0°}{65.1\angle 59.1°}$$

$$= 27.0\angle 36.9° = 21.6 + j16.2\ \Omega.$$

$$\vec{Z}_{Rb} = 0.770 + j3.51 + \frac{1.36}{2 - 0.0417} = 1.47 + j3.51$$

$$= 3.80\angle 67.3°\ \Omega.$$

$$\vec{Z}_b = \frac{j52.4 \times 3.80\angle 67.3°}{j52.4 + 1.47 + j3.51} = \frac{52.4\angle 90° \times 3.80\angle 67.3°}{55.9\angle 88.5°}$$

$$= 3.56\angle 68.8° = 1.29 + j3.32\ \Omega.$$

$$\vec{Z} = \vec{Z}_f + \vec{Z}_b = 21.6 + j16.2 + 1.29 + j3.32$$

$$= 22.9 + j19.5 = 30.1\angle 40.4°\ \Omega.$$

Let

$$\vec{V}_s = 220\angle 0\ \text{V}.$$

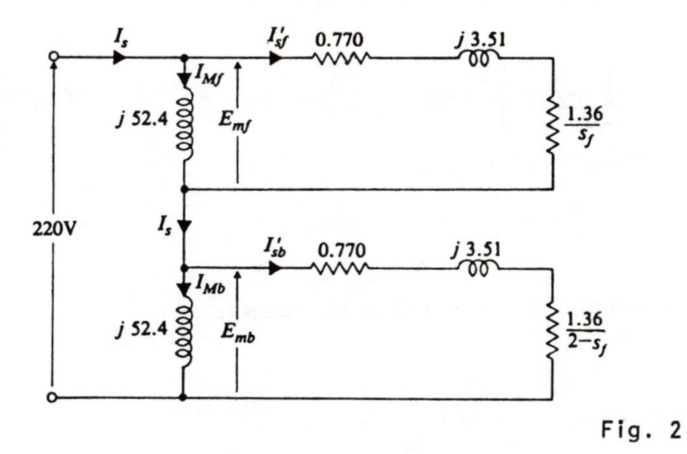

Fig. 2

Then

$$\vec{I}_s = \frac{220\angle 0}{30.1\angle 40.4°} = 7.31\angle -40.4° \text{ A.}$$

$$PF = \cos 40.4° = 0.762.$$

$$\vec{E}_{mf} = \frac{\vec{Z}_f}{\vec{Z}} \times \vec{V}_s = \frac{27.0 \; -\angle 36.9°}{30.1\angle 40.0°} \times 220\angle 0 = 197\angle -3.5° \text{ V.}$$

$$\vec{I}'_{sf} = \frac{\vec{E}_{mf}}{\vec{Z}_{Rf}} = \frac{197\angle -3.5°}{33.6\angle 6.0°} = 5.86\angle -9.5° \text{ A.}$$

$$\vec{E}_{mb} = \frac{\vec{Z}_b}{\vec{Z}} \times \vec{V}_s = \frac{3.56\angle 68.8°}{30.1\angle 40.4°} \times 220\angle 0 = 26.0\angle 28.4° \text{ V.}$$

$$\vec{I}'_{sb} = \frac{\vec{E}_{mb}}{\vec{Z}_{Rb}} = \frac{26.0\angle 28.4°}{3.80\angle 67.3°} = 6.84\angle -38.9° \text{ A.}$$

The internal torque is $T_{av} = T_f + T_b$, where

$$T_f = \frac{p}{2} \cdot \frac{1}{\omega_s} (I'_{s_f})^2 \frac{R'_R}{2s_f} \quad \text{N·m}$$

$$T_b = -\frac{p}{2} \frac{1}{\omega_s} (I'_{sb})^2 \frac{R'_R}{2(2-s_f)} \quad \text{N·m}$$

$$\therefore \quad T_{av} = \frac{4}{2} \frac{1}{120\pi} \left[5.86^2 \times \frac{1.36}{0.0417} - 6.84^2 \times \frac{1.36}{2-0.0417} \right]$$

$$= 5.77 \text{ N·m}$$

Torque absorbed in rotational losses is

$$T_{rot} = \frac{P_{NL}}{\omega_m} = \frac{192}{1725} \times \frac{60}{2\pi} = 1.06 \text{ N·m.}$$

Shaft torque is

$$T_{av} - T_{rot} = 5.77 - 1.06 = 4.71 \text{ N}\cdot\text{m}.$$

Power output is

$$P_0 = 4.71 \times 1725 \times \frac{2\pi}{60} = 851 \text{ W}.$$

Power input is

$$V_s I_s (PF) = 220 \times 7.31 \times 0.762 = 1225 \text{ W}.$$

Efficiency is

$$\eta = \frac{851}{1225} = 0.695.$$

The motor is, thus, overloaded. If the load were reduced to the rated output of 746 W, the speed would increase, and probably the efficiency would increase also.

• PROBLEM 10-21

A 3-phase Y-connected 220-volt (line to line) 10-hp, 60-cps 6-pole induction motor has the following constants in ohms per phase referred to the stator:

$$r_1 = 0.294 \qquad r_2 = 0.144$$

$$x_1 = 0.503 \qquad x_2 = 0.209 \qquad x_\phi = 13.25$$

The total friction, windage, and core losses may be assumed to be constant at 403 watts, independent of load.

For a slip of 2.00 per cent, compute the speed, output torque and power, stator current, power factor, and efficiency when the motor is operated at rated voltage and frequency.

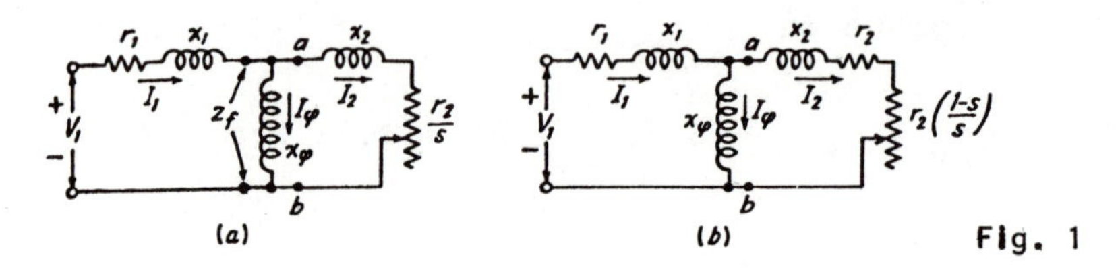

Equivalent circuits.

Solution: The impedance $\vec{Z}_f$ (Fig. 1a) represents physically the per-phase impedance presented to the stator by the air-gap field, both the reflected effect of the rotor and the effect of the exciting current being included therein. From Fig. 1a,

$$\vec{Z}_f = R_f + jX_f = \frac{r_2}{s} + jx_2 \quad \text{in parallel with } jx_\phi$$

Substitution of numerical values gives, for s = 0.0200,

$$R_f + jX_f = 5.41 + j3.11$$

$$r_1 + jx_1 = \underline{0.29 + j0.50}$$

$$\text{Sum} = 5.70 + j3.61 = 6.75\underline{/32.4°} \text{ ohms}$$

$$\text{Applied voltage to neutral} = \frac{220}{\sqrt{3}} = 127 \text{ volts}$$

$$\text{Stator current } I_1 = \frac{127}{6.75} = 18.8 \text{ amp.}$$

$$\text{Power factor} = \cos 32.4° = 0.844$$

$$\text{Synchronous speed} = \frac{2f}{p} = \frac{120}{6} = 20 \text{ rev/sec, or } 1{,}200 \text{ rpm}$$

$$\omega_s = 2\pi(20) = 125.6 \text{ rad/sec}$$

$$\text{Rotor speed} = (1 - s) \times (\text{synchronous speed})$$

$$= (0.98)(1{,}200) = 1{,}176 \text{ rpm}$$

$$P_{g1} = q_1 I_2^2 \frac{r_2}{s} = q_1 I_1^2 R_f$$

$$= (3)(18.8)^2(5.41) = 5{,}740 \text{ watts}$$

From the following equations:

$$P = (1 - s)\omega_s T = q_1 I_2'^2 r_2 \frac{1 - s}{s}$$

$$P_{g1} = q_1 I_2^2 \frac{r_2}{s} ,$$

the internal mechanical power is

$$P = (0.98)(5,740) = 5,630 \text{ watts.}$$

Deducting losses of 403 watts gives

Output power = 5,630 - 403 = 5,230 watts, or 7.00 hp.

$$\text{Output torque} = \frac{\text{output power}}{\omega_{rotor}} = \frac{5,230}{(0.98)(125.6)}$$

$$= 42.5 \text{ newton-m, or } 31.4 \text{ lb-ft.}$$

The efficiency is calculated from the losses.

Total stator copper loss = $(3)(18.8)^2(0.294)$	=	312 watts
Rotor copper loss = $q_1 I_2^2 r_2$ = (0.0200)(5,740)	=	115
Friction, windage, and core losses	=	403
Total losses	=	830 watts
Output	=	5,230 watts
Input	=	6,060 watts

$$\frac{\text{Losses}}{\text{Input}} = \frac{830}{6,060} = 0.137$$

$\therefore$ Efficiency = 1.000 - 0.137 = 0.863.

The complete performance characteristics of the motor can be determined by repeating these calculations for other assumed values of slip.

• PROBLEM 10-22

A single-phase induction motor is rated $\frac{1}{4}$ HP, 100 v, 60 Hz, four poles. The parameters of the equivalent circuit are

$$R_1 = 2\Omega \quad R_2' = 4\Omega \quad X_m = -\frac{1}{B_m} = 66\Omega$$

$$X_1 = 2.4\Omega \quad X_2' = 2.4\Omega$$

The rotational losses, core losses combined with windage and friction, are 37 W. The motor is operating with rated voltage and rated frequency and with a slip of 0.05. Find the input current and the resultant electromagnetic torque.

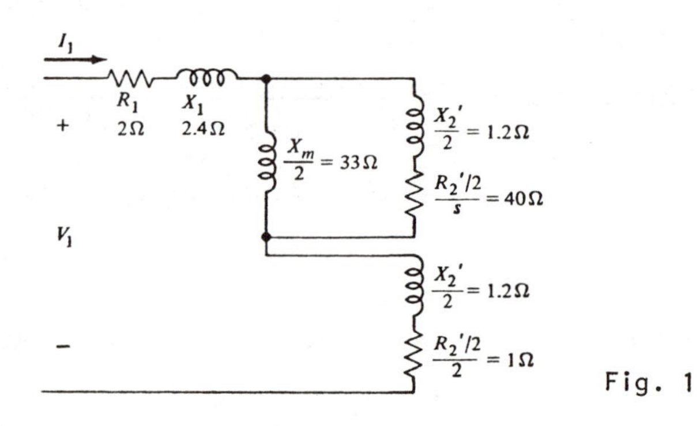

Approximate equivalent circuit of single-phase induction motor.

Solution: Use the equivalent circuit shown in Fig. 1. When numerical values are substituted, one finds that the impedance across E^- is much smaller than that across E^+ to which it is to be added. Therefore, great accuracy is not required in its calculation. Thus

$$0.5\,\vec{Z}_b = 0.5\,\frac{(jX_m)\left[R_2'/(2 - s) + jX_2'\right]}{R_2'/(2 - s) + j(X_m + X_2')}$$

$$\approx 0.5(R_2'/2 + jX_2')$$

This approximation is reasonable for values of slip less than 0.1. The corresponding equivalent circuit is shown in Fig. 1. For this example,

$$0.5\vec{Z}_b = 1 + j1.2\Omega = 0.5R_b + j0.5X_b \quad ,$$

Next, compute $0.5\vec{Z}_f$

$$0.5\vec{Z}_f = \frac{(j33)(40 + j1.2)}{40 + j34.2} = 25.1\underline{/51.2°}$$

$$= 15.7 + j19.6\Omega = 0.5R_f + j0.5X_f$$

$$\vec{Z}_1 = 2 + j2.4\Omega.$$

The total impedance is

$$\vec{Z}_t = \vec{Z}_1 + 0.5\vec{Z}_f + 0.5\vec{Z}_b = 18.7 + j23.2$$

$$= 29.8\underline{/51.1°}\ \Omega.$$

The stator current is

$$\vec{I}_1 = \vec{V}_1/\vec{Z}_t = (110\underline{/0°})/(29.8\underline{/51.1°}) = 3.69\underline{/-51.1°}\ \text{amp.}$$

The positive sequence air gap power is the power in the positive sequence portion of the equivalent circuit.

$$P_g^+ = I_1^2(0.5R_f) = (3.69)^2(15.7) = 213.8\ \text{w.}$$

The negative-sequence air gap power is the power in the negative-sequence portion of the equivalent circuit.

$$P_g^- = I_1^2(0.5R_b) = (3.69)^2(1) = 13.6\ \text{w.}$$

Synchronous speed is $\omega_{s_m} = 4\pi f/p = 188.5$ rad/sec. The

resultant torque is

$$T = (1/\omega_{s_m})(P_g^+ - P_g^-) = 1.06\ \text{nm.}$$

• **PROBLEM 10-23**

(i) An induction motor draws 25 A from a 460-V, three-phase line at a power factor of 0.85, lagging. The stator copper loss is 1000 W, and the rotor copper loss is 500 W. The "rotational" losses are friction and windage = 250 W, core loss = 800 W, and stray load loss = 200 W. Calculate (a) the air-gap power, P_g, (b) the developed mechanical power, DMP, (c) the output horsepower, and (d) the efficiency.

(ii) If the frequency of the source in part (i) is 60 Hz, and the machine has four poles, find (a) the slip, (b) the operating speed, (c) the developed torque, and (d) the output torque.

Solution: (i)

(a) $P_g = P_{in} - SCL$

$$= \sqrt{3} \times 460 \times 25 \times 0.85 - 1000$$

$$= 16,931 - 1000 = 15,931 \text{ W.}$$

(b) $DMP = P_g - RCL = 15,931 - 500 = 15,431$ W.

(c) $P_{out} = DMP - P_{rot}$

$$= DMP - (P_{h+e} + P_{fw} + P_{LL})$$

$$= 15,431 - 1250 = 14,181 \text{ W.}$$

$$\therefore \quad \text{Horsepower} = \frac{P_{out}}{746} = 19.0 \text{ hp}$$

(d) $\eta = \dfrac{P_{out}}{P_{in}} = \dfrac{14,181}{16,931}$ or 83.8%

(ii)

(a) $s = \dfrac{RCL}{P_g} = \dfrac{500}{15,931} = 0.0314.$

(b) $$\omega_s = \frac{4\pi 60}{p} = \frac{754}{4} = 188.5 \text{ rad/s.}$$

$$n_s = \frac{120f}{p} = \frac{120 \cdot 60}{4} = 1800 \text{ rev/min.}$$

$$\omega = \omega_s(1 - s) = 188.5(1 - 0.0314) = 182.6 \text{ rad/s.}$$

$$n = n_s(1 - s) = 1800 \times 0.9689 = 1744 \text{ rev/min.}$$

(c) $$\tau_d = \frac{\text{DMP}}{\omega_s} = \frac{15{,}431}{182.6} = 84.5 \text{ N-m.}$$

(d) $$\tau = \frac{P_{out}}{\omega} = \frac{14{,}181}{182.6} = 77.7 \text{ N-m.}$$

Note: the output torque is 8 percent less than the developed torque.

• PROBLEM 10-24

A 60-Hz, four-pole, Y-connected induction motor is rated 5 HP, 220 v (line to line). The equivalent circuit parameters are:

$R_1 = 0.48\ \Omega$ $\quad R_2' = 0.42\ \Omega$ $\quad B_m = -1/30$ mhos

$X_1 = 0.80\ \Omega$ $\quad X_2' = 0.80\ \Omega$

The motor is operating with slip of 0.04. Find the input current and power factor. Find the air gap power, the mechanical power, and the electromagnetic torque.

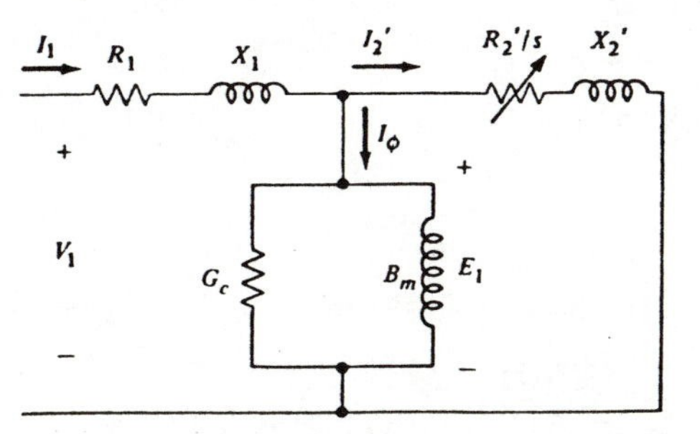

Equivalent circuit.

Fig. 1

Solution: Use the equivalent circuit in Fig. 1. The shunt conductance, G_c, is negligible in determining I_1. Let $\vec{Z}_f$ be the parallel combination of $\vec{Y}_\phi$ and $\vec{Z}_2'$. Let $X_m = -1/B_m = 30\ \Omega$.

$$\vec{Z}_f = R_f + jX_f = \frac{1}{G_c + jB_m + \dfrac{1}{(R_2'/s) + jX_2'}}$$

$$\approx \frac{jX_m[(R_2'/s) + jX_2']}{(R_2'/s) + j(X_2' + X_m)}$$

All of the quantities are given. Substitution yields

$$\vec{Z}_f = 9.71\underline{/23.3°} = 8.92 + j3.84\ \Omega.$$

$$\therefore \quad \vec{Z}_{in} = \vec{Z}_1 + \vec{Z}_f = (0.48 + j0.80) + (8.92 + j3.84)$$
$$= 9.40 + j4.64 = 10.5\underline{/26.3°}\ \Omega.$$

The phase voltage is $220/\sqrt{3} = 127$ v. Use V_1 as the reference. $\vec{V}_1 = 127\underline{/0°}$ V. The input current is

$$\vec{I}_1 = \vec{V}_1/\vec{Z}_{in} = (127\underline{/0°})/(10.5\underline{/26.3°})$$

$$= 12.1\underline{/-26.3°}\ \text{amp}.$$

The power factor is $\cos(-26.3°) = 0.90$. The air gap power is given by

$$P_g = 3(R_2'/s)I_2'^2 = 3R_f I_1^2 = 3(8.92)(12.1)^2 = 3920\ \text{w}.$$

The mechanical power is given by

$$P_{mech} = (1 - s)P_g = (.96)(3920) = 3760\ \text{w}.$$

Synchronous speed is $\omega_{s_m} = 4\pi f/p = 4\pi 60/4 = 188.5$ rad/sec. The operating speed is $\omega_m = (1 - s)\omega_{s_m} = 181$ rad/sec. The electromagnetic torque is

$$T = P_{mech}/\omega_m = 3760/181 = 20.8 \text{ nm.}$$

The electromagnetic torque can also be found from

$$T = P_g/\omega_{s_m} = 3920/188.5 = 20.8 \text{ nm.}$$

• PROBLEM 10-25

With reference to figure 1a, the constants of a 1/4 hp, 230 V, four-pole, 60 Hz one-phase induction motor are: $r_1 = 10.0\Omega$, $r_2 = 11.6\Omega$, $x_1 = 12.8\Omega = x_2$, and $x_m = 258.0\Omega$. For an applied voltage of 210 V, at a 3% slip, calculate (a) input current, (b) power factor, (c) developed power, (d) shaft power (if mechanical losses are 7 W), and (e) efficiency (if iron losses at 210 V are 35.5 W).

Solution: For the given circuit, using the given data,

$$\frac{0.5r_2}{s} = \frac{11.65}{2 \times 0.03} = 194.16\Omega$$

$$\frac{0.5r_2}{2 - s} = \frac{11.65}{2(2 - 0.03)} = 2.96\Omega$$

$$j0.5x_m = j129\Omega$$

and

$$j0.5x_2 = j0.5x_1 = j6.4\Omega.$$

For the forward-field circuit

$$\vec{Z}_f = \frac{194.16 \times j129}{194.16 + j129} = 59.2 + j86\Omega$$

and for the backward-field circuit

$$\vec{Z}_b = \frac{2.96 \times j129}{2.96 + j129} \cong 2.96\Omega$$

The total series impedance, $\vec{Z}_e$, is

$$\vec{Z}_e = \vec{Z}_1 + \vec{Z}_f + \vec{Z}_b = (10 + j12.8) + (59.2 + j89)$$
$$+ 2.96 = 124\angle 55°\Omega$$

(a). Input current

$$\vec{I} = \frac{\vec{V}}{\vec{Z}_e} = \frac{210}{124\angle 55°} = 1.7 \angle -55° \text{ A.}$$

(b). Power factor = cos 55 = 0.573 lagging.

(c). Developed power,

$$P_d = \left(\frac{0.5r_2}{s} I_f^2 - \frac{0.5r_2}{2 - s} I_b^2\right)(1 - s)$$

$$\cong \left[\frac{V_f^2}{0.5r_2/s} - \frac{V_b^2}{0.5r_2/(2 - s)}\right](1 - s)$$

since s = 0.03 (small). But $V_f = IZ_f = 1.7(59.2 + j89) = 182$ V and $V_b = IZ_b = 1.7 \times 2.96 = 5.04$ V. Hence

$$P_d = \left(\frac{182^2}{194} - \frac{5.04^2}{2.96}\right)(1 - 0.03) = 156 \text{ W.}$$

(d). Shaft power, $P_s = P_d - P_{rot} = 156 - 7 = 149$ W.

(e). Input power $= VI \cos\theta = 210 \times 1.7 \times 0.573 = 204$ W,

output power $= P_s - P_{iron} = 149 - 35.5 = 113.5$ W,

efficiency $= \frac{113.5}{204} = 55.6\%$.

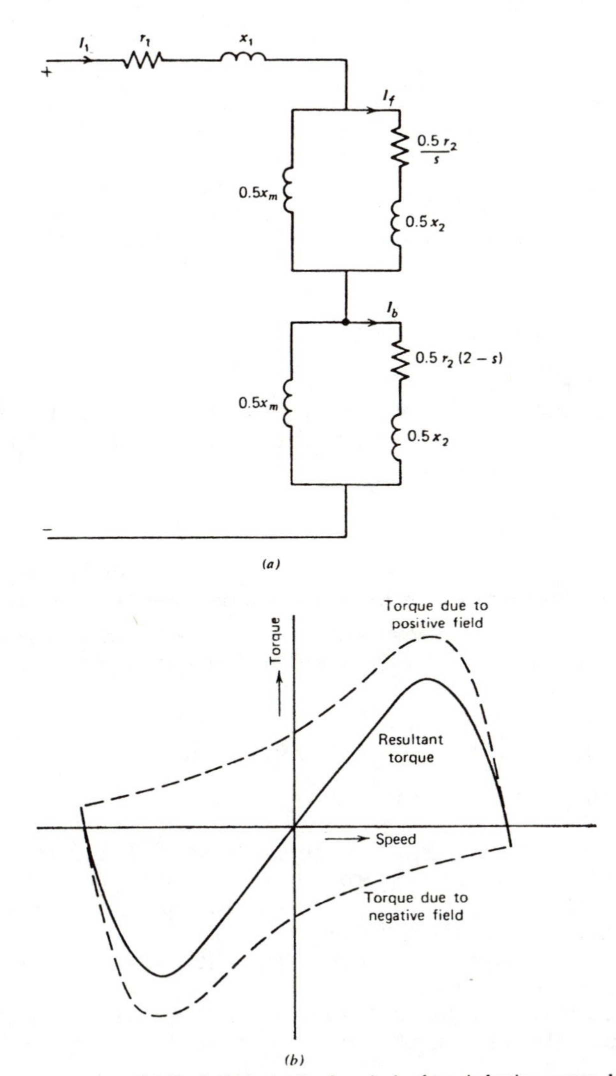

Fig. 1

(*a*) Equivalent circuit of a single-phase induction motor, based on revolving-field theory. (*b*) Torque-speed characteristics of a single-phase induction motor based on revolving field theory.

• **PROBLEM 10-26**

An induction motor carries a rotor current per phase of 10 amperes when loaded until the slip is 5 per cent. If the rotor effective resistance is 0.1 ohm per phase, determine the internal power developed per phase.

Solution:

$$R = r_2 \frac{1 - s}{s}$$

$$= 0.1 \frac{1 - 0.05}{0.05}$$

$$= 1.9 \text{ ohm}$$

$$P_r = I^2R = 10^2 \times 1.9$$

$$= 190 \text{ watts, internal power developed per phase}$$

The true loss $i_2{}^2r_2$ in one phase of the rotor winding is

$$10^2 \times 0.1 = 10 \text{ watts}$$

This, added to the internal power developed, equals the total power transferred across the air gap per phase, or 200 watts, in this case.

• **PROBLEM 10-27**

The equivalent-circuit constants of a 5-hp,220-volt, 60-cps,2-phase,squirrel-cage induction motor are given below, in ohms per phase:

$r_1 = 0.534$ $\quad x_1 = 2.45$ $\quad x_\phi = 70.1$

$r_2 = 0.956$ $\quad x_2 = 2.96$

This motor is operated from an unbalanced 2-phase source whose phase voltages are, respectively, 230 volts and 210 volts, the smaller voltage leading the larger by 80°. For a slip of 0.05, find:

(a) The positive- and negative-sequence components of the applied voltages

(b) The positive- and negative-sequence components of the stator phase currents

(c) The effective values of the phase currents

(d) The internal mechanical power

Solution: (a) Let $\vec{V}_m$ and $\vec{V}_a$ denote the voltages applied to the 2 phases, respectively. Then

$$\vec{V}_m = 230\underline{/0°} = 230 + j0 \text{ volts}$$

$$\vec{V}_a = 210\underline{/80°} = 36.4 + j207 \text{ volts}$$

From the following equations:

$$\vec{V}_{mf} = \frac{1}{2}(\vec{V}_m - j\vec{V}_a)$$

$$\vec{V}_{mb} = \frac{1}{2}(\vec{V}_m + j\vec{V}_a),$$

the forward and backward components of voltages are, respectively,

$$\vec{V}_{mf} = \frac{1}{2}(230 + j0 + 207 - j36.4)$$

$$= 218.5 - j18.2 = 219.5\underline{/-4.8°} \text{ volts}$$

$$\vec{V}_{mb} = \frac{1}{2}(230 + j0 - 207 + j36.4)$$

$$= 11.5 + j18.2 = 21.5\underline{/57.7°} \text{ volts}$$

(b) From the following equation:

$$R_f = \frac{x_\phi^2}{x_{22}} \frac{1}{sQ_2 + (1/sQ_2)},$$

the forward-field impedance is, for a slip of 0.05,

$$\vec{Z}_f = 16.46 + j7.15 \text{ ohms}$$

$$r_1 + jx_1 = \underline{0.53 + j2.45} \text{ ohms}$$

$$16.99 + j9.60 = 19.50\underline{/29.4°} \text{ ohms}$$

Hence the forward component of stator current is

$$\vec{I}_{mf} = \frac{219.5\angle -4.8°}{19.50\angle 29.4°} = 11.26\angle -34.2° \text{ amp}$$

For the same slip, from the following equations:

$$R_b = \frac{r_2}{2 - s}\left(\frac{x_\phi}{x_{22}}\right)^2$$

$$x_b = \frac{x_2 x_\phi}{x_{22}} + \frac{R_b}{(2 - s)Q_2} ,$$

the backward-field impedance is

$$\vec{Z}_b = 0.451 + j2.84 \text{ ohms}$$

$$r_1 + x_1 = \underline{0.534 + j2.45} \text{ ohms}$$

$$0.985 + j5.29 = 5.38\angle 79.5° \text{ ohms}$$

Hence the backward component of stator current is

$$\vec{I}_{mb} = \frac{21.5\angle 57.7°}{5.38\angle 79.5°} = 4.00\angle -21.8° \text{ amp}$$

(c) From the following equations:

$$\vec{I}_m = \vec{I}_{mf} + \vec{I}_{mb}$$

$$\vec{I}_a = j\vec{I}_{mf} - j\vec{I}_{mb} ,$$

the currents in the two phases are, respectively,

$$\vec{I}_m = 13.06 - j7.79 = 15.2\angle -31° \text{ amp}$$

$$\vec{I}_a = 4.81 + j5.64 = 7.40\angle 49.2° \text{ amp}$$

Note that the currents are much more unbalanced than the applied voltages. Even though the motor is not overloaded in so far as shaft load is concerned, the losses are appreciably increased by the current unbalance and the stator winding with the greatest current may overheat.

(d) The power delivered to the forward field by the 2 stator phases is

$$P_{gf} = 2I^2_{mf}R_f = 2 \times 126.8 \times 16.46 = 4,175 \text{ watts}$$

and the power delivered to the backward field is

$$P_{gb} = 2I^2_{mb}R_b = 2 \times 16.0 \times 0.451 = 15 \text{ watts}$$

Thus the internal mechanical power developed is

$$P = 0.95(4,175 - 15) = 3,950 \text{ watts}$$

If the core losses, friction and windage, and stray load losses are known, the shaft output can be found by subtracting them from the internal power. The friction and windage losses depend solely on the speed and are the same as they would be for balanced operation at the same speed. The core and stray load losses, however, are somewhat greater than they would be for balanced operation with the same positive-sequence voltage and current. The increase is caused principally by the (2 - s)-frequency core and stray losses in the rotor caused by the backward field.

• PROBLEM 10-28

A 2-pole, 50 Hz, induction machine has a rotor of diameter 0.20 m and core length 0.12 m. The polyphase stator winding maintains in the gap a sine-distributed travelling wave of flux of peak density 0.54 T. The rotor winding comprises a cage of 33 bars, each of resistance 120 μΩ and leakage inductance 2.50 μH, including the effect of the end-rings. At a given load the rotor runs with a slip of 0.064 corresponding to an angular speed of 294 rad/s (46.8 rev/s). Find (a) the peak current per rotor bar and the rotor I^2R loss; (b) the rotor m.m.f. per pole; (c) the load angle between the gap-flux axis and the axis of the rotor m.m.f.; (d) the electromagnetic torque developed; and (e) the mechanical power developed.

Solution: (a) Rotor bar current: The slip velocity is 314 - 294 = 20 rad/s or 20 × 0.1 = 2.0 m/s peripherally. The maximum e.m.f. induced in a bar is e_{bm} = Blu = 0.54 × 0.12 × 2.0 = 0.13 V. The bar impedance is z_b = (120 + j20 × 2.5) = (120 + j50) = 130 $\angle$ 23° μΩ. The maximum bar current is therefore

$$i_{bm} = e_{bm}/z_b = 0.13/130 \times 10^{-6} = 1{,}000 \text{ A}.$$

and the total rotor I^2R loss is $\frac{1}{2}(1{,}000^2 \times 120 \times 10^{-6} \times 33) = 1{,}980$ W.

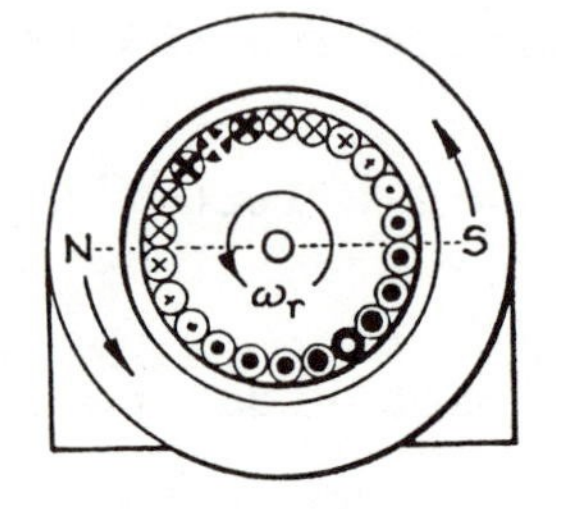

Fig. 1

Elements of the induction machine.

(b) Rotor m.m.f.: The peak linear current density is $A_2 = 1000 \times 33/\pi \times 0.2 = 52{,}500$ A/m. The rotor m.m.f. is

$$F_2 = \frac{1}{2}A_2D = \frac{1}{2} \times 52{,}500 \times 0.2 = 5{,}250 \text{ A-t/pole}.$$

(c) Load angle: If the rotor winding has only resistance, the peak current would occur in the bars at the positions of maximum gap flux density, and the rotor m.m.f. axis would be at right-angles to the gap flux axis. With the actual phase-lag of 23°, the load angle is increased to 90° + 23° = 113° (see Fig. 1).

(d) Torque:

$$M = -\frac{1}{2}\pi D\ell B_1 A_2 \sin\lambda$$

$$= -\frac{1}{2}\pi \times 0.20 \times 0.12 \times 0.54 \times 52{,}500 \times \sin 113°$$

$$= 98.5 \text{ N-m}$$

(e) Power: This is

$$P_m = M\omega_r = 98.5 \times 294 = 29{,}000 \text{ W}.$$

which can be checked from the rotor I^2R loss, using $P_m[s/(1-s)] = 29{,}000(0.064/0.936) = 1{,}980$ W. The total rotor input is $P_2 = 29{,}000 + 1{,}980 = 30{,}980$ W and the efficiency (neglecting stator and mechanical losses) is 0.936 p.u.

• PROBLEM 10-29

Under consideration is 20.0 hp, 440 volt, three-phase, 60 cps, 1735 rpm, four-pole squirrel-cage induction motor. The stator windings are WYE connected and the phase parameters, as determined by open-circuit and short-circuit tests are as follows:

$$R_1 = 0.30 \text{ ohm}$$

$$R_a = 0.36 \text{ ohm}$$

$$j\omega(L_{11} - L'_{1a}) = j\omega(L'_{AA} - L'_{a1}) = j0.63 \text{ ohm}$$

$$j\omega L'_{1a} = j27.3 \text{ ohm}$$

Per phase terminal voltage is $440/\sqrt{3} = 254$.

Determine the following phase values for a relative velocity of rotor to field of 72 rpm.
(a) Input power, current, and power factor.
(b) Air gap power and electromagnetic torque.
(c) The stator and rotor copper losses.

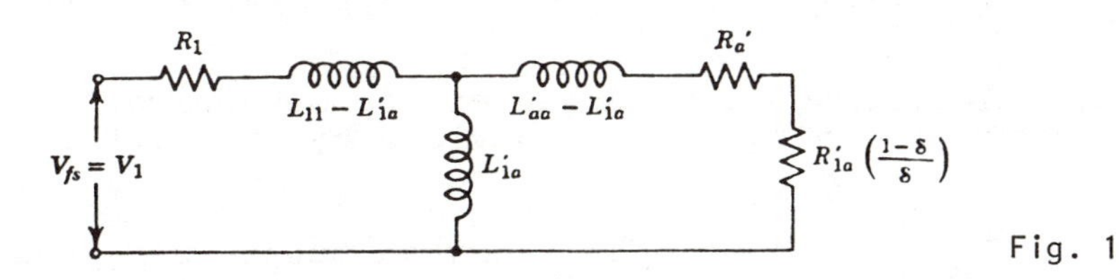

Fig. 1

Equivalent circuits of the symmetrical two-phase induction motor with balanced applied voltages.

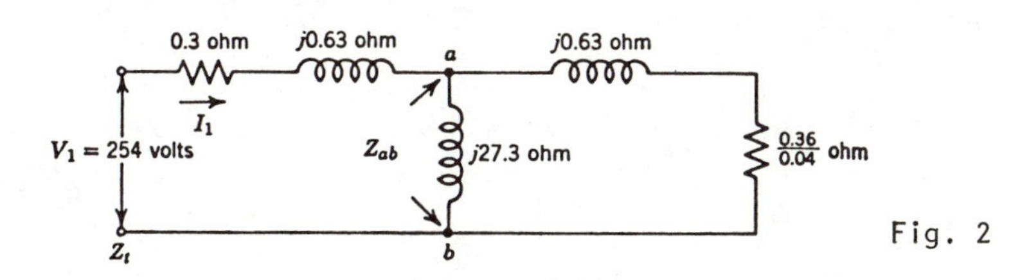

Fig. 2

Solution: Note: for the meaning of the symbols used, refer to Fig. 1 or to Fig. 2.

(a)

$$S = 72/1800 = 0.04$$

$$\vec{Z}_{ab} = j27.3 \text{ in parallel with } (9.0 + j0.63)$$

$$= 7.79 + j3.1\ \Omega$$

$$\vec{Z}_t = 7.79 + j3.1 + 0.3 + j0.63 = 8.09 - j3.73$$

$$= 8.96\underline{/24.8°}\ \Omega$$

With the terminal voltage placed at the reference, the input current in magnitude and phase is

$$\vec{I}_1 = 254\underline{/0°} \div 8.96\underline{/24.8°} = 28.3\underline{/-24.8°} \text{ amp.}$$

The power factor by definition is the cosine of the angle between voltage and current, thus

$$\text{P.F.} = \cos(-24.8°) = 0.905 \text{ lag.}$$

$$P_{in} = VI\cos\theta = 254 \times 28.3 \times 0.905$$

$$= 6540 \text{ watts per phase.}$$

(b) The real power to the air gap per phase is given by the product of the stator current squared times the real part of Z_{ab}:

$$\text{Air gap power} = P_g = 28.3^2 \times 7.79 = 6240 \text{ watts.}$$

$$\text{The field velocity} = \omega = (60 \times 2\pi)/2 = 60\pi \text{ rad/sec.}$$

The electromagnetic torque or the mechanical developed torque is given by power divided by speed, thus

$$T_g = 6240/60\pi = 33.1 \text{ newton-meters per phase.}$$

(c) The stator copper loss is simply the ohmic loss and is given by

$$\text{Stator } (I^2R) \text{ loss} = 28.3^2 \times 0.3 = 240.0 \text{ watts per phase}$$

The rotor copper loss is given by the product of slip and air gap power so

$$\text{Rotor copper loss} = 6240 \times 0.04 = 249.6 \text{ watts per phase}$$

Note that all of the above values of power and torque were calculated for a single phase of a balanced polyphase motor. Thus the total power and torque are obtained by multiplying each of the above powers and torque by the number of phases, in this case three.

• PROBLEM 10-30

Test data on a 3-phase, 7.5-hp, 220-volt, 6-pole, 60-cycle induction motor are

running light: 236 volts, 6.5 amp, 610 watts, 1,197 rpm

under load: 235 volts, 18.4 amp, 6,700 watts, 1,164 rpm

resistance between stator terminals is 0.54 ohm

Determine the power factor, output, and efficiency of the motor when loaded.

Solution: From the above readings, neglecting rotor copper loss at no load,

$$\text{fixed losses} = 610 - 3 \times 6.5^2 \times 0.27 = 576 \text{ watts}$$

$$\text{stator } I^2R \text{ under load} = 3 \times 18.4^2 \times 0.27 = 274 \text{ watts}$$

$$\text{rotor input} = 6{,}700 - (576 + 274) = 5{,}850 \text{ watts}$$

$$\text{rotor } I^2R = 0.03 \times 5850 = 179 \text{ watts (3\% slip)}$$

$$\text{motor output} = 5{,}850 - 179 = 5{,}671 \text{ watts, } 7.59 \text{ hp}$$

$$\text{Efficiency} = \frac{P_{out}}{P_{in}} = \frac{5{,}671}{6{,}700} = 84.7\%$$

$$\text{Power factor} = \frac{\text{Power}}{\sqrt{3}\,EI} = \frac{6{,}700}{\sqrt{3} \times 235 \times 18.4}$$

$$= 0.893.$$

• PROBLEM 10-31

A 5-hp, 60-cycle, 115-volt, eight-pole, three-phase induction motor was tested, and the following data were obtained:

No-load test: $V_{NL} = 115$; $P_1 = 725$; $P_2 = -425$; $I_{NL} = 10$

Load test: $V_L = 115$; $P_1 = 3{,}140$; $P_2 = 1{,}570$; $I_L = 27.3$; $rpm_{rotor} = 810$

D-c stator resistance between terminals = 0.128 ohm

Calculate: (a) the horsepower output; (b) the torque; (c) the per cent efficiency; (d) the power factor of the motor for the given load values.

Solution:

R_{ac} = Effective a-c resistance of stator per phase

$= \frac{0.128}{2} \times 1.25 = 0.08$ ohm

No-load copper loss $= 3I_{NL}^2 R_{ac} = 3 \times (10)^2 \times 0.08$

$= 24$ watts.

Friction + windage + iron losses $= (725 - 425) - 24$

$= 276$ watts.

(It is permissible to charge this loss entirely to the stator with no appreciable error, although the friction and windage portion is actually a part of the rotor loss.)

Load copper loss in stator $= 3I_L^2 R_{ac}$

$= 3 \times (27.3)^2 \times 0.08$

$= 179$ watts.

Total stator loss under load $= 276 + 179 = 455$ watts.

RPI under load $= P_1 + P_2$ - Total stator loss under load

$= (3{,}140 + 1{,}570) - 455 = 4{,}255$ watts

$$s = \frac{rpm_{syn} - rpm_{rotor}}{rpm_{syn}} = \frac{900 - 810}{900} = 0.10.$$

RCL = (RPI)·s = 0.10 × 4,255 = 426 watts.

Rotor power output = RPD = RPI - RCL = 4,255 - 426

= 3,829 watts.

(a) Horsepower output $= \frac{RPD}{746} = \frac{3{,}829}{746} = 5.13.$

(b) $T = 7.04 \times \frac{RPI}{rpm_{sym}} = 7.04 \times \frac{4{,}255}{900} = 33.2$ lb-ft.

(c) Per cent efficiency (η)

$$\eta = \left(1 - \frac{\text{Watts losses}}{\text{Watts output + Watts losses}}\right) \times 100$$

$$= 1 - \frac{455 + 426}{4{,}710} \times 100 = 81.3.$$

(d) Power factor $= \dfrac{\text{Watts output + Watts losses}}{\sqrt{3}\, V_L I_L}$

$$= \frac{4{,}710}{\sqrt{3} \times 115 \times 27.3} = 0.866.$$

• PROBLEM 10-32

A no-load and a blocked test are made on a 220-volt, 25-cycle, 7.5-hp 750-rpm 3-phase wound-rotor induction motor. The transformation ratio of stator to rotor is 3.32. Computations are based on the stator and rotor being Y-connected.

From the following no-load and blocked data, determine for a slip of 0.06 (a) motor output; (b) speed; (c) torque; (d) efficiency; (e) current; (f) power factor.

The data are as follows:

(1) No-load: volts, 220; amperes, 9.63; watts, 390; friction and windage, 150 watts.
(2) Blocked test: volts, 38.2; amperes, 22.7; watts, 1,080.
(3) Average d-c resistance between terminals of stator, 0.375 ohm, or 0.1875 ohm per phase.
(4) Average d-c resistance between terminals of rotor, 0.0726 ohm.
(5) Rotor resistance per phase referred to stator = $(3.32)^2(0.0726/2) = 0.40$ ohm.

Solution: From (1), watts core loss per phase (neglecting loss in stator due to 9.63 amp) = (390 - 150)/3 = 80 watts;

$$\text{energy current } I_e = \frac{390}{(\sqrt{3}\times 220)} = 1.02 \text{ amp;}$$

quadrature current $I_q = \sqrt{(9.63)^2 - (1.02)^2} = 9.58$ amp;

core-loss corrent, $I'_e = 240/(\sqrt{3}\cdot 220) = 0.630$ amp; no-load current with iron only,

$$I_0 = \sqrt{(9.58)^2 + (0.630)^2} = 9.6 \text{ amp};$$

$\cos\theta_0 = 240/(220\sqrt{3}\times 9.6) = 0.0655;\ \theta_0 = 86.2°.$

Voltage to neutral: $V' = 220/\sqrt{3} = 127$ volts; $V'^2 G_0 =$ 80 watts;

$$G_0 = \frac{80}{(127)^2} = 0.00496 \text{ mho};$$

$Y_0 = 9.6/127 = 0.0755$ mho; $B_0 = \sqrt{Y_0^2 - G_0^2} = 0.0754$ mho.

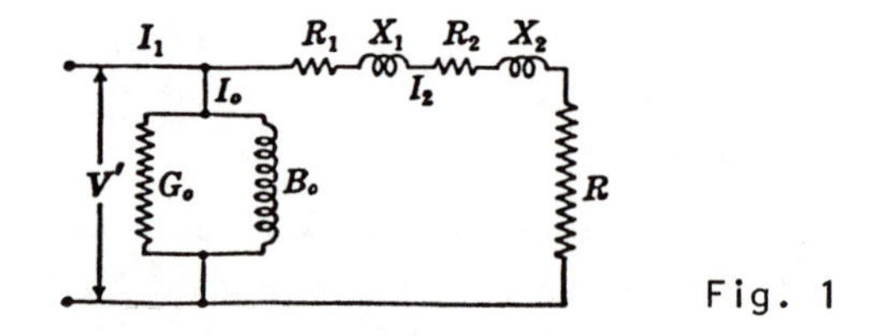

Fig. 1

Approximate equivalent circuit of induction motor.

From (2), since the mechanical power and hence R = 0, Fig. 1,

$$(22.7)^2(R_1 + R_2) = \frac{1{,}080}{3};\quad R_1 + R_2 = 0.70 \text{ ohm}.$$

Effective resistance R_1 of stator [from (3), (5)]

$$= \frac{0.1875}{0.1875 + 0.40}\, 0.70 = 0.223 \text{ ohm}.$$

With low values of slip, the direct-current resistance of rotor (0.40 ohm) is much nearer the operating value than the 60-cycle effective value, because of the low rotor frequency. Hence, R_2=0.40 ohm, Fig. 1.

From (2),

$$Z_B = \frac{38.2}{\sqrt{3}\times 22.7} = 0.972 \text{ ohm},$$

$$X_1 + X_2 = \sqrt{(0.972)^2 - (0.70)^2} = 0.675 \text{ ohm},$$

$$R = R_2 \times \frac{(1 - s)}{s} = 0.40\,\frac{1 - 0.06}{0.06} = 6.26 \text{ ohms}$$

$$I_2 = \frac{127}{\sqrt{(0.223 + 6.26 + 0.40)^2 + (0.675)^2}}$$

$$= \frac{127}{6.91} = 18.4 \text{ amp.}$$

(a) Motor output $= [3 \times (18.4)^2 \times 6.26] - 150$

$= 6{,}210$ watts $= 8.33$ hp.

(b) Speed $= 750(1 - 0.06) = 705$ rpm.

(c) Torque $= \dfrac{8.33 \times 33{,}000}{705 \times 2\pi} = 62.1$ lb-ft.

(d) Total resistance loss $= (18.4)^2(0.223 + 0.40)3$

$= 633$ watts.

From (1), (a), and (d),

$$\text{Efficiency} = \frac{6{,}210}{6{,}210 + 633 + 240 + 150} = 0.859.$$

(e) Total current

$\vec{I}_0 = 127(G_0 - jB_0) = 127(0.00496 - j0.0754)$

$= 0.63 - j9.58$ amp.

$I_0 = \sqrt{0.63^2 + 9.58^2} = 9.58$ amp.

$\vec{I}_2 = \dfrac{127}{6.88 + j0.675} = 18.3 - j1.8$ amp.

$I_2 = \sqrt{18.3^2 + 1.8^2} = 18.4$ amp.

$\vec{I} = I_0 + I_2 = 18.93 - j11.38$ amp.

$I = \sqrt{18.93^2 + 11.38^2} = 22.1$ amp.

(f) $\text{P.F.} = \frac{18.93}{22.1} = 0.857.$

• PROBLEM 10-33

A 25-hp, 230-volt, 60-cycle, six-pole, single-phase induction motor has the following no-load and locked rotor data:

No LOAD				Locked Rotor			
Volts	Frequency	Amperes	Watts	Volts	Frequency	Amperes	Watts
220	60	36.8	938	62.0	60	108.0	2,650

Other data are:

Friction and windage loss	375 watts
Ohmic resistance between stator terminals	0.1261 ohm
Ratio of effective resistance to ohmic resistance:	
Stator at 60 cycles	1.1
Rotor:	
At 60 cycles	1.2
At 120 cycles	1.8

The stator and rotor reactances are assumed to be equal when referred to the stator. Assuming a slip of 2.0 per cent, calculate the torque, power output and efficiency.

Solution:

$$r_{eq}\text{(equivalent)} = \frac{2{,}650}{2 \times (108.0)^2} = 0.1136 \text{ ohm per phase at 60 cycles}$$

$$z_{eq}\text{ (equivalent)} = \frac{62}{2 \times 108.0} = 0.287 \text{ ohm per phase at 60 cycles}$$

$$x_{eq}\text{ (equivalent)} = \sqrt{(0.287)^2 - (0.1136)^2} = 0.2636 \text{ ohm per phase at 60 cycles}$$

Equivalent Three-phase Constants

$$r_1 \text{ (effective at 60 cycles)} = \frac{0.1261}{2} \times 1.1 = 0.06936 \text{ ohm per phase.}$$

r_2 (effective at 60 cycles) $= 0.1136 - 0.06936$

$= 0.0442$ ohm per phase

$$r_2 \text{ (ohmic)} = \frac{0.0442}{1.2} = 0.03683 \text{ ohm per phase}$$

r_2' (effective at 120 cycles) $= 0.03683 \times 1.8$

$= 0.0663$ ohm per phase

$$x_1 = x_2 \text{ (referred to stator)} = \frac{0.2636}{2} = 0.1318 \text{ ohm per phase.}$$

Now $$\vec{z}^{+} = \vec{z}^{-} = \left(2r_1 + \frac{r_2}{s} + \frac{r_2'}{2 - s}\right) + j2(x_1 + x_2)$$

and

$$\vec{z}_n^{\,+} = \left(2r_1 + \frac{r_2'}{2 - s}\right) + j(2x_1 + x_2).$$

$$\vec{z}^{+} = \vec{z}^{-} = \left(2 \times 0.06936 + \frac{0.03683}{0.02} + \frac{0.0663}{1.98}\right) + j2(0.2636)$$

$$= 2.014 + j0.5272 \text{ vector ohms}$$

$$= \sqrt{(2.014)^2 + (0.5272)^2} = 2.082 \text{ ohms.}$$

$$\vec{z}_n^{\,+} = \left(2 \times 0.06936 + \frac{0.0663}{1.98}\right) + j(2 \times 0.1318 + 0.1318)$$

$$= 0.1722 + j0.3954 \text{ vector ohm}$$

$$= \sqrt{(0.1722)^2 + (0.3954)^2} = 0.431 \text{ ohm.}$$

$$\frac{V_L}{\sqrt{3}} = \frac{220}{\sqrt{3}} = 127.1 \text{ volts}$$

$$\frac{V_L'}{\sqrt{3}} = \frac{V_L}{\sqrt{3}} - I_n z_n^{+} = 127.1 - \frac{36.8}{\sqrt{3}} \times 0.431$$

$$= 117.9 \text{ volts}$$

$$I_2^+ = \frac{117.9}{2.082} = 56.6 \text{ amp}$$

Consider the positive-sequence components. The vector diagram for these components is given in Fig. 1.

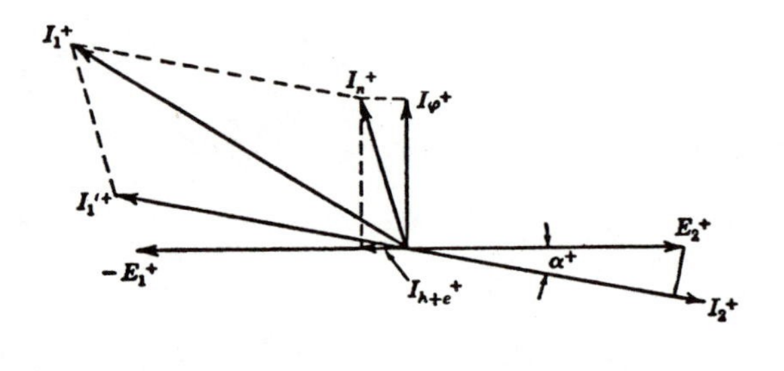

Fig. 1

$$\text{No-load power factor} = \frac{938}{220 \times 36.8} = 0.1156$$

$$\cos\theta_n^+ = 0.1156 \qquad \sin\theta_n^+ = 0.993$$

$$\text{Core loss} = 928 - 375 - (36.8)^2(2 \times 0.06936 + 0.0663)$$

$$= 285 \text{ watts}$$

$$\text{Single-phase } \vec{I}_n = -\frac{285}{220} + j36.8 \times 0.993$$

$$= -1.29 + j36.5 \text{ vector amperes}$$

Neglecting the negative-sequence magnetizing current,

$$\vec{I}_n^+ = \frac{1}{\sqrt{3}}(-1.29 + j36.5) = -0.748 + j21.1 \text{ vector amperes}$$

$$\cos\alpha^+ = \frac{r_2}{\sqrt{(r_2)^2 + (x_2 s)^2}}$$

$$= \frac{0.03683}{\sqrt{(0.03683)^2 + (0.1318 \times 0.02)^2}} = 0.997.$$

$$\sin\alpha^+ = \frac{x_2 s}{\sqrt{(r_2^2) + (x_2 s)^2}} = \frac{0.1318 \times 0.02}{\sqrt{(0.03683)^2 + (0.1318 \times 0.02)^2}}$$

$$= 0.0714.$$

$$\vec{I}_1^{\,+} = \vec{I'}_1^{\,+} + \vec{I}_n^{\,+} = -56.6(0.997 - j0.0714) + $$

$$+ (-0.748 + j21.1)$$

$$= -57.15 + j25.1 \text{ vector amperes}$$

$$= \sqrt{(-57.15)^2 + (25.1)^2} = 63.25 \text{ amp.}$$

If the magnetizing current for the negative-sequence components is neglected, $I_2^- = I_1^+$ in magnitude if referred to the stator. Making this approximation,

$$T = T^+ - T^-$$

$$= 3 \frac{p}{4\pi f} \left[(I_2^+)^2 \frac{r_2}{s} - (I_2^-)^2 \frac{r'_2}{2 - s} \right] \frac{550}{746}$$

$$= 3 \frac{6}{2 \times 377} \left[(56.6)^2 \frac{0.03683}{0.02} - (63.25)^2 \frac{0.0663}{1.98} \right] \frac{550}{746}$$

$$= 0.02386(5{,}900 - 134) \frac{550}{746}$$

$$= 101.5 \text{ lb-ft internal torque.}$$

$$P = \text{pulley output} = 2\pi \text{ speed } T = 2\pi \frac{2f}{p} (1 - s)T - P_{f+w}$$

$$= 2\pi \frac{120}{6} (1.00 - 0.02)101.5 - 375 \times \frac{550}{746}$$

$$= 12{,}230 \text{ ft-lb per sec.}$$

$$= \frac{12{,}230}{550} = 22.2 \text{ hp.}$$

$$\text{Copper loss} = 3[(I_1^+)^2 r_1 + (I_1^-)^2 r_1 + (I_2^+)^2 r_2 + (I_2^-)^2 r'_2]$$

$$= 3[2(63.25)^2 0.0694 + (56.6)^2 0.0368 + (63.25)^2\, 0.0663]$$

$$= 3(555 + 118 + 265)$$

$$= 2{,}814 \text{ watts.}$$

Copper loss	2,814 watts
Core loss	285 watts
Friction and windage loss	375 watts
Total losses	3,474 watts

$$\text{Efficiency} = \frac{22.2 \times 746}{22.2 \times 746 + 3{,}474} = 0.827 \text{ or } 82.7 \text{ per cent}$$

$$I_1 = \sqrt{3}\, I_1^{+} = \sqrt{3} \times 63.25 = 109.5 \text{ amp.}$$

$$\text{Power factor} = \frac{22.2 \times 746 + 3{,}474}{220 \times 109.5} = 0.831.$$

• PROBLEM 10-34

A 15-hp, 440-V, three-phase, 60-Hz, 8-pole, wound-rotor induction motor has its stator and rotor both connected in wye. The ratio of effective stator turns to effective rotor turns is b = 2.4:1. The windage and friction losses are 220 W at rated speed and may be assumed constant from no load to full load. The stator and rotor have the following constants per phase:

Stator (Ω)	Rotor (Ω)
$r_1 = 0.52$	$r_{22} = 0.110$
$x_1 = 1.15$	$x_{22} = 0.20$
$x_M = 40.0$	
$r_{fe} = 360$	

The stray-load loss is 120 W. (a) Use the equivalent circuit in Fig. 1(a) to calculate the following for a slip s = 0.045 with balanced rated voltage and rated frequency applied to the stator and with the rotor slip rings short-circuited: (i) stator current, (ii) power factor, (iii) current in the rotor winding, (iv) output in horsepower, (v) efficiency, and (vi) torque. (b) Repeat part (a) using the approximate equivalent circuit, Fig. 1(b). (c) Compare the results of parts (a) and (b) in tabulated form.

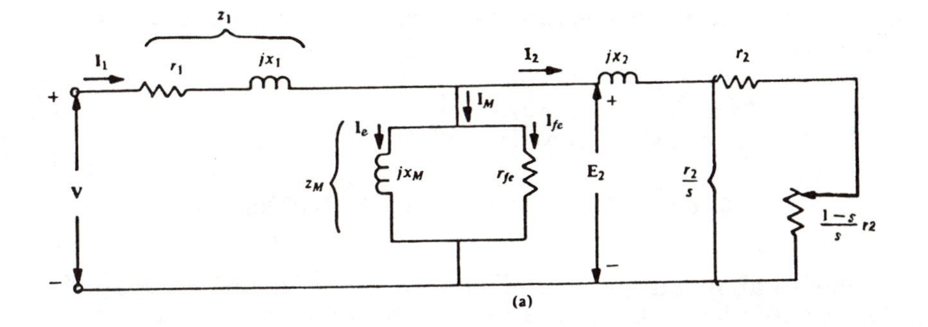

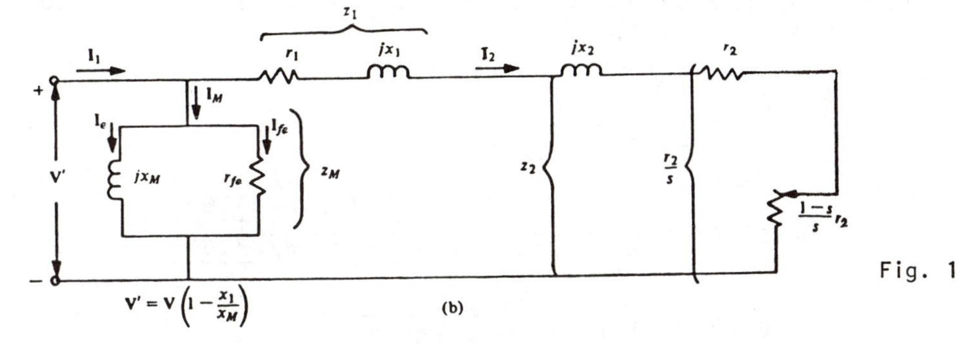

Fig. 1

(a) Equivalent circuit. (b) Approximate equivalent circuit of polyphase motor.

Solution: (a) In both equivalent circuits of Fig. 1, the rotor impedance is referred to the stator by use of the impedance ratio:

$$b^2 = (2.4)^2 = 5.76$$

so that

$$r_2 = b^2 r_{22} = 5.76 \times 0.110 = 0.634 \ \Omega/\text{phase}$$

$$x_2 = b^2 x_{22} = 5.76 \times 0.20 = 1.15 \ \Omega/\text{phase}$$

For a slip s = 0.045, the impedance of the rotor referred to the stator is

$$\vec{z}_2 = \frac{r_2}{s} + jx_2 = \frac{0.634}{0.045} + j1.15$$

$$= 14.10 + j1.15 = 14.13\underline{/4.7°} \ \Omega/\text{phase}$$

The leakage impedance of the stator is

$$\vec{z}_1 = r_1 + jx_1 = 0.52 + j1.15 = 1.26\underline{/65.6°} \ \Omega/\text{phase}$$

and the exciting impedance referred to the stator is

$$\vec{z}_M = \frac{r_{fe} jx_M}{r_{fe} + jx_M} = \frac{(360)(j40)}{360 + j40}$$

$$= 39.8\underline{/83.65°} = 4.40 + j39.5 \ \Omega/\text{phase}$$

(i) The stator current can now be determined simply by dividing the voltage, applied to one phase of the stator by the impedance of the circuit. The impedance is, from Fig. 1(a),

$$\vec{Z} = \vec{z}_1 + \frac{\vec{z}_2 \vec{z}_M}{\vec{z}_2 + \vec{z}_M}$$

$$= 1.26\underline{/65.6°} + \frac{(14.13\underline{/4.7°})(39.8\underline{/83.65°})}{14.13\underline{/4.7°} + 39.8\underline{/83.65°}}$$

$$= 0.52 + j1.15 + \frac{562\underline{/88.35°}}{44.8\underline{/65.6°}}$$

$$= 12.09 + j6.00 = 13.47\underline{/26.4°} \ \Omega/\text{phase}$$

The voltage per phase is

$$V = \frac{440}{\sqrt{3}} = 254$$

which produces the stator current

$$\vec{I}_1 = \frac{\vec{V}}{\vec{Z}} = \frac{254\underline{/-26.4°}}{13.47} = 18.85\underline{/-26.4°} \ \text{A/phase}$$

(ii) The power factor of the motor is

$$\text{P.F.} = \cos\theta = \cos 26.4° = 0.895$$

(iii)

$$\vec{I}_2 = \frac{\vec{E}_2}{\vec{z}_2} = \frac{\vec{I}_1 \vec{z}_2 \vec{z}_M}{(\vec{z}_2 + \vec{z}_M)\vec{z}_2} = \frac{\vec{I}_1 \vec{z}_M}{\vec{z}_2 + \vec{z}_M}$$

$$= \frac{(18.85\underline{/-26.4°})(39.8\underline{/83.65°})}{44.8\underline{/65.6°}}$$

$$= 16.75\underline{/-8.35°} \text{ A/phase}$$

$$\vec{I}_{22} = b\vec{I}_2 = 2.4 \times 16.75\underline{/-8.35°}$$

$$= 40.2\underline{/-8.35°} \text{ A/phase}$$

(iv)

$$P_{em} = mI_2^2 \frac{1 - s}{s} r_2 = 3(16.75)^2 \frac{1 - 0.045}{0.045} 0.634$$

$$= 11{,}300 \text{ W}$$

$$P_{mech} = P_{em} - (P_{fw} + P_{stray})$$

$$= 11{,}300 - (220 + 120) = 10{,}960 \text{ W}$$

$$= 10{,}960 \div 746 = 14.7 \text{ hp}$$

(v) The efficiency is the ratio of output to input. The real power input is

$$P_{in} = mVI_1 \cos \theta$$

$$= 3 \times 254 \times 18.85 \times 0.895 = 12{,}880 \text{ W}$$

$$\text{Efficiency} = \frac{10{,}960}{12{,}880} = 1 - \frac{1920}{12{,}880} = 1 - 0.149$$

$$= 0.851$$

(vi) Torque is the ratio of mechanical power to the mechanical angular velocity of rotation. The angular velocity

$$\omega_m = \frac{2\pi n}{60} = \frac{2\pi(1 - s)n_{syn}}{60}$$

$$= \frac{2\pi(1 - 0.045)[(120 \times 60)/8]}{60} = 90.0 \text{ rad/sec,}$$

and the torque is

$$T = \frac{10{,}960}{90} = 121.8 \text{ N-m/rad}$$

(b) (i) On the basis of the approximate equivalent circuit of Fig. 1(b) and the equation,

$$V' = V\left(1 - \frac{x_1}{x_M}\right) = \left[r_1 + \frac{r_2}{s} + j(x_1 + x_2)\right] I_2 ,$$

the rotor current referred to the stator is found to be

$$\vec{I}_2 = \frac{\vec{V}(1 - x_1/x_M)}{\vec{z}_1 + \vec{z}_2} = \frac{\vec{V}(1 - x_1/x_M)}{r_1 + (r_2/s) + j(x_1 + x_2)}$$

$$= \frac{254(1 - 1.15/40)}{0.52 + 14.1 + j2.30} = \frac{246.7}{14.62 + j2.30}$$

$$= \frac{246.7}{14.80\underline{/8.95°}}$$

$$= 16.7\underline{/-8.95°} = 16.5 - j2.60 \text{ A/phase}$$

The exciting current is

$$\vec{I}_M = \frac{\vec{V}(1 - x_1/x_m)}{\vec{z}_M} = \frac{246.7}{360} + \frac{246.7}{j40}$$

$$= 0.69 - j6.17 \text{ A/phase}$$

and the stator current is

$$\vec{I}_1 = \vec{I}_2 + \vec{I}_M = 16.5 - j2.60 + 0.69 - j6.17$$

$$= 17.19 - j8.77 = 19.30\underline{/-27.0°} \text{ A/phase}$$

(ii) The power factor is

$$\text{P.F.} = \cos\theta = \cos 27.0° = 0.890$$

The actual current in the rotor winding is

$$I_{22} = bI_2 = 2.4 \times 16.7 = 40.1 \text{ A/phase or per slip ring}$$

(iii) The developed mechanical power is

$$P_{em} = mI_2^2 \frac{1 - s}{s} r_2 = 3(16.7)^2 \frac{1 - 0.045}{0.045} 0.634$$

$$= 11{,}260 \text{ W}$$

(iv) The mechanical power is

$$P_{mech} = P_{em} - (P_{fw} + P_{stray})$$

$$= 11{,}260 - (220 + 120) = 10{,}920 \text{ W}$$

with a horsepower output of

$$\text{hp} = 10{,}920 \div 746 = 14.65$$

(v) The power input is

$$P_{in} = 3VI_1 \cos \theta$$

$$= 3(254)(19.30)(0.890) = 13{,}080 \text{ W}$$

and the efficiency is

$$\text{Efficiency} = \frac{P_{mech}}{P_{in}} = \frac{10{,}920}{13{,}080} = 1 - \frac{2160}{13{,}080} = 0.835$$

(vi) $$\text{Torque} = \frac{P_{mech}}{\omega_m} = \frac{11{,}040}{90.0} = 122.7 \text{ N-m/rad}$$

(c) Tabulation of Results

	Circuit	
	Equiv. (*a*)	*Approx. Equiv.* (*b*)
Slip, s	0.045	0.045
Stator current, I_1	18.85	19.3
Rotor current, I_{22}	40.2	40.1
Power factor	0.895	0.890
Power input (kW)	12.88	13.08
Power output (kW)	10.96	10.92
Efficiency	0.851	0.835

PERFORMANCE COMPUTATIONS, CIRCLE DIAGRAM

• PROBLEM 10-35

A 200 W, 240 V, 50 Hz, 4-pole, 1-ph induction motor runs on rated load with a slip of 0.05 p.u. The parameters are: $r_1 = 11.4\ \Omega$, $x_1 = 14.5\ \Omega$, $\frac{1}{2}r_2 = 6.9\ \Omega$, $\frac{1}{2}x_2 = 7.2\ \Omega$, $\frac{1}{2}x_m = 135\ \Omega$; core and mechanical loss, 32 W. Estimate the full-load performance.

Solution: Forward circuit. With s = 0.05, the rotor impedance is

$$\left(\frac{1}{2}\,r_2/s\right) + j\,\frac{1}{2}\,x_2 = \frac{6.9}{0.05} + j7.2 = 138 + j7.2\ \Omega$$

in parallel with

$$j\,\frac{1}{2}\,x_m = j135\ \Omega.$$

This gives

$$\vec{z}_{2f} = 64 + j69 = 94\angle\,47°\ \Omega$$

Backward circuit. Here

$$\frac{1}{2}\,r_2/(2 - s) = 3.5\ \Omega$$

and $$\frac{1}{2}\,jx_2 = j7.2\ \Omega$$

in series. The result of parallel combination with $j\,\frac{1}{2}\,x_m$ is

$$\vec{z}_{2b} = 3 + j7 = 7.6\angle 67°\ \Omega$$

Output. The total series input impedance is $\vec{z}_1 + \vec{z}_{2f} + \vec{z}_{2b} = 78.4 + j90.5 = 120\angle 49°$, the p.f. being cos 49° = 0.66. The forward and backward active rotor currents and torques are found as follows:

$E_{1f} = 2.0 \times 94 = 188$ V $\qquad E_{1b} = 2.0 \times 7.6 = 15.2$ V

$I_{2f} = 188/138 = 1.36$ A $\quad I_{2b} = 15.2/8.0 = 1.9$ A

$M_f = 1.36^2 \times 138 = 255$ W $\quad M_b = 1.9^2 \times 3.5 = 13$ W

The net torque (as a synchronous power) is M = 255 - 13 - 32 = 210 W and the shaft power is 210 × 0.95 = 200 W. The input power is 240 × 2.0 × 0.66 = 316 W and the efficiency is 200/316 = 0.63 p.u.

• PROBLEM 10-36

A 3-Hp, 440/220-volt, 3-phase 60-cycle, 4-pole, 1750-rpm squirrel-cage induction motor has the following data and parameters. Determine the performance at start and during running conditions.

(a) Parameters for starting (p-u)

$r_1 = 0.0311$

$r_2' = 0.0322$

$(x_1 = x_2') = 0.0393$

$x_m = 1.19$

$r_m = 0.0423$

(b) Parameters for running (p-u)

$r_1 = 0.0311$

$r_2' = 0.0248$

$x_1 = 0.0505$

$x_2' = 0.0520$

$x_m = 1.19$

$r_m = 0.0423$

Solution:

$$\text{Unit voltage} = \frac{440}{\sqrt{3}} = 254 \text{ volts}$$

$$\text{Unit current} = \frac{3 \times 746}{3 \times 254} = 2.94 \text{ amp} = I_{HP}$$

$$\text{Unit impedance} = \frac{254}{2.94} = 86.5 \text{ ohms}$$

Unit power = 254 × 2.94 = 746 watts = 1 HP

Unit speed = 1800 rpm

$$\text{Unit torque} = 7.04 \times 3 \times \frac{\text{unit power}}{\text{unit speed}} = 8.80 \text{ lb-ft}$$

(a) Starting performance (p-u)

$$r_2' \left(\frac{1-s}{s}\right) = 0$$

$$\vec{\dot{Z}}_1 = 0.0500\underline{/51.6} \ \Omega$$

$$\vec{\dot{Z}}'_2 = 0.0508\underline{/50.6} \ \Omega$$

$$\vec{\dot{Z}}_m = 1.19\underline{/87.96} \ \Omega$$

From the following equation:

$$\dot{V}_1 = \dot{I}_1 \left[\dot{Z}_1 + \frac{\dot{Z}_m \dot{Z}'_2}{\dot{Z}_m + \dot{Z}'_2}\right],$$

$$1\underline{/0} = \vec{\dot{I}}_1 \left[0.0500\underline{/51.6} + \frac{0.0508\underline{/50.6} \times 1.19\underline{/87.96}}{0.0508\underline{/50.6} + 1.19\underline{/87.96}}\right]$$

$$\dot{I}_1 = 10.05\underline{/-51.8} \text{ (p-u) amp}$$

(check) $I_1(\text{amp}) = 10.05 \times 2.94 = 29.6$

$$\vec{\dot{I}}'_2 = 10.05\underline{/-51.8} \times \frac{1.19\underline{/87.96}}{0.0508\underline{/50.6} + 1.19\underline{/87.96}}$$

$$= 9.66\underline{/-50.4} \text{ amp.}$$

The starting torque on a p-u basis is the same as $P_{rot.f}$ on a p-u basis.

$$P_{rot.f} = m_1 \dot{I}_2'^2 \frac{r_2'}{s} \text{ watts}$$

$$P_{rot.f} = (9.66)^2 \times \frac{0.0322}{1.0} = 3.0$$

$$T_{st}\text{p-u} = 3.0$$

(check) $T_{st}(\text{lb-ft}) = 3.0 \times 8.80 = 26.4$

(b) Running performance $s = 0.03$

$$r_2'\left(\frac{1 - s}{s}\right) = 0.80$$

$$\dot{Z}_1 = 0.0593\underline{/58.3}\ \Omega$$

$$\dot{Z}_2' = 0.825\underline{/3.60}\ \Omega$$

$$\dot{Z}_m = 1.19\underline{/87.96}\ \Omega$$

$$\dot{I}_1 = \frac{\dot{V}_1}{\dot{Z}_1 + [\dot{Z}_m\dot{Z}_2' / (\dot{Z}_m + \dot{Z}_2')]}$$

$$= \frac{1.0\underline{/0}}{\left[0.0593\underline{/58.3} + \dfrac{0.825\underline{/3.60} \times 1.19\underline{/87.96}}{0.825\underline{/3.60} + 1.19\underline{/87.96}}\right]}$$

$$= 1.42\underline{/-38.2}\ \text{amp.}$$

Input power factor = $\cos 38.2° = 0.785$

$$\dot{I}_2' = \dot{I}_1 \times \frac{\dot{Z}_m}{\dot{Z}_m + \dot{Z}_2'}$$

$$= 1.42\underline{/-38.2} \times \frac{1.19\underline{/87.96}}{0.825\underline{/3.60} + 1.19\underline{/87.96}}$$

$$= 1.115\underline{/-5.74}\ \text{amp.}$$

$$P_{rot.f} = m_1 \dot{I}_2'^2 \frac{r_2'}{s}$$

$$= (1.115)^2 \times \frac{0.0248}{0.03} = 1.025\ \text{watt}$$

$$T_{dev} = 1.025$$

(check) $T_{dev}(\text{lb-ft}) = 1.025 \times 8.80 = 9.05$

$$\text{Mech. loss torque (p-u)} = \frac{7.04 \times \frac{(48+44+61)}{1800}}{8.80}$$

$$= 0.068$$

$$T_{del}(\text{p-u}) = 1.025 - 0.068 = 0.957$$

(check) $T_{del}(\text{lb-ft}) = 0.957 \times 8.80 = 8.40$

Power input $= 3 \times 1.0 \times 1.42 \times \cos 38.2° = 3.35$

Losses (in watts)

$m_1 I_1{}^2 r_1 = 0.189$

$m_1 I_2{}'^2 r_2{}' = 0.093$

no-load iron $= 0.163$

stray load $= 0.064$

Friction and windage $= 0.059$

Total $= 0.568$,

$P_{del} = 3.35 - 0.568 = 2.78.$

(check) $P_{del}(\text{watts}) = 2.78 \times 746 = 2076 = 2.78$ HP

$\therefore$ Efficiency $= \frac{2.78}{3.35} = 0.83.$

$$\text{slip} = \frac{0.093}{3.35 - (0.189 - 0.082)} = 0.03. \qquad \text{(check)}$$

$$s_{normal} = \frac{3.0}{2.78} \times 0.03 = 0.0324$$

$$s_{p.o.} \approx \frac{(1 + \tau_1) r_2{}'}{x_1 + (1 + \tau_1) x_2{}'}$$

$$= \frac{1.042 \times 0.0257}{0.0505 + 1.042 \times 0.0520} = 0.246$$

$$r_2{}' \left(\frac{1 - s}{s}\right) = 0.0758$$

$$\dot{Z}_2' = 0.1133\underline{/27.3}\ \Omega$$

$$\dot{I}_1 = 6.12\underline{/-41.4}\ \text{amp}$$

$$\dot{I}_2' = 5.85\underline{/-36.6}\ \text{amp}$$

$$P_{rot.f} = (5.85)^2 \times \frac{0.0248}{0.246} = 3.44 \text{ watts}$$

$$T_{p.o.} = 3.44$$

(check) $\quad T_{p.o.}(\text{lb-ft}) = 3.44 \times 8.80 = 30.2$

The normal torque delivered by this machine at 3.0 HP output, s = 0.0324, is

$$T = \frac{5250}{1800(1 - 0.0324)} \times 3 = 9.05 \text{ lb-ft}$$

$$= 1.025 \text{ p-u}$$

$$\frac{T_{st}}{T_n} = \frac{3.0}{1.025} = 2.92$$

$$\frac{T_{p.o.}}{T_n} = \frac{3.44}{1.025} = 3.36.$$

• PROBLEM 10-37

Calculate the performance characteristics of a 1,000-hp, 2,200-volt, Y-connected, 25-cycle, 12-pole, squirrel-cage induction motor with copper bars in rotor, using the equivalent circuit of Fig. 1. Test data obtained from no-load and blocked-rotor runs and the measured stator resistance are as follows:

Ohmic resistance of stator between terminals at 25°C = 0.210 ohm:

At no load and a temperature of 25°C:

Stator line voltage	2,200 volts
Stator line current	75.1 amp
Total stator input	15.2 kw

With the rotor blocked at approximately full-load current at a temperature of 25°C and a frequency of 15 cycles:

Stator line voltage	185 volts
Stator line current	250 amp
Total stator input	38 kw

Stray load loss = 7.0 kw.

Assume a slip of 0.02 and an operating temperature of 75°C.

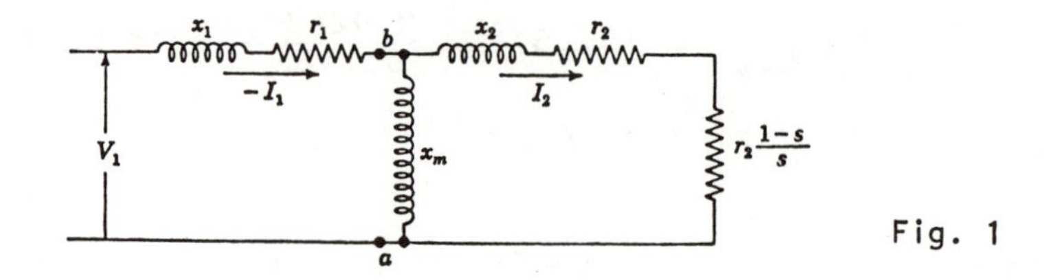

Fig. 1

Solution: The stator resistance at this operating temperature is

$$r_1 \text{ at } 75°C = \frac{0.210}{2} \times \frac{234.5 + 75}{234.5 + 25} = 0.1252 \text{ ohm.}$$

The rotor resistance referred to the stator as found from the blocked-rotor test is

$$r_2 \text{ at } 25°C = \frac{38{,}000}{3 \times (250)^2} - \frac{0.210}{2} = 0.0977 \text{ ohm.}$$

$$r_2 \text{ at } 75°C = 0.0977 \times \frac{234.5 + 75}{234.5 + 25} = 0.1165 \text{ ohm.}$$

The combined reactance of the stator and rotor is

$$x_0 = \frac{25}{15}\sqrt{\left(\frac{185}{\sqrt{3} \times 250}\right)^2 - \left(\frac{38{,}000}{3 \times (250)^2}\right)^2} = 0.626 \text{ ohm.}$$

and

$$x_1 = x_2 = 0.5x_0 = 0.313 \text{ ohm.}$$

The magnetizing reactance as found from the no-load test is

$$x_m = \frac{2{,}200}{\sqrt{3} \times 75.1} - 0.313 = 16.60 \text{ ohms.}$$

$$b_m = \frac{1}{16.60} = 0.0602 \text{ ohm.}$$

The core loss plus the friction and windage loss equals

$$P_c + P_{f+w} = 15{,}200 - 3 \times (75.1)^2 \times 0.105$$
$$= 13{,}420 \text{ watts.}$$

Refer to the equivalent circuit of Fig. 1.

$$\vec{z}_2 = r_2 + r_2\frac{1 - s}{s} + jx_2 = \frac{r_2}{s} + jx_2$$

$$= \frac{0.1165}{0.02} + j0.313$$

$$= 5.825 + j0.313 \ \Omega.$$

$$\vec{y}_2 = \frac{1}{\vec{z}_2} = g_2 - jb_2 = \frac{1}{5.825 + j0.313}$$

$$= 0.1712 - j0.0092 \text{ mho}$$

The total admittance to the right of ab is

$$g_2 - jb_2 - jb_m = 0.1712 - j0.0092 - j0.0602$$
$$= 0.1712 - j0.0694 \text{ mho.}$$

The corresponding impedance is

$$\vec{z}_{ab} = \frac{1}{0.1712 - j0.0694} = 5.02 + j2.04 \ \Omega.$$

The primary impedance is

$$r_1 + jx_1 = 0.1252 + j0.313 \ \Omega.$$

The total impedance of the equivalent circuit is

$$\vec{z}_e = \vec{z}_{ab} + \vec{z}_1 = 5.02 + j2.04 + 0.1252 + j0.313$$

$$= 5.15 + j2.35 \ \Omega$$

The primary current for the assumed slip of 0.02 is

$$I_1 = \frac{2{,}200}{\sqrt{3} \times \sqrt{(5.15)^2 + (2.35)^2}} = \frac{2{,}200}{\sqrt{3} \times 5.66}$$

$$= 224 \text{ amp.}$$

The primary power factor is

$$pf = \frac{r_e}{z_e} = \frac{5.15}{5.66} = 0.910.$$

The power input to the primary equals

$$3P_1 = \frac{\sqrt{3} \times 2,200 \times 224 \times 0.910}{1,000} = 777 \text{ kw}$$

The secondary current referred to the primary is

$$I_2 = I_1 \frac{z_{ab}}{z_2} = 224 \frac{5.41}{5.83} = 208 \text{ amp.}$$

The internal power is

$$3P_2 = I_2^2 r_2 \frac{1 - s}{s} n = (208)^2 \times 0.1165 \times \frac{0.98}{0.02} \times 3$$

$$= 743,000 \text{ watts or } 743 \text{ kw.}$$

The shaft output equals

$$P_0 = 3P_2 - P_c - P_{f+w} - \text{stray-load loss}$$

$$= 743 - 13.4 - 7.0 = 723 \text{ kw or } 970 \text{ hp.}$$

The efficiency is 723/777 = 0.93 or 93 per cent.

The shaft torque is $5,250 \times \frac{970}{250 \times 0.98} = 20,800$ lb-ft.

• PROBLEM 10-38

Test data obtained from no-load and blocked-rotor runs and the measured stator resistance of a 1,000-hp, 2,200-volt, Y-connected, 25-cycle, 12-pole squirrel-cage induction motor, are as follows:

Ohmic resistance of stator between terminals at 25°C = 0.210 ohm:

At no load and a temperature of 25°C:

Stator line voltage	2,200 volts
Stator line current	75.1 amp
Total stator input	15.2 kw

With the rotor blocked at approximately full-load current at a temperature of 25°C and a frequency of 15 cycles:

Stator line voltage	185 volts
Stator line current	250 amp
Total stator input	38 kw

Stray load loss = 7.0 kw

The phase voltage, phase currents, phase resistances, and phase reactances of the motor are

$V_1 = 1{,}270$ volts per phase

$I_n = 75.1$ amp per phase

$r_1 = 0.1252$ ohm at 75°C

$r_2 = 0.1165$ ohm at 75°C

$x_1 = 0.313$ ohm

$x_2 = 0.313$ ohm

$z_1 = \sqrt{(0.1252)^2 + (0.313)^2} = 0.337$ ohm

$x_1 + x_2 = 0.626$ ohm

slip $s = 0.02$

All values are referred to the stator.

Calculate the performance characteristics of the motor using the vector diagram shown in Fig. 1.

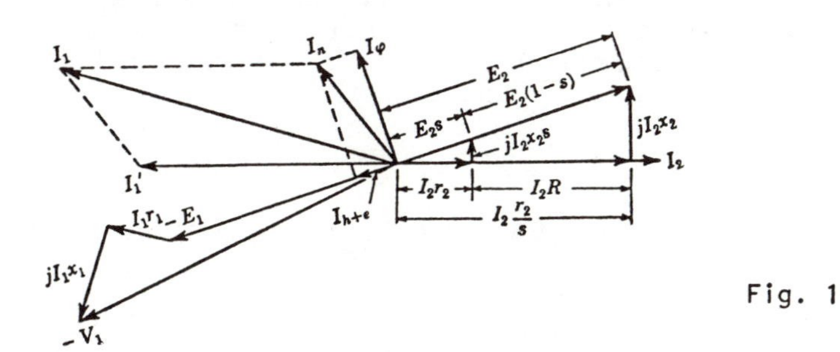

Fig. 1

Solution: The primary impedance drop V_1' is approximately given by

$$V_1' = V_1 - I_n x_1 \quad \text{arithmetically}$$

where

V_1 = Actual stator impressed voltage per phase

I_n = Exciting component of stator current

x_1 = stator reactance

Thus

$$V_1' = 1{,}270 - 75.1 \times 0.313$$
$$= 1{,}246 \text{ V.}$$

The motor internal power is

$$P_2 = \frac{V_1'^2(1 - s)sr_2}{(r_1s + r_2)^2 + s^2(x_1 + x_2)^2}$$

$$(r_1s + r_2)^2 = (0.1252 \times 0.02 + 0.1165)^2$$
$$= 0.01416$$

$$s^2(x_1 + x_2)^2 = (0.02 \times 0.626)^2$$
$$= 0.000157$$

$$3P_2 = 3 \times \frac{(1,246)^2 \times 0.98 \times 0.02 \times 0.1165}{0.01432}$$
$$= 743,000 \text{ watts.}$$

If the I_{h+e} component of the no-load current is omitted in calculating the primary current and instead the core loss is combined with the friction and windage and stray-load losses and subtracted from the internal power in calculating the shaft power, little error is introduced.

The core loss plus the friction and windage loss is

$$P_c + P_{f+w} = 15,200 - 3 \times (75.1)^2 \times 0.105$$
$$= 13,420 \text{ watts}$$

The shaft power then equals

$$P_0 = 3P_2 - P_c - P_{f+w} - \text{stray-load loss}$$
$$= 743 - 13.4 - 7.0 = 723 \text{ kw or } 970 \text{ hp.}$$

The rotor current per phase is

$$I_2 = \frac{V_1's}{\sqrt{(r_1s + r_2)^2 + s^2(x_1 + x_2)^2}}$$

$$= \frac{1,246 \times 0.02}{\sqrt{0.01432}} = 208 \text{ amp.}$$

The no-load power factor is

$$\frac{15.2 \times 1{,}000}{\sqrt{3} \times 2{,}200 \times 75.1} = 0.0531$$

The magnetizing current is

$$I_\phi = I'_n \sqrt{1 - (\text{no-load power factor})^2}$$

where I'_n is the no-load current of the motor at rated voltage.

$$I_\phi = 75.1 \sqrt{1 - (0.0531)^2}$$

$$= 75.0 \text{ amp.}$$

Refer to the vector diagram given in Fig. 1. Use $\vec{E}_2$ as an axis of reference and neglect the I_{h+e} component of the current.

$$\vec{I}_1 = -I_2 \left(\frac{r_2}{\sqrt{r_2{}^2 + x_2{}^2 s^2}} - j \frac{x_2 s}{\sqrt{r_2{}^2 + x_2{}^2 s^2}} \right) + jI_\phi$$

$$= -208 \left[\frac{0.1165}{\sqrt{(0.1165)^2 + (0.313 \times 0.02)^2}} - j \frac{0.313 \times 0.02}{\sqrt{(0.1165)^2 + (0.313 \times 0.02)^2}} \right] + j75.0$$

$$= -208 + j86.2$$

$$I_1 = \sqrt{(208)^2 + (86.2)^2} = 225 \text{ amp.}$$

Primary copper loss = $3 \times (225)^2 \times 0.1252$	= 19,020 watts
Secondary copper loss = $3 \times (208)^2 \times 0.1165$	= 15,140 watts
Core loss plus friction and windage loss	= 13,420 watts
Stray-load loss	= 7,000 watts
Total losses	= 54,580 watts

$$\text{Efficiency} = \frac{\text{output}}{\text{output + losses}} \times 100$$

$$= \frac{723}{723 + 54.6} \times 100$$

$$= 93.0 \text{ per cent.}$$

$$\text{Power factor} = \frac{\text{input}}{3V_1 I_1}$$

$$= \frac{(723 - 54.6) \times 1,000}{3 \times 1,270 \times 225}$$

$$= 0.91 \text{ or } 91 \text{ per cent.}$$

The internal torque in lb-ft is

$$T_2 = \frac{p}{4\pi f_1} I_2{}^2 \frac{r_2}{s} \times 0.7376$$

where p = number of poles, s = slip. Thus

$$T_2 = \frac{12}{4\pi \times 25} \times (208)^2 \times \frac{0.1165}{0.02} \times 0.7376$$

$$= 7,100 \text{ lb-ft}$$

$$\text{The shaft torque} = 3T_2 - (\text{torque corresponding to above losses})$$

$$= 3 \times 7100 - 5,250 \times \frac{(13,420 + 7,000)}{746 \times 250 \times 0.98}$$

$$= 21,300 - 590 = 20,710 \text{ lb-ft.}$$

• PROBLEM 10-39

Consider the following data obtained from tests on a 5-horsepower, 220-volt, 60-cycle, six-pole, three-phase, squirrel-cage induction motor.

Calculate the performance characteristics and plot the performance curves for the above induction motor.

Solution: The running-light values of current and watts input are plotted as a function of line voltage in Fig. 1. Extension of the curve of watts input until it intersects the zero-voltage ordinate shows a value of windage-and-friction loss equal to 100 watts.

Running-light Data

Line Voltage V_0	Line Amperes I_0	Watts Input W_0
225	8.55	357
200	7.00	290
175	5.85	246
180	4.90	215
125	4.06	185
100	3.30	160
75	2.70	141
50	2.15	125

Blocked-rotor Data (60~)

Line Voltage V_s	Line Amperes I_s	Watts Input W_s
110	31.7	2300
100	28.7	1880
90	26.0	1500
80	23.0	1175
70	20.2	860
60	17.2	615
50	14.5	380
40	11.4	245
20	5.75	83

Blocked-rotor Data (40~)

Line Voltage V_s	Line Amperes I_s	Watts Input W_s
88.5	31.7	2160
72.6	26.0	1445
56.4	20.2	885
40.4	14.5	453

Blocked-rotor Data (20~)

Line Voltage V_s	Line Amperes I_s	Watts Input W_s
51.0	31.7	1980
41.9	26.0	1380
31.5	20.2	847
23.3	14.5	435

Stator resistance between terminals = 0.60 ohm (hot).

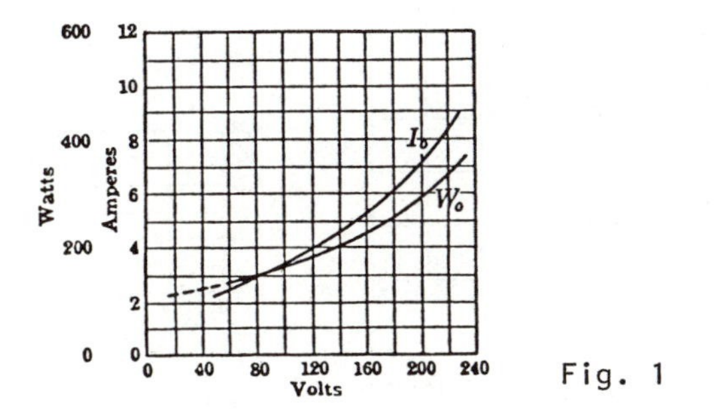

Fig. 1

Running-light Current and Power Versus Voltage.

The blocked-rotor data are secured at reduced values of primary voltage in order to avoid overheating the motor. It is customary to limit the current to about 200% rated. For purposes of computation, however, the values of current and watts input corresponding to rated voltage must be known. This means the curves must be extrapolated. Since the impedance is constant for a given frequency, the values of current vary directly with voltage and plot as straight lines, and extrapolation is easy. A curve of watts as a function of voltage plots as a parabola, and accurate extension of the curve is difficult. To obviate this difficulty, it is customary to plot $I_s \cos\theta_s = W_s/(\sqrt{3}V_s)$ as a function of voltage. Because the impedance is essentially constant for a given frequency, the phase angle θ_s is also constant, and this latter curve is linear and can be extrapolated readily. Curves of I_s and of $I_s \cos\theta_s$ for the present example are shown in Fig. 2. It is seen that at 220 volts and 60 cycles, the value of I_s is 63.5 amperes, and the value of $W_s/(\sqrt{3}V_s)$ is 24. Whence, $W_s = 24 \times \sqrt{3} \times 220 = 9150$ watts and

$$R_0 = \frac{W_s}{3I_s^2} = \frac{9150}{3 \times (63.5)^2} = 0.764 \text{ ohm}$$

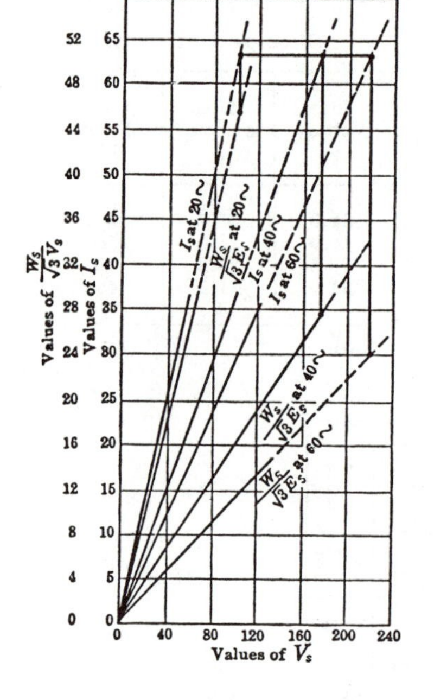

Fig. 2

Extrapolation Curves

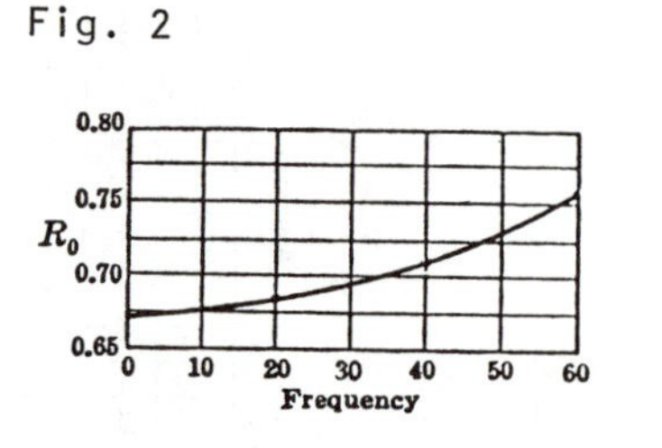

Fig. 3

Equivalent Resistance Versus Frequency

This value of current is also obtained at 178 volts and 40 cycles, and the corresponding value of $W_s/\sqrt{3}V_s)$ equals 27.6. Whence, $W_s = 27.6 \times \sqrt{3} \times 178 = 8510$ watts, and

$$R_0 = \frac{8510}{3 \times (63.5)^2} = 0.71 \text{ ohm}$$

In a similar manner, at 20 cycles $W_s = 8200$ watts and $R_0 = 0.684$ ohm. A plot of the values of R_0 as a function of frequency is shown in Fig. 3. Extension of the curve to the zero-frequency ordinate shows a value of $R_0 = 0.67$. From the test data,

$$r_1 = \frac{0.60}{2} = 0.30 \text{ ohm.}$$

Whence $r_2 = R_0 - r_1 = 0.67 - 0.30 = 0.37$ ohm

assuming the ratio of transformation to be unity.

$$\text{Phase Voltage to Stator Winding} = \frac{220}{\sqrt{3}} = 127 \text{ volts}$$

$$\text{Power Factor with Blocked Rotor} = \cos \theta_s$$

$$= \frac{9150}{3 \times 127 \times 63.5} = 0.38.$$

$$\text{Phase Angle with Blocked Rotor} = \theta_s = \cos^{-1} 0.38$$

$$= 67.6°.$$

$$\text{Winding Reactances} = x_1 = x_2 = \frac{127}{2 \times 63.5} \sin 67.6°$$

$$= 0.923 \text{ ohm.}$$

$$\text{Induced Phase Volts} = E_1 = 127 - (8.1 \times 0.923)$$

$$= 119.5 \text{ volts.}$$

Iron Loss at 220 Volts (line)

$$= 340 - 100 - [(8.1)^2 \times 0.30 \times 3]$$

$$= 240 - 59$$

$$= 181 \text{ watts}$$

$$\text{Exciting Conductance} = g_0 = \frac{181}{3 \times (119.5)^2}$$

$$= 0.00422 \text{ mho.}$$

$$\text{Power Factor, Running Light} = \cos \theta_0$$

$$= \frac{181}{119.5 \times 8.1 \times 3} = 0.0625.$$

$$\text{Phase Angle, Running Light} = \theta_0 = \cos^{-1} 0.0625 = 86.4°.$$

$$\text{Magnetizing Current} = I_m = 8.1 \sin 86.4° = 8.1 \times 0.998$$

$$= 8.09 \text{ amperes.}$$

$$\text{Exciting Susceptance} = b_0 = \frac{I_m}{E_1} = \frac{8.09}{119.5} = 0.0675 \text{ mho.}$$

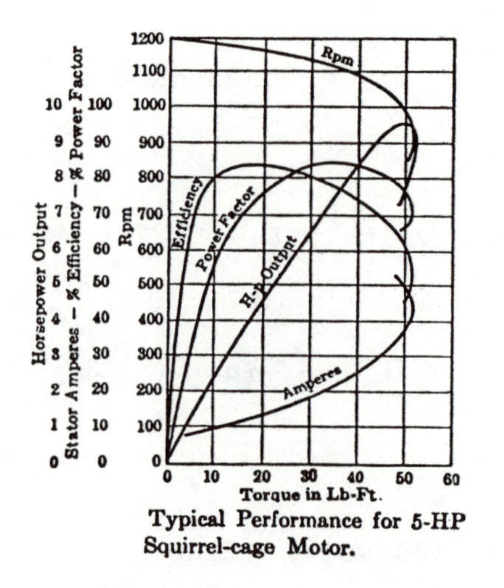

Fig. 4

Typical Performance for 5-HP Squirrel-cage Motor.

The performance characteristics are calculated by using a form sheet, since several values of slip are assumed.

Item	s = Slip	0.01	0.02	0.05	0.10	0.20	0.30
(1)	r_2	0.37	0.37	0.37	0.37	0.37	0.37
(2)	$r_e = r_2/s$	37	18.5	7.4	3.7	1.85	1.23
(3)	$x_1 = a^2x_2$	0.923	0.923	0.923	0.923	0.923	0.923
(4)	$\cot\theta_2 = (2)/(3)$	40	20	8	4	2	1.33
(5)	(2) × (4)	1,480	370	59.2	14.8	3.7	1.64
(6)	(3) + (5)	1,480.92	370.92	60.12	15.72	4.62	2.56
(7)	$B_2 = 1/(6)$	0.000675	0.0027	0.0166	0.0636	0.216	0.39
(8)	b_0	0.0675	0.0675	0.0675	0.0675	0.0675	0.0675
(9)	$B = (7) + (8)$	0.0682	0.0692−	0.0841	0.1311	0.2835	0.4575
(10)	$G_2 = (7) \times (4)$	0.027	0.054	0.1328	0.2544	0.432	0.518
(11)	g_0	0.00422	0.00422	0.00422	0.00422	0.00422	0.00422
(12)	$G = (10) + (11)$	0.0312	0.0582	0.1370	0.2587	0.4362	0.5222
(13)	$\cot\theta = (12)/(9)$	0.455	0.840	1.63	1.975	1.545	1.145
(14)	(12) × (13)	0.0142	0.0490	0.223	0.510	0.675	0.597
(15)	$1/X = (14) + (9)$	0.0824	0.1182	0.3071	0.6411	0.9585	1.055
(16)	$X = 1/(15)$	12.15	8.44	3.25	1.56	1.043	0.947
(17)	$R = (16) \times (13)$	5.54	7.10	5.31	3.08	1.63	1.09
(18)	$X_0 = (16) + (3)$	13.07	9.36	4.18	2.47	1.97	1.87
(19)	$R_0 = (17) + r_1$	5.84	7.40	5.61	3.38	1.93	1.39
(20)	V_1	127	127	127	127	127	127
(21)	$\sqrt{R_0^2 + X_0^2}$	14.3	11.95	7.00	4.18	2.76	2.33
(22)	$I_1 = (20)/(21)$	8.88	10.61	18.15	30.4	46.0	54.5
(23)	I_1^2	78.7	112.5	330	925	2,110	2,970
(24)	$\cos\theta_1 = (19)/(21)$	0.408	0.619	0.802	0.807	0.700	0.595
(25)	$R = m \times (23) \times (19)$	1,380	2,500	5,560	8,960	12,200	12,900
(26)	$G_2/G = (10)/(12)$	0.865	0.963	0.968	0.985	0.990	0.992
(27)	$R/R_0 = (17)/(19)$	0.95	0.96	0.947	0.911	0.844	0.784
(28)	(25) × (26) × (27)	1,133	2,310	5,100	8,050	10,200	10,000
(29)	$P_2 = (28) \times (1 - s)$	1,122	2,260	4,850	7,240	8,150	7,000
(30)	Windage and friction	100	100	100	100	100	100
(31)	$P_0 = (29) - (30)$	1,022	2,160	4,750	7,140	8,050	6,900
(32)	Hp output = (31)/746	1.37	2.90	6.36	9.57	10.8	9.25
(33)	Efficiency = (31)/(25)	74.1	86.4	85.5	79.5	66.0	53.5
(34)	Rpm = $N \times (1 - s)$	1,190	1,176	1,140	1,080	960	840
(35)	Torque (lb-ft) = 7.05 × (31)/(34)	6.05	12.95	29.4	46.5	59.0	58

The performance curves shown in Fig. 4 are plotted from the results of this predetermination, and are typical.

• **PROBLEM 10-40**

The test data for a 1/4-horsepower, four-pole, 60-cycle, 110-volt single-phase induction motor are as follows:

Locked Rotor		Running Light
110	Volts	110
14.8	Amperes	3.2
980	Watts	78

Stator-winding Resistance = 2.35 ohms

Calculate the performance characteristics and plot the performance curves for the above induction motor.

Solution: From these data,

Z_m = magnetizing impedance = $110/(3.2) \times 2 = 68.8$ ohms.

Z_0 = equivalent stator-winding leakage impedance
$= 110/14.8 = 7.43$ ohms.

R_0 = equivalent stator-winding resistance = $980/(14.8)^2$
$= 4.48$ ohms.

X_0 = equivalent stator winding reactance
$= \sqrt{(7.43)^2 - (4.48)^2} = 5.9$ ohms.

$x_1 = x_2 = X_0/2$ = leakage reactances = 2.95 ohms.

The magnetizing reactance is

$$X_m = Z_m - 2x_1 = 68.8 - 5.9 = 62.9 \text{ ohms.}$$

The flux factor is

$$f_\phi = \frac{X_m}{Z_m} = \frac{62.9}{68.8} = 0.913.$$

The magnetizing current is

$$I_m = \frac{2 - f_\phi}{2}(I_0) = \frac{2 - 0.913}{2} \times 3.2 = 1.74 \text{ amperes.}$$

The equivalent rotor-winding resistance is

$$r_2 = \frac{R_0 - r_1}{f_\phi} = \frac{4.48 - 2.35}{0.913} = 2.33 \text{ ohms}$$

Using these constants, the performance characteristics are computed in tabular form as follows.

Item						
(1)	S	1.00	0.98	0.95	0.90	0.80
(2)	$1 - S^2$	0	0.0396	0.0975	0.19	0.36
(3)	$S[1 - S^2]$	0	0.0388	0.0927	0.171	0.288
(4)	K_1	0.0354	0.0354	0.0354	0.0354	0.0354
(5)	$K_1{}^2$	0.001255	0.001255	0.001255	0.001255	0.001255
(6)	K_2	65.85	65.85	65.85	65.85	65.85
(7)	K_3	5.77	5.77	5.77	5.77	5.77
(8)	$K_4 = (2) - (5)$	−0.00125	0.03834	0.09624	0.18874	0.35874
(9)	$K_2 + K_3$	71.62	71.62	71.62	71.62	71.62
(10)	$K_1(K_2 + K_3)$	2.54	2.54	2.54	2.54	2.54
(11)	r_1K_4	−0.00296	0.0900	0.2265	0.444	0.844
(12)	$K_5 = (10) + (11)$	2.537	2.63	2.767	2.984	3.384
(13)	$K_5{}^2$	6.44	6.91	7.67	8.91	11.46
(14)	$(1 - S^2)K_3$	0	0.228	0.562	1.095	2.08
(15)	$K_1{}^2K_2$	0.0826	0.0826	0.0826	0.0826	0.0826
(16)	$2r_1K_1$	0.1665	0.1665	0.1665	0.1665	0.1665
(17)	$K_6 = (14) - (15) - (16)$	−0.2491	−0.0221	0.313	0.846	1.831
(18)	$K_6{}^2$	0.0621	0.0005	0.098	0.718	3.35
(19)	$K_5{}^2 + K_6{}^2$	6.502	6.91	7.77	9.63	14.81
(20)	K_4K_5	−0.00318	0.1007	0.2675	0.5630	1.216
(21)	$2K_1K_6$	−0.0176	−0.00156	0.0222	0.0600	0.1300
(22)	$K_7 = (20) - (21)$	0.0144	0.1023	0.2453	0.503	1.086
(23)	K_4K_6	+0.00031	0.00085	0.0300	0.1595	0.660
(24)	$2K_1K_5$	0.179	0.186	0.196	0.212	0.240
(25)	$K_8 = (23) + (24)$	+0.1793	0.1868	0.2260	0.3815	0.900
(26)	$I_{1a} = E(22)/(19)$	0.243	1.63	3.47	5.75	8.06
(27)	$I_{1b} = E(25)/(19)$	+3.04	2.98	3.19	4.37	6.67
(28)	$I_1 = \sqrt{I_{1a}{}^2 + I_{1b}{}^2}$	3.06	3.39	4.71	7.22	10.43
(29)	$r_2(X_m/X)^2$	2.125	2.125	2.125	2.125	2.125
(30)	$SK_1{}^2 = (1)(5)$	0.00126	0.00123	0.00119	0.00113	0.00100
(31)	$(3) - (30)$	−0.00126	0.03757	0.0915	0.16987	0.287
(32)	$K_9 = (29) \times (31)$	−0.00267	0.0796	0.1948	0.361	0.610
(33)	$T = E^2(32)/(19)$	−4.95	139	304	455	498
(34)	Core Loss and Friction	78	78	78	78	78
(35)	Net Torque	−83	61	226	377	420
(36)	$P_0 = S \times (35)$	−83	59.6	215	340	336
(37)	$P_i = E \times (26)$	−38	116.5	326	588	856
(38)	Efficiency	....	51.1%	65.8%	57.9%	39.2%
(39)	Power Factor		48.1%	72.7%	79.5%	77.5%

The curves shown in the figure are plotted from these data. Power factor is found as the ratio of I_{1a} to I_1.

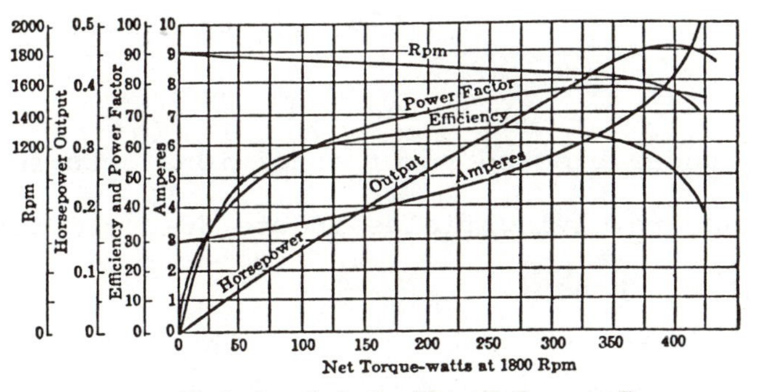

Single-phase Induction Motor Performance Curves.

The rpm curve is obtained by multiplying the values of S by 1800. Horsepower output is equal to watt output divided by 746. If desired, the values of net torque can be converted to pound-feet by multiplying by 7.04 and dividing by 1800.

• PROBLEM 10-41

A 3-phase, Y-connected, 60-cycle, 220-volt, 5-hp, 1800-rpm induction motor has the following per-phase constants: $G_0 = 0.00324$, $B_0 = 0.0351$, $R_1 = 0.545$, $R_2' = 0.538$, $X_1 = X_2' = 0.84$. Windage and friction is 65 watts. This motor is to be operated as a single-phase motor at 220 volts and 60 cycles by leaving one phase open. Calculate its performance characteristics at 4.9 per cent slip by both the double-revolving-field and the cross-field methods, compare the results, and discuss the two methods of solution from the point of view of work involved, accuracy, and design. Assume 60-cycle skin effect to increase the resistance by 50 per cent, and 120 cycles to increase it by 70 per cent.

Solution: 1. Double-revolving field theory

The blocked rotor impedance of the motor on single-phase is twice that per phase of the motor on three-phase. Hence the blocked-rotor, single-phase values are

$$R_1 + jX_1 = 2(0.545 + j0.84) = 1.090 + j1.68$$

$$= 2.04\underline{/57.1^\circ}\ \Omega.$$

$$r_2' + r_2'' + j2x_2' = 2(0.538 + j0.84)$$

$$= 1.076 + j1.68 = 1.994\underline{/57.3^\circ}\ \Omega.$$

Note that under blocked rotor conditions both forward and backward revolving fields induce 60-cycle currents in the rotor. Hence

$$r_2'(60\sim) = r_2''(60\sim) = 0.538$$

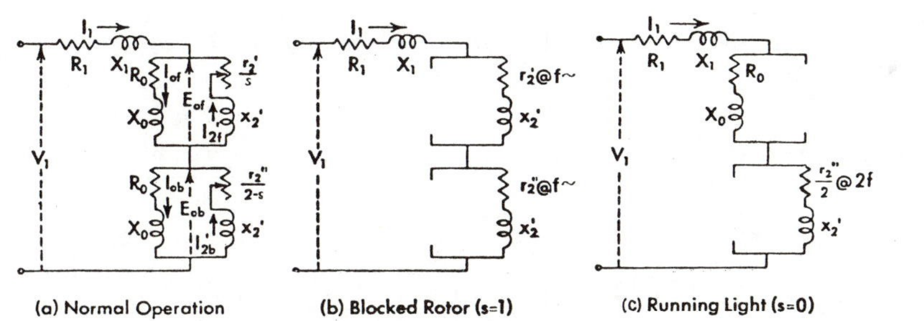

Fig. 1

Equivalent circuit (revolving field theory).

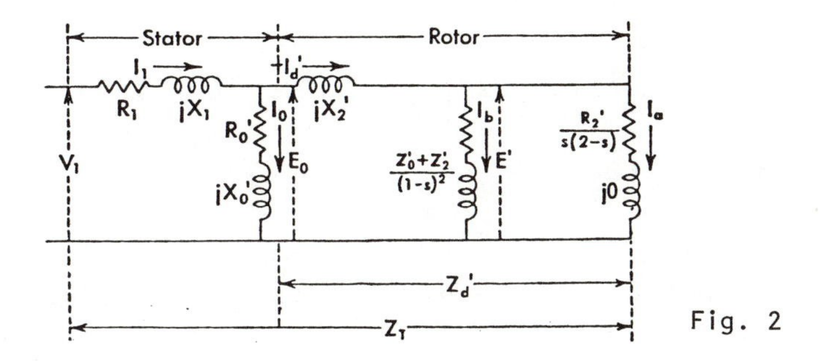

Fig. 2

Equivalent cirucit corresponding to the cross-field theory of the single-phase induction motor.

But under running conditions, r_2' is at only 3 cycles, or practically direct current, while r_2'' is at 117 cycles, or practically double frequency. Hence

$$r_2' = \frac{0.538}{1.50} = 0.359$$

$$r_2'' = 1.70 \times 0.359 = 0.610$$

$$x_2' = 0.84$$

Now under running light conditions the single-phase core loss and magnetizing current, according to Fig. 1(c) are accounted for by $R_0 + jX_0$, and this impedance is twice the three-phase value per phase. Hence

$$R_0 = 2\,\frac{0.00324}{0.0032^2 + 0.0351^2} = 5.23\ \Omega.$$

$$X_0 = 2\,\frac{0.0351}{0.0032^2 + 0.0351^2} = 56.6\ \Omega.$$

Then we have

$$\vec{Z}_0 = R_0 + jX_0 = 5.23 + j56.6 = 56.8\angle 84.7^\circ\ \Omega.$$

$$\vec{Z}_1 = R_1 + jX_1 = 1.090 + j1.680 = 2.04\angle 57.1^\circ\ \Omega.$$

$$\vec{Z}_f = \vec{Z}_0 \| \vec{Z}_1 = \frac{(5.23 + j56.6)(7.33 + j0.84)}{12.56 + j57.44}$$

$$= 6.93 + j1.67 = 7.13\angle 13.5^\circ\ \Omega.$$

$$\vec{Z}_b = \frac{(5.23 + j56.6)(0.31 + j0.84)}{5.54 + j57.44} = 0.306 + j0.826$$

$$= 0.88\angle 69.7^\circ\ \Omega.$$

$$\vec{Z}_T = \vec{Z}_1 + \vec{Z}_f + \vec{Z}_b = 8.326 + j4.170 = 9.31\angle 26.6^\circ\ \Omega.$$

$$\vec{I}_1 = \frac{\vec{V}}{\vec{Z}_T} = \frac{220\angle 0^\circ}{9.31\angle 26.6^\circ} = 21.13 - j10.60$$

$$= 23.62\angle -26.6^\circ \text{ amp.}$$

$$\cos\theta_1 = \cos 26.62^\circ = 0.894$$

$$\vec{E}_{0f} = \vec{Z}_f \cdot \vec{I}_1 = (7.13\angle 13.5^\circ)(23.62\angle -26.6^\circ)$$

$$= 163.6 - j38.1 = 168.5\angle -13.1^\circ \text{ volts.}$$

$$\vec{I}_{0f} = \frac{\vec{E}_{0f}}{\vec{Z}_0} = \frac{168.5\angle -13.1^\circ}{56.8\angle 84.7^\circ} = -0.40 - j2.94$$

$$= 2.97\angle -97.8^\circ \text{ amp.}$$

$$\vec{I}_{2f'} = \frac{-\vec{E}_{0f}}{\vec{Z}_f} = \frac{168.5\angle 166.9^\circ}{7.13\angle 13.5^\circ} = -21.53 + j7.66$$

$$= 22.82\underline{/160.4^\circ} \text{ amp.}$$

$$\vec{E}_{0b} = \vec{Z}_b \cdot \vec{I}_1 = (0.88\underline{/69.7^\circ})(23.62\underline{/-26.6^\circ})$$

$$= 15.18 + j13.28 = 20.8\underline{/43.1^\circ} \text{ volts.}$$

$$\vec{I}_{0b} = \frac{\vec{E}_{0b}}{\vec{Z}_0} = \frac{20.8\underline{/43.1^\circ}}{56.83\underline{/84.7^\circ}} = 0.27 - j0.24$$

$$= 0.37\underline{/-41.6^\circ} \text{ amp.}$$

$$\vec{I}_{2b'} = \frac{-\vec{E}_{0b}}{\vec{Z}_b} = \frac{20.8\underline{/223.1^\circ}}{0.88\underline{/69.7^\circ}} = -20.85 + j10.35$$

$$= 23.27\underline{/153.6^\circ} \text{ amp.}$$

Core loss $= 5.23 \times 2.97^2 + 5.23 \times 0.37^2 \quad = \quad 47$ watts

Copper loss $= 1.09 \times 23.62^2 + 0.359 \times 22.82^2 + 0.610 \times 23.27^2 \quad = 1127$

Windage and friction $= \underline{\quad 65}$

Total losses $= 1239$ watts

Input to stator $= 220 \times 23.62 \times 0.894 = 4650$

$$\text{Gross output} = \frac{1 - s}{s} r_2' I_{2f}'^2 - \frac{1 - s}{2 - s} r_2'' I_{2b}'^2$$

$$= \frac{0.951}{0.049} 187 - \frac{0.951}{1.951} 331 = 3469 \text{ watts}$$

Net output $= 3469 - 65 = 3404$ watts

(check) $\quad = 4650 - 1239 = 3411$

$$\text{Efficiency}' = \frac{3405}{4650} = 0.732.$$

$$\text{Torque} = \frac{33000}{746(2\pi 1800)} \frac{3405}{1 - 0.049} = 14.0 \text{ lb-ft.}$$

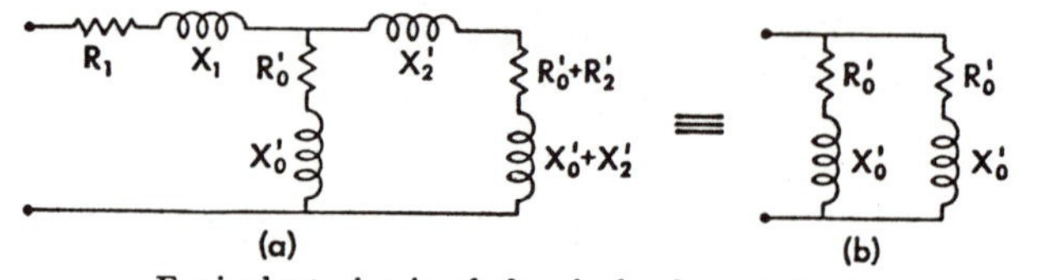

Fig. 3

Equivalent circuit of the single-phase induction motor under running-light conditions.

2. Cross-field theory

Under running-light conditions, $s = 0$, the cross-field equivalent circuit of the single-phase induction motor, Fig. 2, reduces to that shown in Fig. 3(a); and to a first approximation this in turn reduces to Fig. 3(b). That is, the exciting impedance under running-light conditions is $0.5(R_0' + jX_0')$. But the single-phase impedance is twice the three-phase values per phase. Hence,

$$\vec{Z}_0' = R_0' + jX_0' = 4(2.615 + j28.3)$$

$$= 10.46 + j113.2 = 113.7\underline{/84.7°}\ \Omega.$$

$$\vec{Z}_1 = 2(0.545 + j0.84) = 1.090 + j1.68 = 2.0\underline{/56.9°}\ \Omega.$$

Now according to the following equation:

$$e = \frac{K_p N\phi_\omega}{2 \times 10^8}\{[\cos(\omega - \omega_0)t - \cos(\omega + \omega_0)t]$$

$$- \frac{\omega_0}{\omega}[\cos(\omega - \omega_0)t + \cos(\omega + \omega_0)t]\},$$

there are two frequencies of 3 and 117 cycles in each conductor. But unlike the double-revolving field theory, the cross-field theory does not yield an equivalent circuit in which it is possible to state that skin effects corresponding to certain frequencies should be used in certain branches. Actually, the equivalent circuit was derived for the d-axis, in which axis the frequency is 60 cycles. But resistance is in a conductor, not in an axis, and the conductor has two frequencies. However, as seen from double-revolving-field theory, the slip frequency current is principally responsible for the torque. Therefore, in the cross-field theory the slip frequency rotor resistance should be used to secure closest agreement with the numerical values of the revolving field theory. Hence,

$$\vec{Z}_2' = 2(r_2' + jx_2') = 2(0.359 + j0.84)$$

$$= 0.718 + j1.68 = 1.825\underline{/66.5°}\ \Omega.$$

$$\vec{Z}_d' = \left\{\frac{\vec{Z}_0' + \vec{Z}_2'}{(1-s)^2}\right\} \Big\| \left\{\frac{R_2'}{s(2-s)}\right\} + jx_2'$$

$$= \frac{\left(\dfrac{10.46 + j113.2 + 0.72 + j1.68}{0.951^2}\right)\left(\dfrac{0.718}{1.951 \times 0.049}\right)}{19.89 + j127.2}$$

$$+ j168$$

$$= 7.41 + j2.11 = 7.72\underline{/15.9°}\ \Omega.$$

$$\vec{Z}_T = \vec{Z}_1 + (\vec{Z}_0' \| \vec{Z}_d') = (1.09 + j1.68)$$

$$+ \frac{(113.7\underline{/84.7°})(7.72\underline{/15.9°})}{116.7\underline{/81.2°}}$$

$$= 8.18 + j4.17 = 9.18\underline{/27°}\ \Omega.$$

$$\vec{I}_1 = \frac{\vec{V}}{\vec{Z}_T} = \frac{220\underline{/0°}}{9.18\underline{/27°}} = 21.35 - j10.88 = 23.95\underline{/-27°}\ \text{amp.}$$

$$\vec{E} = \vec{V} - \vec{Z}_1\vec{I}_1 = 220 - (2.0\underline{/56.9°})(23.95\underline{/-27°})$$

$$= 178.5 - j23.8 = 180.2\underline{/-7.6°}\ \text{volts.}$$

$$\vec{I}_0 = \frac{\vec{E}}{\vec{Z}_0'} = \frac{180.2\underline{/-7.6°}}{113.7\underline{/84.7°}} = -0.064 - j1.584$$

$$= 1.59\underline{/-92.3°}\ \text{amp.}$$

$$-\vec{I}_d' = \vec{I}_1 - \vec{I}_0 = 21.414 - j9.296 = 23.35\underline{/-23.45°}\ \text{amp}$$

$$\vec{E}' = \vec{E} - j2x_2' \cdot \vec{I}_d' = (178.5 - j23.8) - j1.68$$

$$\times 23.35\underline{/-23.45°} = 162.92 - j59.6$$

$$= 173.4\underline{/-20.1°}\ \text{volts.}$$

$$\vec{I}_a = \frac{\vec{E}' \cdot s(2 - s)}{R_2'} = (1 - 0.951^2)\frac{173.4\underline{/-20.1°}}{0.718}$$

$$= 21.67 - j7.93 = 23.07\underline{/-20.1°} \text{ amp.}$$

$$\vec{I}_b = \frac{\vec{E}' \cdot (1 - s)^2}{\vec{Z}_0' + \vec{Z}_2'} = \frac{0.951^2 \times 173.4\underline{/-20.1°}}{(10.46 + j113.2) + (0.72 + j1.68)}$$

$$= -0.34 - j1.32 = 1.36\underline{/-104.3°} \text{ amp.}$$

Core loss = $10.46 \times 1.59^2 + 10.46 \times 1.36^2$ = 46 watts

Copper loss = $1.09 \times 23.95^2 + 0.718 \times 23.07^2 + 0.718 \times 1.36^2$ = 1008

Windage and friction = 65

Total losses = 1119 watts

Input to stator = $220 \times 23.95 \times 0.891$ = 4700 watts

Net output = 4700 - 1119 = 3581 watts

$$\text{Efficiency} = \frac{3581}{4700} = 0.762.$$

$$\therefore \quad \text{Torque} = \frac{33000}{746(2\pi 1800)}\,\frac{3581}{0.951} = 14.72 \text{ lb-ft.}$$

3. Comparison of calculations

From the point of view of length and time of computation, there is little to choose between the two methods of calculation. There are a few more items in the revolving-field theory, but this is compensated by the greater complexity of some of the items in the cross-field theory.

The revolving-field theory is explicit about the frequencies and corresponding skin-effect corrections to use in its various branches; whereas the 60-cycle frequency required in all branches by the cross-field theory is obviously not appropriate for skin-effect calculations, and it is necessary to arbitrarily use slip frequency for the rotor resistance in order to secure accurate results. In fact, even better numerical agreement between the two theories can be obtained (comparing the equivalent circuits) by putting

Revolving-field		Cross-field
$\frac{r_2'}{s} + \frac{r_2''}{2 - s}$	=	$\frac{R_2'}{s(2 - s)}$

from which

$$R_2' = (2 - s)r_2' + sr_2''$$

$$= 1.951 \times 0.359 + 0.049 \times 0.610$$

$$= 0.730$$

From the point of view of the designer who must juggle design constants to obtain specified or desired characteristics, the revolving-field equivalent circuit is perhaps superior, since it is more symmetrical and the physical significance of its parts easier to see. Also, its derivation is much less involved.

The numerical results of the two methods agree reasonably well, particularly in view of the arbitrary inclusion of skin effect in the cross-field theory. A comparison is given below.

Revolving-field		Cross-field
47	Core loss	46
1127	Copper loss	1008
65	Windage and friction	65
1239	Total losses	1119
4650	Input	4700
3405	Net output	3581
0.732	Efficiency	0.762
0.894	Power factor	0.891
14.00	Torque	14.72

• PROBLEM 10-42

The test-record for an induction-motor is shown in Table 1. (a) Draw the circle diagram for this motor and determine the maximum torque and maximum output. (b) Determine the performance characteristics using the circle diagram.

Solution: (a) For convenience in measuring current vectors, a current scale is laid off on the horizontal axis. The no-load current OA is plotted by using a compass to swing a circular arc with 10.5 amp radius. Point A on this arc is determined by computing the vertical or power component of current,

$$10.5 \times 0.249 = 2.6 \text{ amp}$$

TABLE 1

INDUCTION-MOTOR TEST RECORD

Rating: 35 hp, 1,800 rpm, 3-phase, 60 cycles, 440 volts

Stator resistance 0.238 ohm (between terminals)

Load	Volts	Amperes	Watts	PF	Speed, rpm	Torque, lb-ft
No load	440	10.5	2,000	0.249	1799	0
Locked	440	252	80,000	0.416	0	208
Full load (approx)	440	43.5	31,200	0.942	1738	

where 0.249 is the no-load power factor listed in Table 1. Similarly the locked current

$$OC = 252 \text{ amp}$$

is plotted with a power component of

$$CF = 252 \times 0.416 = 105 \text{ amp.}$$

Now with points A and C determined, a perpendicular bisector of AC will locate the center of the circle at the point V where this bisector intersects the horizontal line through point A, and the circle can be drawn.

$$\text{Total stator resistance} = 3 \times (\tfrac{1}{2}\cdot \text{the value measured between terminals})$$

$$= \frac{3}{2} \times 0.238.$$

$$\text{Stator copper loss at no load} = \frac{3}{2} \times 0.238 \times (10.5)^2$$

$$= 40 \text{ watts.}$$

Therefore, fixed loss = 2000 - 40 = 1960 watts.

Thus on an ampere scale

$$EF = \frac{1,960}{\sqrt{3} \times 440} = 2.6 \text{ amp.}$$

$$\text{The locked stator copper loss} = \frac{3}{2} \times 0.238 \times (252)^2$$

$$= 22670 \text{ watts}$$

which, on an ampere scale, is

$$DE = \frac{22,670}{\sqrt{3} \times 440} = 29.8 \text{ amp}$$

Therefore the rotor loss at standstill, on an ampere scale, is

$$CD = 105 - 2.6 - 29.8 = 72.6 \text{ amp}$$
$$= 55.37 \text{ kw}$$

and

$$T_L = 55.37 \text{ synchronous kw}$$
$$= 216 \text{ lb-ft}$$

The maximum torque is

$$QR = 117 \text{ amp}$$
$$= 89 \text{ synchronous kw}$$
$$= 119.2 \text{ synchronous hp}$$
$$= 342\% \text{ synchronous full-load torque}$$
$$= 348 \text{ lb-ft}$$

which is over 50 per cent higher than the locked value.

The maximum output is

$$NP = 85.6 \text{ amp}$$
$$= 65.2 \text{ kw}$$
$$= 87.5 \text{ hp}$$
$$= 250\% \text{ of full-load rating.}$$

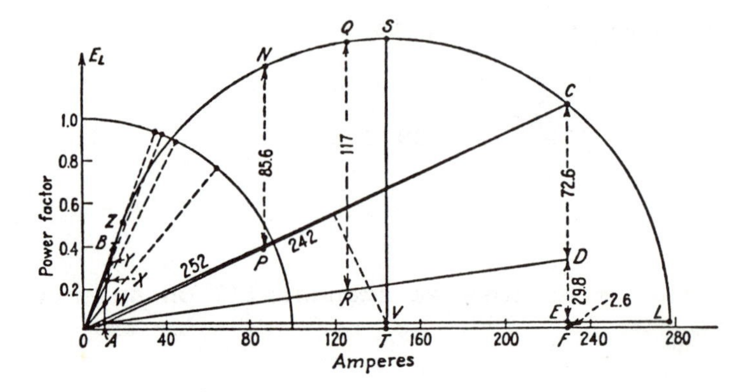

Fig. 1

Circle diagram for motor described in Table 1.

(b) The performance computations are tabulated as shown in Table 2.

TABLE 2
INDUCTION-MOTOR PERFORMANCE SHEET
Rating: 35 hp, 1,800 rpm, 3-phase, 60 cycles, 440 volts

Approximate load	0.25	0.5	0.75	1.00	1.25
Stator current	17	25	34	43.5	54
Fixed loss	1,960	1,960	1,960	1,960	1,960
Stator copper loss	102	223	412	678	1,040
Total stator loss	2,062	2,183	2,372	2,638	3,000
Input to motor	9,850	17,000	23,900	31,200	38,500
Rotor input	7,788	14,817	21,528	28,562	35,500
Rotor current	10.3	18.7	28.8	38.4	48.1
Rotor loss	78	256	607	1,078	1,686
Slip	0.010	0.0173	0.0281	0.0378	0.0475
Output	7,710	14,561	20,921	27,484	33,814
Horsepower	10.3	19.5	28.0	36.8	45.3
Efficiency	0.782	0.856	0.875	0.880	0.879
Power factor	0.758	0.891	0.924	0.942	0.938

The computations are explained below:

Row 1: The values of the stator current for 0.25 load, 0.5 load, etc., are estimated. This can be done accurately enough by considering the fact that the no-load current is

$$OA = 10.5 \text{ amp}$$

and the current at approximately full load is

$$OB = 43.5 \text{ amp}$$

giving a change of current of 33 amp from no load to full load. The increase of current will be roughly proportional to the load and will give some such figures as those given in the table.

Row 2: The fixed loss is assumed constant.

Row 3: Stator copper loss = $\frac{3}{2} \times 0.238 \times (\text{stator current})^2$

Row 4: Total stator loss = fixed loss + stator copper loss

Row 5: The input to the motor may be obtained by laying off points on the circle diagram corresponding to the assumed stator currents and scaling the values of the power components of these currents. On the diagram of Fig. 1 these points are W for 1/4 load, X for 1/2 load, Y for 3/4, and Z for $1\frac{1}{4}$ load. The power components of these currents, multiplied by $\sqrt{3} \times 440$, give the power-input figures.

Row 6: Rotor input = Input to motor - total stator loss

Row 7: Rotor current is scaled directly from Fig. 1.

Row 8: The rotor loss for the motor under approximately full load was computed by multiplying the rotor input by the slip, giving 1,078 watts. To compute the rotor loss at three-fourths load, since the rotor loss varies with the square of the rotor current,

$$\text{Rotor loss} = 1{,}078 \left(\frac{28.8}{38.4}\right)^2 = 607 \text{ watts}$$

The other rotor losses are computed in the same manner.

Row 9: Slip = Rotor loss/Rotor input

Row 10: Output = Rotor input - rotor loss

Row 11: Horsepower = output in watts/746.

Row 12: Efficiency = output/input to motor

Row 13: Power factor = Power component of the stator current scaled from Fig. 1 /stator current.

SOUND INTENSITY COMPUTATION

• PROBLEM 10-43

Compute the sound intensity of the following machine:

Squirrel-cage induction motor, 53.5 hp, 4 poles, 50 cycles.

$D_{outs.} = 13.8$ in. $\Delta = 0.62$

$D = 9.05$ in. $Q_1 = 36$

$D_m = 12.43$ in. Two rotors with $Q_2 = 30$ and 44

$h_c = 1.36$ in.

The slot harmonics of first order of stator and rotor will be considered, because they are the most dangerous with respect to magnetic noise.

Solution: (a) Cylindrical radiation. The slot harmonics are

$$\nu'_{sl} = IcQ_1 + (p/2) \qquad c = 1,2,3....$$

$$\mu'_{sl} = IcQ_2 + (p/2) \qquad c = 1,2,3....$$

or

$$\nu'_{sl} = -34 + 38$$

$$\mu'_{sl} = -28 + 32$$

The smallest value of p' is produced by

$$\nu'_b = -34 \quad \text{and} \quad \mu_a' = +32$$

$$p' = -34 + 32 = -2$$

This is a 4-pole force wave. For this wave

$$B_{\hat{\nu}'b} = 18{,}000 \qquad B_{\hat{\mu}'a} = 17{,}200 \quad \text{lines/in.}^2$$

These values include the influence of the slot openings. From the following equation:

$$F_{r(p'>1)} = 1.385 \times 10^{-8} \frac{1}{p'} D\ell_e B_{\hat{\nu}'b} B_{\hat{\mu}'a} \quad \text{lbs.,}$$

$$\therefore \quad F_r/\ell_e = 1.385 \times 10^{-8} \times \frac{1}{2} \times 0.95 \times 18{,}000 \times 17{,}200$$

$$= 19.4 \text{ lbs./per pole, per inch core length}$$

Now the deflection

$$d = \frac{F_r}{\ell_e} \frac{D_m^3 \times 10^6}{6E' \, h_c^3} \qquad \text{for } p' = 2$$

$$= 19.4 \times 12.43^2 \times \frac{10^6}{6 \times 3.0 \times 10^7 \times 1.36^3}$$

$$= 82.3 \text{ microns.}$$

The natural frequency of the core is

$$f_n = \frac{36{,}700 \; p' \; (p'^2 - 1) h_c}{D_m^2 \sqrt{p'^2 + 1}} \quad \text{cps,}$$

$$= \frac{36{,}700 \times 2 \times 3 \times 1.36}{12.43^2 \sqrt{5}} = 867 \quad \text{c/s.}$$

The frequency of the force wave, since $k_{2a} = +1$, is

$$f'_+ = \left[2 + k_{2a} \frac{Q_2}{p/2} (1 - s)\right] f_1$$

$$= \left[2 + \frac{30}{2} (1 - s)\right] 50 = 825 = f_r.$$

$s = 0.035$. The closeness of f_r and f_n indicates a noisy motor. From $I_d = 7 + 20 \log_{10} (d \times f_r)$ decibels, the sound intensity of a plane wave,

$$I_d = 7 + 20 \log_{10} (82.3 \times 825) = 103.6 \text{ db.}$$

The "effective" radius of the cylinder

$$kr_0 = 0.00559 f_r \times r_0 = k(D_m/24)$$

$$= 0.00559 \times 825 \times (13.8/24) = 2.65$$

The correction factor for cylindrical radiation is found from Fig. 1 equal to +2 db. Thus, the sound intensity of the core is 103.6 + 2 = 105.6. The noise level was measured at a distance of 1.64 feet (= 50 cm.) from the machine. This yields an effective radius

$$kr = 0.00559 \times 825(13.8/24 + 1.64) = 10.2$$

Fig. 2 yields for $kr_0 = 2.65$ the value -6 db and for $kr = 10.2$ the value -12 db. The correction factor for distance is -12 - (-6) = -6. Thus, the computed sound intensity at the distance of 1.64 feet is 105.6 - 6 = 100 db. The tested value is 97 to 110 db.

The same motor was built and tested with $Q_2 = 44$. The slot harmonics of first order are with $Q_2 = 44$

$$\nu'_{sl} = -34 + 38$$

$$\mu'_{sl} = -42 + 46$$

The smallest value of p' is produced by

$$\nu'_b = +38 \quad \text{and} \quad \mu'_a = -42$$

$$p' = +38 - 42 = -4$$

This is an 8-pole force wave. Since $\nu'_b = +38$ is also a slot harmonic,

$$B_{\nu' b} \approx \frac{34}{38}\,18{,}000 = 16{,}100 \text{ lines/in.}^2$$

Considering the following equation:

$$B_{\mu'} = 0.45\,\frac{Q_2}{2}\,\frac{1}{\mu'}\,\frac{3 \cdot 19}{g k_c k_s}\,I_{2\nu'}$$

$I_{2\nu'}$ is inversely proportional to Q_2, and therefore, $B_{\mu' a}$ decreases direction with μ'_a

$$B_{\mu' a} = \frac{32}{42}\,17{,}200 = 13{,}100 \text{ lines/in.}^2$$

$$F_r/\ell_e = 1.385 \times 10^{-8} \times \frac{1}{4} \times 9.05 \times 16{,}100 \times 13{,}100$$

$$= 6.6 \text{ lbs/per pole, per inch - core length}$$

The deflection

$$d = \frac{F_r}{\ell_e}\,\frac{D_m^3 \times 10^6}{75 E' h_c^3} \qquad \text{for } p' = 4$$

$$= 6.6 \times 12.43^3\,\frac{10^6}{75 \times 3 \times 10^7 \times 1.36^3} = 2.24 \text{ microns}$$

$$f_n = \frac{36{,}700 \times 4 \times 15 \times 1.36}{12.43^2\,\sqrt{17}} = 4700 \text{ cps}$$

The frequency of the force wave, since $k_{2a} = -1$

$$f_+ = \left[2 - \frac{44}{2} \times 0.965\right] 50 = 960 = f_r$$

$$I_d = 7 + 20 \log_{10} (2.24 \times 960) = 73 \text{ db}$$

$$k r_0 = 0.00559 \times 960\,\frac{13.8}{24} = 3.08$$

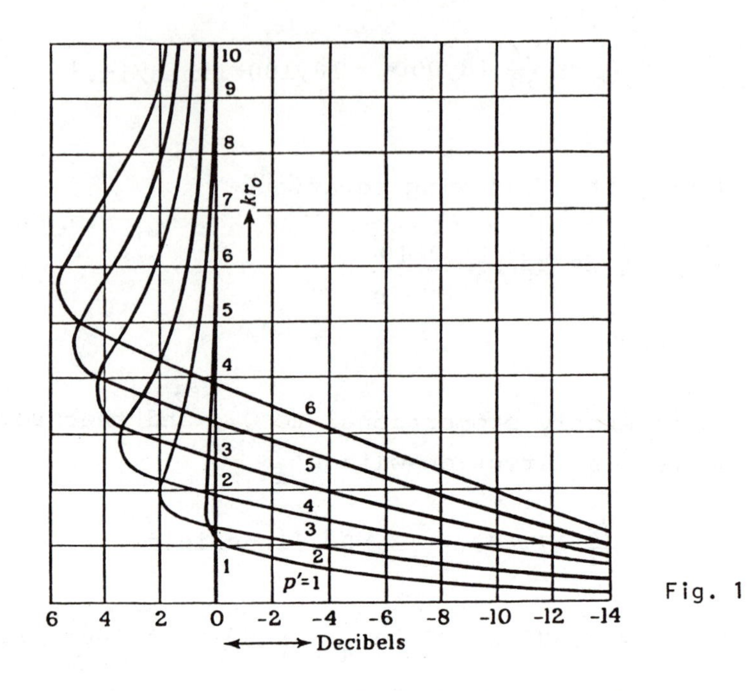

Fig. 1

Correction for an indefinitely long cylinder.

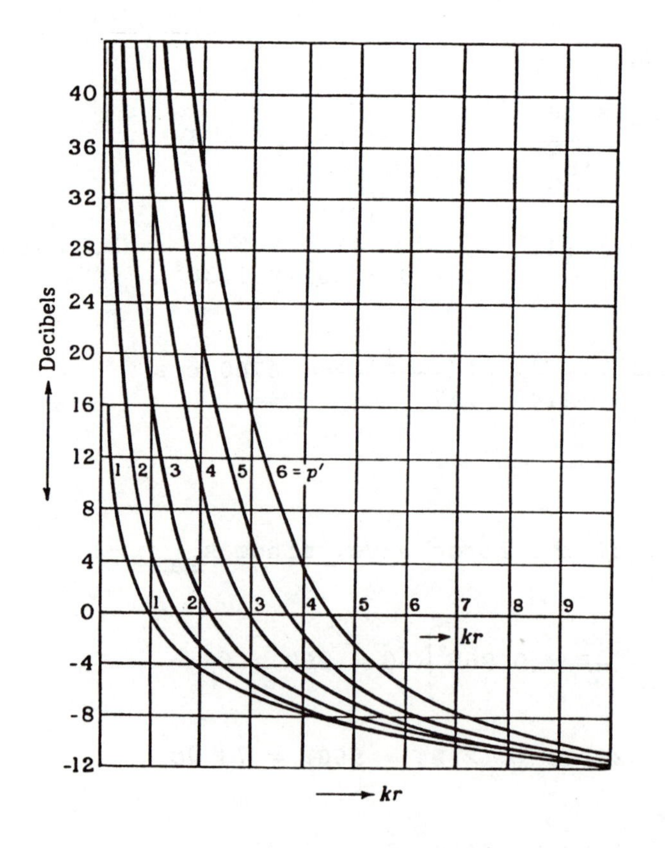

Fig. 2

Correction for distance from the motor surface.

Correction for cylindrical radiation from Fig. 1 is +3 db. The sound intensity at the motor surface us 73 + 3 = 76 db. The test was made at a distance of 1.64 feet from the motor

$$kr = 0.00559 \times 960 \left(\frac{13.8}{24} + 1.64\right) = 11.9$$

From Fig. 2, the correction factor for distance is

$$-12 - (-1) = -11$$

Thus, the computed sound intensity at the distance of 1.64 feet is 76 - 11 = 65. The tested value is 75 to 78 db.

(b) Spherical radiation. Again the slot harmonics will be considered. For the slot combination 36/30

$$p' = -2 \qquad B_{\nu' b} = 18{,}000 \qquad B_{\mu' a} = 17{,}200 \text{ lines/in.}^2$$

$$F_r = 1.385 \times 10^{-8} \times B_{\nu' b} B_{\mu' a} \quad \text{lbs./in.}^2$$

$$= 1.385 + 10^{-8} \times 18{,}000 \times 17{,}200 = 4.27 \text{ lb./in.}^2$$

The static deflection is

$$d_{st} = F_r \frac{D_m/2}{3 \times 10^7} \left(\frac{D_m/2}{h_c}\right)^3 \frac{12}{(p'^2 - 1)^2}$$

$$= 4.27 \frac{12.43/2}{3 \times 10^7} \left(\frac{12.43/2}{1.36}\right)^3 \frac{12}{9} = 114 \text{ microns}$$

The natural frequency of the core is given by the following equations:

$$f_{n(p'=0)} = \frac{83{,}750}{D_m/2} \sqrt{\Delta} \frac{1}{2.54} \quad \text{cps}$$

$$f_{n(p'>1)} = f_{n(p'=0)} \frac{1}{2\sqrt{3}} \frac{h_c}{D_m/2} \frac{p'(p'^2 - 1)}{\sqrt{p'^2 + 1}} \quad \text{cps}$$

$$f_{n(p'=0)} = \frac{83{,}750}{12.43/2} \times \sqrt{0.62} \frac{1}{2.54} = 4180 \text{ cps.}$$

$$f_{n(p'=2)} = 4180 \frac{1}{2\sqrt{3}} \frac{1.36}{12.43/2} \frac{2 \times 3}{\sqrt{5}} = 708 \text{ cps.}$$

The amplification factor is

$$\xi_r = \frac{1}{1 - (f_r/f_n)^2}$$

$$\xi_{r(p'=2)} = \frac{1}{1 - (825/708)^2} = 2.78$$

The forced frequency is the same as under (a) for p' = 2. The radiated wave length is

$$\lambda = \frac{34{,}300}{f_r \times 2.54} \text{ in.}$$

$$= \frac{34{,}300}{825 \times 2.54} = 16.35 \text{ in.}$$

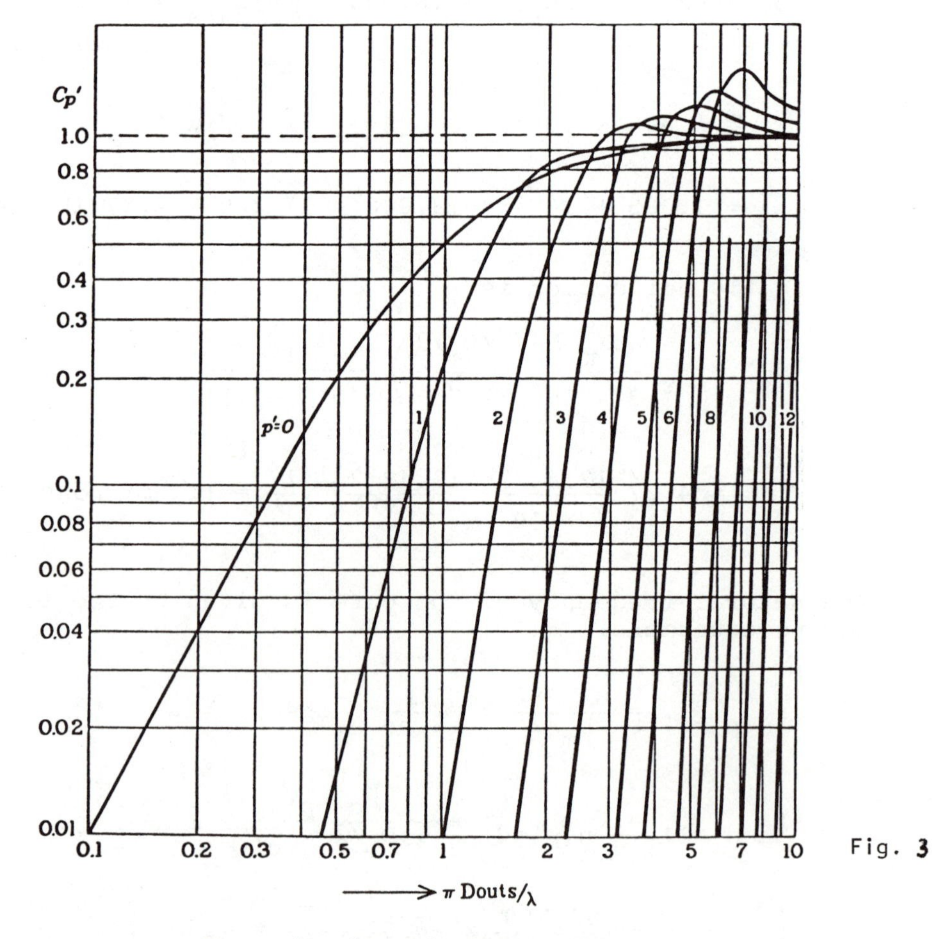

Fig. 3

Correction for spherical radiator.

The quantity $\pi \times D_{outs}\lambda$ is $\pi \times 13.8/16.35 = 2.65$. Fig. 3 yields the correction factor for spherical radiation $C'_{\nu} = 0.85$. From the following equation:

$$L = 20 \log_{10} \left[1900\sqrt{C'}_p \, \xi_r \, \frac{\pi D_{outs}}{\lambda} \left(\frac{D_m/2}{h_c} \right)^3 \frac{F_r}{(p'^2 - 1)^2} \right] ,$$

for the sound intensity at the surface of the motor,

$$L = 20 \log_{10} \left[1900\sqrt{0.85} \times 2.78 \times 2.65 \left(\frac{12.43/2}{1.36} \right)^3 \frac{4.27}{9} \right]$$

$$= 115.5 \text{ phon}$$

Applying the same correction for distance as under (a) for $p' = 2$, the calculated sound intensity at the distance of 1.64 feet from the motor is 118 - 6 = 112 phon. The tested value is 97 to 110 db.

For the slot combination 36/44, $B_{\nu'b}$, $B_{\mu'a}$ and f_r are the same as for this slot combination under (a)

$$B_{\nu'b} = 16{,}100 \qquad B_{\mu'a} = 13{,}100 \qquad f_r = 960 \qquad p' = 4$$

$$F_r = 1.385 \times 10^{-8} \times 16{,}100 \times 13{,}100 = 2.92 \text{ lb./in.}^2$$

$$f_{n(p'=0)} = 4180 \text{ cps}$$

$$f_{n(p'=4)} = 4180 \frac{1}{2\sqrt{3}} \frac{1.36}{12.43/2} \frac{4 \times 15}{\sqrt{17}} = 3840 \text{ cps}$$

$$\xi_{r(p'=4)} = \frac{1}{1 - (960/3840)^2} = 1.065$$

$$\lambda = \frac{34{,}300}{960 \times 2.54} = 14.1 \text{ in.} \quad \therefore \quad \frac{\pi \times 13.8}{14.1} = 3.08.$$

From Fig. 3, $C'_p = 0.14$ and

$$L = 20 \log_{10} \left[1900\sqrt{0.14\times1}.065\times3.08\times\left(\frac{12.43/2}{1.36}\right)^3 \frac{2.92}{225} \right]$$

$$= 69 \text{ phon.}$$

Applying the same correction factor for distance as under (a) for p' = 4 yields 69 - 11 = 58 against the tested value of 75 to 78 phon.

CHAPTER 11

SPECIAL MACHINES

PERMANENT MAGNET GENERATORS

• PROBLEM 11-1

Figure 1 shows the per-phase equivalent circuit of a PM ac generator in which R_s is neglected. The permanent-magnet material is Ferrite D, as described in the B-H characteristic of Fig. 2.

a) Determine the open-circuit terminal potential difference of the generator and find the operating point on the B-H curve of the magnetic material.

b) Determine the short-circuit current for the generator and find the B-H operating point during short circuit.

c) Three capacitors, each having a reactance of 3.75 Ω, are connected in series with the generator to improve its regulation. If the load connected to the generator and series capacitors is short-circuited, will the rotor magnets be demagnetized?

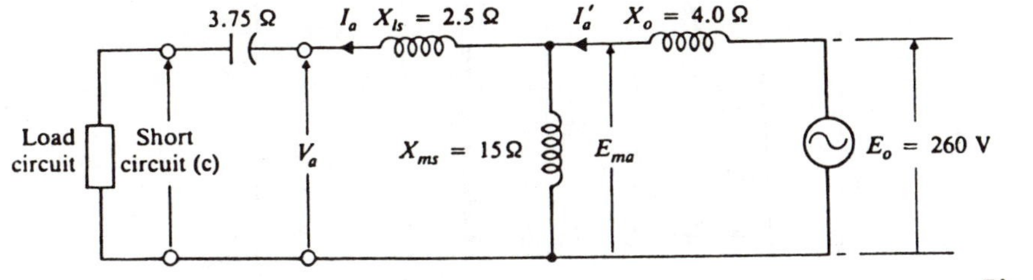

Fig.1

Solution: a) On open circuit

$$V_a = E_{ma} = \frac{15}{15 + 4} \times 260 = 205.3 \text{ V}$$

Thus the line-to-line terminal potential difference is 355.6 V. From Fig. 2, $B_0 = 0.34$ T, which corresponds to $E_0 = 260$ V. Thus $E_{ma} = 205.3$ corresponds to

$$B = \frac{205.3}{260} \times 0.34 = 0.27 \text{ T}$$

From Fig. 2, H = -40 kA/m.

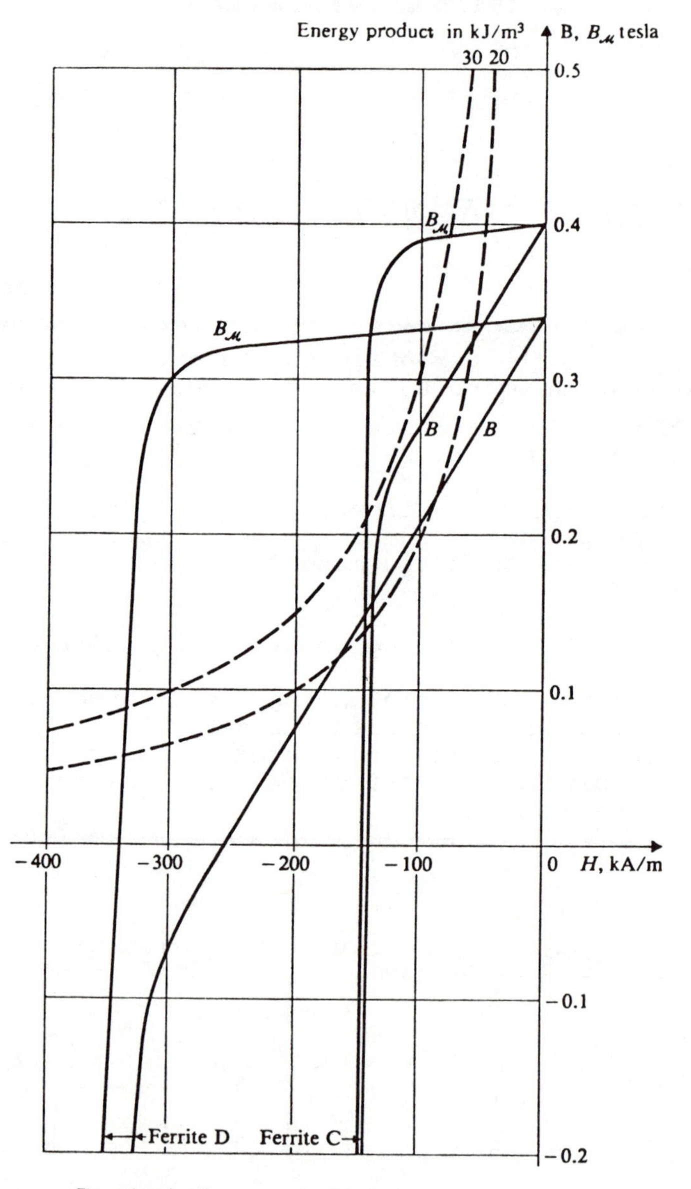

Demagnetization curves for Ferrites C and D.

Fig.2

b) On short circuit, $V_a = 0$

$$I'_a = \frac{260}{[4 + (2.5 \times 15)/(2.5 + 15)]} = 42.33 \text{ A}$$

and

$$I_a = \frac{15}{2.5 + 15} \times 42.33 = 36.28 \text{ A}$$

Thus

$$E_{ma} = 36.28 \times 2.5 = 90.7 \text{ V}$$

and this corresponds to

$$B = \frac{90.7}{260} \times 0.34 = 0.119 \text{ T}$$

From Fig. 2, H =-170 kA/m.

c) With the capacitors in series with the generator

$$X_{\ell s} + X_c = 2.5 - 3.75 = -1.25 \ \Omega$$

and

$$I'_a = \frac{260}{[4 + (-1.25 \times 15)/(-1.25 + 15)]} = 98.6 \quad \text{A}$$

$$I_a = \frac{15}{15 - 1.25} \times 98.6 = 107.6 \quad \text{A}$$

$$E_{ma} = 107.6(2.5 - 3.75) = -134.5 \quad \text{V}$$

Thus

$$B = \frac{-134.5}{260} \times 0.34 = -0.176 \quad \text{T}$$

From Fig. 2, H $\simeq$ -325 kA/m. Since this operating point is beyond the straightline part of the characteristic, the rotor magnets will be demagnetized.

RELUCTANCE MOTORS

• PROBLEM 11-2

When the rotor of a reluctance motor like that shown in the figure is in the direct-axis position, the inductance of its exciting winding is L_d = 1.00 henry. When the rotor is in the quadrature-axis position, the inductance is L_q = 0.50 henry. The exciting winding has N = 1,000 turns. Determine approximately the maximum torque that the motor can develop with 115 volts at 60 cps applied to its exciting winding.

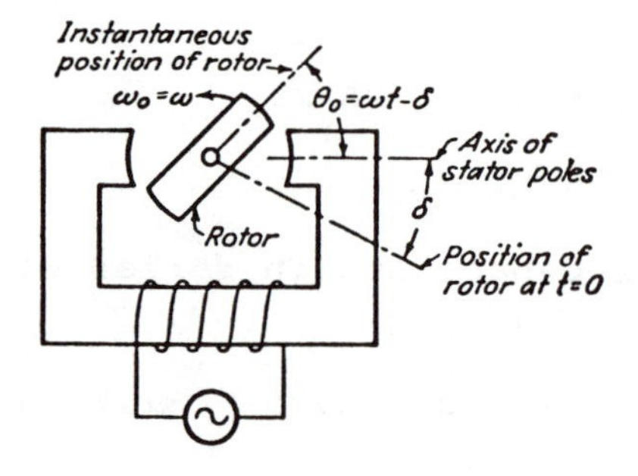

Elementary reluctance motor.

Solution:

$$E_{rms} = 4.44\ fN\phi_{max}\ ,$$

which gives

$$\phi_{max} = E_{rms}/4.44Nf$$

$$= \frac{115}{(4.44)(1{,}000)(60)} = 4.32 \times 10^{-4} \text{ weber}$$

From the definition of inductance

$$L = \frac{N\phi}{I} = \frac{N^2\phi}{NI} = \frac{N^2}{R}$$

or

$$R = \frac{N^2}{L}$$

whence

$$R_q = \frac{10^6}{0.50} = 2.00 \times 10^6 \text{ mks units}$$

$$R_d = \frac{10^6}{1.00} = 1.00 \times 10^6 \text{ mks units}$$

Substitution of numerical values in the following equation:

$$T_{max} = \frac{1}{8}\phi^2_{max}\ (R_q - R_d), \quad \text{gives}$$

$$T_{max} = \frac{1}{8}(4.32)^2(10^{-8})(2.00 - 1.00)(10^6)$$

$$= 2.34 \times 10^{-2} \text{ newton-m}$$

In English units of a convenient size (inch-ounces),

$$T_{max} = (2.34)(10^{-2})(0.738)(16)(12) = 3.31 \text{ in.-oz}$$

• PROBLEM 11-3

A reluctance motor of the type illustrated in Fig. 1 has a stator and rotor of cross section 25 mm × 25 mm, rotor being 50 mm in length. The air gap is approximately 4 mm in length, so that the magnetic circuit reluctances are

Direct-axis reluctance: $R_d = 10 \times 10^6$ A/Wb

Quadrature-axis reluctance: $R_q = 40 \times 10^6$ A/Wb

It may be assumed that reluctance varies with rotor position in the sinusoidal manner illustrated in Fig. 2 and expressed by the following equation:

$$R = R_a - R_b \cos 2\beta \quad \text{A/Wb}$$

The stator coil has 3000 turns and is excited from a 115 V, 25 Hz supply. The coil resistance is negligibly small.

a) Determine the maximum value of average mechanical power which this machine can develop.

b) Determine the rms value of the coil current.

Solution:

a) $R = R_a - R_b \cos 2\beta \quad$ A/Wb

$$= (25 - 15 \cos 2\beta) \times 10^6 \quad \text{A/Wb}$$

from which

$$\frac{dR}{d\beta} = 30 \times 10^6 \sin 2\beta$$

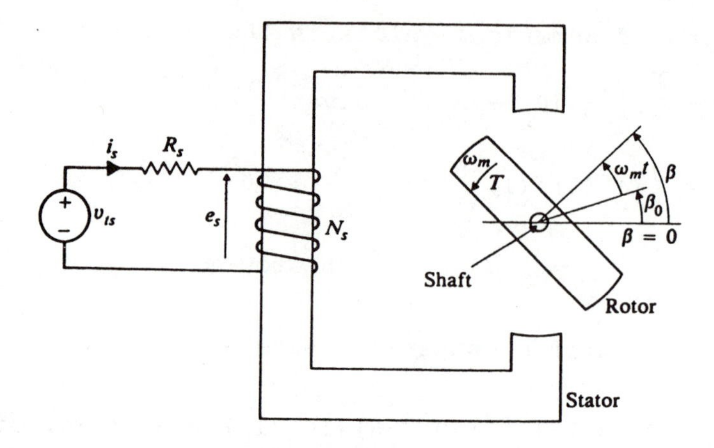

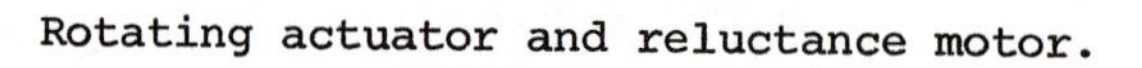

Rotating actuator and reluctance motor.

Fig.1

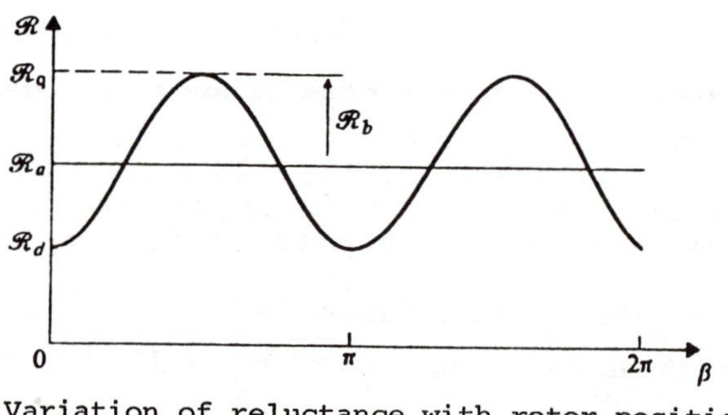

Variation of reluctance with rotor position.

Fig.2

Since the resistance of the coil is negligible,

$$E_s = V_{ts} = \frac{N_s \omega_s \hat{\phi}}{\sqrt{2}}$$

from which

$$= \frac{115\sqrt{2}}{3000 \times 50\pi} = 0.345 \times 10^{-3} \quad \text{Wb}$$

From the following equation:

$$T_{av} = \frac{-R_b \hat{\phi}^2}{4} \sin 2\beta_0 \quad \text{N.m},$$

the maximum average motoring torque is developed when $\beta_0 = -45°$ and

$$T_{av} = \frac{-R_b \hat{\phi}^2}{4} = \frac{15 \times 10^6 \times 0.345^2 \times 10^{-6}}{4} = 0.447 \quad \text{N.m}$$

Necessarily, for the machine to develop torque,

$$|\omega_m| = |\omega_s| = 50\pi \quad \text{rad/s}$$

Thus the power developed is

$$P = T_{av} \times \omega_m = 0.4447 \times 50\pi = 70 \quad W$$

b) For $\beta_0 = -45°$, from the following equation:

$$i = \frac{\phi R_a}{N_s} \cos \omega_s t - \frac{\phi R_b}{2N_s} [\cos (\omega_s t + 2\beta_0) + \cos (3\omega_s t + 2\beta_0)] \quad A,$$

$$i = \frac{0.345 \times 10^{-3} \times 25 \times 10^6}{3000} \cos \omega_s t$$

$$+ \frac{0.345 \times 10^{-3}\ 15 \times 10^6}{2 \times 3000} [\sin \omega_s t + \sin 3\omega_s t]$$

$$= 2.88 \cos \omega_s t + 0.863 \sin \omega_s t + 0.863 \sin 3\ \omega_s t$$

$$= 3.01 \sin (\omega_s t + 73.3°) + 0.863 \sin 3\ \omega_s t$$

The rms value of this current is

$$I = \left[\frac{3.01^2}{2} + \frac{0.863^2}{2}\right]^{\frac{1}{2}} = 2.21 \text{ A.}$$

REPULSION MOTORS

• PROBLEM 11-4

Calculate the starting current and starting torque for the repulsion-start, induction-run motor described below:

Rating: 2 hp 220 volts 4 poles 60 cycles

Full-load torque = 96.5 oz-ft

Maximum torque as induction motor = 228 oz-ft.

$r_1 = 0.765\ \Omega$ $\quad X_0 = 57.0\ \Omega$

$x_1 = 1.88\ \Omega$ $\quad K_r = 0.935$

$x_2 = 1.88\ \Omega$ $\quad r_2 = 1.58\ \Omega$

$a = 4.68$ $\quad$ C (stator conductors) = 576

$k_{w1} = 0.78$ $\qquad$ r_c (short-circuited coils) = 0.00745 Ω

a_3 = ratio of transformation stator to short-circuited coils

$a_3 = 56.25$ $\qquad$ r_b (brush and brush contact) = 0.0140 Ω

Brush-shift angle = 16°

Solution: In stator terms:

$$r_3 = (r_c + r_b) \times a_3{}^2$$

$$= (0.00745 + 0.014) \times 56.25^2 = 67.7 \text{ ohms.}$$

$$\beta = \frac{r_3}{X_0} = 1.18$$

$$G = K_r \sin^2 A \left(\frac{\beta}{\beta + 1}\right) + \frac{r_1}{X_0}$$

$$= 0.935 \times \sin^2 16° \left(\frac{1.18}{1.18 + 1}\right) + \frac{0.765}{57}$$

$$= 0.04785 \text{ mho.}$$

$$H = 1 - K_r\left(\cos^2 A + \frac{1}{\beta^2 + 1} \sin^2 A\right)$$

$$= 1 - 0.935\left(\cos^2 16° + \frac{1}{(1.18)^2 + 1} \sin^2 16°\right)$$

$$= 0.1184$$

$$I_1 = \frac{V}{X_0 \sqrt{G^2 + H^2}} = 30.65 \text{ amperes (32 by test)}$$

Starting torque:

$$T = I_1{}^2 K_r x_0 \left(\frac{\beta^2}{\beta^2 + 1}\right) \cos A \sin A$$

(synchronous watts) =

$$30.65^2 \times 0.935 \times 57.0\left(\frac{1.39}{1.39 + 1}\right) 0.9613 \times 0.2756$$

$$= 7750$$

Starting torque (oz-ft) = $\dfrac{112.5 \times 7750}{1800}$ = 486 oz-ft (472 by test)

Ratio starting torque to full-load torque:

$$\frac{486}{96.5} \times 100 = 504\%.$$

Extensive investigation of this method on a number of different motors indicates that the starting current and torque obtained through a consideration of the effect of the short-circuited commutating coils yields a much higher degree of accuracy than can otherwise be expected.

REACTORS

• PROBLEM 11-5

(i) The following information is given for a saturable reactor: $A = 1$ in.^{2}, $\ell = 10$ in., $N = 1{,}000$ turns, $H_c = 0.4$ A-T/in., and $B_S = 100$ KL/in.^{2} = 0.001 weber/in^2. This reactor is used in the circuit shown in the figure with $R_L = 1{,}000$ ohms and the voltage source a rectangular pulse of 10 volts and 0.1 sec. With an initial flux of zero, find: (a) the exciting current; (b) the flux as a function of time during the exciting interval; (c) the final value of flux; (d) the duration of the exciting interval.

(ii) Repeat Part (i) with the voltage source a rectangular pulse of 20 volts for 0.1 sec.

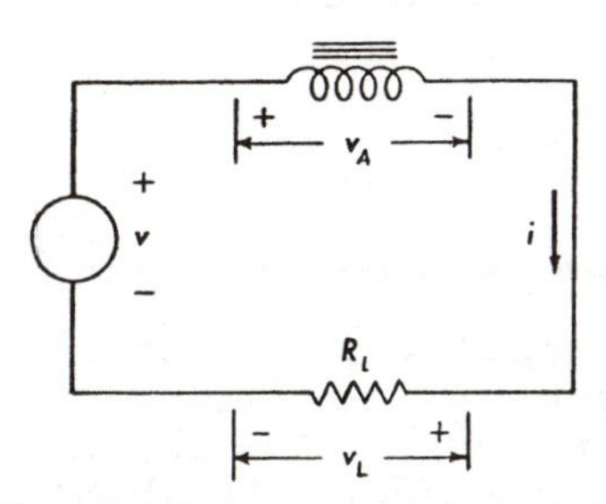

Saturable reactor in series with a resistance.

Solution: (i) (a) $I_x = \dfrac{H_c \ell}{N} = \dfrac{0.4\text{ A-T/in.} \times 10\text{ in.}}{1{,}000\text{ turns}}$

$= 0.004$ amp

(b) $v_A = v - I_x R_L = 10$ volts $-$ (0.004 amp)(1,000 ohms) $= 6$ volts for $0 < t < 0.1$ sec.

$$\phi(t) = \phi_1 + \frac{1}{N}\int_{t_1}^{t} v_A(\tau)\,d\tau$$

$$= 0 + \frac{1}{1,000} \int_0^t 6 \, d\tau = 0.006t \qquad 0 < t < 0.1 \text{ sec.}$$

(c) $\phi_2 = 0.006\ (0.1) = 0.0006$ weber

$\phi_s = B_s A$

$= 0.001 \text{ weber/in.}^2 \times 1 \text{ in.}^2 = 0.001$ weber;

$\phi_2 < \phi_s$

(d) $t_2 = 0.1$ sec

(ii) (a) $I_x = 0.004$ amp

(b) $v_A = v - I_x R_L = 20 - 4 = 16$ volts;

$$\phi = 0 + \frac{1}{1,000} \int_0^t 16 \, d\tau = 0.016t$$

(c) $\phi_2 = 0.016\ (0.1) = 0.0016$ weber

ϕ cannot exceed $\phi_s = 0.001$ weber

$\phi_s = \phi = 0.016t_s = 0.001; \qquad t_s = \frac{0.001}{0.016} = 0.0625$ sec.

The final value of flux is ϕ_s.

(d) The exciting interval ends at t = 0.0625 sec.

• PROBLEM 11-6

A 200-V 3-A 60-Hz shell-type reactor has an air gap 3 mm long in its center leg which is of square cross section. The length of flux path in the iron is 36 cm. The core material is electrical sheet steel for which the magnetization curve is shown in the figure. The maximum flux density in the iron is 1.27 T and the stacking factor of the core is 0.94. The effect of fringing is such as to increase the effective area of the air gap to 1.07 that of the gross area of the core. Calculate (a) the flux density B_m in the air gap, (b) the modified length of the air gap, (c) the number of the turns in the winding, and (d) the cross-sectional area of the center leg of the core.

Solution: a. The flux density in the air gap is

$$B_m = \frac{B_{iron} \times \text{stacking factor}}{\text{correction for fringing}}$$

$$= \frac{1.27 \times 0.94}{1.07} = 1.115 \text{ T}$$

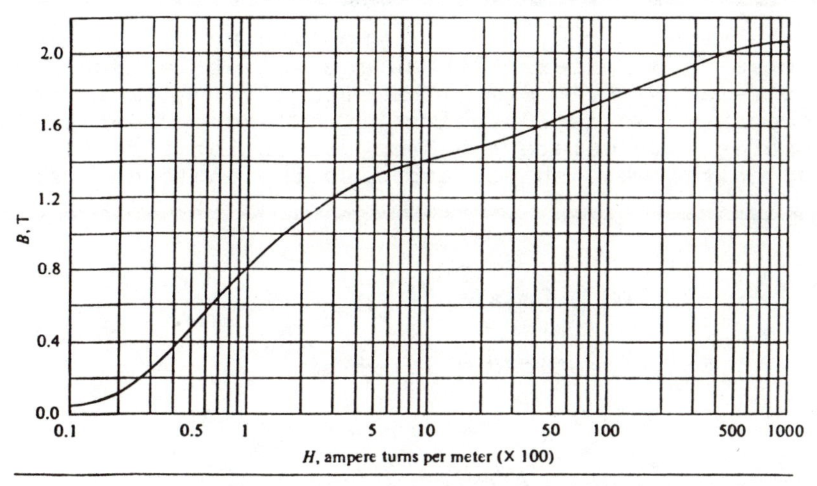

DC magnetization curve for M-19
fully processed 29-guage steel.

b. From the figure, $H_i = 400$, then,

$$g_i = g_0 + \frac{4\pi \times 10^{-7} H_i \ell_i}{B_m},$$

$$g_i = 0.003 + \frac{4\pi \times 10^{-7} \times 400 \times 0.36}{1.115}$$

$$= 0.003163 \text{ m.}$$

c. $$N = \frac{B_m g_1 \times 10^7}{\sqrt{2} \times 4\pi I} = \frac{1.115 \times 0.00317 \times 10^7}{\sqrt{2} \times 4\pi \times 3}$$

$$= 660 \text{ turns}$$

d. The gross area of the core

$$A_g = \frac{4EI \times 10^{-7}}{g_1 f B_m^2} = \frac{4 \times 200 \times 3 \times 10^{-7}}{0.003163 \times 60 \times (1.115)^2}$$

$$= 10.15 \times 10^{-4} \text{ m}^2$$

$$A_{core} = \frac{A_g}{\text{correction for fringing}}$$

$$= \frac{10.15 \times 10^{-4}}{1.07} = 9.48 \times 10^{-4} \text{ m}^2.$$

RECTIFIERS

• PROBLEM 11-7

A three-phase rectifier is supplied by delta-star-connected transformers with an equivalent leakage reactance per Y-leg of 0.25 ohm. The effective voltage from one secondary leg is 100 volts. The load current is 100 amperes. Calculate the drop due to the effect of overlap on the wave shape, and also the angle of overlap.

Solution: $E_{DC} = E_{max} \frac{p}{\pi} \sin \frac{\pi}{p}$

$$\therefore E_{DC} \text{ at no load} = 141 \frac{3}{\pi} \sin \frac{\pi}{3}$$

$$= 117 \text{ volts.}$$

$$\text{The drop} = I_{DC}(p/2\pi)$$

$$= 100 \times 0.25 \times \frac{3}{2\pi}$$

$$= 11.9 \text{ volts}$$

$$\therefore E_{DC} \text{ under load} = 117 - 11.9$$

$$= 105.1 \text{ volts}$$

The angle of overlap:

$$(1 - \cos u) = \frac{I_{DC} X}{E_{max} \sin \frac{\pi}{p}}$$

$$1 - \cos u = \frac{100 \times 0.25}{141 \times 0.866}$$

$$\cos u = 0.795$$

$$\therefore u = 37° 21'.$$

• PROBLEM 11-8

A dc commutator motor is controlled from a single-phase, full-wave bridge controlled rectifier (see figure) energized from a 60-Hz, 120-V power source. The motor has an armature resistance of 0.1 ohm and an armature inductance of 1 mH. The armature is in parallel with a freewheeling diode. The motor is operating at a constant speed such that the armature back emf is 60 V.

At a certain instant when one SCR is gated on (T_α = 0.006 sec.) the armature current is 80 A. Determine

a. The value of armature current at the time T_1 when the freewheeling current begins and the SCR is turned off.

b. The current at the end of the freewheeling current segment.

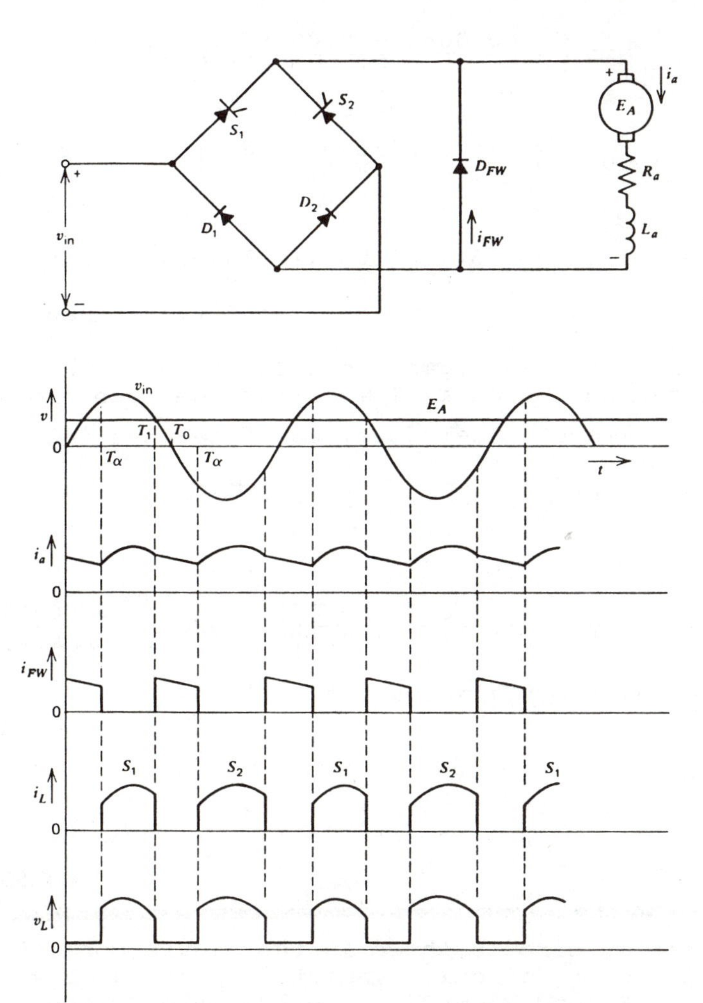

Full wave single-phase controlled rectifier with active load and free-wheeling diode.

<u>Solution</u>: V_m for a 120-V rms source is 169.7 V; $Z = \sqrt{0.1^2 + (377 \times 0.001)^2} = 0.39 \angle 75.2°\Omega$; $\phi = 75.2° = 1.3125$ rad; $V_m \sin \omega T_1 = 60$; $\omega T_1 = \pi - 60/169.7 =$

$2.786 =; \omega T_\alpha 377 \times 0.003 = 1.131$ rad; $T_0 = (1/2)(1/60)$

$= 0.00833$ s; $\omega T_0 = \pi$; $\tau = 0.001/0.1 = 0.01$ s.$i(T\ell)$

$$i(T_1) = \frac{V_m}{Z}[\sin(\omega t - \phi) - e^{-[(t-T_\alpha)/\tau} \sin(\omega T_\alpha - \phi)]$$

$$+ I(T_\alpha) \cdot e^{-[(t-T_\alpha)/\tau}$$

$$= \frac{169.7 - 60}{0.39}\left\{\sin(2.786 - 1.3125)\right.$$

$$\left. - \exp\left[-\left(\frac{0.0083 - 0.003}{0.01}\right)\right] \sin(1.131 - 1.3125)\right\}$$

$$+ 80 \exp\left[-\left(\frac{0.0083 - 0.003}{0.01}\right)\right]$$

$$= 281.3 [0.995 - 0.586 (-0.1805)] + 80 \times 0.586$$

$= 356$ A, at the end of the first segment.

The freewheeling segment continues until the next SCR is gated on, which is T_αs after the zero crossover of the input voltage. Therefore, the free-wheeling current at T_αs is,

$$i_{FW}(T_\alpha) = I(T_1)e^{-t/\tau} - \frac{E}{R}(1 - e^{-t/\tau})$$

$$= 356 \exp\left(\frac{-0.003}{0.01}\right) - \frac{60}{0.1}\left[1 - \exp\left(-\frac{0.003}{0.01}\right)\right]$$

$$= 264 - 155.5 = 108.5 \text{ A.}$$

• PROBLEM 11-9

The series circuit shown in the figure consists of the following circuit elements: R = 0.4, L = 125μH, C = 100μF, and E_b = 72V. Both S_1 and S_2 are to be operated alternately. L_c = 20μh; and assume that, in the steady state, the voltage on C is 180 V each time S_1 and S_2 are turned on. This implies zero resistance in the reversing circuit. Determine the maximum value of the new current pulse through S_1, the pulse width and maximum current of the pulse through S_2. Check to determine that the circuit commutation time, T_Q, is large enough to insure safe commutation of S_1.

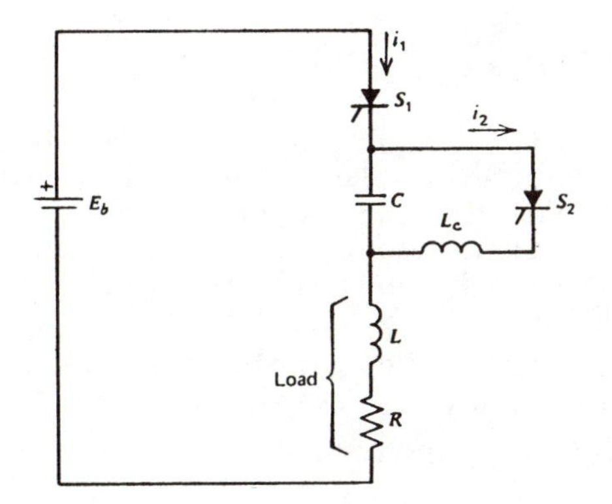

Motor controller illustrating series commutation.

Solution: $z_0 = \sqrt{\frac{L}{C} - \left(\frac{R}{2}\right)^2} = 1.1;$

$$\beta = \sqrt{\frac{1}{LC} - \left(\frac{R}{2L}\right)^2} = 0.88 \times 10^4$$

$$\varepsilon = \tan^{-1} \frac{2Z_0}{R} = 79.7° = 1.39 \text{ rad.}$$

Using these values, we find the current pulse width is

$$T_0 = \frac{\pi}{\beta} = \frac{\pi}{0.88 \times 10^4} = 358\mu s,$$

$$T_m = \frac{\varepsilon}{\beta} = \frac{1.39}{0.89 \times 10^4} = 158\mu s,$$

$$\alpha = \frac{R}{2L} = \frac{0.4}{250 \times 10^{-6}} = 1.6 \times 10^3 .$$

The maximum current can be determined from the following equation:

$$I_{max} = \frac{E}{Z_0} e^{-\alpha T_m} \sin \varepsilon,$$

or

$$I_{max} = \frac{(72 + 180)}{1.1} (e^{-0.252}) (\sin 79.7°) = 176 \text{ A}$$

For the current pulse in the reversing circuit,

$$Z_0 = \sqrt{\frac{L}{C}} = \sqrt{\frac{20}{100}} = 0.448;$$

$$\beta = \sqrt{\frac{1}{LC}} = \sqrt{\frac{10^{12}}{2000}} = 2.24 \times 10^4$$

$$\varepsilon = \frac{\pi}{2}; \quad \alpha = 0$$

$$T_0 = \frac{\pi}{\beta} = 141 \ \mu s; \quad T_m = \frac{\pi/2}{\beta} = 70.5 \ \mu s$$

$$I_{max} = \frac{180}{0.448} = 402 \text{ A}$$

Note that this pulse has a narrow pulse width and high peak current. This is typical of the required current pulses in this type of circuit. The turn-off or circuit commutating time, [the time after S_2 is turned on] is, in this case, the time during which the capacitor voltage is still larger than the source voltage, in order to maintain a reverse bias on S_2. This can be found by setting v_c equal to 72 V in the following equation:

$$v_c = E_b - \frac{E}{\sin \varepsilon} e^{-\alpha t} \sin (\beta t + \varepsilon), \quad (I_{10} = 0),$$

and solving for time. Note that, in the reversing circuit calculation, E_b is zero.

$$72 = 0 - \frac{180}{1.0} (1.0) \sin \left(\beta t_Q + \frac{\pi}{2}\right)$$

Solving for t_Q gives

$$\beta t_Q = 66.4° = 1.14 \text{ rad}; \quad t_Q = \frac{1.14}{2.24 \times 10^4} = 51 \mu s$$

This is adequate for many types of SCRs in this power range.

CONVERTERS

• PROBLEM 11-10

(1) A 1.5-kw single-phase converter operates at full load from a 230-volt a-c source. Neglecting losses, and assuming a power factor of unity, calculate: (a) the d-c voltage and current; (b) the a-c input current.

(2) A 2.5-kw single-phase converter is operated inverted, i.e., it converts direct to alternating current. If the d-c input voltage is 230, calculate: (a) the a-c voltage; (b) the alternating and direct currents at full load.

Solution:

(1) (a) $E_{dc} = \sqrt{2}\, E_{ac} = \sqrt{2} \times 230 = 325$ volts.

$$I_{dc} = \frac{P_{dc}}{E_{dc}} = \frac{1{,}500}{325} = 4.62 \text{ amp.}$$

(b) $I_{ac} = \frac{P_{ac}}{E_{ac}} = \frac{1,500}{230} = 6.52$ amp.

(2) (a) $E_{ac} = \frac{230}{\sqrt{2}} = 162.5$ volts.

$I_{ac} = \frac{2,500}{162.5} = 15.4$ amp.

$I_{dc} = \frac{2,500}{230} = 10.9$ amp.

• **PROBLEM 11-11**

A 1,500-kw six-phase synchronous converter has a full-load d-c voltage of 600. Determine: (a) the a-c voltage between rings; (b) the d-c output; (c) the a-c input per line.

Solution:

(a) $E_{ac} = 0.354\ E_{dc}$

$= 0.354 \times 600 = 212.4$ volts.

(b) $I_{dc} = \frac{P}{E_{dc}} = \frac{1,500,000}{600} = 2,500$ amp.

(c) $I_{ac} = 0.472\ I_{dc}$

$= 0.472 \times 2,500 = 1,180$ amp.

• **PROBLEM 11-12**

A three-phase, 2-pole converter has a d-c output of 100 amperes. Neglecting losses, determine the line current on the a-c side.

Solution: $I_{ac} = 0.943 \times I_{dc}$

$= 94.3$ amperes

The alternating current in the windings will then be

$\frac{94.3}{\sqrt{3}} = 54.6$ amperes

This example considered a 2-pole converter. In general there will usually be as many parallel paths per phase as there are pole pairs, and the following relationship can be shown to exist:

$$I_{a\text{-}c\ coil} = \frac{I_{a\text{-}c\ line}}{\left(2 \sin \frac{\pi}{n}\right) \times \frac{P}{2}} .$$

• PROBLEM 11-13

A 500-kw converter (see figure) has an efficiency of 93 percent at full load and operates at a power factor of 0.94. The d-c voltage is 550 volts.

The converter is supplied from a 6,600-volt 60-cycle 3-phase system. The transformer primaries are connected in delta and the secondaries in 6-phase star. Determine (a) direct current; (b) slip-ring current (c) slip-ring voltage to neutral; (d) diametrical transformer secondary voltage (e) transformer primary currents; (f) a-c line current; (g) kva rating of each transformer. Neglect transformer magnetizing current and losses.

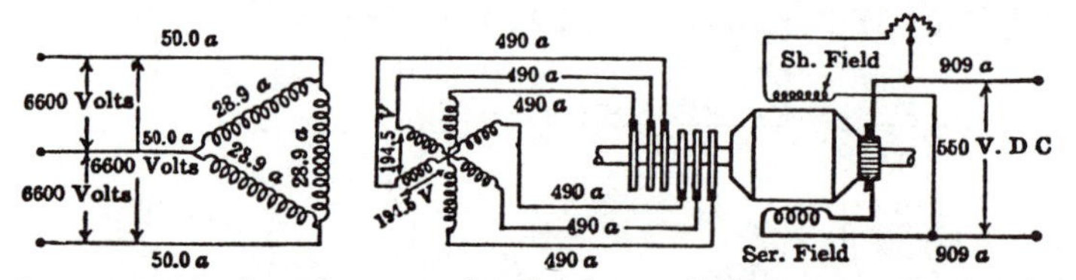

Currents and voltages in 6-phase 500-kw synchronous converter and transformers.

Solution: (a) $V = 550,$

$$I = \frac{500,000}{550} = 909 \text{ amp.}$$

(b) $$I_6 = \frac{0.471\ I}{\eta\,(\text{P.F.})}$$

where P.F. is the power factor.

$$= \frac{0.471 \times 909}{0.93 \times 0.94} = 490 \text{ amp per line.}$$

(c) The voltage between slip rings that are electrically adjacent is (50/141.4)550 = 194.5 volts. Since the secondary connection is 6-phase star, the voltage to neutral is also 194.5 volts.

(d) The diametrical secondary voltage is

$2 \times 194.5 = 389$ volts.

(e) The transformer voltage ratio is 6,600/389. Hence the primary current is

$$490 \frac{389}{6,600} = 28.9 \text{ amp.}$$

(f) The line current is

$$\sqrt{3} \times 28.9 = 50.0 \text{ amp.}$$

(g) The transformer kva rating is

$$\frac{6,600 \times 28.9}{1,000} = 191 \text{ kva.}$$

• PROBLEM 11-14

A 500 kW, 600 V dc, 12 phase synchronous converter operates as a direct converter at a full efficiency of 92 per cent and a power factor of 0.93. Calculate:

(a) The ac voltage between slip-rings

(b) The dc output current

(c) The ac current drawn from a 12-phase transformer-fed supply.

POLYPHASE SYNCHRONOUS CONVERTER RELATIONS

QUANTITY	1-PHASE, 2-RING	3-PHASE, 3-RING	6-PHASE, 6-RING	12-PHASE, 12-RING
E_{ac} between rings	$0.707E_{dc}$	$0.612E_{dc}$	$0.354E_{dc}$	$0.182E_{dc}$
I_{ac}* in rings	$1.414I_{dc}$	$0.943I_{dc}$	$0.472I_{dc}$	$0.236I_{dc}$
δ, electrical degrees between taps	180	120	60	30
T taps per pole pair	2	3	6	12

*At unity power factor 100 per cent efficiency.

Table 1

Solution: Using Table 1,

(a) $E_{ac} = 0.182E_{dc} = 0.182 \times 600\text{ V} = 109\text{ V}$ between slip-rings.

(b) $I_{dc} = \dfrac{\text{kW} \times 1000}{E_{dc}} = \dfrac{500 \times 1000\text{ W}}{600\text{ V}} = 833.3\text{ A}$

(c) $I_{ac} = \frac{0.236 I_{dc}}{PF \times eff.} = \frac{0.236 \times 833.3}{0.93 \times 0.92} = 229.5$ A.

• PROBLEM 11-15

The power for a 100-kw, 250-volt, d-c load is to be supplied by a three-phase rotary converter from a 2300-volt, three-phase supply. If the rotary converter efficiency is 95 per cent for unity power factor operation, determine the rotary converter input and the transformer ratings. (Assume the transformer bank is connected in delta-delta).

Solution:

D-c side:

$$I_{DC} = \frac{100,000}{250} = 400 \text{ amp} \qquad V_{DC} = 250 \text{ volts}$$

A-c side:

$$V_{AC} = 0.612\, V_{DC} = 0.612 \times 250 = 153 \text{ volts}$$

$$I_{AC} = \frac{0.943\, I_{DC}}{\eta \cos\theta} = \frac{0.943 \times 400}{0.95 \times 1.00} = 397 \text{ amp.}$$

Rotary converter input:

$$V_{AC} = 153 \text{ volts} \qquad I_{AC} = 397 \text{ amp.}$$

Current per transformer secondary:

$$I = \frac{I_{AC}}{\sqrt{3}} = \frac{397}{\sqrt{3}} = 229 \text{ amp.}$$

Kilovolt-ampere rating:

$$\frac{V_{AC} I}{1000} = \frac{229 \times 153}{1000} = 35.0 \text{ kva}$$

Three transformers, each having a rating of 2300/153 volt 35.0 kva are required.

This can be checked easily since the efficiency and power factor are given. Input is 100/0.95 = 105 kw at unity power factor. Transformer bank is 105 kva, which is 35 kva per transformer.

• PROBLEM 11-16

The data relating to a 1,000-kw 60-cycle 600-volt (d-c) synchronous converter are as follows:

Rating	1,000 kw
Direct-current voltage	600 volts
Alternating-current voltage (diametrical) . .	424 volts
Direct-current output	1,667 amp
Number of phases	6
Frequency	60 cycles
Poles	12
Speed	600 rpm
Number of armature slots	180
Inductors per slot	6
Armature resistance at 25° C	
Between d-c terminals	0.00589 ohm
Between a-c diametrical terminals	0.00589 ohm
Shunt turns per pole	864
Series turns per pole	2
Resistance at 25° C of shunt field	39.7 ohms
Resistance at 25° C of series winding	0.000610 ohm
Friction and windage loss	8.1 kw

The open-circuit saturation curve and the curve of core loss are plotted in the figure.

Calculate (a) the field excitation and (b) the efficiency of the converter, for a full d-c load and a power factor of 0.95 with a leading current.

Solution: (a) The armature of a synchronous converter carries a current equal to the difference between the components due to the d-c output and the a-c input. As a result the voltage drop in the armature is relatively small and can be neglected when calculating the field excitation and efficiency.

The distorting components of the armature reaction nearly neutralize and need not be considered. The only component of the armature reaction which must be taken into account is that due to the reactive component of the alternating current. This component either strengthens or weakens the field without producing distortion. The ampere-turns corresponding to it add directly to the excitation of the shunt and series fields or subtract directly from it. In a converter delivering direct current, a reactive lagging component of the alternating current strengthens the field. The reactive component of a leading current weakens the field.

The resultant or net ampere-turns of excitation for any terminal voltage under load conditions are apprximately equal to the ampere-turns necessary to produce the required voltage when the converter is driven as a generator at no load.

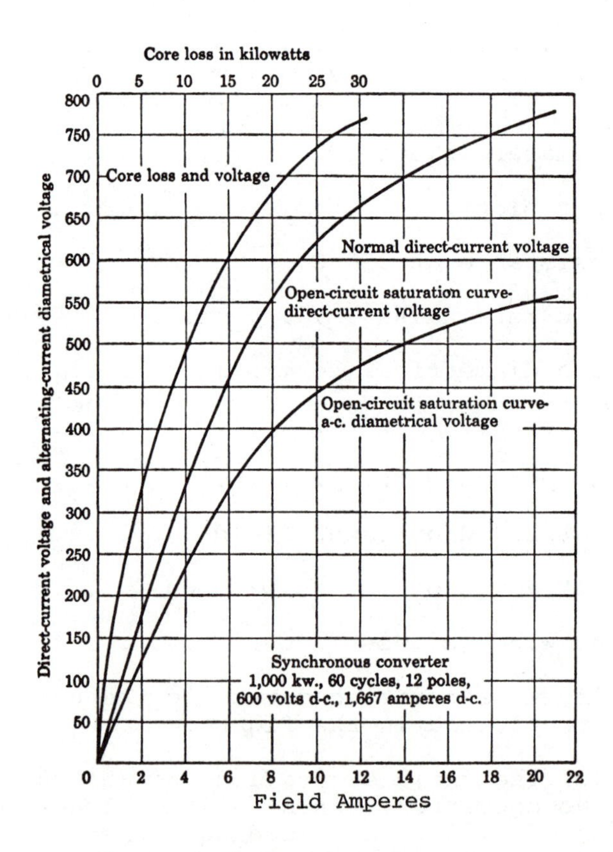

The efficiency of a synchronous converter operating at a power factor in the neighborhood of unity is usually 95 percent or better. On account of this high operating efficiency, it is sufficiently close to assume the efficiency to be 95 percent as a first approximation when calculating the armature reaction caused by the reactive component of the alternating current and the armature copper loss. A second approximation can be made if necessary.

The coil current is the same as the inductor current, and is given by

$$I'_{ac} = I_{dc} \frac{2}{pn(pf)\eta} \frac{V_{dc}}{V_{ac}}$$

$$I_{dc} = \frac{1{,}000 \times 1{,}000}{600} = 1{,}667 \text{ amp.}$$

For an efficiency assumed to be 0.95 as a first approximation and a power factor of 0.95

$$I'_{ac} = 1{,}667 \frac{2}{12 \times 6 \times 0.95 \times 0.95} \frac{\sqrt{2}}{\sin \frac{\pi}{n}} = 145 \text{ amp.}$$

The reactive component of this current is

$$I_x = 145 \sqrt{1 - (0.95)^2} = 45.3 \text{ amp.}$$

The armature reaction A_x per pole for I_x can be found from the equation

$$A_x = 0.75 k_b N I_x .$$

where N is the number of armature turns per pole, I_x is the reactive component of the armature current I'_{ac} and k_b is the breadth factor. The constant 0.75 is used instead of 0.90 in finding A_x because the synchronous converter is a salient-pole machine.

The phase spread of a six-phase converter is 60 deg, or one-third of the pole pitch. The converter has 180 slots and 12 poles, or 180/(12 × 3) = 5 slots per phase belt.

Numner of slots per belt	Breadth Factor				
	Width of phase belt in fractional part of the whole pitch.				
	1/4	1/3	1/2	2/3	Whole
2	0.980	0.966	0.924	0.866	0.707
3	0.977	0.960	0.911	0.844	0.666
4	0.976	0.958	0.906	0.836	0.653
infinite	0.975	0.955	0.901	0.827	0.637

From the table, the breadth factor for a spread of 60 deg and four slots per phase is 0.958. For five slots per phase it is about 0.957.

$$A_x = 0.75 \times 0.957 \times \frac{180 \times 6}{2 \times 12} \times 45.3$$

$$= 1{,}464 \text{ amp-turns per pole}$$

These are demagnetizing ampere-turns since a leading current is assumed. The ampere-turns per pole due to the series field are

$$1{,}667 \times 2 = 3{,}334$$

The field current required for 600 volts when the converter is driven at no load as a d-c generator is 9.25 amp (open-circuit saturation curve, see figure).

This corresponds to

$$9.25 \times 864 = 7{,}990 \text{ amp-turns per pole}$$

The shunt excitation required under full-load conditions at a power factor 0.95, an efficiency 95 percent, and with a leading current is

$$7{,}990 + 1{,}464 - 3{,}334 = 6{,}120 \text{ amp-turns}$$

This corresponds to a shunt-field current of

$$\frac{6{,}120}{864} = 7.08 \text{ amp.}$$

(b) The efficiency is

$$\eta = \frac{I_{dc}V_{dc}}{I_{dc}V_{dc} + HI_{dc}^2 r_{dc} + I_{sh}V_{dc} + I_c^2 r_c + P_c + (F + W)}$$

where I_{dc} = direct current

V_{dc} = d-c voltage

r_{dc} = armature resistance between d-c terminals

I_{sh} = shunt-field current

I_c = compound-field current

r_c = resistance of compound winding

P_c = core loss

$F + W$ = friction and windage loss

The armature copper loss can be found by multiplying the copper loss corresponding to the d-c component of the armature current by the ratio of the copper loss of the converter as a converter to its copper loss at the same output as a d-c generator. This ratio H is given by

$$H = \frac{8}{(pf)^2\eta^2 n^2 \sin^2(\pi/n)} + 1 - \frac{16}{\pi^2\eta}$$

For a power factor of 0.95 and an efficiency to be 95 percent as a first approximation,

$$H = \frac{8}{(0.95)^2(0.95)^2(6)^2(0.5)^2} + 1 - \frac{16}{(3.142)^2(0.95)}$$

$$= 0.385$$

The armature resistance at 75°C between d-c terminals is

$$0.00589(1 + 50 \times 0.00385) = 0.00702 \text{ ohm}$$

The armature copper loss is

$$I_{dc}^{\ 2} \times r_a \times 0.385 = (1{,}667)^2 \times 0.00702 \times 0.385$$

$$= 7{,}510 \text{ watts.}$$

The ohmic resistance is used in finding the armature copper loss. This loss is small and the error introduced by using ohmic resistance in place of effective resistance is probably not great. Since the armature inductors of a converter carry differently shaped current waves, the ratio of the ohmic resistance to effective resistance is not the same for all inductors. It also changes with power factor.

The shunt-field loss including the loss in the field rheostat is equal to the shunt-field current multiplied by the voltage across the d-c brushes. This voltage is equal to the terminal voltage plus the drop in the series field. The drop in the series field is neglected. The shunt-field loss is

$$7.08 \times 600 = 4{,}248 \text{ watts.}$$

The resistance of the series field at 75° C is

$$0.000610 \times (1 + 50 \times 0.00385) = 0.000728 \text{ ohm}$$

The series-field loss is

$$(1{,}667)^2 0.000728 = 2{,}025 \text{ watts.}$$

The core loss corresponding to a d-c voltage of 600 is 14,700 watts, from the figure.

The efficiency is then

$$\eta = \frac{1{,}000}{1{,}000 + 7.5 + 4.2 + 2.0 + 14.7 + 8.1}$$

$$= 0.965 \text{ or } 96.5 \text{ percent.}$$

The difference between the efficiency just found and the efficiency which was assumed to be 95 percent as a first approximation in calculating I_x and H is too small to make a second approximation necessary.

SYNCHRONOUS CONDENSERS

• PROBLEM 11-17

An industrial plant has a load of 1,500 kva at an average power factor of 0.6 lagging. Neglecting all losses, calculate: (a) the kilovolt-ampere input to a synchronous condenser for an over-all power factor of unity; (b) the total kilowatt load.

Solution: Problems of this type are best solved by the use of simple kilovolt-ampere phasor diagrams as shown in the figure. An arbitrary reference voltage phasor is drawn horizontally first. The kilovolt-ampere load is then drawn in a lagging direction at the proper angle (in this case the angle whose cosine is 0.6 = 53°) to a convenient scale.

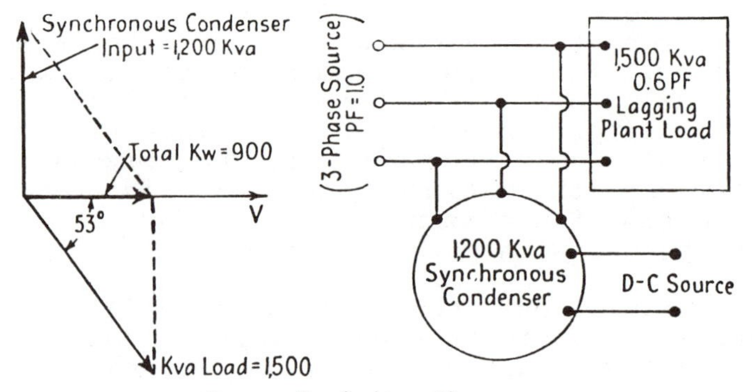

Kva and wiring diagrams.

(a) For an over-all power factor of unity, the synchronous condenser will have to counteract the vertical component of 1,500 kva, which is

1,500 x sin 53° = 1,500 x 0.8 = 1,200 kva

(b) Total load = 1,500 x 0.6 = 900 kw

A wiring diagram for this problem is also shown in the figure.

• PROBLEM 11-18

A factory draws a lagging load of 2000 kW at a power factor of 0.6 from 6000 V mains. A synchronous capacitor is purchased to raise the overall power factor to unity. Assuming that the synchronous capacitor losses are 275 kW, calculate:

(a) Original kilovars of lagging load.

(b) kvars of correction needed to bring the power factor to unity.

(c) kVA rating of synchronous capacitor and its power factor.

Solution: (a) $\text{kVA} = \frac{\text{kW}}{\cos\theta} = \frac{2000}{0.6} = 3333 \text{ kVA.}$

Lagging kvars = kVA sin θ = 3333 × 0.8 = 2667 kvars.

(b) 2667 kvars of correction are required to bring the power factor to unity.

(c) $\tan\theta = \frac{2667 \text{ kvars}}{275 \text{ kW}} = 9.68; \therefore \theta = \text{arc tan } 9.68 = 84.09°$ leading.

$\cos\theta = 0.103$ leading; $\text{kVA} = \frac{\text{kW}}{\cos\theta} = \frac{275}{0.103} = 2755 \text{ kVA.}$

A synchronous capacitor at a power factor of 0.103 leading and a kVA rating of 2755 is required.

• PROBLEM 11-19

Consider a three-phase line about 35 miles long supplying a load of 20,000 kva at a power factor of 0.7 inductive. The receiving-end voltage, between lines, is 66,000 volts. With transformers at each end, the resistance of line and transformers (referred to transmission line voltage) to neutral is 12.5 ohms. Reactance of line and transformers to neutral is 40 ohms. Capacitance of line is neglected. Show the control of line voltage by a synchronous condenser.

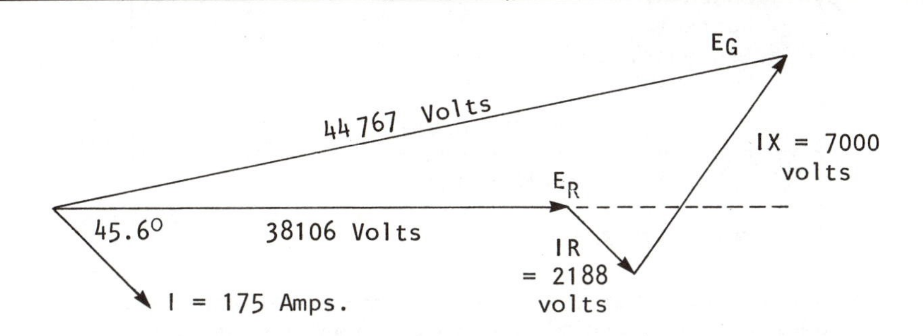

Vector diagram of an inductive load on a transmission line.

Fig.1

Solution: Voltage to neutral = $66,000/\sqrt{3}$ = 38,106 volts.

Full-load current per line = $\frac{20,000,000}{3 \times 38,106}$ = 175 amperes.

If the IR and IX drops are added to the receiving-end voltage, E_R, the voltage at the generating end, E_G, may be calculated (Fig. 1).

Line IR = 17.5 × 12.5 = 2188 volts.

Line IX = 175 × 40 = 7000 volts.

Also cos ϕ = 0.700 and sin ϕ = 0.715.

$$E_G = \left\{(E_R + IR\cos\phi + IX\sin\phi)^2 + (-IR\sin\phi + IX\cos\phi)^2\right\}^{\frac{1}{2}}$$

$$E_G = \left\{(38{,}106 + 2188\cos\phi + 7000\sin\phi) + (-2188\sin\phi + 7000\cos\phi)^2\right\}^{1/2}$$

$$= \left\{(38{,}106 + 1532 + 5005)^2 + (-1564 + 4900)^2\right\}^{1/2} = 44{,}767 \text{ volts.}$$

The drop in phase voltage due to line impedance is 44,767 - 38,106 = 6661 volts. At no load the voltage at the receiving end will rise to 44,767 volts, giving a regulation at the load of 6661/38,106 = 17.5 per cent.

If an over-excited synchronous condenser is used at the load end to draw a leading reactive current equal to the lagging reactive component of the load (175 × 0.715 = 125 amperes), the full-load current over the line would be 175 × 0.7 = 122.5 amperes (neglecting the losses of the synchronous condenser).

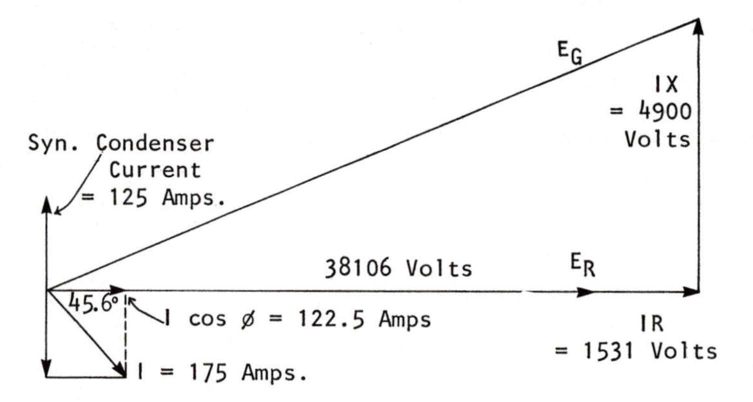

Conditions after a synchronous condenser has been added at the receiving end of the transmission line of Figure 1.

Fig.2

With 38,106 volts at the receiving end, the voltage at the generating end of the line would be (Fig. 2):

$$E_G = \left\{(E_R + I\cos\phi . R)^2 + (I\cos\phi . X)^2\right\}^{1/2}$$

$$= \left\{(38{,}106 + 122.5 \times 12.5)^2 + (122.5 \times 40)^2\right\}^{1/2}$$

$$= \left\{(38{,}106 + 1531)^2 + (4900)^2\right\}^{1/2} = 39{,}939 \text{ volts}$$

Drop due to line is 39,939 - 38,106 = 1833 volts.

Regulation is 1833/38,106 = 4.8 per cent.

The use of the synchronous condenser thus aids in reducing the drop due to line and the losses of the line and improves the regulation.

It is possible, by suitable manipulation of the field current of the synchronous condenser, to hold the voltage at the load constant, as load current varies. If, as the load decreases, the field of the synchronous condenser is weakened, so that, when the load has dropped to zero, the synchronous machine draws a lagging current sufficient to cause a line drop of such value that with 39,939 volts at the generator end, the receiving end voltage is 38,106, the regulation of the line will be zero.

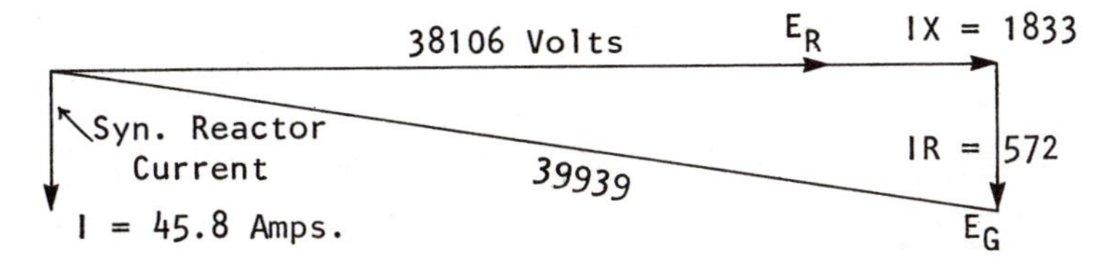

When the load on the transmission line of Fig.2 is removed the synchronous machine is under-excited sufficiently to make it draw such a value of inductive current as to hold the receiving-end voltage constant. The synchronous machine may then be called a synchronous reactor.

Fig.3

Conditions are shown in Fig. 3, from which it will be seen that the line reactance drop is practically equal to the arithmetic difference between the receiving-end voltage and the generator-end voltage. The current to be drawn by the synchronous machine (now properly called a synchronous reactor), is equal to line drop divided by line reactance, or (39,939 - 38,106)/40 = 1833/40 = 45.8 amperes. With this current the generator-end voltage may be checked as

$$E_G = \sqrt{(E_R + IX)^2 + (IR)^2}$$

$$E_G = \sqrt{(38{,}106 + 45.8 \times 40)^2 + (45.8 \times 12.5)^2} = 39{,}942.$$

The proper field adjustment of the synchronous machine, to bring about this condition of zero regulation, may be made automatically by a voltage regulator, so that the receiving-end voltage is held constant.

CHAPTER 12

TRANSMISSION AND DISTRIBUTION OF POWER

TRANSMISSION LINE REACTANCE AND CAPACITANCE

• PROBLEM 12-1

A single-phase transmission line is 40 miles long and consists of two 0000 solid conductors spaced 4 ft on centers.

Determine (a) inductance of entire line and reactance per conductor at 25 cycles per sec; (b) at 60 cycles per sec; (c) total reactance drop with 200 amp, 60 cycles, in line.

Solution: The diameter of a 0000 conductor is 0.460 in the radius r = 0.230 in.

$$\frac{D}{r} = \frac{48}{0.230} = 209.$$

$$\log_{10} 209 = 2.32$$

The inductance per mile is given by

$$L' = 2\ell(0.080 + 0.741 \log_{10} \frac{D}{r}) \times 10^{-3} \text{ henrys}$$

$$L' = 2(0.080 + 0.741 \times 2.32) = 3.60 \text{ milhenrys}$$

(a) The total inductance

L = 3.60 × 40 = 144 mil-henrys, or 72 milhenrys per conductor.

The reactance per conductor at 25 cycles is

$$X_1 = (2\pi) 25 \times 72 \times 10^{-3} = 11.3 \text{ ohms.}$$

(b) The reactance per conductor at 60 cycles is

$$X_2 = (2\pi)\ 60 \times 72 \times 10^{-3} = 27.1 \text{ ohms.}$$

(c) The total reactance drop with 200 amp, 60 cycles, is

$$V = 27.1 \times 200 \times 2 = 10{,}840 \text{ volts.}$$

• PROBLEM 12-2

A 3-phase line consists of three 0000 solid conductors placed at the corners of an equilateral triangle, 4 ft on a side. Determine the reactance drop per conductor per mile with 120-amp 25-cycle alternating current.

Solution: $X = (2\pi)\ f\ (80 + 741 \log_{10}\frac{D}{r})10^{-6}$ ohms per mile.

$$X = (2\pi)\ 25\ (80 + 741 \log_{10}\frac{48}{0.23})10^{-6}$$

$$= 157(80 + 741 \times 2.32)10^{-6}$$

$$= 157 \times 1{,}800 \times 10^{-6} = 0.282 \text{ ohm.}$$

The voltage drop is

$$V = IX$$

$$= 120 \times 0.282 = 33.8 \text{ volts.}$$

• PROBLEM 12-3

(a) A 40-mile 60-cycle single-phase line consists of two 000 conductors spaced 5 ft apart. Determine the charging current if the voltage between wires is 33,000 volts.

(b) Assume that a third wire is added to the system of Part (a) to form a symmetrical spacing and that the system is operated 3-phase, 33,000 volts between conductors. Find the charging current per conductor.

Solution: (a) The diameter of 000 wire is 410 mils. The radius r = 0.205 in.

$$\frac{D}{r} = \frac{60}{0.205} = 293.$$

$$\log_{10} 293 = 2.47.$$

$$C_n = \frac{0.0388}{\log_{10}(D/r)}\ \mu f \text{ per mile to neutral.}$$

$$= 40 \times \frac{0.0388}{2.47} = 0.628 \ \mu f.$$

The charging current is

$$I_c = 2\pi f c_n \ E \ 10^{-6} \text{ amp per mile of line.}$$

$$= 2\pi(60) \times 0.628 \times \frac{33{,}000}{2} \ 10^{-6} = 3.91 \text{ amp.}$$

(b) $r = 0.205$ in.

$$\frac{D}{r} = \frac{60}{0.205} = 293.$$

$$\log_{10} 293 = 2.47.$$

$$C_n = 40 \times \frac{0.0388}{2.47} = 0.628 \ \mu f \text{ to neutral.}$$

Volts to neutral = $33{,}000/\sqrt{3} = 19{,}070$ volts.

Charging current per conductor

$$I_c = 2\pi(60) \times 0.628 \times 19{,}070 \times 10^{-6} = 4.52 \text{ amp.}$$

TRANSMISSION LINE CALCULATIONS

• PROBLEM 12-4

A power station supplies 60 kw to a load over 2,500 ft of 000 2-conductor copper feeder the resistance of which is 0.078 ohm per 1,000 ft. The bus-bar voltage is maintained constant at 600 volts. Determine: (a) current; (b) voltage at load; (c) efficiency of transmission; (d) maximum power which can be transmitted; (e) maximum current which can be supplied.

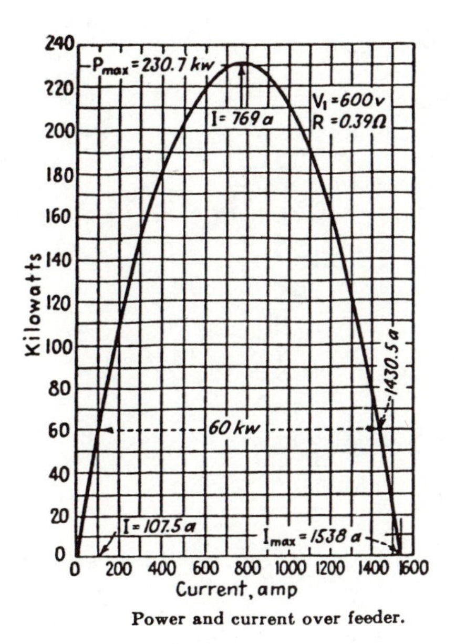

Power and current over feeder.

Solution: (a) Total resistance R = 5 · 0.078 = 0.39 ohm.

Let I be the current and V the voltage at the load:

$$60,000 = VI. \tag{1}$$

$$V = 600 - 0.39I. \tag{2}$$

Substituting V from (2) in (1),

$$60,000 = (600 - 0.39I)I.$$

$$0.39I^2 - 600I = -60,000 \tag{3}$$

Dividing (3) by 0.39,

$$I^2 - 1,538I = -153,800.$$

This is a quadratic equation and must have two roots. Hence, two values of current will satisfy the given conditions. Completing the square and solving,

$$I^2 - 1,538I + (769)^2 = -153,800 + (769)^2.$$

$$(I - 769)^2 = 437,600.$$

$$I - 769 = \pm\sqrt{437,600}.$$

$$I = 769 \pm 661.5$$

$$= 1,430.5, \text{ or } 107.5 \text{ amp.}$$

Both values of current satisfy (1) and (2), but if the larger value of current were used, the efficiency would be too low. Hence, the current

$$I = 107.5 \text{ amp.}$$

(b) $V = 600 - (0.39 \times 107.5) = 600 - 41.9 = 558.1$ volts.

$$P = 558.1 \times 107.5 = 60{,}000 \text{ watts, or } 60 \text{ kw (check).}$$

(c) $\eta = \frac{558.1}{600} = 0.930$, or 93.0%.

(d) Repeating (1), (2), (3), (4) with power represented by P,

$$P = VI = (600 - 0.39I)I.$$

$$0.39I^2 - 600I = -P.$$

Dividing by 0.39 and completing the square ($591{,}400 \simeq 769^2$),

$$I^2 - 1{,}538I + 591{,}400 = 591{,}400 - 2.564P.$$

$$I - 769 = \pm \sqrt{2.564}\ \sqrt{230{,}700 - P}.$$

$$I = 769 \pm 1.6\ \sqrt{230{,}700 - P}.$$

If the power P exceeds 230,700 watts, the quantity under the radical is negative, the radical is an imaginary quantity, and the current is the sum of a real and an imaginary quantity, showing that this condition is impossible. Hence, the maximum value which the power P can have is 230,700 watts, or 230.7 kw.

Since under these conditions the radical is zero, the two roots of the equation are equal. Hence, the current I has but a single value and is equal to 769 amp (see figure).

The voltage at the load,

$$V = 600 - (769 \times 0.39) = 600 - 300, \text{ or } 300 \text{ volts.}$$

The maximum power which can be transmitted,

$$P_m = 300 \times 769 = 230{,}700 \text{ watts, or } 230.7 \text{ kw.}$$

Under these conditions, one-half the power is lost in the line, and the efficiency is 50 percent. This corresponds to the similar condition for batteries.

The value of maximum power is more readily determined by the use of calculus. Let V_1 by the bus-bar voltage, V the load voltage, I the current, r the total line resistance, and P the power to the load. Then,

$$V = V_1 - Ir.$$

$$P = VI = V_1 I - I^2 r.$$

Taking the first derivative of the power with respect to the current, and equating to zero to obtain the condition for maximum power,

$$\frac{dP}{dI} = V_1 - 2Ir = 0,$$

$$Ir = \frac{V_1}{2},$$

and

$$V = V_1 - Ir = V_1 - \frac{V_1}{2} = \frac{V_1}{2}.$$

(e) The maximum current occurs when the load is short-circuited. That is,

$$I = \frac{600}{0.39} = 1{,}538 \text{ amp.}$$

The power supplied to the load under these conditions must be zero since the voltage at the load is zero.

To illustrate further these relations between power and voltage, the power versus current curve is shown in the figure. Current is plotted as the abscissa and power as the ordinate. Different values of current are assumed, substituted in (V), and the equation solved for the power P. The power must be zero when the current is zero (voltage V a maximum) and when the voltage V is zero (current a maximum). It is clear that for any given value of power, except the maximum, there are two different values of current. The maximum power is 230.7 kw and occurs when the current is 769 amp. At this point both roots of the quadratic equation are equal. In practice the power transmitted is usually a small proportion of the maximum power, as can be seen by the position of the 60-kw point on the curve, shown in the figure.

Also the other solution of (4), I = 1,430.5 amp, is found at the second intersection of the characteristic with the 60-kw ordinate.

(i) It is desired to deliver 4,000 kw, single-phase, at a distance of 25 miles, the load voltage being 33,000 volts, 60 cycles, and the power factor of the load being 0.85, lagging current. The conductors are spaced 4 ft on centers. The line loss shall not exceed 10 percent of the power delivered. Determine (a) size of conductor; (b) resistance drop per conductor; (c) reactance drop per conductor; (d) voltage at sending end; (e) line regulation. Neglect capacitive effects.

(ii) Solve part (i), assuming 3-phase transmission, other conditions remaining the same. Power to be delivered, 4,000 kw; load voltage, 33,000 volts between conductors; distance, 25 miles; frequency, 60 cycles; load power factor, 0.85, lagging current; spacing of conductors, 48 in.; allowable line loss, 10 percent of power delivered. Determine (a), (b), (c), (d), (e). (f) Determine sending-end voltage when load power factor is 0.70, leading current.

Table 1

Resistance of Copper Wire, Ohms per Mile, 25°C (77°F)

Size, cir mils or AWG	Number of wires	Outside diameter, mils	Ohms per mile
STRANDED			
1,000,000	61	1.152	0.0571
750,000	61	0.998	0.0760
500,000	37	814	0.1130
450,000	37	772	0.1267
400,000	37	728	0.1426
350,000	37	681	0.1626
300,000	37	630	0.1900
250,000	37	575	0.2278
0000	19	528	0.2690
000	19	470	0.339
00	19	418	0.428
0	19	373	0.538
1	19	332	0.681
2	7	292	0.856
3	7	260	1.083
4	7	232	1.367
SOLID			
0000		460	0.264
000		410	0.333
00		365	0.420
0		325	0.528
1		289	0.665
2		258	0.839
3		229	1.061
4		204	1.335

Table 2

Inductive Reactance per Single Conductor, Ohms per Mile*

Size, cir mils or AWG	No. of stds.	Outside dia., in.	60 cycles per sec — Spacing, ft: 1	2	3	4	5	6	7	8	10	12	15	20
Stranded														
1,000,000	61	1.152	0.400	0.484	0.533	0.568	0.595	0.617	0.636	0.652	0.679	0.702	0.729	0.764
750,000	61	0.998	0.417	0.501	0.550	0.585	0.612	0.634	0.653	0.669	0.696	0.719	0.746	0.781
500,000	37	0.814	0.443	0.527	0.576	0.611	0.638	0.660	0.679	0.695	0.772	0.745	0.772	0.807
400,000	19	0.725	0.458	0.542	0.591	0.626	0.653	0.675	0.694	0.710	0.737	0.760	0.787	0.822
300,000	19	0.628	0.476	0.560	0.609	0.644	0.671	0.693	0.712	0.728	0.755	0.778	0.805	0.840
250,000	19	0.574	0.487	0.571	0.620	0.655	0.682	0.704	0.723	0.739	0.766	0.789	0.816	0.851
0000	19	0.528	0.497	0.581	0.630	0.665	0.692	0.714	0.733	0.749	0.776	0.799	0.826	0.861
000	7	0.464	0.518	0.602	0.651	0.686	0.713	0.735	0.754	0.770	0.797	0.820	0.847	0.882
00	7	0.414	0.532	0.616	0.665	0.700	0.727	0.749	0.768	0.784	0.811	0.834	0.861	0.896
0	7	0.368	0.546	0.630	0.679	0.714	0.741	0.763	0.782	0.798	0.825	0.848	0.875	0.910
1	7	0.328	0.560	0.644	0.693	0.728	0.755	0.777	0.796	0.812	0.839	0.862	0.889	0.924
2	7	0.292	0.574	0.658	0.707	0.742	0.769	0.791	0.810	0.826	0.853	0.876	0.903	0.938
3	7	0.260	0.588	0.672	0.721	0.756	0.783	0.805	0.824	0.840	0.867	0.890	0.917	0.952
4	7	0.232	0.602	0.686	0.735	0.770	0.797	0.819	0.838	0.854	0.881	0.904	0.931	0.966
Solid														
0000		0.4600	0.510	0.594	0.643	0.678	0.705	0.727	0.746	0.762	0.789	0.812	0.839	0.874
000		0.4096	0.524	0.608	0.657	0.692	0.719	0.741	0.760	0.776	0.803	0.826	0.853	0.888
00		0.3648	0.538	0.622	0.671	0.706	0.733	0.755	0.774	0.790	0.817	0.840	0.867	0.902
0		0.3249	0.552	0.636	0.685	0.720	0.747	0.769	0.788	0.804	0.831	0.854	0.881	0.916
1		0.2893	0.566	0.650	0.699	0.734	0.761	0.783	0.802	0.818	0.845	0.868	0.895	0.930
2		0.2576	0.581	0.665	0.714	0.749	0.776	0.798	0.817	0.833	0.860	0.883	0.910	0.945
3		0.2294	0.595	0.679	0.728	0.763	0.790	0.812	0.831	0.847	0.874	0.897	0.924	0.959
4		0.2043	0.609	0.693	0.742	0.777	0.804	0.826	0.845	0.861	0.888	0.911	0.938	0.973
5		0.1819	0.623	0.707	0.756	0.791	0.818	0.840	0.859	0.875	0.902	0.925	0.952	0.987
6		0.1620	0.637	0.721	0.770	0.805	0.832	0.854	0.873	0.889	0.916	0.939	0.966	1.001

* From formula $x = 2\pi f\left(80 + 741.1 \log \frac{D}{r}\right) 10^{-6}$.

Solution: (i)

(a) Line loss $= 4,000 \times 0.10$

$= 400$ kw $= 400,00$ watts.

Loss per conductor $= \frac{400,000}{2} = 200,000$ watts.

Current $I = \frac{4,000,000}{33,000 \times 0.85} = 142.5$ amp.

$I^2R' = (142.5)^2R' = 200,000$ watts.

$R' = \frac{200,000}{(142.5)^2} = 9.85$ ohms.

Resistance (per mile) $= \frac{9.85}{25} = 0.394$ ohm.

From Table 1, the wire having the next lowest resistance per mile is 000 A.W.G., the resistance of which is 0.333 ohm per mile.

(b) Total resistance per conductor

$R = 25 \times 0.333 = 8.33$ ohms.

$IR = 142.5 \times 8.33 = 1{,}188$ volts.

(c) From Table 2, for 000 conductor and 48-in. spacing, the reactance per conductor is 0.692 ohm per mile.

Total reactance per conductor $X = 25 \times 0.0692$

$= 17.3$ ohms.

Reactance drop $IX = 142.5 \times 17.3 = 2{,}470$ volts.

(d) $E_s = \sqrt{(E_R \cos\theta + IR)^2 + (E_R \sin\theta + IX)^2}$

Volts to the neutral, $E_R = 16{,}500$ volts.

$\cos\theta = 0.85, \quad \theta = 31.8° \quad \sin\theta = 0.527.$

$$E_s = \left\{(16{,}500 \times 0.85 + 1{,}188)^2 + (16{,}500 \times 0.527 + 2{,}470)^2\right\}^{\frac{1}{2}}$$

$$= \sqrt{(15{,}220)^2 + (11{,}170)^2} = \sqrt{356.4 \times 10^6}$$

$$= 18{,}870 \text{ volts.}$$

$$E_s = E_R\sqrt{\left(\cos\theta + \frac{IR}{E_R}\right)^2 + \left(\sin\theta + \frac{IX}{E_R}\right)^2}$$

$$E_s = 16{,}500\sqrt{(0.85 + 0.072)^2 + (0.527 + 0.1496)^2}$$

$$= 16{,}500\sqrt{(0.922)^2 + (0.6766)^2}$$

$$= 16{,}500 \times 1.144 = 18{,}870 \text{ volts.}$$

$$\vec{E}_s = E_R + I(\cos\theta - j\sin\theta)(R + jX)$$

$$\vec{E}_s = 16{,}500 + 142.5(0.85 - j0.527)(8.33 + j17.3)$$

$$= 16{,}500 + 1{,}001 + j2{,}095 - j626 + 1{,}299$$

$$= 18{,}800 + j1{,}470.$$

$$|\vec{E}_s| = \sqrt{(18{,}800)^2 + (1{,}470)^2}$$

$= 18{,}870$ volts (check).

Voltage at the sending end $= 2 \times 18{,}870 = 37{,}740$ volts.

(e) The line regulation is defined as the rise in voltage when full load is thrown off the line, divided by the load voltage,

$$\text{Regulation} = \frac{37{,}740 - 33{,}000}{33{,}000}, \text{ or } 14.4\%.$$

(ii)

(a) Power per phase $= \frac{4{,}000}{3} = 1{,}333$ kw.

Load voltage to neutral $E_R = \frac{33{,}000}{\sqrt{3}} = 19{,}070$ volts.

$$\text{Current per conductor } I = \frac{1{,}333{,}000}{19{,}070 \times 0.85}$$

$$= 82.3 \text{ amp.}$$

Allowable loss per conductor $= 1{,}333 \times 0.10$

$= 133.3$ kw $= 133{,}300$ watts.

Resistance per conductor $R' = 133{,}300/(82.3)^2$

$= 19.68$ ohms.

Resistance per mile $= \frac{19.68}{25} = 0.787$ ohm.

From Table 1, the wire having the next lowest resistance per mile is No. 1 A.W.G., the resistance of which is 0.665 ohm per mile.

(b) Total resistance per conductor

$R = 25 \times 0.665 = 16.6$ ohms.

$IR = 82.3 \times 16.6 = 1{,}365$ volts.

(c) From Table 2, for No. 1 A.W.G. wire and 48-in. spacing, the reactance is 0.734 ohm per mile.

Total reactance per conductor

$X = 25 \times 0.734 = 18.35$ ohms.

Reactance drop

$IX = 82.3 \times 18.35 = 1{,}510$ volts.

(d) $E_s = \sqrt{(E_R \cos\theta + IR)^2 + (E_R \sin\theta + IX)^2}$

volts to the neutral, $E_R = 19{,}070$ volts.

$\cos\theta = 0.85, \quad \theta = 31.8°, \quad \sin\theta = 0.527.$

$$E_s = \left\{(19{,}070 \times 0.85 + 1{,}365)^2 + (19{,}070 \times 0.527 + 1{,}510)^2\right\}^{\frac{1}{2}}$$

$$= \sqrt{(17{,}580)^2 + (11{,}560)^2} = \sqrt{443 \times 10^6}$$

$$= 21{,}000 \text{ volts.}$$

$$\vec{E}_s = E_R + I(\cos\theta - j\sin\theta)(R + jX)$$

$$\vec{E}_s = 19{,}070 + 82.3(0.85 - j0.527)(16.6 + j18.35)$$

$$= 21{,}000 + j564.$$

$$|\vec{E}_s| = \sqrt{(21{,}000)^2 + (564)^2} = 21{,}000 \text{ volts}$$

The voltage between conductors at the sending end

$$E'_s = \sqrt{3} \times 21{,}000 = 36{,}400 \text{ volts.}$$

(e) $\text{Regulation} = \dfrac{21{,}000 - 19{,}070}{19{,}070} = \dfrac{1{,}930}{19{,}070} = 0.101,$

or 10.1 per cent.

(f) $I = \dfrac{1{,}333{,}000}{19{,}070 \times 0.70} = 99.6$ amp.

$$\vec{E}_s = E_R + I(\cos\theta + j\sin\theta)(R + jX).$$

$$\vec{E}_s = 19{,}070 + 99.6(0.70 + j0.715)(16.6 + j18.35)$$

$$= 18{,}920 + j2{,}460 \text{ volts.}$$

$$|\vec{E}_s| = \sqrt{(18{,}920)^2 + (2{,}460)^2} = 19{,}070 \text{ volts.}$$

With a leading current at 0.70 power factor, the sending-end and receiving-end voltages are equal. With more line reactance or with a lower power factor, it is possible for the receiving-end voltage to be greater even than the sending-end voltage.

• PROBLEM 12-6

It is required to deliver 40,000 kw, 3-phase, at 0.85 power factor, lagging current, at a distance of 140 miles, with a line loss not exceeding 10 percent of the power delivered. The voltage at the load is 132,000 volts, 60 cycles, and the conductors are arranged at the apexes of an equilateral triangle, 12 ft on a side. Determine (a) voltage between conductors at sending end; (b) line regulation; (c) total power supplied to line; (d) efficiency of transmission.

Table 1

Resistance of Copper Wire, Ohms per Mile, 25°C (77°F)

Size, cir mils or AWG	Number of wires	Outside diameter, mils	Ohms per mile
		STRANDED	
1,000,000	61	1.152	0.0571
750,000	61	0.998	0.0760
500,000	37	814	0.1130
450,000	37	772	0.1267
400,000	37	728	0.1426
350,000	37	681	0.1626
300,000	37	630	0.1900
250,000	37	575	0.2278
0000	19	528	0.2690
000	19	470	0.339
00	19	418	0.428
0	19	373	0.538
1	19	332	0.681
2	7	292	0.856
3	7	260	1.083
4	7	232	1.367
		SOLID	
0000		460	0.264
000		410	0.333
00		365	0.420
0		325	0.528
1		289	0.665
2		258	0.839
3		229	1.061
4		204	1.335

Solution: Power per phase

$$P = \frac{40,000}{3} = 13,330 \text{ kw.}$$

Volts to neutral at load

$$E_R = \frac{132,000}{\sqrt{3}} = 76,200 \text{ volts.}$$

Current per conductor at load

$$I_R = \frac{13,330,000}{76,200 \times 0.85} = 206 \text{ amp.}$$

Power loss per conductor = 13,330 × 0.10 = 1,333 kw = 1,333,000 watts.

$$\text{Conductor resistance } R' = \frac{1,333,000}{(206)^2} = 31.4 \text{ ohms.}$$

$$\text{Resistance per mile} = \frac{31.4}{140} = 0.224 \text{ ohm.}$$

From Table 1, the wire having the nearest resistance per mile is 250,000 cir mils, with resistance per mile of 0.2278 ohm.

Conductor resistance R = 140 × 0.2278 = 31.9 ohms.

Table 2

Inductive Reactance per Single Conductor, Ohms per Mile*

Size, cir mils or AWG	No. of stds.	Outside dia., in.	60 cycles per sec, Spacing, ft: 1	2	3	4	5	6	7	8	10	12	15	20
Stranded														
1,000,000	61	1.152	0.400	0.484	0.533	0.568	0.595	0.617	0.636	0.652	0.679	0.702	0.729	0.764
750,000	61	0.998	0.417	0.501	0.550	0.585	0.612	0.634	0.653	0.669	0.696	0.719	0.746	0.781
500,000	37	0.814	0.443	0.527	0.576	0.611	0.638	0.660	0.679	0.695	0.772	0.745	0.772	0.807
400,000	19	0.725	0.458	0.542	0.591	0.626	0.653	0.675	0.694	0.710	0.737	0.760	0.787	0.822
300,000	19	0.628	0.476	0.560	0.609	0.644	0.671	0.693	0.712	0.728	0.755	0.778	0.805	0.840
250,000	19	0.574	0.487	0.571	0.620	0.655	0.682	0.704	0.723	0.739	0.766	0.789	0.816	0.851
0000	19	0.528	0.497	0.581	0.630	0.665	0.692	0.714	0.733	0.749	0.776	0.799	0.826	0.861
000	7	0.464	0.518	0.602	0.651	0.686	0.713	0.735	0.754	0.770	0.797	0.820	0.847	0.882
00	7	0.414	0.532	0.616	0.665	0.700	0.727	0.749	0.768	0.784	0.811	0.834	0.861	0.896
0	7	0.368	0.546	0.630	0.679	0.714	0.741	0.763	0.782	0.798	0.825	0.848	0.875	0.910
1	7	0.328	0.560	0.644	0.693	0.728	0.755	0.777	0.796	0.812	0.839	0.862	0.889	0.924
2	7	0.292	0.574	0.658	0.707	0.742	0.769	0.791	0.810	0.826	0.853	0.876	0.903	0.938
3	7	0.260	0.588	0.672	0.721	0.756	0.783	0.805	0.824	0.840	0.867	0.890	0.917	0.952
4	7	0.232	0.602	0.686	0.735	0.770	0.797	0.819	0.838	0.854	0.881	0.904	0.931	0.966
Solid														
0000		0.4600	0.510	0.594	0.643	0.678	0.705	0.727	0.746	0.762	0.789	0.812	0.839	0.874
000		0.4096	0.524	0.608	0.657	0.692	0.719	0.741	0.760	0.776	0.803	0.826	0.853	0.888
00		0.3648	0.538	0.622	0.671	0.706	0.733	0.755	0.774	0.790	0.817	0.840	0.867	0.902
0		0.3249	0.552	0.636	0.685	0.720	0.747	0.769	0.788	0.804	0.831	0.854	0.881	0.916
1		0.2893	0.566	0.650	0.699	0.734	0.761	0.783	0.802	0.818	0.845	0.868	0.895	0.930
2		0.2576	0.581	0.665	0.714	0.749	0.776	0.798	0.817	0.833	0.860	0.883	0.910	0.945
3		0.2294	0.595	0.679	0.728	0.763	0.790	0.812	0.831	0.847	0.874	0.897	0.924	0.959
4		0.2043	0.609	0.693	0.742	0.777	0.804	0.826	0.845	0.861	0.888	0.911	0.938	0.973
5		0.1819	0.623	0.707	0.756	0.791	0.818	0.840	0.859	0.875	0.902	0.925	0.952	0.987
6		0.1620	0.637	0.721	0.770	0.805	0.832	0.854	0.873	0.889	0.916	0.939	0.966	1.001

* From formula $x = 2\pi f \left(80 + 741.1 \log \frac{D}{r}\right) 10^{-6}$.

From Table 2, the reactance per conductor per mile for 250,000-cir-mil wire and 12-ft spacing is 0.789 ohm.

Total reactance = 140 × 0.789 = 110.5 ohms.

Table 3

Charging Current per Single Wire, Amperes per Mile per 100,000 Volts from Phase Wire to Neutral*

STRANDED

Size, cir mils or AWG	No. of stds.	Out-side dia., in.	60 cycles per sec											
			Spacing, ft											
			1	2	3	4	5	6	7	8	10	12	15	20
1,000,000	61	1.152	1.110	0.903	0.815	0.762	0.726	0.698	0.677	0.659	0.631	0.611	0.587	0.559
750,000	61	0.998	1.059	0.870	0.787	0.738	0.704	0.678	0.658	0.641	0.615	0.595	0.572	0.546
500,000	37	0.814	0.996	0.826	0.752	0.707	0.679	0.651	0.633	0.617	0.593	0.574	0.553	0.528
400,000	19	0.725	0.963	0.804	0.733	0.690	0.660	0.637	0.619	0.604	0.581	0.563	0.543	0.519
300,000	19	0.628	0.925	0.777	0.711	0.670	0.642	0.620	0.603	0.589	0.567	0.550	0.531	0.508
250,000	19	0.574	0.903	0.762	0.697	0.658	0.631	0.610	0.594	0.580	0.558	0.542	0.523	0.501
0000	19	0.528	0.883	0.747	0.685	0.648	0.621	0.601	0.585	0.571	0.551	0.535	0.517	0.495
000	7	0.464	0.854	0.726	0.668	0.632	0.607	0.588	0.572	0.559	0.539	0.524	0.507	0.485
00	7	0.414	0.830	0.709	0.653	0.619	0.595	0.576	0.561	0.549	0.530	0.515	0.498	0.478
0	7	0.368	0.807	0.692	0.639	0.606	0.582	0.565	0.550	0.539	0.520	0.506	0.490	0.470
1	7	0.328	0.785	0.676	0.625	0.594	0.571	0.554	0.540	0.529	0.511	0.497	0.483	0.463
2	7	0.292	0.765	0.661	0.612	0.582	0.560	0.544	0.531	0.520	0.502	0.489	0.473	0.455
3	7	0.260	0.745	0.646	0.599	0.570	0.550	0.534	0.521	0.510	0.494	0.481	0.466	0.448
4	7	0.232	0.727	0.632	0.588	0.559	0.539	0.524	0.512	0.502	0.485	0.473	0.459	0.442
SOLID														
0000	..	0.4600	0.853	0.725	0.667	0.631	0.606	0.587	0.571	0.559	0.539	0.524	0.506	0.485
000	..	0.4096	0.828	0.707	0.652	0.618	0.593	0.575	0.560	0.548	0.529	0.514	0.497	0.477
00	..	0.3648	0.805	0.691	0.638	0.605	0.581	0.564	0.550	0.538	0.520	0.505	0.489	0.469
0	..	0.3249	0.784	0.675	0.624	0.593	0.570	0.553	0.540	0.528	0.511	0.497	0.483	0.462
1	..	0.2893	0.763	0.659	0.611	0.580	0.559	0.543	0.530	0.519	0.502	0.489	0.473	0.455
2	..	0.2576	0.744	0.645	0.598	0.570	0.549	0.533	0.520	0.510	0.493	0.480	0.465	0.447
3	..	0.2294	0.725	0.631	0.586	0.558	0.538	0.523	0.511	0.501	0.485	0.473	0.458	0.441
4	..	0.2043	0.707	0.617	0.575	0.548	0.529	0.514	0.502	0.492	0.477	0.465	0.451	0.435
5	..	0.1819	0.690	0.604	0.563	0.538	0.519	0.505	0.495	0.484	0.469	0.458	0.444	0.428
6	..	0.1620	0.674	0.592	0.553	0.528	0.510	0.496	0.485	0.476	0.462	0.431	0.437	0.422

* From formula $I = \frac{2\pi f \cdot 38.83 \cdot 10^{-9}}{\log_{10}(D/t)} E$.

The charging current at 60 cycles for 250,000-cir-mil wire with 12 ft spacing and 100,000 volts to neutral is, from Table 3, 0.542 amp per mile.

The total charging current for the line is

$$I_c = 0.542 \times \frac{76,200}{100,000} \times 140 = 57.8 \text{ amp.}$$

As only one-half the line capacitance is assumed at the receiving end, the charging current flowing over the line, $I_c/2 = 57.8/2 = 28.9$ amp.

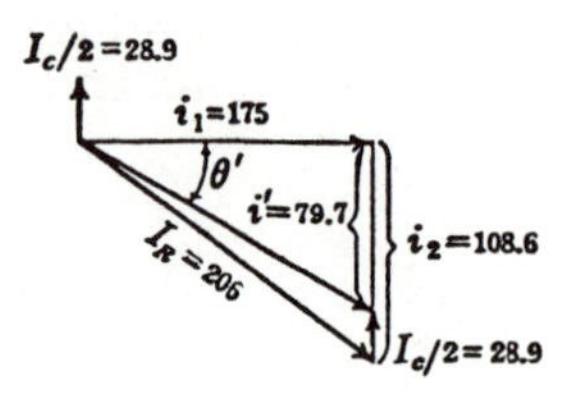

Effect of line-charging current
on total line current.

In order to find the total line current, however, this 28.9 amp must be added vectorially to the 206 amp of load current. The load current, therefore, (see figure), is resolved into an energy component

$$I_R \cos\theta = i_1 = 206 \times 0.85 = 175.0 \text{ amp}$$

and a quadrature component

$$I_R \sin\theta = i_2 = 206 \times 0.527 = 108.6 \text{ amp.}$$

As the quadrature component, 108.6 amp, lags the load voltage by 90°, and the charging current, 28.9 amp, leads the load voltage by 90°, the resulting quadrature component is

$$i' = 108.6 - 28.9 = 79.7 \text{ amp.}$$

Total line current

$$I = \sqrt{(175)^2 + (79.7)^2} = 192.3 \text{ amp.}$$

Let θ' be the angle between this current and the receiving-end voltage.

$$\cos\theta' = \frac{175}{192.3} = 0.910, \qquad \theta' = 24.5°,$$

$$\sin\theta' = \frac{79.7}{192.3} = 0.414.$$

(a) Voltage to neutral at sending end

$$E_S = \sqrt{(76{,}200\times0.910+192.3\times31.9)^2 + (76{,}200\times0.414+192.3\times110.5)^2}$$

$$= \sqrt{(69{,}340 + 6{,}130)^2 + (31{,}550 + 21{,}250)^2}$$

$$= \sqrt{(5{,}700 + 2{,}790)\cdot 10^6} = 92{,}140 \text{ volts.}$$

Voltage between conductors

$$E = \sqrt{3} \times 92{,}140 = 159{,}600 \text{ volts.}$$

(b) Line regulation $= \dfrac{92{,}140 - 76{,}200}{76{,}200} = \dfrac{15{,}940}{76{,}200}$

or 21.0%.

(c) Line loss

$$P_c = 3 \times (192.3)^2 \times 31.9 = 3{,}540{,}000 \text{ watts.}$$

Total sending-end power

$$P_S = 40{,}000 + 3{,}540 = 43{,}540 \text{ kw.}$$

(d) Efficiency

$$\eta = \frac{40{,}000}{43{,}540} \quad \text{or} \quad 91.8\%.$$

• PROBLEM 12-7

Three-phase line; voltage 220,000; 60 cylces; length of line 230 miles; spacing 27 ft; conductors 636,000 cir-mil aluminum cable, steel-reinforced (ACSR); diameter = 0.977 in. Load is 75,000 kw at 0.85 power factor, lag. Determine (a) resistance; (b) inductance; (c) impedance; (d) admittance, neglecting shunt conductance; (e) A; (f) B; (g) C; (h) receiving-end current; (i) sending-end voltage; (j) sending-end current; (k) sending-end power; (ℓ) efficiency.

Table 1

Properties of Aluminum Cable Steel-reinforced (ACSR)

Aluminum Company of America

Cir mils or AWG		No. of wires		Out-side dia., in.	Cross section, sq in.		Total lb per mile	Ohms per mile of single conductor at 25°C				
								0 amp d.c.	200 amp		600 amp	
Alum.	Copper equiv.	Al.	St.		Al.	Total			25 cycles	60 cycles	25 cycles	60 cycles
1,590,000	1,000,000	54	19	1.545	1.249	1.4071	10,735	0.0587	0.0589	0.0594	0.0592	0.0607
1,431,000	900,000	54	19	1.465	1.124	1.2664	9,662	0.0652	0.0654	0.0659	0.0657	0.0671
1,272,000	800,000	54	19	1.382	0.9990	1.1256	8,588	0.0734	0.0736	0.0742	0.0738	0.0752
1,192,500	750,000	54	19	1.338	0.9366	1.0553	8,055	0.0783	0.0785	0.0791	0.0787	0.0801
1,113,000	700,000	54	19	1.293	0.8741	0.9850	7,517	0.0839	0.0841	0.0848	0.0843	0.0857
1,033,500	650,000	54	7	1.246	0.8117	0.9170	7,022	0.0903	0.0906	0.0913	0.0908	0.0922
954,000	600,000	54	7	1.196	0.7493	0.8464	6,481	0.0979	0.0980	0.0985	0.0983	0.0997
874,500	550,000	54	7	1.146	0.6868	0.7759	5,942	0.107	0.107	0.108	0.107	0.109
795,000	500,000	26	7	1.108	0.6244	0.7261	5,776	0.117	0.117	0.117	0.117	0.117
715,500	450,000	54	7	1.036	0.5620	0.6348	4,860	0.131	0.131	0.133	0.131	0.133
636,000	400,000	54	7	0.977	0.4995	0.5642	4,321	0.147	0.147	0.149	0.147	0.149
556,500	350,000	26	7	0.927	0.4371	0.5083	4,044	0.168	0.168	0.168	0.168	0.168
477,000	300,000	26	7	0.858	0.3746	0.4357	3,467	0.196	0.196	0.196	0.196	0.196
397,500	250,000	26	7	0.783	0.3122	0.3630	2,887	0.235	0.235	0.235	0.235	0.235
336,400	0000	26	7	0.721	0.2642	0.3073	2,445	0.278	0.278	0.278	0.278	0.278
266,000	000	26	7	0.633	0.2095	0.2367	1,813	0.350	0.350	0.350	0.350	0.350
0000	00	6	1	0.563	0.1662	0.1939	1,549	0.441	0.443	0.446	0.447	0.464
000	0	6	1	0.502	0.1318	0.1537	1,227	0.556	0.557	0.561	0.562	0.579
00	1	6	1	0.447	0.1045	0.1219	974	0.702	0.703	0.707	0.706	0.718
0	2	6	1	0.398	0.0829	0.0967	773	0.885	0.885	0.889	0.887	0.893
1	3	6	1	0.355	0.0657	0.0767	614	1.12	1.12	1.12	1.12	1.12
2	4	6	1	0.316	0.0521	0.0608	486	1.41	1.41	1.41	1.41	1.41
3	5	6	1	0.281	0.0413	0.0482	386	1.78	1.78	1.78	1.78	1.78
4	6	6	1	0.250	0.0328	0.0383	306	2.24	2.24	2.24	2.24	2.24

Solution: (a) From Table 1, resistance per mile at 25° C, 200 amp, 60 cycles, is 0.149 ohm.

$$R = 230 \times 0.149 = 34.3 \text{ ohms.}$$

(b) The inductance per wire is

$$L = 2\ell(0.080 + 0.741 \log_{10}\frac{D}{r})\ 10^{-3}$$

$$= 230\ (0.080 + 0.741 \log_{10}\frac{27 \cdot 12}{0.4885})\ 10^{-3}$$

$$= 0.4995 \text{ or } 0.500 \text{ henry.}$$

(c) $\vec{\underset{\bullet}{Z}} = 34.3 + j(377 \times 0.500) = 34.3 + j188.5$ ohms.

(d) $C = \dfrac{0.0388}{\log_{10}(D/r)}$ μf per mile to neutral.

$$= 230\ \frac{0.0388}{\log_{10}\dfrac{27 \cdot 12}{0.4885}} = 3.16\ \mu f,$$

$$\vec{\underset{\bullet}{Y}} = +j3.16 \times 10^{-6} \cdot 377 = +j1.191 \times 10^{-3} \text{ mho.}$$

(e) $$\vec{\underset{\bullet}{Z}}\vec{\underset{\bullet}{Y}} = (34.3 + j188.5)(+j1.191 \cdot 10^{-3})$$

$$= -0.225 + j0.0409. \tag{1}$$

$$(\vec{\underset{\bullet}{Z}}\vec{\underset{\bullet}{Y}})^2 = (-0.225 + j0.0409)^2 = 0.0490 - j0.018. \tag{2}$$

$$\vec{\underset{\bullet}{A}} = (1 + \frac{\underset{\bullet}{Z}\underset{\bullet}{Y}}{1 \times 2} + \frac{\underset{\bullet}{Z}^2\underset{\bullet}{Y}^2}{1 \times 2 \times 3 \times 4} + \ldots .)$$

$$= (1 + \frac{-0.225 + j0.0409}{2} + \frac{0.0490 - j0.018}{24} + \ldots)$$

$$= 0.889 + j0.0197.$$

(f) Using (c), (1), (2),

$$\vec{\underset{\bullet}{B}} = \underset{\bullet}{Z}(1 + \frac{\underset{\bullet}{Z}\underset{\bullet}{Y}}{2 \times 3} + \frac{\underset{\bullet}{Z}^2\underset{\bullet}{Y}^2}{2 \times 3 \times 4 \times 5} + \ldots .)$$

$$= (34.3 + j188.5)\Big(1 + \frac{-0.225 + j0.0409}{6}$$

$$+ \frac{0.0490 - j0.018}{120} + \ldots\Big)$$

$$= (34.3 + j188.5)(0.963 + j0.0067)$$

$$= 31.8 + j181.7 \text{ ohms.}$$

(g) Using (d), (1), (2),

$$\vec{\underset{\bullet}{C}} = \underset{\bullet}{Y}(1 + \frac{\underset{\bullet}{Z}\underset{\bullet}{Y}}{2 \times 3} + \frac{\underset{\bullet}{Z}^2\underset{\bullet}{Y}^2}{2 \times 3 \times 4 \times 5} + \ldots .)$$

$$= +j1.191 \times 10^{-3}\left(1 + \frac{-0.225 + j0.0409}{6} + \frac{0.0490 - j0.018}{120} + \ldots\right)$$

$$= +j1.191 \times 10^{-3}(0.963 + j0.0067)$$

$$= (-0.0080 + j1.147)10^{-3}.$$

(h) $$|I_R| = \frac{75{,}000{,}000}{\sqrt{3} \cdot 220{,}000 \cdot 0.85} = 231.5 \text{ amp.}$$

$$I_R = 231.5(0.850 - j0.527)$$

$$= 196.8 - j122.0 \text{ amp,}$$

where 0.850 = cos 31.8° and 0.527 = sin 31.8°.

(i) Volts to neutral = $220{,}000/\sqrt{3}$ = 127,000 volts.

$$\vec{E}_S = \vec{A}\,\vec{E}_R + \vec{B}\,\vec{I}_R$$

$$= (0.889 + j0.0197)(127{,}000) + (31.8 + j181.7)(196.8 - j122.0)$$

$$= 112{,}900 + j2{,}500 + 6{,}260 + j35{,}800 - j3{,}880 + 22{,}200$$

$$= 141{,}400 + j34{,}400 \text{ volts}$$

$$= \sqrt{(141{,}000)^2 + (34{,}400)^2}\ \underline{/13.7°}$$

$$= 145{,}600\ \underline{/13.7°} \text{ volts.}$$

Voltage (absolute) between conductors = $145{,}600\sqrt{3}$ = 252,000 volts.

(j) $$\vec{I}_S = \vec{A}\,\vec{I}_R + \vec{C}\,\vec{E}_R$$

$$= (0.889 + j0.0197)(196.8 - j122.0) + (-0.0080 + j1.147)10^{-3}(127{,}000)$$

$$= 175.0 - j108.5 + j3.9 + 2.4 - 1.0 + j145.7$$

$$= 176.4 + j41.1 \text{ amp}$$

$$I_S = \sqrt{(176.4)^2 + (41.1)^2}\ \underline{/13.1°} = 181.2\ \underline{/13.1°} \text{ amp.}$$

Note that, although the receiving-end current lags its voltage, the sending-end current, due to the line capacitance, is practically in phase with the sending-end voltage.

(k) $P_S = \dot{E}_S \times \dot{I}_S$

$= (141{,}400 \times 176.4) + (34{,}400 \times 41.1)$

$= 24{,}940{,}000 + 1{,}414{,}000$ watts

$= 26{,}350$ kw.

(ℓ) $\eta = \dfrac{25{,}000}{26{,}350} = 0.949$, or 94.9%.

Lines operating at 220 kv almost never operate with considerable voltage difference between receiving and sending ends. The lines almost always operate with receiving- and sending-end voltages substantially equal. This condition is realized by operating synchronous condensers at the receiving end, overexciting them if the receiving-end voltage is low and underexciting them if the receiving-end voltage is high. In this example, the receiving-end, voltage is almost equal to the sending-end voltage if the load power factor is unity.

Thus,

$\dot{I}_R = 196.8 + j0$ amp.

$\dot{E}_S = (0.889 + j0.0197)(127.000) +$

$(31.8 + j181.7)(196.8 + j0)$

$= 112{,}900 + j2{,}500 + 6{,}260 + j35{,}800$

$= 119{,}200 + j38{,}300$ volts

$= \sqrt{(119{,}200)^2 + (38{,}300)^2}\ \underline{/17.8^\circ}$

$= 125{,}200\ \underline{/17.8^\circ}$ volts.

$\dot{I}_S = (0.889 + j0.0197)(196.8) +$

$(-0.0080 + j1.147)10^{-3}(127{,}000)$

$= 175.0 + j3.9 - 1.0 + j145.7$

$= 174.0 + j149.6$ amp

$= \sqrt{(174.0)^2 + (149.6)^2}\ \underline{/40.7^\circ}$

$= 229.4\ \underline{/40.7^\circ}$ amp.

$P_S = (119{,}200 \times 174.0) + (38{,}300 \times 149.6)$

$= 20{,}740 + 5{,}730 = 26{,}470$ kw.

$\eta = \dfrac{25{,}000}{26{,}470} = 0.944$, or 94.4%.

Owing to the line capacitance the sending-end current now leads its voltage by an angle of 22.9°. The current is now greater in magnitude so that the line loss is increased and the efficiency reduced.

DISTRIBUTION CIRCUITS, THREE WIRE SYSTEM

• PROBLEM 12-8

It is desired to deliver 50 kva, 0.80 power factor, lagging current, at 230 volts, single-phase, 60 cycles, from a transformer secondary over a distance of 600 ft, the loss not to exceed 12 percent of the delivered power. The wires are spaced 18 in. between centers. Determine (a) size of conductor; (b) voltage difference between sending and receiving ends; (c) efficiency of transmission.

Table 1

Resistance of Copper Wire, Ohms per Mile, 25°C (77°F)

Size, cir mils or AWG	Number of wires	Outside diameter, mils	Ohms per mile
STRANDED			
1,000,000	61	1.152	0.0571
750,000	61	0.998	0.0760
500,000	37	814	0.1130
450,000	37	772	0.1267
400,000	37	728	0.1426
350,000	37	681	0.1626
300,000	37	630	0.1900
250,000	37	575	0.2278
0000	19	528	0.2690
000	19	470	0.339
00	19	418	0.428
0	19	373	0.538
1	19	332	0.681
2	7	292	0.856
3	7	260	1.083
4	7	232	1.367
SOLID			
0000		460	0.264
000		410	0.333
00		365	0.420
0		325	0.528
1		289	0.665
2		258	0.839
3		229	1.061
4		204	1.335

Table 2

Inductive Reactance per Single Conductor, Ohms per Mile*

Size, cir mils or AWG	No. of stds.	Outside dia., in.	60 cycles per sec — Spacing, ft: 1	2	3	4	5	6	7	8	10	12	15	20
Stranded														
1,000,000	61	1.152	0.400	0.484	0.533	0.568	0.595	0.617	0.636	0.652	0.679	0.702	0.729	0.764
750,000	61	0.998	0.417	0.501	0.550	0.585	0.612	0.634	0.653	0.669	0.696	0.719	0.746	0.781
500,000	37	0.814	0.443	0.527	0.576	0.611	0.638	0.660	0.679	0.695	0.772	0.745	0.772	0.807
400,000	19	0.725	0.458	0.542	0.591	0.626	0.653	0.675	0.694	0.710	0.737	0.760	0.787	0.822
300,000	19	0.628	0.476	0.560	0.609	0.644	0.671	0.693	0.712	0.728	0.755	0.778	0.805	0.840
250,000	19	0.574	0.487	0.571	0.620	0.655	0.682	0.704	0.723	0.739	0.766	0.789	0.816	0.851
0000	19	0.528	0.497	0.581	0.630	0.665	0.692	0.714	0.733	0.749	0.776	0.799	0.826	0.861
000	7	0.464	0.518	0.602	0.651	0.686	0.713	0.735	0.754	0.770	0.797	0.820	0.847	0.882
00	7	0.414	0.532	0.616	0.665	0.700	0.727	0.749	0.768	0.784	0.811	0.834	0.861	0.896
0	7	0.368	0.546	0.630	0.679	0.714	0.741	0.763	0.782	0.798	0.825	0.848	0.875	0.910
1	7	0.328	0.560	0.644	0.693	0.728	0.755	0.777	0.796	0.812	0.839	0.862	0.889	0.924
2	7	0.292	0.574	0.658	0.707	0.742	0.769	0.791	0.810	0.826	0.853	0.876	0.903	0.938
3	7	0.260	0.588	0.672	0.721	0.756	0.783	0.805	0.824	0.840	0.867	0.890	0.917	0.952
4	7	0.232	0.602	0.686	0.735	0.770	0.797	0.819	0.838	0.854	0.881	0.904	0.931	0.966
Solid														
0000		0.4600	0.510	0.594	0.643	0.678	0.705	0.727	0.746	0.762	0.789	0.812	0.839	0.874
000		0.4096	0.524	0.608	0.657	0.692	0.719	0.741	0.760	0.776	0.803	0.826	0.853	0.888
00		0.3648	0.538	0.622	0.671	0.706	0.733	0.755	0.774	0.790	0.817	0.840	0.867	0.902
0		0.3249	0.552	0.636	0.685	0.720	0.747	0.769	0.788	0.804	0.831	0.854	0.881	0.916
1		0.2893	0.566	0.650	0.699	0.734	0.761	0.783	0.802	0.818	0.845	0.868	0.895	0.930
2		0.2576	0.581	0.665	0.714	0.749	0.776	0.798	0.817	0.833	0.860	0.883	0.910	0.945
3		0.2294	0.595	0.679	0.728	0.763	0.790	0.812	0.831	0.847	0.874	0.897	0.924	0.959
4		0.2043	0.609	0.693	0.742	0.777	0.804	0.826	0.845	0.861	0.888	0.911	0.938	0.973
5		0.1819	0.623	0.707	0.756	0.791	0.818	0.840	0.859	0.875	0.902	0.925	0.952	0.987
6		0.1620	0.637	0.721	0.770	0.805	0.832	0.854	0.873	0.889	0.916	0.939	0.966	1.001

* From formula $x = 2\pi f \left(80 + 741.1 \log \frac{D}{r}\right) 10^{-6}$.

Solution: (a) Current $= \frac{50,000}{230} = 217$ amp.

$$(217)^2 R' = 0.12 \times 40,000 = 4,800 \text{ watts.}$$

$$R' = \frac{4,800}{(217)^2} = 0.102 \text{ ohm.}$$

$$\text{Resistance per mile} = \frac{5,280}{1,200}\, 0.102 = 0.449 \text{ ohm.}$$

From Table 1, stranded 00 wire gives the lowest value of resistance. The resistance = 0.428 ohm.

(b) $R = 0.428 \frac{600}{5,280} = 0.0486$ ohm per wire.

Resistance drop

$$IR = 217 \times 0.0486 = 10.55 \text{ volts per wire.}$$

From Table 1, the diameter of 00 stranded wire is 0.418 in.

$$X = 2\pi f \quad 80 + 741 \log_{10}\frac{D}{r}\ 10^{-6} \text{ ohms per mile}$$

$$= 2\pi(60)\left(80 + 741 \log_{10}\frac{18}{0.209}\right) 10^{-6}\left(\frac{600}{5,280}\right)$$

$$= 0.572 \quad \left(\frac{600}{5,280}\right) \quad = 0.065 \text{ ohm.}$$

(Table 2 may also be used.)

$$IX = 217 \times 0.065 = 14.1 \text{ volts.}$$

Volts to neutral = 115; $\cos\theta = 0.80$, $\sin\theta = 0.60$.

$$E_S = \sqrt{(E_R \cos\theta + IR)^2 + (E_R \sin\theta + IX)^2}$$

$$= \sqrt{(115 \times 0.80 + 10.55)^2 + (115 \times 0.60 + 14.1)^2}$$

$$= 132 \text{ volts,}$$

$$2(132 - 115) = 34 \text{ volts.}$$

(c) $I^2R = (217)^2 \cdot 0.0972 = 4,580$ watts,

$$\text{Efficiency} = \frac{50,000 \times 0.80}{50,000 \times 0.80 + 4,580}, \text{ or } 89.8\%.$$

It is interesting to compare the effect on the voltage difference of using 000, the next larger size of wire.

Resistance of 600 ft = 0.0385 ohm.

$$IR = 217 \times 0.0385 = 8.36 \text{ volts.}$$

Reactance of 600 ft = 0.0633 ohm.

$$IX = 217 \times 0.0633 = 13.75 \text{ volts.}$$

$$E_S = \sqrt{(115 \times 0.80 + 8.36)^2 + (115 \times 0.60 + 13.75)^2}$$

$$= 130.1 \text{ volts.}$$

$$2(130.1 - 115) = 260.2 - 230 = 30.2 \text{ volts.}$$

That is, an increase of 26 percent in the cross section of the conductor has almost negligible effect on the value of E_S and small effect on the voltage difference.

• PROBLEM 12-9

A three-wire circuit supplies 500 lamps each taking 1/2 ampere; 275 lamps being on one side of the neutral and 225 on the other side, all at a distance of 1,000 feet from a 231-volt three-wire generator. The voltage between the neutral and either outside main is maintained at 115.5 volts at the generator. The outside wires are of No. 0000 A.w.g. copper wire and the neutral consists of No. 0 wire. Find the voltage across each set of lamps.

<u>Solution</u>: $I_1 = 275 \times 1/2 = 137.5$ amperes.

$I_2 = 225 \times 1/2 = 112.5$ amperes.

$I_0 = 137.5 - 112.5 = 25$ amperes.

$R_1 = 0.05$ ohm at 25° C.

$R_0 = 0.1$ ohm at 25° C.

$R_2 = R_1 = 0.05$ ohm.

Since $I_1 = I_2 + I_0$, one obtains

$$\begin{aligned} I_1 R_A &= E_A - I_1 R_1 - I_0 R_0 \\ &= 115.5 - 137.5 \times 0.05 - 25 \times 0.1 \\ &= 115.5 - 6.875 - 2.5 \\ &= 115.5 - 9.375 \\ &= 106.125 \text{ volts (i.e. voltage across 275 lamps)} \end{aligned}$$

and

$$\begin{aligned} I_2 R_B &= E_B - I_2 R_2 + I_0 R_0 \\ &= 115.5 - 112.5 \times 0.05 + 25 \times 0.1 \\ &= 115.5 - 5.625 + 2.5 \\ &= 115.5 - 3.125 \\ &= 112.375 \text{ volts (voltage across 225 lamps.)} \end{aligned}$$

• PROBLEM 12-10

A three-wire system supplies the load shown in Fig. 1. If the resistance of each lamp is 110 ohms and the motor takes a current of 25 amperes, calculate the voltage across each group of lamps:

(a) When the motor is disconnected.

(b) When the motor is operating.

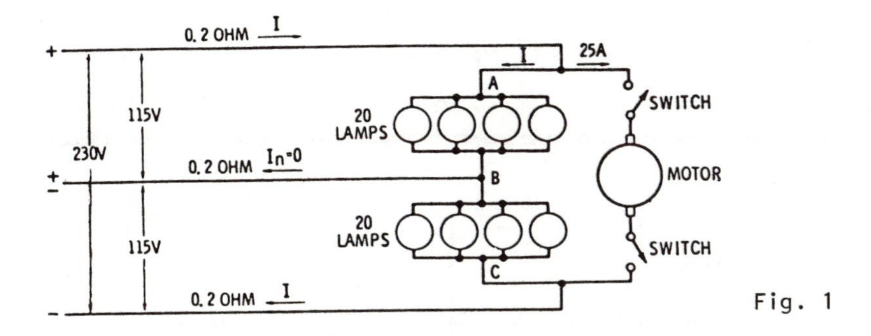

A three-wire system supplying a lamp and motor load.

Solution: The combined resistance of a group of 20 lamps, each having a resistance of 110 ohms, is

$$R = \frac{110}{20} = 5.5 \text{ ohms.}$$

Since the load is a balanced one, it is evident that the current in the neutral is zero.

The current through the lamps with the motor disconnected is

$$I = \frac{V}{(5.5 + 5.5 + 0.2 + 0.2)} = \frac{230}{11.4} = 20.2 \text{ amperes.}$$

(a) Voltage $E_{AB} = E_{BC} = \frac{20.2 \times 11}{2} = 111.1$ volts

The current through the lamps with the motor operating and drawing 25 amperes can be obtained if Kirchhoff's law is applied to the circuit, remembering that the current flowing in the line is now (I + 25) amperes.

$$230 = 0.4(I + 25) + 11\,I = 11.4\,I + 10$$

and

$$I = \frac{220}{11.4} = 19.3 \text{ amperes.}$$

(b) Voltage $E_{AB} = E_{BC} = \frac{19.3 \times 11}{2} = 106.15$ volts

Thus, when the motor is thrown on the line, the voltage across the lamps will fall from 111.10 to 106.15 volts.

• PROBLEM 12-11

In the circuit of Fig. 1, with loads L_1, L_2, and L_3 drawing currents of 25, 8, and 40 amperes, respectively, calculate:

(a) The power supplied by each generator.

(b) The voltages E_1, E_2, and E_3.

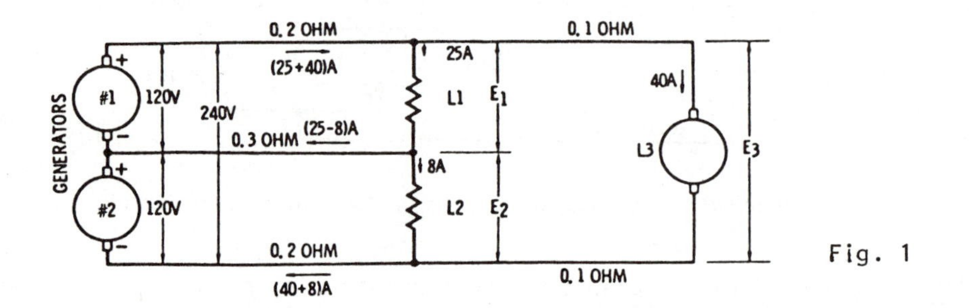

Fig. 1

A three-wire system supplying three loads.

Solution: By inspection, the currents supplied by generators Nos. 1 and 2 are 65 and 48 amperes, respectively.

(a) $P_{G1} = 120 \times 65 = 7800$ watts.

$P_{G2} = 120 \times 48 = 5760$ watts.

(b) According to Kirchhoff's law

$E_1 = 120 - (65 \times 0.2) - [(25 - 8) \times 0.3]$

$= 120 - (65 \times 0.2) - (17 \times 0.3) = 101.9$ volts.

$E_2 = 120 + [(25 - 8) \times 0.3] - [(40 + 8) \times 0.2]$

$= 120 + (17 \times 0.3) - (48 \times 0.2) = 115.5$ volts.

and

$E_1 + E_2 = 217.4$ volts.

Similarly

$E_3 = [(E_1 + E_2)] - [40 \times (0.1 + 0.1)]$

$217.4 - (0.2 \times 40) = 209.4$ volts.

• PROBLEM 12-12

A 125/250-volt three-wire system has a load on the positive side of 500 amp, on the negative side of 450 amp, the neutral current therefore being 50 amp. If each machine of the balancer set has an efficiency of 86 percent, calculate the current for each of these machines and for the main generator.

Solution: Efficiency of balancer set

$$\eta = \frac{\text{output of generator}}{\text{input of motor}} = \frac{I_g \times 125}{I_m \times 125}$$

$$= \frac{I_g}{I_m} = \frac{I_g}{50 - I_g} = 0.86 \times 0.86 = 0.74$$

Therefore

$$I_g = 0.74(50 - I_g)$$

from which

$$I_g = 21.26 \text{ amp}$$

$$I_m = 50 - I_g = 28.74 \text{ amp}$$

$$\text{I of main generator} = 500 - 21.26 = 450 + 28.74$$

$$= 478.74 \text{ amp.}$$

SECTION II

SUMMARY OF ELECTRICAL MACHINERY

FUNDAMENTALS OF ELECTRICITY

1–3. Nature of Electric Current

Although the repairman is primarily concerned with the practical application of electricity and not with theory, some understanding of the fundamentals is necessary.

a. Definition and Direction of Electric Current. Electric current can be simply defined as the movement of free electrons within an electrical conductor caused by a potential difference between the ends of the conductor. Since the electron carries a negative charge, it will be attracted to the positive end of the conductor, i.e., to the point of higher potential. Hence, an electric current actually moves in a direction from negative to positive. When the phenomenon of electricity was first discovered, it was thought that an electric current was the flow of positive charges from points of higher to points of lower potential, i.e., from positive to negative. Consequently, all of the early texts on electricity and many of the current texts used in the civilian industry are based upon the assumption that electric current moved in a direction from positive to negative. All explanations in this manual are based upon the fact that an electric current is the flow of electrons from negative to positive.

b. Direct Current. A direct current is a current which flows in one direction only. A pure direct current has a constant magnitude (value) with respect to time (①, fig 1–1). A varying direct current varies in magnitude with respect to time (②, fig 1–1). When such variations occur at regular intervals, the direct current is called a "pulsating direct current" (③, fig 1–1).

c. Alternating Current. Alternating current is an electric current which moves first in one direction for a fixed period of time and then in the opposite direction for an equal period of time, constantly changing in magnitude. From a zero value, alternating current builds up to a maximum in a positive direction, then falls off to zero value again before building up to a maximum in the opposite or negative direction, and then

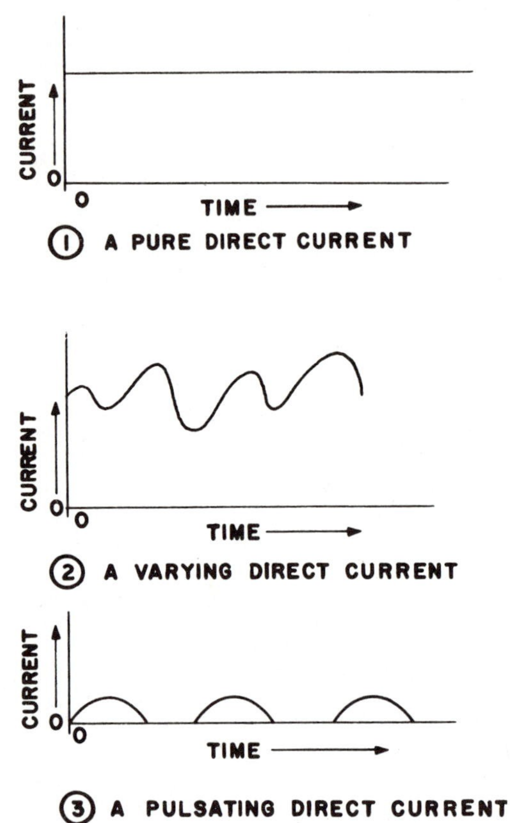

Figure 1–1. Types of direct currents.

finally returning to zero. For this reason, alternating current may be further defined as a current which is constantly changing in magnitude (either building up or falling off) and periodically (at set intervals of time) changing direction. The shape of such alternating flow is that of a wave.

(1) The most common and most important alternating current wave shape is the sine wave, so named after the trigonometric function whose graphic pattern it follows. However, due to the presence of electrical noise (random voltages generated within an electric circuit), sine waves are often distorted. Consequently, many different kinds of alternating wave shapes can occur; these alternating wave shapes are referred to as complex waves. Figure 1–2 illustrates a sine wave and two types of complex waves.

(2) In each instance, one cycle or complete pattern from positive to negative is shown (fig 1–2). Assuming that each wave shape starts at

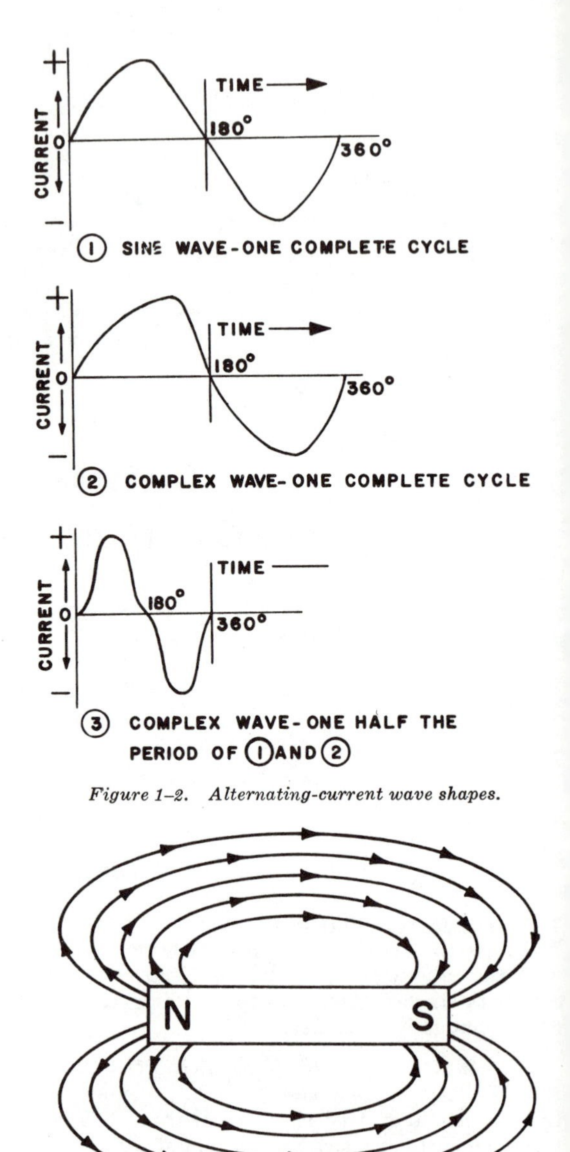

Figure 1–2. Alternating-current wave shapes.

Figure 1–3. Magnetic lines of force about a bar magnet.

the same instant of time, it can be seen that the wave shape of ③ completes its cycle in half the time of either ① or ②. Therefore, the frequency of ③ is twice that of ① and ②.

(3) It is apparent then, that the proper recognition and definition of an alternating current depends upon:

(*a*) The wave shape of one cycle.

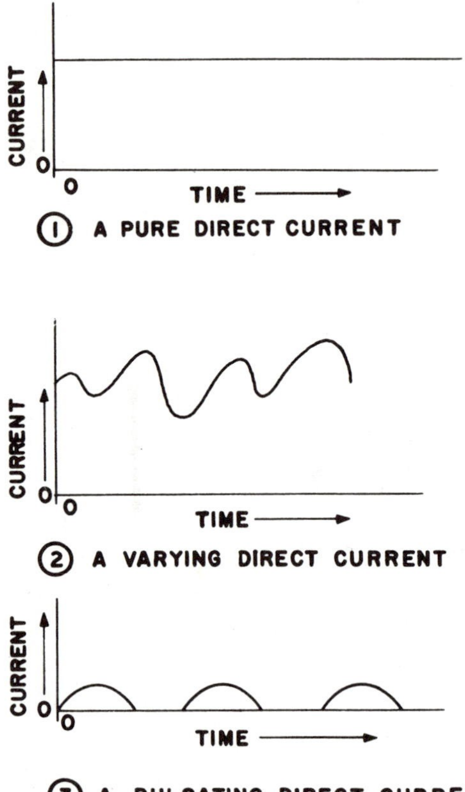

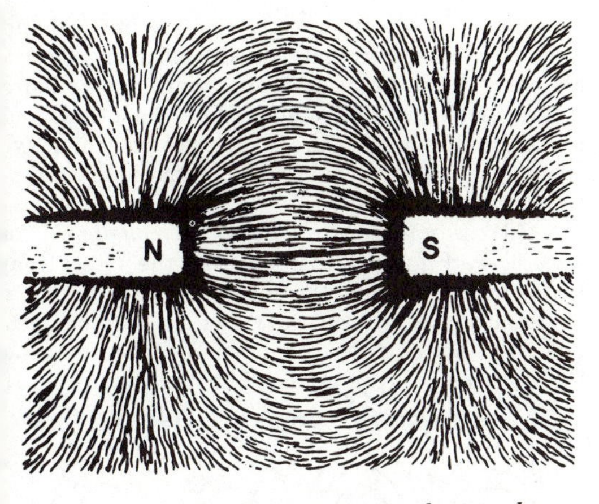

Figure 1-4. Magnetic lines of force about two bar magnets of opposite polarity.

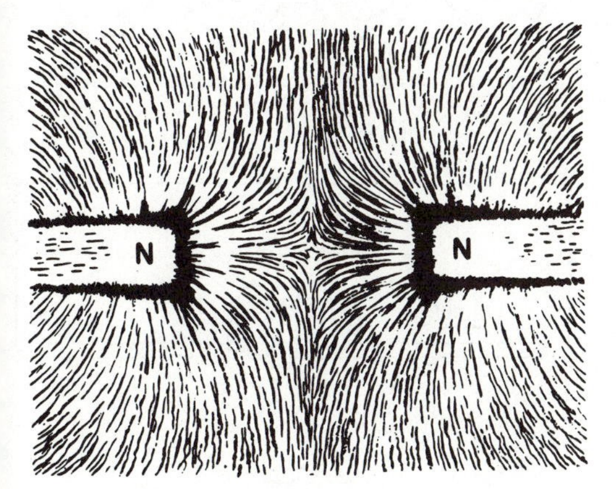

Figure 1-5. Magnetic lines of force about two bar magnets of like polarity.

(*b*) The value at some specified point in the cycle.

(*c*) The length of time to complete one cycle (the period).

1-4. Magnetism

Magnetism is a power of attraction or repulsion which can be introduced most pronouncedly in certain metals by means of an electric current. Magnets formed in this way are referred to as artificial magnets. Magnets may also be classified as either permanent magnets or electromagnets, depending upon their ability to retain their magnetic properties. Permanent magnets are usually made of steel or steel alloys since *steel* has a tendency to *retain* its magnetic properties once it has been magnetized by an electric current. Electromagnets are usually made of soft iron or iron alloys since *iron* tends to *lose* most of its magnetic properties once the magnetizing influence of an electric current has been removed. Hence, the effect of magnetism can be controlled by an electric current. Such is the case in many types of electric machines, wherein an electromagnet is a major component.

a. Magnetic Influence. The influence of magnets in the space surrounding them may be detected in many ways. Experiments have shown that this influence, i.e., the force of attraction and repulsion, varies inversely as the square of the distance from the magnet. To account for this influence, a magnet is said to establish a magnetic field around itself which is represented pictorially by directed lines. Consider the permanent bar magnet of figure 1-3. For convenience, the ends of a magnet are arbitrarily referred to as poles. The north pole is the end from which the magnetic lines of force leave the magnet and the south pole is the end into which the magnetic lines of force reenter the magnet. The magnetic lines of force pass through the magnet from its south to its north pole. Lines of magnetic force are always closed loops.

b. Lines of Force. If a second bar magnet is placed near the first bar magnet (fig 1-4) so that the unlike poles are adjacent, a force of attraction results and some of the magnetic lines of force from the first bar magnet are diverted towards the pole of the second bar magnet. However, if one magnet is reversed so that the like poles are now adjacent, a force of mutual repulsion will result, tending to separate the magnets (fig 1-5). The magnetic lines of force leaving from each of the adjacent like poles of the two bar magnets will tend to repel each other while simultaneously seeking to reenter the opposite end (opposite pole) of their respective magnets. From these observations it can be seen that unlike poles of magnets attract each other and like poles repel each other. This phenomenon of magnetic attraction and repulsion produces the torque action in many types of electric motors.

c. Electromagnetism. When an electric current is passed through a wire, a magnetic field is produced around the wire (fig 1-6). The direction of the magnetic lines of force, which form concentric circles around the wire, depends upon the direction of the electric current through the wire. The direction of the magnetic lines of force about a current-carrying conductor can be determined by using the left-hand rule for a current-carrying conductor. This rule states that if a current-carrying conductor is grasped in the left hand with the thumb pointing in the direction of current flow (negative to positive), the fingers will encircle the wire in the direction of the magnetic lines of force, as illustrated in figure 1–7.

(1) *Magnetic field about a coil.* The magnetic field resulting under the conditions shown in figure 1–6, even with high currents, is relatively weak. But if the electrical conductor is formed into a coil, a relatively stronger magnetic field is created with magnetic lines of force running through the center of and perpendicular to each turn of the coil as shown in figure 1–8. The coil current sets up circular magnetic lines of force around each coil turn in accordance with the left-hand rule. Consequently, the magnetic lines of force encircling the upper part of each coil turn in a counterclockwise direction and those encircling the lower part of each coil turn in a clockwise direction. In the center of the coil, all of the magnetic lines of force run in the same direction, thereby aiding each other to produce a net positive effect. Between adjacent coil turns, the magnetic lines of force cancel each other. Consequently, magnetic lines of force travel the entire length of the coil in order to complete their loops. This makes the coil behave as a magnet with a north and a south pole. By using the left hand, again, the north pole end of a current-carrying coil can be determined. This time, if the coil is grasped so that the fingers of the left hand encircle the coil in the direction of the current flowing through the coil, the thumb will point to the north pole of the coil.

(2) *Electromagnets.* If a piece of magnetic material, usually soft iron, is placed within a coil through which current is flowing, the magnetic properties of the coil are tremendously increased. This increase in magnetic strength is due to the greater permeability of the soft iron. Permeability is the ease with which magnetic lines of force pass through a substance. A coil wound around a core of magnetic material is called an electromagnet. The coil may be wound with one or more layers of wire from one end to the other and back, providing, of course, that the current flows around the core continuously in the same direction.

(3) *Magnetic field about a coil with an iron core.* Figure 1–9 illustrates the effect that an iron core has on the magnetic lines of force surrounding a current-carrying coil. In ①, figure 1–9, notice that the lines of force passing through the coil are confined to the iron core. If

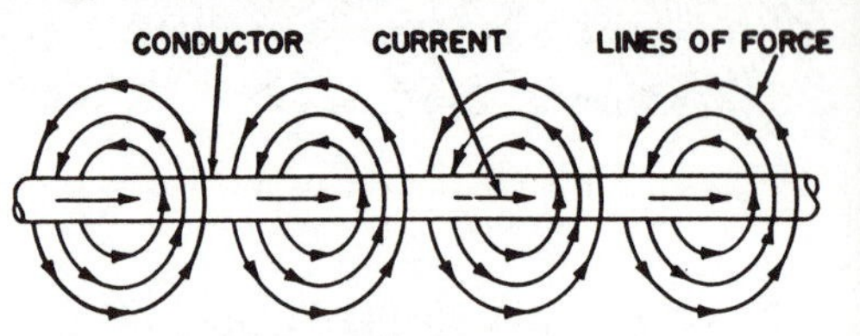

Figure 1–6. Magnetic lines of force about a current-carrying conductor.

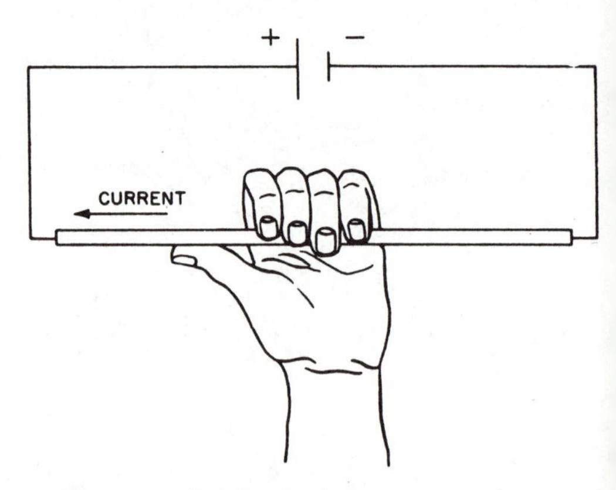

Figure 1–7. Left-hand rule for a current-carrying conductor.

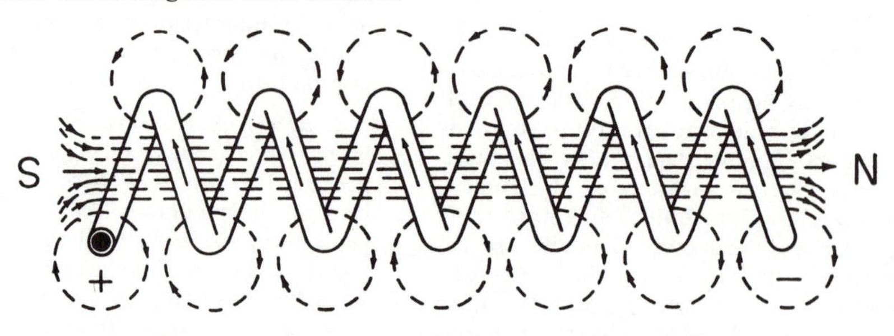

Figure 1–8. Magnetic field about a current-carrying coil.

the iron core is pulled partially out of the coil as shown in ②, figure 1–9, the magnetic lines of force will be extended in order to enter the end of the iron core which is outside of the coil. Once the lines have established themselves in the core, they tend to shorten, thereby exerting a force on the core. This force tends to pull the core until its center coincides with the center of the coil as shown by the dotted lines in figure 1–9. This action has many practical applications in the various types of electrical controlling devices which are used in industry and in the home.

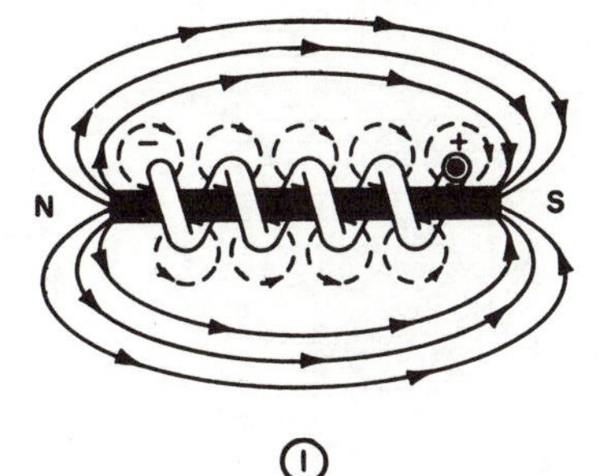

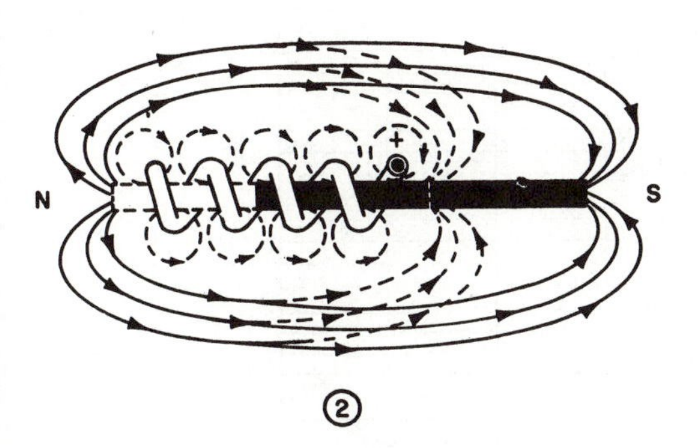

Figure 1–9. The effect of an iron core on the magnetic field of a coil.

1–5. Induced Electromotive Force

Having determined the existence of magnetic lines of force around any wire due to the flow of an electric current, one may conceive of establishing an electric current by means of a magnetic field. Such is the case when an electrical conductor is moved across a magnetic field. As the conductor cuts the lines of magnetic force, an electromotive force is induced in the conductor which causes an electric current to flow only if the conductor is part of a closed loop or circuit. If the conductor is not part of a closed loop or circuit, then current will not flow through the conductor. However, there will still be an electromotive force induced in the conductor as long as it cuts across magnetic lines of force. Electromotive force (simply referred to as emf) is defined as the force or pressure necessary to cause the flow of an electric current. An emf can still be present in a circuit without the flow of an electric current. An emf can also be induced in a stationary electrical conductor if the magnetic field is made to move so that its lines of force cut across the conductor. In each of these instances, i.e., where a magnetic field and an electrical conductor are moving relative to each other causing an induced emf in the conductor, the magnetic field is assumed to be constant in magnitude. However, a magnetic field which does not move relative to a conductor within its boundaries can also induce an emf in the conductor. This is done by varying the magnitude of the magnetic field with respect to time.

1–6. EMF Produced by Cutting Magnetic Lines of Force

All of the methods mentioned above for inducing an emf in an electrical conductor find their practical application in the various types of electric motors and generators.

a. Figure 1–10 shows a loop wire revolving in a magnetic field. The magnetic field is created by field poles of the kind found in the most elementary type of electric machine. The ends of the loop are connected to sliprings which revolve with the loop. Stationary brushes are used to collect the current from the rings and deliver it to an external circuit. In ①, figure 1–10, the white conductor moves to the left while the black conductor moves to the right and the induced emf's in both conductors may be added together. The total emf will depend upon the position of the loop in the field. This is true since only that portion of the motion perpendicular to the field is effective in producing emf. The wave shape which results from one complete revolution of the loop is also shown in figure 1–10.

(1) To trace the development of the wave shape, let us start with the loop as shown in ①, figure 1–10. In this position each conductor is moving parallel to the magnetic field, the loop is in a neutral position, and the generated emf is zero.

(2) As the loop continues in a clockwise direction the emf increases, due to the loops cutting more lines in a given period of time, until it reaches a maximum at which time the loop is parallel to the field (②, fig 1–10).

(3) As the loop continues to rotate to the

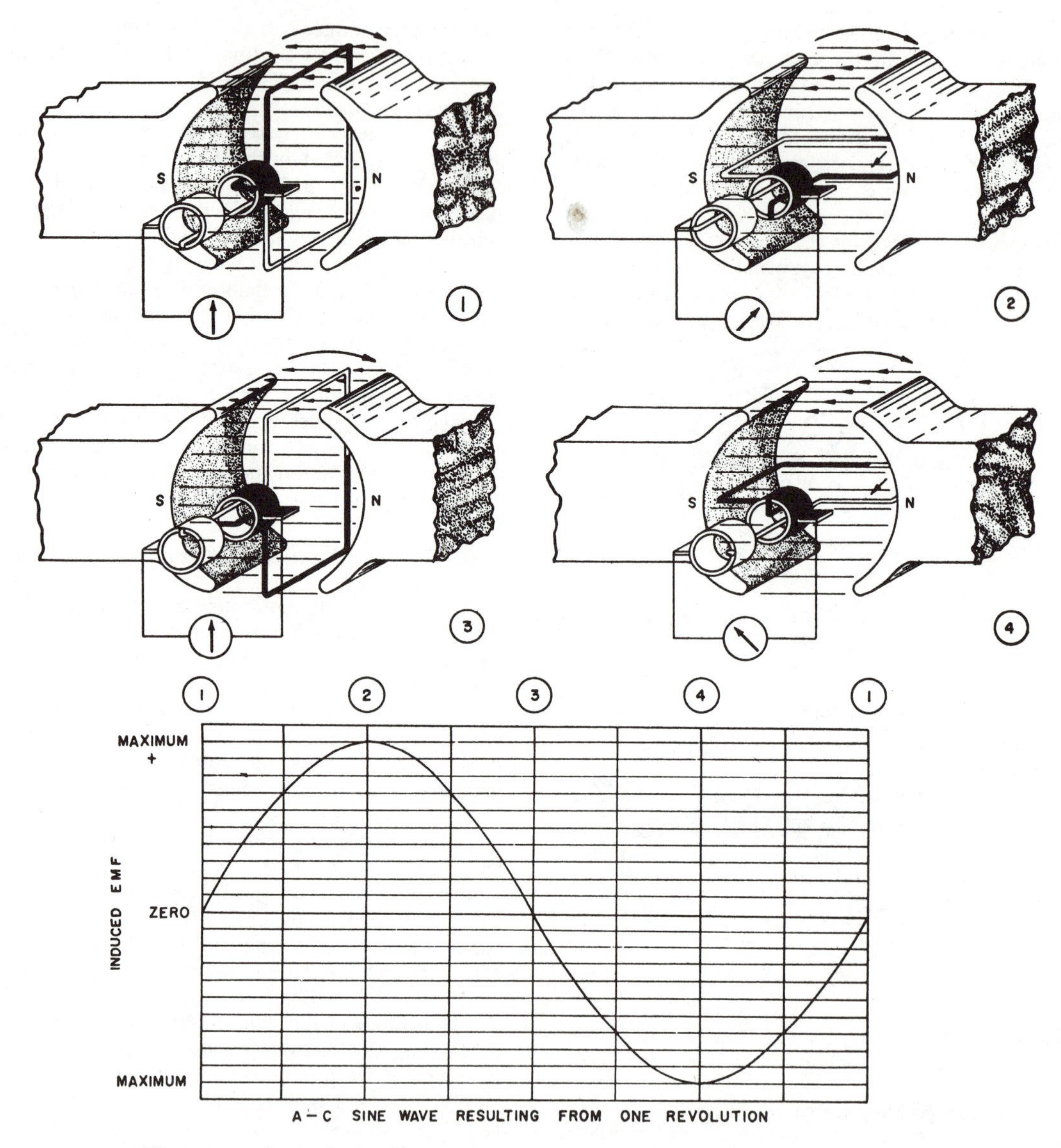

Figure 1–10. Loop of wire rotating in a magnetic field. The encircled arrow simulates a galvanometer for indicating the strength and direction of the induced emf.

position shown in ③, figure 1–10, the emf decreases until it is again zero.

(4) If the loop is turned further through an angle of 90° in the same direction (④, fig 1–10), it will again be cutting lines of force at a maximum rate; however, the emf and resulting current will be reversed with respect to the loop. The reversal occurs because of the change in the direction in which the conductor passes through the field.

(5) As the conductor continues to rotate, the generated emf decreases until, at the starting point, it is again zero. The wave shape of figure 1–10 is a sine wave.

b. Let us see what would happen if the sides of the loop of figure 1–10 were connected to split rings instead of sliprings (fig 1–11). Notice that one brush is shown as being black and the other white.

(1) Starting with the loop in a neutral po-

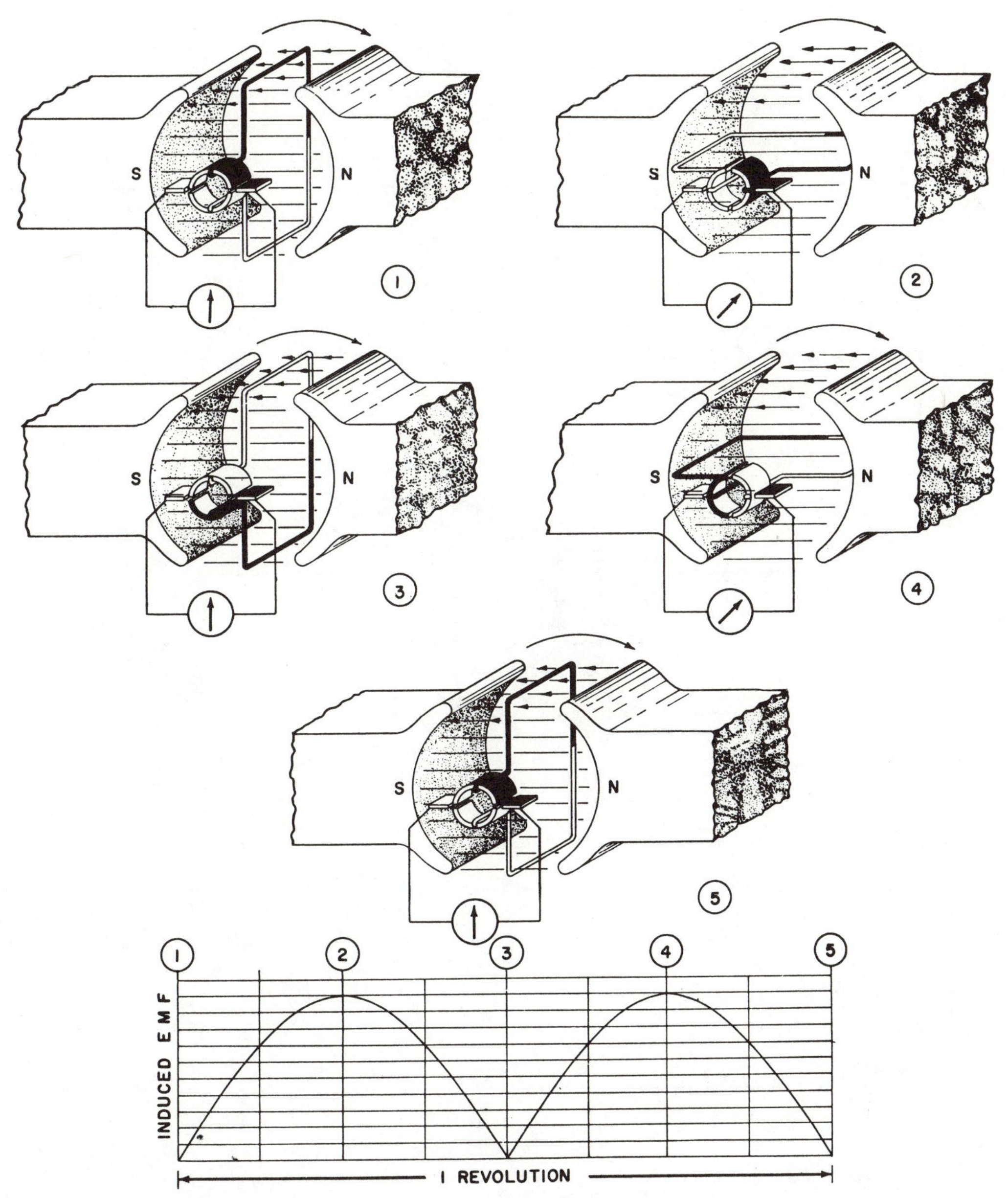

Figure 1–11. Loop of wire connected to a split ring, rotating in a magnetic field.

sition (①, fig 1–11), the generated emf rises from zero to a maximum at the position shown in ②, figure 1–11. This corresponds to the first portion of the wave shape of ②, figure 1–10. Notice that the black brush is in contact with the black side of the coil.

(2) As the loop continues to rotate in a clockwise direction, the induced emf again decreases to zero, position ③, figure 1–11, as it did in ③, figure 1–10. However, the black brush is directly over the split portion of the ring. This means the white portion of the loop will be in contact with the black brush while the loop is rotated through the next 180° (④ and ⑤, fig 1–11).

(3) As the loop is rotated through the re-

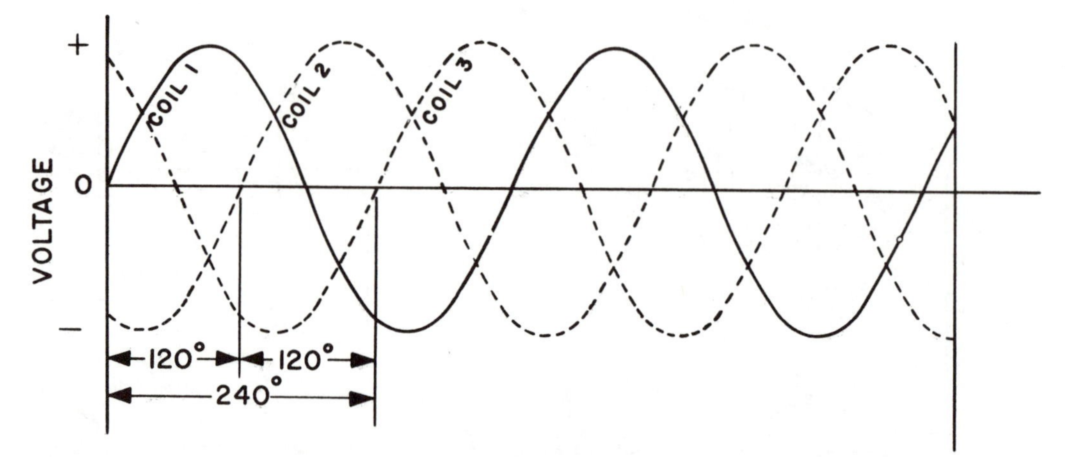

Figure 1–12. Wave shapes of three separate loops of wires connected to sliprings.

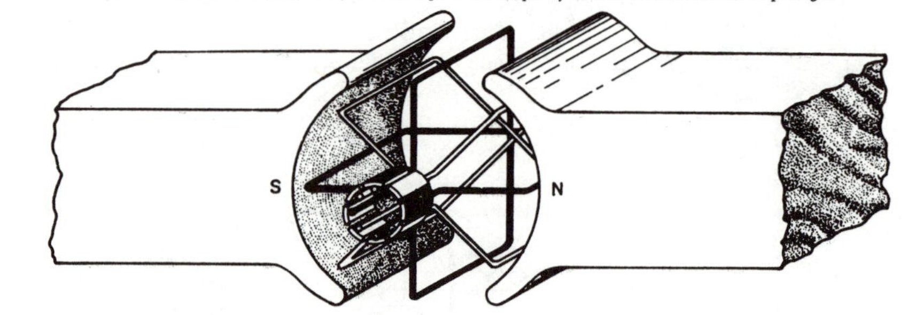

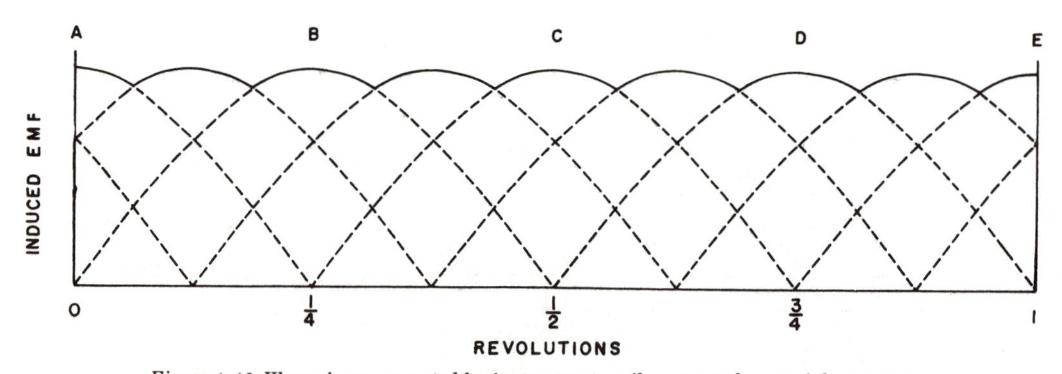

Figure 1–13. Wave shape generated by four separate coils connected to an eight-segment commutator.

maining 180° (④ and ⑤, fig 1–11) to complete one revolution, the generated emf once again rises to a maximum at the position shown in ④, figure 1–11 and decreases to a minimum at the position shown in ⑤, figure 1–11. However, since the black brush is always in contact with the side of the loop passing from left to right through the field, the current through the brush will always be in the same direction. This is is shown by the reversal of current in the second half of the wave shape in figure 1–11. The action of collecting the induced current in the same direction is called commutation and the split ring by which it is accomplished is called a commutator.

(4) Commutation can be explained as fol-

lows: At the instant that each brush is contacting two segments on the commutator (positions ①, ③, and ⑤, fig 1–11), a direct short circuit is produced. If an emf is generated at this time, a high current will flow in the short circuit, causing an arc, and thus damaging the commutator and the brushes. For this reason, the brushes must be positioned so that the short between the commutator segments occurs when the generated emf is zero. This position is referred to as the *neutral plane* of the brushes. The generated emf is zero, of course, at the instant when the coil is not cutting magnetic lines of force. This instant occurs at the positions ①, ③, and ⑤ in figure 1–11.

c. Figure 1–12 shows the wave shape generated by three separate loops which are connected to sliprings and are rotating in a magnetic field. Notice that the voltage generated by the second and third coils follows the first by 120° and 240° respectively. This displacement, or phase angle, results from the loops being mutually spaced at 120°; that is to say, the distance between the first and second loops is one third the periphery of the armature core as is the distance between the second and third loops. The resultant wave shape is similar to that of a three-phase generator.

d. Figure 1–13 shows the wave shape generated by four separate coils connected to an eight-segment commutator, rotating in the magnetic field established by two poles. A comparison of figures 1–11 and 1–13 will show that the variation between maximum and minimum values of emf decreases with the addition of loops or coils. This variation, called *ripple,* is present in all emf's generated in the manner described above.

e. The maximum value of the emf is not affected by the number of loops, but rather by the number of turns of wire per loop, the field strength, and the speed of rotation. This is sometimes expressed as being the rate of change of flux linkages. Flux is merely another name for lines of magnetic force.

1–7. Left-Hand Rule for Generators

The relationship found to exist between the direction of the magnetic field, the direction of motion of the conductor, and the direction of the induced current is illustrated in figure 1–14. This relationship, referred to as "left-hand rule for generators," may be stated as follows: Extend the thumb, forefinger, and middle finger of the left hand so that they are at right angles to each other. Place the forefinger in the direction of the magnetic field (north to south pole) with the thumb in the direction of the motion; then the middle finger will indicate the direction of current flow through the conductor.

1–8. Motor Effect

If current is generated in a wire moving through a magnetic field, one might logically assume it possible to cause a conductor, located in a magnetic field, to move by passing a current through it. Such is the case. In fact, this action, called the motor effect, is the basis of all electric motors. Consideration of the conductor shown in figure 1–15 will aid in understanding how this comes about.

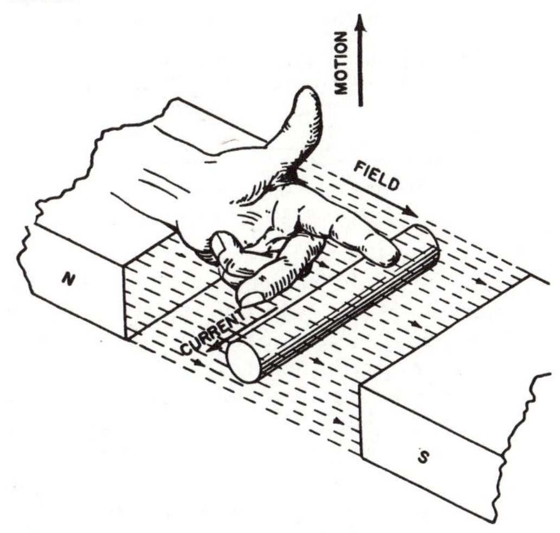

Figure 1–14. Left-hand rule for generators.

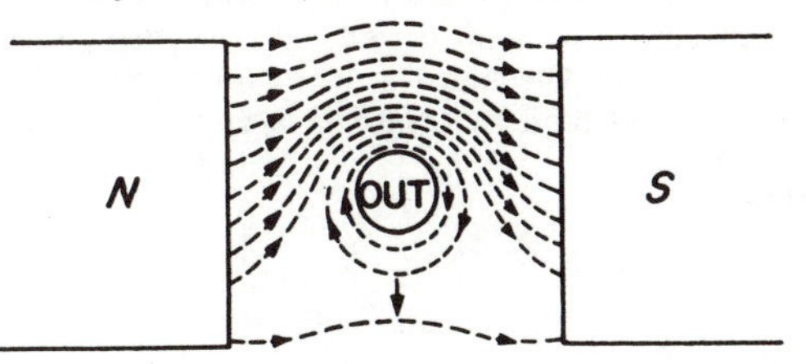

Figure 1–15. Expulsion of current-carrying wire from a magnetic field.

a. The conductor in the figure is assumed to carry an electric current coming out of the page. This current establishes lines of force about the wire in a clockwise direction according to the left-hand rule. With the magnetic field in the position shown, some of the lines which would normally pass through, or immediately beneath, the area occupied by the conductor may be thought of as being deflected over it. In doing so, they are stretched somewhat and crowded into the area above the conductor. The natural tendency of the lines to straighten exerts a force on the conductor which would tend to push it downward entirely out of the magnetic field.

b. Let us see what would happen if a loop of wire, rather than a single conductor, were placed in a magnetic field and current passed through it. Figure 1–16 shows a loop, with the direction of current flow indicated, located in a magnetic

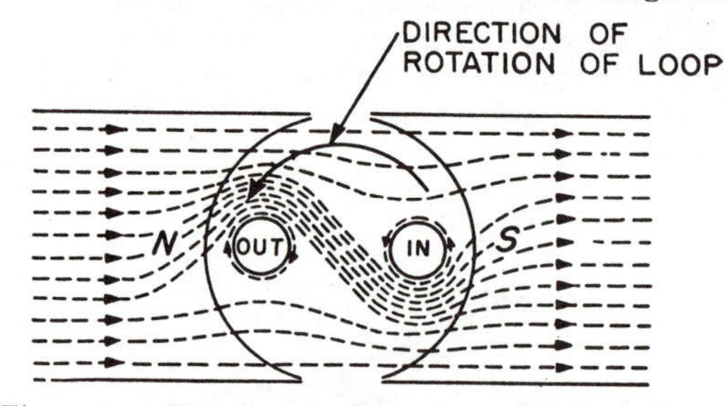

Figure 1–16. Rotation of a current-carrying loop of wire in a magnetic field.

field. The direction of rotation of the lines of force set up by the current is clockwise about the left half of the loop and counterclockwise about the right half (left-hand rule). This distorts the lines of the magnetic field by deflecting the lines in such a manner that they pass over the left conductor and beneath the right conductor. The left half of the loop experiences a downward force, as in figure 1–15, while the right half is forced upward. If the loop were free to move about an axis located midway between the sides of the coil, it would rotate in a counterclockwise direction. If the direction of current flow through the loop were reversed, the loop would tend to rotate in a reverse direction, or clockwise.

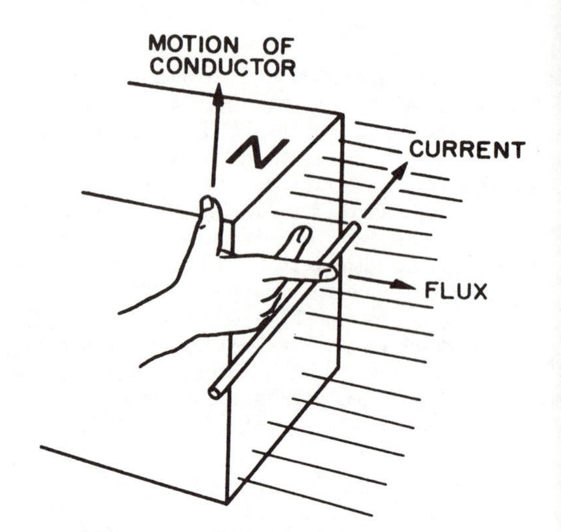

Figure 1–17. Right-hand rule for motors.

1–9. Right-Hand Rule for Motors

Since an electric motor performs the reverse function of a generator, in that it changes electrical energy into mechanical energy, the opposite hand, or right hand, is used in the motor rule (fig 1–17). The rule, referred to as the right-hand rule for motors, can be performed as follows: Extend the thumb, forefinger, and middle finger of the right hand at right angles to each other. Then place the forefinger in the direction of the magnetic field (north to south pole), with the middle finger in the direction of current flow in the conductor; the thumb will then indicate the direction of motion. Notice that, as in the case of the generator rule, the forefinger is associated with the field, the middle finger with the current, and the thumb with the direction of motion. When you know any two of the following—motion, magnetic field, or current—you can determine the direction of the third.

Section I. DC GENERATORS

2–1. Principles of Operation

A generator is a machine which converts mechanical energy into electrical energy. This is done by rotating an armature, which contains and moves conductors, through a magnetic field, thus inducing an emf in the moving conductors. In any generator, a relative motion between the conductors and the magnetic field must always exist by the application of a constant mechanical force or twist on the shaft. In the direct current generator (fig 2–1), the magnetic lines of force (called the field) are stationary. The armature, which contains the conductors, rotates through the stationary field.

2–2. Major Components of DC Generators

The principal components of a dc generator are the armature, the commutator, the field poles, the brushes and brush rigging, the yoke or frame, and the end bells or end frames, as shown in figure 2–1. Other components found on some dc generators are interpoles, compensating windings, and various controls and devices for regulating the generator output.

a. Armature. The armature (fig 2–2) is the structure upon which are mounted the coils which cut the magnetic lines of force. It is fixed on a shaft which is suspended at each end of the machine by bearings set in the end bells

Figure 2–1. Typical dc generator.

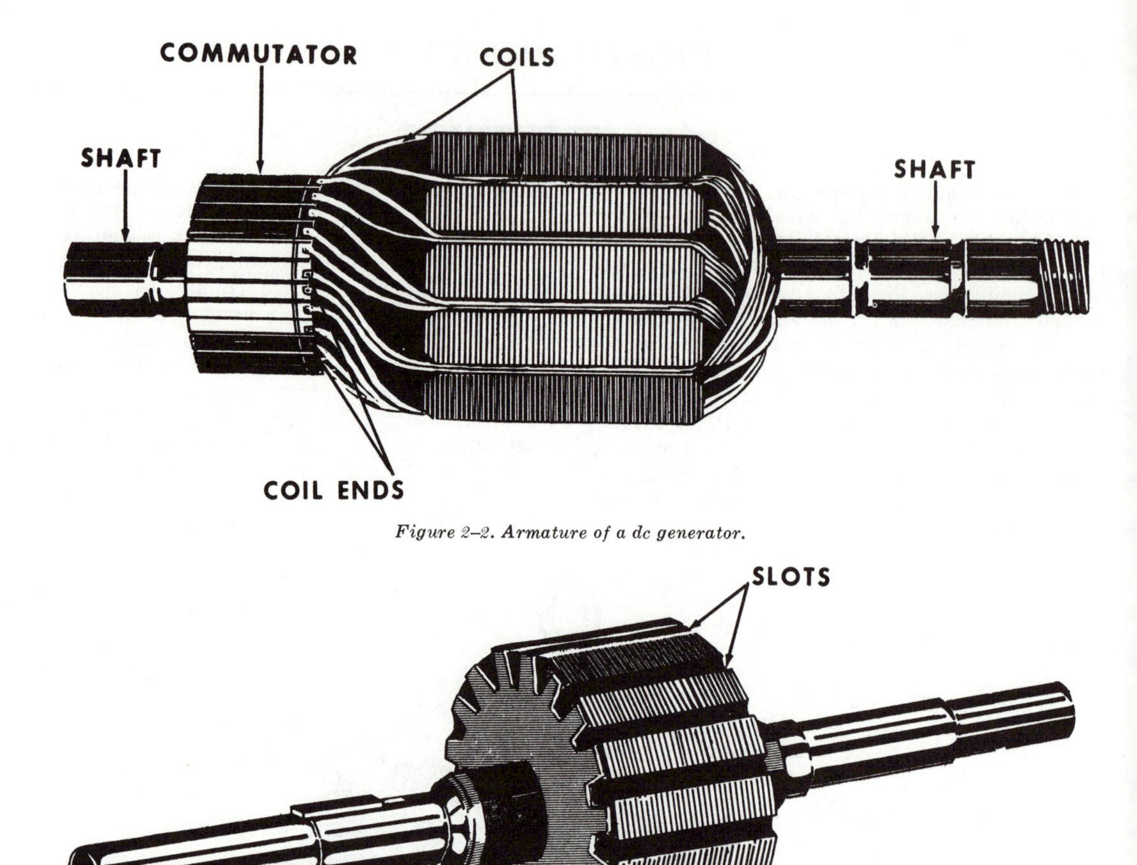

Figure 2–2. Armature of a dc generator.

Figure 2–3. Unwound armature core on a shaft.

or end frames. The armature core is circular in cross section and is built up from sheets of soft iron (fig 2–3). The circumferential edge of the laminated core is slotted in order to receive the coil windings. The windings are held in place and in their slots by wooden or fiber wedges. Steel bands are sometimes wrapped around the completed armature and, with the wedges, hold the coil windings in place. On small machines the laminations of the armature core are usually pressed onto the armature shaft. In comparison, on large machines (usually ac generators) where the rotating member is the field winding, a spider is secured to the rotating shaft and forms the base for the cores of the field windings. The spider, as shown in figure 2–4, consists of a hub and projecting arms to which the field pole laminations may be rigidly fastened. This type of rotor construction permits air to flow freely through the rotating member thus keeping it ventilated and cooled.

b. Commutator. The commutator is that component of the generator which rectifies the generated alternating current to provide direct current output and connects the stationary output terminals to the rotating armature. A typical commutator (fig 2–5) consists of commutator bars, which are wedged-shaped segments of hard-drawn copper, insulated from each other by thin strips of mica. These commutator bars are held in place by steel V-rings or clamping flanges which are bolted to the commutator sleeve by hexagonal cap screws. The commuator sleeve is keyed to the shaft which rotates the armature. A mica collar or ring insulates the commutator bars from the commutator sleeve. The commuta-

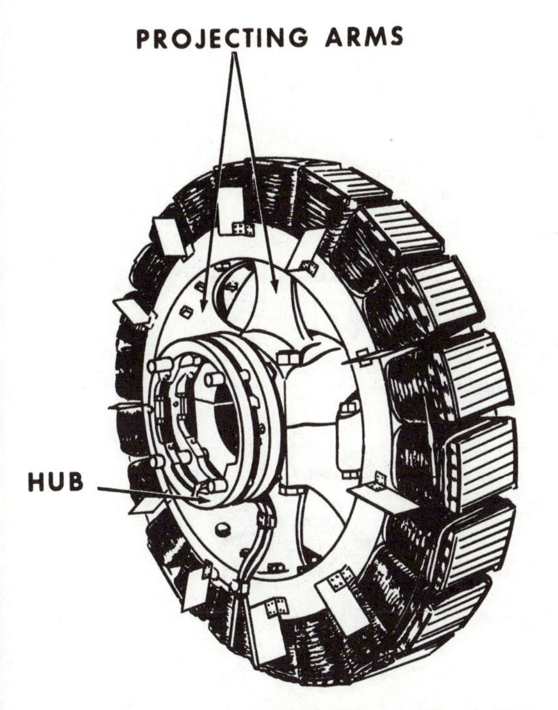

Figure 2–4. Field pole cores and spider assembly of a large ac generator.

tor bars usually have risers or flanges to which the leads from the associated armature coils are soldered (fig 2–5 and 3–25). These risers serve as a shield for the soldered connections when the commuator bars are turned down due to wear. When risers are not provided, it is necessary to solder the leads from the armature coils to short slits in the ends of the commutator bars. The brushes, which make contact with the commutator bars, collect the current generated by the armature coils and, through the brush holders, pass the current to the main terminals. As the commutator bars are insulated from each other, each set of brushes, as it makes contact with the commutator bars, collects current of the same polarity, resulting in a continuous flow of direct current. The finer the division of the commutator bars, the less ripples will be present and the smoother will be the flow of the dc output.

c. Field Pole and Frame. The frame, or yoke, of a generator serves both as a mechanical support for the machine and as a path for the completion of the magnetic circuit. The lines of force which pass from the north to the south pole through the armature are returned to the north pole through the frame. Frames are made of electrical-grade steel. The method of construction of field poles and frames varies with the manufacturer.

(1) Field poles (fig 2–6) are required to produce the magnetic field, or flux, which passes through the conductors of the armature. The minimum number of field poles required to complete the magnetic circuit is two, a north pole and a south pole; but generators with only two poles are inefficient due to leakage of flux. Most commercial generators are made with four or more field poles, depending on the speed of the generator. The slower the generator, the more poles are needed to produce the same output. The number of field poles must always be an even number, each set consisting of a north pole and a south pole. Multipole generators are in common use. In dc generators, each set of field poles requires a set of brushes. Thus, a four-pole generator requires four brushes, a six-pole generator requires six brushes, and so on.

(2) Field poles are usually fastened to the inside of the frame with screws or bolts, except for some small units which are made with the field poles cast as part of the frame. Most field poles consist of rectangular laminations held together by bolts or rivets. That section of the pole away from the frame is generally flared. These flared sections, called shoes, distribute the flux beneath the pole and hold the coil on the pole.

(3) The field coils are made up of turns of wire taped together before assembly on the core. Coils which are to be connected in series with the armature are wound to have a few turns of heavy wire. Those to be connected for shunt, or separate excitation, are made up of many turns of finer wire. Thus, the series winding offers little resistance and can carry more current, while shunt windings present a high resistance which passes little current.

(4) Field poles of compound-wound generators contain both a series and a shunt winding. These windings are generally placed one inside the other on smaller units and side by side on larger ones. On small units it is customary to wind and insulate the series winding; wind and insulate the shunt winding; and finally tape the two windings together before placing them on the pole piece. On machines with interpoles, the interpole coils are treated simply as series windings.

d. Brushes. The brushes of a generator are the points of contact between the external circuit conductors and the commutator. These points "brush" the commutator in such a way as to

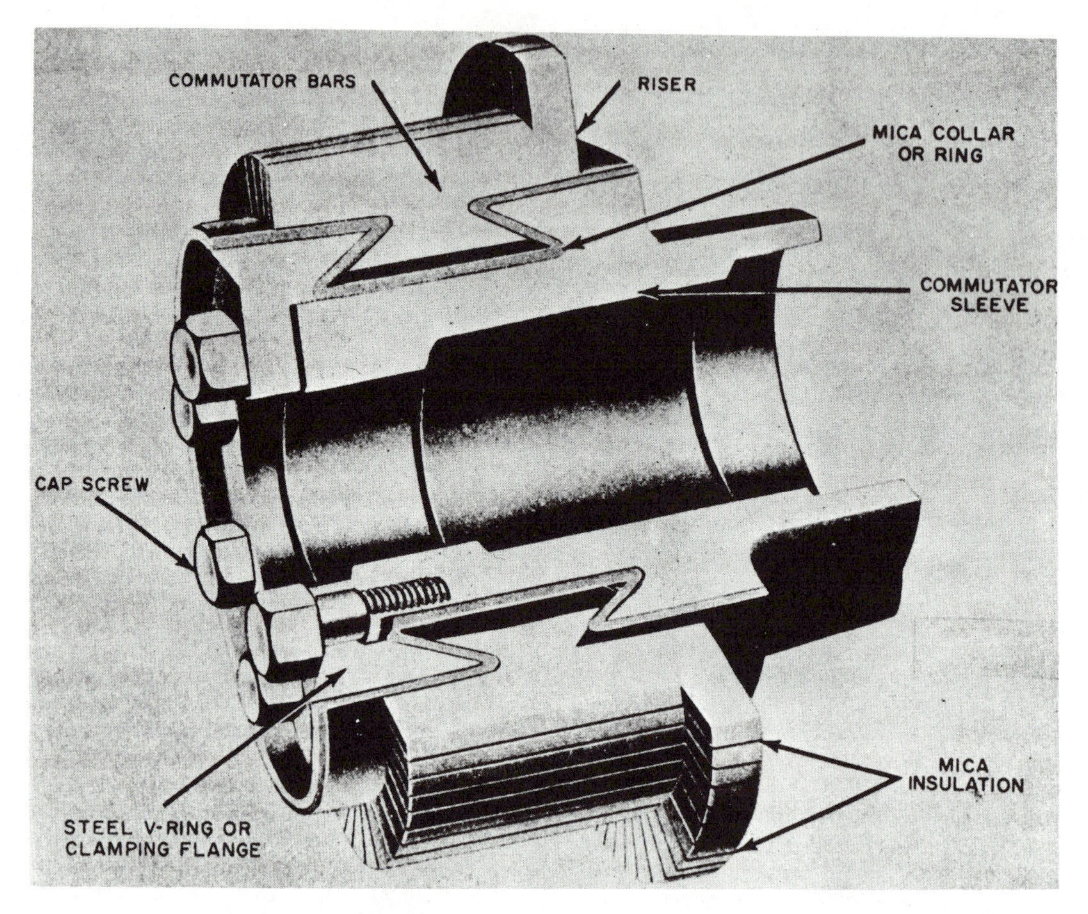

Figure 2–5. Typical commutator.

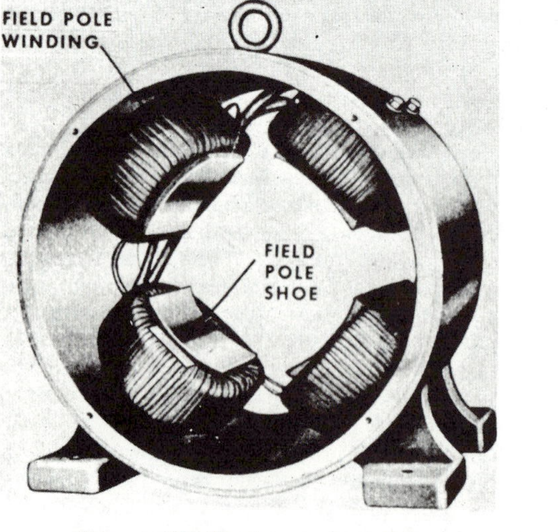

Figure 2–6. Field poles on a dc machine.

take off the generated emf. Several types of brushes may be used, depending on the application. It is of utmost importance that substituted brushes are always of the same type as the original brush on the equipment. The types of brushes are as follows:

(1) Brushes are usually made of high grade carbon. These brushes are primarily restricted in use to low-speed machines with low current densities (amperes per unit area of brush face) and where economy in brush cost is a major factor.

(2) Electrographite brushes are made from carbon which has been processed at high temperatures in an electric graphitizing furnace. This treatment lowers its hardness and increases its electrical and thermal conductivity and toughness. This results in a brush that has low friction, great resilience (hugs commutator without bouncing), and high current capacity, and that is nonabrasive and cool running.

(3) Natural graphite brushes can be identified by their silvery appearance and soft flaky structure. They cannot be made as mechanically strong as the carbon or the electrographite brush. They are also more prone to selective action and shunt burning at high current densities. Their application is therefore rather restricted, but with proper allowances, these brushes can give long brush life with minimum maintenance.

(4) Copper-graphite brushes are made from a mixture of powdered copper and graphite pressed together and baked at relatively low temperatures. They are primarily used on rings and commutators where high current densities are being carried.

e. Brush Rigging. A flexible braided-copper conductor, commonly called a pigtail, connects each brush to the external circuit. The brush rigging (fig 2–7) consists of brushes set in brush holders fastened to a rocker arm which in turn is connected to the yoke or frame of the generator. The brush holders hold the brushes in place as the brushes ride over the surface of the commutator. Each brush is free to slide up and down in its holder so that it may follow irregularities in the surface of the commutator. Each brush is also insulated from its holder. A spring on each brush holder forces each brush to bear on the commutator with from 1 1/2 to 2 pounds of pressure for every square inch of brush surface riding on the commutator. These springs are usually mounted so that the brush pressure is adjustable. The rocker arm to which the brush holders are fastened permits shifting the brush positions about the commutator without changing the relative position of the brushes. On large machines, several brush holders are usually connected to a single brush arm (fig 2–8). Each brush arm is then bolted to the machine frame.

f. End Bell or End Frames. The end bells or end frames inclose, support, and protect the armature, frame, and field poles. They also house the bearings in which the shaft rotates (fig 2–9).

g. Interpoles and Compensating Windings. Interpoles (fig 2–10), sometimes called commutating poles, are small auxiliary poles placed midway between the main poles and provided with a winding in series with the armature. Their

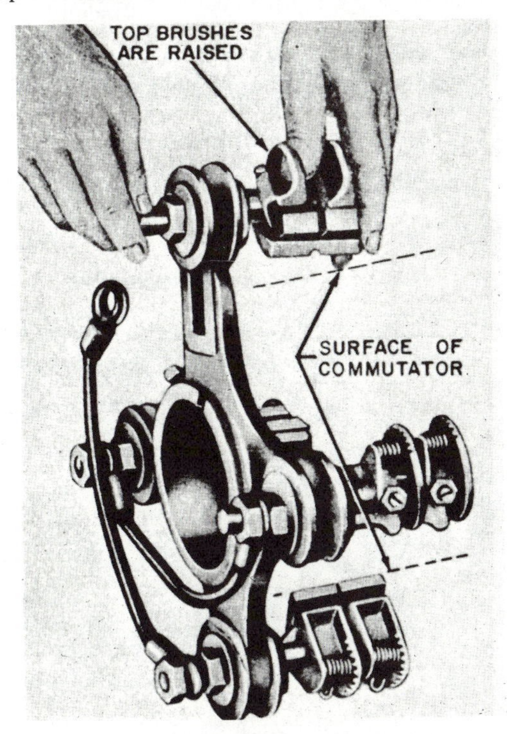

Figure 2–7. Typical brush rigging.

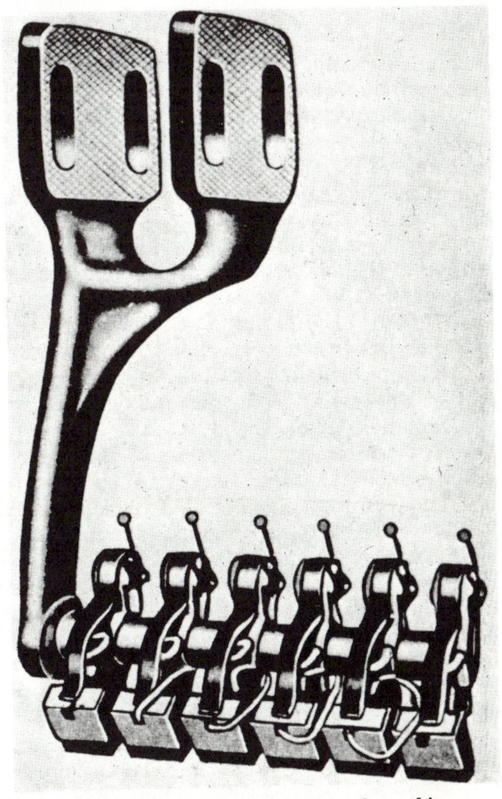

Figure 2–8. Brush arm from a large dc machine.

272-646 O - 78 - 2

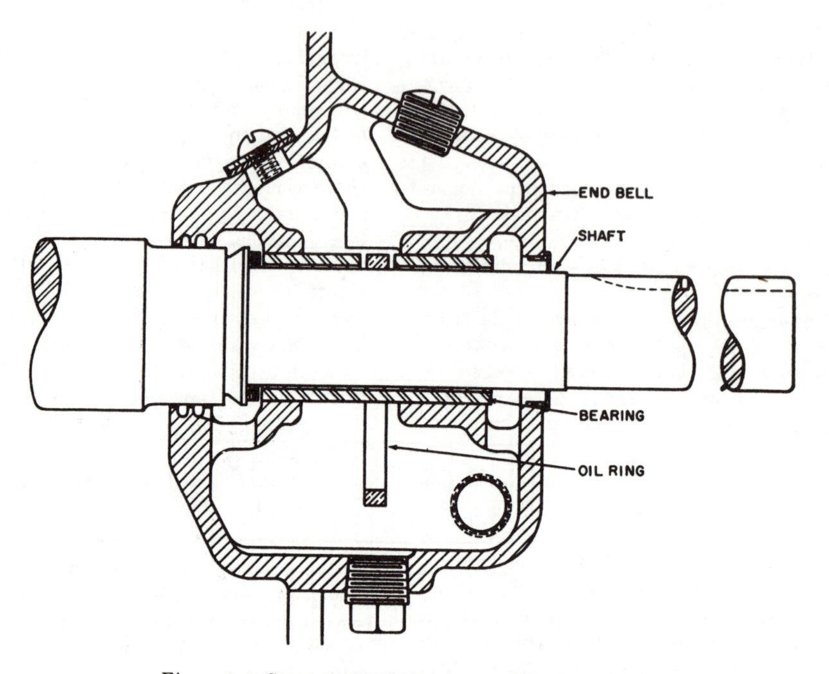

Figure 2–9. Cross-sectional view of an end bell assembly.

function is to improve commutation and to reduce sparking at the brushes to a minimum. In order to understand this, one should first understand the phenomenon of armature reaction. This is the effect that the armature flux has upon the field flux. The current flowing in the armature coils sets up a magnetic field of its own which opposes the magnetic field set up by the field structure. The presence of this opposing field causes a shift in the neutral plane of the brushes (para 1–6*b*(4)). Figure 2–11 illustrates how this effect, called armature reaction, is produced. ①, figure 2–11 shows the flux lines produced by the pair of poles when no current is flowing through the armature coils. The curvature in the flux lines is produced by the presence of the iron-core armature between the poles and is a normal distortion of the magnetic field in a generator. The line *ab* indicates the zero axis (neutral plane) of the field. ②, figure 2–11 shows the flux lines produced by current flowing through the armature coils alone; that is, it is assumed that the field coils are unexcited and that no field is produced by the poles. ③, figure 2–11 shows the resultant field of two fluxes: the flux of ① produced by the poles and the flux of ② produced by the current-carrying armature. Note that the zero axis of the resultant field is displaced as indicated by the line *a′b′*. This displacement results in a shift in the position of the neutral plane. The shift of the main field brings the armature coils, which are being short circuited by the brushes, under the influence of an additional slight field, which induces a low voltage in them. This low voltage is short-circuited by the brushes, causing sparking, which results in the burning and pitting of the commutator. As

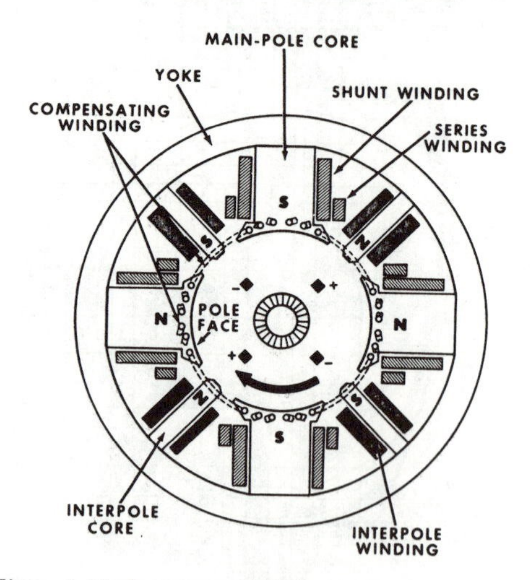

Figure 2–10. Arrangement of the various field windings on a dc machine.

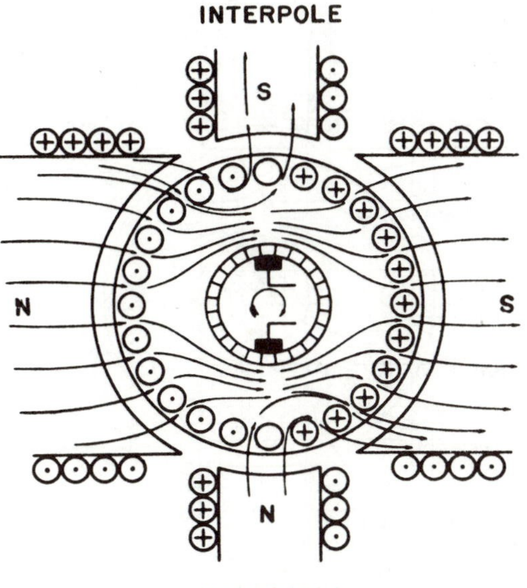

Figure 2–11. Armature reaction.

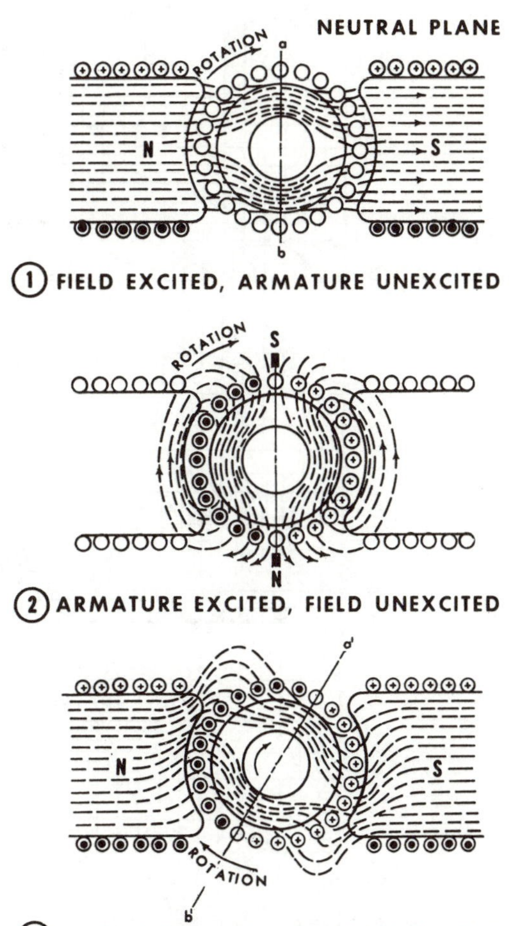

Figure 2–12. Effect of interpoles.

the load current and the resulting armature reaction increase, this effect becomes more pronounced. Various methods are used to counteract this effect.

(1) The entire brush assembly can be adjusted to bring the brushes in line with the shifted neutral plane. Since the neutral plane shifts with the load, it follows that the brush assembly must be rotated or shifted each time the load on the machine is changed. Such a procedure is impractical in most cases, and other means must be used for reducing the effects of armature reaction. In some generators without interpoles, the brushes are located a little ahead of the neutral plane. This allows for a slight increase in armature reaction (due to a change in the load condition) without causing serious damage to the brushes and the commutator.

(2) Commutating poles can be placed between the main field poles. These poles are smaller and narrower than the main field poles and the winding on them is in series with the armature and so connected that the field set up by them opposes the field caused by armature reaction. This can be seen by reference to ③, figure 2–11 and figure 2–12. Notice how the armature flux in ③, figure 2–11 combines with the field flux to cause distortion in the resultant flux passing through the armature. The distortion occurs to the top left and to the bottom right of the armature vertical centerline. Now, suppose that interpoles are placed between the main field poles with the polarities as shown in figure 2–12. Note how the flux lines created by the interpoles oppose the distorted flux caused by armature reaction. This, in effect, results in a flux which is almost as evenly distributed as the flux produced by the field poles with the armature unexcited as shown in ①, figure 2–11. Hence the neutral plane is not affected in a machine that has interpoles. In generators with interpoles, the brushes are centered on the interpole axis as shown in figure 2–12.

(3) The faces of the main field poles may be slotted longitudinally and windings placed in the slots. These windings are then connected in such a way that the field produced by them opposes the field set up by the armature. These

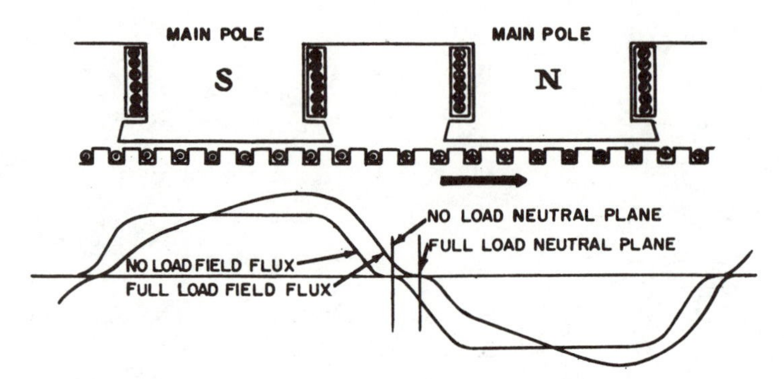

① WAVE FORM OF FIELD FLUX IN GENERATOR WITHOUT INTERPOLES

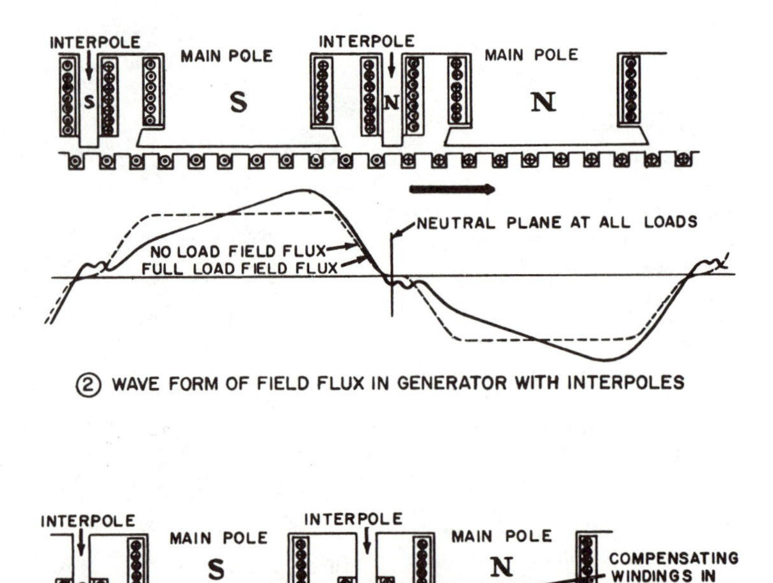

② WAVE FORM OF FIELD FLUX IN GENERATOR WITH INTERPOLES

INTERPOLE
MAIN POLE
INTERPOLE
MAIN POLE
S
S
N
COMPENSATING WINDINGS IN SLOTS
NEUTRAL PLANE AT ALL LOADS
NO LOAD FIELD FLUX
FULL LOAD FIELD FLUX

③ WAVE FORM OF FIELD FLUX IN GENERATOR WITH INTERPOLES AND COMPENSATING WINDINGS

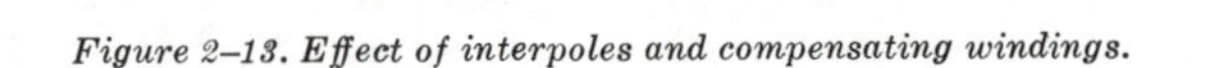
Figure 2–13. Effect of interpoles and compensating windings.

windings, referred to as compensating windings, help to produce a smoother dc output as well as to improve commutation. This effect and the effect of interpoles are shown in figure 2–13 ①, figure 2–13 shows the form of the field flux in a generator, without interpoles, at no load

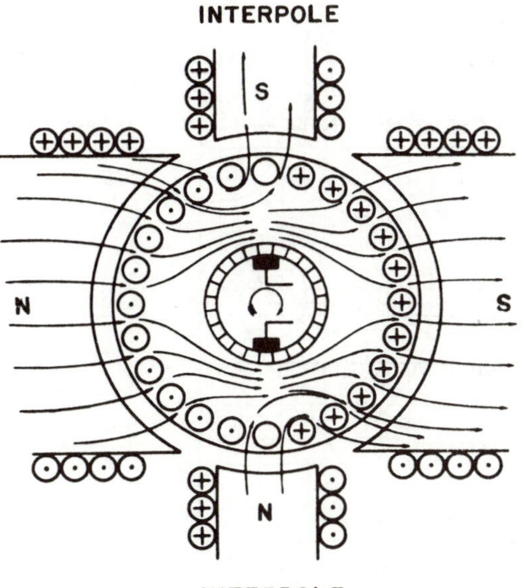

Figure 2–11. Armature reaction.

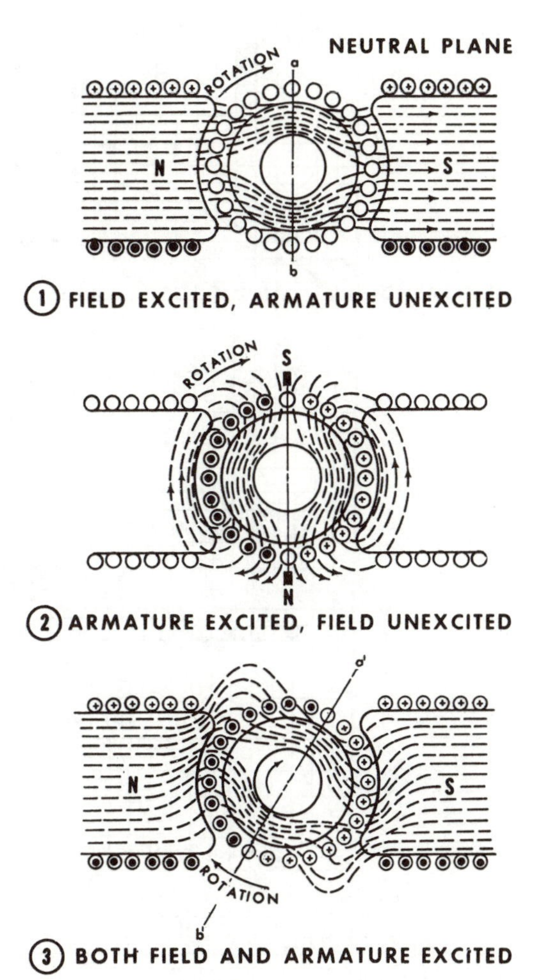

Figure 2–12. Effect of interpoles.

the load current and the resulting armature reaction increase, this effect becomes more pronounced. Various methods are used to counteract this effect.

(1) The entire brush assembly can be adjusted to bring the brushes in line with the shifted neutral plane. Since the neutral plane shifts with the load, it follows that the brush assembly must be rotated or shifted each time the load on the machine is changed. Such a procedure is impractical in most cases, and other means must be used for reducing the effects of armature reaction. In some generators without interpoles, the brushes are located a little ahead of the neutral plane. This allows for a slight increase in armature reaction (due to a change in the load condition) without causing serious damage to the brushes and the commutator.

(2) Commutating poles can be placed between the main field poles. These poles are smaller and narrower than the main field poles and the winding on them is in series with the armature and so connected that the field set up by them opposes the field caused by armature reaction. This can be seen by reference to ③, figure 2–11 and figure 2–12. Notice how the armature flux in ③, figure 2–11 combines with the field flux to cause distortion in the resultant flux passing through the armature. The distortion occurs to the top left and to the bottom right of the armature vertical centerline. Now, suppose that interpoles are placed between the main field poles with the polarities as shown in figure 2–12. Note how the flux lines created by the interpoles oppose the distorted flux caused by armature reaction. This, in effect, results in a flux which is almost as evenly distributed as the flux produced by the field poles with the armature unexcited as shown in ①, figure 2–11. Hence the neutral plane is not affected in a machine that has interpoles. In generators with interpoles, the brushes are centered on the interpole axis as shown in figure 2–12.

(3) The faces of the main field poles may be slotted longitudinally and windings placed in the slots. These windings are then connected in such a way that the field produced by them opposes the field set up by the armature. These

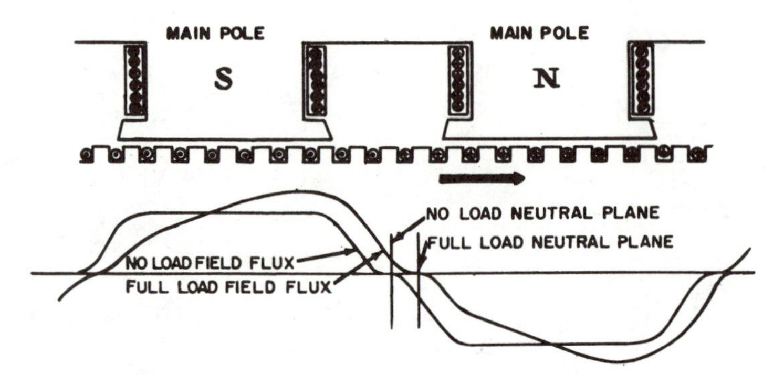

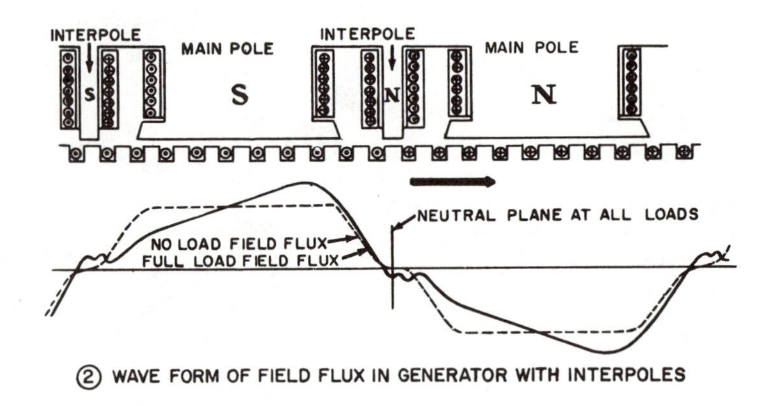

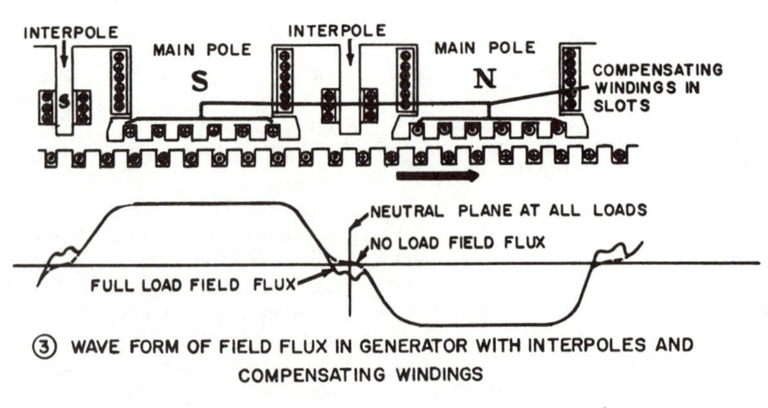

Figure 2–13. Effect of interpoles and compensating windings.

windings, referred to as compensating windings, help to produce a smoother dc output as well as to improve commutation. This effect and the effect of interpoles are shown in figure 2–13 ①, figure 2–13 shows the form of the field flux in a generator, without interpoles, at no load and at full load. At full load, the flux is distorted and the neutral plane, which is the commutating zone, is shifted. ②, figure 2–13 shows the form of the field flux in a generator with interpoles. At full load, almost the same flux distortion takes place as in ①, figure 2–13 but the shifting of the commutating zone is eliminated. By distributing, in slots on the pole faces, a compensating winding in series with the armature, but with current in the opposite direction to that in the adjacent armature conductors, the flux distribution and the commutating zone remain constant at both no load and full load (③, fig 2–13). Compensating windings are not used in many generators because of their high cost of manufacture. However, in high speed and high voltage generators where armature reaction is prevalent, compensating windings are used.

2–3. Types of DC Generators

DC generators are classified by the method of supplying excitation current to the field coils. The two major classifications are separately excited and self-excited generators. Self-excited generators are further classified by the method of connecting the field coils, as series-connected, shunt-connected, and compound-connected generators.

a. Separately Excited Generator. A dc generator which has its field supplied by another generator, batteries, or some other outside source is referred to as a separately excited generator (fig 2–14). Figure 2–15 shows the voltage characteristics of a separately excited generator. When operated at constant speed with constant field excitation but not supplying current, the terminal voltage of this type of generator will equal the generated voltage. When the unit is delivering current, the terminal voltage will be less than the generated voltage by an amount equal to the drop due to armature reaction (para 2–2*g*) plus the voltage drop due to the resistance of the armature and the brushes. Separately

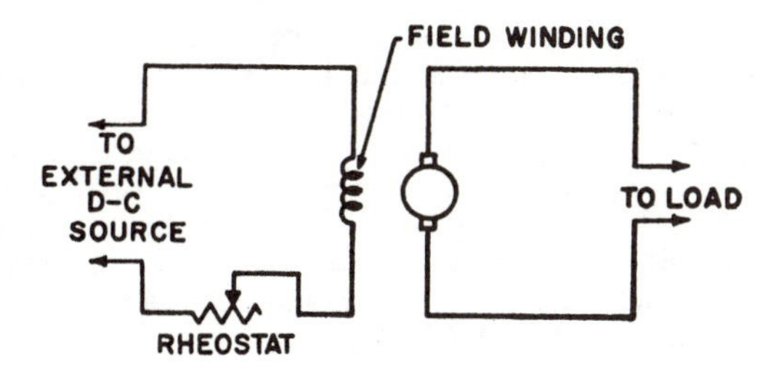

Figure 2–14. Connection of a separately excited dc generator.

excited generators, however, are seldom used for

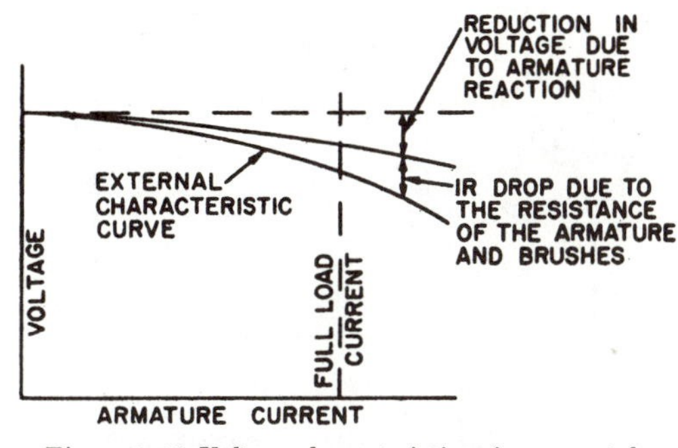

Figure 2–15. Voltage characteristics of a separately excited dc generator.

b. Series Generators. When all of the windings are connected directly in series with the armature, the generator is series-connected (fig 2–16). Figure 2–17 shows the voltage characteristics of a series generator. At no load, the only voltage present is that due to the cutting of the flux established by residual magnetism. (Residual magnetism is that which is retained by the poles of a generator when it is not in operation.) However, as the load is applied, the current through the field coil increases the flux and, therefore, the generated voltage. The voltage generated tends to increase directly as the current, but is prevented from doing so by three factors. The first of these is saturation of the field core. If the field excitation is increased beyond the point at which the flux produced no longer increases directly as the exciting current, the core is said to be saturated. The second factor is armature reaction, the effect of which can be seen to increase as the load of current increases. The third, loss in terminal voltage, is caused by ohmic resistance of the armature winding, the brushes, and the series field. This loss also increases as the unit is loaded. Since the terminal voltage of series generators varies under changing load conditions, they are generally connected in a circuit that demands constant current. When so used, they are sometimes referred to as constant-current generators, despite the fact they do not in themselves tend to maintain a constant current. This is done by connecting a variable resistance, either manually or automatically controlled, in parallel with the series field. Thus, as the load is increased, the resistance of the shunt path is decreased, permitting more of the current to pass through it, thereby maintaining a relatively constant field.

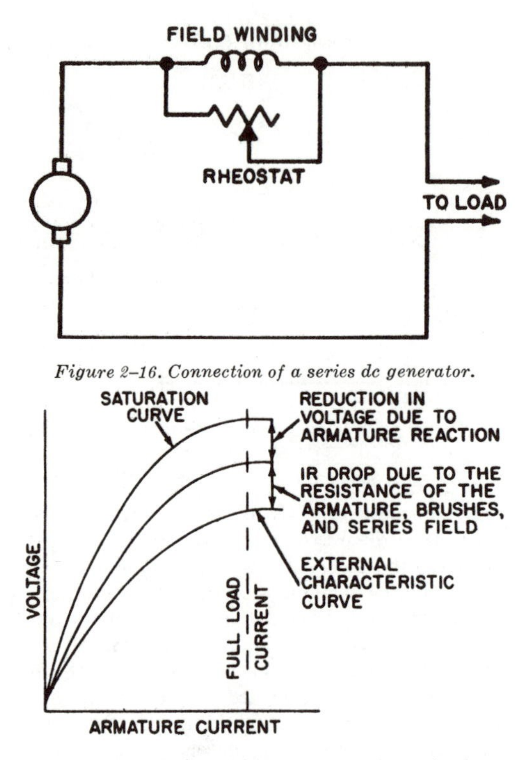

Figure 2–16. Connection of a series dc generator.

Figure 2–17. Voltage characteristics of a series dc generator.

c. Shunt Generator. When the field windings are connected in parallel with the armature, the generator is shunt-connected (fig 2–18). Figure 2–19 shows the voltage characteristics of a shunt generator. A comparison of the voltage characteristics of a shunt generator (fig 2–19) shows a similarity in behavior with that of a separately excited generator (fig 2–15). In both instances the terminal voltage drops from the no-load value as the load is increased, but the terminal voltage of the shunt generator remains fairly constant until it approaches full load. This is so, even though the graph of the shunt generator has a third factor, other than the armature reaction and the ohmic resistance, which causes the terminal voltage to decrease. This is the weakening of the field as the current approaches full load. It is, therefore, advantageous to use a shunt generator in place of a separately excited or a series generator, where a constant voltage at varying load is required. Shunt generators are readily adaptable to applications where the speed of the prime mover cannot be held constant, as in the aircraft and automotive fields. When so used, it is not desirable to have a constantly fluctuating terminal voltage. It is necessary, therefore, to control field current by varying the shunt field resistance to compensate for changes in speed of the prime mover.

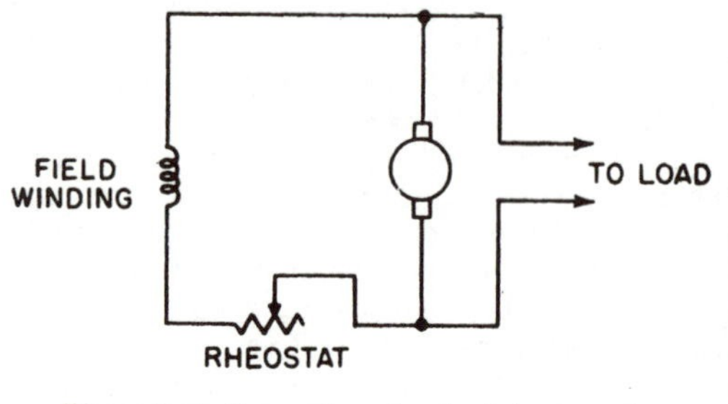

Figure 2–18. Connection of a shunt dc generator.

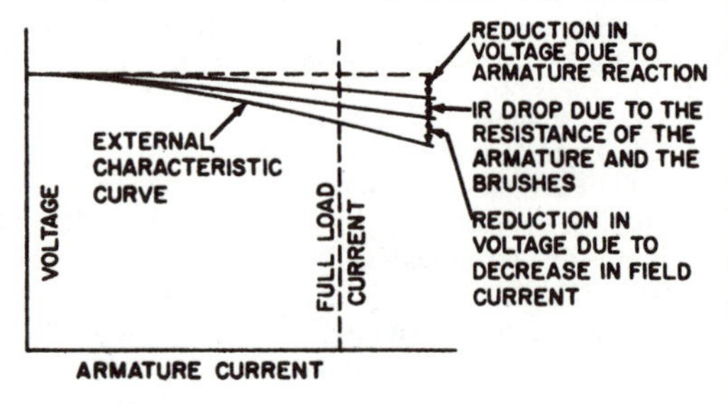

Figure 2–19. Voltage characteristics of a shunt dc generator.

d. Compound Generators. As described in *b* and *c* above, terminal voltages associated with series- and shunt-connected generators vary in opposite directions, with load-series connections increasing and shunt connections decreasing slightly with an increase in load. Thus, as shown in figure 2–20, if both a series and a shunt field were included in the same unit, it would be possible, by proper design of the respective fields, to obtain a generator with a voltage-load characteristic somewhere between that of either previous type (fig 2–21). By changing the number of turns in the series field it is possible to obtain three distinct types of compound generators.

(1) *Overcompound.* If the turns of the series field are more in number than is necessary to give approximately the same voltage at all loads, the generator is overcompounded. Thus the terminal voltage at full load will be higher than the no-load voltage. This is desirable where the power must be transmitted some distance. The rise in generated voltage compensates for the drop in the transmission line.

(2) *Flat-compound.* If the relationship between the turns of the series and shunt fields is

such that the terminal voltage is approximately the same over the entire load range, the unit is flat-compounded.

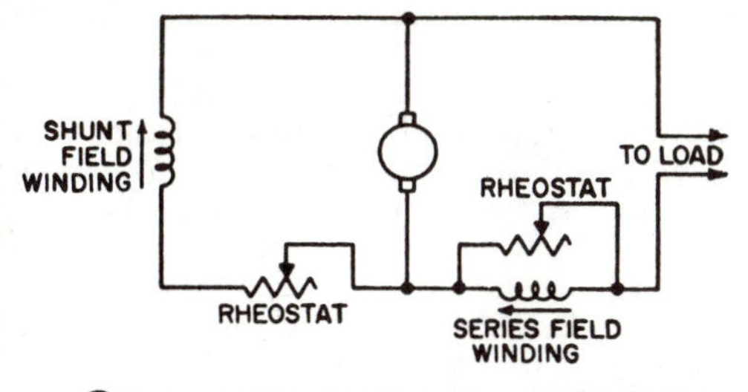

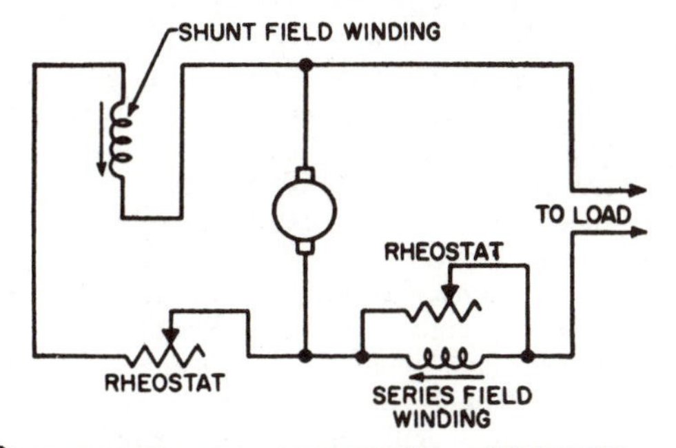

Figure 2–20. Connection of compound-wound dc generators.

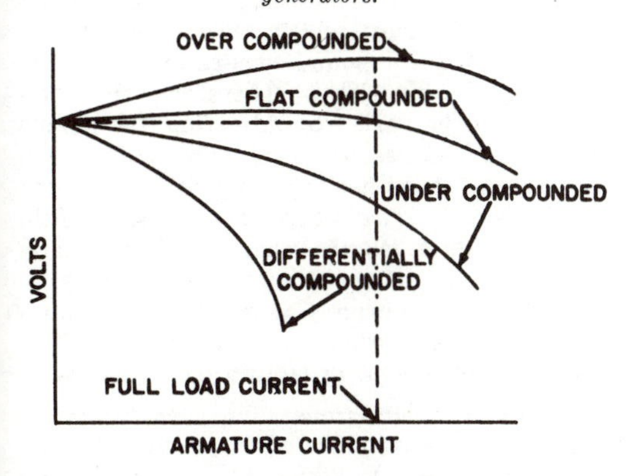

Figure 2–21. Voltage characteristics of compound-wound dc generators.

(3) *Undercompound.* If the series field is wound of so few turns that it does not compensate entirely for the voltage drop associated with the shunt field, the generator is undercompounded. In this type, the voltage at full load is less than the no-load voltage. An undercompounded generator, in which the series and shunt fields are connected so as to oppose rather than aid one another, is referred to as a differentially compounded generator. With this type of generator the terminal voltage decreases rapidly as the load increases. Undercompounded generators are used in applications where a short might occur, such as in welding machines.

e. Third-Brush Generators. This type of generator (fig 2–22) receives its field excitation from the armature by means of a third brush on the commutator. The output is controlled by the brush position using the principle of armature reaction. The third brush is set to take advantage of the reduction in generated voltage over part of the windings as generator speed is increased. Because of the action of the third brush, the output current increases until a maximum is reached and then is reduced as the speed is increased further. These generators find their largest application in the automotive field. A regulator is sometimes used to reduce the output current when the maximum is not needed.

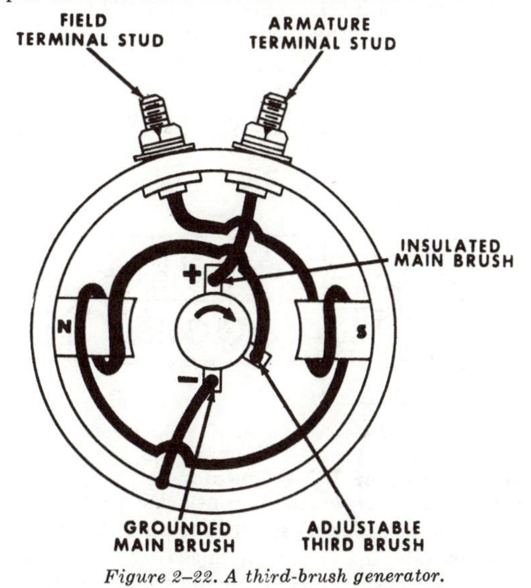

Figure 2–22. A third-brush generator.

f. Three-Wire Generators. The three-wire generator is similar to the two-wire generator in operation. It develops 240 volts across the armature terminals. The generator is arranged so a third wire, or neutral wire, is brought out from a point midway in potential between the positive and negative terminals. This provides for a lead at half generator voltage. This midvoltage is obtained by connecting a reactance coil, called a stationary compensator, across the armature winding through the collector rings. The neutral wire is connected to the midpoint of the compensator. One type of three-wire generator has a low-resistance compensator coil of heavy wire wound on an iron core. The coil is mounted on the armature and connected to the regular armature winding at diametrically opposite points (a and b, fig 2–23). The neutral wire is joined to the midpoint (c, fig 2–23) of the coil by a slip-ring and brush. The emf across the terminals a and b is alternating. Consequently, an alternating current will flow through the coil, but its value is small because of the large inductance of the coil. The midpoint (c, fig 2–23) has a potential midway between the potential of the brushes connected to the outside wires. If the load taken from one side of the circuit is equal to the load taken from the other, no direct current will flow in the neutral wire nor through the coil. But if the loads are unbalanced, the neutral wire and coil will carry a direct current equal to the difference in currents, or the amount of unbalanced current in the two circuits. The chief advantage of the three-wire system is a saving in copper, because the neutral wire carries less than 25 percent of the rated current output of the generator. Therefore, it can be much smaller than either of the two outer wires.

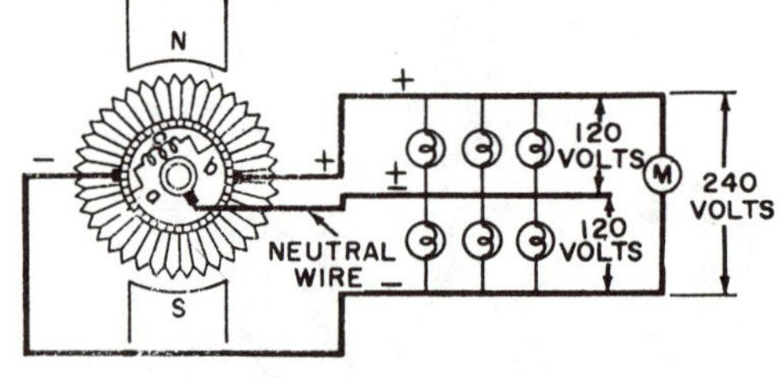

Figure 2–23. Diagram of a three-wire generator.

2–4. Control of DC Generators

Generally, a dc generator is controlled by a variable resistance called a rheostat, after the generator is brought up to proper speed by the prime mover. The rheostat may be manually or automatically operated. The adjustment of the rheostat controls the amount of exciter current fed to the field coils. Metering requires the use of a dc voltmeter and ammeter of appropriate ranges in the generator output circuit. Matched sets of shunt-wound or compound-wound generators with series-field equalizer connections are used for parallel operation. Precautions must be observed when connecting the machines to generator buses.

a. Voltage Regulation.

(1) *Series generator.* The series generator may be classified as a constant-current generator and, as such, may be used to supply series motors, series arc-lighting systems, and voltage boosting on long dc feeders. The series generator is excited entirely by low-resistance field coils connected in series with the armature terminals and the load (fig 2–16). The voltage increases with load since the load current provides the necessary additional field excitation. Low-resistance shunts may be used across series field coils to obtain the desired voltage characteristic. The series field of the generator is adjusted so the output voltage may be maintained at a constant value. Because series generators have poor regulation, only a few are in actual use.

(2) *Shunt generator.* The shunt dc generator may be classified as a constant-potential generator although it is seldom used for lighting and power because of its poor voltage regulation. The field coils have a comparatively high resistance in this type of generator and are connected across the armature terminals in series with the rheostat (fig 2–18). Shunt generators sometimes have separate excitation to prevent reversal of the generator polarity and to obtain better voltage regulation. Shunt generators are frequently used in conjunction with automatic voltage regulators as exciters for ac generators.

(3) *Compound generator.* The compound generator is the most widely used dc generator. The speed of a compound generator affects its generating characteristics. Therefore, the compounding can be varied by adjusting the engine governor for higher or lower speeds and then adjusting the shunt-field rheostat for the proper no-load voltage. The range of the shunt-field rheostat and the engine characteristic usually limit the amount of speed variation that may be obtained for this purpose. Compound generators can be connected either cumulatively or differentially, as illustrated in ① and ②, figure 2–20.

b. Parallel Connections. The generators in power stations are connected in parallel through common bus lines, to which the output feeders are also connected.

(1) *Shunt generators.* Figure 2–24 shows two shunt generators connected in parallel. The load is kept in balance by adjustment of the rheostats. The voltage can be raised, while keeping the currents equal, by increasing the field excitation of both generators. For generators of unequal capacities, the loads are generally divided in proportion to their ratings.

(2) *Compound generators.* Undercompounded generators connected in parallel will operate satisfactorily, but overcompounded generators will not operate satisfactorily in parallel unless their series fields are also parallel. This is done by bringing the negative brush connections of each generator to a common point as shown in figure 2–25. The conductor or bus used to connect these brushes is called an equalizer. When the equalizer is used, a stabilizing action takes place. If generator A (fig 2–25) takes more than its proper share of the load, the increased current will flow through the series field of generator A, but some ot it will also flow through the equalizer and through the series field of generator B. Thus, both generators are affected in a similar manner and neither machine takes the entire load.

c. Bus Connections. The procedure for paralleling a generator on the bus is as follows:

(1) Refer to figure 2–25 and bring generator A up to rated speed and voltage.

(2) Close the negative switch of generator A.

(3) Close the positive switch of generator A.

(4) Load generator A.

(5) Bring generator B up to rated speed and voltage.

(6) Close the equalizer switch on generator A.

(7) Close the negative and equalizer switch

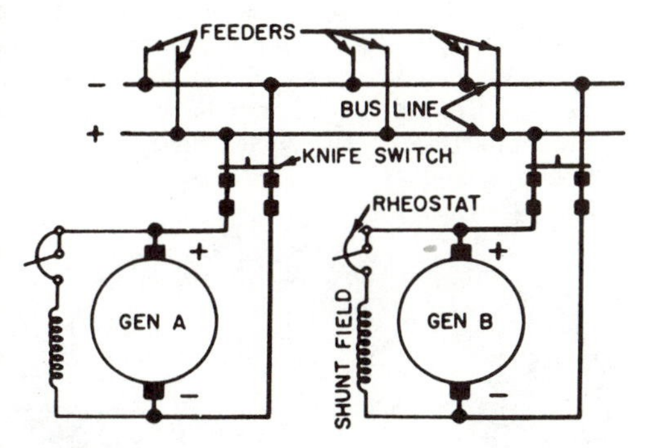

Figure 2–24. Parallel connection of shunt dc generators.

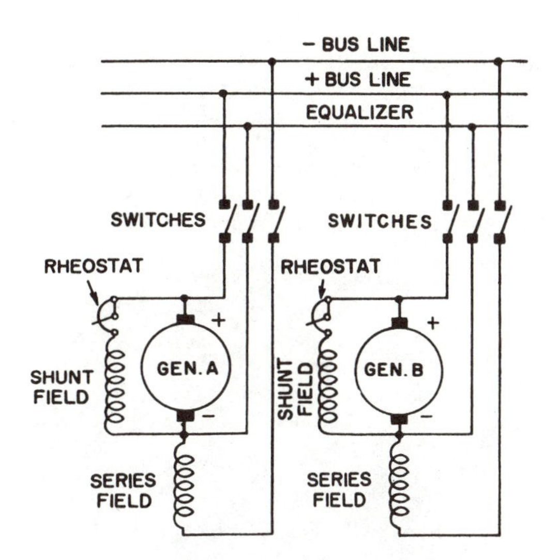

Figure 2–25. Parallel connection of overcompounded dc generators with equalizer.

on generator B, causing some of the load current to pass through the series field of machine B, by adjusting the field rheostat.

(8) Adjust the voltage of generator B to equal that of the bus line.

(9) Close the positive switch of generator B.

(10) Adjust or balance out the loads of both machines by increasing the field strength of machine B with the field rheostat.

d. Shutting Down a Generator Connected in Parallel.

(1) Refer to figure 2–25 and bring load current to zero on generator to be shut down and shift load to generator to be kept in service by adjusting the corresponding field rheostats, meanwhile keeping the output voltage of both generators equal to the bus voltage.

(2) Open the positive switch.

(3) Open the negative and equalizer switches of the generator.

(4) Shut down and secure the generator.

Section II. DC MOTORS

2–5. Basic Considerations

Fundamentally, there is little difference in the construction of dc generators and dc motors, except in shunt-field windings and the size of the field-coil resistance, which must be somewhat less for a generator. The brush angle, too, may be different, since motors are often required to reverse rotation. It is quite possible to convert a dc motor into a dc generator by means of a prime mover connected to the shaft, providing the direction of rotation of the armature remains the same. Otherwise it is necessary to reverse the polarity of the residual magnetism in the field poles. Usually the residual magnetism in the field poles of a motor will, as in generators, provide sufficient voltage to create a field. A typical dc motor is shown in figure 2–26.

Figure 2–26. Typical dc motor.

2–6. Principles of Operation

Paragraph 1–9 showed that a definite relation exists between the direction of the lines in a magnetic field, the direction of the current in the conductor, and the direction in which the conductor tends to move. Furthermore, a conductor carrying current in a magnetic field tends to move at right angles across the field. Therefore, if an electromotive force is impressed across the windings of an armature by means of brushes and commutator, and also the field poles, a force is created sufficient to turn the shaft of the motor. The overall turning effect on the armature, that is, the combined force on the two sides of the loop times the distance of the conductors from the axis (shaft center), is called the torque.

2–7. Types of DC Motors

The connections of fields and armatures in motors are the same as in generators. The only differences between motors and generators are mechanical. A motor may be series-wound, shunt-wound, or compound-wound. Ther are also other ways of classifying dc motors—by speed rating and mechanical modifications. Physical characteristics of motors serve to classify as well. For example, motors are identified as being open, enclosed (fig 2–27), dripproof, and so on. All of the foregoing types of motors are available with interpoles to improve commutation.

a. Series Motors. The field of a series motor is obtained by a low-resistance coil connected in series with the armature. Any additional load placed on a series motor will cause more current to flow through the armature to produce the necessary torque. Since this increased current must pass through the series field, there will be a greater flux, which in turn will produce a large counterelectromotive force at low speed. The motor speed will therefore greatly decrease. The characteristic of a series motor is that speed changes rapidly with torque and when torque is high, speed is low; or if torque is low, speed is high. Never start a series motor without a load or remove the load while the motor is in operation. The motor will gain speed until it goes so fast the motor is damaged. The connection of a series dc motor is illustrated in figure 2–28.

b. Shunt Motors. In a shunt motor the field is across the line or in parallel with the armature (fig 2–29). The rheostat in series with the field winding is used for speed control. The field current stays the same, regardless of changes in armature current. Therefore, when the armature current is doubled, the torque is doubled. The speed of a shunt motor changes very little with change of load, the speed increasing when the load decreases. The characteristic of the shunt motor is an almost constant speed for all reasonable loads.

c. Compound Motors. Compound motors differ from the stabilized shunt types by having a more predominant series field. Like compound generators, compound motors can be divided into two classes, differential and cumulative, depending on the connection of the series field in relation to the shunt field.

Figure 2–27. Totally enclosed, fan-cooled, dc motor.

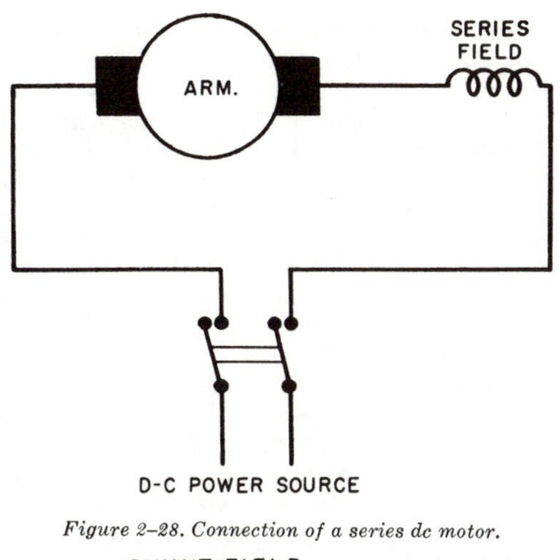

Figure 2–28. Connection of a series dc motor.

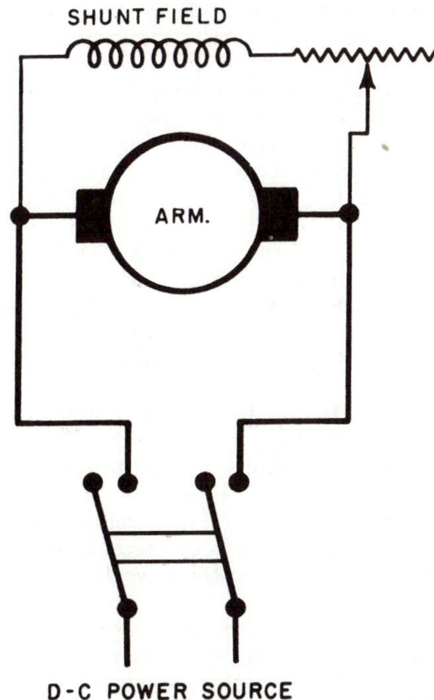

Figure 2–29. Connection of a shunt dc motor.

(1) *Differential-compound motors.* A diagram of a differential-compound motor is shown in figure 2–30. In this type, the series field opposes the connected shunt field. Therefore, this motor operates at practically a constant speed. As the load increases, the armature current increases to provide more torque. The series magneto motive force increases, thus weakening the shunt field and reducing the counterelectromotive force in the armature without causing a reduction in speed.

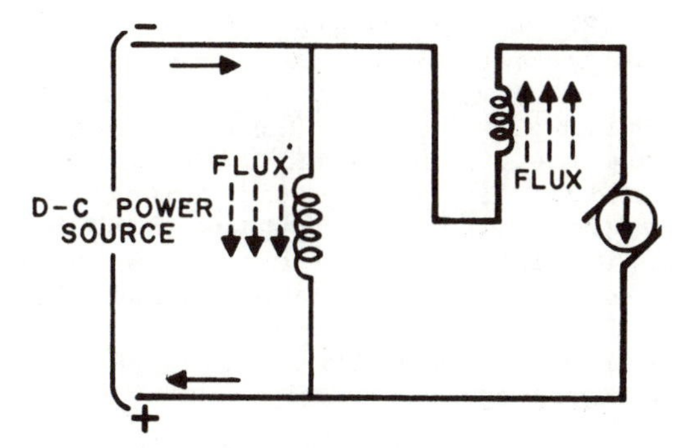

Figure 2–30. Connection of a differential-compounded motor.

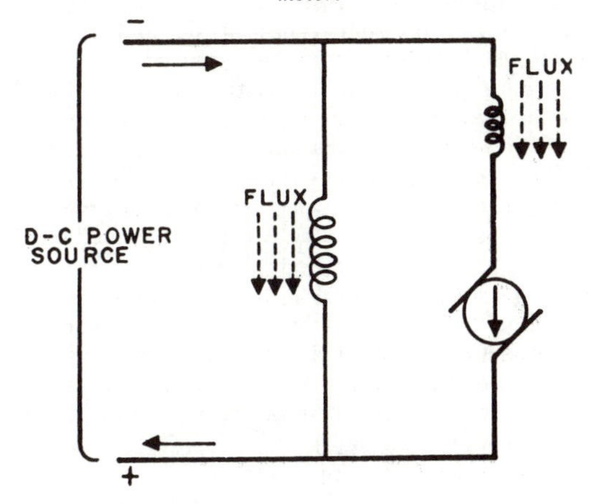

Figure 2–31. Connection of a cumulative-compound motor.

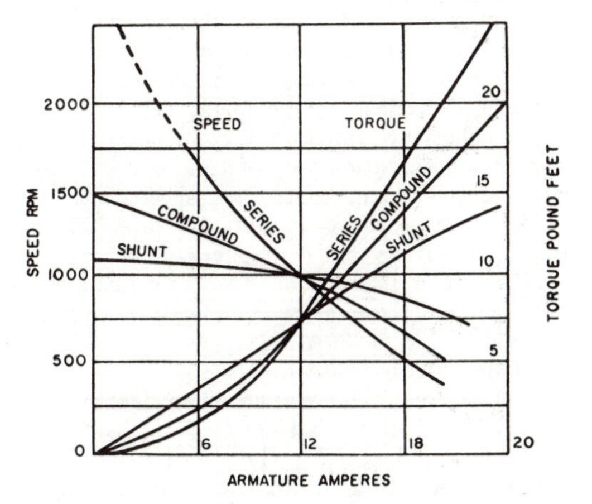

Figure 2–32. Operating characteristics of series, shunt, and compound motors.

(2) *Cumulative-compound motor.* The cumulative-compound motor diagrammed in figure 2–31 is connected so that its series and shunt fields aid each other. From this comes the name, cumulative-compound motor. A motor thus connected will have a very strong starting torque, but poor speed regulation. Motors of this type are used for machinery where speed regulation is not necessary but where great torque is desired to overcome sudden application of heavy loads. The operating characteristics of series, shunt, and compound motors are shown in figure 2–32.

2–8. Application Data

The several different types of dc motors that have just been discussed have varied particular uses depending on their construction. Continued use of any dc motor depends on the kind of load to be carried, the speed, and the torque. DC motors not adapted to the load or of improper rating should not be used even for short periods due to possible damage by overheating, inefficient operation, or improper speed control. The location in which the motor is to be operated is also a factor to consider when selecting a machine for a particular job. A dusty atmosphere requires an enclosed motor; high temperatures necessitate a motor with special insulation, and so on. Table F–1 in appendix F provides information on motor selection and application.

2–9. Controllers

A controller is a device for regulating the operation of electrical equipment. A controller for electric motors is simply a mechanism which conveniently and safely performs several or all of the following functions: connection to the power line, limitation of starting current, control of acceleration, control of speed, and disconnection from the power line. Because of this functional variety in controllers, they can be classified in several ways. This paragraph discusses some of these. More complete coverage can be found in TM 5–680A.

a. Types of Controllers.

(1) *Manual.* The manual type controller is one having all of its basic functions performed by hand. The basic functions are usually line closing, acceleration, retardation, and reversing. Manual control permits regulation of machines from only one position and is limited in the size and capacity of the equipment that can be so controlled. Figures 2–33, 2–35, and 2–36 are good examples of manual controllers.

(2) *Semimagnetic.* This controller has part of its basic function performed by electromagnets, and part by other means (fig 2–33).

(3) *Full magnetic.* The full magnetic controller performs all of its basic functions by electromagnets. The power circuits to the motor are closed and opened by magnetic contactors (fig 2–34). The contactors are controlled by a pilot device which has small current capacity. The pilot device may be manually operated, by push-button or master switch, or it may be automatically operated by a float switch or thermostat. Magnetic controllers make it possible to control motors automatically. This has advantages over the manual types. For example, the operator may accelerate the motor too rapidly with the manual type controller, with the result that the motor may take excessive current. This causes fuses to blow or circuit breakers to open. Furthermore, the starting resistance or resistances may burn out. With automatic controllers, the starting resistances may be cut out at the maximum safe rate by magnetically operated contacts. The operator need only press a button and the electromagnetic relays start the motor and bring it up to speed automatically in the proper time sequence. The motor is stopped by merely pressing the stop button. Magnetic controllers are used principally for—

(*a*) Smooth acceleration, retardation, and reversing of motors without damage.

(*b*) Control of a motor from one or more stations.

(*c*) Automatic control where an attendant is not present.

(*d*) Operation of a motor at a distant point.

(*e*) Operation of high-voltage equipment.

(*f*) Conservation of space, by locating the controller in an out-of-the-way place and operating it by a pilot device. These types of controllers may be further divided into general classes, starters and speed regulators. The starter is designed for accelerating a motor to normal speed in one direction of rotation. (If it is designed for starting a motor in either direction of rotation, it is usually called a reversing controller.) The speed regulator is designed for operating a motor at a speed either below or above normal.

b. Faceplate Controllers. The difference in appearance of a faceplate starter (fig 2–33) and faceplate controller (fig 2–35) is the arrangement of contact segments on the face of the control. On a starter, the contact segments are mounted in an arc of a circle. On a controller,

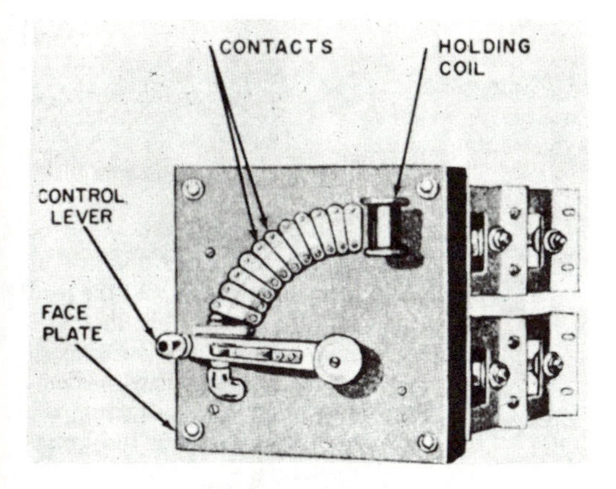

Figure 2–33. Open type motor starter.

the contact segments are mounted in a complete outer circle and another set (fewer in number) in a complete inner circle. On a faceplate controller, the segments are contacted by brushes attached to (but insulated from) the horizontal arm, which is actuated by a handle. The two sets of brushes at each end of the arm are connected together, thus forming a circuit between the inner contact segments and outer contact segments. When the controller is in the OFF position, the brushes rest on insulation pieces. When the handle is moved in the FORWARD direction, sections of starting resistance are cut out in steps, thus causing the motor to accelerate. When the handle is moved in the REVERSE direction, the motor reverses.

c. Drum Controllers. A drum controller consists essentially of a drum cylinder insulated from a control shaft to which an operating handle is keyed (fig 2–36). Copper contact segments are attached to the drum, and are connected to, or insulated from, one another as the situation requires. A series of stationary fingers is arranged to contact with the segments. The fingers are insulated from one another but connect to the starting resistance and to the motor circuit. Just under the handle, a notched wheel is keyed to the shaft. A spring forces a roller into one of the notches when contacts are properly made; this indicates to the operator the correct position of the handle and also that a step has been completed.

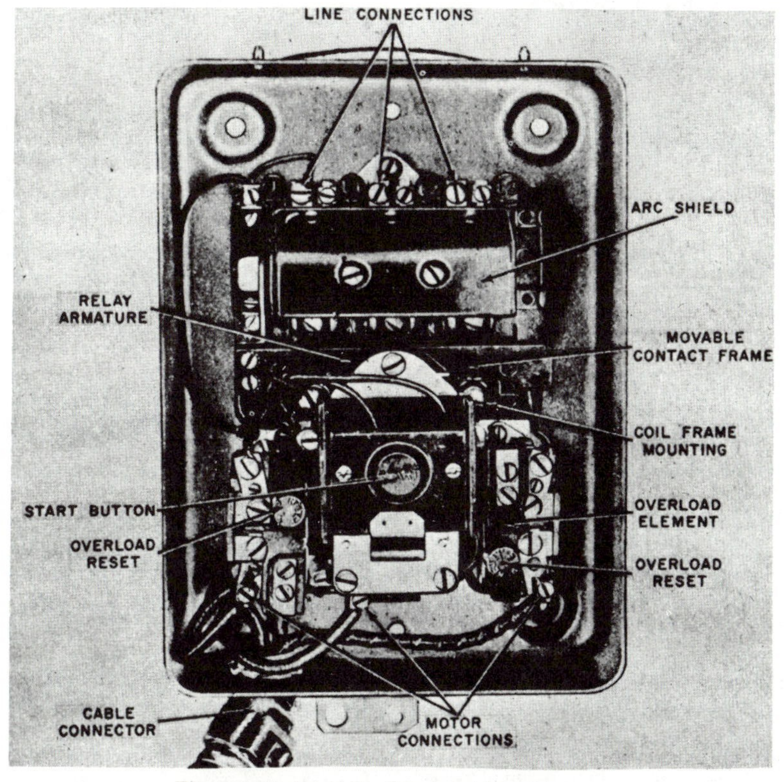

Figure 2–34. Across-the-line magnetic starter.

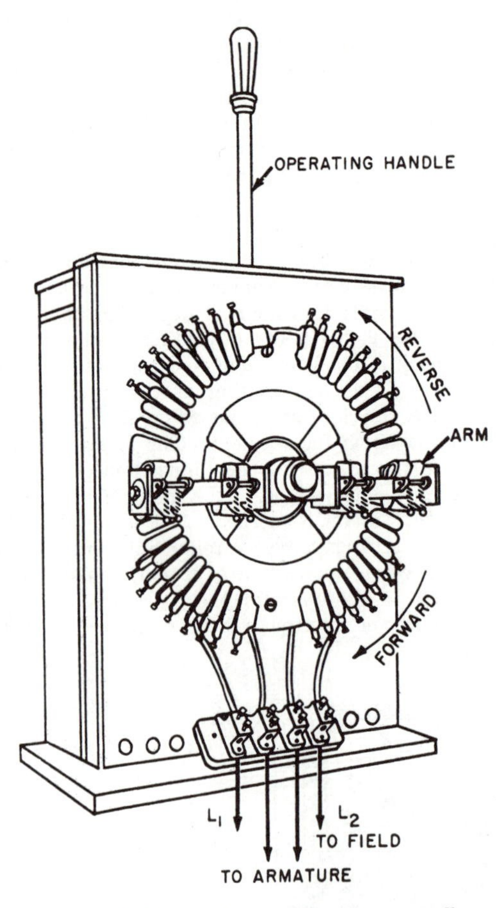

Figure 2–35. A forward-reverse faceplate controller.

When the controller handle is moved forward one notch, the motor starts, with resistance in its armature circuit. As the handle is turned further (notch by notch), the resistance is cut out of the armature circuit in steps and is inserted in the field circuit. When the handle reaches its limit, all the resistance is cut out of the armature circuit and the motor is operating at its maximum speed. By moving the handle from the OFF position in the opposite direction, the motor will be caused to run in the reverse direction. Figure 2–37 is a diagrammatic representation of the same drum controller (fig 2–36) with its contact fingers rolled out flat.

d. Starting and Accelerating.

(1) There are two methods of starting dc motors—on full voltage and on reduced voltage. The full-voltage controller connects the motor directly to the power lines. The reduced-voltage controller impresses, at first, less than full line voltage on the armature terminals, and then,

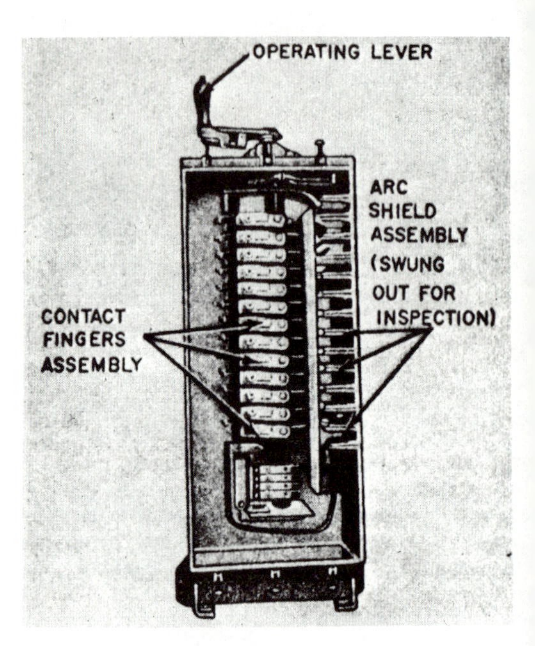

Figure 2–36. A drum type controller.

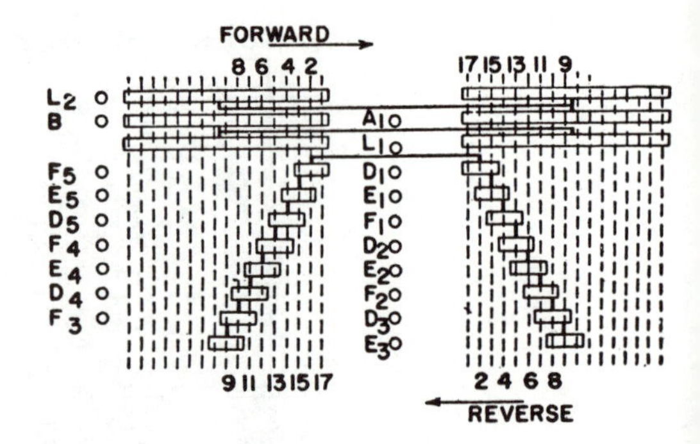

Figure 2–37. Diagrammatic representation of a drum controller with its contact fingers rolled out flat.

by one or more steps, increases the voltage at the armature to full-line voltage. Good practice dictates that a dc motor of over 1/3 horsepower be started by a reduced-voltage controller, either manually or magnetically operated. Acceleration of dc motors is obtained by manual control, with a faceplate type starter (fig 2–33) or drum controllers and resistors (fig 2–36), and by magnetic control, applying the principle of counterelectromotive force, current limit, or definite time. By counterelectromotive force (cemf) is meant the countervoltage built up by the rotation of the motor armature, acting as a generator, in the magnetic field. This counterelectromotive force always opposes the applied electromotive force and is therefore defined as countervoltage or back voltage.

(2) The counterelectromotive force (cemf) principle is the simplest form of acceleration for magnetic starters with dc motors of up to about 10 horsepower. The acceleration depends on the increasing value of the armature voltage or cemf as the motor accelerates. This cemf is impressed across accelerating contactor coils, which operate contactors cutting out resistors in the armature circuit. Starters of this type usually have not more than three points of acceleration, and are satisfactory where the motor load and line voltage are fairly constant.

(3) Current-limit acceleration requires current relays which are adjusted to close the contacts at a value of current higher than that drawn by the load to be accelerated. The high current peaks hold the current relay contacts open. When the armature current has decreased to a value for which the relay is adjusted, the relay contacts close and energize the coil of the contactor. The contactor, in turn, cuts out one step of resistance, so that the motor accelerates more.

(4) In definite-time acceleration, time-element starters cut out the accelerating resistors within a fixed time limit, regardless of the load. If the motor is heavily overloaded and stalled on the first point, the starting and acceleration are automatically forced on a later point. The overload relay will protect the motor from very high currents as well as normal overloads.

e. Speed Regulation.

(1) Faceplate type starters are made and used for various services, such as starting duty with speed regulation by armature resistance, and starting duty with speed regulation by both armature and field resistance.

(2) Drum controllers, standard, listed types, and resistors can be used for speed regulation by either armature resistance or field resistance, but not a combination of both.

(3) Other methods for regulating speed are by the variable-voltage or Ward-Leonard system, by multivoltage control, and by shunting the armature with resistance.

f. Retardation. Deceleration is obtained by the following means:

(1) The resistance in the armature circuit is increased, or in the case of the adjustable speed dc motor, resistance is reduced in the field circuit.

(2) Dynamic braking is applied for quick-stopping of shunt and compound motors, and sometimes as a step in reversing service. When the motor circuit is opened, a dynamic-braking resistor is instantly connected in parallel with the armature. The shunt field remains energized. The motor, running by inertia, now acts as a generator, and the braking resistor acts as a load on the generator. Braking resistors are seldom designed to permit more than 150 percent load current at the instant the braking circuit is closed. Dynamic braking is seldom applied to series motors because the series field must be reversed to be effective. This complicates the control.

(3) Another method of bringing motors to a quick stop is plugging. This is done by reversing the motor connections to the power line. It is used most often with series motors or heavily compounded motors.

g. Application.

(1) Full-voltage controllers, either manual or magnetic, can be used to connect small dc motors of up to about 2 horsepower directly to the line.

(2) Reduced-voltage controllers, either manual or magnetic, are used to operate dc motors of 1/2 horsepower (rated) and larger.

(3) Manual reduced-voltage starters and speed regulators of the faceplate type, with self-contained resistors, are nonreversing and are used with motors of up to about 25 horsepower. Separate devices are necessary for reversing service.

(4) Drum controllers with separately mounted resistors for starting or speed regulating duty may be used with motors of up to about 50 horsepower.

(5) Magnetic controllers are used for any size of motor and can be obtained for nonreversing of reversing service and other basic functions.

(6) Table F–3 provides further information on the application of deceleration controllers.

Section III. AC GENERATORS

2–10. Salient-Pole Type Alternators and Synchronous Motors

An alternator is an ac generator in which the armature is stationary and the magnetic field is rotated inside the armature. This is the reversal of the dc and ac generators, previously described, in which the armature rotates inside the magnetic field. In alternators, the armature coils are held in place in slots in the frame and the field coils are wound on poles or slots in the rotating part. A salient-pole type alternator is merely an alternator with salient-type (projecting) field poles as contrasted with the turbo-type (slotted) field poles. Just as the usual dc generator can be energized to operate as a dc motor, so can an ac generator be operated as an ac motor. Similarly, an alternator will operate as a motor without any changes in construction. When so used, it is called a synchronous motor. The so-called induction generator is simply an alternator of the salient-pole type construction. It is used in traction work, as an induction motor on the level and up grades. As a traction units goes down a steep grade, the motor may speed beyond synchronism and generate power which is returned to the source. This is known as regenerative braking. An application of this method is the automatic control of the downgrade speed of electric trains which are equipped with induction motors. Refer to TM 5–680G for further information.

2–11. Principles of Operation of AC Generators

a. The fundamental principle for the generation of an emf in an ac generator is the same as in a dc generator. The generation of an emf in an armature conductor depends solely on a relative motion between the conductor and the magnetic field. The magnetic field may be stationary, in which case it is called the stator; and the armature is rotated, in which case it is called the rotor. When the armature is stationary, it is called the stator, and when the magnetic field is rotated, it is called the rotor. In either case, the part of the machine which rotates is called the rotor and the part which remains stationary is called the stator. In practically all dc generators, the field is stationary and the armature is rotated. But in practically all ac generators, the armature is stationary and the field is rotated. This construction has several advantages. A rotating armature requires sliprings for carrying current to the external load. Such rings are difficult to insulate and are a frequent source of trouble, often causing open and short circuits. A stationary armature needs no sliprings. The armature leads can be continuously insulated conductors from the armature coils to the bus bars. It is also more difficult to insulate conductors in a rotating armature than in a stationary armature, because of the centrifugal force resulting from rotation in the former type. The stationary armature enables alternators to operate at voltages that are impossible in dc generators.

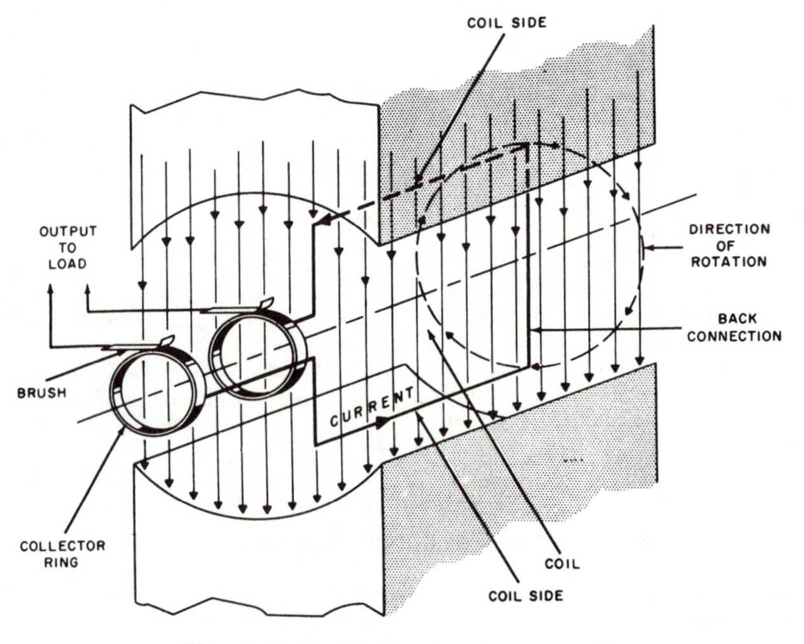

Figure 2–38. Simplified drawing of an ac generator.

b. A simplified drawing of an ac generator is shown in figure 2–38. Certain features of this generator are basic to the design of all ac generators:

(1) Some means of providing a magnetic field is necessary. In this case, the north and south poles of a permanent magnet are used.

(2) The coil must be rotated through the magnetic field.

(3) A coil is composed of two coil sides in which emf's are induced.

(4) A coil of more than one turn presents a coil side of more than one conductor. The total emf induced in a coil side is equal to the emf induced in one conductor multiplied by the number of conductors.

(5) At any instant, the emf induced in one coil side is equal and opposite in direction to that induced in the other coil side. The two emf's appear in series between the collector rings because of the back connection.

(6) Each coil side is connected to a metallic collector ring.

(7) A brush is continually in contact with

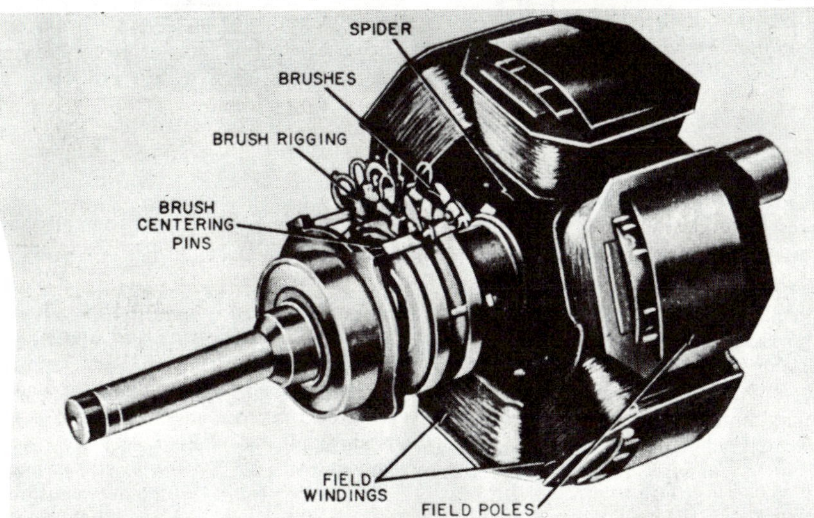

① SALIENT-POLE ROTOR (6 POLES) FOR SLOW SPEED ALTERNATOR

② SALIENT-POLE ROTOR SECTION FOR SLOW SPEED ALTERNATOR

Figure 2–39. Salient-pole type rotor.

272-646 O - 78 - 3

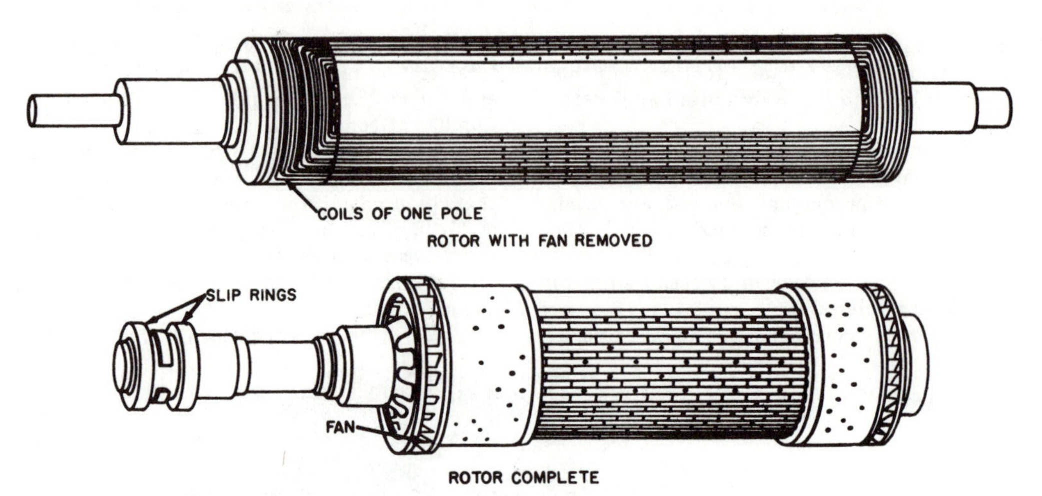

Figure 2–40. Turbo type rotors.

each collector ring. The brushes conduct the current in the coil to the load.

c. The voltage produced by this generator is an alternating voltage. One complete revolution of the coil will produce one cycle of voltage. That is, the voltage builds up from zero to a maximum, falls off to zero, builds up again in the opposite direction to a maximum, and falls off to zero to complete the cycle. Such a cycle of alternating current or voltage is usually represented as a sine wave.

2–12. Components of an AC Generator

a. Rotor. As previously stated, the rotating field in a generator is called the rotor and, when the field of the ac generator is placed upon the rotor, it is either of the salient-pole type (fig 2–39) or the turbo type (fig 2–40). When the ac generator is to be driven by a slow-speed diesel engine or by a water turbine (from about 720 rpm down), the salient-pole or projecting pole rotor (fig 2–39) is used. The field poles are formed by fastening a number of steel laminations to a spoked frame or spider. The heavy pole pieces produce a flywheel effect on the slow-speed rotor. This helps to keep the angular speed constant and reduce variation in the voltage and frequency of the generator output. In high-speed alternators (up to 3,600 rpm), the smooth surface turbo type rotor (fig 2–40) is used because it has less air-friction (heating) loss and the windings can be so placed that they can withstand the centrifugal forces developed at high speeds. The turbo type rotors are made from a solid-steel forging, or a number of steel disks fastened together with the field coils locked in slots. These field coils are usually distributed so they distribute the field flux evenly around the rotor.

b. Stators. In a rotating-field, ac generator, the armature windings are stationary, and are called the stator. The armature iron, being in a moving magnetic field, is laminated in order to reduce eddy-current losses. A typical ac generator stator is shown in figure 2–41. In high-speed turbo type generators (fig 2–42) the stator laminations are ribbed to provide sufficient ventilation because the high temperature developed in the windings cannot be dissipated in the small air gap be-

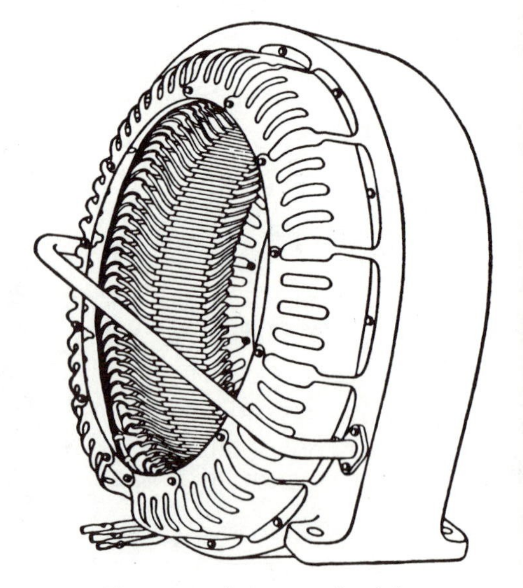

Figure 2–41. An ac generator stator.

tween the rotor and the stator. In some larger installations the alternators are totally enclosed and cooled by hydrogen gas under pressure, which has greater heat-dissipating properties than air. Stator coils in high-speed alternators must be well braced to prevent their being pulled out of place when the alternator is operating at heavy load.

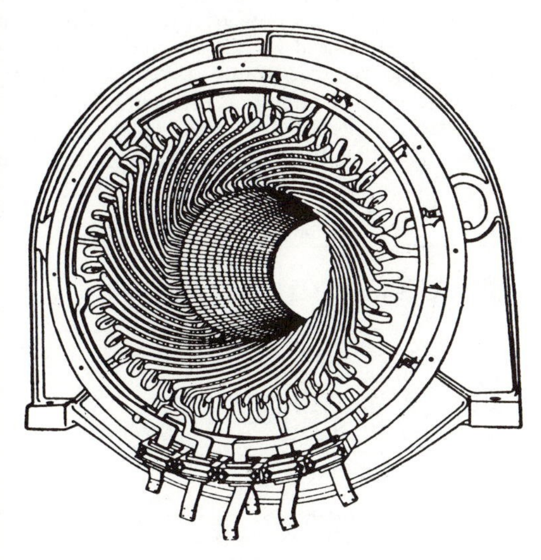

Figure 2–42. A turbo type ac generator stator.

c. Exciters. Like many dc generators, ac generators need a separate dc source for their fields. This dc field current must be obtained from an external source called an exciter. The exciter used to supply this current is usually a flat, compound-wound dc generator designed to furnish 125 to 250 volts. The exciter armature may be mounted directly on the rotor shaft of the ac generator (fig 2–43) or it may be belt-driven. Brushless exciters are also used to provide the dc fields. The brushless exciter is an ac generator that converts its ac power to dc by means of a diode rectifier assembly which is attached to but insulated from the generator shaft. The brushless exciter has no friction-producing parts such as brushes, brush holders, commutators, and sliprings (fig 2–44), so it needs very little maintenance.

d. Static Exciters. Another method of field excitation commonly in use is the static exciter, so called because it contains no moving parts. In this method, a portion of the ac current from each phase of the generator output is fed back to the field windings as dc excitation current through a system of transformers, rectifiers and reactors. With this system an external source of dc current is necessary for initial excitation of the field windings. On engine driven generators, the initial "field flash" may be obtained from the storage batteries which are also used to start the engine. Repair of static exciters is not covered in this manual due to the numerous designs in use. For troubleshooting or repair, always refer to the manufacturer's service manual or technical manual covering the specific generator under repair.

e. Frame and Shaft. The frame and shaft of an ac generator serve the same purpose as in dc generators. The frame completes the magnetic circuit of the field and supports the component parts and windings. The shaft, upon which the rotor turns, is supported on the end bells or end frames.

2–13. Types of AC Generators

a. Single-Phase AC Generators. Single-phase ac generators are seldom used except for special applications. As a rule, this type of ac generator

Figure 2–43. Exciter armature and alternator field mounted on the same shaft.

Figure 2–44. Brushless rotor.

is low-powered and self-excited. Actually, in construction it may be likened to a dc generator with an auxiliary ac winding on the dc armature. The dc winding on the rotating member is of the conventional lap or wave type, and is connected to the commutator bars in the usual manner. The dc winding output provides the current for dc field excitation and other dc power applications. A second winding of the open wave or lap type is laid in the slots of the rotating member, on top of the dc winding. This second winding is connected to sliprings and supplies the ac output.

b. Single-Phase Revolving Field Alternators. Alternators are designed to produce voltages which have as perfect a sine wave as possible, because this output characteristic is the most desirable for general applications. The waveform of a voltage produced by an alternator is determined by the type of stator armature winding used. Figure 2–45 shows a revolving field with the faces of the field poles (rotor poles) rounded. In this case, the flux density of the rotor field

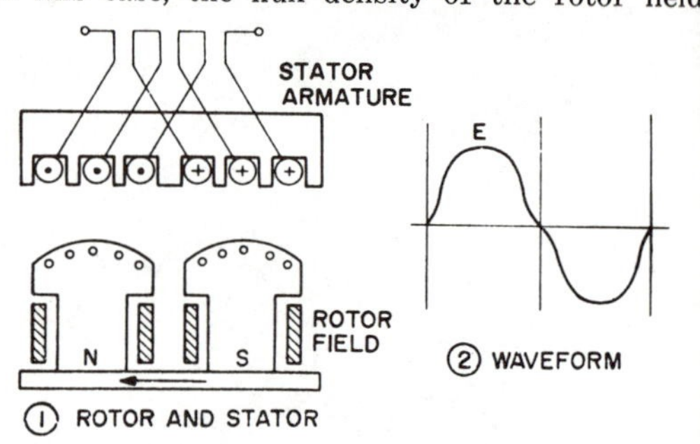

Figure 2–45. Alternator, with rotor having rounded pole faces and stator having three coils in series.

decreases as the air gap increases. The induced stator voltage gradually decreases in amplitude as the stator conductor cuts the less dense field flux at the pole edges. This gradual decrease in the amplitude changes the waveform so that it becomes trapezoidal. To further improve the waveform, three coils can be connected in series under the same stator field pole, as shown in ① of figure 2–45. When more than one coil is connected together under the same stator field pole, the group of conductors is called a "phase belt." Each individual coil produces a trapezoidal wave, and the sum of these waves closely resembles that of the waveform shown in ② of figure 2–45. Single-phase generators are usually limited to 25 kw or less and generate ac power at utilization voltages. Figure 2–46 shows a simplified diagram of a two-wire, single-phase generator.

c. Two-Phase Revolving Field Alternators. Assume that two coils are located at right angles to each other on a stator field. When the field flux of the rotor moves past these coils, each coil has induced in it a voltage of its own; when the output voltage is at its peak in one coil, the other is at minimum. These alternations are constantly changing in amplitude and direction;

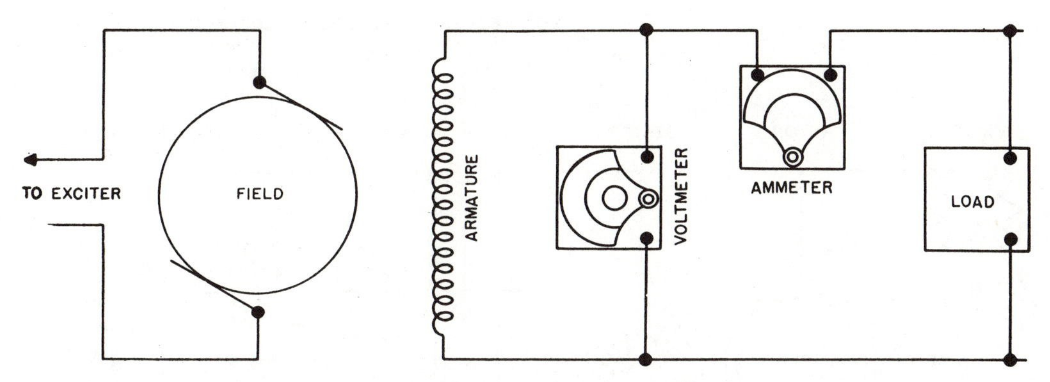

Figure 2–46. Two wire single phase generator, simplified drawing.

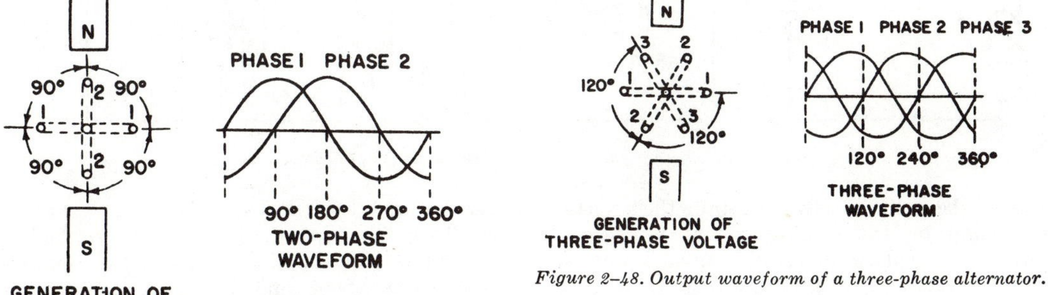

Figure 2–48. Output waveform of a three-phase alternator.

Figure 2–47. Output waveform of a two-phase alternator.

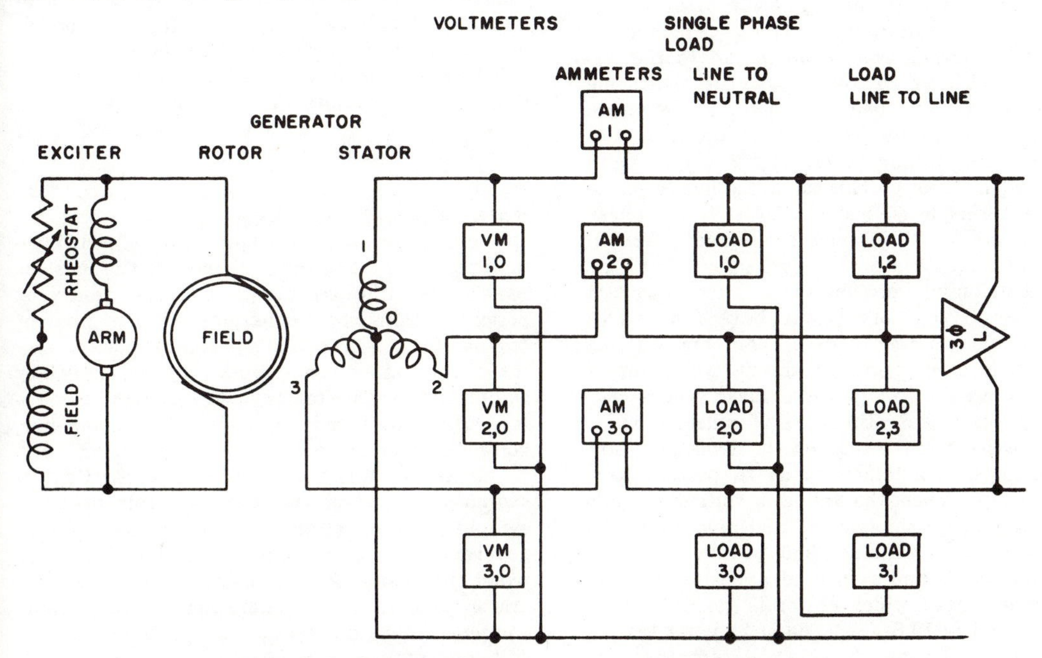

Figure 2–49. Four-wire, three-phase generator, simplified drawing.

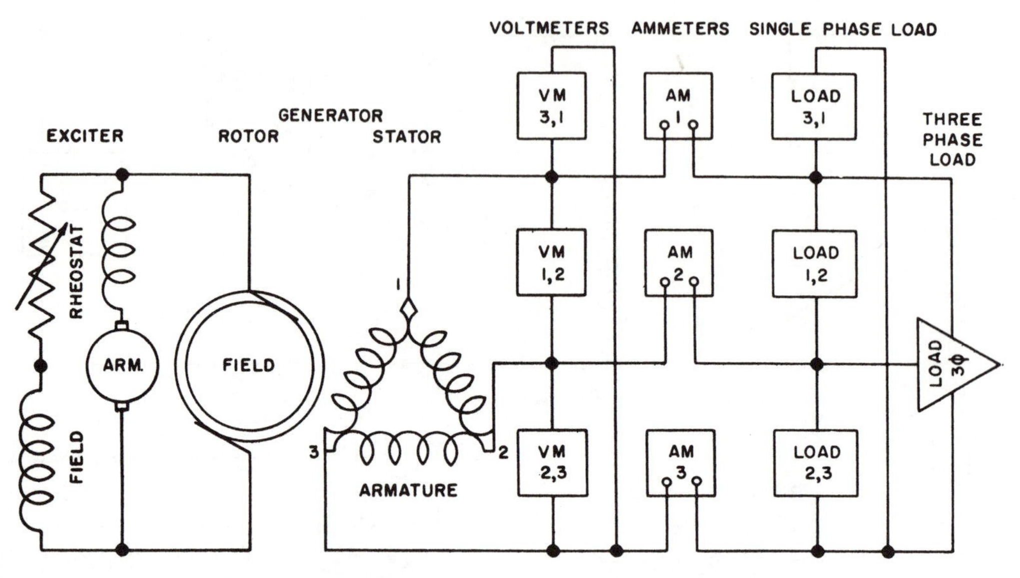

Figure 2–50. Three-wire, three-phase generator, simplified drawing.

however, the voltages always retain their phase relationship by the same number of degrees. Therefore, two-phase power will always have a phase difference of 90 electrical degrees when a two-pole rotor is used, because this is the angle between the armature (stator) coils (fig 2–47).

d. Three-Phase Revolving Field Alternators. Most three-phase alternators (fig 2–48) have six armature (stator) leads brought out to the terminal box. There are three pairs of leads, one pair for each coil or phase winding. The separate leads make it possible to connect the armature (stator) coils in star (wye) (fig 2–49) or delta network (fig 2–50). This is a method used to balance three-phase systems so that all phase voltages are equal in magnitude and differ in phase by equal angles. The three phases are usually distributed between three lines, although some star (wye) connections have four wires when a common line is used. The common line is generally used to ground the neutral point of the windings. To keep a three-phase system balanced, a phase separation of 120° must be maintained; therefore, the coils on the armature (stator) must be positioned so as to produce this relationship. When the induced voltage of one phase is 0, the other two induced phase voltages each have an amplitude 86.6 percent of their maximum value and each is of opposite polarity. The frequency at which power is generally supplied in the United States is 60 or 25 hertz (cycles per second). Twenty five-hertz alternators are mostly used for railroads. There are very few localities in the United States where 50-hertz current is still supplied. In European countries, frequencies of 50 and 25 hertz are commonly used. The speed and number of poles of an ac generator determine its frequency. Since the most common frequency is 60 hertz, a two-pole machine must operate at 3,600 rpm, a four-pole machine at 1,800 rpm, and a six-pole machine at 1,200 rpm to maintain the standard frequency. Polyphase generators are built in capacities of from 3 to 250,000 kw, and in voltages of from 110 to 13,800 volts.

2–14. Control and Metering

The generator and distribution switchboard usually consists of generator panels, the bus tie panels, and the feeder panels. The generator panel contains the generator circuit breaker (manually or electrically operated), the generator disconnect switch or links, and the necessary equipment for satisfactory control of the generator. This equipment consists of an ammeter, voltmeter, wattmeter, power-factor meter, and dc ammeter for the exciter circuit. Control switches to control the above instruments, and switches to control the engine or turbine-speed governor, are also mounted on the generator panel. Rheostats for the exciter and main generator field circuits complete the generator-panel equipment. The automatic voltage regulator controls and their equipment are generally mounted beside the generator panel. Mounted on either the bus tie panel or the generator panel are synchronizing lamps and a frequency meter. The necessary switches to cut the synchronizing equipment in or out at will, and to transfer the frequency meter from the generator to the bus tie lines, are also mounted on the panel. The bus tie panel usually contains the bus tie breaker (manually or electrically operated) and also a voltmeter, an ammeter, and a wattmeter for indicating the power in the bus tie circuit. On large installations, for efficiency of operation, all the meters, switches, and similar control equipment of a unit are set up remotely from the switchboard. This is known as a control switchboard of the benchboard type. Refer to TM 5-680G for further information.

a. Regulation of Alternators. AC generators usually have automatic voltage regulators which overcome voltage drop within the generator by changing the field excitation automatically as the voltage varies with the load. The internal impedance of the usually small and medium size generators is on the order of 40 percent more than its ohmic resistance. In addition to the voltage drop in the armature (stator), a considerable amount of armature (stator) reaction exists, particularly with lagging power factor. There is also an appreciable decrease in voltage with increasing load as the result of the slight slowing down of the prime mover. As the generator unit slows down, the generated ac voltage decreases. Since the exciter is usually driven by the same prime mover, the field excitation also decreases, causing the generated ac voltage to decrease even more. Without an automatic voltage regulator, ac generators would have very poor voltage regulation in all but the small sizes.

b. Measure of Voltage and Current. Refer to figures 2-46, 2-49, and 2-50 for connections of meters into the generator output circuits. In figure 2-46, an ammeter and a voltmeter are shown connected into the output of a single-phase ac generator. If a wattmeter and power-factor meter are added to this circuit (not shown), connection is made by connecting the current coil of each meter in series with the output of the line and the voltage coil across the line. Figures 2-49 and 2-50 show ammeters and voltmeters connected into four-wire, three-phase and three-wire, three-phase output circuits.

(1) *Ammeter and voltmeter.* In practice, a single ammeter and voltmeter is used with a selector switch for switching the meters in and out of the phase circuit.

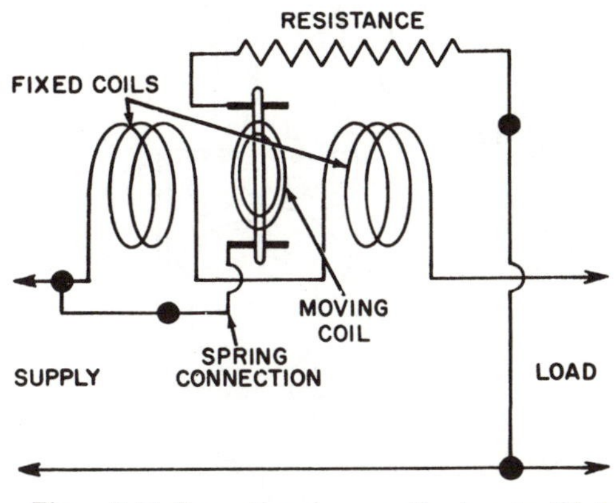

Figure 2-51. Connections for ac wattmeter, simplified diagram.

(2) *Wattmeter.* A wattmeter must be used to measure the power in watts on an ac circuit. The ordinary voltmeter and ammeter method, which is used for measuring watts on dc circuits, cannot be used for ac circuits due to the tendency of the current and the voltage to differ in phase rotation. Figure 2-51 shows the connections of a typical ac wattmeter, using a moving coil similar to that in any meter that operates on the principle of a dynamometer. The moving coil is wound with fine wire and is connected directly across the line in series with high resistance. Connection between the meter terminals and the moving coil is made through springs. The two fixed coils are called the current coils and are wound with a few turns of heavy wire which can carry the load current. The magnetic field in the current coils is therefore proportional to the load current at every instant, and the current in the moving coil is proportional to the voltage at every instant. Thus, every position of the moving coil, as reflected by the position of the needle, is proportional to the instantaneous power of the ac circuit. ①, figure 2-52 shows connections for measuring power in a singe-phase two-wire ac circuit, using an ordinary ac wattmeter. ②, figure 2-52 shows connections for a two-phase four-wire circuit and ③, figure 2-52 for two-phase and three-phase, three-wire circuits. These all require the use of polyphase wattmeters. ④, figure 2-52 shows connections for measuring power in a three-phase, four-wire circuit. This requires the use of three single-phase wattmeters, one for each phase, and the total power is the arithmetical sum of the watt

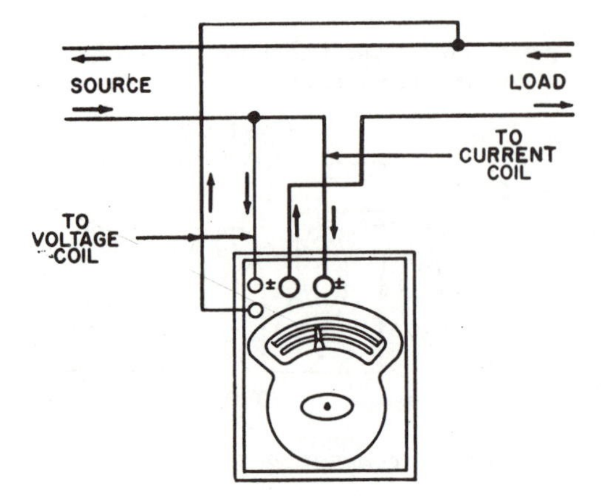

① CONNECTIONS FOR MEASURING WATTS IN SINGLE PHASE A-C CIRCUIT

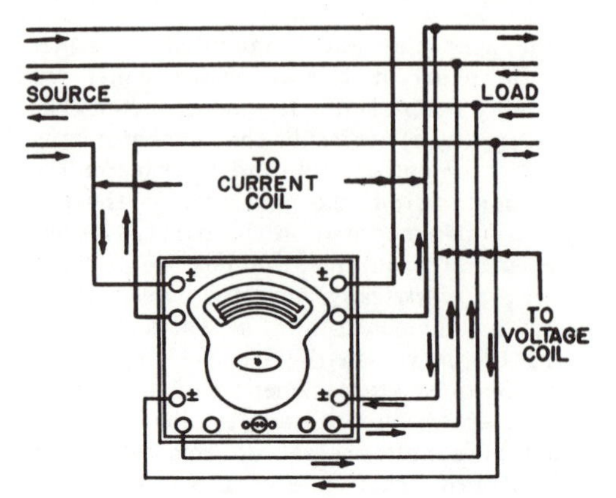

② CONNECTIONS FOR MEASURING WATTS IN TWO PHASE FOUR WIRE A-C CIRCUIT

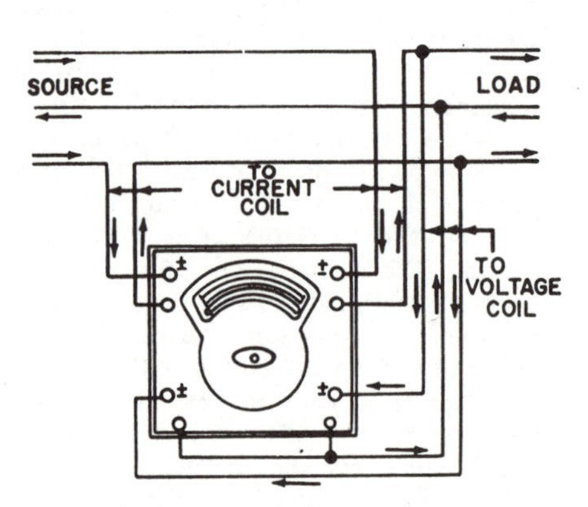

③ CONNECTIONS FOR MEASURING WATTS IN TWO PHASE AND THREE PHASE THREE WIRE A-C CIRCUITS

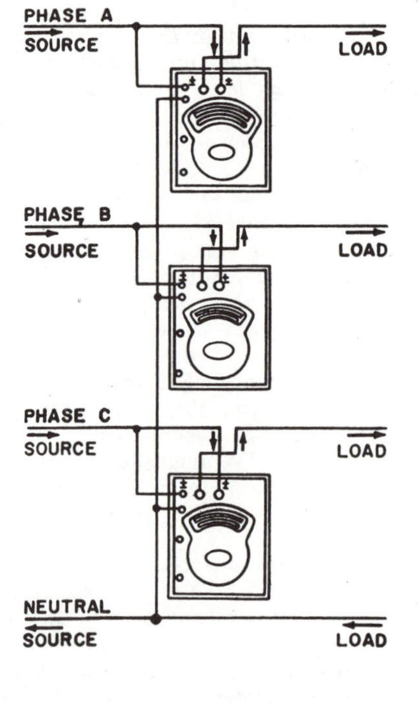

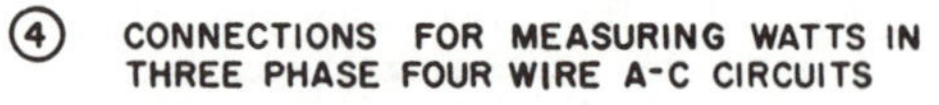
④ CONNECTIONS FOR MEASURING WATTS IN THREE PHASE FOUR WIRE A-C CIRCUITS

Figure 2–52. Connections for typical ac wattmeters.

readings on the three meters. Although all of these circuits (fig 2–52) show the wattmeters connected directly across the line, it is the practice to insert current transformers where the current is in excess of 10 amperes, and to use voltage transformers where line potential exceeds 600 volts.

(3) *Power-factor meter.* A polyphase

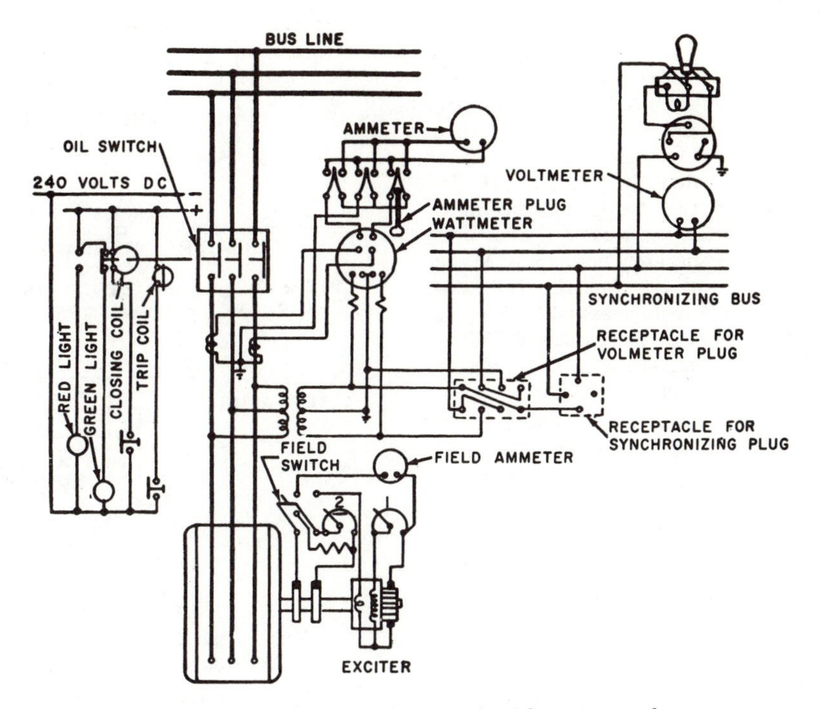

Figure 2–53. Typical wiring diagram of an alternator panel.

power-factor meter is connected (not shown) by connecting each of the voltage-coil leads to phases one, two, and three, respectively, and the current coil leads to phases one and three.

(4) *Alternator panel.* Figure 2–53 shows a typical wiring diagram of an alternator panel. Since it is unsafe to connect measuring instruments directly to the alternator leads, transformers are used.

c. *Parallel Operation of Alternators.* To connect one alternator is parallel with another machine or system, the output voltage of the two must be equal and the machines must be synchronized as to speed and phase rotation. By speed is meant the electrical speed or frequency. For example, a six-pole alternator operating at 1,200 rpm has the same electrical speed, or frequency, as an eight-pole alternator operating at 900 rpm. Assuming that two alternators are to be operated in parallel, the same instantaneous voltage of the two machines is obtained by adjusting the excitation voltage of dc supply to the alternator fields. To regulate the electrical speed or frequency, adjust the governor of the prime mover, this usually being steam-, water-, or fuel oil-driven. To synchronize the phase rotation of the two machines, perform the operation called "phasing out" as described in (1) and (2) below. When these conditions are fulfilled, the potential differences between corresponding terminals of the two machines are zero. If the speeds of the two machines are not equal, the voltages between corresponding terminals vary from zero to about twice the terminal voltage at a frequency equal to the difference in speeds (electrical) between the two machines.

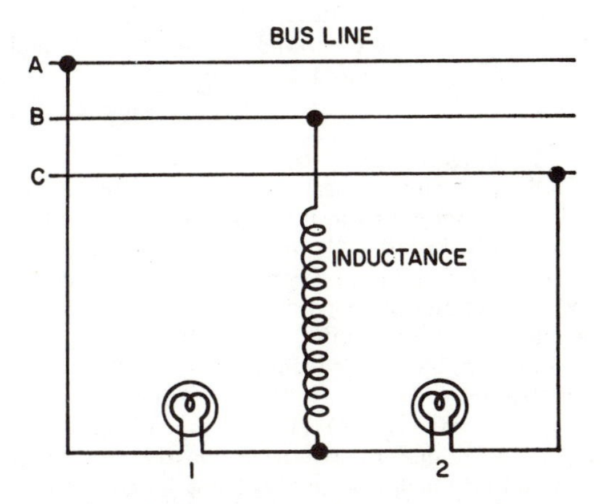

Figure 2–54. Diagram of a phase-sequence indicator.

(1) *Phase rotation.* In polyphase machines, the coils for each phase are equally spaced about the armature and the maximum voltage in the coils occurs at different times. When the ends of the coils are brought out to terminals without markings, phase rotation may be determined with a phase-sequence indicator (fig 2–54). The indicator consists of two lamps and a highly reactive coil, such as the potential coil of a watt-hour meter. If the generator voltage is higher than the allowable voltage for the indicator, voltage and current transformers should be inserted between the line and the indicator. If lamp 1 is bright, the phase sequence is A, B, C; if lamp 2 is bright, the phase sequence is C, B, A.

(2) *Synchronizing lamps.* The simplest system uses a bank of lamps connected across each pole of the switch that connects the generator to the bus, the bus being live from another generator or system ①, fig 2–55). This is known as the "all dark" method of synchronizing two alternators. The lamps should have a voltage rating 15 percent higher than the generator terminal voltage, although lamps of standard rating are ordinarily used. If the lamps are connected correctly, they should become bright and dim together. If they brighten and dim in sequence, the phase rotation of the two machines is opposite and one phase must be reversed. The lamps flicker at a frequency equal to the difference in frequency between the two machines. As the machines approach equal frequency, the flicker becomes slower. When the lamps are dark, the switch may be closed and the generator will be placed in parallel with the bus. The system works very well for small and low-speed generators. A more sensitive arrangement for the larger, high-speed generators may be obtained by connecting the lamps as shown in ②, figure 2–55. This is known as the "one dark and two bright" method

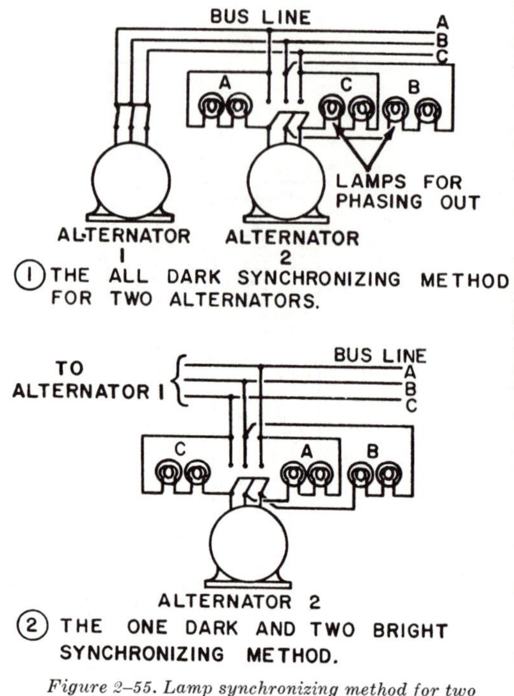

Figure 2–55. Lamp synchronizing method for two alternators.

of synchronizing two alternators. When the machine is in synchronism, lamps A and B are bright and C is dark. As one of the bright lamps is increasing the other is decreasing in brilliancy and the instant of synchronism can be determined quite accurately. By noticing the sequence of brightness of the lamps, it can be determined whether the incoming machine is fast or slow. Standard practice requires closing the generator switch when the incoming machine is gaining speed slowly and is just below synchronous speed.

Section IV. AC MOTORS

2–15. Introduction

Of all the various types of ac motors, the induction type motor is the most popular. The induction motor is divided into types according to methods of starting and construction. The majority of fractional-horsepower motors are operated on single-phase circuits. The universal type ac motors can be divided into three classes: (1) the straight-series type; (2) the series type with shifted brushes; (3) the compensated-series type. These motors are used mostly for drills and adding machines and will not be covered in this paragraph. The popularity and wide application of the ac induction motor are principally due to its simplicity of construction. Rugged and reliable, it has constant-speed characteristics. For example, the speed is substantially independent of the load, within the normal working range. The polyphase induction motor is comparatively simple. Current is established by an emf across the stator. By transformer action, this induces an emf in the rotor. The secondary windings are short-circuited upon themselves, either directly or through an external resistance, and are not

connected to an external power source. The secondary winding is usually put on the rotor, and commutator or collector rings and brushes are not required except in the case of special applications.

2–16. Principles of Operation

An induction motor consists essentially of two units, called stator and rotor. The induction motor stator is like the stator used on alternators. The motor consists of a laminated cylinder with slots in its surface. The winding placed in these slots may be one of two types. One type, called a squirrel-cage winding, ssually consists of heavy copper bars, connected together at each end by a conducting end ring made of copper or brass. The joints between the bars and the end rings are commonly made by an electric weld. In some cases the rotor, including the bars, is placed in a mold and the bar ends are cast together with copper. In small squirrel-cage motors the bars are of aluminum, cast in one piece. Some industrial applications require a motor with a wound rotor instead of the squirrel-cage type. In this second type, the rotor is provided with a winding similar to that used on the stator. The winding is generally connected in star (wye), and its open ends are fastened to sliprings mounted on the shaft. Brushes mounted on the rings are attached to a star-connected rheostat. The purpose of this arrangement is to provide a means of varying the rotor resistance.

a. Number of Poles. It is usually not possible to determine the number of poles on an induction motor by visual inspection, but the information can be obtained from the nameplate of the machine. The nameplate gives the speed of the machine and the frequency of the applied current. To determine the number of poles per phase, divide the product of 120 times the frequency by the rated speed. This, written as an equation is—

$$P = \frac{120 \times f}{N}$$

where—*P* is the number of poles per phase.
F is the frequency in hertz.
N is the rated speed in rpm and 120 is a constant.

The result will be very nearly equal to the number of poles per phase. For example, consider a 60-hertz, three-phase machine with a rated speed of 1,750 rpm. In this case—

$$P = \frac{120 \times 60}{1{,}750} = \frac{7{,}200}{1{,}750} = 4.1$$

Therefore, the machine has four poles per phase.

b. Slip. When the rotor of an induction motor is subjected to the revolving magnetic field produced by the stator windings, a voltage is induced in the longitudinal bars. The induced voltage causes a current to flow through the bars. This current, in turn, produces its own magnetic field, which combines with the revolving field to cause the rotor to assume a position in which the induced voltage is minimized. As a result, the rotor revolves at very nearly the synchronous speed of the stator field. The difference in speed is just sufficient to produce enough current in the rotor to overcome the mechanical and electrical losses. If the rotor were to turn at the same speed as the rotating field, the rotor conductors would not be cut by any magnetic line of force, no emf would be induced in them, no current could flow, and there would be no torque. The rotor would then slow down. For this reason, there must always be a difference in speed between the rotor and the rotating field. This difference in speed is called slip and is expressed as a percentage of the synchronous speed. For example, if the rotor turns at 1,750 rpm and the synchronous speed is 1,800 rpm, the difference in speed is 50 rpm. The slip is then equal to 50/1800, or 2.78 percent.

c. Single-Phase Machines. In a single-phase ac machine, the field, instead of rotating as in two- and three-phase, merely pulsates. No rotation of the rotor takes place. However, a single-phase pulsating field may be visualized as two rotating fields revolving at the same speed but in opposite directions. It follows, therefore, that the rotor will revolve in either direction at nearly synchronous speed, provided it is given an initial impetus in either one direction or the other. The exact value of this initial rotational velocity varies widely with different machines but a velocity higher than 15 percent of the synchronous speed is usually sufficient to cause the rotor to accelerate to rated speed. A single-phase motor can be made self-starting if means can be provided to give the effect of a rotating field.

2–17. Types of Induction Motors

a. Shaded-Pole Starting. The first development of a self-starting, single-phase motor was the shaded-pole induction motor (fig 2–56). This machine has salient stator poles, a portion of each pole being encircled by a heavy copper ring. The presence of the copper ring causes the magnetic field through the ringed portion of the pole faced to lag appreciably behind the field through the other part of the pole face. This produces a slight component of rotation of the field, enough to cause the rotor to revolve. As the rotor accelerates, the torque increases until the rated speed is obtained. Such motors have low starting torque and find their greatest application in small fan motors where the initial torque required is low.

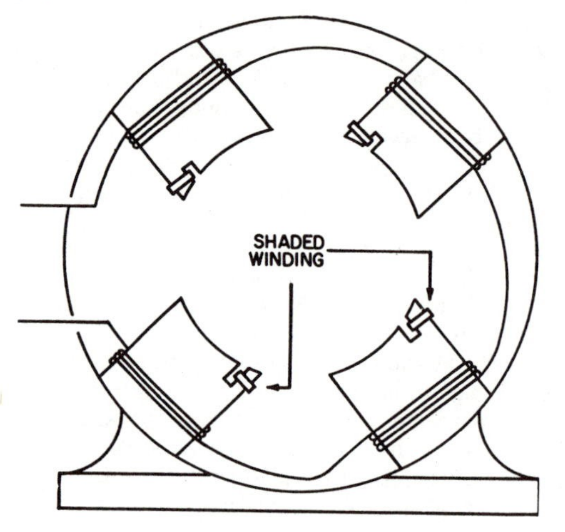

Figure 2–56. Shaded-pole motor.

b. Split-Phase Motors. Various types of self-starting, single-phase ac motors, known as split-phase motors, are manuafctured. They are usually of fractional horsepower and are used to operate such devices as washing machines, small pumps, and blowers. The motor has four main parts, the rotor, the stator, the mounting frame, and the centrifugal switch (fig 2–57). The winding of the rotor is usually of the squirrel-cage type, consisting of copper bars placed in slots in the laminated iron core and connected to each other by copper rings. The stator has two windings, the main or running winding and the starting winding, which are wound into slots in the iron core. The main winding is connected across the line in the usual manner and produces the magnetic field for the main poles. Between the main poles are auxiliary poles, the windings of which have a resistance greater than that of the main winding. Sometimes additional resistance is put in series with the auxiliary winding. As the ratio of resistance to impedance in the auxiliary winding is greater than that in the main winding, the voltages in the auxiliary and main windings differ by about 90° electrical. These two sets of poles produce a sort of rotating field which starts the motor. When the motor reaches a predetermined speed, the starting winding is automatically disconnected by a centrifugal switch inside the motor.

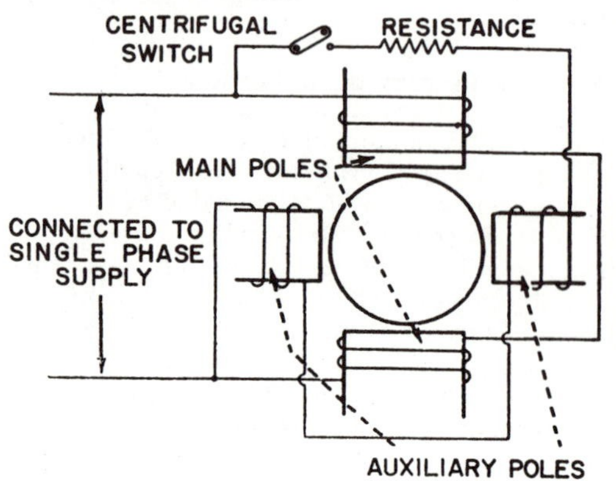

Figure 2–57. Split-phase motor with auxiliary starting winding.

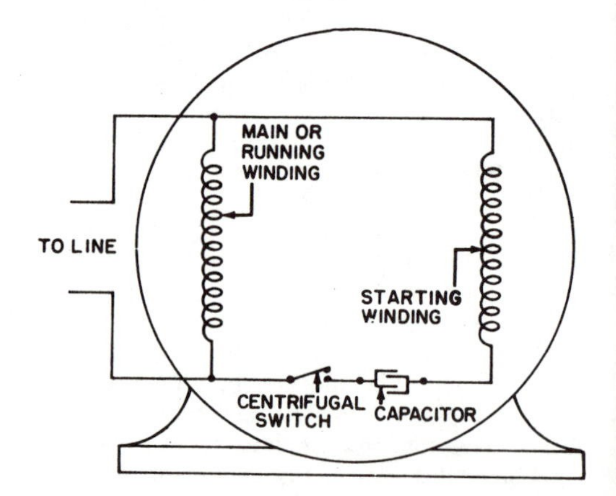

Figure 2–58. Typical capacitor-start motor connections.

c. Capacitor-Start Motors. With the development of high-capacity, electrolytic capacitors, a variation of the split-phase motor, known as capacitor-start motor, was made (fig 2–58). Nearly all fractional-horsepower motors in use today on refrigerators, oil burners, and other similar appliances are of this type. In this adaptation, the starting winding and running winding have the same size wire of the same resistance value. The phase shift of 90° electrical between currents of the two windings is obtained by means of the capacitor connected in series with the starting winding. Capacitor-start motors have a starting torque comparable to their torque at rated speed and can be used in applications where

the initial load is heavy. Again, a centrifugal switch is necessary for disconnecting the starting winding when the rotor is at approximately 75 percent of rated speed.

d. Capacitor Start-and-Run Motor. A variation of the capacitor-start motor is the capacitor start-and-run motor (fig 2–59). This is similar to the capacitor-start motor except that the starting winding and capacitor are connected in the circuit all the time. This type of motor is quiet and smooth running in operation but can be used only where medium starting torque is required. Such applications include blowers, fans, and oil burners. The capacitor is usually a paper-insulated, oil-filled type.

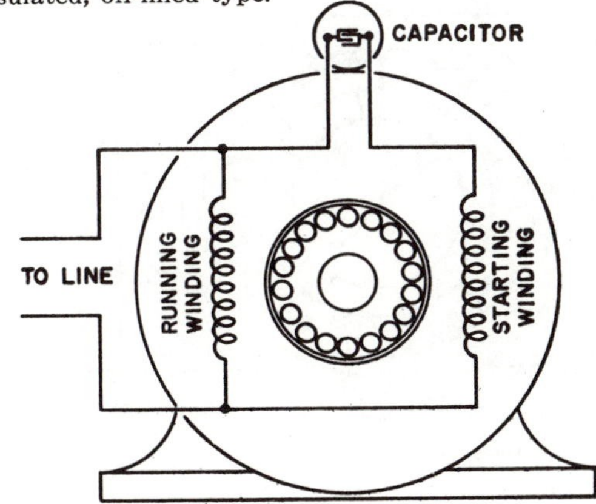

Figure 2–59. Typical capacitor start-and-run motor connections.

e. Reversing Direction of Rotation. The direction of rotation of a three-phase, three-wire motor can be reversed by simply interchanging the connections of any two motor leads A, B, C, (①, fig 2–60). The same effect can be produced in a two-phase, four-wire motor by reversing the motor leads of either phase A or B (②, fig 2–60). To reverse the direction of rotation of a two-phase, three-wire motor, simply interchange the outer two motor leads, 1 and 2 (③, fig 2–60). Three-phase, four-wire motors are not generally used. In a single-phase motor, reversing the connections of the starting winding in relation to that of the running winding will reverse the rotation of the motor. Nothing can be done to many shaded-pole motors to reverse the direction of rotation because the direction is determined by the physical location of the copper shading ring. In others, removal and reversal of the entire stator field will change the direction of rotation.

f. Repulsion-Induction Motor. This type of motor consists of a wound rotor with commutator and brushes. The brushes are not connected to the supply line, but are short-circuited. The principle involved in this method of starting may be understood by a consideration of figure 2–61. Although the current through the stator is alternating, the arrows indicate the direction of the current during a half cycle. In ①, figure 2–61, suppose the stator current is increasing

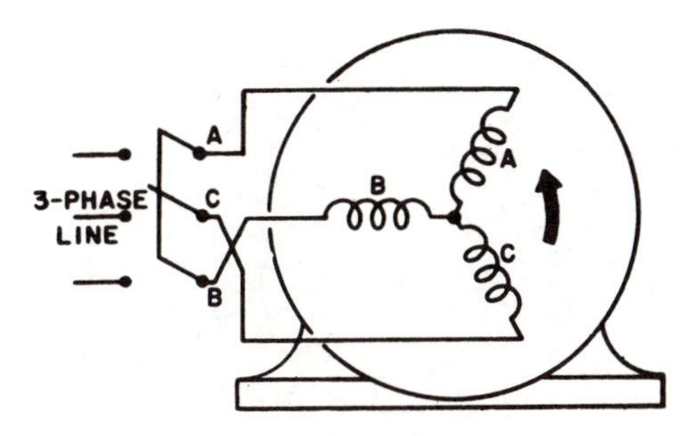

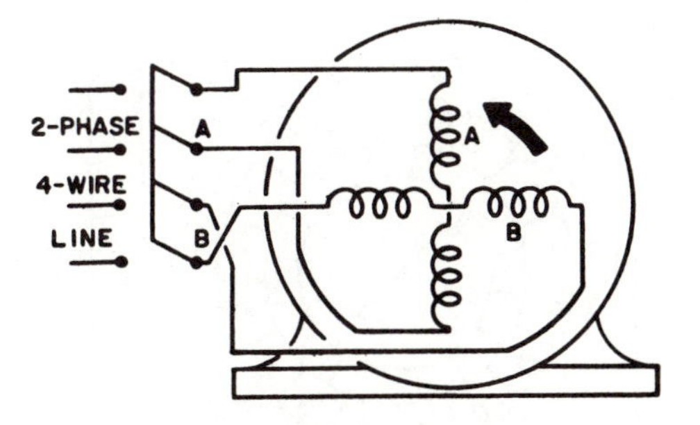

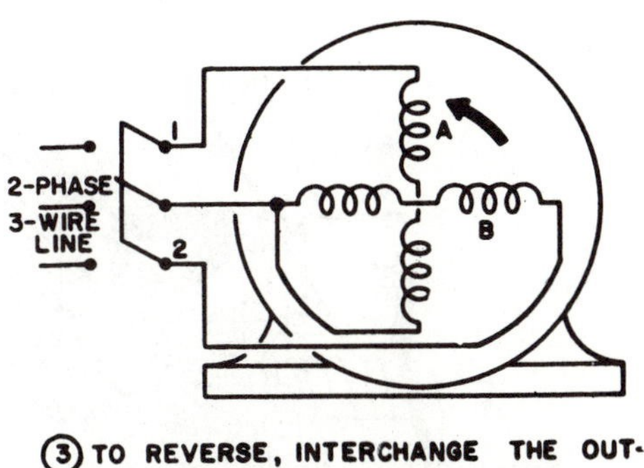

Figure 2–60. Reversing direction of rotation of ac motors.

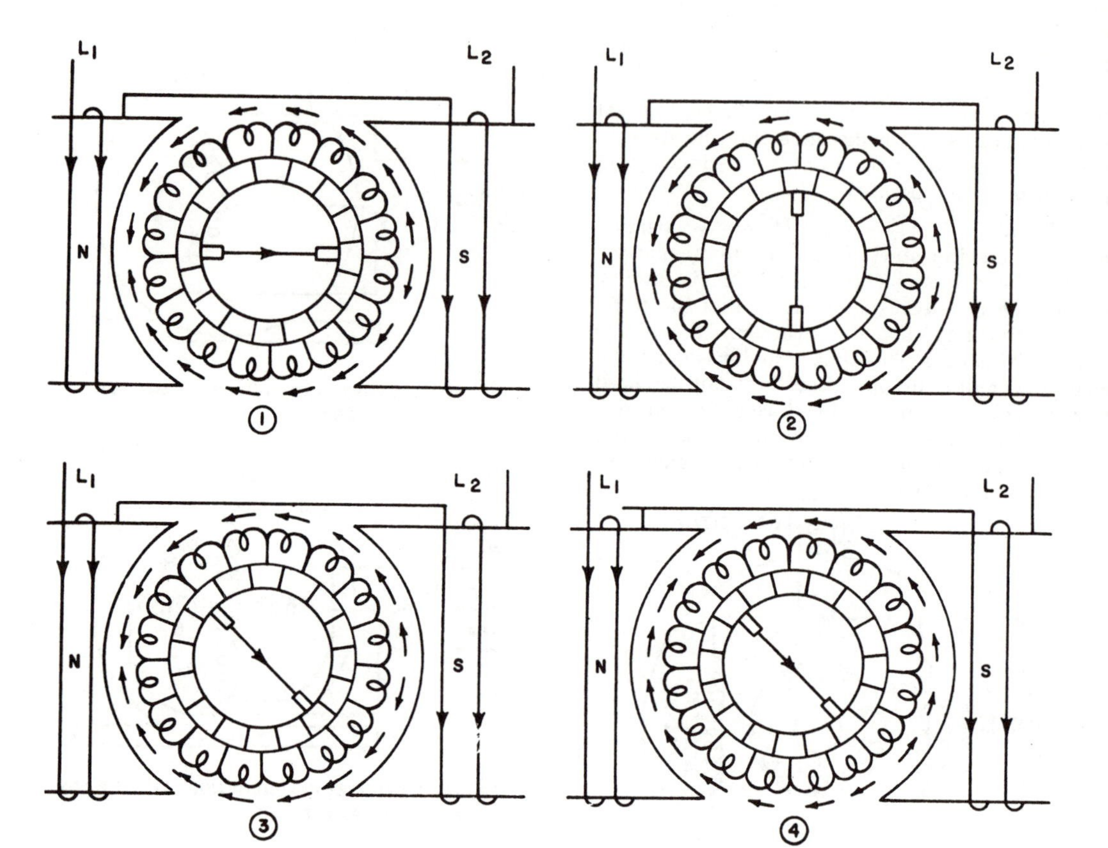

Figure 2–61. Principle of repulsion starting.

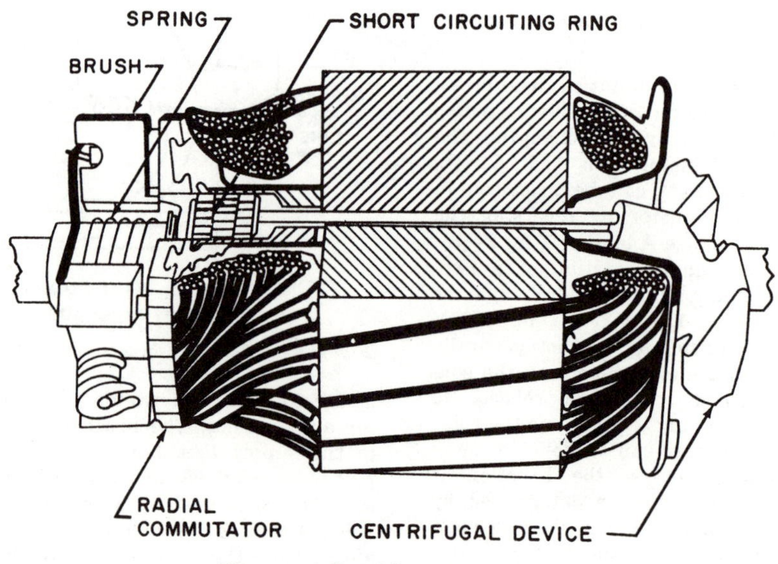

Figure 2–62. Repulsion-induction rotor.

and is in the direction indicated. The flux produced will induce voltages in the armature conductors as shown. These voltages are additive on each side of the brushes and, therefore, send a high current through the rotor and short-circuited brushes. No torque will be developed, however, because half of the conductors under each pole carry current in one direction and the other half carry current in the opposite direction. If the brushes are shifted 90°, as shown in ②, figure 2–61, the emf in each path will be neutralized. Then no voltage exists at the brushes; consequently, no current flows through the rotor, and no torque is produced. If the brushes are shifted to the position shown in ③, figure 2–61, a resultant voltage will exist in each path, sending current through the rotor as indicated in ④, figure 2–61. Now all conductors under one pole carry current in one direction, and all conductors under the opposite pole carry current in the opposite direction. Then a torque is developed. If the brushes were shifted to the opposite direction from the position shown in ③, figure 2–61, the rotor would turn in the opposite direction. This method is used for starting single-phase motors. For this purpose, the rotor is provided with a commutator that has either a radial commutator as shown in figure 2–62 or a cylindrical commutator as in dc machines. At starting, the machine functions as a repulsion motor developing a high starting torque. As soon as the rotor, assumed to have a structure as shown in figure 2–62, reaches nearly full speed, the centrifugal device forces the short-circuiting ring into contact with the inner surface of the commutator. This short-circuits the segments (bars), and therefore converts the machine into an induction motor. At the same time, the centrifugal device rasises the brushes, which reduces the wear of the brushes and commutator. This machine is used where a high starting torque is desired.

g. The Brush-Shifting Repulsion Motor. The brush-shifting repulsion motor is applicable for varying types of service. Speeds can be reduced to one-third the motor rating. Horsepower decreases as the speed is reduced. This type of motor is used on applications where adjustable speed is desired, as on small job printing presses.

h. Polyphase Squirrel-Cage Motors. Refer to paragraph 2–16 for principles of operation. Polyphase motors are available for many service classes. For details, see table F–2 in appendix F.

i. Wound-Rotor Motors. Wound-rotor or slipring motors differ from the squirrel-cage motors only in the rotor winding, which consists of insulated coil windings, usually three-phase, star (wye) connected. The leads are connected to sliprings. This secondary circuit is extended to an adjustable star (wye) connected external resistance. By inserting this resistance, with suitable control, during the starting period, the motor can be made to develop a high starting torque with comparatively low starting current. As the motor speed increases, the resistance is gradually reduced, until at full speed the rotor windings are short-circuited. The motor then operates at constant speed as a squirrel-cage machine with good efficiency and low slip. Wound-rotor or slipring motors are commonly used where the excessive starting current of a large squirrel-cage motor would be objectionable. They are also used where the capacity of the motor is large with respect to the capacity of the generating plant, transformers, or lines from which the current is taken. With correctly designed external resistors in the rotor circuit, the motor may be used for speed control. Speed can be reduced approximately 50 percent from normal, but when it is regulated in this manner, the speed depends largely on the load.

2–18. Synchronous Motors

Synchronous motors are not self-starting, and must first be brought up to speed before they operate.

a. Starting Torque. More care is required in starting a synchronous motor than an induction motor. There are several methods of starting it:

(1) Where the dc exciter is coupled directly to the shaft of the synchronous motor and enough dc power is available, the dc exciter may be operated as a motor to bring the synchronous motor up to speed. Normally there is enough residual magnetism in the field poles of the synchronous motor to do this. The field of the synchronous motor is then excited and the motor synchronized, as with an alternator.

(2) If a separate dc motor with enough dc power is available, it may be used to bring the synchronous motor up to speed, after which the motor is synchronized as before.

(3) If a separate induction motor is available, it may be used to start the synchronous motor as in (2) above.

(4) A synchronous motor may be started as an induction motor. The starting torque to bring the motor from rest to synchronous speed

is obtained by means of damper winding (*b* below). These are separate windings, placed in the slots of the rotating field poles, which make it possible to insert external resistance and produce high starting torque, as with a wound-rotor induction motor. First the field circuit is opened, but the field poles retain enough residual magnetism to maintain their polarity. Then the stator is connected to a source of ac supply at full voltage or at a reduced voltage by means of an autotransformer. The stator winding develops a rotating magnetic field which induces an emf in the damper windings. The reaction between the rotating field of the stator winding and the magnetic field of the damper winding exerts a torque that accelerates the motor. The stator winding also develops a voltage in the rotor field winding. The rotor field-winding circuit is closed through a voltage dropping resistor. This prevents the development of excessive voltage in these windings when starting. The excessive voltage might puncture the insulation of the windings. After the motor comes up to speed, which is slightly less than synchronous speed, the rotor field resistor is shunted out at the same time dc voltage is applied. Figure 2–63 shows a simplified diagram of this method. If the synchronous motor should lose its dc voltage, it would immediately stop.

(5) Automatic controllers are used to start synchronous motors of appreciable size. Figure 2–64 shows an automatic controller for starting a synchronous motor directly across the line. The circuit of coil M (fig 2–64) is completed when the START button is depressed. This closes the main contact M, which connects the motor terminals to the line. After the START button is released, the circuit is maintained by the auxiliary contact 1. Current is then supplied to the stator, producing a rotating magnetic field. This field establishes a current in the squirrel-cage bars and the field winding, which is closed through the discharge resistance R and coil C_1. The current in these two windings is effective in producing a starting torque. When the motor is just starting, a high voltage is induced in the field at a frequency which is the same as that in the stator. As the motor gains speed, the induced voltage and frequency decrease. The reactance of the circuit decreases with the frequency and, as a result, the field current remains practically constant until the rotor attains about 75 percent of its synchronous speed. About this point, the field current decreases rapidly. The field current flows through C_1 (fig 2–64), and while the current in this coil is constant, it produces enough force to hold contact 2 open. The current, through coil C_2, which is connected across the line when contact 1 closes, establishes a constant force which tends to close contact 2. When the rotor reaches about 75 percent of synchronous speed, the current through coil C_1 starts to decrease and coil C_2 closes contact 2, energizing coil F. This coil closes contacts 3 and 4 and opens contacts 5 and 6. The closing of contacts 3 and 4 energizes the field with direct current, completing the starting period of the motor. On opening, contact 5 disconnects the discharge resistance. With contact 6 open, contact 7 will disconnect the motor from the line if the dc voltage fails. The synchronous motor is stopped merely by pressing the STOP button, which allows the contacts to return to their normal position.

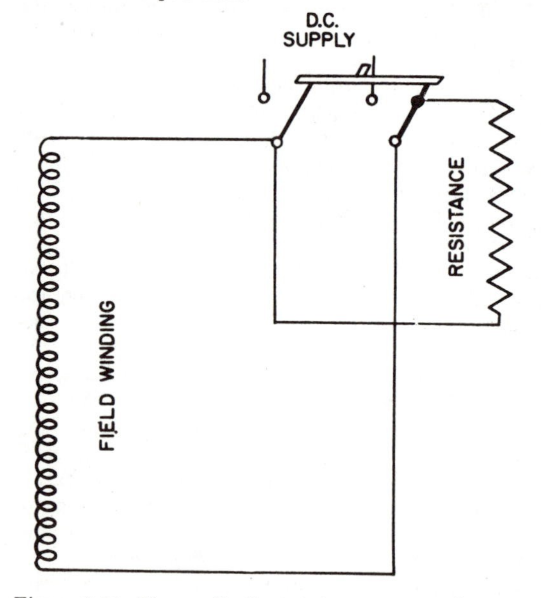

Figure 2–63. The application of dc current to the rotor field.

b. Hunting. Hunting is the tendency of the speed of the synchronous motor to oscillate above or below synchronism. The damper windings are useful in stabilizing the operation of the motor. When the motor is operating at uniform speed, the damper winding is linked by the steady flux of the combined emf of the field poles and the armature. This normally has no effect on the operation of the motor. The synchronous motor, however, is very sensitive to changes in phase. A slight angular shift of the rotor, with respect to the armature field, produces large changes in

frequency, and decreased with an increase in frequency.

b. Mechanical Modifications. Induction motors with mechanical modifications are made for special applications. The important ones are:

(1) The splashproof motor is so constructed that dripping or splashing liquids will not enter the motor. However, since the motor is self-ventilated, and since moisture-saturated air may be circulated through the motor, the windings are made moisture-resistant. Motors of this construction are used to drive pumps, and are used as components of machinery in such places as dairies, paper mills, and chemical plants.

(2) The totally inclosed, fan-cooled motor has totally inclosed windings and rotor. Cooling air is circulated over the inclosure to carry away the heat. This motor is used widely where the air contains dust, as in grain elevators, coal-handling plants, large foundries, and machine shops.

(3) The explosion-proof motor is similar to the totally inclosed, fan-cooled motor, but is constructed to prevent any explosion within the motor from igniting gases or dust in the surrounding air.

(4) The gearmotor is usually built with a gear mechanism and the motor as an integral unit. In some cases, the gear unit is designed to permit the use of motors having the other mechanical modifications, as well as the electrical characteristics. This construction is often used to connect motors directly to low-speed or high-speed loads.

(5) The vertical motor has the shaft vertical to the working plane. Mountings are flanged, ring-based, tripod-based, or with machined end shields for mounting on the bolt circle of a machine.

c. Requirements. The selection of a motor for a job is usually relatively simple and, with minor variations, can be based on similar jobs. However, to insure correct application, these important requirements should be kept in mind:

(1) Starting or breakaway torque.
(2) Accelerating time.
(3) Frequency of starting.
(4) Speed requirements.
(5) Variation of load.
(6) Special limitations imposed by the power supply.
(7) Type of drive (direct or belted).
(8) Position of motor with respect to driven load.
(9) Condition of surrounding atmosphere (temperature, humidity, clean or dust laden, or presence of gas).
(10) Noise.

d. Motor Selection. After determining the characteristics of the load, a suitable type of polyphase motor can usually be selected from the motor characteristics and selection chart in appendix F. In the majority of cases, standard motors can be selected. However, there will always be some applications which are not covered by the standard types. Important special conditions may affect the choice of motor drives. High voltage may be a consideration. Reversing duty and dynamic braking may be required.

e. Classes of Polyphase AC Motors. The NEMA (National Electrical Manufacturers Association) code identification of polyphase ac motors is listed below:

(1) *Squirrel-cage, constant speed.*
(*a*) Normal starting torque, normal starting current.
(*b*) Sizes for (*a*): 1 to 1,000 horsepower for low voltages, and up to 15,000 horsepower for high voltages, direct connection.
(*c*) Normal starting torque, low starting current.
(*d*) Sizes for (*c*): 7 1/2 to 500 horsepower.
(*e*) High starting torque, low starting current.
(*f*) Sizes for (*e*): 7 1/2 to 500 horsepower.
(*g*) High torque, high slip.
(*h*) Sizes for (*g*): 1 to 50 horsepower.
(*i*) High torque, medium slip.
(*j*) Sizes for (*i*): 3 to 50 horsepower.

(2) *Squirrel-cage, multispeed.*
(*a*) Constant horsepower.
(*b*) Constant torque.
(*c*) Variable torque.
(*d*) Sizes: 1 to 200 horsepower.

(3) *Wound-rotor.*
(*a*) General purpose.
(*b*) Crane and hoist.
(*c*) Sizes: 1 to 1,000 horsepower in low voltages, and up to 15,000 horsepower for high voltage, direct connection.

(4) *Synchronous.*
(*a*) General purpose.
(*b*) Low speed.
(*c*) Sizes: 25 to 5,000 horsepower.

f. Classes of Single-Phase, AC Motors. The NEMA code identification of single-phase, ac motors is listed below:

(1) *Repulsion-induction.*

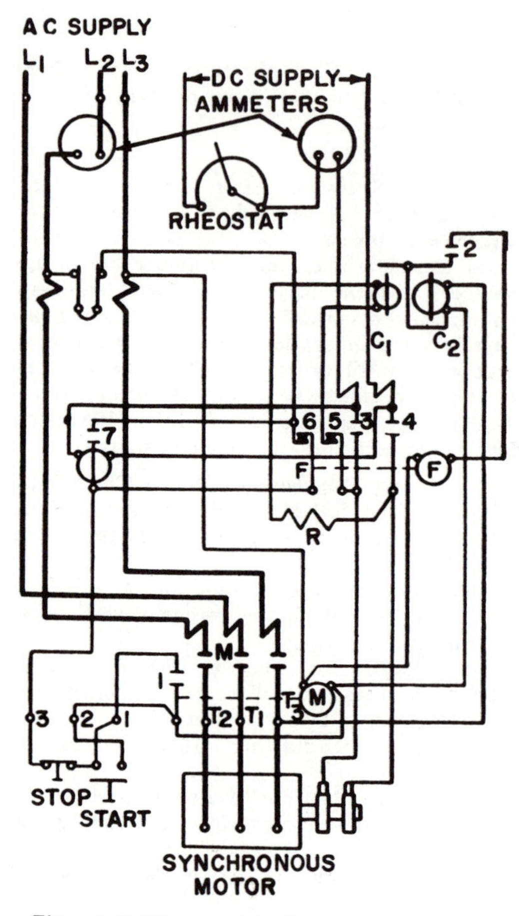

Figure 2–64. Diagram of synchronous motor controller.

the armature current. Disturbances in the source of power may cause the rotor to oscillate about its normal position, and the resulting phase shift of the rotor produces pulsations in the power system which, in turn, may accentuate the oscillations. The purpose of the damper winding is to oppose and dampen these tendencies of the rotor to oscillate while rotating. Again, if the rotor slows down momentarily, the rotating field in the armature gets out of line with the rotating field in the field structure. The resultant induction produces currents in the dampers in such direction that they tend to pull the rotor back into synchronism. If the field poles, for some reason, swing ahead of their normal position, the rotor tends to slow down. The dampers will always tend to pull the motor back into synchronism.

2–19. Selection of AC Motors

It is very important to select and use the proper motor for the application, service, and location. Refer to table F–2 for ac motor characteristics and selection charts.

a. Characteristics. Important ac motor characteristics are listed below:

(1) *Starting torque.* For squirrel-cage motors of various operating characteristics, minimum values of starting torques, which are known as normal, have been established. With full voltage applied, the starting torques range from 150 percent of full-load torque for 2- and 4-pole motors down to 105 percent for 16-pole motors.

(2) *Starting current.* Normal starting currents were established for squirrel-cage motors. These range from 450 to 650 percent of full-load current, depending on the size and speed of the motor. Check regulations of the local power company or other source of primary power to make sure the starting-current characteristics do not overload or unbalance the line.

(3) *Slip.* Slip is often important in selecting a motor for a specific application. It is the difference between synchronous speed and actual rotor speed under load, expressed as a percent of synchronous speed.

(4) *Voltage and frequency.* Motor performance guarantees are based on normal or rated voltage and frequency applied at the motor terminals. Satisfactory operation can be obtained at rated load and frequency, if the applied voltage is within the limits of 10 percent above or below rated voltage. Motors for 220-volt circuits will operate satisfactorily on network systems of 208 or 199 volts, but not in accordance with the performance established at rated voltage. Output (operating), starting, and maximum torque of an induction motor will vary as the square of the applied voltage. Reduction in torques should not affect the usual application of polyphase induction motors, because starting and maximum torques are not usually the limiting factors. However, where standard motors are to be used on torque applications, such as hoists, cranes, elevators and valves, the factor of reduction in torque should be given careful consideration. Standard motors will operate successfully on frequencies within the limits of 5 percent above or below normal. The output starting and maximum torques will be increased slightly with a drop in

(*a*) General purpose.
(*b*) High starting torque.
(*c*) Constant-speed regulation.
(*d*) Sizes: 1/8 to 10 horsepower.

(2) *Capacitor.*
(*a*) General purpose.
(*b*) Fan service.
(*c*) Sizes: 1/4 to 10 horsepower.

(3) *Brush-shifting repulsion.*
(*a*) Adjustable speed.
(*b*) Sizes: 1/4 to 2 horsepower.

(4) *Fractional horsepower.*
(*a*) Shaded-pole.
(*b*) Split-phase.
(*c*) Capacitor.
(*d*) Series.
(*e*) Sizes: 1/200 to 3/4 horsepower.

Properties of Copper Conductors

Size A W G	Area Cir Mils.	Concentric lay stranded conductors		Bare conductors		D–C resistance ohms/M ft at 25° C. 77° F.	
		No. wires	Diameter each wire, inches	Diameter inches	Area,* square inches	Bare conductor	Tinned conductor
18	1624	Solid	0.0403	0.0403	0.0013	6.510	6.77
16	2583	Solid	.0508	.0508	.0020	4.094	4.25
14	4107	Solid	.0641	.0641	.0032	2.575	2.68
12	6530	Solid	.0808	.0808	.0051	1.619	1.69
10	10380	Solid	.1019	.1019	.0081	1.018	1.06
8	16510	Solid	.1285	.1285	.0130	.641	.660
6	26250	7	.0612	.184	.027	.410	.426
4	41740	7	.0772	.232	.042	.259	.269
3	52640	7	.0867	.260	.053	.205	.213
2	66370	7	.0974	.292	.067	.162	.169
1	83690	19	.0664	.332	.087	.129	.134
0	105500	19	.0745	.373	.109	.102	.106
00	133100	19	.0837	.418	.137	.0811	.0844
000	167800	19	.0940	.470	.173	.0642	.0668
0000	211600	19	.1055	.528	.219	.0509	.0524
2500	250000	37	.0822	.575	.260	.0431	.0444
	300000	37	.0900	.630	.312	.0360	.0371
	350000	37	.0973	.681	.364	.0308	.0318
	400000	37	.1040	.728	.416	.0270	.0278
	500000	37	.1162	.814	.520	.0216	.0225
	600000	61	.0992	.893	.626	.0180	.0185
	700000	61	.1071	.964	.730	.0154	.0159
	750000	61	.1109	.998	.782	.0144	.0148
	800000	61	.1145	1.031	.835	.0135	.0139
	900000	61	.1215	1.093	.938	.0120	.0124
	1000000	61	.1280	1.152	1.042	.0108	.0111
	1250000	91	.1172	1.289	1.305	.00864	.00890
	1500000	91	.1284	1.412	1.566	.00719	.00740
	1750000	127	.1174	1.526	1.829	.00617	.00636
	2000000	127	.1255	1.631	2.089	.00539	.00555

* Area given is that of a circle having a diameter equal to the overall diameter of a stranded conductor.

Other values given in the table are those given in Circular 31 of the National Bureau of Standards, except that those shown in the last column are those given in specifications B33 of the American Society

The resistance values given in the last 2 columns are applicable only to direct current. When conductors larger than No. 4/0 are used with alternating current, the multiplying factors in table B-2 should be used to compensate for skin effect.

Multiplying Factors for Converting DC Resistance to AC Resistance

Size C M	Multiplying factor		Size CM	Multiplying factor	
	25 Hertz	60 Hertz		25 Hertz	60 Hertz
250000		1.005	750000	1.007	1.039
300000		1.006	800000	1.008	1.044
350000		1.009	900000	1.010	1.055
400000		1.011	1000000	1.012	1.067
500000		1.018	1250000	1.019	1.102
600000	1.005	1.025	1500000	1.027	1.142
700000	1.006	1.034	1750000	1.037	1.185
			2000000	1.048	1.233

Table B–3. Conductor Sizes and Overcurrent Protection for Motors

These values are in accordance with the National Electrical Code. They can be used for all installations except those intended for commercial or industrial use. For commercial or industrial use, consult pertinent sections of the National Electrical Code.

					Maximum allowable rating or setting of branch circuit protective devices			
(1)	(2)	(3)	(5)	(6)	(7) With code letters	(8) With code letters	(9) With code letters	(10) With code letters
Full load current rating of motor in amperes.	Minimum-sized conductor in raceways. For conductors in air, or for other insulations, see tables B–4 and B–5. AWG and MCM		For running protection of motors.[a]		Single-phase and squirrel-cage and synchronous. Full voltage, resistor, or reactor starting. Code letters F to V inc.	Single-phase and squirrel-cage and synchronous. Full voltage, resistor, or reactor starting. Code letters B to E inc.	Squirrel-cage and synchronous. Autotransformer starting. Code letters B to E inc.	All motors, Code letter A
	Rubber Type R Type T	Rubber Type RH	Maximum rating of non-adjustable protective devices. Amperes	Maximum setting of adjustable protective devices. Amperes	Without code letters	Without code letters	Without code letters	Without code letters
					Same as above.	Squirrel-cage and synchronous, autotransformer starting. High reactance squirrel-cage.[b] Both not more than 30 amperes.	Squirrel-cage and synchronous, autotransformer starting. High reactance squirrel-cage.[b] Both more than 30 amperes.	DC and wound-rotor motors.
1	14	14	2	1.25	15	15	15	15
2	14	14	3	2.50	15	15	15	15
3	14	14	4	3.75	15	15	15	15
4	14	14	6	5.00	15	15	15	15
5	14	14	8	6.25	15	15	15	15
6	14	14	8	7.50	20	15	15	15
7	14	14	10	8.75	25	20	15	15
8	14	14	10	10.00	25	20	20	15
9	14	14	12	11.25	30	25	20	15
10	14	14	15	12.50	30	25	20	15
11	14	14	15	13.75	35	30	25	20
12	14	14	15	15.00	40	30	25	20

Conductor Sizes and Overcurrent Protection for Motors—Continued

(1)	(2)	(3)	(5)	(6)	Maximum allowable rating or setting of branch circuit protective devices (7) With code letters	(8) With code letters	(9) With code letters	(10) With code letters
13	12	12	20	16.25	40	35	30	25
14	12	12	20	17.50	45	35	30	25
15	12	12	20	18.75	45	40	30	25
16	12	12	20	20.00	50	40	35	25
17	10	10	25	21.35	60	45	35	30
18	10	10	25	22.50	60	45	40	30
19	10	10	25	23.75	60	50	40	30
20	10	10	25	25.00	60	50	40	30
22	10	10	30	27.50	70	60	45	35
24	10	10	30	30.00	80	60	50	40
26	8	8	35	32.50	80	70	60	40
28	8	8	35	35.00	90	70	60	45
30	8	8	40	37.50	90	70	60	45
32	8	8	40	40.00	100	80	70	50
34	6	8	45	42.50	110	90	70	60
36	6	8	45	45.00	110	90	80	60
38	6	6	50	47.50	125	100	80	60
40	6	6	50	50.00	125	100	80	60
42	6	6	50	52.50	125	110	90	70
44	6	6	60	55.00	125	110	90	70
46	4	6	60	57.50	150	125	100	70
48	4	6	60	60.00	150	125	100	80
50	4	6	60	62.50	150	125	100	80
52	4	6	70	65.00	175	150	110	80
54	4	4	70	67.50	175	150	110	90
56	4	4	70	70.00	175	150	120	90
58	3	4	70	72.50	175	150	120	90
60	3	4	80	75.00	200	150	120	90
62	3	4	80	77.50	200	175	125	100
64	3	4	80	80.00	200	175	150	100
66	2	4	80	82.50	200	175	150	100
68	2	4	90	85.00	225	175	150	110
70	2	3	90	87.50	225	175	150	110
72	2	3	90	90.00	225	200	150	110
74	2	3	90	92.50	225	200	150	125
76	2	3	100	95.00	250	200	175	125
78	1	3	100	97.50	250	200	175	125
80	1	3	100	100.00	250	200	175	125
82	1	2	110	102.50	250	225	175	125
84	1	2	110	105.00	250	225	175	150
86	1	2	110	107.50	300	225	175	150
88	1	2	110	110.00	200	225	200	150
90	0	2	110	112.50	300	225	200	150
92	0	2	125	115.00	300	250	200	150
94	0	1	125	117.50	300	250	200	150
96	0	1	125	120.00	300	250	200	150
98	0	1	125	122.50	300	250	200	150
100	0	1	125	125.00	300	250	200	150
105	00	1	150	131.50	350	300	225	175
110	00	0	150	137.50	350	300	225	175
115	00	0	150	144.00	350	300	250	175
120	000	0	150	150.00	400	300	250	200
125	000	00	175	156.50	400	350	250	200

Conductor Sizes and Overcurrent Protection for Motors—Continued

					Maximum allowable rating or setting of branch circuit protective devices with code letters			
(1)	(2)	(3)	(5)	(6)	(7)	(8)	(9)	(10)
130	000	00	175	162.50	400	350	300	200
135	0000	00	175	169.00	450	350	300	225
140	0000	00	175	175.00	450	350	300	225
145	0000	000	200	181.50	450	400	300	225
150	0000	000	200	187.50	450	400	300	225
155	0000	000	200	194.00	500	400	350	250
160	250	000	200	200.00	500	400	350	250
165	250	0000	225	206.00	500	450	350	250
170	250	0000	225	213.00	500	450	350	300
175	300	0000	225	219.00	600	450	350	300
180	300	0000	225	225.00	600	450	400	300
185	300	0000	250	231.00	600	500	400	300
190	300	250	250	238.00	600	500	400	300
195	350	250	250	244.00	600	500	400	300
200	350	250	250	250.00	600	500	400	300
210	400	300	250	263.00	------	600	450	350
220	400	300	300	275.00	------	600	450	350
230	500	300	300	288.00	------	600	500	350
240	500	350	300	300.00	------	600	500	400
250	500	350	300	313.00	------	------	500	400
260	600	400	350	325.00	------	------	600	400
270	600	400	350	338.00	------	------	600	450
280	600	400	350	338.00	------	------	600	450
290	700	500	350	363.00	------	------	600	450
300	700	500	400	375.00	------	------	600	450
320	750	600	400	400.00	------	------	------	500
340	900	600	450	425.00	------	------	------	600
360	1000	700	450	450.00	------	------	------	600
380	1250	750	500	475.00	------	------	------	600
400	1500	900	500	500.00	------	------	------	600
420	1750	1000	600	525.00				
440	2000	1250	600	550.00				
460	------	1250	600	575.00				
480	------	1500	600	600.00				
500	------	1500	------	625.00				

running protection of motors, notify values in columns 5 and 6. [b] High-reactance, squirrel-cage motors are those gned to limit if nameplate-full-load current values different than those shown in the starting current by means eep-slot secondaries or double-wound table. Reduce current values shown in columns 5 and 6 by 8 percent for secondaries are generally started on full voltage.

otors other than open-type motors marked to have a temperature of over 40° C. (72° F.)

Allowable Current-Carrying Capacity in Amperes of Insulated Conductors (not more th three) in Raceways or Cables or Direct Burial

(1) Size AWG MCM	(2) Rubber Type R Type RW Type RU Type RUW (14-2) — Type RH-RW — Thermoplastic Type T Type TW	(3) Rubber Type RH — Type RH-RW — Type RHW	(4) Paper — thermoplastic Asbestos Type TA — Var-Cam Type V — Asbestos Var-Cam Type AVB — MI Cable	(5) Asbestos Var-Cam Type AVA Type AVL	(6) Impregnated Asbestos Type AI (14-8) Type AIA	(7) Asbestos Type A (14-8) Type AA
14	15	15	25	30	30	30
12	20	20	30	35	40	40
10	30	30	40	45	50	55
8	40	45	50	60	65	70
6	55	65	70	80	85	95
4	70	85	90	105	115	120
3	80	100	105	120	130	145
2	95	115	120	135	145	165
1	110	130	140	160	170	190
0	125	150	155	190	200	225
00	145	175	185	215	230	250
000	165	200	210	245	265	285
0000	195	230	235	275	310	340
250	215	255	270	315	335	
300	240	285	300	345	380	
350	260	310	325	390	420	
400	280	335	360	420	450	
500	320	380	405	470	500	
600	355	420	455	525	545	
700	385	460	490	560	600	
750	400	475	500	580	620	
800	410	490	515	600	640	
900	435	520	555			
1,000	455	545	585	680	730	
1,250	495	590	645			
1,500	520	625	700	785		
1,750	545	650	735			
2,000	560	665	775	840		

These values are in accordance with the National Electrical Code.

Allowable Current-Carrying Capacity in Amperes of Single, Insulated Conductor in Free Air

(1) Size AWG MCM	(2) Rubber Type R Type RW Type RU Type RUW (14-2) Type RH-RW Thermoplastic Type T Type TW	(3) Rubber Type RH Type RH-RW Type RHW	(4) Thermoplastic Asbestos Type TA Var-Cam Type V Asbestos Var-Cam Type AVB MI Cable	(5) Asbestos Var-Cam Type AVA Type AVL	(6) Impregnated Asbestos Type AI (14-8) Type AIA	(7) Asbestos Type A (14-8) Type AA	(8) Slow- burning Type SB Weather- proof Type WP Type SBW
14	20	20	30	40	40	45	30
12	25	25	40	50	50	55	40
10	40	40	55	65	70	75	55
8	55	65	70	85	90	100	70
6	80	95	100	120	125	135	100
4	105	125	135	160	170	180	130
3	120	145	155	180	195	210	150
2	140	170	180	210	225	240	175
1	165	195	210	245	265	280	205
0	195	230	245	285	305	325	235
00	225	265	285	330	355	370	275
000	260	310	330	385	410	430	320
0000	300	360	385	445	475	510	370
250	340	405	425	495	539	—	410
300	375	445	480	555	590	—	460
350	420	505	530	610	655	—	510
400	455	545	575	665	710	—	555
500	515	620	660	765	815	—	630
600	575	690	740	855	910	—	710
700	630	755	815	940	1005	—	780
750	655	785	845	980	1045	—	810
800	680	815	880	1020	1085	—	845
900	730	870	940	—	—	—	905
1000	780	935	1000	1165	1240	—	965
1250	890	1065	1130	—	—	—	—
1500	980	1175	1260	1450	—	—	1215
1750	1070	1280	1370	—	—	—	—
2000	1155	1385	1470	1715	—	—	1405

hese values are in accordance with the National Electrical Code.

*Full-Load Current for Direct-Current Motors**

HP	120V	240V
¼	2.9	1.5
⅓	3.6	1.8
½	5.2	2.6
¾	7.4	3.7
1	9.4	4.7
1½	13.2	6.6
2	17	8.5
3	25	12.2
5	40	20
7½	58	29
10	76	38
15		55
20		72
25		89
30		106
40		140
50		173
60		206
75		255
100		341
125		425
150		506
200		675

*These values of full-load current are average for all speeds, and are in accordance with the National Electrical Code.

Full-Load Current for Single-Phase AC Motors [a]

HP	115V	230V
⅙	4.4	2.2
¼	5.8	2.9
⅓	7.2	3.6
½	9.8	4.9
¾	13.8	6.9
1	16	8
1½	20	10
2	24	12
3	34	17
5	56	28
7½	80	40
10	100	50

[a] These values of full-load current are in accordance with the National Electrical Code, and are for motors running at speeds usual for belted motors and motors with normal torque characteristics. Motors built for especially low speeds or high torques may require more running current, in which case the nameplate current rating should be used.

[b] For full-load currents of 208- and 200-volt motors, increase corresponding 230-volt motor full-load current by 10 and 15 percent, respectively.

Full-Load Current for Three-Phase AC Motors [a]

	Induction type squirrel-cage and wound motor Amperes				Synchronous type unity power factor† Amperes				
HP	110v	220v [b]	440v	550v	2300v	220v [b]	440v	550v	230v
½	4.0	2.0	1.0	0.8	----	----	----	----	----
¾	5.6	2.8	1.4	1.1	----	----	----	----	----
1	7.0	3.5	1.8	1.4	----	----	----	----	----
1½	10.0	5.0	2.5	2.0	----	----	----	----	----
2	13.0	6.5	3.3	2.6	----	----	----	----	----
3	----	9.0	4.5	4.0	----	----	----	----	----
5	----	15.0	7.5	6.0	----	----	----	----	----
7½	----	22.0	11.0	9.0	----	----	----	----	----
10	----	27.0	14.0	11.0	----	----	----	----	----
15	----	40.0	20.0	16.0	----	----	----	----	----
20	----	52.0	26.0	21.0	----	----	----	----	----
25	----	64.0	32.0	26.0	7.0	54.0	27.0	22.0	5.4
30	----	78.0	39.0	31.0	8.5	65.0	33.0	26.0	6.5
40	----	104.0	52.0	41.0	10.5	86.0	43.0	35.0	8.0
50	----	125.0	63.0	50.0	13.0	108.0	54.0	44.0	10.0
60	----	150.0	75.0	60.0	16.0	128.0	64.0	51.0	12.0
75	----	185.0	93.0	74.0	19.0	161.0	81.0	65.0	15.0
100	----	246.0	123.0	98.0	25.0	211.0	106.0	85.0	20.0
125	----	310.0	155.0	124.0	31.0	264.0	132.0	106.0	25.0
150	----	360.0	180.0	144.0	37.0	----	158.0	127.0	30.0
200	----	480.0	240.0	192.0	48.0	----	210.0	168.0	40.0

[a] These values of full-load current are in accordance with the National Electrical Code, and are motors running at speeds usual for belted motors and motors with normal torque characteristics. Motors built for especially low speeds or high torques may require more running current, in which case, the nameplate current rating should be used.

[b] For full-load currents of 208- and 200-volt motors, increase the corresponding 220-volt motor full-load current by 6 and 10 percent, respectively.

†For 90 and 80 percent power factor the above figures should be multiplied by 1.1 and 1.25, respectively.

INDEX

Numbers on this page refer to PROBLEM NUMBERS, not page numbers

Numbers on this page refer to **PROBLEM NUMBERS**, not page numbers

Numbers on this page refer to PROBLEM NUMBERS, not page numbers

Numbers on this page refer to **PROBLEM NUMBERS**, not page numbers

Numbers on this page refer to PROBLEM NUMBERS, not page numbers

Numbers on this page refer to PROBLEM NUMBERS, not page numbers